T0186150

Frontiers in Mathematics

Frontiers in Elliptic and Parabolic Problems

Series Editor

Michel Chipot, Institute of Mathematics, Zürich, Switzerland

The goal of this series is to reflect the impressive and ongoing evolution of dealing with initial and boundary value problems in elliptic and parabolic PDEs. Recent developments include fully nonlinear elliptic equations, viscosity solutions, maximal regularity, and applications in finance, fluid mechanics, and biology, to name a few. Many very classical notions have been revisited, such as degree theory or Sobolev spaces. Books in this series present the state of the art keeping applications in mind wherever possible.

The series is curated by the Series Editor.

More information about this subseries at http://www.springer.com/series/16595

Vincenzo Ambrosio

Nonlinear Fractional Schrödinger Equations in $\mathbb{R}^N$

 Birkhäuser

Vincenzo Ambrosio
Dipartimento di Ingegneria Industriale
e Scienze Matematiche
Università Politecnica delle Marche
Ancona, Italy

ISSN 1660-8046   ISSN 1660-8054   (electronic)
Frontiers in Mathematics
ISSN 2730-549X   ISSN 2730-5503   (electronic)
Frontiers in Elliptic and Parabolic Problems
ISBN 978-3-030-60219-2   ISBN 978-3-030-60220-8   (eBook)
https://doi.org/10.1007/978-3-030-60220-8

Mathematics Subject Classification: 35R11, 35A15, 35B33, 35S05, 35J60, 35B09, 35B65, 35B40, 47G20, 58E05

This book is published under the imprint Birkhäuser, www.birkhauser-science.com, by the registered company
Springer Nature Switzerland AG.
The registered company address is: Gewerbestrasse 11, 6330 Cham, Switzerland

Preface

The aim of this book is to collect a set of results concerning nonlinear Schrödinger equations in the whole space driven by fractional operators. The material presented here was mainly taken from some papers carried out by the author and some joint works with his collaborators in these last years. It concerns some existence, multiplicity, and regularity results and qualitative properties of solutions established by means of suitable variational and topological methods which take care of the nonlocal character of the involved fractional operator. We deal with fractional Schrödinger equations involving different types of potentials and nonlinearities satisfying certain growth assumptions. Moreover, fractional Kirchhoff problems, fractional Choquard equations, fractional Schrödinger–Poisson systems, and fractional Schrödinger equations with magnetic field are considered. The book is principally addressed to researchers interested in pure and applied mathematics, physics, mechanics, and engineering. It is also appropriate for graduate students in mathematical and applied sciences, who will find updated information and a systematic exposition of important parts of modern mathematics. In particular, the proofs are given in detail and some of them are presented in a more different and elegant way with respect to the original ones in order to make the exposition more accessible and attractive for the interested reader.

The author would like to express his sincere gratitude to some dear friends and colleagues, including C. O. Alves, V. Coti Zelati, P. d'Avenia, G. M. Figueiredo, T. Isernia, G. Molica Bisci, H.-M. Nguyen, and E. Valdinoci, for their friendship, encouragement, and discussions on mathematics of common interest. Finally, the author would like to thank Prof. M. Chipot for giving him the opportunity to write this book.

Ancona, Italy
31 July 2020

Vincenzo Ambrosio

Introduction

Recently, a great attention has been devoted to the study of nonlinear problems involving fractional elliptic operators, whose prototype is given by the fractional Laplacian operator $(-\Delta)^s$ defined for any $u : \mathbb{R}^N \to \mathbb{R}$ sufficiently smooth by

$$(-\Delta)^s u(x) = C(N, s) P.V. \int_{\mathbb{R}^N} \frac{u(x) - u(y)}{|x - y|^{N+2s}} \, dy$$

where $s \in (0, 1)$, $N \in \mathbb{N}$, $P.V.$ stands for the Cauchy principal value and $C(N, s)$ is a positive constant depending only on N and s. Indeed, this operator has a great interest both for pure mathematical research, due to its intriguing structure, and in view of several concrete real-world applications, such as phase transition phenomena, crystal dislocation, population dynamics, flame propagation, chemical reactions of liquids, conservation laws, ultra-relativistic limits of quantum mechanics, quasi-geostrophic flows, minimal surface, and game theory. Moreover, from a probabilistic point of view, the fractional Laplacian is the infinitesimal generator of a (rotationally) symmetric $2s$-stable Lévy process. For a very nice introduction on fractional nonlocal operators and their applications, we refer to [106, 115, 168, 203, 273].

In this book we focus on the following nonlinear fractional Schrödinger equation:

$$(-\Delta)^s u + V(x)u = f(x, u) \text{ in } \mathbb{R}^N, \tag{1}$$

where $V : \mathbb{R}^N \to \mathbb{R}$ is a continuous potential and $f : \mathbb{R}^N \times \mathbb{R} \to \mathbb{R}$ is a suitable nonlinear term. Equation (1) arises in the study of the time-dependent fractional Schrödinger equation

$$\iota \frac{\partial \Psi}{\partial t} = (-\Delta)^s \Psi + V(x)\Psi - F(x, \Psi) \text{ in } \mathbb{R}^N \times \mathbb{R}, \tag{2}$$

when we look for standing wave solutions, which is solutions of the form $\Psi(x, t) = u(x)e^{-\iota ct}$, where c is a constant. We recall that (2) plays a fundamental role in the study of fractional quantum mechanics. It has been discovered by Laskin [245, 246] as a result of expanding the Feynman path integral, from the Brownian-like to the Lévy-like quantum mechanical paths; see also [158, 247] for more physical background. We stress that for $N > 1$, when $s \to 1^-$, $(-\Delta)^s u \to -\Delta u$ for all $u \in C_c^\infty(\mathbb{R}^N)$, and Eq. (1) boils down to the well-known nonlinear Schrödinger equation

$$-\Delta u + V(x)u = f(x, u) \text{ in } \mathbb{R}^N,$$

which has been widely investigated in these last 30 years by many authors [1, 13, 28, 90, 146, 154, 165, 197, 284, 299, 330] due to its relevance in several physical problems arising in nonlinear optics, plasma physics, and condensed matter physics, only to cite a few.

Motivated by the interest shared by the mathematical community in nonlocal fractional problems, the goal of this monograph is to present some recent results concerning the existence, multiplicity, and qualitative properties of different types of solutions to nonlinear fractional Schrödinger equations by applying suitable variational and topological methods. Indeed, due to the variational nature of (1), we will look for critical points of the functional

$$J(u) = \frac{1}{2} \left[\frac{C(N, s)}{2} \iint_{\mathbb{R}^{2N}} \frac{|u(x) - u(y)|^2}{|x - y|^{N+2s}} \, dx \, dy + \int_{\mathbb{R}^N} V(x)u^2(x) \, dx \right] - \int_{\mathbb{R}^N} F(x, u) \, dx$$

where $F(x, t) = \int_0^t f(x, \tau) \, d\tau$. For our purpose, the constant $\frac{C(N,s)}{2}$ appearing in the above definition does not play an essential role, so, along the book, we will omit it.

The book consists of 17 chapters and is organized as follows. In Chap. 1, we recall some basic properties of fractional Sobolev spaces, we give different definitions of the fractional Laplacian, and we present some regularity theorems, maximum principles, technical lemmas, and a concentration-compactness principle which will be frequently used along the book.

Chapter 2 addresses variational and topological principles and critical point theory, including minimax theorems, genus and category theory, and the recent approach developed by Szulkin and Weth [321, 322] concerning the Nehari method for non-differentiable manifolds.

In Chap. 3, we consider fractional scalar field equations of the type

$$(-\Delta)^s u = g(u) \text{ in } \mathbb{R}^N,$$

where g is a Berestycki-Lions type nonlinearity with subcritical or critical growth. We use the mountain pass theorem [29] to deduce the main existence and multiplicity results. We further analyze regularity, decay, and symmetry properties of ground state solutions.

In Chap. 4, we focus on the existence and qualitative properties of ground state solutions for a fractional Schrödinger equation driven by the pseudo-relativistic operator $(-\Delta + m^2)^s$ used in the study of stability of relativistic matter; see [250, 251].

In Chap. 5, we introduce fractional Schrödinger equations with periodic potentials and bounded potentials, involving nonlinearities with subcritical or critical growth. A compactness lemma in this last case is proved.

In Chap. 6, we study existence, multiplicity, and concentration of positive solutions for

$$\varepsilon^{2s}(-\Delta)^s u + V(x)u = f(u) + \gamma |u|^{2_s^*-2}u \text{ in } \mathbb{R}^N, \tag{3}$$

where $\varepsilon > 0$ is a small parameter, $\gamma \in \{0, 1\}$, f is a continuous nonlinearity with subcritical growth, and the potential V satisfies the following global condition due to Rabinowitz [299]:

$$\inf_{x \in \mathbb{R}^N} V(x) < \liminf_{|x| \to \infty} V(x) \in (0, \infty].$$

In Chap. 7, we construct a family of positive solutions to (3), which concentrates around a local minimum of the potential V that fulfills the following local condition introduced by del Pino and Felmer [165]: there exists a bounded open set $\Lambda \subset \mathbb{R}^N$ such that

$$\inf_{x \in \Lambda} V(x) < \min_{x \in \partial\Lambda} V(x).$$

Here, we consider superlinear nonlinearities satisfying the Ambrosetti-Rabinowitz condition [29] (shortly (AR)), and the monotonicity assumption $t \mapsto \frac{f(t)}{t}$ is increasing for $t > 0$, which plays a fundamental role in applying Nehari method. An existence result for a supercritical fractional Schrödinger equation is also discussed.

In Chap. 8, we investigate the previous problem in the subcritical case with asymptotically linear nonlinearities as well as superlinear nonlinearities which do not satisfy (AR). We point out that no monotonicity assumption on the function $\frac{f(t)}{t}$ is required.

In Chap. 9, we study the multiplicity and concentration of positive solutions for the following fractional Choquard equation:

$$\varepsilon^{2s}(-\Delta)^s u + V(x)u = \varepsilon^{\mu-N}\left(\frac{1}{|x|^\mu} * F(u)\right)f(u) \text{ in } \mathbb{R}^N,$$

where $0 < \mu < 2s$ and f is a continuous nonlinearity with subcritical growth. The main results will be obtained by applying suitable variational methods and Lusternik-Schnirelman theory.

In Chap. 10, we examine the following fractional Kirchhoff equation [196]

$$M \left(\iint_{\mathbb{R}^{2N}} \frac{|u(x) - u(y)|^2}{|x - y|^{N+2s}} \, dx dy \right) (-\Delta)^s u = g(u) \text{ in } \mathbb{R}^N,$$

where $M(t) = p + qt$ and g fulfills Berestycki-Lions type assumptions. A multiplicity result is obtained provided that $q > 0$ is small enough.

Chapter 11 focuses on the multiplicity and concentration of positive solutions for a fractional Kirchhoff problem in $\mathbb{R}^3$ under del Pino and Felmer assumptions on the potential.

In Chap. 12, we show the existence of positive concentrating solutions for a fractional Kirchhoff equation with critical growth.

Chapter 13 concerns the study of a fractional Schrödinger–Poisson system with critical growth

$$\begin{cases} \varepsilon^{2s} (-\Delta)^s u + V(x)u + \phi u = f(u) + |u|^{2^*_s - 2} u \text{ in } \mathbb{R}^3, \\ \varepsilon^{2t} (-\Delta)^t \phi = u^2 \hspace{4.5cm} \text{ in } \mathbb{R}^3, \end{cases}$$

where $s \in (\frac{3}{4}, 1), t \in (0, 1)$, V satisfies local conditions, and f is a subcritical nonlinearity. Applying variational and topological arguments, we relate the number of positive solutions with the topology of the set where the potential attains its minimum value.

In Chap. 14, we obtain an existence result for the fractional Kirchhoff-Schrödinger–Poisson system

$$\begin{cases} \left(p + q \iint_{\mathbb{R}^6} \frac{|u(x) - u(y)|^2}{|x-y|^{N+2s}} \, dx dy \right) (-\Delta)^s u + \mu \phi u = g(u) \text{ in } \mathbb{R}^3, \\ (-\Delta)^t \phi = \mu u^2 \hspace{5.5cm} \text{ in } \mathbb{R}^3, \end{cases}$$

as long as $\mu > 0$ is sufficiently small, by means of the Struwe-Jeanjean monotonicity trick.

Chapter 15 is devoted to the multiplicity of positive solutions for the non-homogeneous fractional Schrödinger equation

$$(-\Delta)^s u + V(x)u = k(x)f(u) + h(x) \text{ in } \mathbb{R}^N,$$

where $k(x)$ is a bounded positive function and h is a L^2-small perturbation.

In Chap. 16, we use a minimization argument and a quantitative deformation lemma to deal with the existence of sign-changing solutions for a fractional Schrödinger equation with vanishing potentials.

In Chap. 17 we consider the following fractional magnetic Schrödinger equation:

$$\varepsilon^{2s} (-\Delta)^s_A u + V(x)u = f(|u|^2)u \text{ in } \mathbb{R}^N,$$

where $A : \mathbb{R}^N \to \mathbb{R}^N$ is a smooth magnetic potential and

$$(-\Delta)_A^s u(x) = C(N, s) P.V. \int_{\mathbb{R}^N} \frac{u(x) - u(y) e^{\iota A(\frac{x+y}{2}) \cdot (x-y)}}{|x - y|^{N+2s}} \, dy$$

is the magnetic fractional Laplacian introduced in [157, 225]. We present a multiplicity result when the potential V satisfies the global Rabinowitz condition and an existence result under local del Pino–Felmer assumptions. We also prove a multiplicity result when $s = \frac{1}{2}$, $N = 1$ and f has exponential critical growth at infinity. A list of references and an index conclude the book.

Contents

Preliminaries

1

In this chapter we collect some useful facts about the fractional Sobolev spaces, define the fractional Laplacian operator and give some its interesting properties. For more details, we refer to [2, 78, 166, 209, 244, 258, 268, 315, 326] and the recent works [115, 168, 203, 273].

1.1   Function Spaces

1.1.1   Fractional Sobolev Spaces

Let Ω be an open set in $\mathbb{R}^N$. For any $s \in (0, 1)$ and $p \in [1, \infty)$, we define the fractional Sobolev space $W^{s,p}(\Omega)$ as

$$W^{s,p}(\Omega) = \left\{ u \in L^p(\Omega) : \frac{|u(x) - u(y)|}{|x - y|^{\frac{N+sp}{p}}} \in L^p(\Omega^2) \right\},$$

endowed with the norm

$$\|u\|_{W^{s,p}(\Omega)} = \left(\int_\Omega |u|^p \, dx + \iint_{\Omega^2} \frac{|u(x) - u(y)|^p}{|x - y|^{N+sp}} \, dxdy \right)^{\frac{1}{p}},$$

where the term

$$[u]_{W^{s,p}(\Omega)} = \left(\iint_{\Omega^2} \frac{|u(x) - u(y)|^p}{|x - y|^{N+sp}} \, dxdy \right)^{\frac{1}{p}}$$

is the so-called Gagliardo (semi)norm of u. Here we used the notation $\Omega^2 = \Omega \times \Omega$.

© The Author(s), under exclusive license to Springer Nature Switzerland AG 2021
V. Ambrosio, *Nonlinear Fractional Schrödinger Equations in* $\mathbb{R}^N$,
Frontiers in Mathematics, https://doi.org/10.1007/978-3-030-60220-8_1

Next we list some useful properties on $W^{s,p}$ spaces.

Proposition 1.1.1 ([168]) *Let $p \in [1, \infty)$ and $0 < s \leq s' < 1$. Let $\Omega \subset \mathbb{R}^N$ be an open set and $u : \Omega \to \mathbb{R}$ be a measurable function. Then $\|u\|_{W^{s,p}(\Omega)} \leq C\|u\|_{W^{s',p}(\Omega)}$ for some positive constant $C = C(N, s, p)$, and $W^{s',p}(\Omega) \subset W^{s,p}(\Omega)$.*

Proposition 1.1.2 ([168]) *Let $p \in [1, \infty)$ and $0 < s \leq s' < 1$. Let $\Omega \subset \mathbb{R}^N$ be an open set of class $C^{0,1}$ with bounded boundary and $u : \Omega \to \mathbb{R}$ be a measurable function. Then $\|u\|_{W^{s,p}(\Omega)} \leq C\|u\|_{W^{1,p}(\Omega)}$ for some positive constant $C = C(N, s, p)$, and $W^{1,p}(\Omega) \subset W^{s,p}(\Omega)$.*

Remark 1.1.3 As proved in [334], the previous result holds true if one assumes that Ω has the $W^{1,p}$-extension property.

Lemma 1.1.4 ([334]) *Let $p \in [1, \infty)$ and $s \in (0, 1)$. Let $u \in W^{s,p}(\Omega)$ and $\varphi \in C^{0,1}(\overline{\Omega}) \cap L^\infty(\Omega)$. Then $\varphi u \in W^{s,p}(\Omega)$ and there exists a positive constant $C = C(N, s, p, \|\varphi\|_{L^\infty(\Omega)})$ such that $\|\varphi u\|_{W^{s,p}(\Omega)} \leq C\|u\|_{W^{s,p}(\Omega)}$.*

Lemma 1.1.5 ([334]) *Let $p \in [1, \infty)$ and $s \in (0, 1)$. Then the following assertions hold:*

(i) *If $u \in W^{s,p}(\Omega)$, then $|u|, u^+, u_- \in W^{s,p}(\Omega)$ and*

$$\||u|\|_{W^{s,p}(\Omega)} \leq \|u\|_{W^{s,p}(\Omega)}, \quad \|u^+\|_{W^{s,p}(\Omega)} \leq \|u\|_{W^{s,p}(\Omega)}, \quad \|u_-\|_{W^{s,p}(\Omega)} \leq \|u\|_{W^{s,p}(\Omega)},$$

where $u^+ = \max\{u, 0\}$ and $u_- = \max\{-u, 0\}$.

(ii) *If $k \in \mathbb{R}$, $k \geq 0$ and $u \in W^{s,p}(\Omega)$ is nonnegative, then $u \wedge k = \min\{u, k\} \in W^{s,p}(\Omega)$ and $\|u \wedge k\|_{W^{s,p}(\Omega)} \leq \|u\|_{W^{s,p}(\Omega)}$.*

(iii) *If $u_1, u_2 \in W^{s,p}(\Omega)$ are non-negative, then $u = u_1 \vee u_2 = \max\{u_1, u_2\}$ and $v = u_1 \wedge u_2 = \min\{u_1, u_2\}$ are in $W^{s,p}(\Omega)$ and*

$$\|u\|^p_{W^{s,p}(\Omega)} + \|u\|^p_{W^{s,p}(\Omega)} \leq \|u_1\|^p_{W^{s,p}(\Omega)} + \|u_2\|^2_{W^{s,p}(\Omega)}.$$

(iv) *If $u \in W^{s,p}(\Omega)$ and $n \in \mathbb{N}$, then $(u \wedge n) \vee (-n) \in W^{s,p}(\Omega)$.*

When $s > 1$ and $s \notin \mathbb{N}$, we can write $s = m + \sigma$, where $m \in \mathbb{N}$ and $\sigma \in (0, 1)$. Then $W^{s,p}(\Omega)$ consists of the equivalence classes of functions $u \in W^{m,p}(\Omega)$ whose distributional derivatives $D^\alpha u$, with $|\alpha| = m$, belong to $W^{\sigma,p}(\Omega)$. When $s = m \in \mathbb{N}$, the space $W^{s,p}(\Omega)$ coincides with the classical Sobolev space $W^{m,p}(\Omega)$.

Corollary 1.1.6 ([168]) *Let $p \in [1, \infty)$ and $t \geq s > 1$. Let Ω be an open set in $\mathbb{R}^N$ of class $C^{0,1}$. Then, $W^{t,p}(\Omega) \subset W^{s,p}(\Omega)$.*

We also have the following density result.

Theorem 1.1.7 ([2, 166, 258]) *For any* $s \geq 0$, *the space* $C_c^\infty(\mathbb{R}^N)$ *of smooth functions with compact support is dense in* $W^{s,p}(\mathbb{R}^N)$.

In general, if $\Omega \neq \mathbb{R}^N$, the space $C_c^\infty(\Omega)$ is not dense in $W^{s,p}(\Omega)$. Hence, for $s \geq 0$, we denote by $W_0^{s,p}(\Omega)$ the closure of $C_c^\infty(\Omega)$ with respect to the $\|\cdot\|_{W^{s,p}(\Omega)}$-norm. When $s < 0$ and $p \in (1, \infty)$, we can define $W^{s,p}(\Omega)$ as the dual space of $W_0^{-s,p'}(\Omega)$, where p' is the conjugate exponent of p, i.e., $\frac{1}{p} + \frac{1}{p'} = 1$. It follows by the reflexivity of $W^{s,p}(\Omega)$ spaces, with $s \in \mathbb{R}$ and $p \in (1, \infty)$, that the dual of $W^{s,p}(\mathbb{R}^N)$ is isometrically isomorphic to $W^{-s,p'}(\mathbb{R}^N)$, with p' the conjugate exponent of p.

Now, we focus on the case $\Omega = \mathbb{R}^N$, $s \in (0, 1)$ and $p = 2$. We denote by $\mathcal{D}^{s,2}(\mathbb{R}^N)$ the completion of $C_c^\infty(\mathbb{R}^N)$ with respect to the Gagliardo seminorm

$$[u]_s = [u]_{W^{s,2}(\mathbb{R}^N)} = \left(\iint_{\mathbb{R}^{2N}} \frac{|u(x) - u(y)|^2}{|x - y|^{N+2s}} \, dx dy \right)^{\frac{1}{2}},$$

or equivalently (see [164]), when $N > 2s$,

$$\mathcal{D}^{s,2}(\mathbb{R}^N) = \{u \in L^{2_s^*}(\mathbb{R}^N) : [u]_s < \infty\},$$

where

$$2_s^* = \frac{2N}{N - 2s}$$

is the fractional critical Sobolev exponent. Occasionally, we also use the notation

$$\langle u, v \rangle_{\mathcal{D}^{s,2}(\mathbb{R}^N)} = \iint_{\mathbb{R}^{2N}} \frac{(u(x) - u(y))(v(x) - v(y))}{|x - y|^{N+2s}} \, dx dy$$

for all $u, v \in \mathcal{D}^{s,2}(\mathbb{R}^N)$. We let $H^s(\mathbb{R}^N)$ denote the space $W^{s,2}(\mathbb{R}^N)$ endowed with the natural norm

$$\|u\|_{H^s(\mathbb{R}^N)} = \left([u]_s^2 + \|u\|_{L^2(\mathbb{R}^N)}^2 \right)^{\frac{1}{2}}.$$

Clearly, $H^s(\mathbb{R}^N)$ is a Hilbert space with the inner product

$$\langle u, v \rangle_{H^s(\mathbb{R}^N)} = \langle u, v \rangle_{\mathcal{D}^{s,2}(\mathbb{R}^N)} + \int_{\mathbb{R}^N} uv \, dx$$

for any $u, v \in H^s(\mathbb{R}^N)$.

1.1.2 Sobolev Embeddings

We review the main embedding results for the fractional Sobolev space $H^s(\mathbb{R}^N)$.

Theorem 1.1.8 ([2, 166, 268]) *Let $s \in (0, 1)$ and $N \geq 1$.*

- *If $N > 2s$, then there exists a sharp constant $S_* = S(N, s) > 0$ such that*

$$S_* \|u\|^2_{L^{2^*_s}(\mathbb{R}^N)} \leq [u]^2_s \tag{1.1.1}$$

  for all $u \in \mathcal{D}^{s,2}(\mathbb{R}^N)$. Moreover, $H^s(\mathbb{R}^N)$ is continuously embedded in $L^q(\mathbb{R}^N)$ for every $q \in [2, 2^*_s]$ and compactly embedded in $L^q_{loc}(\mathbb{R}^N)$ for any $q \in [1, 2^*_s)$.
- *When $N = 2s$, then $H^s(\mathbb{R}^N)$ is continuously embedded in $L^q(\mathbb{R}^N)$ for every $q \in [2, \infty)$ and compactly embedded in $L^q_{loc}(\mathbb{R}^N)$ for any $q \in [1, \infty)$.*
- *If $N < 2s$, then $H^s(\mathbb{R}^N)$ is continuously embedded in $C^{0,s-\frac{N}{2}}(\mathbb{R}^N)$ and compactly embedded in $C^{0,\lambda}_{loc}(\mathbb{R}^N)$ for all $0 < \lambda < s - \frac{N}{2}$.*

Remark 1.1.9 As proved in [155], (1.1.1) becomes an equality if and only if

$$u(x) = c(\mu^2 + (x - x_0)^2)^{-\frac{N-2s}{2}}, \ x \in \mathbb{R}^N,$$

where $c \in \mathbb{R}$, $\mu > 0$ and $x_0 \in \mathbb{R}^N$ are fixed constants. Moreover, the value of the best constant S_* is given by

$$S_*^{-1} = 2^{-\frac{2s}{N}} \pi^{-\frac{s(N+1)}{N}} \frac{\Gamma(\frac{N-2s}{2})}{\Gamma(\frac{N+2s}{2})} \left[\Gamma\left(\frac{N+1}{2}\right) \right]^{\frac{2s}{N}}.$$

We also have the following fractional Hardy inequality:

Theorem 1.1.10 ([200, 217]) *Let $s \in (0, 1)$ and $N > 2s$. Then for all $u \in \mathcal{D}^{s,2}(\mathbb{R}^N)$,*

$$\int_{\mathbb{R}^N} \frac{|u(x)|^2}{|x|^{2s}} \, dx \leq H_{N,s}[u]^2_s, \tag{1.1.2}$$

with

$$H_{N,s} = 2\pi^{\frac{N}{2}} \frac{\Gamma^2(\frac{N+2s}{4}) \ \Gamma(\frac{N+2s}{2})}{\Gamma^2(\frac{N-2s}{4}) \ |\Gamma(-s)|}.$$

Now we introduce the space of radial functions in $H^s(\mathbb{R}^N)$, namely

$$H^s_{\text{rad}}(\mathbb{R}^N) = \{u \in H^s(\mathbb{R}^N) : u(x) = u(|x|)\}.$$

We recall the following compactness result due to Lions [255]:

Theorem 1.1.11 ([255]) *Let* $s \in (0, 1)$ *and* $N \geq 2$. *Then* $H^s_{\text{rad}}(\mathbb{R}^N)$ *is compactly embedded in* $L^q(\mathbb{R}^N)$ *for every* $q \in (2, 2^*_s)$.

1.2   The Fractional Laplacian Operator

1.2.1   Definition via Singular Integrals

Consider the Schwartz space $\mathcal{S}(\mathbb{R}^N)$ of rapidly decreasing functions in $\mathbb{R}^N$, that is, the set of functions $\phi \in C^\infty(\mathbb{R}^N)$ such that

$$\|\phi\|_{\alpha,\beta} = \sup_{x \in \mathbb{R}^N} |x^\alpha D^\beta \phi(x)| < \infty,$$

for all multi-indices α and β. The vector space $\mathcal{S}(\mathbb{R}^N)$ with the natural topology given by the seminorms $\|\cdot\|_{\alpha,\beta}$ is a Fréchet space, i.e., a complete metrizable locally convex space. As it is customary, if $\alpha = (\alpha_1, \ldots, \alpha_N)$ is a multi-index (that is an N-tuple of nonnegative integers), then $|\alpha| = \alpha_1 + \cdots + \alpha_N$, the monomial x^α (with $x = (x_1, \ldots, x_N) \in \mathbb{R}^N$) is defined by $x^\alpha = x_1^{\alpha_1} \cdots x_N^{\alpha_N}$, and D^α denotes the differential operator $\frac{\partial^{|\alpha|}}{\partial_{x_1}^{\alpha_1} \ldots \partial_{x_N}^{\alpha_N}}$, with the convention that $D^{(0,\ldots,0)}$ is the identity operator.

Let $\mathcal{S}'(\mathbb{R}^N)$ be the space of tempered distributions, that is, the topological dual of $\mathcal{S}(\mathbb{R}^N)$. For any $\phi \in \mathcal{S}(\mathbb{R}^N)$, we denote by

$$\mathcal{F}\phi(\xi) = \frac{1}{(2\pi)^{\frac{N}{2}}} \int_{\mathbb{R}^N} e^{-\iota \xi \cdot x} \phi(x) \, dx$$

the Fourier transform of ϕ. We recall that $\mathcal{F}$ is an isomorphism from $\mathcal{S}(\mathbb{R}^N)$ onto itself, with inverse given by

$$\mathcal{F}^{-1}\phi(x) = \frac{1}{(2\pi)^{\frac{N}{2}}} \int_{\mathbb{R}^N} e^{\iota x \cdot \xi} \phi(\xi) \, d\xi,$$

which is called the inverse Fourier transform of $\phi \in \mathcal{S}(\mathbb{R}^N)$. Moreover, one can extend the Fourier transformation to $\mathcal{S}'(\mathbb{R}^N)$ so that $\mathcal{F}$ is an isomorphism from $\mathcal{S}'(\mathbb{R}^N)$ onto itself, with inverse $\mathcal{F}^{-1}$.

Fix $s \in (0, 1)$ and $u \in \mathcal{S}(\mathbb{R}^N)$. Then we define the fractional Laplace operator by

$$(-\Delta)^s u(x) = C(N, s) \, \text{P.V.} \int_{\mathbb{R}^N} \frac{u(x) - u(y)}{|x - y|^{N+2s}} \, dy$$

$$= C(N, s) \lim_{r \to 0} \int_{\mathbb{R}^N \setminus B_r(x)} \frac{u(x) - u(y)}{|x - y|^{N+2s}} \, dy$$

(1.2.1)

where P.V. stands for the Cauchy principal value, and $C(N, s)$ is a positive dimensional constant that depends on N and s, precisely given by (see [124, 317])

$$C(N, s) = \left(\int_{\mathbb{R}^N} \frac{1 - \cos(\xi_1)}{|\xi|^{N+2s}} \, d\xi \right)^{-1} = \pi^{-\frac{N}{2}} 2^{2s} \frac{\Gamma(\frac{N+2s}{2})}{-\Gamma(-s)}$$

$$= \pi^{-\frac{N}{2}} 2^{2s} \frac{\Gamma(\frac{N+2s}{2})}{\Gamma(1 - s)} s$$

$$= \pi^{-\frac{N}{2}} 2^{2s} \frac{\Gamma(\frac{N+2s}{2})}{\Gamma(2 - s)} s(1 - s).$$

We recall the following useful integral representation formula for the fractional Laplacian.

Lemma 1.2.1 ([168]) *For all* $u \in \mathcal{S}(\mathbb{R}^N)$,

$$(-\Delta)^s u(x) = -\frac{1}{2} C(N, s) \int_{\mathbb{R}^N} \frac{u(x + y) + u(x - y) - 2u(x)}{|y|^{N+2s}} \, dy \quad \forall x \in \mathbb{R}^N.$$

For $u \in \mathcal{S}(\mathbb{R}^N)$ it is not true in general that $(-\Delta)^s u \in \mathcal{S}(\mathbb{R}^N)$. However, one can verify that $(-\Delta)^s u \in C^\infty(\mathbb{R}^N)$ and decays at infinity according to the following result.

Lemma 1.2.2 ([203]) *Let* $u \in \mathcal{S}(\mathbb{R}^N)$. *Then,* $(-\Delta)^s u \in C^\infty(\mathbb{R}^N) \cap L^1(\mathbb{R}^N)$. *Moreover, for all* $x \in \mathbb{R}^N$ *such that* $|x| > 1$, *we have*

$$|(-\Delta)^s u(x)| \leq C_{u,N,s} |x|^{-(N+2s)},$$

where

$$C_{u,N,s} = C_{N,s} \left(\|u\|_{N+2,\mathcal{S}} + \|u\|_{N,\mathcal{S}} + \|u\|_{L^1(\mathbb{R}^N)} \right),$$

$C_{N,s}$ *is a positive constant depending only on* N *and* s, *and*

$$\|u\|_{k,\mathcal{S}} = \sup_{|\alpha| \leq k} \sup_{x \in \mathbb{R}^N} (1 + |x|^2)^{\frac{k}{2}} |D^\alpha u(x)| \quad \text{for } k \in \mathbb{N} \cup \{0\}.$$

One can also define the fractional Laplacian acting on spaces of functions with weaker regularity. Indeed, following [244, 313], one considers the space $\bar{\mathcal{S}}_s(\mathbb{R}^N)$ of C^∞ functions u such that, for every $k \in \mathbb{N} \cup \{0\}$, the quantity $(1 + |x|^{N+2s})D^k u$ is bounded. The class $\bar{\mathcal{S}}_s(\mathbb{R}^N)$ is endowed with the topology induced by the countable family of seminorms

$$\rho_k(u) = \sup_{x \in \mathbb{R}^N} |(1 + |x|^{N+2s})D^k u|.$$

Denote by $\bar{\mathcal{S}}'_s(\mathbb{R}^N)$ the dual space of $\mathcal{S}_s(\mathbb{R}^N)$. One can then verify that $(-\Delta)^s$ maps $\mathcal{S}(\mathbb{R}^N)$ into $\bar{\mathcal{S}}_s(\mathbb{R}^N)$. The symmetry of $(-\Delta)^s$ allows us to extend its definition to $\bar{\mathcal{S}}'_s(\mathbb{R}^N)$ by duality; i.e., if $u \in \bar{\mathcal{S}}'_s(\mathbb{R}^N)$,

$$\langle (-\Delta)^s u, f \rangle = \langle u, (-\Delta)^s f \rangle \quad \text{for every } f \in \mathcal{S}(\mathbb{R}^N).$$

This definition coincides with the previous ones in the case where $u \in \mathcal{S}(\mathbb{R}^N)$, and $(-\Delta)^s$ is a continuous operator from $\bar{\mathcal{S}}'_s(\mathbb{R}^N)$ to $\mathcal{S}'(\mathbb{R}^N)$. Anyway, it is more convenient to define $(-\Delta)^s$ for functions in the space

$$\mathcal{L}_s(\mathbb{R}^N) = \left\{ u : \mathbb{R}^N \to \mathbb{R} \text{ measurable} : \int_{\mathbb{R}^N} \frac{|u(x)|}{1 + |x|^{N+2s}} \, dx < \infty \right\} = L^1_{\text{loc}}(\mathbb{R}^N) \cap \bar{\mathcal{S}}'_s(\mathbb{R}^N).$$

Then we have the following result:

Proposition 1.2.3 ([313]) *Let Ω be an open subset of $\mathbb{R}^N$ and let $u \in \mathcal{L}_s(\mathbb{R}^N)$, $s \in (0, 1)$. If $u \in C^{0,2s+\varepsilon}(\Omega)$ (or $u \in C^{1,2s+\varepsilon-1}(\Omega)$ if $s \in [\frac{1}{2}, 1)$) for some $\varepsilon > 0$, then $(-\Delta)^s u$ is a continuous function in Ω and $(-\Delta)^s u(x)$ is given by the integral in (1.2.1), for every $x \in \Omega$.*

1.2.2 Definition via Fourier Transform: Riesz and Bessel Potentials

The fractional Laplacian $(-\Delta)^s$ can be also defined via the Fourier transform and viewed as a pseudo-differential operator of symbol $|\xi|^{2s}$ (see [220, 221]). More precisely, for any $u \in \mathcal{S}(\mathbb{R}^N)$, we have

$$\mathcal{F}((-\Delta)^s u)(\xi) = |\xi|^{2s} \mathcal{F}u(\xi).$$

This last identity is motivated by the fact that, for $u \in \mathcal{S}(\mathbb{R}^N)$, $\mathcal{F}(-\Delta u)(\xi) = |\xi|^2 \mathcal{F}u(\xi)$, and thus one can define $(-\Delta)^s$ as the operator given by multiplication with the function $|\xi|^{2s}$ on the Fourier transform.

Using Plancherel's formula we see that

$$[u]_s^2 = 2C(N, s)^{-1} \int_{\mathbb{R}^N} |\xi|^{2s} |\mathcal{F}u(\xi)|^2 \, d\xi \quad \text{for all } u \in \mathcal{S}(\mathbb{R}^N), \tag{1.2.2}$$

which combined with the definition of $(-\Delta)^s$ via the Fourier transform implies that

$$[u]_s^2 = 2C(N,s)^{-1}\|(-\Delta)^{\frac{s}{2}}u\|_{L^2(\mathbb{R}^N)}^2 \quad \text{for all } u \in \mathcal{S}(\mathbb{R}^N).$$

In the light of these relations, we can see that for $s \in (0,1)$, $H^s(\mathbb{R}^N)$ is equivalent to the following space:

$$\widehat{H}^s(\mathbb{R}^N) = \left\{u \in L^2(\mathbb{R}^N) : \int_{\mathbb{R}^N}(1+|\xi|^2)^s|\mathcal{F}u(\xi)|^2\,d\xi < \infty\right\}.$$

We note that the above space, unlike the spaces defined via the Gagliardo seminorm, is defined for any real $s \geq 0$, whereas if $s < 0$ we set

$$\widehat{H}^s(\mathbb{R}^N) = \left\{u \in \mathcal{S}'(\mathbb{R}^N) : \int_{\mathbb{R}^N}(1+|\xi|^2)^s|\mathcal{F}u(\xi)|^2\,d\xi < \infty\right\}.$$

Moreover, for any $s \in \mathbb{R}$, $C_c^\infty(\mathbb{R}^N)$ is dense in $\widehat{H}^s(\mathbb{R}^N)$, and $\widehat{H}^{-s}(\mathbb{R}^N)$ coincides with the dual of $\widehat{H}^s(\mathbb{R}^N)$; see [2,166,258]. From now on, whenever $s \in (0,1)$, we will use the same notation for $H^s(\mathbb{R}^N)$ and $\widehat{H}^s(\mathbb{R}^N)$, and will denote by $H^{-s}(\mathbb{R}^N)$ the dual of $H^s(\mathbb{R}^N)$.

Now, we note that when $N > 2s$, the operator $(-\Delta)^{-s}$ is defined as the inverse of $(-\Delta)^s$ and is given by

$$(-\Delta)^{-s}u(x) = c(N,-s)\int_{\mathbb{R}^N}\frac{u(y)}{|x-y|^{N-2s}}\,dy, \quad \text{for all } u \in \mathcal{S}(\mathbb{R}^N),$$

which is a Riesz potential of order $2s$. More generally, we have the following definition.

Definition 1.2.4 Let $N \in \mathbb{N}$ and $0 < s < N$. The Riesz potential of order s is the operator whose action on a function $u \in \mathcal{S}(\mathbb{R}^N)$ is given by

$$\mathcal{I}_s u(x) = (-\Delta)^{-\frac{s}{2}}u(x) = (I_s * u)(x) = \frac{\Gamma\left(\frac{N-s}{2}\right)}{\pi^{\frac{N}{2}}2^s\Gamma\left(\frac{s}{2}\right)}\int_{\mathbb{R}^N}\frac{u(y)}{|x-y|^{N-s}}\,dy, \quad (1.2.3)$$

where $I_s(x) = \dfrac{\Gamma\left(\frac{N-s}{2}\right)}{\pi^{\frac{N}{2}}2^s\Gamma\left(\frac{s}{2}\right)}|x|^{s-N}$ is called the Riesz kernel.

The following identities exhibit essential properties of the Riesz operators $\mathcal{I}_s$ (see [315]):

$$\mathcal{I}_s(\mathcal{I}_t u) = \mathcal{I}_{s+t}u, \quad \text{for all } u \in \mathcal{S}(\mathbb{R}^N),\ s,t > 0,\ s+t < N,$$

$$\Delta(\mathcal{I}_s u) = \mathcal{I}_s(\Delta u) = -\mathcal{I}_{s-2}u, \quad \text{for all } u \in \mathcal{S}(\mathbb{R}^N),\ N \geq 3,\ 2 \leq s \leq N.$$

The importance of the Riesz potentials lies in the fact that they are indeed smoothing operators. This is the essence of the Hardy–Littlewood–Sobolev theorem on fractional integration.

Theorem 1.2.5 ([315]) *Let $s \in (0, N)$ and let $1 \le p < q < \infty$ satisfy*

$$\frac{s}{N} = \frac{1}{p} - \frac{1}{q}.$$

- *If $u \in L^p(\mathbb{R}^N)$, then the integral defined by (1.2.3) is absolutely convergent for a.e. $x \in \mathbb{R}^N$.*
- *If $p > 1$, then there exists a positive constant $C = C(N, p, s)$ such that for all $u \in L^p(\mathbb{R}^N)$,*

$$\|\mathcal{I}_s u\|_{L^q(\mathbb{R}^N)} \le C \|u\|_{L^p(\mathbb{R}^N)}.$$

- *If $p = 1$, then for some positive constant $C = C(N, s)$ and for all $u \in L^1(\mathbb{R}^N)$,*

$$\left| \{x \in \mathbb{R}^N : |\mathcal{I}_s u(x)| > \lambda \} \right| \le \left(\frac{C \|u\|_{L^1(\mathbb{R}^N)}}{\lambda} \right)^{\frac{N}{N-s}}$$

for all $\lambda > 0$.

While the behavior of the kernels $|x|^{s-N}$ as $|x| \to 0$ is well suited to their smoothing properties, their decay as $|x| \to \infty$ gets worse as s increases. A way out of this dilemma is to slightly adjust the Riesz potentials so that we maintain their essential behavior near zero, but achieve exponential decay at infinity. The simplest way to accomplish this goal is to replace the "nonnegative" operator $-\Delta$ by the "strictly positive" operator $1 - \Delta$. This fact justifies the introduction of Bessel potentials (see [78, 128, 209, 315]).

Definition 1.2.6 Let $s > 0$. The Bessel potential of order s of $u \in \mathcal{S}(\mathbb{R}^N)$ is defined as

$$\mathcal{J}_s u(x) = (1 - \Delta)^{-\frac{s}{2}} u(x) = (G_s * u)(x) = \int_{\mathbb{R}^N} G_s(x - y) u(y) \, dy,$$

where G_s is called the Bessel kernel and its Fourier transform satisfies

$$\mathcal{F} G_s(\xi) = (2\pi)^{-\frac{N}{2}} (1 + |\xi|^2)^{-\frac{s}{2}}.$$

Remark 1.2.7 If $s \in \mathbb{R}$ (or $s \in \mathbb{C}$), then we may define the Bessel potential of a temperate distribution $u \in \mathcal{S}'(\mathbb{R}^N)$ (see [128]) by setting

$$\mathcal{F}\mathcal{J}_s u(\xi) = (2\pi)^{-\frac{N}{2}}(1 + |\xi|^2)^{-\frac{s}{2}}\mathcal{F}u(\xi).$$

It is possible to prove (see [78]) that

$$G_s(x) = \frac{1}{2^{\frac{N+s-2}{2}}\pi^{\frac{N}{2}}\Gamma(\frac{s}{2})}K_{\frac{N-s}{2}}(|x|)|x|^{\frac{s-N}{2}},$$

where K_ν denotes the modified Bessel function of the third kind (or Macdonald function) and order ν (see [178, 336]), and satisfies the following asymptotic formulas for $\nu \in \mathbb{R}$ and $r > 0$:

$$K_\nu(r) \sim \frac{\Gamma(\nu)}{2}\left(\frac{r}{2}\right)^{-\nu} \quad \text{as } r \to 0, \text{ for } \nu > 0, \tag{1.2.4}$$

$$K_\nu(r) \sim \sqrt{\frac{\pi}{2}}r^{-\frac{1}{2}}e^{-r} \quad \text{as } r \to \infty, \text{ for } \nu \in \mathbb{R}. \tag{1.2.5}$$

Thus $G_s(x)$ is a positive, decreasing function of $|x|$, analytic except at $x = 0$, and for $x \in \mathbb{R}^N \setminus \{0\}$, $G_s(x)$ is an entire function of s. Moreover,

$$G_s(x) \sim \frac{\Gamma\left(\frac{N-s}{2}\right)}{\pi^{\frac{N}{2}}2^s\Gamma\left(\frac{s}{2}\right)}|x|^{s-N} \quad \text{as } |x| \to 0, \text{ if } 0 < s < N,$$

$$G_s(x) \sim \frac{1}{2^{\frac{N+s-1}{2}}\pi^{\frac{N-1}{2}}\Gamma(\frac{s}{2})}|x|^{\frac{s-N-1}{2}}e^{-|x|} \quad \text{as } |x| \to \infty,$$

$G_s \in L^1(\mathbb{R}^N)$ for all $s > 0$, and $\int_{\mathbb{R}^N} G_s(x)\,dx = 1$. We also have the composition formula $G_{s+t} = G_s * G_t$ for all $s, t > 0$.

One the most interesting facts concerning Bessel potentials is they can be employed to define the Bessel potential spaces; see [2, 78, 128, 315]. For $p \in [1, \infty]$ and $\alpha \in \mathbb{R}$ we define the Banach space

$$\mathcal{L}_\alpha^p = G_\alpha(L^p(\mathbb{R}^N)) = \{u : u = G_\alpha * f, \quad f \in L^p(\mathbb{R}^N)\}$$

endowed with the norm

$$\|u\|_{\mathcal{L}_\alpha^p} = \|f\|_{L^p(\mathbb{R}^N)} \quad \text{if } u = G_\alpha * f.$$

Thus $\mathcal{L}_\alpha^p$ is a subspace of $L^p(\mathbb{R}^N)$ for all $\alpha \geq 0$. We also have the following useful result:

Theorem 1.2.8 ([2,128])

(i) *If $\alpha \geq 0$ and $1 \leq p < \infty$, then $\mathcal{D}(\mathbb{R}^N)$ is dense in $\mathcal{L}_\alpha^p$.*

(ii) *If $1 < p < \infty$ and p' its conjugate exponent, then the dual of $\mathcal{L}_\alpha^p$ is isometrically isomorphic to $\mathcal{L}_{-\alpha}^{p'}$.*

(iii) *If $\beta < \alpha$, then $\mathcal{L}_\alpha^p$ is continuously embedded in $\mathcal{L}_\beta^p$.*

(iv) *If $\beta \leq \alpha$ and if either $1 < p \leq q \leq \frac{Np}{N-(\alpha-\beta)p} < \infty$, or $p = 1$ and $1 \leq q < \frac{N}{N-\alpha+\beta}$, then $\mathcal{L}_\alpha^p$ is continuously embedded in $\mathcal{L}_\beta^q$.*

(v) *If $0 < \mu \leq \alpha - \frac{N}{p} < 1$, then $\mathcal{L}_\alpha^p$ is continuously embedded in $C^{0,\mu}(\mathbb{R}^N)$.*

(vi) *$\mathcal{L}_k^p = W^{k,p}(\mathbb{R}^N)$ for all $k \in \mathbb{N}$ and $1 < p < \infty$, $\mathcal{L}_\alpha^2 = W^{\alpha,2}(\mathbb{R}^N)$ for any α.*

(vii) *If $1 < p < \infty$ and $\varepsilon > 0$, then for every α we have the following continuous embeddings:*

$$\mathcal{L}_{\alpha+\varepsilon}^p \subset W^{\alpha,p}(\mathbb{R}^N) \subset \mathcal{L}_{\alpha-\varepsilon}^p.$$

1.2.3   Definition via Caffarelli-Silvestre Extension

Caffarelli and Silvestre [127] showed that $(-\Delta)^s$ can be characterized as the Dirichlet-to-Neumann operator for a suitable local degenerate elliptic problem posed on the upper half-space $\mathbb{R}_+^{N+1} = \{(x, y) \in \mathbb{R}^{N+1} : y > 0\}$.

Let $s \in (0, 1)$ be given. For a measurable function $u : \mathbb{R}^N \to \mathbb{R}$, we first formally define its s-harmonic extension to $\mathbb{R}_+^{N+1}$ by setting

$$\text{Ext}(u)(x, y) = (P_s(\cdot, y) * u)(x) = \int_{\mathbb{R}^N} P_s(x - z, y)u(z)\,dz,$$

where the Poisson kernel $P_s(x, y)$ is given by

$$P_s(x, y) = p_{N,s} \frac{y^{2s}}{(|x|^2 + y^2)^{\frac{N+2s}{2}}}, \qquad p_{N,s} = \pi^{-\frac{N}{2}} \frac{\Gamma(\frac{N+2s}{2})}{\Gamma(s)},$$

and the constant $p_{N,s}$ is such that $\int_{\mathbb{R}^N} P_s(x, y)\,dx = 1$ for all $y > 0$.

If $u \in \mathcal{S}(\mathbb{R}^N)$, then $U = \text{Ext}(u) \in C^\infty(\mathbb{R}_+^{N+1}) \cap C(\overline{\mathbb{R}_+^{N+1}})$, $U(\cdot, y) \to 0$ as $y \to \infty$, and U solves the following problem

$$\begin{cases} -\,\text{div}(y^{1-2s}\nabla U) = 0 & \text{in } \mathbb{R}_+^{N+1}, \\ U(\cdot, 0) = u & \text{on } \partial\mathbb{R}_+^{N+1} \cong \mathbb{R}^N. \end{cases} \qquad (1.2.6)$$

Moreover, for any $x \in \mathbb{R}^N$,

$$\frac{\partial U}{\partial v^{1-2s}}(x, 0) = -\lim_{y \to 0} y^{1-2s} \frac{\partial U}{\partial y}(x, y) = \kappa_s (-\Delta)^s u(x) \tag{1.2.7}$$

where

$$\kappa_s = \frac{\Gamma(1-s)}{2^{2s-1}\Gamma(s)}.$$

Then (1.2.7) expresses the fact that $(-\Delta)^s$ can be regarded as the Dirichlet-to-Neumann map for the extension problem (1.2.6) with the weight y^{1-2s}. Qualitatively, the result in [127] states that one can localize the fractional Laplacian by adding the new variable y.

We now consider a more general setting. Let us define the spaces $\dot{H}^s(\mathbb{R}^N)$ and $X_0^s(\mathbb{R}_+^{N+1})$ as the completion of $C_c^\infty(\mathbb{R}^N)$ and $C_c^\infty(\overline{\mathbb{R}_+^{N+1}})$, respectively, under the norms

$$\|u\|_{\dot{H}^s(\mathbb{R}^N)} = \left(\int_{\mathbb{R}^N} |\xi|^{2s} |\mathcal{F}u(\xi)|^2 \, d\xi \right)^{\frac{1}{2}},$$

$$\|U\|_{X_0^s(\mathbb{R}_+^{N+1})} = \left(\iint_{\mathbb{R}_+^{N+1}} y^{1-2s} |\nabla U|^2 \, dx dy \right)^{\frac{1}{2}}.$$

If $u \in \dot{H}^s(\mathbb{R}^N)$, then $\mathrm{Ext}(u) \in X_0^s(\mathbb{R}_+^{N+1})$ and

$$\|\mathrm{Ext}(u)\|_{X_0^s(\mathbb{R}_+^{N+1})}^2 = \kappa_s \|u\|_{\dot{H}^s(\mathbb{R}^N)}^2.$$

Furthermore, the function $U = \mathrm{Ext}(u)$ is a weak solution to $\mathrm{div}(y^{1-2s}\nabla U) = 0$ in $\mathbb{R}_+^{N+1}$, and, as $\varepsilon \to 0$, we have

$$U(\cdot, \varepsilon) \to u \text{ in } \dot{H}^s(\mathbb{R}^N) \quad \text{and} \quad -\varepsilon^{1-2s} \frac{\partial U}{\partial y}(\cdot, \varepsilon) \to \kappa_s (-\Delta)^s u \text{ in } \dot{H}^{-s}(\mathbb{R}^N),$$

where $\dot{H}^{-s}(\mathbb{R}^N)$ is the dual of $\dot{H}^s(\mathbb{R}^N)$. Now, we recall (see [109, 198, 199]) that there exists a linear trace operator $\mathrm{Tr} : X_0^s(\mathbb{R}_+^{N+1}) \to \dot{H}^s(\mathbb{R}^N)$ such that

$$\sqrt{\kappa_s} \|\mathrm{Tr}(v)\|_{\dot{H}^s(\mathbb{R}^N)} \le \|v\|_{X_0^s(\mathbb{R}_+^{N+1})} \quad \text{for any } v \in X_0^s(\mathbb{R}_+^{N+1}), \tag{1.2.8}$$

and the equality is attained if and only if $v = \mathrm{Ext}(u)$ for some $u \in \dot{H}^s(\mathbb{R}^N)$. Combining Theorem 1.1.8 with (1.2.8), we derive the following Sobolev inequality:

$$\|\mathrm{Tr}(v)\|_{L^{2_s^*}(\mathbb{R}^N)} \le C \|v\|_{X_0^s(\mathbb{R}_+^{N+1})} \tag{1.2.9}$$

for any $v \in X_0^s(\mathbb{R}_+^{N+1})$. In what follows, for $v \in X_0^s(\mathbb{R}_+^{N+1})$, we denote $\mathrm{Tr}(v)$ by $v(\cdot, 0)$.

The extension technique is commonly used in the recent literature since it allows to transform a given nonlocal problem into a local one and use classical PDE tools and ideas that are available for these kind of problems. Further results about the extension method can be found in [31, 109, 124, 126, 131, 181, 198, 317].

Remark 1.2.9 As shown in [127], if $u \in \mathcal{S}(\mathbb{R}^N)$, a way to prove (1.2.7) is to use a partial Fourier transform with respect to $x \in \mathbb{R}^N$. Indeed, it suffices to verify that $\|U\|^2_{X_0^s(\mathbb{R}_+^{N+1})} = \kappa_s \|u\|^2_{\dot{H}^s(\mathbb{R}^N)}$. First, one observes that $\mathcal{F}U(\xi, y) = \mathcal{F}u(\xi)\theta(|\xi|y)$, where $\theta(y)$ solves the problem

$$\begin{cases} \theta'' + \frac{1-2s}{y}\theta' - \theta = 0 \text{ in } (0, \infty), \\ \theta(0) = 1, \\ \lim_{y \to \infty} \theta(y) = 0. \end{cases} \tag{1.2.10}$$

We note that (see [131, 181])

$$\theta(y) = \frac{2}{\Gamma(s)} \left(\frac{y}{2}\right)^s K_s(y)$$

and that $K_s(y)$ solves (see [78, 178, 336]) the following differential equation of second order:

$$y^2 K_s'' + y K_s' - (y^2 + s^2)K_s = 0.$$

Then we obtain

$$\iint_{\mathbb{R}_+^{N+1}} y^{1-2s}|\nabla U|^2 \, dx\, dy = \int_{\mathbb{R}^N} \int_0^\infty \left(|\xi|^2|\mathcal{F}U(\xi, y)|^2 + |\mathcal{F}U_y(\xi, y)|^2\right)y^{1-2s}\, d\xi\, dy$$

$$= \int_{\mathbb{R}^N} \int_0^\infty |\xi|^2|\mathcal{F}u(\xi)|^2\left(|\theta(|\xi|y)|^2 + |\theta'(|\xi|y)|^2\right)y^{1-2s}\, d\xi\, dy$$

$$= \left(\int_0^\infty y^{1-2s}\left(|\theta(y)|^2 + |\theta'(y)|^2\right) dy\right)\left(\int_{\mathbb{R}^N} |\xi|^{2s}|\mathcal{F}u(\xi)|^2\, d\xi\right).$$

Now, we observe that (see [78, 178, 336])

$$(y^s K_s(y))' = -y^s K_{s-1}(y) \quad \text{and} \quad K_{s-1}(y) = K_{1-s}(y)$$

yield $(y^s K_s(y))' = -y^s K_{1-s}(y)$. Moreover, by (1.2.4), $K_{1-s}(y) \sim \frac{\Gamma(1-s)}{2} \left(\frac{y}{2}\right)^{s-1}$ as $y \to 0$. Recalling the expression of $\theta(y)$, $\theta(0) = 1$, and the precise value of κ_s, it follows from $(y^{1-2s}\theta'(y))' = y^{1-2s}\theta(y)$ (by (1.2.10)) and an integration by parts that

$$\int_0^\infty y^{1-2s}\left(|\theta(y)|^2 + |\theta'(y)|^2\right) dy = -\lim_{y\to 0} y^{1-2s}\theta'(y) = \kappa_s. \tag{1.2.11}$$

1.3   Regularity Results and Maximum Principles

The following propositions obtained in [313] recall how the fractional Laplace operators interact with the Hölder norms. For more details see also [316].

Proposition 1.3.1 ([313]) *Let* $w = (-\Delta)^s u$. *Assume* $w \in C^{0,\alpha}(\mathbb{R}^N)$ *and* $u \in L^\infty(\mathbb{R}^N)$ *for* $\alpha \in (0, 1]$ *and* $s > 0$.

- *If* $\alpha + 2s \leq 1$, *then* $u \in C^{0,\alpha+2s}(\mathbb{R}^N)$. *Moreover,*

$$\|u\|_{C^{0,\alpha+2s}(\mathbb{R}^N)} \leq C\left(\|u\|_{L^\infty(\mathbb{R}^N)} + \|w\|_{C^{0,\alpha}(\mathbb{R}^N)}\right)$$

  for a constant C that depends only on n, α, and s.
- *If* $\alpha + 2s > 1$, *then* $u \in C^{1,\alpha+2s-1}(\mathbb{R}^N)$. *Moreover,*

$$\|u\|_{C^{1,\alpha+2s-1}(\mathbb{R}^N)} \leq C\left(\|u\|_{L^\infty(\mathbb{R}^N)} + \|w\|_{C^{0,\alpha}(\mathbb{R}^N)}\right)$$

  for a constant C that depends only on n, α, and s.

Proposition 1.3.2 ([313]) *Let* $w = (-\Delta)^s u$. *Assume* $w \in L^\infty(\mathbb{R}^N)$ *and* $u \in L^\infty(\mathbb{R}^N)$ *for* $s > 0$.

- *If* $2s \leq 1$, *then* $u \in C^{0,\alpha}(\mathbb{R}^N)$ *for any* $\alpha < 2s$. *Moreover,*

$$\|u\|_{C^{0,\alpha}(\mathbb{R}^N)} \leq C\left(\|u\|_{L^\infty(\mathbb{R}^N)} + \|w\|_{L^\infty(\mathbb{R}^N)}\right)$$

  for a constant C depending only on n, α, and s.
- *If* $2s > 1$, *then* $u \in C^{1,\alpha}(\mathbb{R}^N)$ *for any* $\alpha < 2s - 1$. *Moreover,*

$$\|u\|_{C^{1,\alpha}(\mathbb{R}^N)} \leq C\left(\|u\|_{L^\infty(\mathbb{R}^N)} + \|w\|_{L^\infty(\mathbb{R}^N)}\right)$$

  for a constant C depending only on n, α, and s.

Next, we collect some maximum principles for $(-\Delta)^s$. For more details we refer to [106, 163, 222, 228, 282, 313]. We start by recalling the following maximum principle for distributional supersolutions.

Proposition 1.3.3 ([313]) *Let $\Omega \Subset \mathbb{R}^N$ be an open set, and let $u \in \mathcal{L}_s(\mathbb{R}^N)$ be a lower-semicontinuous function in $\bar{\Omega}$ such that $(-\Delta)^s u \geq 0$ in Ω and $u \geq 0$ in $\mathbb{R}^N \setminus \Omega$. Then $u \geq 0$ in $\mathbb{R}^N$. Moreover, if $u(x) = 0$ for one point x inside Ω, then $u \equiv 0$ in all $\mathbb{R}^N$.*

Next we introduce several definitions. Let Ω be an open set in $\mathbb{R}^N$. Set

$$\widetilde{H}^s(\Omega) = \{u \in W^{s,2}(\Omega) : \tilde{u} \in W^{s,2}(\mathbb{R}^N)\},$$

where $\tilde{u}$ is the extension of u by zero outside Ω. Denote by $\widetilde{H}^{-s}(\Omega)$ the dual space of $\widetilde{H}^s(\Omega)$, and denote by $\langle \cdot, \cdot \rangle$ the duality pairing between $\widetilde{H}^s(\Omega)$ and $\widetilde{H}^{-s}(\Omega)$. Let $\mathcal{H}^s(\Omega)$ be the space of functions $u \in L^2_{\text{loc}}(\mathbb{R}^N)$ such that for any bounded $\Omega' \subseteq \Omega$ there is an open set $U \supseteq \Omega'$ so that $u \in W^{s,2}(U)$, and

$$\int_{\mathbb{R}^N} \frac{|u(x)|}{(1+|x|)^{N+2s}} \, dx < \infty.$$

When Ω is bounded, then $u \in \mathcal{H}^s(\Omega)$ if and only if there is an open set $U \supseteq \Omega$ such that $u \in W^{s,2}(U)$ and

$$\int_{\mathbb{R}^N} \frac{|u(x)|}{(1+|x|)^{N+2s}} \, dx < \infty.$$

Given $f \in \widetilde{H}^{-s}(\Omega)$, we say that $f \geq (\leq)0$ if for any $\phi \in \widetilde{H}^s(\Omega)$, $\phi \geq 0$, it holds that $\langle f, \phi \rangle \geq (\leq)0$. When Ω is bounded, we say that $u \in \mathcal{H}^s(\Omega)$ is a weak supersolution (subsolution) of $(-\Delta)^s u = f$ in Ω if

$$\langle u, \phi \rangle_{\mathcal{D}^{s,2}(\mathbb{R}^N)} \geq (\leq)\langle f, \phi \rangle$$

for all $\phi \in \widetilde{H}^s(\Omega)$, $\phi \geq 0$.

When Ω is unbounded, we say that $u \in \mathcal{H}^s(\Omega)$ is a weak supersolution (subsolution) of $(-\Delta)^s u = f$ in Ω if u is a weak supersolution (subsolution) of $(-\Delta)^s u = f$ in Ω' for every bounded set $\Omega' \subset \Omega$. In both cases, u is a weak solution of $(-\Delta)^s u = f$ in Ω if u is a supersolution and subsolution of $(-\Delta)^s u = f$ in Ω.

Given $c \in L^1_{\text{loc}}(\Omega)$, we say that $u \in \mathcal{H}^s(\Omega)$ is a weak supersolution (subsolution) of $(-\Delta)^s u = c(x)u$ in Ω if $f = c(x)u \in \widetilde{H}^{-s}(\Omega)$ and u is a weak supersolution (subsolution) of $(-\Delta)^s u = f$ in Ω. Finally, $u \in \mathcal{H}^s(\Omega)$ is a weak solution of $(-\Delta)^s u = c(x)u$ in Ω if u is both a supersolution and a subsolution of $(-\Delta)^s u = c(x)u$ in Ω.

Theorem 1.3.4 ([163] Strong Maximum Principle (Variant 1)) *Let $c \in L^1_{loc}(\Omega)$ be a non-positive function and $u \in \mathcal{H}^s(\Omega)$ be a weak supersolution of*

$$(-\Delta)^s u = c(x)u \quad in \; \Omega. \tag{1.3.1}$$

- *If Ω is bounded and $u \geq 0$ a.e. in $\mathbb{R}^N \setminus \Omega$, then either $u > 0$ a.e. in Ω, or $u = 0$ a.e. in $\mathbb{R}^N$.*
- *If $u \geq 0$ a.e. in $\mathbb{R}^N$, then either $u > 0$ a.e. in Ω, or $u = 0$ a.e. in $\mathbb{R}^N$.*

Theorem 1.3.5 ([163] Strong Maximum Principle (Variant 2)) *Let $c \in C(\overline{\Omega})$ be a non-positive function and $u \in \mathcal{H}^s(\Omega) \cap C(\overline{\Omega})$ be a weak supersolution of (1.3.1).*

- *If Ω is bounded and $u \geq 0$ a.e. in $\mathbb{R}^N \setminus \Omega$, then either $u > 0$ in Ω, or $u = 0$ a.e. in $\mathbb{R}^N$.*
- *If $u \geq 0$ a.e. in $\mathbb{R}^N$, then either $u > 0$ in Ω, or $u = 0$ a.e. in $\mathbb{R}^N$.*

Remark 1.3.6 In the case where $u \geq 0$ a.e. in $\mathbb{R}^N$, the non-positivity assumption on the function $c(x)$ in the previous theorems is not necessary.

Theorem 1.3.7 ([163] Hopf Lemma) *Let Ω satisfy the interior ball condition in $x_0 \in \partial\Omega$, namely there exists a ball $B_R \subseteq \Omega$ such that $x_0 \in \partial B_R$. Let $c \in C(\overline{\Omega})$ and $u \in \mathcal{H}^s(\Omega) \cap C(\overline{\Omega})$ be a weak supersolution of (1.3.1).*

- *If Ω is bounded, $c(x) \leq 0$ in Ω and $u \geq 0$ a.e. in $\mathbb{R}^N \setminus \Omega$, then either $u = 0$ a.e. in $\mathbb{R}^N$, or*

$$\liminf_{x \in B_R, x \to x_0} \frac{u(x)}{\delta_R(x)^s} > 0 \tag{1.3.2}$$

  where $B_R \subseteq \Omega$ and $x_0 \in \partial B_R$ and $\delta_R(x) = \mathrm{dist}(x, \mathbb{R}^N \setminus B_R)$.
- *If $u \geq 0$ a.e. in $\mathbb{R}^N$, then either $u = 0$ a.e. in $\mathbb{R}^N$, or (1.3.2) holds.*

Lemma 1.3.8 (Comparison Principle [164]) *Let $\Omega \subset \mathbb{R}^N$ be an open set, $\mu \in \mathbb{R}_+$, $u_1, u_2 \in H^s(\mathbb{R}^N)$ such that $u_1 \leq u_2$ in $\mathbb{R}^N \setminus \Omega$ and*

$$(-\Delta)^s u_1 + \mu u_1 \leq (-\Delta)^s u_2 + \mu u_2 \quad in \; \Omega,$$

that is

$$\iint_{\mathbb{R}^{2N}} \frac{(u_1(x) - u_1(y))(\phi(x) - \phi(y))}{|x - y|^{N+2s}} \, dxdy + \int_{\mathbb{R}^N} \mu u_1 \phi \, dx$$

$$\leq \iint_{\mathbb{R}^{2N}} \frac{(u_2(x) - u_2(y))(\phi(x) - \phi(y))}{|x - y|^{N+2s}} \, dxdy + \int_{\mathbb{R}^N} \mu u_2 \phi \, dx$$

for all $\phi \in \widetilde{H}^s(\Omega)$, $\phi \geq 0$. Then $u_1 \leq u_2$ in Ω.

Due to the strong connection between the fractional Laplacian and degenerate elliptic problems with Neumann boundary conditions, we also recall some useful regularity results and maximum principle established in [124, 237] (see also [180]). First we introduce some notations. We let $|x| = \sqrt{\sum_{i=1}^N x_i^2}$ denote the euclidean norm of $x = (x_1, \ldots, x_N) \in \mathbb{R}^N$ and let $|(x, y)| = \sqrt{|x|^2 + y^2}$ denote the euclidean norm of $(x, y) \in \mathbb{R}_+^{N+1}$. Let $D \subset \mathbb{R}^{N+1}$ be a bounded domain, that is, a bounded connected open set, with boundary ∂D. We denote by $\partial'D$ the interior of $\overline{D} \cap \partial \mathbb{R}_+^{N+1}$ in $\mathbb{R}^N$, and we set $\partial''D = \partial D \setminus \partial'D$. For $R > 0$, we set

$$B_R^+ = \{(x, y) \in \mathbb{R}_+^{N+1} : |(x, y)| < R\},$$

$$\Gamma_R^0 = \partial'B_R^+ = \{(x, 0) \in \partial \mathbb{R}_+^{N+1} : |x| < R\},$$

$$\Gamma_R^+ = \partial''B_R^+ = \{(x, y) \in \mathbb{R}^{N+1} : y \geq 0, |(x, y)| = R\}.$$

Now, we introduce the weighted Lebesgue space (see [180, 237, 328] for more details). Let $D \subset \mathbb{R}_+^{N+1}$ be an open set, $s \in (0, 1)$ and $r \in [1, \infty)$. Denote by $L^r(D, y^{1-2s})$ the weighted Lebesgue space of all measurable functions $v : D \to \mathbb{R}$ such that

$$\|v\|_{L^r(D, y^{1-2s})} = \left(\iint_D y^{1-2s} |v|^r \, dx dy \right)^{\frac{1}{r}} < \infty.$$

We say that $v \in H^1(D, y^{1-2s})$ if $v \in L^2(D, y^{1-2s})$ and its weak derivatives, collectively denoted by ∇v, exist and belong to $L^2(D, y^{1-2s})$. The norm of v in $H^1(D, y^{1-2s})$ is given by

$$\|v\|_{H^1(D, y^{1-2s})} = \left(\iint_D y^{1-2s} (|\nabla v|^2 + v^2) \, dx dy \right)^{\frac{1}{2}} < \infty.$$

It is clear that $H^1(D, y^{1-2s})$ is a Hilbert space with the inner product

$$\iint_D y^{1-2s} (\nabla v \cdot \nabla w + vw) \, dx dy.$$

We note that when $D = \mathbb{R}_+^{N+1}$ then we have the following weighted Sobolev embeddings.

Lemma 1.3.9 ([170, 181])

(i) *There exists a constant $\hat{S} > 0$ such that*

$$\left(\iint_{\mathbb{R}^{N+1}_+} y^{1-2s} |u|^{2\gamma}\, dxdy\right)^{\frac{1}{2\gamma}} \leq \hat{S} \left(\iint_{\mathbb{R}^{N+1}_+} y^{1-2s} |\nabla u|^2\, dxdy\right)^{\frac{1}{2}}$$

for all $u \in X_0^s(\mathbb{R}^{N+1}_+)$, where $\gamma = 1 + \dfrac{2}{N-2s}$.

(ii) *Let $R > 0$ and let $\mathcal{T}$ be a subset of $X_0^s(\mathbb{R}^{N+1}_+)$ such that*

$$\sup_{u \in \mathcal{T}} \iint_{\mathbb{R}^{N+1}_+} y^{1-2s} |\nabla u|^2\, dxdy < \infty.$$

Then $\mathcal{T}$ is pre-compact in $L^2(B_R^+, y^{1-2s})$.

Now we give the following definitions.

Definition 1.3.10 Let $D \subset \mathbb{R}^{N+1}_+$ be a bounded domain, with $\partial' D \neq \emptyset$. Let $a \in L_{loc}^{\frac{2N}{N+2s}}(\partial' D)$ and $b \in L^1_{loc}(\partial' D)$. Consider the problem

$$\begin{cases} -\text{div}(y^{1-2s}\nabla v) = 0 & \text{in } D, \\ \dfrac{\partial v}{\partial v^{1-2s}} = a(x)v + b(x) & \text{on } \partial' D. \end{cases} \tag{1.3.3}$$

We say that $v \in H^1(D, y^{1-2s})$ is a weak supersolution (resp. subsolution) to (1.3.3) in D if for any nonnegative $\varphi \in C_c^\infty(D \cup \partial' D)$,

$$\iint_D y^{1-2s}\, \nabla v(x, y) \cdot \nabla\phi(x, y)\, dxdy \geq (\leq) \int_{\partial' D} [a(x)v(x, 0) + b(x)]\varphi(x, 0)\, dx.$$

We say that $v \in H^1(D, y^{1-2s})$ is a weak solution to (1.3.3) in D if it is both a weak supersolution and a weak subsolution.

Set $Q_R = B_R \times (0, R) \subset \mathbb{R}^{N+1}_+$, where $B_R \subset \mathbb{R}^N$ is the ball in $\mathbb{R}^N$ of radius R centered at 0. We recall the following De Giorgi–Nash–Moser type theorems.

Proposition 1.3.11 ([237]) *Let $a, b \in L^q(B_1)$ for some $q > \dfrac{N}{2s}$.*

(i) *Let $v \in H^1(Q_1, y^{1-2s})$ be a weak subsolution to (1.3.3) in Q_1. Then for all $v > 0$,*

$$\sup_{Q_{1/2}} v^+ \leq C\left(\|v^+\|_{L^v(Q_1, y^{1-2s})} + \|b\|_{L^q(B_1)}\right),$$

where $C > 0$ depends only on N, s, q, v and $\|a^+\|_{L^q(B_1)}$.

(ii) *(Weak Harnack inequality) Let $v \in H^1(Q_1, y^{1-2s})$ be a nonnegative weak superso-lution to (1.3.3) in Q_1. Then for any $0 < \mu < \tau < 1$ and $0 < v \leq \frac{N+1}{N}$,*

$$\inf_{Q_\mu} v + \|b_-\|_{L^q(B_1)} \geq C\|v\|_{L^v(Q_\tau, y^{1-2s})},$$

where $C > 0$ depends only on N, s, q, v, μ, τ and $\|a_-\|_{L^q(B_1)}$.

(iii) *Let $v \in H^1(Q_1, y^{1-2s})$ be a non-negative weak solution to (1.3.3) in Q_1. Then*

$$\sup_{Q_{1/2}} v \leq C\left(\inf_{Q_{1/2}} v + \|b\|_{L^q(B_1)}\right),$$

where $C > 0$ depends only on N, s, q, $\|a\|_{L^q(B_1)}$. Consequently, there exists $\alpha \in (0, 1)$ depending only on N, s, q, $\|a\|_{L^q(B_1)}$ such that any weak solution v of (1.3.3) is of class $C^{0,\alpha}(\overline{Q_{1/2}})$. Moreover,

$$\|v\|_{C^{0,\alpha}(\overline{Q_{1/2}})} \leq C\left(\|v\|_{L^\infty(Q_1)} + \|b\|_{L^q(B_1)}\right),$$

where $C > 0$ depends only on N, s, q, $\|a\|_{L^q(B_1)}$.

The next results are very useful for obtaining local Schauder estimates for non-negative solutions of fractional Laplace equations.

Proposition 1.3.12 ([237]) *Let $a, b \in C^k(B_1)$ for some positive integer $k \geq 1$. Let $v \in H^1(Q_1, y^{1-2s})$ be a weak solution to (1.3.3) in Q_1. Then we have*

$$\sum_{i=0}^{k} \|\nabla_x^i v\|_{L^\infty(Q_{1/2})} \leq C\left(\|v\|_{L^2(Q_1, y^{1-2s})} + \|b\|_{C^k(B_1)}\right),$$

where $C > 0$ depends only on N, s, k, $\|a\|_{C^k(B_1)}$.

Lemma 1.3.13 ([237] (Schauder Estimate)) *Let $v \in H^1(Q_1, y^{1-2s})$ be a weak solution to (1.3.3), where $a, b \in C^\alpha(B_1)$ for some $0 < \alpha \notin \mathbb{N}$. If $\alpha + 2s$ is not an integer, then $v(\cdot, 0) \in C^{\alpha+2s}(B_{1/2})$. Moreover, we have*

$$\|v(\cdot, 0)\|_{C^{\alpha+2s}(B_{1/2})} \leq C(\|v\|_{L^\infty(Q_1)} + \|b\|_{C^\alpha(B_1)}),$$

where $C > 0$ depends only on N, s, α, $\|a\|_{C^\alpha(B_1)}$. Here, for simplicity, we used the symbol $C^\alpha(\Omega)$ to denote the standard Hölder space $C^{0,\alpha}(\Omega)$ over a bounded domain Ω whenever $\alpha \in (0, 1)$, and the space $C^{[\alpha],\alpha-[\alpha]}(\Omega)$ when $1 < \alpha \notin \mathbb{N}$.

Let Ω be a bounded domain in $\mathbb{R}^N$, $a \in L_{\text{loc}}^{\frac{2N}{N+2s}}(\Omega)$ and $b \in L_{loc}^1(\Omega)$. We say $u \in \mathcal{D}^{s,2}(\mathbb{R}^N)$ is a weak solution of

$$(-\Delta)^s u = a(x)u + b(x) \quad \text{in } \Omega, \tag{1.3.4}$$

if for every non-negative $v \in C^\infty(\mathbb{R}^N)$ supported in Ω,

$$\langle u, v \rangle_{\mathcal{D}^{s,2}(\mathbb{R}^N)} = \int_\Omega [a(x)u(x) + b(x)]v(x)\,dx.$$

Then $u \in \mathcal{D}^{s,2}(\mathbb{R}^N)$ is a weak solution of

$$(-\Delta)^s u = \frac{1}{\kappa_s}(a(x)u + b(x)) \text{ in } B_1,$$

if and only if the s-harmonic extension $U(x, y) = P_s(x, y) * u(x)$ of u is a weak solution of (1.3.3) in Q_1.

Theorem 1.3.14 ([237]) *Let* $a, b \in C^\alpha(B_1)$ *with* $0 < \alpha \notin \mathbb{N}$. *Let* $u \in \mathcal{D}^{s,2}(\mathbb{R}^N)$ *and* $u \geq 0$ *in* $\mathbb{R}^N$ *be a weak solution of*

$$(-\Delta)^s u = a(x)u + b(x) \text{ in } B_1.$$

Suppose that $2s + \alpha$ *is not an integer. Then* $u \in C^{2s+\alpha}(B_{1/2})$. *Moreover,*

$$\|u\|_{C^{2s+\alpha}(B_{1/2})} \leq C\left(\inf_{B_{3/4}} u + \|b\|_{C^\alpha(B_{3/4})}\right),$$

where $C > 0$ *depends only on* N, s, α, $\|a\|_{C^\alpha(B_{3/4})}$.

Finally, we collect some maximum principles for (1.3.3). The following (weak) maximum principle holds true.

Lemma 1.3.15 ((Weak Maximum Principle)[124]) *Let* $v \in H^1(B_R^+, y^{1-2s})$ *be a weak supersolution to (1.3.3) in* B_R^+ *with* $a \equiv b \equiv 0$. *If* $v \geq 0$ *on* Γ_R^+ *in the trace sense, then* $v \geq 0$ *in* B_R^+.

Remark 1.3.16 In addition, one has the strong maximum principle: either $v \equiv 0$, or $v > 0$ in $B_R^+ \cup \Gamma_R^0$. That v cannot vanish at an interior point follows from the classical strong maximum principle for strictly elliptic operators. That v cannot vanish at a point in Γ_R^0 follows from the Hopf principle below or by the strong maximum principle of [180]. Note that the same weak and strong maximum principles hold in other bounded domains of $\mathbb{R}^{N+1}_+$ different than B^+_R. It also holds for the Dirichlet problem in B^+_R.

The next lemma provides estimates for solutions of the Neumann problem in a half-ball.

Lemma 1.3.17 ([124]) *Let $g \in C^{0,\gamma}(\Gamma^0_{2R})$ for some $\gamma \in (0,1)$. Let $v \in H^1(B^+_{2R}, y^{1-2s}) \cap L^\infty(B^+_{2R})$ be a weak solution to*

$$\begin{cases} -\mathrm{div}(y^{1-2s}\nabla v) = 0 & \text{in } B^+_{2R}, \\ \dfrac{\partial v}{\partial\nu^{1-2s}} = g & \text{on } \Gamma^0_{2R}. \end{cases}$$

Then there exists $\alpha \in (0,1)$ depending only on N, s, γ such that $v \in C^{0,\alpha}(\overline{B^+_R})$ and $y^{1-2s}\frac{\partial v}{\partial y} \in C^{0,\alpha}(\overline{B^+_R})$. Moreover, there exists positive constants C^1_R and C^2_R depending only on N, s, R, $\|u\|_{L^\infty(B^+_{2R})}$ and also on $\|g\|_{L^\infty(\Gamma^0_{2R})}$ (for C^1_R) and $\|g\|_{C^{0,\gamma}(\Gamma^0_{2R})}$ (for C^2_R) such that

$$\|v\|_{C^{0,\alpha}(\overline{B^+_R})} \leq C^1_R$$

and

$$\left\| y^{1-2s}\frac{\partial v}{\partial y} \right\|_{C^{0,\alpha}(\overline{B^+_R})} \leq C^2_R.$$

The following proposition provides a Hopf boundary lemma.

Proposition 1.3.18 ([124] Hopf Boundary Lemma) *Let $\mathcal{C}_{R,1} = B_R \times (0,1) \subset \mathbb{R}^{N+1}_+$ and let $v \in H^1(\mathcal{C}_{R,1}, y^{1-2s}) \cap C(\overline{\mathcal{C}_{R,1}})$ be a weak solution to*

$$\begin{cases} -\mathrm{div}(y^{1-2s}\nabla v) \geq 0 & \text{in } \mathcal{C}_{R,1}, \\ v > 0 & \text{in } \mathcal{C}_{R,1}, \\ v(0,0) = 0. \end{cases}$$

Then,

$$\limsup_{y\to 0} -y^{1-2s}\frac{v(0,y)}{y} < 0.$$

In addition, if $y^{1-2s}\frac{\partial v}{\partial y} \in C(\overline{\mathcal{C}_{R,1}})$, then

$$\frac{\partial v}{\partial\nu^{1-2s}}(0,0) < 0.$$

Corollary 1.3.19 ([124]) *Let d be a Hölder continuous function in Γ_R^0 and $v \in L^\infty(B_R^+) \cap H^1(B_R^+, y^{1-2s})$ be a weak solution to*

$$
\begin{cases}
-\operatorname{div}(y^{1-2s}\nabla v) = 0 & \text{in } B_R^+, \\
v \geq 0 & \text{on } B_R^+, \\
\dfrac{\partial v}{\partial v^{1-2s}} + d(x)v = 0 & \text{on } \Gamma_R^0.
\end{cases}
$$

Then, $v > 0$ in $B_R^+ \cup \Gamma_R^0$ unless $v \equiv 0$ in B_R^+.

1.4   Some Useful Lemmas

We recall some useful technical lemmas which will be frequently used later. We start with two well-known results that can be found in [100].

Lemma 1.4.1 ([100] Radial Lemma) *Let $u \in L^t(\mathbb{R}^N)$, $1 \leq t < \infty$ be a non-negative radially decreasing function (that is, $0 \leq u(x) \leq u(y)$ if $|x| \geq |y|$). Then*

$$
|u(x)| \leq \left(\frac{N}{\omega_{N-1}}\right)^{\frac{1}{t}} |x|^{-\frac{N}{t}} \|u\|_{L^t(\mathbb{R}^N)}, \quad \text{for any } x \in \mathbb{R}^N \setminus \{0\}, \tag{1.4.1}
$$

where ω_{N-1} is the surface area of the unit sphere $\mathbb{S}^{N-1} = \{x \in \mathbb{R}^N : |x| = 1\}$.

Lemma 1.4.2 ([83,100,318] The Compactness Lemma of Strauss) *Let P and $Q : \mathbb{R} \to \mathbb{R}$ be two continuous functions satisfying*

$$
\lim_{|t|\to\infty} \frac{P(t)}{Q(t)} = 0.
$$

Let (v_n), v and w be real measurable functions on $\mathbb{R}^N$, with w bounded, such that

$$
\sup_{n\in\mathbb{N}} \int_{\mathbb{R}^N} |Q(v_n(x))w(x)|\, dx < \infty,
$$

$$
P(v_n(x)) \to v(x) \quad \text{a.e. } x \in \mathbb{R}^N, \text{ as } n \to \infty.
$$

Then $\|(P(v_n) - v)w\|_{L^1(B)} \to 0$ as $n \to \infty$, for any bounded Borel set $B \subset \mathbb{R}^N$.
Moreover, if we also have

$$
\lim_{t\to 0} \frac{P(t)}{Q(t)} = 0,
$$

and

$$\lim_{\substack{|x|\to\infty}} \sup_{n\in\mathbb{N}} |v_n(x)| = 0,$$

then $\|(P(v_n) - v)w\|_{L^1(\mathbb{R}^N)} \to 0$ *as* $n \to \infty$.

The next result will be useful for obtaining compactness when the property of uniform decay at infinity, such as in the Strauss type radial lemma, is not clear.

Lemma 1.4.3 ([139]) *Let* $(X, \|\cdot\|)$ *be a Banach space such that* X *is embedded continuously and compactly in* $L^q(\mathbb{R}^N)$ *for* $q \in [q_1, q_2]$ *and* $q \in (q_1, q_2)$, *respectively, where* $q_1, q_2 \in (0, \infty)$. *Let* $(u_n) \subset X$, *let* $u : \mathbb{R}^N \to \mathbb{R}$ *be a measurable function and let* $P \in C(\mathbb{R}, \mathbb{R})$. *Assume that*

(i) $\lim_{|t|\to 0} \dfrac{P(t)}{|t|^{q_1}} = 0,$

(ii) $\lim_{|t|\to\infty} \dfrac{P(t)}{|t|^{q_2}} = 0,$

(iii) $\sup_{n\in\mathbb{N}} \|u_n\| < \infty,$

(iv) $\lim_{n\to\infty} P(u_n(x)) = u(x)$ *for a.e.* $x \in \mathbb{R}^N$.

Then, up to a subsequence,

$$\lim_{n\to\infty} \|P(u_n) - u\|_{L^1(\mathbb{R}^N)} = 0.$$

We recall the following version of vanishing Lions-type result (see [256]) for $H^s(\mathbb{R}^N)$.

Lemma 1.4.4 *Let* $N > 2s$ *and* $r \in [2, 2_s^*)$. *If* (u_n) *is a bounded sequence in* $H^s(\mathbb{R}^N)$ *and if*

$$\lim_{n\to\infty} \sup_{y\in\mathbb{R}^N} \int_{B_R(y)} |u_n|^r \, dx = 0 \tag{1.4.2}$$

for some $R > 0$, *then* $u_n \to 0$ *in* $L^\tau(\mathbb{R}^N)$ *for all* $\tau \in (r, 2_s^*)$.

Proof Let $\tau \in (r, 2_s^*)$. Given $y \in \mathbb{R}^N$, the Hölder inequality shows that

$$\|u_n\|_{L^\tau(B_R(y))} \leq \|u_n\|_{L^r(B_R(y))}^{1-\alpha} \|u_n\|_{L^{2_s^*}(B_R(y))}^{\alpha} \qquad \text{for all } n \in \mathbb{N},$$

where

$$\frac{1}{t} = \frac{1-\alpha}{r} + \frac{\alpha}{2_s^*}.$$

Now, covering $\mathbb{R}^N$ by balls of radius R, in such a way that each point of $\mathbb{R}^N$ is contained in at most $N + 1$ balls, and applying Theorem 1.1.8, we find that

$$\|u_n\|_{L^\tau(\mathbb{R}^N)}^\tau \leq (N+1) \sup_{y \in \mathbb{R}^N} \left(\int_{B_R(y)} |u_n|^r \, dx \right)^{(1-\alpha)\tau} \|u_n\|_{L^{2_s^*}(\mathbb{R}^N)}^{\alpha\tau}$$

$$\leq C \sup_{y \in \mathbb{R}^N} \left(\int_{B_R(y)} |u_n|^r \, dx \right)^{(1-\alpha)\tau} \qquad \text{for all } n \in \mathbb{N}.$$

Using (1.4.2), we conclude that $u_n \to 0$ in $L^\tau(\mathbb{R}^N)$. $\square$

In what follows we give some technical results whose proofs can be obtained arguing as in [40, 47, 196, 344]. However, we present different proofs.

Lemma 1.4.5 *Let* $(u_n) \subset \mathcal{D}^{s,2}(\mathbb{R}^N)$ *be a bounded sequence in* $\mathcal{D}^{s,2}(\mathbb{R}^N)$, *and let* $\eta \in C^\infty(\mathbb{R}^N)$ *be a function such that* $0 \leq \eta \leq 1$ *in* $\mathbb{R}^N$, $\eta = 0$ *in* B_1 *and* $\eta = 1$ *in* $\mathbb{R}^N \setminus B_2$. *Set* $\eta_R(x) = \eta(\frac{x}{R})$ *for all* $R > 0$. *Then*

$$\lim_{R \to \infty} \limsup_{n \to \infty} \iint_{\mathbb{R}^{2N}} |u_n(x)|^2 \frac{|\eta_R(x) - \eta_R(y)|^2}{|x-y|^{N+2s}} \, dx \, dy = 0.$$

Proof Fix $R > 0$. We start by proving that

$$\limsup_{n \to \infty} \iint_{\mathbb{R}^{2N}} |u_n(x)|^2 \frac{|\eta_R(x) - \eta_R(y)|^2}{|x-y|^{N+2s}} \, dx \, dy \leq \iint_{\mathbb{R}^{2N}} |u(x)|^2 \frac{|\eta_R(x) - \eta_R(y)|^2}{|x-y|^{N+2s}} \, dx \, dy.$$

$$(1.4.3)$$

Set $\psi_R = 1 - \eta_R$ and note that $\psi_R \in C_c^\infty(\mathbb{R}^N)$, $0 \leq \psi_R \leq 1$, $\psi_R = 1$ in B_R and $\psi_R = 0$ in $\mathbb{R}^N \setminus B_{2R}$. Define

$$W_R(x) = \int_{\mathbb{R}^N} \frac{|\eta_R(x) - \eta_R(y)|^2}{|x-y|^{N+2s}} \, dy = \int_{\mathbb{R}^N} \frac{|\psi_R(x) - \psi_R(y)|^2}{|x-y|^{N+2s}} \, dy.$$

We observe that $W_R \in L^\infty(\mathbb{R}^N)$, because

$$W_R(x) = \int_{|y-x| \leq R} \frac{|\psi_R(x) - \psi_R(y)|^2}{|x-y|^{N+2s}} \, dy + \int_{|y-x| > R} \frac{|\psi_R(x) - \psi_R(y)|^2}{|x-y|^{N+2s}} \, dy$$

$$\leq R^{-2} \|\nabla \eta\|^2_{L^\infty(\mathbb{R}^N)} \int_{|y-x| \leq R} \frac{1}{|x-y|^{N+2s-2}} \, dy + 4 \int_{|y-x| > R} \frac{1}{|x-y|^{N+2s}} \, dy$$

$$= R^{-2} \|\nabla \eta\|^2_{L^\infty(\mathbb{R}^N)} \omega_{N-1} \int_0^R \frac{1}{r^{2s-1}} \, dr + 4\omega_{N-1} \int_R^\infty \frac{1}{r^{2s+1}} \, dr$$

$$\leq \frac{C_1}{R^{2s}}. \tag{1.4.4}$$

On the other hand, for all $|x| > 4R$, we have

$$W_R(x) = \int_{|y| \leq 2R} \frac{|\psi_R(y)|^2}{|x-y|^{N+2s}} \, dy \leq \frac{2^{N+2s}|B_{2R}|}{|x|^{N+2s}} = \frac{C_2 R^N}{|x|^{N+2s}}, \tag{1.4.5}$$

where we used that $\psi_R = 0$ in $\mathbb{R}^N \setminus B_{2R}$, $0 \leq \psi_R \leq 1$, and

$$|x-y| \geq |x| - |y| \geq |x| - 2R \geq \frac{|x|}{2} \quad \text{for } |x| > 4R, \ |y| \leq 2R.$$

We note that the constants C_1 and C_2 in (1.4.4) and (1.4.5) are independent of R. Take $M > 4R$. Using (1.4.5) and the boundedness of (u_n) in $L^{2^*_s}(\mathbb{R}^N)$, and applying the Hölder inequality, we see that

$$\int_{\mathbb{R}^N} |u_n(x)|^2 W_R(x) \, dx = \int_{|x| \leq M} |u_n(x)|^2 W_R(x) \, dx + \int_{|x| > M} |u_n(x)|^2 W_R(x) \, dx$$

$$\leq \int_{|x| \leq M} |u_n(x)|^2 W_R(x) \, dx + \|u_n\|^2_{L^{2^*_s}(\mathbb{R}^N)} \left(\int_{|x| > M} |W_R(x)|^{\frac{2^*_s}{2^*_s-2}} \, dx \right)^{\frac{2^*_s-2}{2^*_s}}$$

$$\leq \int_{|x| \leq M} |u_n(x)|^2 W_R(x) \, dx + C_3 R^N \left(\int_{|x| > M} \frac{1}{|x|^{\frac{N2}{2s}+N}} \, dx \right)^{\frac{2^*_s-2}{2^*_s}}.$$

Since $u_n \to u$ in $L^2_{loc}(\mathbb{R}^N)$ and $W_R \in L^\infty(\mathbb{R}^N)$, we have

$$\lim_{n \to \infty} \int_{|x| \leq M} |u_n(x)|^2 W_R(x) \, dx = \int_{|x| \leq M} |u(x)|^2 W_R(x) \, dx,$$

and consequently

$$\limsup_{n\to\infty} \int_{\mathbb{R}^N} |u_n(x)|^2 W_R(x)\, dx$$

$$\leq \int_{|x|\leq M} |u(x)|^2 W_R(x)\, dx + C_3 R^N \left(\int_{|x|>M} \frac{1}{|x|^{\frac{N^2}{2s}+N}}\, dx \right)^{\frac{2_s^*-2}{2_s^*}} \quad \forall M > 4R.$$

Letting here $M \to \infty$, we obtain

$$\limsup_{n\to\infty} \int\!\!\int_{\mathbb{R}^{2N}} |u_n(x)|^2 \frac{|\eta_R(x) - \eta_R(y)|^2}{|x-y|^{N+2s}}\, dx\, dy \leq \int_{\mathbb{R}^N} |u(x)|^2 W_R(x)\, dx,$$

that is, (1.4.3) holds true. Since (1.4.4) and (1.4.5) imply that

$$W_R(x) \leq C_1 R^{-2s} \text{ for all } |x| \leq 4R, \quad W_R(x) \leq C_2(R|x|^{-1})^N |x|^{-2s} \text{ for all } |x| > 4R,$$

we deduce that, for some $\bar{C} > 0$ independent of R,

$$0 \leq W_R(x) \leq \bar{C}|x|^{-2s} \quad \text{for all } x \neq 0.$$

This fact combined with the fractional Hardy–Sobolev inequality (1.1.2) yields

$$|u(x)|^2 W_R(x) \leq \bar{C} \frac{|u(x)|^2}{|x|^{2s}} \in L^1(\mathbb{R}^N).$$

On the other hand, (1.4.4) implies feat $W_R \to 0$ as $R \to \infty$. Then, by the dominated convergence theorem,

$$\limsup_{R\to\infty} \limsup_{n\to\infty} \int\!\!\int_{\mathbb{R}^{2N}} |u_n(x)|^2 \frac{|\eta_R(x) - \eta_R(y)|^2}{|x-y|^{N+2s}}\, dx\, dy \leq \lim_{R\to\infty} \int_{\mathbb{R}^N} |u(x)|^2 W_R(x)\, dx = 0.$$

$$\square$$

Remark 1.4.6 If $(u_n) \subset H^s(\mathbb{R}^N)$, then the proof of Lemma 1.4.5 can be simplified. Indeed, using the boundedness of (u_n) in $L^2(\mathbb{R}^N)$ and polar coordinates, we have

$$\int\!\!\int_{\mathbb{R}^{2N}} |u_n(x)|^2 \frac{|\eta_R(x) - \eta_R(y)|^2}{|x-y|^{N+2s}}\, dx\, dy$$

$$= \int_{\mathbb{R}^N} |u_n(x)|^2 \left(\int_{|y-x|>R} \frac{|\eta_R(x) - \eta_R(y)|^2}{|x-y|^{N+2s}}\, dy + \int_{|y-x|\leq R} \frac{|\eta_R(x) - \eta_R(y)|^2}{|x-y|^{N+2s}}\, dy \right) dx$$

$$\leq \int_{\mathbb{R}^N} |u_n(x)|^2 \left(\int_{|y-x|>R} \frac{4\|\eta\|^2_{L^\infty(\mathbb{R}^N)}}{|x-y|^{N+2s}} \, dy + R^{-2} \int_{|y-x|\leq R} \frac{\|\nabla\eta\|^2_{L^\infty(\mathbb{R}^N)}}{|x-y|^{N+2s-2}} \, dy \right) dx$$

$$\leq C \int_{\mathbb{R}^N} |u_n(x)|^2 dx \left(\int_R^\infty \frac{1}{\rho^{2s+1}} \, d\rho + R^{-2} \int_0^R \frac{1}{\rho^{2s-1}} \, d\rho \right)$$

$$\leq \frac{C}{R^{2s}},$$

and letting first $n \to \infty$ and then $R \to \infty$ we get the thesis.

Lemma 1.4.7 *Let* $(u_n) \subset \mathcal{D}^{s,2}(\mathbb{R}^N)$ *be a bounded sequence in* $\mathcal{D}^{s,2}(\mathbb{R}^N)$ *and* $\psi \in C_c^\infty(\mathbb{R}^N)$ *be such that* $0 \leq \psi \leq 1$ *in* $\mathbb{R}^N$, $\psi = 1$ *in* B_1, $\psi = 0$ *in* $\mathbb{R}^N \setminus B_2$ *and* $\|\nabla\psi\|_{L^\infty(\mathbb{R}^N)} \leq 2$. *Set* $\psi_\rho(x) = \psi(\frac{x-x_0}{\rho})$ *for all* $\rho > 0$, *where* $x_0 \in \mathbb{R}^N$ *is a fixed point. Then*

$$\lim_{\rho \to 0} \limsup_{n \to \infty} \iint_{\mathbb{R}^{2N}} |u_n(x)|^2 \frac{|\psi_\rho(x) - \psi_\rho(y)|^2}{|x-y|^{N+2s}} \, dx dy = 0.$$

Proof We modify the proof of Lemma 1.4.5 in a convenient way. Without loss of generality we may assume that $x_0 = 0$. Fix $\rho > 0$ and set

$$W_\rho(x) = \int_{\mathbb{R}^N} \frac{|\psi_\rho(x) - \psi_\rho(y)|^2}{|x-y|^{N+2s}} \, dy.$$

It is easy to verify that

$$W_\rho(x) \leq C_1 \rho^{2s} \quad \text{for all } x \in \mathbb{R}^N, \quad W_\rho(x) \leq C_2 \rho^{-N} |x|^{-(N+2s)} \quad \text{for all } |x| > \frac{4}{\rho}.$$

Then, arguing as in the proof of Lemma 1.4.5, we get

$$\limsup_{n \to \infty} \iint_{\mathbb{R}^{2N}} |u_n(x)|^2 \frac{|\psi_\rho(x) - \psi_\rho(y)|^2}{|x-y|^{N+2s}} \, dx dy \leq \int_{\mathbb{R}^N} |u(x)|^2 W_\rho(x) \, dx.$$

Since $0 \leq W_\rho(x) \leq \bar{C} |x|^{-2s}$ and (1.1.2) imply that $|u(x)|^2 W_\rho(x) \leq \bar{C} \frac{|u(x)|^2}{|x|^{2s}} \in L^1(\mathbb{R}^N)$, and $W_\rho \to 0$ as $\rho \to 0$, we can apply the dominated convergence theorem to deduce that

$$\limsup_{\rho \to 0} \limsup_{n \to \infty} \iint_{\mathbb{R}^{2N}} |u_n(x)|^2 \frac{|\psi_\rho(x) - \psi_\rho(y)|^2}{|x-y|^{N+2s}} \, dx dy \leq \lim_{\rho \to 0} \int_{\mathbb{R}^N} |u(x)|^2 W_\rho(x) \, dx = 0.$$

$\square$

The next result was established in [288] and provides a way to manipulate smooth truncations for the fractional Laplacian. Here we give an alternative proof.

Lemma 1.4.8 ([288]) *Let $u \in \mathcal{D}^{s,2}(\mathbb{R}^N)$ and $\phi \in C_c^\infty(\mathbb{R}^N)$ be such that $0 \le \phi \le 1$ in $\mathbb{R}^N$, $\phi = 1$ in B_1 and $\phi = 0$ in $\mathbb{R}^N \setminus B_2$. Set $\phi_R(x) = \phi(\frac{x}{R})$ for $R > 0$. Then*

$$\lim_{R \to \infty} [u\phi_R - u]_s = 0.$$

Proof We have that

$$[u\phi_R - u]_s^2 \le 2\left[\iint_{\mathbb{R}^{2N}} |u(x)|^2 \frac{|\phi_R(x) - \phi_R(y)|^2}{|x - y|^{N+2s}} \, dx \, dy + \iint_{\mathbb{R}^{2N}} \frac{|\phi_R(x) - 1|^2 |u(x) - u(y)|^2}{|x - y|^{N+2s}} \, dx \, dy\right]$$

$$= 2[A_R + B_R].$$

Since $|\phi_R(x) - 1| \le 2$, $|\phi_R(x) - 1| \to 0$ a.e. in $\mathbb{R}^N$ and $u \in \mathcal{D}^{s,2}(\mathbb{R}^N)$, the dominated convergence theorem implies that

$$B_R \to 0 \quad \text{as } R \to \infty.$$

On the other hand, setting

$$W_R(x) = \int_{\mathbb{R}^N} \frac{|\phi_R(x) - \phi_R(y)|^2}{|x - y|^{N+2s}} \, dy,$$

we can argue as in the second part of the proof of Lemma 1.4.5 to deduce that

$$A_R \to 0 \quad \text{as } R \to \infty.$$

This ends the proof of lemma. $\square$

1.5   A Concentration-Compactness Principle

Here we present a concentration-compactness lemma in the spirit of Lions [256, 257] and Chabrowski [135] (see also [94, 104, 340]). For more details and comments on this topic we refer to [40, 61, 170, 288, 344].

Lemma 1.5.1 ([61]) *Let (u_n) be a sequence that converges weakly to u in $\mathcal{D}^{s,2}(\mathbb{R}^N)$ and such that*

$$|(-\Delta)^{\frac{s}{2}} u_n|^2 \rightharpoonup \mu \quad \text{and} \quad |u_n|^{2_s^*} \rightharpoonup \nu, \tag{1.5.1}$$

in the sense of measures, where μ and ν are two bounded nonnegative measures on $\mathbb{R}^N$. Then, there exist an at most countable index set I and three families of distinct points, $(x_i)_{i\in I}$ in $\mathbb{R}^N$ and $(\mu_i)_{i\in I}$ and $(\nu_i)_{i\in I}$ in $(0, \infty)$, such that

$$\nu = |u|^{2_s^*} + \sum_{i\in I} \nu_i \delta_{x_i}, \tag{1.5.2}$$

$$\mu \geq |(-\Delta)^{\frac{s}{2}} u|^2 + \sum_{i\in I} \mu_i \delta_{x_i}, \tag{1.5.3}$$

$$\mu_i \geq S_* \nu_i^{\frac{2}{2_s^*}} \quad \forall i \in I, \tag{1.5.4}$$

where δ_x is the Dirac mass at $x \in \mathbb{R}^N$. Moreover, if we define

$$\mu_\infty = \lim_{R\to\infty} \limsup_{n\to\infty} \int_{|x|>R} |(-\Delta)^{\frac{s}{2}} u_n|^2 \, dx, \tag{1.5.5}$$

and

$$\nu_\infty = \lim_{R\to\infty} \limsup_{n\to\infty} \int_{|x|>R} |u_n|^{2_s^*} \, dx, \tag{1.5.6}$$

then

$$\limsup_{n\to\infty} \int_{\mathbb{R}^N} |(-\Delta)^{\frac{s}{2}} u_n|^2 \, dx = \int_{\mathbb{R}^N} d\mu + \mu_\infty, \tag{1.5.7}$$

$$\limsup_{n\to\infty} \int_{\mathbb{R}^N} |u_n|^{2_s^*} \, dx = \int_{\mathbb{R}^N} d\nu + \nu_\infty, \tag{1.5.8}$$

$$\mu_\infty \geq S_* \nu_\infty^{\frac{2}{2_s^*}}. \tag{1.5.9}$$

Proof   In order to prove (1.5.2), we aim to pass to the limit in the following relation which holds in view of the Brezis-Lieb lemma [113]:

$$\int_{\mathbb{R}^N} |\psi|^{2_s^*} |u_n|^{2_s^*} \, dx = \int_{\mathbb{R}^N} |\psi|^{2_s^*} |u|^{2_s^*} \, dx + \int_{\mathbb{R}^N} |\psi|^{2_s^*} |u_n - u|^{2_s^*} \, dx + o_n(1), \tag{1.5.10}$$

where $\psi \in C_c^\infty(\mathbb{R}^N)$. Set $\tilde{u}_n = u_n - u$. Then, by Theorem 1.1.8, $\tilde{u}_n \to 0$ in $L^2_{\mathrm{loc}}(\mathbb{R}^N)$ and a.e. in $\mathbb{R}^N$. Up to subsequence, we may assume that

$$|(-\Delta)^{\frac{s}{2}} \tilde{u}_n|^2 \rightharpoonup \tilde{\mu} \quad \text{and} \quad |\tilde{u}_n|^{2_s^*} \rightharpoonup \tilde{\nu} \tag{1.5.11}$$

in the sense of measures. Fix $\psi \in C_c^\infty(\mathbb{R}^N)$. Using the fractional Sobolev inequality (1.1.1) in Theorem 1.1.8 we have

$$
\left(\int_{\mathbb{R}^N} |\psi|^{2_s^*} |u_n - u|^{2_s^*} \, dx \right)^{\frac{1}{2_s^*}} = \left(\int_{\mathbb{R}^N} |\psi \tilde{u}_n|^{2_s^*} \, dx \right)^{\frac{1}{2_s^*}}
$$

$$
\leq S_*^{-\frac{1}{2}} \left(\int_{\mathbb{R}^N} |(-\Delta)^{\frac{s}{2}} (\psi \tilde{u}_n)|^2 \, dx \right)^{\frac{1}{2}}
$$

$$
= S_*^{-\frac{1}{2}} \left(\iint_{\mathbb{R}^{2N}} \frac{|(\psi \tilde{u}_n)(x) - (\psi \tilde{u}_n)(y)|^2}{|x - y|^{N+2s}} \, dx dy \right)^{\frac{1}{2}}.
$$

$$\tag{1.5.12}$$

Next, by the Minkowski inequality,

$$
\left(\iint_{\mathbb{R}^{2N}} \frac{|\psi(x)\tilde{u}_n(x) - \psi(y)\tilde{u}_n(y)|^2}{|x - y|^{N+2s}} \, dx dy \right)^{\frac{1}{2}}
$$

$$
\leq \left(\iint_{\mathbb{R}^{2N}} |\psi(y)|^2 \frac{|\tilde{u}_n(x) - \tilde{u}_n(y)|^2}{|x - y|^{N+2s}} \, dx dy \right)^{\frac{1}{2}} + \left(\iint_{\mathbb{R}^{2N}} |\tilde{u}_n(x)|^2 \frac{|\psi(x) - \psi(y)|^2}{|x - y|^{N+2s}} \, dx dy \right)^{\frac{1}{2}}
$$

$$
= \left(\int_{\mathbb{R}^N} |\psi|^2 |(-\Delta)^{\frac{s}{2}} \tilde{u}_n|^2 dx \right)^{\frac{1}{2}} + \left(\int_{\mathbb{R}^N} |\tilde{u}_n|^2 |(-\Delta)^{\frac{s}{2}} \psi|^2 dx \right)^{\frac{1}{2}},
$$

where in the last equality we used a simple change of variable. Now we prove that

$$
\int_{\mathbb{R}^N} |\tilde{u}_n|^2 |(-\Delta)^{\frac{s}{2}} \psi|^2 \, dx = \iint_{\mathbb{R}^{2N}} |\tilde{u}_n(x)|^2 \frac{|\psi(x) - \psi(y)|^2}{|x - y|^{N+2s}} \, dx dy = o_n(1). \tag{1.5.13}
$$

Assume for simplicity that $\text{supp}(\psi) = \bar{B}_1$ and set

$$
W(x) = \int_{\mathbb{R}^N} \frac{|\psi(x) - \psi(y)|^2}{|x - y|^{N+2s}} \, dy.
$$

Note that $W \in L^\infty(\mathbb{R}^N)$, because

$$
W(x) = \int_{|y-x| \leq 1} \frac{|\psi(x) - \psi(y)|^2}{|x - y|^{N+2s}} \, dy + \int_{|y-x| > 1} \frac{|\psi(x) - \psi(y)|^2}{|x - y|^{N+2s}} \, dy
$$

$$
\leq \|\nabla \psi\|_{L^\infty(\mathbb{R}^N)}^2 \int_{|y-x| \leq 1} \frac{1}{|x - y|^{N+2s-2}} \, dy + 4\|\psi\|_{L^\infty(\mathbb{R}^N)}^2 \int_{|y-x| > 1} \frac{1}{|x - y|^{N+2s}} \, dy
$$

$$
\leq C. \tag{1.5.14}
$$

Moreover, for $|x| > 2$, we get

$$W(x) = \int_{|y| \le 1} \frac{|\psi(y)|^2}{|x - y|^{N+2s}} \, dy \le \frac{2^{N+2s} \omega_N \|\psi\|^2_{L^\infty(\mathbb{R}^N)}}{|x|^{N+2s}} = \frac{C}{|x|^{N+2s}}, \qquad (1.5.15)$$

where we used the fact that, for all $|x| > 2$ and $|y| \le 1$,

$$|x - y| \ge |x| - |y| \ge |x| - 1 \ge \frac{|x|}{2}.$$

Fix $R > 2$. In view of (1.5.14), (1.5.15) and using the Hölder inequality we see that

$$\int_{\mathbb{R}^N} |\tilde{u}_n(x)|^2 W(x) \, dx = \int_{|x| \le R} |\tilde{u}_n(x)|^2 W(x) \, dx + \int_{|x| > R} |\tilde{u}_n(x)|^2 W(x) \, dx$$

$$\le \|W\|_{L^\infty(\mathbb{R}^N)} \int_{|x| \le R} |\tilde{u}_n(x)|^2 \, dx + \|\tilde{u}_n\|^2_{L^{2^*_s}(\mathbb{R}^N)} \left(\int_{|x| > R} |W(x)|^{\frac{2^*_s}{2^*_s - 2}} \, dx \right)^{\frac{2^*_s - 2}{2^*_s}}$$

$$\le C \int_{|x| \le R} |\tilde{u}_n(x)|^2 \, dx + C \left(\int_{|x| > R} \frac{1}{|x|^{\frac{N^2}{2s} + N}} \, dx \right)^{\frac{2^*_s - 2}{2^*_s}},$$

and recalling that $\tilde{u}_n \to 0$ in $L^2(B_R)$ we have

$$\limsup_{n \to \infty} \int_{\mathbb{R}^N} |\tilde{u}_n(x)|^2 W(x) \, dx \le C \left(\int_{|x| > R} \frac{1}{|x|^{\frac{N^2}{2s} + N}} \, dx \right)^{\frac{2^*_s - 2}{2^*_s}} \qquad \forall R > 2.$$

Taking here the limit as $R \to \infty$, we obtain (1.5.13). On the other hand, by the first limit relation in (1.5.11), we have

$$\int_{\mathbb{R}^N} |\psi|^2 |(-\Delta)^{\frac{s}{2}} \tilde{u}_n|^2 \, dx \to \int_{\mathbb{R}^N} |\psi|^2 \, d\tilde{\mu}.$$

Since the second limit relation in (1.5.11) implies that

$$\int_{\mathbb{R}^N} |\psi|^{2^*_s} |u_n - u|^{2^*_s} \, dx \to \int_{\mathbb{R}^N} |\psi|^{2^*_s} \, d\tilde{\nu},$$

from (1.5.12) we obtain that

$$\left(\int_{\mathbb{R}^N} |\psi|^{2^*_s} \, d\tilde{\nu} \right)^{\frac{1}{2^*_s}} \le S_*^{-\frac{1}{2}} \left(\int_{\mathbb{R}^N} |\psi|^2 \, d\tilde{\mu} \right)^{\frac{1}{2}}, \quad \text{for all } \psi \in C_c^\infty(\mathbb{R}^N).$$

Then, using Lemma 1.2 in [257], there exist an at most a countable set I, families $(x_i)_{i \in I}$ of distinct points in $\mathbb{R}^N$, $(v_i)_{i \in I}$ in $(0, \infty)$ such that

$$\tilde{v} = \sum_{i \in I} v_i \delta_{x_i}, \quad \tilde{\mu} \geq S_* \sum_{i \in I} v_i^{\frac{2}{2_s^*}} \delta_{x_i}. \tag{1.5.16}$$

In view of (1.5.10), we deduce that $v = |u|^{2_s^*} + \tilde{v}$, which together with (1.5.16) implies that

$$v = |u|^{2_s^*} + \sum_{i \in I} v_i \delta_{x_i},$$

that is, (1.5.2) is satisfied.

Now, we prove that (1.5.4) holds true. Fix $i \in I$ and take $\psi_\rho(x) = \psi(\frac{x - x_i}{\rho})$, where $\psi \in C_c^\infty(\mathbb{R}^N)$, $0 \leq \psi \leq 1$ in $\mathbb{R}^N$, $\psi(0) = 1$ and $\text{supp}(\psi) = \bar{B}_1$. As before, we obtain

$$S_*^{\frac{1}{2}} \left(\int_{\mathbb{R}^N} |\psi_\rho|^{2_s^*} |u_n|^{2_s^*} \, dx \right)^{\frac{1}{2_s^*}} \leq \left(\iint_{\mathbb{R}^{2N}} \frac{|\psi_\rho(x) u_n(x) - \psi_\rho(y) u_n(y)|^2}{|x - y|^{N + 2s}} \, dx dy \right)^{\frac{1}{2}}$$

$$\leq \left(\int_{\mathbb{R}^N} |u_n|^2 |(-\Delta)^{\frac{s}{2}} \psi_\rho|^2 \, dx \right)^{\frac{1}{2}} + \left(\int_{\mathbb{R}^N} |\psi_\rho|^2 |(-\Delta)^{\frac{s}{2}} u_n|^2 \, dx \right)^{\frac{1}{2}}. \tag{1.5.17}$$

Now, taking into account (1.5.1) and (1.5.2), we have

$$\lim_{n \to \infty} \int_{\mathbb{R}^N} |\psi_\rho|^{2_s^*} |u_n|^{2_s^*} \, dx = \int_{B_\rho(x_i)} |\psi_\rho|^{2_s^*} |u|^{2_s^*} \, dx + v_i.$$

Since $0 \leq \psi_\rho \leq 1$ implies

$$\left| \int_{B_\rho(x_i)} |\psi_\rho|^{2_s^*} |u|^{2_s^*} \, dx \right| \leq C \int_{B_\rho(x_i)} |u|^{2_s^*} \, dx \to 0 \text{ as } \rho \to 0,$$

we deduce that

$$\lim_{\rho \to 0} \lim_{n \to \infty} \int_{\mathbb{R}^N} |\psi_\rho|^{2_s^*} |u_n|^{2_s^*} \, dx = v_i. \tag{1.5.18}$$

On the other hand, (1.5.1) gives

$$\lim_{n \to \infty} \int_{\mathbb{R}^{2N}} |\psi_\rho|^2 |(-\Delta)^{\frac{s}{2}} u_n|^2 \, dx dy = \int_{\mathbb{R}^N} |\psi_\rho|^2 \, d\mu, \tag{1.5.19}$$

and using Lemma 1.4.7 we see that

$$\lim_{\rho \to 0} \limsup_{n \to \infty} \int_{\mathbb{R}^N} |u_n|^2 |(-\Delta)^{\frac{s}{2}} \psi_\rho|^2 \, dx = 0. \tag{1.5.20}$$

Combining (1.5.17), (1.5.18), (1.5.19) and (1.5.20), we get

$$S_* \nu_i^{\frac{2}{2_s^*}} \leq \lim_{\rho \to 0} \mu(B_\rho(x_i)).$$

Setting

$$\mu_i = \lim_{\rho \to 0} \mu(B_\rho(x_i)),$$

we deduce that (1.5.4) holds true. Now, we note that

$$\mu \geq \sum_{i \in I} \mu_i \delta_{x_i},$$

and that the weak lower semi-continuity implies that $\mu \geq |(-\Delta)^{\frac{s}{2}} u|^2$. Then, since $|(-\Delta)^{\frac{s}{2}} u|^2$ is orthogonal to $\sum_{i \in I} \mu_i \delta_{x_i}$, we infer that (1.5.3) is satisfied. Finally, we show the validity of (1.5.7)–(1.5.9). Let η_R be defined as in Lemma 1.4.5. Then we have

$$\int_{\mathbb{R}^N} |(-\Delta)^{\frac{s}{2}} u_n|^2 \, dx = \int_{\mathbb{R}^N} |(-\Delta)^{\frac{s}{2}} u_n|^2 \eta_R^2 \, dx + \int_{\mathbb{R}^N} |(-\Delta)^{\frac{s}{2}} u_n|^2 (1 - \eta_R^2) \, dx. \tag{1.5.21}$$

Since

$$\int_{|x|>2R} |(-\Delta)^{\frac{s}{2}} u_n|^2 \, dx \leq \int_{\mathbb{R}^N} |(-\Delta)^{\frac{s}{2}} u_n|^2 \eta_R^2 \, dx \leq \int_{|x|>R} |(-\Delta)^{\frac{s}{2}} u_n|^2 \, dx,$$

we obtain that

$$\mu_\infty = \lim_{R \to \infty} \limsup_{n \to \infty} \int_{\mathbb{R}^N} |(-\Delta)^{\frac{s}{2}} u_n|^2 \eta_R^2 \, dx. \tag{1.5.22}$$

Now, using the fact that μ is finite and $1 - \eta_R^2$ is smooth and compactly supported, we can apply the dominated convergence theorem to get

$$\lim_{R \to \infty} \limsup_{n \to \infty} \int_{\mathbb{R}^N} |(-\Delta)^{\frac{s}{2}} u_n|^2 (1 - \eta_R^2) \, dx = \lim_{R \to \infty} \int_{\mathbb{R}^N} (1 - \eta_R^2) \, d\mu = \int_{\mathbb{R}^N} d\mu. \tag{1.5.23}$$

Combining (1.5.21), (1.5.22) and (1.5.23), we obtain (1.5.7). In a similar fashion, we can prove that

$$v_\infty = \lim_{R\to\infty} \limsup_{n\to\infty} \int_{\mathbb{R}^N} |u_n|^{2_s^*} \eta_R^{2_s^*} \, dx, \tag{1.5.24}$$

and arguing as before we deduce that (1.5.8) is verified. To show that (1.5.9) is satisfied, we observe that

$$S_*^{\frac{1}{2}} \left(\int_{\mathbb{R}^N} |\eta_R u_n|^{2_s^*} \, dx \right)^{\frac{1}{2_s^*}} \leq \left(\int_{\mathbb{R}^N} |(-\Delta)^{\frac{s}{2}} (\eta_R u_n)|^2 \, dx \right)^{\frac{1}{2}}$$

$$\leq \left(\int_{\mathbb{R}^N} \eta_R^2 |(-\Delta)^{\frac{s}{2}} u_n|^2 \, dx \right)^{\frac{1}{2}} + \left(\int_{\mathbb{R}^N} |u_n|^2 |(-\Delta)^{\frac{s}{2}} \eta_R|^2 \, dx \right)^{\frac{1}{2}}. \tag{1.5.25}$$

By Lemma 1.4.5,

$$\lim_{R\to\infty} \limsup_{n\to\infty} \int_{\mathbb{R}^N} |u_n|^2 |(-\Delta)^{\frac{s}{2}} \eta_R|^2 \, dx = 0,$$

which together with (1.5.22), (1.5.24), and (1.5.25) yields (1.5.9). This ends the proof of lemma. $\qquad\square$

Remark 1.5.2 In the case $\Omega = \mathbb{R}^N$, the concentration-compactness principle established in [288] does not provide any information about a possible loss of mass at infinity. Then Lemma 1.5.1 expresses this fact in quantitative terms.

Some Abstract Results

In this section we review some useful results on critical point theory, minimax methods and Nehari manifold arguments. For more details we refer to [27, 137, 267, 298, 319, 321, 322, 340].

2.1   Critical Point Theory

We start with the definitions of derivatives in Banach spaces. Let X be a Banach space equipped with the norm $\| \cdot \|$, and denote by X^* its topological dual, that is, the space of all continuous linear functionals on X. The (dual) norm on X^* is defined by

$$\|f\|_{X^*} = \sup_{u \in X, \|u\| \leq 1} |f(u)| = \sup_{u \in X, \|u\| \leq 1} f(u).$$

We denote by $\langle \cdot, \cdot \rangle$ the duality pairing between X and X^*, and by $\mathcal{L}(X, X^*)$ the vector space of all continuous linear operators from X into X^*.

Definition 2.1.1 Let $U \subset X$ be an open set of X and $\varphi : U \to \mathbb{R}$ be a functional. We say that φ has a Gateaux derivative $f \in X^*$ at $u \in X$ if, for every $h \in X$,

$$\lim_{t \to 0} \frac{\varphi(u + th) - \varphi(u) - \langle f, th \rangle}{t} = 0.$$

The Gateaux derivative at u is denoted by $\varphi'(u)$ and it holds that

$$\langle \varphi'(u), h \rangle = \lim_{t \to 0} \frac{\varphi(u + th) - \varphi(u)}{t}.$$

V. Ambrosio, *Nonlinear Fractional Schrödinger Equations in* $\mathbb{R}^N$,
Frontiers in Mathematics, https://doi.org/10.1007/978-3-030-60220-8_2

The functional φ has a Fréchet derivative $f \in X^*$ at $u \in U$ if

$$\lim_{h \to 0} \frac{\varphi(u+h) - \varphi(u) - \langle f, h \rangle}{\|h\|} = 0.$$

Evidently, any Fréchet derivative is a Gateaux derivative. The functional φ belongs to $C^1(U, \mathbb{R})$ if the Fréchet derivative of φ exists and is continuous on U. Using the mean value theorem, it is easy to see that if φ has a continuous Gateaux derivative on U, then $\varphi \in C^1(U, \mathbb{R})$.

If X is a Hilbert space with inner product $(\cdot, \cdot)$ and φ has a Gateaux derivative at $u \in U$, the gradient $\nabla \varphi(u)$ of φ at u is defined by

$$(\nabla \varphi(u), h) = \langle \varphi'(u), h \rangle.$$

A point $u \in U$ is called a critical point of φ if $\varphi'(u) = 0$; otherwise u is called a regular point. A number $c \in \mathbb{R}$ is a critical value of φ if there exists a critical point u of φ with $\varphi(u) = c$. Otherwise, c is called regular.

Definition 2.1.2 Let $\varphi \in C^1(U, \mathbb{R})$. The functional φ has a second Gateaux derivative $L \in \mathcal{L}(X, X^*)$ at $u \in U$ if, for every $h, v \in X$,

$$\lim_{t \to 0} \frac{\langle \varphi'(u+th) - \varphi'(u) - Lth, v \rangle}{t} = 0.$$

The second Gateaux derivative at u is denoted by $\varphi''(u)$.

The functional φ has a second Fréchet derivative $L \in \mathcal{L}(X, X^*)$ at $u \in U$ if

$$\lim_{h \to 0} \frac{\varphi'(u+h) - \varphi'(u) - Lh}{\|h\|} = 0.$$

The functional φ belongs to $C^2(X, \mathbb{R})$ if the second Fréchet derivative of φ exists and is continuous on U.

The second Gateaux derivative is given by

$$\langle \varphi''(u)h, k \rangle = \lim_{t \to 0} \frac{\langle \varphi'(u+th) - \varphi'(u), k \rangle}{t}.$$

Clearly, any second Fréchet derivative of φ is a second Gateaux derivative. Using the mean value theorem, if φ has a continuous second Gateaux derivative on U, then $\varphi \in C^2(U, \mathbb{R})$.

Let $f(x, t)$ be a function on $\Omega \times \mathbb{R}$, where Ω is either bounded or unbounded. We say that f is a Carathéodory function if $f(x, t)$ is continuous in t for a.e. $x \in \Omega$ and measurable in x for every $t \in \mathbb{R}$.

Lemma 2.1.3 ([351]) *Assume $p, q \geq 1$. Let $f(x, t)$ be a Carathéodory function on $\Omega \times \mathbb{R}$ satisfying*

$$|f(x, t)| \leq a + b|t|^{\frac{p}{q}}, \quad \forall (x, t) \in \Omega \times \mathbb{R},$$

where $a, b > 0$ and Ω is either bounded or unbounded. Define a Carathéodory operator by setting for all $u \in L^p(\Omega)$

$$Bu = f(x, u(x)).$$

Let $(u_n) \subset L^p(\Omega)$ be a sequence. If $u_n \to u$ in $L^p(\Omega)$ as $n \to \infty$, then $Bu_n \to Bu$ in $L^q(\Omega)$ as $n \to \infty$. In particular, if Ω is bounded, then B is a continuous and bounded mapping from $L^p(\Omega)$ to $L^q(\Omega)$ and the same conclusion holds true if Ω is unbounded and $a = 0$.

Lemma 2.1.4 ([351]) *Assume $p_1, p_2, q_1, q_2 \geq 1$. Let $f(x, t)$ be a Carathéodory function on $\Omega \times \mathbb{R}$ satisfying*

$$|f(x, t)| \leq a|t|^{\frac{p_1}{q_1}} + b|t|^{\frac{p_2}{q_2}}, \quad \forall (x, t) \in \Omega \times \mathbb{R},$$

where $a, b \geq 0$ and Ω is bounded or unbounded. Define a Carathéodory operator by setting for every $u \in \mathcal{D} = L^{p_1}(\Omega) \cap L^{p_2}(\Omega)$

$$Bu = f(x, u(x)).$$

Define the space

$$\mathcal{E} = L^{q_1}(\Omega) + L^{q_2}(\Omega),$$

equipped with the norm

$$\|u\|_{\mathcal{E}} = \inf\{\|v\|_{L^{q_1}(\Omega)} + \|w\|_{L^{q_2}(\Omega)} : u = v + w \in \mathcal{E}, v \in L^{q_1}(\Omega), w \in L^{q_2}(\Omega)\}.$$

Then $B = B_1 + B_2$, where B_i is a bounded and continuous mapping from $L^{p_i}(\Omega)$ to $L^{q_i}(\Omega)$, $i = 1, 2$. In particular, B is a bounded continuous mapping from $\mathcal{D}$ to $\mathcal{E}$.

Theorem 2.1.5 ([351]) *Assume $\sigma, p \geq 0$. Let $f(x, t)$ be a Carathéodory function on $\Omega \times \mathbb{R}$ satisfying*

$$|f(x, t)| \leq a|t|^{\sigma} + b|t|^{p}, \quad \forall (x, t) \in \Omega \times \mathbb{R},$$

where $a, b > 0$ and Ω is either bounded or unbounded. Define a functional I by

$$I(u) = \int_\Omega F(x, u)\, dx, \quad where \quad F(x, t) = \int_0^t f(x, \tau)\, d\tau.$$

Assume that $(X, \|\cdot\|)$ is a Sobolev Banach space such that E is continuously embedded in $L^{p+1}(\Omega)$ and $L^{\sigma+1}(\Omega)$. Then, $I \in C^1(X, \mathbb{R})$ and

$$\langle I'(u), h \rangle = \int_\Omega f(x, u) h\, dx \quad \forall h \in X.$$

Moreover, if E is compactly embedded in $L^{p+1}(\Omega)$ and $L^{\sigma+1}(\Omega)$, then $I' : X \to X^*$ is compact.

2.2   Minimax Methods

This subsection is devoted to some classical minimax theorems of the critical point theory. We start by recalling Ekeland's variational principle [176]:

Theorem 2.2.1 ([176] Ekeland's Variational Principle) *Let (X, d) be a a complete metric space with metric d, and let $\varphi : X \to \mathbb{R} \cup \{\infty\}$ be a lower semi-continuous function, bounded from below and $\varphi \not\equiv \infty$. Let $\varepsilon > 0$ and $\lambda > 0$ be given and $u \in X$ be such that $\varphi(u) \leq \inf_X \varphi + \varepsilon$. Then there exists $v \in X$ such that*

- $\varphi(v) \leq \varphi(u)$,
- $d(u, v) \leq \lambda$,
- $\varphi(w) > \varphi(v) - \frac{\varepsilon}{\lambda} d(v, w)$ *for all* $w \neq v$.

Let us introduce some useful definitions (see [111, 267, 287, 319] for more details).

Definition 2.2.2 Let X be a Banach space and $\varphi \in C^1(X, \mathbb{R})$.

We say that φ satisfies the Palais–Smale condition *(shortly, (PS) condition)* if any sequence $(u_n) \subset X$ such that

$$(\varphi(u_n)) \text{ is bounded and } \varphi'(u_n) \to 0 \text{ in } X^* \tag{2.2.1}$$

admits a strongly convergent subsequence. Any sequence satisfying (2.2.1) is called a Palais–Smale sequence.

Let $c \in \mathbb{R}$. The functional φ is said to satisfy the Palais–Smale condition at the level $c \in \mathbb{R}$ (shortly, $(PS)_c$ condition), if every sequence $(u_n) \subset X$ such that

$$\varphi(u_n) \to c \text{ and } \varphi'(u_n) \to 0 \text{ in } X^*$$

admits a strongly convergent subsequence.

Now we give the notion of pseudogradient defined in [286].

Definition 2.2.3 Let E be a metric space, X a normed space and $h : E \to X^* \setminus \{0\}$ a continuous mapping. A pseudogradient vector field for h on E is a locally Lipschitz continuous vector field $g : E \to X$ such that, for every $u \in E$,

$$\|g(u)\| \leq 2\|h(u)\|_{X^*},$$

$$\langle h(u), g(u) \rangle \geq \|h(u)\|_{X^*}^2.$$

Lemma 2.2.4 *Under the assumptions of the preceding definition, there exists a pseudogradient vector field for h on E.*

Before stating the next result, we introduce some notations. Let $\varphi : X \to \mathbb{R}$ be a functional and $S \subset X$. Then, for all $d \in \mathbb{R}$, we set

$$\varphi^d = \{v \in X : \varphi(v) \leq d\},$$

and for all $\delta > 0$

$$S_\delta = \{u \in X : \text{dist}(u, S) \leq \delta\}.$$

At this point we present the following quantitative deformation lemma for continuously differentiable functions defined on a Banach space.

Lemma 2.2.5 ([340]) *Let X be a Banach space, $\varphi \in C^1(X, \mathbb{R})$, $S \subset X$, $c \in \mathbb{R}$, $\varepsilon, \delta > 0$ such that*

$$\|\varphi'(u)\| \geq \frac{8\varepsilon}{\delta} \quad \forall u \in \varphi^{-1}([c - 2\varepsilon, c + 2\varepsilon]) \cap S_{2\delta}.$$

Then there exists $\eta \in C([0, 1] \times X, X)$ such that

(1) $\eta(t, u) = u$, *if $t = 0$ or if $u \notin \varphi^{-1}([c - 2\varepsilon, c + 2\varepsilon]) \cap S_{2\delta}$,*
(2) $\eta(1, \varphi^{c+\varepsilon} \cap S) \subset \varphi^{c-\varepsilon}$,
(3) $\eta(t, \cdot)$ *is an homeomorphism of X, $\forall t \in [0, 1]$,*

(4) $\|\eta(t, u) - u\| \leq \delta, \forall u \in X, \forall t \in [0, 1],$

(5) $\varphi(\eta(\cdot, u))$ *is nonincreasing,* $\forall u \in X,$

(6) $\varphi(\eta(t, u)) < c, \forall u \in \varphi^c \cap S_\delta, \forall t \in (0, 1].$

Remark 2.2.6 As pointed out in [340], a simple application of Lemma 2.2.5 gives a version of Ekeland variational principle for Banach spaces.

A very important application of Lemma 2.2.5 is the following general minimax principle.

Theorem 2.2.7 ([340]) *Let X be a Banach space. Let M_0 be a closed subspace of the metric space M and let $\Gamma_0 \subset C(M_0, X)$. Define*

$$\Gamma = \{\gamma \in C(M, X) : \gamma|_{M_0} \in \Gamma_0\}.$$

If $\varphi \in C^1(X, \mathbb{R})$ satisfies

$$\infty > c = \inf_{\gamma \in \Gamma} \sup_{u \in M} \varphi(\gamma(u)) > a = \sup_{\gamma_0 \in \Gamma_0} \sup_{u \in M_0} \varphi(\gamma_0(u)) \tag{2.2.2}$$

then, for every $\varepsilon \in (0, \frac{c-a}{2})$, $\delta > 0$ and $\gamma \in \Gamma$ such that

$$\sup_M \varphi \circ \gamma \leq c + \varepsilon,$$

there exists $u \in X$ such that

(a) $c - 2\varepsilon \leq \varphi(u) \leq c + 2\varepsilon,$

(b) $\text{dist}(u, \gamma(M)) \leq 2\delta,$

(c) $\|\varphi'(u)\|_{X^*} \leq \frac{8\varepsilon}{\delta}.$

As a consequence of Theorem 2.2.7 we have:

Theorem 2.2.8 ([340]) *Under assumption (2.2.2), there exists a sequence $(u_n) \subset X$ satisfying*

$$\varphi(u_n) \to c \quad and \quad \varphi'(u_n) \to 0 \text{ in } X^*.$$

In particular, if φ satisfies the $(PS)_c$ condition, then c is a critical value of J.

With the aid of Theorem 2.2.8 we obtain the next renowned minimax theorems.

Theorem 2.2.9 (Mountain Pass Theorem [29, 298, 340]) *Let X be a Banach space, $\varphi \in C^1(X, \mathbb{R})$, $e \in X$ and $r > 0$ be such that $\|e\| > r$ and*

$$b = \inf_{\|u\|=r} \varphi(u) > \varphi(0) \geq \varphi(e).$$

If φ satisfies the (PS)$_c$ condition with

$$c = \inf_{\gamma \in \Gamma} \max_{t \in [0,1]} \varphi(\gamma(t)),$$

$$\Gamma = \{\gamma \in C([0, 1], X) : \gamma(0) = 0, \gamma(1) = e\},$$

then c is a critical value of φ.

Remark 2.2.10 If φ satisfies the geometric assumptions of the mountain pass theorem, then Theorem 2.2.8 guarantees the existence of a Palais–Smale sequence at the mountain pass level c. In the literature, this result is sometimes referred to as a variant of the mountain pass theorem without the Palais–Smale condition (see [111, 114, 177]). More precisely, Brezis and Nirenberg [114] observed that if X is a Banach space, $\varphi \in C^1(X, \mathbb{R})$, and there exist a neighborhood U of 0 in X and a constant ρ such that $\varphi(u) \geq \rho$ for every u in the boundary of U, $\varphi(0) < \rho$ and $\varphi(v) < \rho$ for some $v \notin U$, then there exists a Palais–Smale sequence at the level

$$c = \inf_{P \in \mathcal{P}} \max_{w \in P} \varphi(w) \geq \rho,$$

where $\mathcal{P}$ denotes the class of paths joining 0 to v.

Theorem 2.2.11 (Saddle-Point Theorem [298, 340]) *Let $X = Y \oplus Z$ be a Banach space with dim $Y < \infty$. Define, for $\rho > 0$,*

$$M = \{u \in Y : \|u\| \leq \rho\}, \quad M_0 = \{u \in Y : \|u\| = \rho\}.$$

Let $\varphi \in C^1(X, \mathbb{R})$ be such that

$$b = \inf_Z \varphi > a = \max_{M_0} \varphi.$$

If φ satisfies the (PS)$_c$ condition with

$$c = \inf_{\gamma \in \Gamma} \max_{u \in M} \varphi(\gamma(u)),$$

$$\Gamma = \{\gamma \in C(M, X) : \gamma = \text{id on } M_0\},$$

then c is a critical value of φ.

Theorem 2.2.12 (Linking Theorem [298, 340]) *Let $X = Y \oplus Z$ be a Banach space with $\dim Y < \infty$. Let $\rho > r > 0$ and let $z \in Z$ be such that $\|z\| = r$. Define*

$$M = \{u = y + \lambda z : \|u\| \leq \rho, \, \lambda \geq 0, \, y \in Y\},$$

$$M_0 = \{u = y + \lambda z : y \in Y, \, \|u\| = \rho \text{ and } \lambda \geq 0 \text{ or } \|u\| \leq \rho \text{ and } \lambda = 0\},$$

$$N = \{u \in Z : \|u\| = r\}.$$

Let $\varphi \in C^1(X, \mathbb{R})$ be such that

$$b = \inf_N \varphi > a = \max_{M_0} \varphi.$$

If φ satisfies the $(PS)_c$ condition with

$$c = \inf_{\gamma \in \Gamma} \max_{u \in M} \varphi(\gamma(u)),$$

$$\Gamma = \{\gamma \in C(M, X) : \gamma = \text{id on } M_0\},$$

then c is a critical value of φ.

We also recall the following compactness-type condition proposed in [134].

Definition 2.2.13 Let X be a Banach space and $\varphi \in C^1(X, \mathbb{R})$.

We say that φ satisfies the Cerami condition (shortly, (C) condition) if any sequence $(u_n) \subset X$ such that

$$(\varphi(u_n)) \text{ is bounded and } (1 + \|u_n\|)\varphi'(u_n) \to 0 \text{ in } X^* \tag{2.2.3}$$

admits a strongly convergent subsequence. Any sequence satisfying (2.2.3) is called a Cerami sequence.

Let $c \in \mathbb{R}$. The functional φ is said to satisfy the Cerami condition at the level $c \in \mathbb{R}$ (shortly, $(C)_c$ condition), if every sequence $(u_n) \subset X$ such that

$$\varphi(u_n) \to c \text{ and } (1 + \|u_n\|)\varphi'(u_n) \to 0 \text{ in } X^*$$

admits a strongly convergent subsequence.

Remark 2.2.14 The Cerami condition is weaker than the Palais–Smale condition.

When a C^1-functional possesses a mountain pass geometry, we can use Ekeland's variational principle to obtain the following variant of the mountain pass lemma with the Cerami condition.

Theorem 2.2.15 ([177]) *Let X be a Banach space and suppose that $\varphi \in C^1(X, \mathbb{R})$ satisfies*

$$\max\{\varphi(0), \varphi(e)\} \leq \mu < \alpha \leq \inf_{\|u\|=\rho} \varphi(u),$$

for some $\mu < \alpha$, $\rho > 0$ and $e \in X$ with $\|e\| > \rho$. Let $c \geq \alpha$ be characterized by

$$c = \inf_{\gamma \in \Gamma} \max_{t \in [0,1]} \varphi(\gamma(t)),$$

where

$$\Gamma = \{\gamma \in C([0, 1], X) : \gamma(0) = 0, \gamma(1) = e\}.$$

Then there exists a Cerami sequence $(u_n) \subset X$ at the level c.

2.3 Genus and Category

Topological tools play a central role in the study of variational problems. In this subsection we recall the notions of genus and category as well as some of their basic properties.

Definition 2.3.1 Let X be a real Banach space and denote

$$\Sigma(X) = \{A \subset X \setminus \{0\}| \quad A \text{ is closed and symmetric with respect to } 0\}.$$

We say that the positive integer k is the Krasnoselski genus of $A \in \Sigma(X)$, if there exists an odd map $\varphi : A \to \mathbb{R}^k \setminus \{0\}$ and k is the smallest integer with this property. The genus of the set A is denoted by $\gamma(A) = k$. When there is no finite such k, set $\gamma(A) = \infty$. Finally, set $\gamma(\emptyset) = 0$.

Remark 2.3.2 The notion of genus is due to Krasnoselskii [242]; here we use the equivalent definition due to Coffman [149].

The genus has the following well-known properties:

Proposition 2.3.3 ([29, 298]) *Let $A, B \in \Sigma(X)$. Then:*

  (i) *if there exists an odd map $f \in C(A, B)$, then $\gamma(A) \leq \gamma(B)$;*
 (ii) *if $A \subset B$, then $\gamma(A) \leq \gamma(B)$;*
(iii) *if there exists an odd homeomorphism between A and B, then $\gamma(A) = \gamma(B)$;*
(iv) *$\gamma(A \cup B) \leq \gamma(A) + \gamma(B)$;*

(v) *if $\gamma(B) < \infty$, then $\gamma\left(\overline{A \setminus B}\right) \geq \gamma(A) - \gamma(B)$;*

(vi) *if A is compact, then $\gamma(A) < \infty$, and there exists $\delta > 0$ such that $\gamma(N_\delta(A)) = \gamma(A)$, where $N_\delta(A) = \{x \in X : \text{dist}(x, A) \leq \delta\}$;*

(vii) *if A is homeomorphic by an odd homeomorphism to the boundary of a symmetric bounded open neighborhood of 0 in $\mathbb{R}^m$, then $\gamma(A) = m$;*

(viii) *let $A \in \Sigma(X)$, V be a k-dimensional subspace of X, and $V^\perp$ an algebraically and topologically complementary subspace. If $\gamma(A) > k$, then $A \cap V^\perp \neq \emptyset$.*

The genus can be used to get infinitely many distinct pairs of critical points for even functionals:

Theorem 2.3.4 ([298]) *Let X be an infinite-dimensional Hilbert space and let $\varphi \in C^1(X, \mathbb{R})$ be even. Suppose $r > 0$, $\varphi|_{\partial B_r}$ satisfies the (PS) condition, and $\varphi|_{\partial B_r}$ is bounded from below. Then $\varphi|_{\partial B_r}$ possesses infinitely many distinct pairs of critical points.*

We also recall the following symmetric mountain pass theorem due to Ambrosetti and Rabinowitz [29].

Theorem 2.3.5 (Symmetric Mountain Pass Theorem [29]) *Let X be an infinite-dimensional Banach space and let $\varphi \in C^1(X, \mathbb{R})$ be even, satisfy the (PS) condition, and $\varphi(0) = 0$. If $X = V \oplus W$, where V is finite-dimensional, and φ satisfies*

(i) *there are constants $\rho, \alpha > 0$ such that $\varphi(u) \geq \alpha$ for all $u \in W$ such that $\|u\| = \rho$, and*

(ii) *for each finite-dimensional subspace $E \subset X$, there exists an $R = R(E)$ such that $\varphi(u) \leq 0$ for any $u \in E$ such that $\|u\| \geq R$,*

then φ possesses an unbounded sequence of critical values.

Next we present the notion of category introduced by Lusternik and Schnirelman [265].

Definition 2.3.6 Let M be a topological space and $A \subset M$ a subset. The continuous map $h : [0, 1] \times A \to M$ is called a deformation of A in M if $h(0, u) = 0$ for all $u \in A$. The set A is said be contractible in M if there exist a deformation $h : [0, 1] \times A \to M$ and $p \in M$ such that $h(1, u) = \{p\}$ for all $u \in A$.

Definition 2.3.7 Let M be a topological space. A set $A \subset M$ is said to be of Lusternik-Schnirelman category k in M (denoted $\text{cat}_M(A)$) if it can be covered by k but not by $k - 1$ closed sets that are contractible to a point in M. If such k does not exist, then $\text{cat}_M(A) = \infty$.

The Lusternik-Schnirelman category has the following basic properties.

Proposition 2.3.8 ([137, 298, 319, 340]) *Let A, $B \subset M$.*

(i) $\mathrm{cat}_M(A) = 0 \Longleftrightarrow A = \emptyset$;
(ii) $A \subset B \Longrightarrow \mathrm{cat}_M(A) \leq \mathrm{cat}_M(B)$;
(iii) $\mathrm{cat}_M(A \cup B) \leq \mathrm{cat}_M(A) + \mathrm{cat}_M(B)$;
(iv) *if A is closed and $\eta : [0, 1] \times A \to M$ is a deformation of A in M, then $\mathrm{cat}_M(A) \leq$*
   $\mathrm{cat}_M(\overline{\eta(1, A)})$;
(v) *if M is a Banach-Finsler manifold of class C^1 and $A \subset M$, then there is a closed*
   neighborhood N of A in M such that $\mathrm{cat}_M(A) = \mathrm{cat}_M(N)$.

Let M be a Banach-Finsler manifold of class C^1 (see [137, 286, 319] for the definition)
and $\varphi \in C^1(M, \mathbb{R})$. For any $k \leq \mathrm{cat}(M)$ we define

$$c_k = \inf_{A \in A_k} \max_{u \in A} \varphi(u),$$

$$A_k = \{A \subset M : A \text{ is compact}, \mathrm{cat}_M(A) \geq k\}.$$

With the aid of the category we can state the following celebrated multiplicity result (see
[137] for a proof).

Theorem 2.3.9 (Lusternik-Schnirelman Theorem [265]) *Let M be a complete Banach-*
Finsler manifold of class C^1, and let $\varphi : M \to \mathbb{R}$ be a functional of class C^1 which is
bounded from below on M. If φ satisfies the (PS) condition, then φ has at least $\mathrm{cat}_M(M)$
critical points.

In what follows, we assume that

(A) X is a real Banach space, $\psi \in C^2(X, \mathbb{R})$, $V = \{u \in X : \psi(v) = 1\} \neq \emptyset$, $\psi'(v) \neq 0$
   for every $v \in V$.

The set V is a differentiable manifold of class C^2. The norm on X induces a metric on V
and so V becomes a metric manifold, i.e., a metric space and a manifold. We denote by
$T_v V$ the tangent space of V at v, i.e.,

$$T_v V = \{w \in X : \langle \psi'(v), w \rangle = 0\}.$$

Let $\varphi \in C^1(X, \mathbb{R})$ and $v \in \mathcal{V}$. The norm of the derivative of the restriction of φ to $\mathcal{V}$ is defined by

$$\|\varphi'(v)\|_* = \sup_{w \in T_v \mathcal{V},\, \|w\|=1} \langle \psi'(v), w \rangle.$$

The point v is a critical point of the restriction of φ to v if the restriction of $\varphi'(v)$ to $T_v \mathcal{V}$ is equal to 0.

Proposition 2.3.10 ([340]) *If $\varphi \in C^1(X, \mathbb{R})$ and $u \in \mathcal{V}$, then*

$$\|\varphi'(u)\|_* = \min_{\lambda \in \mathbb{R}} \|\varphi'(u) - \lambda \psi'(u)\|.$$

In particular, u is a critical point of $\varphi|_\mathcal{V}$ if and only if there exists $\lambda \in \mathbb{R}$ such that $\varphi'(u) = \lambda \psi'(u)$.

Definition 2.3.11 The functional $\varphi|_\mathcal{V}$ satisfies the $(PS)_c$ condition if any sequence $(u_n) \subset \mathcal{V}$ such that $\varphi(u_n) \to c$ and $\|\varphi'(u_n)\|_* \to 0$ has a convergent subsequence.

Theorem 2.3.12 ([340]) *If $\varphi|_\mathcal{V}$ is bounded from below, $d \geq \inf_\mathcal{V} \varphi$ and φ satisfies the $(PS)_c$ condition for any $c \in [\inf_\mathcal{V} \varphi, d]$, then $\varphi|_\mathcal{V}$ has a minimum and φ^d contains at least $\mathrm{cat}_{\varphi^d}(\varphi^d)$ critical points of $\varphi|_\mathcal{V}$.*

2.4   The Method of Nehari Manifold

Let X be a Banach space and let $\varphi \in C^1(X, \mathbb{R})$ be a functional such that $\varphi'(0) = 0$. Suppose that $u \in X$ is a nontrivial critical point of φ. Then necessarily u is contained in the set

$$\mathcal{N} = \{u \in X \setminus \{0\} : \langle \varphi'(u), u \rangle = 0\}$$

which is a natural constraint for the problem of finding nontrivial critical points of φ. $\mathcal{N}$ is called the Nehari manifold, though in general it may not be a manifold. Set

$$c = \inf_{u \in \mathcal{N}} \varphi(u).$$

Under suitable conditions on φ one can prove that c is attained at some $u_0 \in \mathcal{N}$ and that u_0 is a critical point. For instance, if φ is bounded below on $\mathcal{N}$ and $u \in \mathcal{N}$ satisfies $\varphi(u) = c$, then u is a critical point of φ different from 0. We recall the following result:

Proposition 2.4.1 ([280]) *Let X be a Banach space and $\varphi \in C^2(X, \mathbb{R})$. Assume that $\mathcal{N} \neq \emptyset$, $\langle \varphi''(u)u, u \rangle \neq 0$ for all $u \in \mathcal{N}$, and there exists $r > 0$ such that $B_r \cap \mathcal{N} = \emptyset$. Then $\mathcal{N}$ is a complete C^1-Banach submanifold of X of codimension 1, and it is a natural constraint of φ.*

In what follows, we present the approach developed in [321, 322] to the method of Nehari manifold. It does not need to make customary assumptions which imply that $\varphi \in C^2(X, \mathbb{R})$ and $\langle \varphi''(u)u, u \rangle \neq 0$ on $\mathcal{N}$.

Let X be a uniformly convex real Banach space, $\varphi \in C^1(X, \mathbb{R})$ and $\varphi(0) = 0$. Let $S = S_1(0) = \{x \in X : \|x\| = 1\}$ be the unit sphere in X. A function $v \in C(\mathbb{R}_+, \mathbb{R}_+)$ is said to be a normalization function if $v(0) = 0$, v is increasing and $v(t) \to \infty$ as $t \to \infty$. We also introduce the following assumptions:

(A_1) There exists a normalization function v such that

$$u \mapsto \psi(u) = \int_0^{\|u\|} v(t)\, dt \in C^1(X \setminus \{0\}, \mathbb{R}),$$

$J = \psi'$ is bounded on bounded sets and $\langle J(w), w \rangle = 1$ for all $w \in S$.

(A_2) For each $w \in X \setminus \{0\}$ there exists s_w such that if $\alpha_w(s) = \varphi(sw)$, then $\alpha'_w(s) > 0$ for $0 < s < s_w$ and $\alpha'_w(s) < 0$ for $s > s_w$.

(A_3) There exists $\delta > 0$ such that $s_w \geq \delta$ for all $w \in S$ and for each compact subset $\mathcal{W} \subset S$ there exists a constant $C_{\mathcal{W}}$ such that $s_w \leq C_{\mathcal{W}}$ for all $w \in \mathcal{W}$.

The map J in (A_1) is called the duality mapping corresponding to φ. It follows from (A_1) that S is a C^1-submanifold of X. Let us note that (A_2) implies that $sw \in \mathcal{N}$ if and only if $s = s_w$. Moreover, by the first part of (A_3), $\mathcal{N}$ is closed in X and bounded away from 0. Define the mappings $\hat{m} : X \setminus \{0\} \to \mathcal{N}$ and $m : S \to \mathcal{N}$ by setting

$$\hat{m}(w) = s_w w \quad \text{and} \quad m = \hat{m}|_S.$$

Proposition 2.4.2 ([322]) *Suppose v satisfies (A_2) and (A_3). Then:*

(a) *The mapping $\hat{m}$ is continuous.*
(b) *The mapping m is a homeomorphism between S and $\mathcal{N}$, and the inverse of m is given by $m^{-1}(u) = u/\|u\|$.*

Let us consider the functionals $\hat{\Psi} : X \setminus \{0\} \to \mathbb{R}$ and $\Psi : S \to \mathbb{R}$ defined by

$$\hat{\Psi}(w) = \varphi(\hat{m}(w)) \quad \text{and} \quad \Psi = \hat{\Psi}|_S.$$

Proposition 2.4.3 ([322]) *Suppose X is a Banach space satisfying (A_1). If φ satisfies (A_2) and (A_3), then $\hat{\Psi} \in C^1(X \setminus \{0\}, \mathbb{R})$ and*

$$\langle \hat{\Psi}'(w), z \rangle = \frac{\|\hat{m}(w)\|}{\|w\|} \langle \varphi'(\hat{m}(w)), z \rangle \quad \text{for all } w, z \in X, w \neq 0.$$

Corollary 2.4.4 ([322]) *Suppose X is a Banach space satisfying (A_1). If φ satisfies (A_2) and (A_3), then:*

(a) $\Psi \in C^1(S, \mathbb{R})$ *and*

$$\langle \Psi'(w), z \rangle = \|m(w)\| \langle \varphi'(m(w)), z \rangle \quad \text{for all } z \in T_w S.$$

(b) *If (w_n) is a Palais–Smale sequence for Ψ, then $(m(w_n))$ is a Palais–Smale sequence for φ. If $(u_n) \subset \mathcal{N}$ is a bounded Palais–Smale sequence for φ, then $(m^{-1}(u_n))$ is a Palais–Smale sequence for Ψ.*

(c) *w is a critical point of Ψ if and only if $m(w)$ is a nontrivial critical point of φ. Moreover, the corresponding values of Ψ and φ coincide and $\inf_S \Psi = \inf_{\mathcal{N}} \varphi$.*

(d) *If φ is even, then so is Ψ.*

We note that the infimum of φ over $\mathcal{N}$ has the following minimax characterization:

$$c = \inf_{u \in \mathcal{N}} \varphi(u) = \inf_{w \in X \setminus \{0\}} \max_{s > 0} \varphi(sw) = \inf_{w \in S} \max_{s > 0} \varphi(sw). \tag{2.4.1}$$

Now, we introduce the subset $S_+ = \{u \in X : \|u\| = 1, u^+ \neq 0\}$ of the unit sphere S in X, and consider the following assumptions:

(I) For each $w \in X$ with $w^+ \neq 0$ there exists s_w such that if $\alpha_w(s) = \varphi(sw)$, then $\alpha_w'(s) > 0$ for $0 < s < s_w$ and $\alpha_w'(s) < 0$ for $s > s_w$.

(II) There exists $\delta > 0$ such that $s_w \geq \delta$ for all $w \in S_+$ and for each compact subset $\mathcal{W} \subset S_+$ there exists a constant $C_{\mathcal{W}}$ such that $s_w \leq C_{\mathcal{W}}$ for all $w \in \mathcal{W}$.

(III) The map $m : S_+ \to \mathcal{N}$ defined by $m(u) = s_u u$ is a homeomorphism between S_+ and $\mathcal{N}$, and the inverse of m is given by $m^{-1}(u) = u/\|u\|$.

As before, $\mathcal{N} = \{u \in X \setminus \{0\} : \langle \varphi'(u), u \rangle = 0\}$ is the Nehari manifold. Let $\Psi : S_+ \to \mathbb{R}$ be defined by $\Psi(w) = \varphi(\hat{m}(w))$. Then Ψ is a C^1-functional on the open subset S_+ of S

and it holds

$$\langle \Psi'(w), z \rangle = \|m(w)\| \langle \varphi'(m(w)), z \rangle \text{ for all } w \in S_+ \text{ and } z \in T_w S_+.$$

Hence, the nontrivial critical points of Ψ are in a one-to-one correspondence with the nontrivial critical points of φ. Moreover, the following result holds true:

Lemma 2.4.5 ([322])

(i) *Let $(u_n) \subset S_+$ be a sequence such that* $\text{dist}(u_n, \partial S_+) \to 0$ *as* $n \to \infty$ *(where the distance is taken with respect to the norm $\| \cdot \|$). Then $\Psi(u_n) \to \infty$.*
(ii) *Ψ satisfies the Palais–Smale condition in S_+.*

Similarly as before, we see that

$$c = \inf_{u \in \mathcal{N}} \varphi(u) = \inf_{w \in X \setminus \{0\}} \max_{s > 0} \varphi(sw) = \inf_{w \in S_+} \max_{s > 0} \varphi(sw) = \inf_{w \in S_+} \Psi(w).$$

By standard deformation arguments with respect to the flow of a pseudogradient vector field of Ψ on S_+, we have an abstract multiplicity result for critical points of Ψ in terms of the Lusternik-Schnirelman category with respect to sublevel sets.

Theorem 2.4.6 ([322]) *If there exist $d \geq c$ and a compact set $K \subset S_+^d$ such that $\text{cat}_{S_+^d}(K) \geq k$ for some $k \in \mathbb{N}$, where $S_+^d = \{u \in S_+ : \Psi(u) \leq d\}$, then S_+^d contains at least k critical points of Ψ. If furthermore $k \geq 2$ and there exists $e > d$ such that K is contractible in S_+^e, then there exists another critical point of Ψ in $S_+^e \setminus S_+^d$.*

Using the definitions of Ψ and m and property (III) above, we obtain the following corollary.

Corollary 2.4.7 ([322]) *If there exist $d \geq c$ and a compact set $K \subset S_+^d$ such that $\text{cat}_{\mathcal{N}^d}(K) \geq k$ for some $k \in \mathbb{N}$, where $\mathcal{N}^d = \{u \in \mathcal{N} : \varphi(u) \leq d\}$, then $\mathcal{N}^d$ contains at least k critical points of φ. If furthermore $k \geq 2$ and there exists $e > d$ such that K is contractible in $\mathcal{N}^e$, then there exists another critical point of φ in $\mathcal{N}^e \setminus \mathcal{N}^d$.*

In the next sections we will see that the approach presented above is fundamental for obtaining multiplicity results for some nonlinear fractional elliptic problems for which the corresponding Nehari manifolds are not differentiable.

Fractional Scalar Field Equations

3.1 Introduction

In the fundamental papers [100, 101], Berestycki and Lions studied the existence and the multiplicity of nontrivial solutions to the equation

$$- \Delta u = g(u) \text{ in } \mathbb{R}^N, \qquad (3.1.1)$$

where $N \geq 3$ and $g : \mathbb{R} \to \mathbb{R}$ is an odd continuous function such that

(BL1) $-\infty < \liminf\limits_{t \to 0} \dfrac{g(t)}{t} \leq \limsup\limits_{t \to 0} \dfrac{g(t)}{t} = -m < 0$;

(BL2) $-\infty < \limsup\limits_{t \to \infty} \dfrac{g(t)}{t^{2^*-1}} \leq 0$, where $2^* = \frac{2N}{N-2}$ is the critical exponent;

(BL3) there exists $\xi_0 > 0$ such that $G(\xi_0) > 0$, where $G(t) = \displaystyle\int_0^t g(\tau) \, d\tau$.

To obtain the existence of a positive solution to (3.1.1), the authors in [100] developed a subtle Lagrange multiplier procedure which ultimately relies on the Pohozaev identity [293] for (3.1.1), namely

$$\frac{N-2}{2} \int_{\mathbb{R}^N} |\nabla u|^2 \, dx = N \int_{\mathbb{R}^N} G(u) \, dx.$$

The lack of compactness due to the translational invariance of (3.1.1) is regained by working in the subspace $H^1_{\text{rad}}(\mathbb{R}^N)$ of $H^1(\mathbb{R}^N)$ of radially symmetric functions. They also proved the existence of a positive solution to (3.1.1) when $m = 0$, the so called zero mass case. In [101] the authors showed that (3.1.1) possesses infinitely many distinct

© The Author(s), under exclusive license to Springer Nature Switzerland AG 2021
V. Ambrosio, *Nonlinear Fractional Schrödinger Equations in* $\mathbb{R}^N$,
Frontiers in Mathematics, https://doi.org/10.1007/978-3-030-60220-8_3

bound states by applying the genus theory to the even functional $V(u) = \int_{\mathbb{R}^N} G(u)\,dx$ defined on the symmetric manifold $M = \{u \in H^1_{\text{rad}}(\mathbb{R}^N) : \|\nabla u\|_{L^2(\mathbb{R}^N)} = 1\}$. In [234] Jeanjean and Tanaka studied a mountain pass characterization of least energy solutions of (3.1.1). Indeed, without the assumption of monotonicity of $t \mapsto \frac{g(t)}{t}$, they showed that the mountain pass value gives the least energy level. Subsequently, the existence and multiplicity of solutions of closely related problems to (3.1.1) have been extensively studied by many authors; see for instance [23, 83, 99, 218, 233].

Later, Zhang and Zou [345] extended the existence result in [100] under the following critical growth assumptions on g:

(ZZ1) $g \in C(\mathbb{R}, \mathbb{R})$ and g is odd;

(ZZ2) $\lim\limits_{t \to 0} \dfrac{g(t)}{t} = -a < 0$;

(ZZ3) $\lim\limits_{t \to \infty} \dfrac{g(t)}{t^{2^*-1}} = b > 0$;

(ZZ4) there exist $C > 0$ and $\max\left\{2, \dfrac{4}{N-2}\right\} < q < 2^*$ such that

$$g(t) - bt^{2^*-1} + at \geq Ct^{q-1} \text{ for all } t > 0.$$

In the spirit of [234], they also proved a mountain pass characterization of least energy solutions of (3.1.1) in the critical case.

In this chapter we deal with the following fractional nonlinear scalar field equation with fractional diffusion

$$(-\Delta)^s u = g(u) \text{ in } \mathbb{R}^N, \quad u \in H^s(\mathbb{R}^N), \tag{3.1.2}$$

with $s \in (0, 1)$ and $N \geq 2$. A naturally arising question is whether or not the above classical existence and multiplicity results for (3.1.1) still hold in the nonlocal framework of (3.1.2). In [139] the authors obtained the existence of a positive solution to (3.1.2) when g is a subcritical nonlinearity, by using the Struwe–Jeanjean monotonicity trick as in [307], and by proving the following Pohozaev identity for the fractional Laplacian (see also [198, 302, 307]):

$$\frac{N - 2s}{2} \int_{\mathbb{R}^N} |(-\Delta)^{\frac{s}{2}} u|^2 \, dx = N \int_{\mathbb{R}^N} G(u) \, dx. \tag{3.1.3}$$

The first aim of this chapter is to answer the question concerning the existence of infinitely many solutions for (3.1.2) when g is a general nonlinearity with subcritical growth. We also provide a mountain pass characterization of least energy solutions to (3.1.2) as in [234]. In this way, we are able to complement and improve the result proved in [139].

Now we state our main assumptions. We will assume that $g : \mathbb{R} \to \mathbb{R}$ is an odd function of class $C^{1,\alpha}$, for some $\alpha > \max\{0, 1 - 2s\}$, with the following properties:

(g1) $-\infty < \liminf\limits_{t \to 0} \dfrac{g(t)}{t} \leq \limsup\limits_{t \to 0} \dfrac{g(t)}{t} = -m < 0$;

(g2) $-\infty < \limsup\limits_{t \to \infty} \dfrac{g(t)}{t^{2^*_s - 1}} \leq 0$;

(g3) there exists $\xi_0 > 0$ such that $G(\xi_0) = \displaystyle\int_0^{\xi_0} g(\tau)\, d\tau > 0$.

We emphasize that the regularity of g is higher than in [100, 101], and this seems to be due to the more demanding assumptions for elliptic regularity in the framework of fractional operators; see [124, 313].

Under assumptions (g1)–(g3), problem (3.1.2) admits a variational formulation, so its weak solutions can be found as critical points of the energy functional $I : H^s(\mathbb{R}^N) \to \mathbb{R}$ defined as

$$I(u) = \frac{1}{2} \iint_{\mathbb{R}^{2N}} \frac{|u(x) - u(y)|^2}{|x - y|^{N+2s}}\, dx\, dy - \int_{\mathbb{R}^N} G(u)\, dx.$$

We recall that by weak solution of problem (3.1.2) we mean a function $u \in H^s(\mathbb{R}^N)$ such that

$$\iint_{\mathbb{R}^{2N}} \frac{(u(x) - u(y))(\varphi(x) - \varphi(y))}{|x - y|^{N+2s}}\, dx\, dy = \int_{\mathbb{R}^N} g(u)\varphi\, dx$$

for all $\varphi \in H^s(\mathbb{R}^N)$, and we say that u is a least energy solution (or ground state solution) to (3.1.2) if

$$I(u) = \mathfrak{m} \quad \text{and} \quad \mathfrak{m} = \inf\left\{I(u) : u \in H^s(\mathbb{R}^N) \setminus \{0\} \text{ is a solution of (3.1.2)}\right\}.$$

Our first main result is the following:

Theorem 3.1.1 ([39]) *Let $s \in (0, 1)$ and $N \geq 2$ and let $g \in C^{1,\alpha}(\mathbb{R}, \mathbb{R})$, with $\alpha > \max\{0, 1 - 2s\}$, be an odd function satisfying (g1)–(g3). Then (3.1.2) possesses a positive least energy solution and infinitely many (possibly sign-changing) radially-symmetric solutions (u_n) such that $I(u_n) \to \infty$ as $n \to \infty$. Moreover, these solutions are of class $C^{2,\beta}(\mathbb{R}^N)$, for some $\beta \in (0, 1)$, and they are characterized by a mountain pass and a symmetric mountain pass arguments, respectively.*

In order to prove our main result, we borrow some arguments developed in [120, 218, 234]. Clearly, due to the nonlocal nature of the fractional Laplacian, some additional difficulties arise in the study of (3.1.2). We would like to point out that our results are in clear agreement with those for the classical local counterpart. For this reason, Theorem 3.1.1 can be seen as the fractional version of the existence and multiplicity results given in [100, 101, 218].

Secondly, we use the mountain pass approach in [218, 234] to prove the existence of a positive solution to (3.1.2) in the null mass case, that is, when g satisfies the following properties:

(h_1) $\displaystyle\limsup_{t \to 0} \frac{g(t)}{t^{2_s^*-1}} \leq 0$;

(h_2) $\displaystyle\lim_{t \to \infty} \frac{g(t)}{t^{2_s^*-1}} = 0$;

(h_3) there exists $\xi_0 > 0$ such that $G(\xi_0) = \displaystyle\int_0^{\xi_0} g(\tau)\, d\tau > 0$.

We point out that the main difficulty related to the zero mass case is due to the fact that the energy of solutions of (3.1.2) can be infinite. In the local setting, i.e., when $s = 1$, several results for zero mass problems were established in [24,26,84,97,98,100,102,319]. We also recall that from a physical point of view, this type of problem is related to the Yang–Mills equations; see for instance [205,206]. However, differently from the classic literature, there is only one work [38] dealing with zero mass problems in nonlocal setting. Motivated by this fact, here we study (3.1.2) when g is a general nonlinearity such that $g'(0) = 0$. Our main second result can be stated as follows:

Theorem 3.1.2 ([39]) *Let $s \in (0,1)$ and $N \geq 2$ and let $g \in C^{1,\alpha}(\mathbb{R}, \mathbb{R})$, with $\alpha > \max\{0, 1 - 2s\}$, be an odd function satisfying (h_1)–(h_3). Then (3.1.2) possesses a positive radially decreasing solution.*

The proof of Theorem 3.1.2 is obtained by combining the mountain pass approach and an approximation argument. Indeed, we show that a solution of (3.1.2) can be approximated by a sequence of positive radially-symmetric solutions (u_ε) in $H^s(\mathbb{R}^N)$, each of which solves an approximate "positive mass" problem $(-\Delta)^s u = g(u) - \varepsilon u$ in $\mathbb{R}^N$. Taking into account the mountain pass characterization of least energy solution of (3.1.2) given in Theorem 3.1.1, we are able to obtain lower and upper bounds for the mountain pass critical levels b_{mp}^ε, which can be estimated independently of ε, when ε is sufficiently small. This allows us to pass to the limit in $(-\Delta)^s u_\varepsilon = g(u_\varepsilon) - \varepsilon u_\varepsilon$ in $\mathbb{R}^N$ as $\varepsilon \to 0$, and to find a nontrivial solution u to (3.1.2). We emphasize that our proof is different from the one given in [100], and that it works even when $s = 1$.

In the second part of this chapter, we consider (3.1.2) with a general critical nonlinearity. More precisely, we assume that $g : \mathbb{R} \to \mathbb{R}$ satisfies the following conditions:

(g_1) $g \in C^{1,\beta}(\mathbb{R}, \mathbb{R})$ for some $\beta > \max\{0, 1 - 2s\}$, and g is odd;

(g_2) $\displaystyle\lim_{t \to 0} \frac{g(t)}{t} = -a < 0$;

(g_3) $\displaystyle\lim_{t\to\infty} \frac{g(t)}{t^{2_s^*-1}} = b > 0$;

(g_4) there exist $C > 0$ and $\max\{2_s^* - 2, 2\} < q < 2_s^*$ such that

$$g(t) - bt^{2_s^*-1} + at \geq Ct^{q-1} \text{ for all } t > 0.$$

Then we are able to prove the following result:

Theorem 3.1.3 ([36]) *Let* $s \in (0,1)$ *and* $N \geq 2$. *Assume that* g *satisfies* (g_1)–(g_4). *Then* (3.1.2) *possesses a radial positive least energy solution* $\omega \in H^s_{\mathrm{rad}}(\mathbb{R}^N)$.

Remark 3.1.4 Let us observe that if $g(t) = b|t|^{2_s^*-2}t - at$, then g satisfies (g_1)–(g_3), but not (g_4). Moreover, in view of the Pohozaev identity, it is not difficult to prove that (3.1.2) does not admit solution. Hence, without (g_4), assumptions (g_1)–(g_3) are not sufficient to guarantee the existence of a ground state to (3.1.2).

To establish the existence of a solution for (3.1.2), we set

$$\mathcal{I}(u) = \frac{1}{2} \iint_{\mathbb{R}^{2N}} \frac{|u(x) - u(y)|^2}{|x-y|^{N+2s}} \, dx dy - \int_{\mathbb{R}^N} G(u) \, dx = \mathcal{T}(u) - \mathcal{V}(u),$$

and we consider the constrained minimization problem

$$M = \inf\left\{\mathcal{T}(u) : \mathcal{V}(u) = 1, u \in H^s(\mathbb{R}^N)\right\}.$$

Motivated by Zhang and Zou [345], we use some crucial estimates obtained in [310] and prove that M satisfies the following bounds:

$$0 < M < \frac{1}{2}(2_s^*)^{\frac{N-2s}{N}} S_*,$$

where S_* is the best constant in the fractional Sobolev embedding $\mathcal{D}^{s,2}(\mathbb{R}^N) \to L^{2_s^*}(\mathbb{R}^N)$. Thanks to this information, we can see that the above minimization problem admits a minimizer in $H^s(\mathbb{R}^N)$ which is positive and radially symmetric. Then we show the existence of a least energy solution $\omega \in H^s(\mathbb{R}^N)$ to (3.1.2). Finally, observing that $\mathcal{I}$ has a mountain pass geometry, we also obtain the following result in the spirit of [234]:

Theorem 3.1.5 ([36]) *Assume that* g *satisfies* (g_1)–(g_4). *Then*

$$c = \mathfrak{m},$$

where

$$c = \inf_{\gamma \in \Gamma} \max_{t \in [0,1]} \mathcal{I}(\gamma(t)), \quad \Gamma = \{\gamma \in C([0,1], H^s(\mathbb{R}^N)) : \gamma(0) = 0 \text{ and } \mathcal{I}(\gamma(1)) < 0\},$$

(3.1.4)

and

$$\mathfrak{m} = \inf \left\{ \mathcal{I}(u) : u \in H^s(\mathbb{R}^N) \setminus \{0\} \text{ is a solution of } (3.1.2) \right\}.$$

That is, the mountain pass value gives the least energy level. Moreover, for any least energy solution ω of (3.1.2), there exists a path $\gamma \in \Gamma$ such that $\omega \in \gamma([0,1])$ and $\max_{t \in [0,1]} \mathcal{I}(\gamma(t)) = \mathcal{I}(\omega)$.

In Sect. 3.2 we prove the existence of infinitely many solutions to (3.1.2) by using an auxiliary functional $\tilde{I}(\theta, u)$ on the augmented space $\mathbb{R} \times H^s_{\text{rad}}(\mathbb{R}^N)$. Then we obtain the existence of a least energy solution to (3.1.2) characterized by a mountain pass argument, and we investigate the regularity and symmetry of this solution. In Sect. 3.3, by using an approximation argument, we get a positive radially symmetric solution to (3.1.2) in the null mass case. In Sect. 3.4 we give the proofs of Theorems 3.1.3 and 3.1.5. In Section 3.5 we give a proof of the Pohozaev identity for $(-\Delta)^s$.

3.2   The Subcritical Case

3.2.1   Introduction of a Penalty Functional

Following [100], we redefine the nonlinearity g in a convenient way.

(*i*) If $g(t) > 0$ for all $t \geq \xi_0$, we simply extend g to the negative axis:

$$\tilde{g}(t) = \begin{cases} g(t), & \text{for } t \geq 0, \\ -g(-t), & \text{for } t < 0. \end{cases}$$

(*ii*) If there exists $t_0 > \xi_0$ such that $g(t_0) = 0$, we put

$$\tilde{g}(t) = \begin{cases} g(t), & \text{for } t \in [0, t_0], \\ 0, & \text{for } t > t_0, \\ -\tilde{g}(-t), & \text{for } t < 0. \end{cases}$$

Then $\tilde{g}$ satisfies $(g1)$, $(g3)$ and

$$\lim_{t \to \infty} \frac{\tilde{g}(t)}{|t|^{2_s^* - 1}} = 0. \tag{$g2'$}$$

Clearly, any solution to $(-\Delta)^s u = \tilde{g}(u)$ in $\mathbb{R}^N$ is also a solution to (3.1.2). Indeed, in the case (ii) above, any solution to $(-\Delta)^s u = \tilde{g}(u)$ in $\mathbb{R}^N$ satisfies $-t_0 \leq u \leq t_0$ in $\mathbb{R}^N$. To prove this, we use $(u - t_0)^+ \in H^s(\mathbb{R}^N)$ as test function in the weak formulation of $(-\Delta)^s u = \tilde{g}(u)$ in $\mathbb{R}^N$, and we get

$$\iint_{\mathbb{R}^{2N}} \frac{(u(x) - u(y))}{|x - y|^{N+2s}} ((u(x) - t_0)^+ - (u(y) - t_0)^+) \, dx dy = \int_{\mathbb{R}^N} \tilde{g}(u)(u(x) - t_0)^+ \, dx.$$

By the definition of $\tilde{g}$, it follows that $\int_{\mathbb{R}^N} \tilde{g}(u)(u(x) - t_0)^+ \, dx = 0$. Then, recalling that $(x - y)(x^+ - y^+) \geq |x^+ - y^+|^2$ for all $x, y \in \mathbb{R}$, we see that

$$\iint_{\mathbb{R}^{2N}} \frac{|(u(x) - t_0)^+ - (u(y) - t_0)^+|^2}{|x - y|^{N+2s}} \, dx dy \leq 0,$$

which implies that $[(u - t_0)^+]_s^2 \leq 0$. Since $u \in H^s(\mathbb{R}^N)$, we deduce that $u \leq t_0$ in $\mathbb{R}^N$. The other inequality is obtained in a similar way by using $(u + t_0)^-$ as test function. We recall that $x^+ = \max\{x, 0\}$ and $x^- = \min\{x, 0\}$. Hereafter, we tacitly write g instead of $\tilde{g}$, and we will assume that g satisfies $(g1)$, $(g2')$ and $(g3)$.

Now we introduce a penalty function to construct an auxiliary functional. For $t \geq 0$ we define

$$f(t) = \max\left\{0, \frac{1}{2}mt + g(t)\right\},$$

and

$$h(t) = t^p \sup_{0 < \tau \leq t} \frac{f(\tau)}{\tau^p},$$

where p is a positive number such that $1 < p < 2_s^* - 1$. Note that f and h are well defined in view of $(g1)$. We extend h as an odd function on $\mathbb{R}$ and we set

$$H(t) = \int_0^t h(\tau) \, d\tau.$$

The next result (except the last property) can be obtained following [218]. For the reader's convenience we provide the details.

Lemma 3.2.1

(h1) $h \in C(\mathbb{R}, \mathbb{R})$, $h(t) \geq 0$ and $h(-t) = -h(t)$ for all $t \geq 0$.
(h2) There exists $\beta > 0$ such that $h = 0 = H$ on $[-\beta, \beta]$.
(h3) For all $t \in \mathbb{R}$

$$\frac{1}{2}mt^2 + g(t)t \leq h(t)t \quad and \quad \frac{1}{4}mt^2 + G(t) \leq H(t).$$

(h4) $\lim\limits_{|t| \to \infty} \dfrac{h(t)}{|t|^{2_s^* - 1}} = 0.$
(h5) h satisfies the following Ambrosetti-Rabinowitz condition:

$$0 \leq (p+1)H(t) \leq h(t)t \text{ for all } t \in \mathbb{R}.$$

(h6) If (u_k) is a bounded sequence in $H_{\mathrm{rad}}^s(\mathbb{R}^N)$, then

$$\lim_{k \to \infty} \int_{\mathbb{R}^N} h(u_k)u_k \, dx = \int_{\mathbb{R}^N} h(u)u \, dx.$$

Proof Clearly, (h1), (h2), and (h3) follow from (g1) and the definitions of f, h and H. Concerning (h4), we first note that

$$\frac{h(t)}{t^{2_s^* - 1}} = t^{-(2_s^* - 1 - p)} \sup_{0 < \tau \leq t} \frac{f(\tau)}{\tau^p} = \sup_{0 < \tau \leq t} \frac{f(\tau)}{\tau^{2_s^* - 1}} \frac{\tau^{2_s^* - 1 - p}}{t^{2_s^* - 1 - p}}.$$

Since f satisfies (g2'), for any $\varepsilon > 0$ there exists $t_\varepsilon > 0$ such that

$$\left| \frac{f(\tau)}{\tau^{2_s^* - 1}} \right| < \varepsilon \text{ for all } \tau \geq t_\varepsilon.$$

Set $C_\varepsilon = \sup\limits_{0 < \tau \leq t_\varepsilon} \left| \dfrac{f(\tau)}{\tau^{2_s^* - 1}} \right|$. Hence we obtain

$$\frac{h(t)}{t^{2_s^* - 1}} \leq \max \left\{ \sup_{0 < \tau \leq t_\varepsilon} \left| \frac{f(\tau)}{\tau^{2_s^* - 1}} \right| \frac{t_\varepsilon^{2_s^* - 1 - p}}{t^{2_s^* - 1 - p}}, \sup_{t_\varepsilon \leq \tau \leq t_\varepsilon} \left| \frac{f(\tau)}{\tau^{2_s^* - 1}} \right| \right\} \leq \max \left\{ \frac{C_\varepsilon t_\varepsilon^{2_s^* - 1 - p}}{t^{2_s^* - 1 - p}}, \varepsilon \right\}$$

which implies that

$$0 \leq \limsup_{t \to \infty} \frac{h(t)}{t^{2_s^* - 1}} \leq \varepsilon.$$

The arbitrariness of ε shows that $(h4)$ holds. Property $(h5)$ follows from the definitions of h and H:

$$(p+1)H(t) - h(t)t = \int_0^t [(p+1)h(\tau) - h(t)] \, d\tau$$

$$= \int_0^t \left[(p+1)\tau^p \sup_{0<\xi\leq\tau} \frac{f(\xi)}{\xi^p} - t^p \sup_{0<\xi\leq t} \frac{f(\xi)}{\xi^p} \right] d\tau$$

$$\leq \sup_{0<\xi\leq t} \frac{f(\xi)}{\xi^p} \int_0^t (p+1)\tau^p - t^p \, d\tau = 0.$$

Finally, we prove $(h6)$. Let (u_k) be a bounded sequence in $H^s_{\text{rad}}(\mathbb{R}^N)$. Using $(h2)$ and $(h4)$ we know that

$$\lim_{|t|\to 0} \frac{h(t)t}{|t|^2} = 0 \quad \text{and} \quad \lim_{|t|\to\infty} \frac{h(t)t}{|t|^{2^*_s}} = 0.$$

Then we can apply Lemma 1.4.3 with $P(t) = h(t)t$, $q_1 = 2$ and $q_2 = 2^*_s$ to deduce that

$$\lim_{k\to\infty} \int_{\mathbb{R}^N} h(u_k)u_k \, dx = \int_{\mathbb{R}^N} h(u)u \, dx. \qquad \square$$

3.2.2 Comparison between Functionals

Let us consider the following norm on $H^s(\mathbb{R}^N)$

$$\|u\| = \left([u]_s^2 + \frac{m}{2}\|u\|^2_{L^2(\mathbb{R}^N)} \right)^{\frac{1}{2}}$$

which is equivalent to the standard norm $\|\cdot\|_{H^s(\mathbb{R}^N)}$ defined in Chap. 1. Let us define the functionals $I : H^s_{\text{rad}}(\mathbb{R}^N) \to \mathbb{R}$ and $J : H^s_{\text{rad}}(\mathbb{R}^N) \to \mathbb{R}$ by setting

$$I(u) = \frac{1}{2}\|u\|^2 - \int_{\mathbb{R}^N} \left[\frac{m}{4}u^2 + G(u) \right] dx$$

and

$$J(u) = \frac{1}{2}\|u\|^2 - 2\int_{\mathbb{R}^N} H(u) \, dx.$$

By the growth assumptions on g, it is easy to check that I and J are well defined and that they are $C^1(H^s_{\text{rad}}(\mathbb{R}^N))$-functionals (to prove that $\int_{\mathbb{R}^N} G(u) \, dx$ and $\int_{\mathbb{R}^N} H(u) \, dx$ are of class C^1 in $H^s(\mathbb{R}^N)$ one can argue as in [100]; see also Theorem 2.1.5 with $p = 1$,

$\sigma = 2_s^* - 1$, $\Omega = \mathbb{R}^N$ and $X = H^s(\mathbb{R}^N)$). Clearly, critical points of I and J are weak solutions to (3.1.2) and $(-\Delta)^s u + \dfrac{m}{2} u = 2h(u)$ in $\mathbb{R}^N$, respectively. In what follows, we show that I and J have a symmetric mountain pass geometry [29].

Lemma 3.2.2

(i) $I(u) \geq J(u)$ for all $u \in H_{\mathrm{rad}}^s(\mathbb{R}^N)$.

(ii) There are $\delta > 0$ and $\rho > 0$ such that

$$I(u), J(u) \geq \delta > 0 \quad \text{for } \|u\| = \rho,$$
$$I(u), J(u) \geq 0 \qquad \text{for } \|u\| \leq \rho.$$

(iii) For any $n \in \mathbb{N}$ there exists an odd continuous map $\gamma_n : \mathbb{S}^{n-1} \to H_{\mathrm{rad}}^s(\mathbb{R}^N)$ such that

$$I(\gamma_n(\sigma)), J(\gamma_n(\sigma)) < 0 \quad \text{for all } \sigma \in \mathbb{S}^{n-1}.$$

Proof

(i) is a consequence of Lemma 3.2.1-(h3).

(ii) By Lemma 3.2.1, we know that there exists $C > 0$ such that

$$H(t) \leq C|t|^{2_s^*} \quad \text{for all } t \in \mathbb{R}.$$

Then, by Theorem 1.1.8, it follows that

$$J(u) \geq \frac{1}{2}\|u\|^2 - C\|u\|_{L^{2_s^*}(\mathbb{R}^N)}^{2_s^*} \geq \|u\|^2 \left[\frac{1}{2} - C_*'\|u\|^{2_s^* - 2}\right].$$

Since $2_s^* > 2$, we can find $\delta, \rho > 0$ such that $J(u) \geq \delta$ for $\|u\| = \rho$, and $J(u) \geq 0$ if $\|u\| \leq \rho$. Now (i) implies (ii).

(iii) Arguing as in the proof of Theorem 10 in [101], for every $n \in \mathbb{N}$ there exists an odd continuous map $\pi_n : \mathbb{S}^{n-1} \to H_{\mathrm{rad}}^1(\mathbb{R}^N)$ such that

$$0 \notin \pi_n(\mathbb{S}^{n-1}), \quad \int_{\mathbb{R}^N} G(\pi_n(\sigma)) \, dx \geq 1 \quad \text{for all } \sigma \in \mathbb{S}^{n-1}.$$

Since $H^1(\mathbb{R}^N) \subset H^s(\mathbb{R}^N)$ and $\pi_n(\mathbb{S}^{n-1})$ is compact, there exists $M > 0$ such that

$$\|\pi_n(\sigma)\|_{H^s(\mathbb{R}^N)} \leq M \quad \text{for any } \sigma \in \mathbb{S}^{n-1}.$$

For $t > 0$, define $\psi_n^t(\sigma)(x) = \pi_n(\sigma)(\frac{x}{t})$. Hence,

$$I(\psi_n^t(\sigma)) = \frac{t^{N-2s}}{2}[\pi_n(\sigma)]_{H^s(\mathbb{R}^N)}^2 - t^N \int_{\mathbb{R}^N} G(\pi_n(\sigma))\,dx$$

$$\leq t^{N-2s}\left[\frac{M}{2} - t^{2s}\right] \to -\infty \quad \text{as } t \to \infty,$$

and thus we can find $\bar{t} > 0$ sufficiently large such that $I(\psi_n^{\bar{t}}(\sigma)) < 0$ for all $\sigma \in \mathbb{S}^{n-1}$. By setting $\gamma_n(\sigma)(x) = \psi_n^{\bar{t}}(\sigma)(x)$, we infer that γ_n satisfies the required properties for I. In view of (i), we deduce that γ_n has the desired properties for J.   □

Differently from $I(u)$, the comparison functional $J(u)$ enjoys the following compactness property:

Theorem 3.2.3 *The functional J satisfies the Palais–Smale condition.*

Proof Let $c \in \mathbb{R}$ and let $(u_k) \subset H_{\mathrm{rad}}^s(\mathbb{R}^N)$ be a sequence such that

$$J(u_k) \to c \quad \text{and} \quad J'(u_k) \to 0 \text{ in } (H_{\mathrm{rad}}^s(\mathbb{R}^N))^*. \tag{3.2.1}$$

Using $(h5)$ of Lemma 3.2.1, we get

$$C(1 + \|u_k\|) \geq J(u_k) - \frac{1}{p+1}\langle J'(u_k), u_k\rangle$$

$$= \left(\frac{1}{2} - \frac{1}{p+1}\right)\|u_k\|^2 - 2\int_{\mathbb{R}^N}\left[H(u_k) - \frac{1}{p+1}h(u_k)u_k\right]dx$$

$$\geq \left(\frac{1}{2} - \frac{1}{p+1}\right)\|u_k\|^2$$

so we deduce that (u_k) is bounded in $H_{\mathrm{rad}}^s(\mathbb{R}^N)$. Then, in light of Theorem 1.1.11, up to a subsequence, we may assume that

$$u_k \rightharpoonup u \text{ in } H_{\mathrm{rad}}^s(\mathbb{R}^N),$$

$$u_k \to u \text{ in } L^q(\mathbb{R}^N) \quad \forall q \in (2, 2_s^*), \tag{3.2.2}$$

$$u_k \to u \text{ a.e. } \mathbb{R}^N.$$

Using $(h1)$ and $(h4)$ in Lemma 3.2.1 and (3.2.2), we can apply the first part of Lemma 1.4.2 with $P(t) = h(t)$ and $Q(t) = |t|^{2_s^* - 1}$ to infer that

$$\lim_{k \to \infty} \int_{\mathbb{R}^N} h(u_k) \varphi \, dx = \int_{\mathbb{R}^N} h(u) \varphi \, dx \tag{3.2.3}$$

for all $\varphi \in C_c^\infty(\mathbb{R}^N)$. Putting together (3.2.1), (3.2.2), and (3.2.3) we obtain

$$\langle J'(u), \varphi \rangle = \langle J'(u_k), \varphi \rangle - \langle u_k - u, \varphi \rangle_{\mathcal{D}^{s,2}(\mathbb{R}^N)} - \frac{m}{2} \int_{\mathbb{R}^N} (u_k - u) \varphi \, dx$$

$$+ 2 \int_{\mathbb{R}^N} (h(u_k) - h(u)) \varphi \, dx \to 0$$

for all $\varphi \in C_c^\infty(\mathbb{R}^N)$. Since $C_c^\infty(\mathbb{R}^N)$ is dense in $H_{\mathrm{rad}}^s(\mathbb{R}^N)$, it follows that

$$\langle J'(u), \varphi \rangle = 0 \quad \text{for all } \varphi \in H_{\mathrm{rad}}^s(\mathbb{R}^N).$$

Moreover, $\langle J'(u), u \rangle = 0$, which implies that $\|u\|^2 = 2 \int_{\mathbb{R}^N} h(u) u \, dx$.

On the other hand, by $(h6)$ of Lemma 3.2.1, we know that

$$\lim_{k \to \infty} \int_{\mathbb{R}^N} h(u_k) u_k \, dx = \int_{\mathbb{R}^N} h(u) u \, dx. \tag{3.2.4}$$

Taking into account the boundedness of (u_k) and (3.2.1), we get

$$\langle J'(u_k), u_k \rangle \to 0,$$

which together with (3.2.4) implies that $\|u_k\| \to \|u\|$ as $k \to \infty$. Therefore, $u_k \to u$ in $H_{\mathrm{rad}}^s(\mathbb{R}^N)$ as $k \to \infty$. $\qquad \square$

Now we define minimax values of I and J by using the maps (γ_n) in Lemma 3.2.2. For any $n \in \mathbb{N}$, we set

$$b_n = \inf_{\gamma \in \Gamma_n} \max_{\sigma \in D_n} I(\gamma(\sigma)), \quad c_n = \inf_{\gamma \in \Gamma_n} \max_{\sigma \in D_n} J(\gamma(\sigma)),$$

where $D_n = \{\sigma \in \mathbb{R}^n : |\sigma| \le 1\}$ and

$$\Gamma_n = \left\{ \gamma \in C(D_n, H_{\mathrm{rad}}^s(\mathbb{R}^N)) : \gamma \text{ is odd and } \gamma = \gamma_n \text{ on } \mathbb{S}^{n-1} \right\}.$$

The values b_n and c_n have the following properties.

Lemma 3.2.4

(i) $\Gamma_n \neq \emptyset$ for any $n \in \mathbb{N}$.
(ii) $0 < \delta \leq c_n \leq b_n$ for any $n \in \mathbb{N}$, where δ appears in Lemma 3.2.2.

Proof

(i) Let us define

$$\tilde{\gamma}_n(\sigma) = \begin{cases} |\sigma| \gamma_n(\frac{\sigma}{|\sigma|}), & \text{for } \sigma \in D_n \setminus \{0\}, \\ 0, & \text{for } \sigma = 0. \end{cases}$$

Clearly, $\tilde{\gamma}_n \in \Gamma_n$.

(ii) By (i) of Lemma 3.2.2, $c_n \leq b_n$ for any $n \in \mathbb{N}$. The property $\delta \leq c_n$ follows from the fact that

$$\{u \in H^s_{\mathrm{rad}}(\mathbb{R}^N) : \|u\| = \rho\} \cap \gamma(D_n) \neq \emptyset \text{ for all } \gamma \in \Gamma_n. \qquad \square$$

Lemma 3.2.5

(i) *The value c_n is a critical value of J.*
(ii) $c_n \to \infty$ *as $n \to \infty$.*

Proof

(i) Follows from Theorem 3.2.3.
(ii) Set

$$\Sigma_n = \left\{ h \in C(\overline{D_m \setminus Y}) : h \in \Gamma_m, m \geq n, Y \in \mathcal{E}_m \text{ and genus}(Y) \leq m - n \right\}$$

where $\mathcal{E}_m$ is the family of closed sets $A \subset \mathbb{R}^m \setminus \{0\}$ such that $-A = A$ and genus(A) is the Krasnoselskii genus of A. Now we define another sequence of minimax values by setting

$$d_n = \inf_{A \in \Sigma_n} \max_{u \in A} J(u).$$

Then $d_n \leq d_{n+1}$ for all $n \in \mathbb{N}$, and $d_n \leq c_n$ for all $n \in \mathbb{N}$. Since J satisfies the Palais–Smale condition, we can proceed as in the proof of Proposition 9.33 in [298] to infer that $d_n \to \infty$ as $n \to \infty$. Consequently, $c_n \to \infty$ as $n \to \infty$. $\qquad \square$

We observe that Lemmas 3.2.4-(ii) and 3.2.5-(ii) yield

$$b_n > 0 \text{ for all } n \in \mathbb{N} \quad \text{and} \quad \lim_{n \to \infty} b_n = \infty. \tag{3.2.5}$$

In the next section we will prove that b_n are critical values of $I(u)$.

3.2.3   An Auxiliary Functional on the Augmented Space $\mathbb{R} \times H^s_{\mathrm{rad}}(\mathbb{R}^N)$

Let us introduce the following functional

$$\tilde{I}(\theta, u) = \frac{1}{2} e^{(N-2s)\theta}[u]^2_s - e^{N\theta} \int_{\mathbb{R}^N} G(u)\, dx$$

for $(\theta, u) \in \mathbb{R} \times H^s_{\mathrm{rad}}(\mathbb{R}^N)$. We endow $\mathbb{R} \times H^s_{\mathrm{rad}}(\mathbb{R}^N)$ with the norm

$$\|(\theta, u)\|_{\mathbb{R} \times H^s(\mathbb{R}^N)} = \sqrt{|\theta|^2 + \|u\|^2}.$$

Let us point out that $\tilde{I} \in C^1(\mathbb{R} \times H^s_{\mathrm{rad}}(\mathbb{R}^N), \mathbb{R})$, $\tilde{I}(0, u) = I(u)$ and that

$$\iint_{\mathbb{R}^{2N}} \frac{|u(e^{-\theta}x) - u(e^{-\theta}y)|^2}{|x - y|^{N+2s}}\, dxdy = e^{(N-2s)\theta}[u]^2_s$$

$$\int_{\mathbb{R}^N} G(u(e^{-\theta}x))\, dx = e^{N\theta} \int_{\mathbb{R}^N} G(u(x))\, dx.$$

In particular, we have

$$\tilde{I}(\theta, u(x)) = I(u(e^{-\theta}x)) \quad \text{for all } \theta \in \mathbb{R}, u \in H^s_{\mathrm{rad}}(\mathbb{R}^N). \tag{3.2.6}$$

We also define minimax values $\tilde{b}_n$ for $\tilde{I}(\theta, u)$ by

$$\tilde{b}_n = \inf_{\tilde{\gamma} \in \tilde{\Gamma}_n} \max_{\sigma \in D_n} \tilde{I}(\tilde{\gamma}(\sigma)),$$

where

$$\tilde{\Gamma}_n = \{\tilde{\gamma} \in C(D_n, \mathbb{R} \times H^s_{\mathrm{rad}}(\mathbb{R}^N)) : \tilde{\gamma}(\sigma) = (\theta(\sigma), \eta(\sigma)) \text{ satisfies}$$

$$(\theta(-\sigma), \eta(-\sigma)) = (\theta(\sigma), -\eta(\sigma)) \text{ for all } \sigma \in D_n,$$

$$(\theta(\sigma), \eta(\sigma)) = (0, \gamma_n(\sigma)) \text{ for all } \sigma \in \mathbb{S}^{n-1}\}.$$

Through the modified functional $\tilde{I}(\theta, u)$ we aim to show that b_n are critical values for $I(u)$. We begin by proving that

Lemma 3.2.6 $\tilde{b}_n = b_n$ for all $n \in \mathbb{N}$.

Proof Observe that $(0, \gamma(\sigma)) \in \tilde{\Gamma}_n$ for any $\gamma \in \Gamma_n$, and so $\Gamma_n \subset \tilde{\Gamma}_n$. Since $\tilde{I}(0, u) = I(u)$, by the definitions of b_n and $\tilde{b}_n$, it follows that $\tilde{b}_n \leq b_n$ for all $n \in \mathbb{N}$.

Now, take $\tilde{\gamma}(\sigma) = (\theta(\sigma), \eta(\sigma)) \in \tilde{\Gamma}_n$ and put $\gamma(\sigma) = \eta(\sigma)(e^{-\theta(\sigma)}x)$. Then it is easily verified that $\gamma \in \Gamma_n$ and, by using (3.2.6), $I(\gamma(\sigma)) = \tilde{I}(\tilde{\gamma}(\sigma))$ for all $\sigma \in D_n$. Consequently, we have $\tilde{b}_n \geq b_n$ for all $n \in \mathbb{N}$. $\qquad\square$

Next we show that the functional $\tilde{I}(\theta, u)$ admits a Palais–Smale sequence in $\mathbb{R} \times H^s_{\text{rad}}(\mathbb{R}^N)$ with a property related to the Pohozaev identity (3.1.3). First, we give a version of Ekeland's variational principle [176].

Lemma 3.2.7 *Let* $n \in \mathbb{N}$ *and* $\varepsilon > 0$. *Assume that* $\tilde{\gamma}_n \in \tilde{\Gamma}_n$ *satisfies*

$$\max_{\sigma \in D_n} \tilde{I}(\tilde{\gamma}(\sigma)) \leq \tilde{b}_n + \varepsilon.$$

Then there exists $(\theta, u) \subset \mathbb{R} \times H^s_{\text{rad}}(\mathbb{R}^N)$ *such that:*

(i) $\text{dist}_{\mathbb{R} \times H^s_{\text{rad}}(\mathbb{R}^N)}((\theta, u), \tilde{\gamma}(D_n)) \leq 2\sqrt{\varepsilon}$;
(ii) $\tilde{I}(\theta, u) \in [b_n - \varepsilon, b_n + \varepsilon]$;
(iii) $\|\nabla \tilde{I}(\theta, u)\|_{\mathbb{R} \times (H^s_{\text{rad}}(\mathbb{R}^N))^*} \leq 2\sqrt{\varepsilon}$.

Here we used the notations

$$\text{dist}_{\mathbb{R} \times H^s_{\text{rad}}(\mathbb{R}^N)}((\theta, u), A) = \inf_{(t, v) \in A} \sqrt{|\theta - t|^2 + \|u - v\|^2}$$

for $A \subset \mathbb{R} \times H^s_{\text{rad}}(\mathbb{R}^N)$, *and*

$$\nabla \tilde{I}(\theta, u) = \left(\frac{\partial}{\partial \theta} \tilde{I}(\theta, u), \tilde{I}'(\theta, u) \right).$$

Proof Since $\tilde{I}(\theta, -u) = \tilde{I}(\theta, u)$ for all $(\theta, u) \in \mathbb{R} \times H^s_{\text{rad}}(\mathbb{R}^N)$, we deduce that $\tilde{\Gamma}_n$ is stable under the pseudo-deformation flow generated by $\tilde{I}(\theta, u)$. Since $\tilde{b}_n = b_n > 0$, $I(0) = 0$ and $\max_{\sigma \in D_n} \tilde{I}(0, \gamma_n(\sigma)) < 0$, the proof of Lemma 3.2.7 goes as in [230]. $\qquad\square$

Then we can deduce the following result:

Theorem 3.2.8 *For every $n \in \mathbb{N}$ there exists a sequence $((\theta_k, u_k)) \subset \mathbb{R} \times H^s_{\mathrm{rad}}(\mathbb{R}^N)$ such that*

(i) $\theta_k \to 0$;
(ii) $\tilde{I}(\theta_k, u_k) \to b_n$;
(iii) $\tilde{I}'(\theta_k, u_k) \to 0$ *strongly in* $(H^s_{\mathrm{rad}}(\mathbb{R}^N))^*$;
(iv) $\frac{\partial}{\partial \theta} \tilde{I}(\theta_k, u_k) \to 0$.

Proof For every $k \in \mathbb{N}$ there exists $\gamma_k \in \Gamma_n$ such that

$$\max_{\sigma \in D_n} I(\gamma_k(\sigma)) \le b_n + \frac{1}{k}.$$

Since $\tilde{\gamma}_k(\sigma) = (0, \gamma_k(\sigma)) \in \tilde{\Gamma}_n$ and $\tilde{b}_n = b_n$, we have

$$\max_{\sigma \in D_n} \tilde{I}(\tilde{\gamma}_k(\sigma)) \le \tilde{b}_n + \frac{1}{k}.$$

It follows from Lemma 3.2.7 that there exists $((\theta_k, u_k)) \subset \mathbb{R} \times H^s_{\mathrm{rad}}(\mathbb{R}^N)$ such that

$$\mathrm{dist}_{\mathbb{R} \times H^s_{\mathrm{rad}}(\mathbb{R}^N)}((\theta_k, u_k), \tilde{\gamma}_k(D_n)) \le \frac{2}{\sqrt{k}}, \tag{3.2.7}$$

$$\tilde{I}(\theta_k, u_k) \in \left[b_n - \frac{1}{k}, b_n + \frac{1}{k} \right], \tag{3.2.8}$$

$$\|\nabla \tilde{I}(\theta_k, u_k)\|_{\mathbb{R} \times (H^s_{\mathrm{rad}}(\mathbb{R}^N))^*} \le \frac{2}{\sqrt{k}}. \tag{3.2.9}$$

Then, (3.2.7) and $\tilde{\gamma}_k(D_n) \subset \{0\} \times H^s_{\mathrm{rad}}(\mathbb{R}^N)$ yield (i). Clearly, (ii) follows from (3.2.8), while (iii) and (iv) are consequences of (3.2.9). $\qquad \square$

Next, we investigate the boundedness and the compactness properties of the sequence $(\theta_k, u_k) \subset \mathbb{R} \times H^s_{\mathrm{rad}}(\mathbb{R}^N)$ obtained in Theorem 3.2.8. More precisely, we are able to prove

Theorem 3.2.9 *Let $((\theta_k, u_k)) \subset \mathbb{R} \times H^s_{\mathrm{rad}}(\mathbb{R}^N)$ be a sequence satisfying (i)–(iv) of Theorem 3.2.8. Then we have*

(a) (u_k) *is bounded in* $H^s_{\mathrm{rad}}(\mathbb{R}^N)$.
(b) (θ_k, u_k) *has a strongly convergent subsequence in* $\mathbb{R} \times H^s_{\mathrm{rad}}(\mathbb{R}^N)$.

Proof

(a) Using (ii) and (iv) of Theorem 3.2.8, we can see that

$$\frac{1}{2}e^{(N-2s)\theta_k}[u_k]_s^2 - e^{N\theta_k}\int_{\mathbb{R}^N} G(u_k)\,dx \to b_n$$

and

$$\frac{N-2s}{2}e^{(N-2s)\theta_k}[u_k]_s^2 - Ne^{N\theta_k}\int_{\mathbb{R}^N} G(u_k)\,dx \to 0$$

as $k \to \infty$. Consequently,

$$[u_k]_s^2 \to \frac{Nb_n}{s} \tag{3.2.10}$$

and

$$\int_{\mathbb{R}^N} G(u_k)\,dx \to \frac{N-2s}{2s}b_n$$

as $k \to \infty$. Thanks to Lemma 3.2.1, there exists $C > 0$ such that

$$|h(t)| \le C|t|^{2_s^*-1} \text{ for all } t \in \mathbb{R}. \tag{3.2.11}$$

Set

$$\varepsilon_k = \|\tilde{I}'(\theta_k, u_k)\|_{(H_{\mathrm{rad}}^s(\mathbb{R}^N))^*}.$$

By item (iii) of Theorem 3.2.8, we deduce that $\varepsilon_k \to 0$ as $k \to \infty$, so we get

$$|\langle \tilde{I}'(\theta_k, u_k), u_k\rangle| \le \varepsilon_k \|u_k\|. \tag{3.2.12}$$

Taking into account (3.2.10), (3.2.11), and (3.2.12), $(h3)$ of Lemma 3.2.1 and invoking Theorem 1.1.8, we have

$$e^{(N-2s)\theta_k}[u_k]_s^2 + \frac{m}{2}e^{N\theta_k}\|u_k\|_{L^2(\mathbb{R}^N)}^2 \le e^{N\theta_k}\int_{\mathbb{R}^N} \frac{m}{2}u_k^2 + g(u_k)u_k\,dx + \varepsilon_k\|u_k\|$$

$$\le e^{N\theta_k}\int_{\mathbb{R}^N} h(u_k)u_k\,dx + \varepsilon_k\|u_k\|$$

$$\le Ce^{N\theta_k}\|u_k\|_{L^{2_s^*}(\mathbb{R}^N)}^{2_s^*} + \varepsilon_k\|u_k\|$$

$$\leq CC_* e^{N\theta_k} [u_k]_s^{2_s^*} + \varepsilon_k \|u_k\|$$

$$\leq CC_* C' e^{N\theta_k} + \varepsilon_k \|u_k\|,$$

which implies the boundedness of $\|u_k\|_{L^2(\mathbb{R}^N)}$. Now using (3.2.10), we infer that (u_k) is bounded in $H_{\mathrm{rad}}^s(\mathbb{R}^N)$.

(b) In view of (a), passing to a subsequence if necessary, we may assume that

$$u_k \rightharpoonup u \text{ in } H_{\mathrm{rad}}^s(\mathbb{R}^N),$$

$$u_k \to u \text{ in } L^q(\mathbb{R}^N) \quad \forall q \in (2, 2_s^*), \tag{3.2.13}$$

$$u_k \to u \text{ a.e. } \mathbb{R}^N.$$

By item (iii) of Theorem 3.2.8,

$$e^{(N-2s)\theta_k} \iint_{\mathbb{R}^{2N}} \frac{(u_k(x) - u_k(y))(\varphi(x) - \varphi(y))}{|x-y|^{N+2s}} \, dx dy - e^{N\theta_k} \int_{\mathbb{R}^N} g(u_k)\varphi \, dx \to 0 \tag{3.2.14}$$

for all $\varphi \in C_c^\infty(\mathbb{R}^N)$. Then $\langle I'(u), \varphi \rangle = 0$ for any $\varphi \in C_c^\infty(\mathbb{R}^N)$, and by the denseness of $C_c^\infty(\mathbb{R}^N)$ in $H_{\mathrm{rad}}^s(\mathbb{R}^N)$, we get that $\langle I'(u), \varphi \rangle = 0$ for any $\varphi \in H_{\mathrm{rad}}^s(\mathbb{R}^N)$. In particular,

$$[u]_s^2 = \int_{\mathbb{R}^N} g(u)u \, dx. \tag{3.2.15}$$

Now, we observe that (3.2.14) holds for all $\varphi \in H_{\mathrm{rad}}^s(\mathbb{R}^N)$ and (u_k) is bounded in $H^s(\mathbb{R}^N)$. Therefore, taking $\varphi = u_k$ in (3.2.14), we have

$$e^{(N-2s)\theta_k} [u_k]_s^2 - e^{N\theta_k} \int_{\mathbb{R}^N} g(u_k)u_k \, dx = o(1) \text{ as } k \to \infty.$$

Consequently,

$$e^{(N-2s)\theta_k} [u_k]_s^2 + \frac{m}{2} e^{N\theta_k} \|u_k\|_{L^2(\mathbb{R}^N)}^2$$

$$= e^{N\theta_k} \int_{\mathbb{R}^N} \left[\frac{m}{2} u_k^2 + g(u_k)u_k \right] dx + o(1)$$

$$= e^{N\theta_k} \int_{\mathbb{R}^N} h(u_k)u_k \, dx - e^{N\theta_k} \int_{\mathbb{R}^N} \left[h(u_k)u_k - \frac{m}{2} u_k^2 - g(u_k)u_k \right] dx + o(1)$$

$$= e^{N\theta_k} A_k - e^{N\theta_k} B_k + o(1). \tag{3.2.16}$$

Now, item ($h6$) of Lemma 3.2.1 implies that

$$\lim_{k\to\infty} A_k = \int_{\mathbb{R}^N} h(u)u\,dx, \tag{3.2.17}$$

and by ($h3$) of Lemma 3.2.1 and Fatou's lemma we get

$$\liminf_{k\to\infty} B_k \geq \int_{\mathbb{R}^N} \left[h(u)u - \frac{m}{2}u^2 - g(u)u \right] dx. \tag{3.2.18}$$

Finally, using (3.2.13), (3.2.15), (3.2.16), (3.2.17), and (3.2.18), we infer that

$$\|u\|^2 \leq \limsup_{k\to\infty} \|u_k\|^2 = \limsup_{k\to\infty} \left[e^{(N-2s)\theta_k}[u_k]_s^2 + \frac{m}{2}e^{N\theta_k}\|u_k\|_{L^2(\mathbb{R}^N)}^2 \right]$$

$$\leq \int_{\mathbb{R}^N} \left[\frac{m}{2}u^2 + g(u)u \right] dx = \|u\|^2$$

which gives $u_k \to u$ in $H_{\mathrm{rad}}^s(\mathbb{R}^N)$ as $k \to \infty$. $\square$

Combining Theorem 3.2.9, Lemma 3.2.6 and (3.2.5), we can deduce the following multiplicity result.

Theorem 3.2.10 *Under the assumptions of Theorem* 3.1.1, *there exist infinitely many radially symmetric solutions to* (3.1.2).

Proof Fix $n \in \mathbb{N}$, and let $((\theta_k, u_k)) \subset \mathbb{R} \times H_{\mathrm{rad}}^s(\mathbb{R}^N)$ be a sequence satisfying (i)–(iv) of Theorem 3.2.8. Using Theorem 3.2.9, we know that there exists $u_n \in H_{\mathrm{rad}}^s(\mathbb{R}^N)$ such that $u_k \to u_n$ in $H_{\mathrm{rad}}^s(\mathbb{R}^N)$ as $k \to \infty$. Then u_n satisfies

$$I(u_n) = \tilde{I}(0, u_n) = b_n \text{ and } I'(u_n) = \tilde{I}'(0, u_n) = 0.$$

Since $b_n \to \infty$ as $n \to \infty$, we conclude that (3.1.2) admits infinitely many distinct solutions in $H_{\mathrm{rad}}^s(\mathbb{R}^N)$. $\square$

3.2.4 Mountain Pass Value Gives the Least Energy Level

In this section we prove the existence of a positive solution to (3.1.2) by using the mountain pass approach developed in the previous section. We recall that the existence of a positive ground state solution to (3.1.2) was obtained in [139]. Here, we give an alternative proof of this fact and, in addition, we show that the least energy solution of (3.1.2) coincides with the mountain pass value. This is in clear agreement with the result obtained in the classic framework by Jeanjean and Tanaka in [234]. Moreover, we investigate regularity, decay and symmetry properties of solutions to (3.1.2). The main result of this section can be stated as follows.

Theorem 3.2.11 *Under assumptions* (g1)–(g3), *there exists a classical positive solution u of* (3.1.2). *Moreover, u is radially decreasing, has a polynomial decay at infinity, and can be characterized by a mountain pass argument in* $H_{rad}^s(\mathbb{R}^N)$.

Let us consider the functionals $I : H_{rad}^s(\mathbb{R}^N) \to \mathbb{R}$ and $\tilde{I} : \mathbb{R} \times H_{rad}^s(\mathbb{R}^N) \to \mathbb{R}$ introduced in the previous subsections:

$$I(u) = \frac{1}{2}[u]_s^2 - \int_{\mathbb{R}^N} G(u)\, dx$$

and

$$\tilde{I}(\theta, u) = \frac{1}{2}e^{(N-2s)\theta}[u]_s^2 - e^{N\theta}\int_{\mathbb{R}^N} G(u)\, dx.$$

We define the following minimax values:

$$b_{\mathrm{mp},r} = \inf_{\gamma \in \Gamma_r} \max_{t \in [0,1]} I(\gamma(t)),$$

$$b_{\mathrm{mp}} = \inf_{\gamma \in \Gamma} \max_{t \in [0,1]} I(\gamma(t)),$$

$$\tilde{b}_{\mathrm{mp},r} = \inf_{\tilde{\gamma} \in \tilde{\Gamma}_r} \max_{t \in [0,1]} \tilde{I}(\tilde{\gamma}(t)),$$

where

$$\Gamma_r = \{\gamma \in C([0,1], H_{rad}^s(\mathbb{R}^N)) : \gamma(0) = 0 \text{ and } I(\gamma(1)) < 0\},$$

$$\Gamma = \{\gamma \in C([0,1], H^s(\mathbb{R}^N)) : \gamma(0) = 0 \text{ and } I(\gamma(1)) < 0\},$$

$$\tilde{\Gamma}_r = \{\tilde{\gamma} = (\theta, \gamma) \in C([0,1], \mathbb{R} \times H_{rad}^s(\mathbb{R}^N)) : \gamma \in \Gamma_r \text{ and } \theta(0) = 0 = \theta(1)\}.$$

Then, we show that the following result holds:

Lemma 3.2.12

$$b_{\mathrm{mp}} = b_{\mathrm{mp},r} = b_1 = \tilde{b}_{\mathrm{mp},r}.$$

Proof Arguing as in the proof of Lemma 3.2.6, we have that $b_{\mathrm{mp},r} = \tilde{b}_{\mathrm{mp},r}$. Furthermore, it follows from the above definitions that $b_{\mathrm{mp}} \leq b_{\mathrm{mp},r} \leq b_1$. To prove the lemma, it is enough to show that $b_1 \leq b_{mp}$. Firstly, we show that

$$b_{mp} = \bar{b}, \tag{3.2.19}$$

where

$$\bar{b} = \inf_{\gamma \in \bar{\Gamma}} \max_{t \in [0,1]} I(\gamma(t))$$

and

$$\bar{\Gamma} = \left\{ \gamma \in C([0,1], H^s(\mathbb{R}^N)) : \gamma(0) = 0 \text{ and } \gamma(1) = \gamma_1(1) \right\}.$$

Here γ_1 is the path appearing in Lemma 3.2.2 (with $n = 1$). As observed in [100, 101], we may assume that γ_1 satisfies the following properties: $\gamma(1)(x) \geq 0$ for all $x \in \mathbb{R}^N$, $\gamma_1(|x|) = \gamma(1)(x)$ and $r \mapsto \gamma_1(1)(r)$ is piecewise linear and nonincreasing.

Since $I(\gamma_1(1)) < 0$, it is clear that $\bar{\Gamma} \subset \Gamma$, which gives $b_{mp} \leq \bar{b}$. Now, we show that $b_{mp} \geq \bar{b}$. For this purpose, it is suffices to prove that $B = \{u \in H^s(\mathbb{R}^N) : I(u) < 0\}$ is path connected. We proceed as in [120]. Let $u_1, u_2 \in B$. By Theorem 1.1.7, we may assume that $u_1, u_2 \in C_c^\infty(\mathbb{R}^N)$. Set $u_i^t(x) = u_i(\frac{x}{t})$ for $t > 0$ and $i = 1, 2$. Then

$$I(u_i^t) = \frac{t^{N-2s}}{2}[u_i]_s^2 - t^N \int_{\mathbb{R}^N} G(u_i)\,dx,$$

and so

$$\frac{d}{dt} I(u_i^t) = N t^{N-1} \left(t^{-2s} \frac{1}{2}[u_i]_s^2 - \int_{\mathbb{R}^N} G(u_i)\,dx \right) - s t^{N-2s-1}[u_i]_s^2.$$

In particular, for $t \geq 1$,

$$\frac{d}{dt} I(u_i^t) \leq N t^{N-1} I(u_i) - s t^{N-2s-1}[u_i]_s^2 < 0.$$

Since $u_i \in B$, we deduce that $\int_{\mathbb{R}^N} G(u_i)\,dx > 0$, so we see that $I(u_i^t) \to -\infty$ as $t \to \infty$. Hence, we can choose $t_0 > 0$ such that

$$I(u_i^{t_0}) \leq -2 \max \left\{ \max_{t \in [0,1]} I(t u_1), \max_{t \in [0,1]} I(t u_2) \right\} < 0, \quad i = 1, 2. \tag{3.2.20}$$

Noting that $u_i^{t_0}, u_i \in C_c^\infty(\mathbb{R}^N)$ and

$$\langle u_1^{t_0}, tu_2(\cdot - Re_1)\rangle_{\mathcal{D}^{s,2}(\mathbb{R}^N)} + \int_{\mathbb{R}^N} u_1^{t_0} tu_2(\cdot - Re_1)\, dx$$

$$= t \int_{\mathbb{R}^N} (|\xi|^{2s} + 1)(t_0^N \mathcal{F}u_1(t_0\xi))(\overline{\mathcal{F}u_2(\xi)e^{-i\xi_1 R}})\, d\xi \to 0,$$

$$\int_{\mathbb{R}^N} G(u_1^{t_0}(x) + tu_2(x - Re_1))\, dx \to \int_{\mathbb{R}^N} G(u_1^{t_0}(x))\, dx + \int_{\mathbb{R}^N} G(tu_2(x))\, dx$$

uniformly with respect to $t \in [0, 1]$ as $R \to \infty$, where $e_1 = (1, 0, \ldots, 0)$, it follows from the choice of t_0 that as $R \to \infty$

$$\max_{t \in [0,1]} I(u_1^{t_0} + tu_2(\cdot - Re_1)) \to I(u_1^{t_0}) + \max_{t \in [0,1]} I(tu_2) < -\max_{t \in [0,1]} I(tu_2^{t_0}) < 0.$$

Hence, choosing $R_0 > 0$ sufficiently large, we have

$$\text{supp}(u_1^t) \cap \text{supp}(u_2(\cdot - R_0e_1)) = \emptyset \quad \text{for } t \in [1, t_0],$$

$$\max_{t \in [0,1]} I(u_1^{t_0} + tu_2(\cdot - R_0e_1)) < 0,$$

$$I(u_1^t + u_2(\cdot - R_0e_1)) = I(u_1^t) + I(u_2) < 0 \quad \text{for } t \in [1, t_0].$$

Then, considering the following paths

$$\begin{aligned}
\gamma_1(t) &= u_1^t && \text{for } t \in [1, t_0], \\
\gamma_2(t) &= u_1^{t_0} + tu_2(\cdot - R_0e_1) && \text{for } t \in [0, 1], \\
\gamma_3(t) &= u_1^{t_0-t} + u_2(\cdot - R_0e_1) && \text{for } t \in [0, t_0 - 1],
\end{aligned}$$

we can connect u_1 and $u_1 + u_2(\cdot - R_0e_1)$ in B. In a similar fashion, we can find a path between u_2 and $u_1 + u_2(\cdot - R_0e_1)$ in B. Therefore, B is path connected and $b_{\text{mp}} \geq \bar{b}$. This implies that (3.2.19) holds.

Now, observing that $I(u) = I(-u)$, $\gamma_1(1) \in H_{\text{rad}}^s(\mathbb{R}^N)$ and $\gamma(-t) = -\gamma(t)$ for any $\gamma \in \Gamma_1$, we aim to show that

$$\bar{b} = \inf_{\gamma \in \bar{\Gamma}_r} \max_{t \in [0,1]} I(\gamma(t)) (= b_1) \tag{3.2.21}$$

where

$$\bar{\Gamma}_r = \{\gamma \in C([0, 1], H_{\text{rad}}^s(\mathbb{R}^N)) : \gamma(0) = 0 \text{ and } \gamma(1) = \gamma_1(1)\}.$$

By definition, it is clear that $\bar{b} \leq b_1$. In what follows, we show that $\bar{b} \geq b_1$. Take $\eta \in \bar{\Gamma}$ and we set $\gamma(t) = |\eta(t)|$. Obviously, $\gamma \in C([0, 1], H^s(\mathbb{R}^N))$. Moreover, recalling that G is even, and using the fact that $[|u|]_s \leq [u]_s$ for any $u \in H^s(\mathbb{R}^N)$ (this inequality follows from $||x| - |y|| \leq |x - y|$ for any $x, y \in \mathbb{R}$), we can see that for any $t \in [0, 1]$,

$$
\begin{aligned}
I(\gamma(t)) &= \frac{1}{2}[|\eta(t)|]_s^2 - \int_{\mathbb{R}^N} G(|\eta(t)|) \, dx \\
&= \frac{1}{2}[|\eta(t)|]_s^2 - \int_{\mathbb{R}^N} G(\eta(t)) \, dx \qquad\qquad (3.2.22) \\
&\leq \frac{1}{2}[\eta(t)]_s^2 - \int_{\mathbb{R}^N} G(\eta(t)) \, dx = I(\eta(t)).
\end{aligned}
$$

Since $\gamma_1(1) \geq 0$, we have that $\gamma(1) = |\eta(1)| = |\gamma_1(1)| = \gamma_1(1)$, and thus $\gamma \in \bar{\Gamma}$. Let $\gamma^*(t)$ denote the Schwartz symmetrization (symmetric-decreasing rearrangement) of $\gamma(t)$. By Almgren and Lieb [3], $\gamma^* \in C([0, 1], H^s_{\mathrm{rad}}(\mathbb{R}^N))$ and there holds the fractional Polya-Szegö inequality (see also [200])

$$
[u^*]_s \leq [u]_s \quad \text{for any } u \in \mathcal{D}^{s,2}(\mathbb{R}^N), \qquad\qquad (3.2.23)
$$

which yields $[\gamma^*(t)]_s \leq [\gamma(t)]_s$ for any $t \in [0, 1]$. On the other hand, observing that $G \in C(\mathbb{R})$ is even, $G(0) = 0$ and $G(u) \in L^1(\mathbb{R}^N)$ for all $u \in H^s(\mathbb{R}^N)$, we deduce that (see [100, 249])

$$
\int_{\mathbb{R}^N} G(u^*) \, dx = \int_{\mathbb{R}^N} G(u) \, dx \quad \text{for all } u \in H^s(\mathbb{R}^N),
$$

and thus

$$
\int_{\mathbb{R}^N} G(\gamma^*(t)) \, dx = \int_{\mathbb{R}^N} G(\gamma(t)) \, dx \quad \text{for all } t \in [0, 1].
$$

Consequently, $I(\gamma^*(t)) \leq I(\gamma(t))$ for all $t \in [0, 1]$. By using the properties of γ_1, we see that $(\gamma_1(1))^* = \gamma_1(1)$, and consequently $\gamma^* \in \bar{\Gamma}_r$. Therefore, in view of (3.2.22), we have

$$
b_1 \leq \max_{t \in [0,1]} I(\gamma^*(t)) \leq \max_{t \in [0,1]} I(\gamma(t)) \leq \max_{t \in [0,1]} I(\eta(t)),
$$

which implies that $b_1 \leq \bar{b}$. This shows that (3.2.21) is satisfied. Combining (3.2.19) and (3.2.21), we infer that $b_{\mathrm{mp}} = \bar{b} = b_1$. This ends the proof of lemma. $\square$

Then we are able to prove the following result.

Theorem 3.2.13 *There exists a positive solution to (3.1.2) such that $I(u) = b_{\text{mp},r}$. Moreover, for any non-trivial solution v to (3.1.2), we have $b_{\text{mp},r} \leq I(v)$. This means that u is the least energy solution to (3.1.2) and that $b_{\text{mp},r}$ is the least energy level.*

Proof Let $(\gamma_k) \subset \Gamma_r$ be a sequence such that

$$\max_{t \in [0,1]} I(\gamma_k(t)) \leq b_{\text{mp},r} + \frac{1}{k}. \tag{3.2.24}$$

Since $I(|u|) \leq I(u)$ for all $u \in H^s_{\text{rad}}(\mathbb{R}^N)$, we may assume that γ_k satisfies

$$\gamma_k(t)(x) \geq 0 \quad \text{for all } t \in [0, 1] \text{ and } x \in \mathbb{R}^N.$$

In view of Lemma 3.2.12 and the equality $\tilde{I}(0, u) = I(u)$, we deduce that

$$\max_{t \in [0,1]} \tilde{I}(0, \gamma_k(t)) \leq \tilde{b}_{\text{mp},r} + \frac{1}{k} = b_1 + \frac{1}{k}. \tag{3.2.25}$$

Therefore, applying Lemma 3.2.7 to $\tilde{I}$ and $(0, \gamma_k)$, we can find a sequence $((\theta_k, u_k)) \subset \mathbb{R} \times H^s_{\text{rad}}(\mathbb{R}^N)$ such that as $k \to \infty$

(i) $\text{dist}_{\mathbb{R} \times H^s_{\text{rad}}(\mathbb{R}^N)}((\theta_k, u_k), \{0\} \times \gamma_k([0, 1])) \to 0$;
(ii) $\tilde{I}(\theta_k, u_k) \to b_1$;
(iii) $\|\nabla \tilde{I}(\theta_k, u_k)\|_{\mathbb{R} \times (H^s_{\text{rad}}(\mathbb{R}^N))^*} \to 0$.

Taking into account item (i), we see that $\theta_k \to 0$ as $k \to \infty$, and

$$\|u_k^-\|_{H^s(\mathbb{R}^N)} \leq \text{dist}_{\mathbb{R} \times H^s_{\text{rad}}(\mathbb{R}^N)}((\theta_k, u_k), \{0\} \times \gamma_k([0, 1])) \to 0.$$

Proceeding as in the proof of Theorem 3.2.9, we can show that $u_k \to u$ in $H^s(\mathbb{R}^N)$, $I'(u) = 0$ and $I(u) = b_1$, for some $u \in H^s_{\text{rad}}(\mathbb{R}^N)$ such that $u \geq 0$, $u \not\equiv 0$. By Proposition 3.2.14 below and Proposition 1.3.2, we deduce that $u \in C^{0,\beta}(\mathbb{R}^N)$, and then applying Proposition 1.3.11-(ii), we conclude that $u > 0$ in $\mathbb{R}^N$. This ends the proof of the first statement of theorem.

Now, let $v \in H^s(\mathbb{R}^N)$ be a nontrivial solution to (3.1.2) and consider the curve $\gamma : [0, \infty) \to H^s(\mathbb{R}^N)$ defined by

$$\gamma(t)(x) = \begin{cases} v(\frac{x}{t}), & \text{for } t > 0, \\ 0, & \text{for } t = 0. \end{cases}$$

Then,

(1) $\|\gamma(t)\|^2_{H^s(\mathbb{R}^N)} = t^{N-2s}[v]^2_s + t^N \|v\|^2_{L^2(\mathbb{R}^N)}$,

(2) $I(\gamma(t)) = \dfrac{t^{N-2s}}{2}[v]^2_s - t^N \displaystyle\int_{\mathbb{R}^N} G(v)\,dx = \dfrac{t^{N-2s}}{2}[v]^2_s - t^N \left(\dfrac{N-2s}{2N}\right)[v]^2_s$,

where in (2) we used the Pohozaev identity (3.1.3) to deduce that $\int_{\mathbb{R}^N} G(v)\,dx = \frac{N-2s}{2N}[v]^2_s > 0$. By (1), we have that $\gamma \in C([0, \infty), H^s(\mathbb{R}^N))$. Using (2), we have

$$\frac{d}{dt}I(\gamma(t)) = \frac{N-2s}{2}[v]^2_s\, t^{N-2s-1}(1 - t^{2s}),$$

whence

$$\frac{d}{dt}I(\gamma(t)) > 0 \text{ for } t \in (0, 1) \quad \text{and} \quad \frac{d}{dt}I(\gamma(t)) < 0 \text{ for } t > 1.$$

Therefore, $\max_{t\geq 0} I(\gamma(t)) = I(\gamma(1)) = I(v)$. On the other hand, by (2) and $N \geq 2 > 2s$, we obtain that

$$I(\gamma(t)) = \frac{t^N}{2}[v]^2_s \left(\frac{1}{t^{2s}} - \frac{N-2s}{N}\right) \to -\infty \quad \text{as } t \to \infty,$$

so there exists $\mu > 0$ such that $I(\gamma(\mu)) < 0$. Then we consider the rescaled curve $\tilde{\gamma} \in C([0, 1], H^s(\mathbb{R}^N))$ defined by $\tilde{\gamma}(t)(x) = \gamma(\mu t)(x)$. Then, $\tilde{\gamma} \in \Gamma_r$ and satisfies

$$v \in \tilde{\gamma}([0, 1]) \quad \text{and} \quad \max_{t\in[0,1]} I(\tilde{\gamma}(t)) = I(v).$$

By using Lemma 3.2.12, we deduce that $b_{mp,r} \leq I(v)$ and this completes the proof. $\qquad\square$

3.2.5 Regularity, Symmetry and Asymptotic Behavior of Ground States

We start by proving some interesting results for fractional elliptic PDEs in $\mathbb{R}^N$.

Proposition 3.2.14 (Regularity of Solutions) *Let $s \in (0, 1)$ and $N > 2s$. Let $u \in H^s(\mathbb{R}^N)$ be a weak solution to $(-\Delta)^s u = g(x, u)$ in $\mathbb{R}^N$, where $g : \mathbb{R}^N \times \mathbb{R} \to \mathbb{R}$ is a Carathéodory function such that*

$$|g(x, t)| \leq C(|t| + |t|^p) \quad \text{for a.e. } x \in \mathbb{R}^N, \text{ for all } t \in \mathbb{R},$$

for some $1 \leq p \leq 2^*_s - 1$ and $C > 0$. Then, $u \in L^q(\mathbb{R}^N)$ for all $q \in [2, \infty]$.

Proof We combine a Brezis-Kato type argument [112] with a Moser iteration scheme [278]. For simplicity, we assume that $u \geq 0$ in $\mathbb{R}^N$. The general case can be dealt with in a similar way by considering u^+ and u_-. For any $L > 0$ and $\beta > 0$, we consider the Lipschitz function $\gamma(t) = t t_L^{2\beta}$ for $t \geq 0$, where $t_L = \min\{t, L\}$. Since γ is a nondecreasing function,

$$(a - b)(\gamma(a) - \gamma(b)) \geq 0 \quad \text{for any } a, b \geq 0.$$

Define the function

$$\Gamma(t) = \int_0^t (\gamma'(\tau))^{\frac{1}{2}} \, d\tau.$$

Fix $a, b \geq 0$ such that $a > b$. Then, the Schwartz inequality yields,

$$(a - b)(\gamma(a) - \gamma(b)) = (a - b) \int_b^a \gamma'(t) \, dt$$

$$= (a - b) \int_b^a (\Gamma'(t))^2 \, dt$$

$$\geq \left(\int_b^a \Gamma'(t) \, dt \right)^2 = (\Gamma(a) - \Gamma(b))^2.$$

In a similar fashion, we can prove that this inequality is valid for any $a \leq b$. Thus we can infer that

$$(a - b)(\gamma(a) - \gamma(b)) \geq |\Gamma(a) - \Gamma(b)|^2 \quad \text{for any } a, b \geq 0. \tag{3.2.26}$$

In particular, this implies that

$$|\Gamma(u(x)) - \Gamma(u(y))|^2 \leq (u(x) - u(y))((uu_L^{2\beta})(x) - (uu_L^{2\beta})(y)). \tag{3.2.27}$$

Taking $\gamma(u) = uu_L^{2\beta} \in H^s(\mathbb{R}^N)$ as a test function in (3.1.2), in view of (3.2.27) we have

$$[\Gamma(u)]_s^2 \leq \iint_{\mathbb{R}^{2N}} \frac{(u(x) - u(y))}{|x - y|^{N+2s}} ((uu_L^{2\beta})(x) - (uu_L^{2\beta})(y)) \, dx \, dy$$

$$= \int_{\mathbb{R}^N} g(x, u) uu_L^{2\beta} \, dx. \tag{3.2.28}$$

Since

$$\Gamma(t) \geq \frac{1}{\beta + 1} t t_L^{\beta} \quad \text{for any } t \geq 0,$$

the Sobolev inequality (1.1.1) implies that

$$[\Gamma(u)]_s^2 \geq S_* \|\Gamma(u)\|_{L^{2_s^*}(\mathbb{R}^N)}^2 \geq \left(\frac{1}{\beta+1}\right)^2 S_* \|uu_L^\beta\|_{L^{2_s^*}(\mathbb{R}^N)}^2. \tag{3.2.29}$$

In view of (3.2.29) and the growth assumption on g, we see that (3.2.28) yields

$$\|w_L\|_{L^{2_s^*}(\mathbb{R}^N)}^2 \leq C(\beta+1)^2 \int_{\mathbb{R}^N} (u^2 u_L^{2\beta} + u^{p+1} u_L^{2\beta})\, dx, \tag{3.2.30}$$

where $w_L = uu_L^\beta$. We claim that there exist a constant $c > 0$ and a function $h \in L^{N/2s}(\mathbb{R}^N)$, $h \geq 0$, and independent of L and β, such that

$$u^2 u_L^{2\beta} + u^{p+1} u_L^{2\beta} \leq (c+h) u^2 u_L^{2\beta} \quad \text{on } \mathbb{R}^N. \tag{3.2.31}$$

To show this, notice first that

$$u^2 u_L^{2\beta} + u^{p+1} u_L^{2\beta} = u^2 u_L^{2\beta} + u^{p-1} u^2 u_L^{2\beta} \quad \text{on } \mathbb{R}^N.$$

Moreover,

$$u^{p-1} \leq 1 + h \quad \text{on } \mathbb{R}^N,$$

for some $h \in L^{N/2s}(\mathbb{R}^N)$. Indeed,

$$u^{p-1} = \chi_{\{0 \leq u \leq 1\}} u^{p-1} + \chi_{\{u>1\}} u^{p-1} \leq 1 + \chi_{\{u>1\}} u^{p-1} \quad \text{on } \mathbb{R}^N,$$

and if $(p-1)\frac{N}{2s} < 2$, then

$$\int_{\mathbb{R}^N} \chi_{\{u>1\}} u^{\frac{N}{2s}(p-1)}\, dx \leq \int_{\mathbb{R}^N} \chi_{\{u>1\}} u^2\, dx \leq \int_{\mathbb{R}^N} u^2\, dx < \infty,$$

while if $2 \leq (p-1)\frac{N}{2s}$ we deduce that $\frac{N}{2s}(p-1) \in [2, 2_s^*]$. Taking into account (3.2.30) and (3.2.31) we obtain that

$$\|w_L\|_{L^{2_s^*}(\mathbb{R}^N)}^2 \leq C(\beta+1)^2 \int_{\mathbb{R}^N} (c+h) u^2 u_L^{2\beta}\, dx,$$

and by the monotone convergence theorem (u_L is nondecreasing with respect to L) we have as $L \to \infty$

$$\|u^{\beta+1}\|_{L^{2_s^*}(\mathbb{R}^N)}^2 \leq Cc(\beta+1)^2 \int_{\mathbb{R}^N} u^{2(\beta+1)}\, dx + C(\beta+1)^2 \int_{\mathbb{R}^N} hu^{2(\beta+1)}\, dx. \tag{3.2.32}$$

Fix $M > 0$ and let $A_1 = \{h \leq M\}$ and $A_2 = \{h > M\}$. Then,

$$\int_{\mathbb{R}^N} h u^{2(\beta+1)}\, dx \leq M \|u^{\beta+1}\|^2_{L^2(\mathbb{R}^N)} + \varepsilon(M) \|u^{\beta+1}\|^2_{L^{2^*_s}(\mathbb{R}^N)} \tag{3.2.33}$$

where

$$\varepsilon(M) = \left(\int_{A_2} h^{N/2s}\, dx \right)^{\frac{2s}{N}} \to 0 \quad \text{as } M \to \infty.$$

In view of (3.2.32) and (3.2.33) we get

$$\|u^{\beta+1}\|^2_{L^{2^*_s}(\mathbb{R}^N)} \leq (\beta+1)^2 (Cc + M) \|u^{\beta+1}\|^2_{L^2(\mathbb{R}^N)} + C(\beta+1)^2 \varepsilon(M) \|u^{\beta+1}\|^2_{L^{2^*_s}(\mathbb{R}^N)}. \tag{3.2.34}$$

Choosing $M > 0$ sufficiently large so that

$$\varepsilon(M) C (\beta+1)^2 < \frac{1}{2},$$

and using (3.2.34) we obtain

$$\|u^{\beta+1}\|^2_{L^{2^*_s}(\mathbb{R}^N)} \leq 2C(\beta+1)^2 (c + M) \|u^{\beta+1}\|^2_{L^2(\mathbb{R}^N)}. \tag{3.2.35}$$

Then we can start a bootstrap argument: since $u \in L^{2^*_s}(\mathbb{R}^N)$, we can apply (3.2.35) with $\beta_1 + 1 = \frac{N}{N-2s}$ to deduce that $u \in L^{\frac{(\beta_1+1)2N}{N-2s}}(\mathbb{R}^N) = L^{\frac{2N^2}{(N-2s)^2}}(\mathbb{R}^N)$. Applying again (3.2.35), after k iterations, we find that $u \in L^{\frac{2N^k}{(N-2s)^k}}(\mathbb{R}^N)$, and so $u \in L^q(\mathbb{R}^N)$ for all $q \in [2, \infty)$.

Now we prove that $u \in L^\infty(\mathbb{R}^N)$. Since $u \in L^q(\mathbb{R}^N)$ for all $q \in [2, \infty)$, we have that $h \in L^{\frac{N}{s}}(\mathbb{R}^N)$. By the generalized Hölder inequality, we can verify that, for all $\lambda > 0$,

$$\int_{\mathbb{R}^N} h u^{2(\beta+1)}\, dx \leq \|h\|_{L^{\frac{N}{s}}(\mathbb{R}^N)} \|u^{\beta+1}\|_{L^2(\mathbb{R}^N)} \|u^{\beta+1}\|_{L^{2^*_s}(\mathbb{R}^N)}$$

$$\leq \|h\|_{L^{\frac{N}{s}}(\mathbb{R}^N)} \left(\lambda \|u^{\beta+1}\|^2_{L^2(\mathbb{R}^N)} + \frac{1}{\lambda} \|u^{\beta+1}\|^2_{L^{2^*_s}(\mathbb{R}^N)} \right).$$

Then, using (3.2.32), we deduce that

$$\|u^{\beta+1}\|^2_{L^{2^*_s}(\mathbb{R}^N)} \leq (\beta+1)^2 (Cc + \|h\|_{L^{\frac{N}{s}}(\mathbb{R}^N)} \lambda) \|u^{\beta+1}\|^2_{L^2(\mathbb{R}^N)}$$

$$+ \frac{C(\beta+1)^2 \|h\|_{L^{\frac{N}{s}}(\mathbb{R}^N)}}{\lambda} \|u^{\beta+1}\|^2_{L^{2^*_s}(\mathbb{R}^N)}. \qquad (3.2.36)$$

Taking $\lambda > 0$ such that

$$\frac{C(\beta+1)^2 \|h\|_{L^{\frac{N}{s}}(\mathbb{R}^N)}}{\lambda} = \frac{1}{2}$$

obtain

$$\|u^{\beta+1}\|^2_{L^{2^*_s}(\mathbb{R}^N)} \leq 2(\beta+1)^2 (Cc + \|h\|_{L^{\frac{N}{s}}(\mathbb{R}^N)} \lambda) \|u^{\beta+1}\|^2_{L^2(\mathbb{R}^N)} = M_\beta \|u^{\beta+1}\|^2_{L^2(\mathbb{R}^N)},$$

and the advantage with respect to (3.2.35) is that now we control the dependence of the constant M_β on β. Indeed, for some $M_0 > 0$ independent of β,

$$M_\beta \leq C(1+\beta)^4 \leq M_0^2 e^{2\sqrt{\beta+1}},$$

which implies that

$$\|u\|_{L^{2^*_s(\beta+1)}(\mathbb{R}^N)} \leq M_0^{\frac{1}{\beta+1}} e^{\frac{1}{\sqrt{\beta+1}}} \|u\|_{L^{2(\beta+1)}(\mathbb{R}^N)}.$$

Iterating this last relation and choosing $\beta_0 = 0$ and $2(\beta_{n+1} + 1) = 2^*_s(\beta_n + 1)$, we get

$$\|u\|_{L^{2^*_s(\beta_n+1)}(\mathbb{R}^N)} \leq M_0^{\sum_{i=0}^n \frac{1}{\beta_i+1}} e^{\sum_{i=0}^n \frac{1}{\sqrt{\beta_i+1}}} \|u\|_{L^{2(\beta_0+1)}(\mathbb{R}^N)}.$$

Since $1 + \beta_n = (\frac{N}{N-2s})^n$, we have that

$$\sum_{i=0}^\infty \frac{1}{\beta_i+1} < \infty \quad \text{and} \quad \sum_{i=0}^\infty \frac{1}{\sqrt{\beta_i+1}} < \infty,$$

which implies that

$$\|u\|_{L^\infty(\mathbb{R}^N)} = \lim_{n\to\infty} \|u\|_{L^{2^*_s(\beta_n+1)}(\mathbb{R}^N)} < \infty.$$

An inspection of the above arguments shows that if $u \in H^s(\mathbb{R}^N)$ is a non-negative weak subsolution of $(-\Delta)^s u = g(x, u)$ in $\mathbb{R}^N$, then u is bounded in $\mathbb{R}^N$. □

Remark 3.2.15 In [170] the authors proved that if $u \in \mathcal{D}^{s,2}(\mathbb{R}^N)$ is a non-negative weak solution to $(-\Delta)^s u = f(x, u)$ in $\mathbb{R}^N$, where $|f(x, t)| \leq C(1 + |t|^p)$ for some $1 \leq p \leq 2_s^* - 1$ and $C > 0$, then $u \in L^\infty(\mathbb{R}^N)$.

Next we prove a symmetry result for classical positive solutions to (3.1.2) by using the method of moving planes [206, 207] for the fractional Laplacian; see for instance [126, 138, 171, 183, 185]. We recall that given $g \in C(\mathbb{R}^N)$, we say that a function $u \in C(\mathbb{R}^N)$ is a classical solution of

$$(-\Delta)^s u + u = g \quad \text{in } \mathbb{R}^N \tag{3.2.37}$$

if $(-\Delta)^s u$ is defined at any point of $\mathbb{R}^N$, according the definition given in (1.2.1), and Eq. (3.2.37) is satisfied pointwise in all $\mathbb{R}^N$.

Theorem 3.2.16 (Symmetry of Solutions) *Let $s \in (0, 1)$ and $N \geq 2$. Let u be a positive classical solution of*

$$\begin{cases} (-\Delta)^s u = g(u) \ \text{in } \mathbb{R}^N, \\ u > 0 \ \text{in } \mathbb{R}^N, \quad \lim_{|x| \to \infty} u(x) = 0. \end{cases}$$

If $g(t)$ is a locally Lipschitz function in $[0, \infty)$, nonincreasing for positive and small values of t, then u is radially symmetric and radially decreasing with respect to a point of $\mathbb{R}^N$.

Proof Here we modify in a convenient way some arguments in [185]. We point out that, compared with [185], we do not require any condition on the decay of u at infinity.

We write points $x \in \mathbb{R}^N$ as $x = (x_1, x_2, \ldots, x_N) = (x_1, x')$ and, for $\lambda \in \mathbb{R}$, we set

$$\Sigma_\lambda = \{(x_1, x') \in \mathbb{R}^N : x_1 > \lambda\},$$

$$T_\lambda = \{(x_1, x') \in \mathbb{R}^N : x_1 = \lambda\},$$

$$x_\lambda = (2\lambda - x_1, x') \text{ and } u_\lambda(x) = u(x_\lambda).$$

We divide the proof into three steps.

Step 1 We show that $\lambda_0 = \sup\{\lambda : u_\lambda \leq u \text{ in } \Sigma_\lambda\}$ is finite.

Let us define

$$w(x) = \begin{cases} (u_\lambda - u)^+(x) \ \text{if } x \in \Sigma_\lambda, \\ (u_\lambda - u)^-(x) \ \text{if } x \in \Sigma_\lambda^c. \end{cases}$$

Note that for $x \in \Sigma_\lambda$

$$w(x_\lambda) = \min\{(u_\lambda - u)(x_\lambda), 0\} = \min\{(u - u_\lambda)(x), 0\} = -(u_\lambda - u)^+(x) = -w(x),$$

and similarly $w(x) = -w(x_\lambda)$ for $x \in \Sigma_\lambda^c$. Thus $w(x) = -w(x_\lambda)$ for all $x \in \mathbb{R}^N$, and this implies that

$$\int_{\mathbb{R}^N} |w|^{2_s^*} \, dx = \int_{\Sigma_\lambda} |w|^{2_s^*} \, dx + \int_{\Sigma_\lambda^c} |w|^{2_s^*} \, dx = 2 \int_{\Sigma_\lambda} |w|^{2_s^*} \, dx. \qquad (3.2.38)$$

We also see that for every $x \in \Sigma_\lambda \cap \mathrm{supp}(w)$ we have that $w(x) = (u_\lambda - u)(x)$ and

$$(-\Delta)^s w(x) \le (-\Delta)^s (u_\lambda - u)(x), \qquad \forall x \in \Sigma_\lambda \cap \mathrm{supp}(w), \qquad (3.2.39)$$

since

$$(-\Delta)^s w(x) - (-\Delta)^s (u_\lambda - u)(x)$$

$$= \int_{\mathbb{R}^N} \frac{(u_\lambda - u)(y) - w(y)}{|x - y|^{N+2s}} \, dy$$

$$= \int_{\Sigma_\lambda \cap (\mathrm{supp}(w))^c} \frac{(u_\lambda - u)(y)}{|x - y|^{N+2s}} \, dy + \int_{\Sigma_\lambda^c \cap (\mathrm{supp}(w))^c} \frac{(u_\lambda - u)(y)}{|x - y|^{N+2s}} \, dy$$

$$= \int_{\Sigma_\lambda \cap (\mathrm{supp}(w))^c} (u_\lambda(y) - u(y)) \left(\frac{1}{|x - y|^{N+2s}} - \frac{1}{|x - y_\lambda|^{N+2s}} \right) dy \le 0,$$

where we used that

$$u_\lambda - u \le 0 \text{ in } \Sigma_\lambda \cap (\mathrm{supp}(w))^c, \qquad \text{and } |x - y| \le |x - y_\lambda| \text{ for } x, y \in \Sigma_\lambda.$$

Choosing $(u_\lambda - u)^+$ as test function in the equations for u and u_λ in Σ_λ, respectively, we obtain

$$\int_{\Sigma_\lambda} (-\Delta)^s (u_\lambda - u)(u_\lambda - u)^+ \, dx = \int_{\Sigma_\lambda} (g(u_\lambda) - g(u))(u_\lambda - u)^+ \, dx.$$

Since $g(t)$ is nonincreasing for positive and small values of t (say $0 < t < \varepsilon_0$), and $u(x) \to 0$ as $|x| \to \infty$, there exists $R > 0$ such that for all $\lambda < -R$ we have $u_\lambda < \varepsilon_0$ on Σ_λ, and hence $g(u_\lambda) - g(u) \le 0$ on $\Sigma_\lambda \cap \{u_\lambda > u\}$. Therefore, for $\lambda < -R$,

$$\int_{\Sigma_\lambda} (-\Delta)^s (u_\lambda - u)(u_\lambda - u)^+ \, dx \le 0.$$

Using (3.2.39), we deduce that

$$\int_{\Sigma_\lambda} (-\Delta)^s w \, w \, dx \le \int_{\Sigma_\lambda} (-\Delta)^s (u_\lambda - u)(u_\lambda - u)^+ \, dx \le 0.$$

Then, by (3.2.38) and Theorem 1.1.8, we see that

$$
\begin{aligned}
0 \ge \int_{\Sigma_\lambda} (-\Delta)^s w \, w \, dx &= \frac{1}{2} \int_{\mathbb{R}^N} (-\Delta)^s w \, w \, dx \\
&= \frac{1}{2} \int_{\mathbb{R}^N} |(-\Delta)^{\frac{s}{2}} w|^2 \, dx \\
&\ge C \left(\int_{\mathbb{R}^N} |w|^{2_s^*} \, dx \right)^{\frac{2}{2_s^*}} \\
&= C \left(2 \int_{\Sigma_\lambda} |w|^{2_s^*} \, dx \right)^{\frac{2}{2_s^*}},
\end{aligned}
$$

which shows that $w = 0$ on Σ_λ for all $\lambda < -R$. Thus $u_\lambda \le u$ in Σ_λ for all $\lambda < -R$, and this implies that $\lambda_0 \ge -R$. Since u decays at infinity, there exists λ_1 such that $u(x) < u_{\lambda_1}(x)$ for some $x \in \Sigma_{\lambda_1}$. Hence λ_0 is finite.

Step 2 We prove that $u \equiv u_{\lambda_0}$ in Σ_{λ_0}.

By continuity, $u \ge u_{\lambda_0}$ in Σ_{λ_0}. Assume by contradiction that $u \not\equiv u_{\lambda_0}$ in Σ_{λ_0}. Note that if $x_0 \in \Sigma_{\lambda_0}$ is such that $u_{\lambda_0}(x_0) = u(x_0)$, then

$$(-\Delta)^s u_{\lambda_0}(x_0) - (-\Delta)^s u(x_0) = g(u_{\lambda_0}(x_0)) - g(u(x_0)) = 0,$$

while since $|x_0 - y| < |x_0 - y_{\lambda_0}|$ for all $y \in \Sigma_{\lambda_0}$, $u \ge u_{\lambda_0}$ and $u \not\equiv u_{\lambda_0}$ in Σ_{λ_0}, we have

$$
\begin{aligned}
&(-\Delta)^s u_{\lambda_0}(x_0) - (-\Delta)^s u(x_0) \\
&= - \int_{\mathbb{R}^N} \frac{u_{\lambda_0}(y) - u(y)}{|x_0 - y|^{N+2s}} \, dy \\
&= - \int_{\Sigma_{\lambda_0}} (u_{\lambda_0}(y) - u(y)) \left(\frac{1}{|x_0 - y|^{N+2s}} - \frac{1}{|x_0 - y_{\lambda_0}|^{N+2s}} \right) dy > 0.
\end{aligned}
$$

Therefore, $u > u_{\lambda_0}$ in Σ_{λ_0}. In order to complete Step 2, we only need to prove that the inequality $u \ge u_\lambda$ in Σ_λ continues to hold when $\lambda > \lambda_0$ is close to λ_0, contradicting the definition of λ_0. Fix $\delta > 0$ whose value will be chosen later, and take $\lambda \in (\lambda_0, \lambda_0 + \delta)$. Let $P = (\lambda, 0)$ and consider $\tilde{B} = \Sigma_\lambda \cap B(P, R)$, where $B(P, R)$ is the ball centered at P and with radius $R > 0$ such that $u(x) < \varepsilon_0$ for x

outside $B(P, R)$. Choosing $(u_\lambda - u)^+$ as test function in the equation for u and u_λ in Σ_λ, we have

$$\int_{\Sigma_\lambda} (-\Delta)^s (u_\lambda - u)(u_\lambda - u)^+ \, dx$$

$$= \int_{\Sigma_\lambda} (g(u_\lambda) - g(u))(u_\lambda - u)^+ \, dx$$

$$= \int_{\tilde{B}} (g(u_\lambda) - g(u))(u_\lambda - u)^+ \, dx + \int_{\Sigma_\lambda \setminus \tilde{B}} (g(u_\lambda) - g(u))(u_\lambda - u)^+ \, dx$$

$$\leq \int_{\tilde{B}} (g(u_\lambda) - g(u))(u_\lambda - u)^+ \, dx \tag{3.2.40}$$

where we used that $g(u_\lambda) - g(u) \leq 0$ on $\Sigma_\lambda \setminus \tilde{B}$ (note that $u < u_\lambda < \varepsilon_0$ on that set). Since g is Lipschitz on the compact set $\tilde{B}$, we deduce that

$$\int_{\Sigma_\lambda} (-\Delta)^s (u_\lambda - u)(u_\lambda - u)^+ \, dx \leq L \int_{\tilde{B}} |(u_\lambda - u)^+|^2 \, dx.$$

On the other hand, arguing as in Step 1, we get

$$\int_{\Sigma_\lambda} (-\Delta)^s (u_\lambda - u)(u_\lambda - u)^+ \, dx \geq \int_{\Sigma_\lambda} (-\Delta)^s w \, w \, dx \geq C \left(2 \int_{\Sigma_\lambda} |w|^{2_s^*} \, dx \right)^{\frac{2}{2_s^*}}.$$

Consequently,

$$\|w\|^2_{L^{2_s^*}(\Sigma_\lambda)} \leq C' \int_{\tilde{B}} |(u_\lambda - u)^+|^2 \, dx. \tag{3.2.41}$$

Using Hölder's inequality, we have that

$$C' \int_{\tilde{B}} |(u_\lambda - u)^+|^2 \, dx \leq C'' \int_{\tilde{B}} |w|^2 \chi_{\mathrm{supp}(u_\lambda - u)^+} \, dx$$

$$\leq C'' |\tilde{B} \cap \mathrm{supp}(u_\lambda - u)^+|^{\frac{2s}{N}} \left(\int_{\tilde{B}} |w|^{2_s^*} dx \right)^{\frac{2}{2_s^*}}. \tag{3.2.42}$$

Taking into account that $u > u_{\lambda_0}$ in Σ_{λ_0} and the continuity of u, we see that $u > u_\lambda$ on any compact $K \subset \Sigma_\lambda$, for λ close to λ_0. This means that $\mathrm{supp}(u_\lambda - u)^+$ is small in $\tilde{B}$ for $\lambda > \lambda_0$, λ close to λ_0, so we can choose $\delta > 0$ such that, for all $\lambda \in (\lambda_0, \lambda_0 + \delta)$,

$$C |\tilde{B} \cap \mathrm{supp}(u_\lambda - u)^+|^{\frac{2s}{N}} < \frac{1}{2}. \tag{3.2.43}$$

Combining (3.2.41), (3.2.42), and (3.2.43), we deduce that $w = 0$ in Σ_λ, which gives a contradiction. This completes the proof of Step 2.

Step 3 Conclusion.

By translation, we may say that $\lambda_0 = 0$. The symmetry (and monotonicity) in the x_1-direction follows by repeating the argument in the $(-x_1)$-direction. The radial symmetry result (and the monotonicity of the solution) follows as well by applying the moving plane method in any direction $\nu \in \mathbb{S}^{N-1}$.   □

Finally, we recall the following useful lemma proved in [183], which allows us to deduce some asymptotic estimates of solutions for $|x|$ large.

Lemma 3.2.17 ([183]) *Let $N \geq 2$ and $s \in (0, 1)$. There exists a continuous function w in $\mathbb{R}^N$ such that*

$$(-\Delta)^s w(x) + \frac{1}{2} w(x) = 0 \text{ for } |x| > 1$$

in the classical sense, and $0 < w(x) \leq C/|x|^{N+2s}$ for $|x| > 1$, for some $C > 0$.

Remark 3.2.18 This result holds true for all $N \geq 1$ and $s \in (0, 1)$. Indeed, in [199] it is proved that if $\mathcal{G}_{s,\lambda} \in \mathcal{S}'(\mathbb{R}^N)$, with $\lambda > 0$, denotes the fundamental solution of $(-\Delta)^s \mathcal{G}_{s,\lambda} + \lambda \mathcal{G}_{s,\lambda} = \delta_0$ in $\mathbb{R}^N$, then $\mathcal{G}_{s,\lambda}$ is radial, positive, strictly decreasing in $|x|$, smooth for $|x| \neq 0$, $\mathcal{G}_{s,\lambda} \in L^r(\mathbb{R})$ for all $r \in [1, \infty]$ with $1 - \frac{1}{r} < \frac{2s}{N}$, and $|x|^{N+2s}\mathcal{G}_{s,\lambda}(x) \to \lambda^{-2}\gamma_{N,s} > 0$ as $|x| \to \infty$. Take $\tilde{w} = \mathcal{G}_{s,1} * \chi_{B_a}$, where χ_{B_a} is the characteristic function of the ball B_a with radius $a = 2^{-\frac{1}{2s}}$. Then $w(x) = \tilde{w}(ax)$ satisfies all conclusions in Lemma 3.2.17.

Now we are ready to give the proofs of Theorems 3.2.11 and 3.1.1.

Proof of Theorem 3.2.11 Using Theorem 3.2.13 we know that there exists a nontrivial non-negative solution u to (3.1.2). By Proposition 3.2.14, $u \in L^\infty(\mathbb{R}^N)$. Now we argue as in the proof of Lemma 4.4 in [124]. Since $g(u)$ is bounded, we can apply Proposition 1.3.2 to deduce that

- If $s \leq \frac{1}{2}$, then for any $\alpha < 2s$, $u \in C^{0,\alpha}(\mathbb{R}^N)$.
- If $s > \frac{1}{2}$, then for any $\alpha < 2s - 1$, $u \in C^{1,\alpha}(\mathbb{R}^N)$.

This implies in particular that $g(u) \in C^\alpha(\mathbb{R}^N)$. Applying now Proposition 1.3.1 we have

- If $\alpha + 2s \leq 1$, then $u \in C^{0,\alpha+2s}(\mathbb{R}^N)$.
- If $\alpha + 2s > 1$, then $u \in C^{1,\alpha+2s-1}(\mathbb{R}^N)$.

Therefore, iterating the procedure a finite number of times, one gets that $u \in C^{1,\sigma}(\mathbb{R}^N)$ for some $\sigma \in (0, 1)$ that depends only on s. Indeed, if $\alpha + 2s > 1$, then one can take $\sigma = \alpha + 2s - 1$. On the other hand, if $\alpha + 2s \leq 1$, then $g(u) \in C^{0,\alpha+2s}(\mathbb{R}^N)$. Consequently, $u \in C^{0,\alpha+4s}(\mathbb{R}^N)$, and so iterating a finite number of times, we end up with $\alpha + 2sk > 1$ for some integer k. This gives the $C^{1,\sigma}$ regularity.

We now differentiate the equation to obtain

$$(-\Delta)^s u_{x_i} = g'(u)u_{x_i} \text{ in } \mathbb{R}^N,$$

for $i = 1, \ldots, N$, with u_{x_i} and $g'(u)$ belonging to $C^{0,\sigma}(\mathbb{R}^N)$ provided we take $\sigma < \gamma$. Therefore, applying Proposition 1.3.1 we get $u_{x_i} \in C^{0,\sigma+2s}(\mathbb{R}^N)$. We iterate this procedure a finite number of times (as long as the Hölder exponent is smaller than γ). Now, since $\gamma + 2s > 1$ by assumption, we finally conclude that $u_{x_i} \in C^{1,\delta}(\mathbb{R}^N)$ for all $i = 1, \ldots, N$, and thus $u \in C^{2,\delta}(\mathbb{R}^N)$ for some $\delta > 0$ depending only on s and γ. By Proposition 1.3.11-(ii), we deduce that $u > 0$ in $\mathbb{R}^N$. Note that, by Proposition 1.2.3, u is a classical solution of (3.1.2).

Observing that $u \in C^{0,\delta}(\mathbb{R}^N) \cap L^{2^*_s}(\mathbb{R}^N)$, we have that $u(x) \to 0$ as $|x| \to \infty$. Let us show that

$$0 < u(x) \leq \frac{C}{|x|^{N+2s}} \text{ for } |x| \gg 1.$$

For $t \geq 0$ we define $g_1(t) = (g(t) + mt)^+$ and $g_2(t) = g_1(t) - g(t)$. Extend g_1 and g_2 as odd functions for $t \leq 0$. Then $g = g_1 - g_2$, $g_1, g_2 \geq 0$ in $[0, \infty)$, $g_1(t) = o(t)$ as $t \to 0$, $\lim_{t \to \infty} \frac{g_1(t)}{t^{2^*_s - 1}} = 0$, $g_2(t) \geq mt$ for all $t \geq 0$.

Hence, $(-\Delta)^s u + g_2(u) = g_1(u)$ in $\mathbb{R}^N$. Since $u(x) \to 0$ as $|x| \to \infty$, there exists $R > 0$ such that $g_1(u(x)) \leq \frac{m}{2}u(x)$ for all $|x| > R$. Consequently, $(-\Delta)^s u + \frac{m}{2}u \leq 0$ for $|x| > R$. By Lemma 3.2.17, there exist a positive continuous function w and a constant $C > 0$ such that for large $|x| > R$ (taking R larger if necessary), it holds that $w(x) \leq C|x|^{-(N+2s)}$ and $(-\Delta)^s w + \frac{m}{2}w = 0$. In view of the continuity of u and w, there exists $C_1 = \frac{\max_{|x| \leq R} u}{\min_{|x| \leq R} w} > 0$ such that $\eta(x) = u(x) - C_1 w(x) \leq 0$ for $|x| \leq R$. Moreover, $(-\Delta)^s \eta + \frac{m}{2}\eta \leq 0$ in $\mathbb{R}^N \setminus \bar{B}_R$. Applying Lemma 1.3.8 with $\Omega = \mathbb{R}^N \setminus \bar{B}_R$, $\mu = \frac{m}{2}$, $u_1 = \eta \in H^s(\mathbb{R}^N)$ and $u_2 = 0$, we conclude that $\eta(x) \leq 0$ for $|x| > R$, that is $0 < u(x) \leq C|x|^{-(N+2s)}$ for all $|x| > R$.

Finally, in view of (g2), we can use Theorem 3.2.16 to deduce that u is radially symmetric and radially decreasing with respect to a point of $\mathbb{R}^N$. □

Remark 3.2.19 The positivity of u can be also deduced by using the integral representation of $(-\Delta)^s$ given in Lemma 1.2.1, which holds for functions in $H^{2s}(\mathbb{R}^N) \cap C^\tau(\mathbb{R}^N)$ with $\tau > 2s$; see Lemma 2.4 in [122]. Hence, recalling that $u \geq 0$ in $\mathbb{R}^N$ and $u \not\equiv 0$, if by contradiction there exists $x_0 \in \mathbb{R}^N$ such that $u(x_0) = 0$, then

$$0 = g(u(x_0)) = (-\Delta)^s u(x_0) = -\frac{1}{2} C(N, s) \int_{\mathbb{R}^N} \frac{u(x_0 + y) + u(x_0 - y)}{|y|^{N+2s}} \, dy \leq 0$$

that is $u \equiv 0$, which is absurd. Therefore, $u > 0$ in $\mathbb{R}^N$.

Proof of Theorem 3.1.1 The result is a consequence of Theorems 3.2.10 and 3.2.11. □

We conclude this subsection by proving that if $\mathcal{L}$ is the set of least energy positive solutions to (3.1.2) satisfying $u(0) = \max_{\mathbb{R}^N} u(x)$, then the following compactness result holds true.

Proposition 3.2.20 *$\mathcal{L}$ is compact in $H^s(\mathbb{R}^N)$. Moreover, there exists $C > 0$ independent of $u \in \mathcal{L}$, such that*

$$0 < u(x) \leq \frac{C}{1 + |x|^{N+2s}} \quad \text{for all } x \in \mathbb{R}^N.$$

Proof The proof is inspired by Byeon and Jeanjean [121]. From (3.1.3), we see that

$$I(u) = \frac{s}{N} [u]_s^2$$

for all $u \in \mathcal{L}$, thus $\{[u]_s^2 : u \in \mathcal{L}\}$ is bounded in $\mathbb{R}$. Note that for any $u \in \mathcal{L}$

$$[u]_s^2 = \int_{\mathbb{R}^N} g(u)u \, dx = \int_{\mathbb{R}^N} [g_1(u) - g_2(u)]u \, dx,$$

where g_1 and g_2 are defined as in the proof of Theorem 3.2.11. Observe that for all $\varepsilon > 0$ there exists $c_\varepsilon > 0$ such that $g_1(t) \leq \varepsilon g_2(t) + c_\varepsilon t^{2_s^* - 1}$ for all $t \geq 0$, and $g_2(t) \geq mt$ for all $t \geq 0$. Hence, fixing $\varepsilon \in (0, 1)$, we have

$$[u]_s^2 + \int_{\mathbb{R}^N} g_2(u)u \, dx = \int_{\mathbb{R}^N} g_1(u)u \, dx \leq \varepsilon \int_{\mathbb{R}^N} g_2(u)u \, dx + c_\varepsilon \|u\|_{L^{2_s^*}(\mathbb{R}^N)}^{2_s^*}$$

and thus

$$(1 - \varepsilon)\frac{m}{2} \|u\|_{L^2(\mathbb{R}^N)}^2 \leq [u]_s^2 + (1 - \varepsilon) \int_{\mathbb{R}^N} g_2(u)u \, dx \leq c_\varepsilon \|u\|_{L^{2_s^*}(\mathbb{R}^N)}^{2_s^*}.$$

Using the Sobolev inequality (1.1.1) and the boundedness of $\{[u]_s^2 : u \in \mathcal{L}\}$, we deduce that $\{\|u\|_{L^2(\mathbb{R}^N)}^2 : u \in \mathcal{L}\}$ is bounded in $\mathbb{R}$. Consequently, $\mathcal{L}$ is bounded in $H^s(\mathbb{R}^N)$. Using Propositions 3.2.14 and 1.3.2, we deduce that $\mathcal{L}$ is bounded in $C^{0,\alpha}(\mathbb{R}^N)$.

Let $\mathfrak{m}$ be the least energy level corresponding to (3.1.2). Then, for any $u \in \mathcal{L}$, we have $\mathfrak{m} = I(u) = \frac{s}{N}[u]_s^2$. Moreover, from the above calculations (with $\varepsilon = \frac{1}{2}$), and recalling that $\|u\|_{L^2(\mathbb{R}^N)} \leq \bar{C}$ for all $u \in \mathcal{L}$, we see that

$$\frac{N}{s}\mathfrak{m} \leq [u]_s^2 + \frac{\mathfrak{m}}{4}\|u\|_{L^2(\mathbb{R}^N)}^2 \leq c_{\frac{1}{2}}\int_{\mathbb{R}^N} u^{2_s^*}\, dx$$

$$\leq c_{\frac{1}{2}}\|u\|_{L^\infty(\mathbb{R}^N)}^{2_s^*-2}\int_{\mathbb{R}^N} u^2\, dx$$

$$\leq c_{\frac{1}{2}}\bar{C}^2\|u\|_{L^\infty(\mathbb{R}^N)}^{2_s^*-2}$$

which implies that there exists $\sigma > 0$ such that $u(0) = \|u\|_{L^\infty(\mathbb{R}^N)} \geq \sigma > 0$ for all $u \in \mathcal{L}$, i.e., $\mathcal{L}$ is bounded away from zero in $L^\infty(\mathbb{R}^N)$. We now claim that $\lim_{|x|\to\infty} u(x) = 0$ uniformly for $u \in \mathcal{L}$. Suppose, by contradiction, that there exist sequences $(u_k) \subset \mathcal{L}$ and $(x_k) \subset \mathbb{R}^N$ such that $\lim\inf_{k\to\infty}|x_k| = \infty$ and $\lim\inf_{k\to\infty} u_k(x_k) > 0$. Define $v_k(x) = u_k(x + x_k)$. Then we may assume that $u_k \rightharpoonup u$ and $v_k \rightharpoonup v$ in $H^s(\mathbb{R}^N)$ and uniformly on compact sets of $\mathbb{R}^N$, for some $u, v \in H^s(\mathbb{R}^N)$. Since $u_k(0) \geq \sigma$ and $v_k(0) = u_k(x_k)$, u and v are nontrivial solutions to (3.1.2). Consequently,

$$I(u), I(v) \geq I(w) \quad \text{for all } w \in \mathcal{L}.$$

Take $R > 0$. Thus, for k large enough, we have $|x_k| \geq 2R$. For these values of k,

$$I(u_k) = \frac{s}{N}[u_k]_s^2 \geq \frac{s}{N}[u_k]_{W^{s,2}(B_R)}^2 + \frac{s}{N}[u_k]_{W^{s,2}(B_R^c)}^2$$

$$\geq \frac{s}{N}[u_k]_{W^{s,2}(B_R)}^2 + \frac{s}{N}[u_k]_{W^{s,2}(B_R(x_k))}^2$$

$$= \frac{s}{N}[u_k]_{W^{s,2}(B_R)}^2 + \frac{s}{N}[v_k]_{W^{s,2}(B_R)}^2$$

whence

$$\lim\inf_{k\to\infty} I(u_k) \geq \frac{s}{N}[u]_{W^{s,2}(B_R)}^2 + \frac{s}{N}[v]_{W^{s,2}(B_R)}^2.$$

Letting $R \to \infty$ we find that

$$\lim\inf_{k\to\infty} I(u_k) \geq \frac{s}{N}[u]_s^2 + \frac{s}{N}[v]_s^2 \geq 2I(w) \quad \text{for all } w \in \mathcal{L},$$

and this is in contrast with $I(u_k) = I(w) = \mathfrak{m} > 0$ for all $w \in \mathcal{L}$. Therefore, $\lim_{|x|\to\infty} u(x) = 0$ uniformly for $u \in \mathcal{L}$, and using the fact that $\mathcal{L}$ is bounded in $C^{0,\alpha}(\mathbb{R}^N)$, we can argue as in the last part of the proof of Theorem 3.2.11 (in this case we take $C_1 = \frac{M}{\min_{|x|\leq R} w}$, where $M > 0$ is such that $\|u\|_{L^\infty(\mathbb{R}^N)} \leq M$ for all $u \in \mathcal{L}$) to deduce, by comparison, that there exists $C > 0$ such that for any $u \in \mathcal{L}$ and $x \in \mathbb{R}^N$,

$$u(x) \leq \frac{C}{1 + |x|^{N+2s}}. \tag{3.2.44}$$

Let now $(u_k) \subset \mathcal{L}$. Taking a subsequence if necessary, we can assume that $u_k \rightharpoonup u$ in $H^s(\mathbb{R}^N)$, $u_k \to u$ in $L^p_{loc}(\mathbb{R}^N)$ for all $p \in [1, 2^*_s)$, and $u_k \to u$ a.e. in $\mathbb{R}^N$. Clearly, u is a weak solution of (3.1.2). Moreover, by (3.2.44),

$$u_k(x) \leq \frac{C}{1 + |x|^{N+2s}} \quad \text{for all } x \in \mathbb{R}^N, \ k \in \mathbb{N},$$

and consequently $u_k \to u$ in $L^2(\mathbb{R}^N) \cap L^{2^*_s}(\mathbb{R}^N)$. Thus, using the growth assumptions on g, it follows by the dominated convergence theorem that

$$\int_{\mathbb{R}^N} g(u_k) u_k \, dx \to \int_{\mathbb{R}^N} g(u) u \, dx \quad \text{as } k \to \infty.$$

Since $[u_k]_s^2 = \int_{\mathbb{R}^N} g(u_k) u_k \, dx$ and $[u]_s^2 = \int_{\mathbb{R}^N} g(u) u \, dx$, we deduce that $[u_k]_s \to [u]_s^2$ as $k \to \infty$, which combined with the strong convergence in $L^2(\mathbb{R}^N)$ shows that $u_k \to u$ in $H^s(\mathbb{R}^N)$. This completes the proof of the proposition. $\square$

3.3   The Zero Mass Case

This last section is devoted to the proof of the existence of a positive solution of (3.1.2) in the zero mass case.

Let $\varepsilon_0 = \frac{G(\xi_0)}{\xi_0^2} > 0$ and define $g_\varepsilon(t) = g(t) - \varepsilon t$ with $\varepsilon \in (0, \varepsilon_0]$. Then g_ε satisfies the assumption of Theorem 3.2.11, so we know that for every $\varepsilon \in (0, \varepsilon_0]$ there exists $u_\varepsilon \in H^s_{rad}(\mathbb{R}^N)$ positive and radially decreasing in $r = |x|$, such that $I_\varepsilon(u_\varepsilon) = b_\varepsilon$ and $I'_\varepsilon(u_\varepsilon) = 0$, where the mountain pass minimax value b_ε is defined by

$$b_\varepsilon = \inf_{\gamma \in \Gamma_\varepsilon} \max_{t \in [0,1]} I_\varepsilon(\gamma(t)),$$

with

$$\Gamma_\varepsilon = \{\gamma \in C([0, 1], H^s_{rad}(\mathbb{R}^N)) : \gamma(0) = 0, I_\varepsilon(\gamma(1)) < 0\}.$$

Since u_ε satisfies the Pohozaev identity (3.1.3), we deduce that $\frac{s}{N}[u_\varepsilon]_s^2 = b_\varepsilon$. Now we prove that it is possible to estimate b_ε from above independently of ε:

Lemma 3.3.1 *There exists $b_0 > 0$ such that $0 < b_\varepsilon \leq b_0$ for all $\varepsilon \in (0, \varepsilon_0]$.*

Proof For $R > 1$, we define

$$w_R(x) = \begin{cases} \xi_0, & \text{for } |x| \leq R, \\ \xi_0(R + 1 - |x|), & \text{for } |x| \in [R, R + 1], \\ 0, & \text{for } |x| \geq R + 1. \end{cases}$$

It is clear that $w_R \in H^1(\mathbb{R}^N) \subset H^s(\mathbb{R}^N)$. By the definitions of w_R and g_ε, we obtain that for all $\varepsilon \in (0, \varepsilon_0]$

$$\int_{\mathbb{R}^N} G_\varepsilon(w_R) \, dx \geq \int_{\mathbb{R}^N} G_{\varepsilon_0}(w_R) \, dx$$

$$= G_{\varepsilon_0}(\xi_0)|B_R| + \int_{\{R \leq |x| \leq R+1\}} G_{\varepsilon_0}(\xi_0(R + 1 - |x|)) \, dx$$

$$\geq G_{\varepsilon_0}(\xi_0)|B_R| - |B_{R+1} - B_R| \max_{t \in [0, \xi_0]} |G_{\varepsilon_0}(t)|$$

$$= \frac{\pi^{\frac{N}{2}}}{\Gamma(\frac{N}{2} + 1)} [G_{\varepsilon_0}(\xi_0) R^N - \max_{t \in [0, \xi_0]} |G_{\varepsilon_0}(t)|((R + 1)^N - R^N)]$$

$$\geq \frac{\pi^{\frac{N}{2}}}{\Gamma(\frac{N}{2} + 1)} \left[G_{\varepsilon_0}(\xi_0) - \max_{t \in [0, \xi_0]} |G_{\varepsilon_0}(t)| \left(\left(1 + \frac{1}{R} \right)^N - 1 \right) \right] R^N$$

so there exists $\bar{R} > 0$ (independent of ε) such that $\int_{\mathbb{R}^N} G_\varepsilon(w_R) \, dx > 0$ for all $R \geq \bar{R}$ and $\varepsilon \in (0, \varepsilon_0]$. Then, upon setting $w_t(x) = w_{\bar{R}}(\frac{x}{t})$, we have that for every $\varepsilon \in (0, \varepsilon_0]$,

$$I_\varepsilon(w_t) = \frac{t^{N-2s}}{2} [w_{\bar{R}}]_s^2 - t^N \int_{\mathbb{R}^N} G_\varepsilon(w_{\bar{R}}) \, dx \to -\infty \tag{3.3.1}$$

as $t \to \infty$. Hence, there exists $\tau > 0$ such that $I_\varepsilon(w_\tau) < 0$ for any $\varepsilon \in (0, \varepsilon_0]$. We put $\bar{e}(x) = w_\tau(x)$. Therefore,

$$I_\varepsilon(u_\varepsilon) = b_\varepsilon \leq \sup_{\varepsilon \in (0, \varepsilon_0]} \max_{t \in [0, 1]} I_\varepsilon(t\bar{e}) = b_0$$

for any $\varepsilon \in (0, \varepsilon_0]$. $\qquad \square$

In view of Lemma 3.3.1, we infer that

$$[u_\varepsilon]_s^2 = \frac{N}{s} b_\varepsilon \leq \frac{N}{s} b_0 \quad \text{for all } \varepsilon \in (0, \varepsilon_0], \tag{3.3.2}$$

and consequently, we may assume that as $\varepsilon \to 0$

$$u_\varepsilon \rightharpoonup u \text{ in } \mathcal{D}^{s,2}_{rad}(\mathbb{R}^N),$$

$$u_\varepsilon \to u \text{ in } L^q_{loc}(\mathbb{R}^N), \quad \text{for any } q \in [1, 2^*_s), \quad (3.3.3)$$

$$u_\varepsilon \to u \text{ a.e. in } \mathbb{R}^N.$$

Since u_ε is a weak solution to (3.1.2), we know that

$$\langle u_\varepsilon, \varphi \rangle_{\mathcal{D}^{s,2}(\mathbb{R}^N)} = \int_{\mathbb{R}^N} (g(u_\varepsilon) - \varepsilon u_\varepsilon)\varphi \, dx, \quad (3.3.4)$$

for all $\varphi \in C^\infty_c(\mathbb{R}^N)$. Taking into account (3.3.2), (3.3.3), (h1)–(h2), and the fact that $\mathcal{D}^{s,2}(\mathbb{R}^N) \subset L^{2^*_s}(\mathbb{R}^N)$ we apply the first part of Lemma 1.4.2 with $P(t) = g(t)$, $Q(t) = |t|^{2^*_s-1}$, $v_\varepsilon = u_\varepsilon$, $v = u$ and $w = \varphi$ to pass to the limit in (3.3.4) as $\varepsilon \to 0$. Then we obtain

$$\langle u, \varphi \rangle_{\mathcal{D}^{s,2}(\mathbb{R}^N)} = \int_{\mathbb{R}^N} g(u)\varphi \, dx,$$

for all $\varphi \in C^\infty_c(\mathbb{R}^N)$. This means that u is a weak nonnegative solution to $(-\Delta)^s u = g(u)$ in $\mathbb{R}^N$. Now, we only need to prove that $u \not\equiv 0$. We recall that $u_\varepsilon > 0$ and satisfies the Pohozaev identity (3.1.3) with G replaced by G_ε, that is

$$\frac{N-2s}{2}[u_\varepsilon]^2_s = N \int_{\mathbb{R}^N} \left(G(u_\varepsilon) - \frac{\varepsilon}{2}u^2_\varepsilon \right) dx. \quad (3.3.5)$$

Now, using (h_1)–(h_2), there exists $c > 0$ such that

$$G(t) \le c|t|^{2^*_s} \quad \text{for all } t \in \mathbb{R}. \quad (3.3.6)$$

Then, in view of (3.3.5) and (3.3.6), we get

$$[u_\varepsilon]^2_s \le \frac{2N}{N-2s}\frac{\varepsilon}{2}\|u_\varepsilon\|^2_{L^2(\mathbb{R}^N)} + [u_\varepsilon]^2_s = \frac{2N}{N-2s}\int_{\mathbb{R}^N} G(u_\varepsilon) \, dx$$

$$\le \frac{2Nc}{N-2s}\int_{\mathbb{R}^N} |u_\varepsilon|^{2^*_s} \, dx \le C[u_\varepsilon]^{2^*_s}_s$$

which implies that

$$[u_\varepsilon]_s \ge c > 0, \quad (3.3.7)$$

for some c independent of ε.

Next, we argue by contradiction, and we assume that $u = 0$. We begin by proving that

$$\int_{\mathbb{R}^N} G^+(u_\varepsilon)\, dx \to 0 \quad \text{as } \varepsilon \to 0. \tag{3.3.8}$$

Using Lemma 1.4.1 with $t = 2_s^*$, Theorem 1.1.8 and (3.3.2), we can see that, for any $x \in \mathbb{R}^N \setminus \{0\}$ and $\varepsilon \in (0, \varepsilon_0]$,

$$|u_\varepsilon(x)| \leq \left(\frac{N}{\omega_{N-1}}\right)^{\frac{1}{2_s^*}} |x|^{-\frac{N}{2_s^*}} \|u_\varepsilon\|_{L^{2_s^*}(\mathbb{R}^N)}$$

$$\leq \left(\frac{N}{\omega_{N-1}}\right)^{\frac{1}{2_s^*}} |x|^{-\frac{N}{2_s^*}} S_*[u_\varepsilon]_s$$

$$\leq \left(\frac{N}{\omega_{N-1}}\right)^{\frac{1}{2_s^*}} |x|^{-\frac{N}{2_s^*}} S_*\sqrt{\frac{N}{s} b_0} = C|x|^{-\frac{N}{2_s^*}}, \tag{3.3.9}$$

where C is independent of ε. By (3.3.9), for any $\delta > 0$ there exists $R > 0$ such that $|u_\varepsilon(x)| \leq \delta$ for $|x| \geq R$. Moreover, by assumption (h_1), for every $\eta > 0$ there exists $\delta > 0$ such that

$$G^+(t) \leq \eta |t|^{2_s^*} \quad \text{for } |t| \leq \delta.$$

Hence, for large R, we obtain the estimate

$$\int_{\mathbb{R}^N \setminus B_R} G^+(u_\varepsilon)\, dx \leq \eta \int_{\mathbb{R}^N \setminus B_R} |u_\varepsilon|^{2_s^*} dx \leq C\eta[u_\varepsilon]_s^{2_s^*} \leq C\eta,$$

uniformly in $\varepsilon \in (0, \varepsilon_0]$. On the other hand, by assumption (h_2), for every $\eta > 0$ there exists a constant C_η such that

$$G^+(t) \leq \eta |t|^{2_s^*} + C_\eta \quad \text{for all } t \in \mathbb{R}.$$

Now, we fix $\Omega \subset \mathbb{R}^N$ with sufficiently small measure $|\Omega| < \frac{\eta}{C_\eta}$. Then, for any $\varepsilon \in (0, \varepsilon_0]$

$$\int_\Omega G^+(u_\varepsilon)\, dx \leq \eta \int_\Omega |u_\varepsilon|^{2_s^*} dx + C_\eta |\Omega| \leq C\eta.$$

Applying Vitali's convergence theorem, we deduce that (3.3.8) is satisfied.

Therefore, using (3.3.5) and the facts $G = G^+ + G^-$ and $G^- \leq 0$, we have

$$[u_\varepsilon]_s^2 \leq [u_\varepsilon]_s^2 - \frac{2N}{N-2s} \int_{\mathbb{R}^N} G^-(u_\varepsilon) \, dx + \frac{2N}{N-2s} \int_{\mathbb{R}^N} \frac{\varepsilon}{2} u_\varepsilon^2 \, dx$$

$$= \frac{2N}{N-2s} \int_{\mathbb{R}^N} G^+(u_\varepsilon) \, dx,$$

which together with (3.3.8) shows that $[u_\varepsilon]_s \to 0$ as $\varepsilon \to 0$. This gives a contradiction in view of (3.3.7). Then, $u \in \mathcal{D}_{\mathrm{rad}}^{s,2}(\mathbb{R}^N)$ is a nontrivial non-negative weak solution to (3.1.2). By combining Remark 3.2.15 and Proposition 1.3.2, we deduce that $u \in C^{0,\beta}(\mathbb{R}^N)$. From Proposition 1.3.11-(ii), we conclude that $u > 0$ in $\mathbb{R}^N$.

3.4   The Critical Case

3.4.1   Ground State for the Critical Case

This section is devoted to the proof of Theorem 3.1.3. Without loss of generality, we will assume that $b = 1$ in (g_3). Let us consider the functional $\mathcal{I} : H^s(\mathbb{R}^N) \to \mathbb{R}$ given by

$$\mathcal{I}(u) = \mathcal{T}(u) - \mathcal{V}(u) \text{ for any } u \in H^s(\mathbb{R}^N),$$

where

$$\mathcal{T}(u) = \frac{1}{2}[u]_s^2 \quad \text{and} \quad \mathcal{V}(u) = \int_{\mathbb{R}^N} G(u) \, dx.$$

By Theorem 1.1.8 and the assumptions on g, it is clear that $\mathcal{I}$ is well defined on $H^s(\mathbb{R}^N)$ and that $\mathcal{I} \in C^1(H^s(\mathbb{R}^N), \mathbb{R})$ (to prove that $\mathcal{V}(u)$ is of class C^1 in $H^s(\mathbb{R}^N)$ one can argue as in [100]; see also Theorem 2.1.5). We begin by proving the following result:

Lemma 3.4.1 *Consider the constrained minimization problem*

$$M = \inf \left\{ \mathcal{T}(u) : \mathcal{V}(u) = 1, u \in H^s(\mathbb{R}^N) \right\}. \tag{3.4.1}$$

Then, $0 < M < \frac{1}{2}(2_s^*)^{\frac{N-2s}{N}} S_*$.

Proof First, we show that $\{u \in H^s(\mathbb{R}^N) : \mathcal{V}(u) = 1\}$ is not empty. We know (see Remark 1.1.9) that S_* is achieved by the extremal functions

$$U_\varepsilon(x) = \frac{\kappa \, \varepsilon^{-\frac{N-2s}{2}}}{\left(\mu^2 + \left|\dfrac{x}{\varepsilon \, S_*^{\frac{1}{2s}}}\right|^2\right)^{\frac{N-2s}{2}}},$$

for any $\varepsilon > 0$, where $\kappa \in \mathbb{R}$, $\mu > 0$ are fixed constants. Let $\eta \in C_c^\infty(\mathbb{R}^N)$ be a cut-off function with support in B_2 and such that $0 \leq \eta \leq 1$, and $\eta = 1$ on B_1. For $\varepsilon > 0$, we define $\psi_\varepsilon(x) = \eta(x)U_\varepsilon(x)$, and we set

$$v_\varepsilon(x) = \frac{\psi_\varepsilon}{\|\psi_\varepsilon\|_{L^{2_s^*}(\mathbb{R}^N)}}.$$

By performing similar calculations to those in Proposition 21 and 22 in [310] we can see that

$$\|\psi_\varepsilon\|_{L^{2_s^*}(\mathbb{R}^N)}^{2_s^*} = S_*^{\frac{N}{2s}} + O(\varepsilon^N) \quad \text{and} \quad [\psi_\varepsilon]_s^2 \leq S_*^{\frac{N}{2s}} + O(\varepsilon^{N-2s}), \tag{3.4.2}$$

so, in particular, we deduce that

$$[v_\varepsilon]_s^2 \leq S_* + O(\varepsilon^{N-2s}). \tag{3.4.3}$$

By assumption (g_4), we get $\mathcal{V}(v_\epsilon) \geq \dfrac{1}{2_s^*} + \Gamma_\varepsilon$, where

$$\Gamma_\varepsilon = \frac{C}{q}\|v_\varepsilon\|_{L^q(\mathbb{R}^N)}^q - \frac{a}{2}\|v_\varepsilon\|_{L^2(\mathbb{R}^N)}^2.$$

Our aim is to prove that

$$\lim_{\varepsilon \to 0} \frac{\Gamma_\varepsilon}{\varepsilon^{N-2s}} = \infty. \tag{3.4.4}$$

Since $\{2_s^* - 2, 2\} < q < 2_s^*$, we know that $(N - 2s)q > N$. Then we can find two positive constants $C_1(N, s)$ and $C_2(N, s)$ such that

$$\|v_\varepsilon\|_{L^q(\mathbb{R}^N)}^q \geq \frac{1}{\|\psi_\varepsilon\|_{L^{2_s^*}(\mathbb{R}^N)}^q} \int_{B_1} |U_\varepsilon(x)|^q \, dx$$

$$\geq C_1(N, s) \, \varepsilon^{N - \frac{(N-2s)q}{2}} \int_0^{1/(\varepsilon \, S_*^{\frac{1}{2s}})} \frac{r^{N-1}}{(\mu^2 + r^2)^{\frac{(N-2s)q}{2}}} \, dr$$

$$= O(\varepsilon^{N - \frac{(N-2s)q}{2}})$$

and

$$\|v_\varepsilon\|_{L^2(\mathbb{R}^N)}^2 \le \frac{1}{\|\psi_\varepsilon\|_{L^{2^*_s}(\mathbb{R}^N)}^2} \int_{B_2} |U_\varepsilon(x)|^2 \, dx \le C_2(N,s)\,\varepsilon^{2s} \int_0^{2/(\varepsilon\, S_*^{\frac{1}{2s}})} \frac{r^{N-1}}{(\mu^2 + r^2)^{N-2s}} \, dr$$

$$= \begin{cases} O(\varepsilon^{2s}), & \text{if } N > 4s, \\ O(\varepsilon^{2s}\log(\frac{1}{\varepsilon})), & \text{if } N = 4s, \\ O(\varepsilon^{N-2s}), & \text{if } N < 4s. \end{cases}$$

Thus we deduce that $\Gamma_\varepsilon \ge O(\varepsilon^{N-\frac{(N-2s)q}{2}})$. Using the fact that $\{2^*_s - 2, 2\} < q < 2^*_s$, it is easy to deduce that (3.4.4) holds true. Consequently, there exists $\varepsilon_0 > 0$ such that

$$\mathcal{V}(v_\varepsilon) \ge \frac{1}{2^*_s} \quad \text{for all } 0 < \varepsilon < \varepsilon_0\,.$$

Set $\omega(x) = v_\varepsilon\left(\frac{x}{\sigma}\right)$, where $\sigma = (\mathcal{V}(v_\varepsilon))^{-\frac{1}{N}}$. Then, $\mathcal{V}(\omega) = 1$ and we have that $\{u \in H^s(\mathbb{R}^N) : \mathcal{V}(u) = 1\}$ is not empty.

Now we show that $0 < M < \frac{1}{2}(2^*_s)^{\frac{N-2s}{N}} S_*$. Clearly, $0 \le M < \infty$. Since $\mathcal{T}(v_\varepsilon) = \sigma^{N-2s}\mathcal{T}(\omega)$ and $\mathcal{V}(\omega) = 1$, we deduce that

$$M \le \mathcal{T}(\omega) = \frac{\mathcal{T}(v_\varepsilon)}{(\mathcal{V}(v_\varepsilon))^{\frac{2}{2^*_s}}}.$$

Then, by (3.4.3), we can see that for every ε such that $0 < \varepsilon < \varepsilon_0$

$$0 \le M \le \frac{\frac{1}{2}[v_\varepsilon]_s^2}{\left(\frac{1}{2^*_s} + \Gamma_\varepsilon\right)^{\frac{2}{2^*_s}}} \le \frac{1}{2}(2^*_s)^{\frac{2}{2^*_s}} S_* \frac{1 + O(\varepsilon^{N-2s})}{(1 + 2^*_s\Gamma_\varepsilon)^{\frac{2}{2^*_s}}}. \tag{3.4.5}$$

Noting that for $p \ge 1$

$$(1+y)^p \le 1 + p(1+y)^{p+1}y \quad \text{for all } y \ge -1,$$

by (3.4.4) we deduce that for all $\varepsilon > 0$ sufficiently small

$$(1 + O(\varepsilon^{N-2s}))^{\frac{2^*_s}{2}} - 1 \le \frac{2^*_s}{2}(1 + O(\varepsilon^{N-2s}))^{1+\frac{2^*_s}{2}} O(\varepsilon^{N-2s}) < 2^*_s\Gamma_\varepsilon,$$

that is,

$$(1 + O(\varepsilon^{N-2s}))^{\frac{2^*_s}{2}} < 1 + 2^*_s\Gamma_\varepsilon. \tag{3.4.6}$$

Hence, in view of (3.4.5) and (3.4.6),

$$M < \frac{1}{2}(2_s^*)^{\frac{N-2s}{N}} S_*.$$

Finally we prove that $M > 0$. If, by contradiction, $M = 0$, then we can find a sequence $(u_n) \subset H^s(\mathbb{R}^N)$ such that $\mathcal{V}(u_n) = 1$ and $\mathcal{T}(u_n) \to 0$ as $n \to \infty$. By (1.1.1), we see that $\|u_n\|_{L^{2_s^*}(\mathbb{R}^N)} \to 0$ as $n \to \infty$. On the other hand, using (g_2) and (g_3), we know that there is a constant $K > 0$ such that $G(t) \leq K|t|^{2_s^*}$ for all $t \in \mathbb{R}$, which implies that $1 = \mathcal{V}(u_n) \leq K\|u_n\|_{L^{2_s^*}(\mathbb{R}^N)}^{2_s^*} \to 0$ as $n \to \infty$, a contradiction. $\quad\square$

Using Lemma 3.4.1, we show that:

Lemma 3.4.2 *Under the assumptions of Theorem* 3.1.3, *there exists a solution* $u \in H^s(\mathbb{R}^N)$ *of the minimization problem* (3.4.1).

Proof By Lemma 3.4.1, $\{\mathcal{T}(u) : \mathcal{V}(u) = 1, u \in H^s(\mathbb{R}^N)\}$ is not empty. Now, we show the existence of a radial minimizing sequence. Let $(u_n) \subset H^s(\mathbb{R}^N)$ be a sequence such that $\mathcal{V}(u_n) = 1$ for all $n \in \mathbb{N}$, and $\mathcal{T}(u_n) \to M$ as $n \to \infty$. Denote by u_n^* the Schwarz spherical rearrangement of $|u_n|$. Using the fractional Polya-Szegő inequality (3.2.23), we have

$$[u_n^*]_s \leq [u_n]_s \quad \text{for all } n \in \mathbb{N}.$$

On the other hand, observing that $G \in C(\mathbb{R})$ is even, $G(0) = 0$, and $G(u) \in L^1(\mathbb{R}^N)$ for all $u \in H^s(\mathbb{R}^N)$, we deduce that

$$\int_{\mathbb{R}^N} G(u_n^*)\,dx = \int_{\mathbb{R}^N} G(u_n)\,dx \quad \text{for all } n \in \mathbb{N}.$$

Hence, $u_n^* \in H_{\text{rad}}^s(\mathbb{R}^N)$, $M \leq \mathcal{T}(u_n^*) \leq \mathcal{T}(u_n)$ and $\mathcal{V}(u_n^*) = 1$ for all $n \in \mathbb{N}$, that is, (u_n^*) is also a minimizing sequence. Therefore, we can assume that for every $n \in \mathbb{N}$, u_n is non-negative and radially symmetric. Using (1.1.1), we see that (u_n) is bounded in $L^{2_s^*}(\mathbb{R}^N)$. By (g_2) and (g_3), there exists $K > 0$ such that

$$G(t) \leq K|t|^{2^*} - \frac{a}{4}t^2 \quad \text{for all } t \in \mathbb{R}. \tag{3.4.7}$$

Then, combining the equality $\mathcal{V}(u_n) = 1$, (3.4.7) and the fact that (u_n) is bounded in $L^{2_s^*}(\mathbb{R}^N)$, we obtain that (u_n) is bounded in $L^2(\mathbb{R}^N)$. Consequently, (u_n) is bounded in $H^s(\mathbb{R}^N)$, so $u_n \rightharpoonup u$ in $H^s(\mathbb{R}^N)$, and by Theorem 1.1.11 we have $u_n \to u$ in $L^q(\mathbb{R}^N)$ and a.e. in $\mathbb{R}^N$.

Let $v_n = u_n - u$. Then, the weak convergence in $H^s(\mathbb{R}^N)$ implies that

$$\mathcal{T}(u_n) = \mathcal{T}(v_n) + \mathcal{T}(u) + o(1), \tag{3.4.8}$$

as $n \to \infty$, while the Brezis-Lieb lemma [113] yields

$$\|u_n\|_{L^{2_s^*}(\mathbb{R}^N)}^{2_s^*} = \|v_n\|_{L^{2_s^*}(\mathbb{R}^N)}^{2_s^*} + \|u\|_{L^{2_s^*}(\mathbb{R}^N)}^{2_s^*} + o(1) \tag{3.4.9}$$

and

$$\|u_n\|_{L^2(\mathbb{R}^N)}^2 = \|v_n\|_{L^2(\mathbb{R}^N)}^2 + \|u\|_{L^2(\mathbb{R}^N)}^2 + o(1), \tag{3.4.10}$$

as $n \to \infty$. Let $f(t) = g(t) - (t^+)^{2_s^* - 1} + at$ and $F(t) = \int_0^t f(\xi)\, d\xi$. By (g_2) and (g_3),

$$\lim_{|t| \to 0} \frac{F(t)}{t^2} = 0 \quad \text{and} \quad \lim_{|t| \to \infty} \frac{F(t)}{|t|^{2_s^*}} = 0,$$

and by the boundedness of (u_n) in $H^s(\mathbb{R}^N)$, Lemma 1.4.3 implies that

$$\int_{\mathbb{R}^N} F(v_n)\, dx = o(1) \quad \text{and} \quad \int_{\mathbb{R}^N} F(u_n)\, dx = \int_{\mathbb{R}^N} F(u)\, dx + o(1) \tag{3.4.11}$$

as $n \to \infty$. Combining (3.4.9), (3.4.10), and (3.4.11) we get

$$\mathcal{V}(u_n) = \mathcal{V}(v_n) + \mathcal{V}(u) + o(1) \quad \text{as } n \to \infty. \tag{3.4.12}$$

Set $\tau_n = \mathcal{T}(v_n)$, $\tau = \mathcal{T}(u)$, $\nu_n = \mathcal{V}(v_n)$ and $\nu = \mathcal{V}(u)$. Then, with these notations, (3.4.8) and (3.4.12) read

$$\tau_n = M - \tau + o(1) \quad \text{and} \quad \nu_n = 1 - \nu + o(1). \tag{3.4.13}$$

In order to prove the existence of a minimizer of (3.4.1), it is enough to prove that $\nu = 1$. Firstly, we note that, if $u_\sigma(x) = u(\frac{x}{\sigma})$, then $\mathcal{T}(u_\sigma) = \sigma^{N-2s}\mathcal{T}(u)$ and $\mathcal{V}(u_\sigma) = \sigma^N \mathcal{V}(u)$, so we have

$$\mathcal{T}(u) \geq M(\mathcal{V}(u))^{\frac{N-2s}{N}}, \tag{3.4.14}$$

for all $u \in H^s(\mathbb{R}^N)$ and $\mathcal{V}(u) \geq 0$.

If, by contradiction, $\nu > 1$, then $\tau \geq M\nu^{\frac{N-2s}{N}} > M$, which is in contrast with the fact that $\tau \leq M$. Therefore, $\nu \leq 1$. If $\nu < 0$, then $\nu_n > 1 - \frac{\nu}{2} > 1$ for n sufficiently large.

Using (3.4.14) we see that

$$\tau_n \geq M v_n^{\frac{N-2s}{N}} > M \left(1 - \frac{v}{2} \right)^{\frac{N-2s}{N}}$$

for n sufficiently large. This is impossible, since $\tau_n \leq M + o(1)$ by (3.4.13). Therefore, $v \in [0, 1]$. Let us show that $v = 1$. If, by contradiction, $v \in [0, 1)$, then $v_n > 0$ for all n big enough. By (3.4.14) we have

$$\tau_n \geq M v_n^{\frac{N-2s}{N}} \quad \text{and} \quad \tau \geq M v^{\frac{N-2s}{N}}.$$

This yields

$$M = \lim_{n \to \infty} (\tau + \tau_n)$$

$$\geq \lim_{n \to \infty} M \left(v^{\frac{N-2s}{N}} + v_n^{\frac{N-2s}{N}} \right)$$

$$= M \left(v^{\frac{N-2s}{N}} + (1 - v)^{\frac{N-2s}{N}} \right)$$

$$\geq M(v + 1 - v) = M.$$

Consequently, $v^{\frac{N-2s}{N}} + (1 - v)^{\frac{N-2s}{N}} = 1$, and since $v \in [0, 1)$, we get $v = 0$. Then $u = 0$ and $\tau_n \to M$ as $n \to \infty$. Moreover, by (3.4.13), $v_n \to 1$ as $n \to \infty$, and by the definition of F and (3.4.11) we can infer that

$$\| v_n \|_{L^{2_s^*}(\mathbb{R}^N)}^{2_s^*} = 2_s^* + \frac{2_s^*}{2} a \| v_n \|_{L^2(\mathbb{R}^N)}^2 + o(1),$$

which implies that

$$\limsup_{n \to \infty} \| v_n \|_{L^{2_s^*}(\mathbb{R}^N)}^2 \geq (2_s^*)^{\frac{2}{2_s^*}}.$$

Therefore, recalling that $S_* = \inf\limits_{u \in H^s(\mathbb{R}^N) \setminus \{0\}} \dfrac{[u]_s^2}{\| u \|_{L^{2_s^*}(\mathbb{R}^N)}^2}$, we get

$$M = \frac{1}{2} \lim_{n \to \infty} [v_n]_s^2 \geq \frac{1}{2} (2_s^*)^{\frac{2}{2_s^*}} \liminf_{n \to \infty} \frac{[v_n]_s^2}{\| v_n \|_{L^{2_s^*}(\mathbb{R}^N)}^2} \geq \frac{1}{2} (2_s^*)^{\frac{2}{2_s^*}} S_*,$$

which contradicts Lemma 3.4.1. Consequently, $v = 1$ and M is achieved by $u \in H^s_{\mathrm{rad}}(\mathbb{R}^N)$. $\qquad\square$

Before giving the proof of Theorem 3.1.3, we recall that any solution u of (3.1.2) satisfies the following Pohozaev identity (see [36]):

$$\frac{N-2s}{2}[u]_s^2 = N\int_{\mathbb{R}^N} G(u)\,dx. \tag{3.4.15}$$

Now we are able to prove the main result of this section:

Proof of Theorem 3.1.3 Let us show that there exists a least energy solution $\omega(x)$ of (3.1.2) and that $\mathrm{m} = \dfrac{s}{N}\left(\dfrac{N-2s}{2N}\right)^{\frac{N-2s}{2}}(2M)^{\frac{N}{2s}}$, where m is the least energy level of (3.1.2). We introduce the sets

$$\mathcal{S} = \left\{v \in H^s(\mathbb{R}^N) : \mathcal{V}(v) = 1\right\}$$

and

$$\mathcal{P} = \left\{v \in H^s(\mathbb{R}^N)\setminus\{0\} : \mathcal{J}(v) = 0\right\},$$

where

$$\mathcal{J}(v) = \frac{1}{2}[v]_s^2 - \frac{N}{N-2s}\int_{\mathbb{R}^N} G(v)\,dx \in C^1(H^s(\mathbb{R}^N),\mathbb{R}).$$

Then we have a one-to-one correspondence $\Phi : \mathcal{S} \to \mathcal{P}$ between $\mathcal{S}$ and $\mathcal{P}$, defined by

$$(\Phi(v))(x) = v\left(\frac{x}{\tau_u}\right), \quad \text{where } \tau_u = \left(\frac{N-2s}{2N}\right)^{\frac{1}{2s}}[v]_s^{\frac{1}{s}}.$$

Let us notice that for any $v \in \mathcal{S}$

$$\mathcal{I}(\Phi(v)) = \tau_u^{N-2s}\mathcal{T}(v) - \tau_u^N\mathcal{V}(v) = \frac{s}{N}\left(\frac{N-2s}{2N}\right)^{\frac{N-2s}{2s}}[v]_s^{\frac{N}{s}},$$

and thus

$$\inf_{v\in\mathcal{P}}\mathcal{I}(v) = \inf_{v\in\mathcal{S}}\mathcal{I}(\Phi(v)) = \frac{s}{N}\left(\frac{N-2s}{2N}\right)^{\frac{N-2s}{2s}}\inf_{v\in\mathcal{S}}[v]_s^{\frac{N}{s}}.$$

Using Lemma 3.4.2, we know that there exists $u \in \mathcal{S}$ such that

$$\inf_{v\in\mathcal{S}}[v]_s^2 = [u]_s^2 = 2M.$$

Consequently,

$$\inf_{v \in \mathcal{P}} \mathcal{I}(v) = \mathcal{I}(\Phi(u)) = \frac{s}{N} \left(\frac{N - 2s}{2N} \right)^{\frac{N-2s}{2s}} (2M)^{\frac{N}{2s}}.$$

Let $\omega = \Phi(u)$. By the theorem on Lagrange multipliers, there exists $\lambda \in \mathbb{R}$ such that

$$\langle \mathcal{I}'(\omega), \varphi \rangle = \lambda \langle \mathcal{J}'(\omega), \varphi \rangle \quad \text{for any } \varphi \in H^s(\mathbb{R}^N).$$

This means that ω is a weak solution to

$$(1 - \lambda)(-\Delta)^s \omega = \left(1 - \lambda \frac{N}{N - 2s} \right) g(\omega) \quad \text{in } \mathbb{R}^N,$$

and, in view of (3.4.15), ω satisfies the Pohozaev identity

$$(1 - \lambda) \frac{(N - 2s)}{2} [\omega]_s^2 = N \left(1 - \lambda \frac{N}{N - 2s} \right) \int_{\mathbb{R}^N} G(\omega) \, dx. \tag{3.4.16}$$

Since $\omega = \Phi(u) \in \mathcal{P}$, we also know that

$$\frac{N - 2s}{2} [\omega]_s^2 = N \int_{\mathbb{R}^N} G(\omega) \, dx. \tag{3.4.17}$$

Putting together (3.4.16) and (3.4.17) we deduce that

$$\lambda \left(1 - \frac{N}{N - 2s} \right) \int_{\mathbb{R}^N} G(\omega) \, dx = 0,$$

so we have $\lambda = 0$ and $\langle \mathcal{I}'(\omega), \varphi \rangle = 0$ for all $\varphi \in H^s(\mathbb{R}^N)$. Then, ω is a least energy solution to (3.1.2) and

$$\mathfrak{m} = \frac{s}{N} \left(\frac{N - 2s}{2N} \right)^{\frac{N-2s}{2}} (2M)^{\frac{N}{2s}}.$$

Arguing as in the proof of Theorem 3.2.11, we conclude that $\omega \in C^{2,\beta}(\mathbb{R}^N)$, for some $\beta > 0$, and $\omega > 0$ in $\mathbb{R}^N$. $\qquad\square$

Remark 3.4.3 Since $\omega(x) \to 0$ as $|x| \to \infty$, there exists $R > 0$ such that $g(\omega(x)) \leq -\frac{a}{2}\omega(x)$ for all $|x| > R$. Consequently, $(-\Delta)^s \omega + \frac{a}{2}\omega \leq 0$ for $|x| > R$. Using

Lemma 3.2.17 and arguing as in the proof of Theorem 3.2.11, we deduce that $0 < \omega(x) \leq C|x|^{-(N+2s)}$ for $|x| > R$. Moreover, we can see that ω is radially decreasing with respect to some point of $\mathbb{R}^N$.

3.4.2 Mountain Pass Characterization of Least Energy Solutions

In this section we give the proof of Theorem 3.1.5. First we prove

Lemma 3.4.4 $\mathcal{I}$ *has a mountain pass geometry, that is:*

  (i) $\mathcal{I}(0) = 0$;
 (ii) *there exist $\rho > 0$, and $\eta > 0$ such that $\mathcal{I}(u) \geq \eta$ for all $\|u\|_{H^s(\mathbb{R}^N)} = \rho$;*
(iii) *there exists $u_0 \in H^s(\mathbb{R}^N)$ such that $\|u_0\|_{H^s(\mathbb{R}^N)} > \rho$ and $\mathcal{I}(u_0) < 0$.*

Then

$$c = \inf_{\gamma \in \Gamma} \max_{t \in [0,1]} \mathcal{I}(\gamma(t))$$

where

$$\Gamma = \left\{ \gamma \in C([0,1], H^s(\mathbb{R}^N)) : \gamma(0) = 0 \text{ and } \mathcal{I}(\gamma(1)) < 0 \right\},$$

is well defined.

Proof Clearly, $\mathcal{I}(0) = 0$. By assumptions (g_2) and (g_3), there exists a positive constant C_a such that

$$G(t) \leq -\frac{a}{4}t^2 + C_a|t|^{2_s^*} \text{ for all } t \in \mathbb{R}. \tag{3.4.18}$$

Then, by (1.1.1) and (3.4.18), we get

$$\mathcal{I}(u) \geq \frac{1}{2}[u]_s^2 + \frac{a}{4}\|u\|_{L^2(\mathbb{R}^N)}^2 - C_a\|u\|_{L^{2_s^*}(\mathbb{R}^N)}^{2_s^*}$$

$$\geq \min\left\{\frac{1}{2}, \frac{a}{4}\right\} \|u\|_{H^s(\mathbb{R}^N)}^2 - C_a\|u\|_{L^{2_s^*}(\mathbb{R}^N)}^{2_s^*}$$

$$\geq \min\left\{\frac{1}{2}, \frac{a}{4}\right\} \|u\|_{H^s(\mathbb{R}^N)}^2 - C_a S_*^{-\frac{2_s^*}{2}} \|u\|_{H^s(\mathbb{R}^N)}^{2_s^*},$$

which implies that there exist $\rho > 0$ and $\eta > 0$ such that $\mathcal{I}(u) \geq \eta$ for all $\|u\|_{H^s(\mathbb{R}^N)} = \rho$. Using again (g_2) and (g_3), there exists $C_a' > 0$ such that

$$G(t) \geq -\frac{C_a'}{2} t^2 + \frac{1}{22_s^*} |t|^{2_s^*} \quad \text{for all } t \in \mathbb{R}.$$

Then, for any $u \in H^s(\mathbb{R}^N) \setminus \{0\}$ and $t > 0$,

$$\mathcal{I}(tu) \leq \frac{t^2}{2} [u]_s^2 + \frac{C_a' t^2}{2} \|u\|_{L^2(\mathbb{R}^N)}^2 - \frac{t^{2_s^*}}{22_s^*} \|u\|_{L^{2_s^*}(\mathbb{R}^N)}^{2_s^*} \to -\infty \quad \text{as } t \to \infty,$$

so we can find $u_0 \in H^s(\mathbb{R}^N)$ such that $\|u_0\|_{H^s(\mathbb{R}^N)} > \rho$ and $\mathcal{I}(u_0) < 0$.   □

Proof of Theorem 3.1.5 Let ω be a least energy solution to (3.1.2). We begin by proving that there exists $\gamma \in C([0, 1], H^s(\mathbb{R}^N))$ such that $\gamma(0) = 0$, $\mathcal{I}(\gamma(1)) < 0$, $\omega \in \gamma([0, 1])$ and

$$\max_{t \in [0, 1]} \mathcal{I}(\gamma(t)) = \mathfrak{m}.$$

Let

$$\gamma(t)(x) = \begin{cases} \omega(\frac{x}{t}), & \text{for } t > 0, \\ 0, & \text{for } t = 0. \end{cases}$$

Then,

$$\|\gamma(t)\|_{H^s(\mathbb{R}^N)}^2 = t^{N-2s} [\omega]_s^2 + t^N \|\omega\|_{L^2(\mathbb{R}^N)}^2,$$

$$\mathcal{I}(\gamma(t)) = \frac{t^{N-2s}}{2} [\omega]_s^2 - t^N \int_{\mathbb{R}^N} G(\omega) \, dx. \tag{3.4.19}$$

Hence, $\gamma \in C([0, \infty), H^s(\mathbb{R}^N))$. By the Pohozaev identity (3.4.15),

$$\int_{\mathbb{R}^N} G(\omega) \, dx = \frac{N - 2s}{2N} [\omega]_s^2 > 0. \tag{3.4.20}$$

Combining (3.4.19) and (3.4.20) we have

$$\frac{d}{dt} \mathcal{I}(\gamma(t)) = t^{N-2s-1} (1 - t^{2s}) \frac{N - 2s}{2} [\omega]_s^2,$$

and, in particular,

$$\frac{d}{dt}\mathcal{I}(\gamma(t)) > 0 \quad \text{for } t \in (0, 1) \quad \text{and} \quad \frac{d}{dt}\mathcal{I}(\gamma(t)) < 0 \quad \text{for } t > 1.$$

Thus, for $L > 1$ sufficiently large, there exists a path $\gamma : [0, L] \to H^s(\mathbb{R}^N)$ such that $\gamma(0) = 0, \mathcal{I}(\gamma(L)) < 0, \omega \in \gamma([0, L])$ and

$$\max_{t \in [0, L]} \mathcal{I}(\gamma(t)) = \mathfrak{m}.$$

After a suitable scale change in t, we get the desired path $\gamma \in \Gamma$. Hence, $c \leq \mathfrak{m}$. From the proof of Theorem 3.1.3, we deduce that $\mathfrak{m} = \inf_{v \in \mathcal{P}} \mathcal{I}(v)$.

Now, let

$$\mathcal{H}(u) = \frac{N - 2s}{2}[u]_s^2 - N \int_{\mathbb{R}^N} G(u)\, dx = N\mathcal{I}(u) - s[u]_s^2.$$

Modifying slightly the arguments of the proof of Lemma 3.4.4, one can show that there exists $\rho_0 > 0$ such that $\mathcal{H}(u) > 0$ for all $0 < \|u\|_{H^s(\mathbb{R}^N)} \leq \rho_0$. For any $\gamma \in \Gamma$, we have $\gamma(0) = 0$ and $\mathcal{H}(\gamma(1)) \leq N\mathcal{I}(\gamma(1)) < 0$. Consequently, there exists $t_0 \in [0, 1]$ such that

$$\|\gamma(t_0)\|_{H^s(\mathbb{R}^N)} > \rho_0 \quad \text{and} \quad \mathcal{H}(\gamma(t_0)) = 0.$$

Since $\gamma(t_0) \in \gamma([0, 1]) \cap \mathcal{P}$, we have $\gamma([0, 1]) \cap \mathcal{P} \neq \emptyset$ and thus $c \geq \mathfrak{m}$. Therefore, $c = \mathfrak{m}$. □

3.5 The Pohozaev Identity for $(-\Delta)^s$

In this section we give a proof of the Pohozaev identity for the fractional Laplacian in $\mathbb{R}^N$. For more details we refer to [36, 122, 139, 302, 307].

Theorem 3.5.1 *Let g satisfy (g_1) and either $(g1)$–$(g3)$ or (g_2)–(g_4). If $u \in H^s(\mathbb{R}^N)$ is a weak solution to (3.1.2), then u satisfies the Pohozaev identity*

$$\frac{N - 2s}{2} \int_{\mathbb{R}^N} |(-\Delta)^{\frac{s}{2}} u|^2\, dx = N \int_{\mathbb{R}^N} G(u)\, dx.$$

Proof Transforming (3.1.2) into a local problem via the extension method, we see that the s-harmonic extension $v(x, y) = P_s(x, y) * u(x)$ of u solves

$$\begin{cases} -\operatorname{div}(y^{1-2s}\nabla v) = 0 & \text{in } \mathbb{R}^{N+1}_+, \\ v(\cdot, 0) = u & \text{on } \mathbb{R}^N, \\ \frac{\partial v}{\partial \nu^{1-2s}} = \kappa_s g(u) & \text{on } \mathbb{R}^N, \end{cases} \tag{3.5.1}$$

and satisfies (see Sect. 1.2.3)

$$\iint_{\mathbb{R}^{N+1}_+} y^{1-2s} |\nabla v|^2 \, dx dy = \kappa_s \int_{\mathbb{R}^N} |(-\Delta)^{\frac{s}{2}} u|^2 \, dx = \kappa_s \frac{C(N,s)}{2} [u]_s^2 < \infty. \quad (3.5.2)$$

Using Proposition 3.2.14 and the bootstrap argument in Theorem 3.2.11, we infer that $u \in C^{2,\beta}(\mathbb{R}^N)$ and thus $v \in C^2(\overline{\mathbb{R}^{N+1}_+})$. For any $R > 0$ and $\delta \in (0, R)$, denote

$$D^+_{R,\delta} = \{(x, y) \in \mathbb{R}^N \times [\delta, \infty) : |(x, y)| \le R\},$$

and

$$\partial D^1_{R,\delta} = \{(x, y) \in \mathbb{R}^N \times \{y = \delta\} : |x|^2 \le R^2 - \delta^2\},$$

$$\partial D^2_{R,\delta} = \{(x, y) \in \mathbb{R}^N \times [\delta, \infty) : |(x, y)| = R\}.$$

Denote by n the unit outward normal vector on $\partial D^+_{R,\delta}$. Then,

$$n = \begin{cases} (0, \ldots, 0, -1), & \text{on } \partial D^1_{R,\delta}, \\ (\frac{x}{R}, \frac{y}{R}), & \text{on } \partial D^2_{R,\delta}. \end{cases}$$

Note that

$$\text{div}(y^{1-2s} \nabla v)((x, y) \cdot \nabla v) = \text{div}\left[y^{1-2s} \nabla v((x, y) \cdot \nabla v) - y^{1-2s}(x, y)\frac{|\nabla v|^2}{2}\right]$$
$$+ \frac{N-2s}{2} y^{1-2s} |\nabla v|^2.$$

Multiplying (3.5.1) by $(x, y) \cdot \nabla v$, integrating on $D^+_{R,\delta}$ and applying the divergence theorem we get

$$0 = \iint_{D^+_{R,\delta}} \text{div}(y^{1-2s} \nabla v)((x, y) \cdot \nabla v) \, dx dy$$

$$= \int_{\partial D^1_{R,\delta}} y^{1-2s} \left[((x, y) \cdot \nabla v)(-v_y) + \frac{y}{2}|\nabla v|^2\right] d\sigma$$

$$+ \int_{\partial D^2_{R,\delta}} y^{1-2s} \left[\frac{1}{R}|(x, y) \cdot \nabla v|^2 - \frac{R}{2}|\nabla v|^2\right] d\sigma + \frac{N-2s}{2} \iint_{D^+_{R,\delta}} y^{1-2s} |\nabla v|^2 \, dx dy$$

$$= A_1(R, \delta) + A_2(R, \delta) + A_3(R, \delta). \quad (3.5.3)$$

Standard computations imply the following identity:

$$g(u)(x \cdot \nabla u) = \operatorname{div}(x\, G(u)) - N G(u) \text{ on } \mathbb{R}^N.$$

Then, using (3.5.1) and the divergence theorem, we see that

$$\lim_{\delta \to 0} \int_{\partial D^1_{R,\delta}} y^{1-2s}((x, y) \cdot \nabla v)(-v_y)\, d\sigma = \kappa_s \int_{\mathcal{B}_R} g(u)(x \cdot \nabla u)\, dx$$

$$= R\kappa_s \int_{\partial \mathcal{B}_R} G(u)\, d\sigma - N\kappa_s \int_{\mathcal{B}_R} G(u)\, dx,$$

$$(3.5.4)$$

where $\mathcal{B}_R = \{(x, 0) \in \partial \mathbb{R}^{N+1}_+ : |x|^2 \le R^2\}$. Clearly,

$$\lim_{\delta \to 0} \int_{\partial D^1_{R,\delta}} y^{1-2s} y |\nabla v|^2\, d\sigma = 0. \qquad (3.5.5)$$

Taking into account that $G(u) \in L^1(\mathbb{R}^N)$ and (3.5.2), we can find a sequence (R_n) with $R_n \to \infty$ as $n \to \infty$, such that

$$\lim_{n \to \infty} R_n \int_{\partial \mathcal{B}_{R_n}} G(u)\, d\sigma = 0, \qquad (3.5.6)$$

$$\lim_{n \to \infty} A_2(R_n, \delta) = 0 \text{ for all } \delta > 0. \qquad (3.5.7)$$

Indeed, suppose, by contradiction, that there exist $\tau, R_1 > 0$ such that, for all $R \ge R_1$,

$$\int_{\partial \mathcal{B}_{R_1}} |G(u)|\, d\sigma \ge \frac{\tau}{R}.$$

Then

$$\int_{\mathbb{R}^N} |G(u)|\, dx \ge \int_{R_1}^{\infty} \int_{\partial \mathcal{B}_{R_1}} |G(u)|\, d\sigma\, dr$$

$$\ge \int_{R_1}^{\infty} \int_{\partial \mathcal{B}_{R_1}} |G(u)|\, d\sigma\, dr$$

$$\ge \int_{R_1}^{\infty} \frac{\tau}{r}\, dr = \infty$$

that is absurd. The same argument works for proving (3.5.7) and this shows that the claim holds true. Combining (3.5.4)–(3.5.7), we deduce that

$$\lim_{n\to\infty}\lim_{\delta\to 0} A_1(R_n,\delta) = -N \int_{\mathbb{R}^N} G(u)\,dx,$$

$$\lim_{n\to\infty}\lim_{\delta\to 0} A_2(R_n,\delta) = 0. \tag{3.5.8}$$

Then, using (3.5.8) and the fact that

$$\lim_{n\to\infty}\lim_{\delta\to 0} A_3(R_n,\delta) = \frac{N-2s}{2} \iint_{\mathbb{R}^{N+1}_+} y^{1-2s}|\nabla v|^2\,dxdy,$$

we can see that (3.5.3) yields

$$\frac{N-2s}{2} \iint_{\mathbb{R}^{N+1}_+} y^{1-2s}|\nabla v|^2\,dxdy = \kappa_s \int_{\mathbb{R}^N} G(v(x,0))\,dx. \tag{3.5.9}$$

Now (3.5.2) and (3.5.9) yield the desired result. $\qquad\square$

Remark 3.5.2 As observed in [122], the regularity assumptions on g needed to obtain the Pohozaev identity can be weakened (in fact, it is enough to require the C^1 regularity for weak solutions of (3.1.2) to prove Theorem 3.5.1).

Ground States for a Pseudo-Relativistic Schrödinger Equation

4

4.1 Introduction

In this chapter we consider the nonlinear fractional equation

$$[(-\Delta + m^2)^s - m^{2s}]u + \mu u = |u|^{p-2}u \quad \text{in } \mathbb{R}^N, \tag{4.1.1}$$

where $s \in (0, 1)$, $N \geq 2$, $p \in (2, 2_s^*)$, $m \geq 0$ and $\mu > 0$. The fractional operator

$$(-\Delta + m^2)^s \tag{4.1.2}$$

which appears in (4.1.1) is defined in the Fourier space by setting

$$\mathcal{F}(-\Delta + m^2)^s u(\xi) = (|\xi|^2 + m^2)^s \mathcal{F}u(\xi),$$

or via the Bessel potential, that is, for every $u \in C_c^\infty(\mathbb{R}^N)$

$$(-\Delta + m^2)^s u(x) = c_{N,s} m^{\frac{N+2s}{2}} \text{P.V.} \int_{\mathbb{R}^N} \frac{u(x) - u(y)}{|x - y|^{\frac{N+2s}{2}}} K_{\frac{N+2s}{2}}(m|x - y|)\,dy + m^{2s}u(x) \tag{4.1.3}$$

for every $x \in \mathbb{R}^N$; see [181, 249]. Here P.V. stands for the Cauchy principal value,

$$c_{N,s} = 2^{-\frac{N+2s}{2}+1} \pi^{-\frac{N}{2}} 2^{2s} \frac{s(1 - s)}{\Gamma(2 - s)}$$

and K_ν denotes the modified Bessel function of the third kind and order ν; see [78, 178, 336] for more details.

© The Author(s), under exclusive license to Springer Nature Switzerland AG 2021
V. Ambrosio, *Nonlinear Fractional Schrödinger Equations in* $\mathbb{R}^N$,
Frontiers in Mathematics, https://doi.org/10.1007/978-3-030-60220-8_4

When $s = 1/2$ the operator (4.1.2) has a clear meaning in quantum mechanics: it corresponds to the free Hamiltonian of a free relativistic particle of mass m. We stress that the study of $\sqrt{-\Delta + m^2}$ has been strongly influenced by the study of Lieb and Yau [250, 251] on the stability of relativistic matter. For more details on this topic one can consult [152, 202, 217, 249, 337, 338].

On the other hand, the operator (4.1.2) has a deep connection with the theory of stochastic processes: in this context, $-[(-\Delta + m^2)^s - m^{2s}]$ is the infinitesimal generator of a Lévy process called the $2s$-stable relativistic process: see [76, 119, 132, 304] for more details.

We remark that the hardest issue in dealing with this operator is the lack of scaling properties: there is no standard group action under which $(-\Delta + m^2)^s$ behaves as a local differential operator. Indeed, the most striking difference between $(-\Delta + m^2)^s$ and $(-\Delta)^s$ is that the latter enjoys some scaling properties that the former does not have. As should be clear from (4.1.3), the operator $(-\Delta + m^2)^s$ is not compatible with the semigroup $\mathbb{R}_+$ acting on functions as $t * u : x \mapsto u(t^{-1}x)$ for $t > 0$. In simpler words, the operator $(-\Delta + m^2)^s$ does not scale. For this reason, some interesting papers studied fractional nonlocal problems involving (4.1.2); see for instance [31, 34, 143, 152, 181, 184, 308] and the references therein.

In the present chapter we are interested in the study of the ground states of (4.1.1). Such problems are motivated in particular by the search of standing wave solutions, namely $\psi(x, t) = u(x)e^{-\iota \omega t}$, where ω is a constant, for the fractional Schrödinger-Klein-Gordon equation

$$\iota \frac{\partial \psi}{\partial t} = [(-\Delta + m^2)^s - m^{2s}]\psi + F(x, \psi) \quad \text{in } \mathbb{R}^N,$$

which describes dynamics of boson systems.

Our first result can be stated as follows:

Theorem 4.1.1 ([34]) *Let $m > 0$ and $\mu > 0$. Then there exists a ground state solution $u \in H_m^s(\mathbb{R}^N)$ to (4.1.1) which is positive, radially symmetric and decreasing.*

The main difficulty in the study of (4.1.1) is related to the nonlocal character of the operator (4.1.2). To overcome this difficulty, we use a variant of the Caffarelli-Silvestre extension method [127], which consists in writing a given nonlocal problem in a local form via the Dirichlet-Neumann map. This allow us to apply known variational techniques to this kind of problems; see for instance [31, 152, 181, 317]. More precisely, for any $u \in H_m^s(\mathbb{R}^N)$ there exists a unique $v \in H_m^1(\mathbb{R}_+^{N+1}, y^{1-2s})$ that solves in the weak sense

$$\begin{cases} -\operatorname{div}(y^{1-2s} \nabla v) + m^2 y^{1-2s} v = 0 & \text{in } \mathbb{R}_+^{N+1}, \\ v(\cdot, 0) = u & \text{on } \partial\mathbb{R}_+^{N+1}, \end{cases}$$

and such that

$$\frac{\partial v}{\partial v^{1-2s}}(x, 0) = - \lim_{y \to 0} y^{1-2s} \frac{\partial v}{\partial y}(x, y) = (-\Delta + m^2)^s u(x) \quad \text{in } H_m^{-s}(\mathbb{R}^N).$$

We exploit this fact and we study the existence, regularity and qualitative properties of solutions to the problem

$$\begin{cases} -\text{div}(y^{1-2s}\nabla v) + m^2 y^{1-2s} v = 0 & \text{in } \mathbb{R}_+^{N+1}, \\ \dfrac{\partial v}{\partial v^{1-2s}} = \kappa_s[m^{2s} v - \mu v + |v|^{p-2} v] & \text{on } \partial \mathbb{R}_+^{N+1}. \end{cases}$$

Finally, we are able to pass in (4.1.1) to the limit as $m \to 0$ and find a nontrivial solution to

$$(-\Delta)^s u + \mu u = |u|^{p-2} u \quad \text{in } \mathbb{R}^N. \tag{4.1.4}$$

In this way, we rediscover the existence result contained in Theorem 3.1.1 of Chap. 3 with $g(u) = -\mu u + |u|^{p-2} u$. More precisely, we obtain the following result.

Theorem 4.1.2 ([34]) *There exists a solution $u \in H^s(\mathbb{R}^N)$ to (4.1.4) which is positive and radially symmetric.*

In Sect. 4.2 we give some useful results needed to deal with (4.1.2). In Sect. 4.3 we obtain the existence of a ground state solution to (4.1.1). In Sect. 4.4 we investigate the qualitative properties of solutions to (4.1.1). Finally, we give the proof of Theorem 4.1.2.

4.2 Preliminaries

4.2.1 The Extension Method for $(-\Delta + m^2)^s$

Let $s \in (0, 1)$ and $m > 0$. We define $H_m^s(\mathbb{R}^N)$ as the completion of $C_c^\infty(\mathbb{R}^N)$ with respect to the norm

$$|u|_{H_m^s(\mathbb{R}^N)} = \left(\int_{\mathbb{R}^N} (|\xi|^2 + m^2)^s |\mathcal{F}u(\xi)|^2 d\xi \right)^{\frac{1}{2}} < \infty,$$

where $\mathcal{F}u(k)$ is the Fourier transform of u. We note that

$$\int_{\mathbb{R}^N} |(-\Delta + m^2)^{\frac{s}{2}} u|^2 dx = \int_{\mathbb{R}^N} (|\xi|^2 + m^2)^s |\mathcal{F}u(\xi)|^2 d\xi$$

for all $u \in H_m^s(\mathbb{R}^N)$. When $m = 1$, we use the notation $H^s(\mathbb{R}^N) = H_1^s(\mathbb{R}^N)$ and $|\cdot|_{H^s(\mathbb{R}^N)} = |\cdot|_{H_1^s(\mathbb{R}^N)}$. In [181], it is also proved that

$$
\int_{\mathbb{R}^N} [|(-\Delta + m^2)^{\frac{s}{2}} u|^2 - m^{2s} u^2] \, dx
$$

$$
= \frac{c_{N,s}}{2} m^{\frac{N+2s}{2}} \iint_{\mathbb{R}^{2N}} \frac{|u(x) - u(y)|^2}{|x-y|^{N+2s}} K_{\frac{N+2s}{2}}(m|x-y|) \, dx \, dy
$$

(4.2.1)

for all $u \in H_m^s(\mathbb{R}^N)$. We denote by $H_m^1(\mathbb{R}_+^{N+1}, y^{1-2s})$ the completion of $C_c^\infty(\overline{\mathbb{R}_+^{N+1}})$ with respect to the norm

$$
\|v\|_{H_m^1(\mathbb{R}_+^{N+1}, y^{1-2s})} = \left(\iint_{\mathbb{R}_+^{N+1}} y^{1-2s}(|\nabla v|^2 + m^2 v^2) \, dx \, dy \right)^{\frac{1}{2}} < \infty.
$$

When $m = 1$, we set $H^1(\mathbb{R}_+^{N+1}, y^{1-2s}) = H_1^1(\mathbb{R}_+^{N+1}, y^{1-2s})$ and $\|\cdot\|_{H^1(\mathbb{R}_+^{N+1}, y^{1-2s})} = \|\cdot\|_{H_1^1(\mathbb{R}_+^{N+1}, y^{1-2s})}$.

We recall that there exists a trace operator which relates the spaces $H_m^1(\mathbb{R}_+^{N+1}, y^{1-2s})$ and $H_m^s(\mathbb{R}^N)$:

Theorem 4.2.1 ([181]) *There exists a trace operator* $\mathrm{Tr} : H_m^1(\mathbb{R}_+^{N+1}, y^{1-2s}) \to H_m^s(\mathbb{R}^N)$ *such that:*

(i) $\mathrm{Tr}(v) = v(\cdot, 0)$ *for every* $v \in C_c^\infty(\overline{\mathbb{R}_+^{N+1}})$.
(ii) $\kappa_s |\mathrm{Tr}(v)|_{H_m^s(\mathbb{R}^N)}^2 \leq \|v\|_{H_m^1(\mathbb{R}_+^{N+1}, y^{1-2s})}^2$ *for every* $v \in H_m^1(\mathbb{R}_+^{N+1}, y^{1-2s})$, *where* $\kappa_s = 2^{1-2s} \frac{\Gamma(1-s)}{\Gamma(s)}$.

Equality holds in (ii) for a function $v \in H_m^1(\mathbb{R}_+^{N+1}, y^{1-2s})$ *if and only if* v *is a weak solution to*

$$
-\mathrm{div}(y^{1-2s} \nabla v) + m^2 y^{1-2s} v = 0 \quad \text{in } \mathbb{R}_+^{N+1}.
$$

Remark 4.2.2 By item (ii) in Theorem 4.2.1 we deduce that

$$
m^{2s} \|\mathrm{Tr}(v)\|_{L^2(\mathbb{R}^N)}^2 \leq \iint_{\mathbb{R}_+^{N+1}} y^{1-2s}(|\nabla v|^2 + m^2 v^2) \, dx \, dy
$$

(4.2.2)

for all $v \in H_m^1(\mathbb{R}_+^{N+1}, y^{1-2s})$. Since $H_m^s(\mathbb{R}^N) \subset L^q(\mathbb{R}^N)$ for any $2 \leq q \leq 2_s^*$, we deduce that for any $v \in H_m^1(\mathbb{R}_+^{N+1}, y^{1-2s})$ and for any $q \in [2, 2_s^*]$

$$
\begin{aligned}
C_{q,s,N} \|u\|_{L^q(\mathbb{R}^N)}^2 &\leq \kappa_s \int_{\mathbb{R}^N} (|\xi|^2 + m^2)^s |\mathcal{F}u(\xi)|^2 \, d\xi \\
&\leq \iint_{\mathbb{R}_+^{N+1}} y^{1-2s}(|\nabla v|^2 + m^2 v^2) \, dx \, dy,
\end{aligned} \tag{4.2.3}
$$

where $u(x) = v(x, 0)$ is the trace of v on $\partial \mathbb{R}_+^{N+1}$.

Remark 4.2.3 In view of (1.2.8) and (1.2.2),

$$
\kappa_s \frac{C(N, s)}{2} [\mathrm{Tr}(v)]_s^2 \leq \|\nabla v\|_{L^2(\mathbb{R}_+^{N+1}, y^{1-2s})}^2
$$

for all $v \in X_0^s(\mathbb{R}_+^{N+1})$.

The following result holds true (see also [317]):

Theorem 4.2.4 ([181]) *Let* $u \in H_m^s(\mathbb{R}^N)$. *Then there exists a unique* $v \in H_m^1(\mathbb{R}_+^{N+1}, y^{1-2s})$ *which solves the problem*

$$
\begin{cases}
-\mathrm{div}(y^{1-2s} \nabla v) + m^2 y^{1-2s} v = 0 & \text{in } \mathbb{R}_+^{N+1}, \\
v(\cdot, 0) = u & \text{on } \partial \mathbb{R}_+^{N+1}.
\end{cases} \tag{4.2.4}
$$

In addition,

$$
-\lim_{y \to 0} y^{1-2s} \frac{\partial v}{\partial y}(x, y) = \kappa_s(-\Delta + m^2)^s u(x) \quad \text{in } H_m^{-s}(\mathbb{R}^N),
$$

where $H_m^{-s}(\mathbb{R}^N)$ *denotes the dual of* $H_m^s(\mathbb{R}^N)$.

Remark 4.2.5 As proved in [181], the extension $v = \mathrm{Ext}_m(u)$ of $u \in C_c^\infty(\mathbb{R}^N)$ can be defined via the Fourier transform with respect to the variable x as

$$
\mathcal{F}v(\xi, y) = \mathcal{F}u(\xi)\theta(y\sqrt{|\xi|^2 + m^2}),
$$

where θ is the Bessel function defined as in Remark 1.2.9. More precisely,

$$
v(x, y) = (\tilde{P}_m(\cdot, y) * u)(x) = \int_{\mathbb{R}^N} \tilde{P}_m(x - z, y) u(z) \, dz,
$$

and $\widetilde{P}_m(x, y)$ is the Fourier transform of $\xi \mapsto \theta(y\sqrt{|\xi|^2 + m^2})$, namely

$$\widetilde{P}_m(x, y) = C'_{N,s} y^{2s} m^{\frac{N+2s}{2}} (|x|^2 + y^2)^{-\frac{N+2s}{4}} K_{\frac{N+2s}{2}} (m\sqrt{|x|^2 + y^2}), \qquad (4.2.5)$$

for some constant $C'_{N,s} > 0$. Furthermore, $\int_{\mathbb{R}^N} \widetilde{P}_m(x, y)\, dx = \theta(my)$ for all $y > 0$.

In the light of Theorem 4.2.4, we will look for positive solutions of the following problem:

$$\begin{cases} -\operatorname{div}(y^{1-2s}\nabla v) + m^2 y^{1-2s} v = 0 & \text{in } \mathbb{R}^{N+1}_+, \\ \frac{\partial v}{\partial v^{1-2s}} = \kappa_s [m^{2s} v - \mu v + |v|^{p-2} v] & \text{on } \partial \mathbb{R}^{N+1}_+, \end{cases} \qquad (4.2.6)$$

where

$$\frac{\partial v}{\partial v^{1-2s}}(x, 0) = -\lim_{y \to 0} y^{1-2s} \frac{\partial v}{\partial y}(x, y).$$

For simplicity we will assume that $\kappa_s = 1$. Finally, by using Theorem 4.2.1 and Theorem 1.1.11, we deduce the following result:

Theorem 4.2.6 *Let*

$$X^m_{\mathrm{rad}} = \left\{ v \in H^1_m(\mathbb{R}^{N+1}_+, y^{1-2s}) : v \text{ is radially symmetric with respect to } x \right\}.$$

Then, X^m_{rad} is compactly embedded in $L^q(\mathbb{R}^N)$ for any $q \in (2, 2^*_s)$.

4.2.2   Local Schauder Estimates and Maximum Principles

Here we collect some results about local Schauder estimates and maximum principles for problems involving the operator

$$-\operatorname{div}(y^{1-2s}\nabla v) + m^2 y^{1-2s} v.$$

We begin with the following definition:

Definition 4.2.7 Let $R > 0$ and $h \in L^1(\Gamma^0_R)$. We say that $v \in H^1_m(B^+_R)$ is a weak solution to

$$\begin{cases} -\operatorname{div}(y^{1-2s}\nabla v) + m^2 y^{1-2s} v = 0 & \text{in } B^+_R, \\ \frac{\partial v}{\partial v^{1-2s}} = h & \text{on } \Gamma^0_R, \end{cases}$$

if

$$\iint_{B_R^+} y^{1-2s}[\nabla v \cdot \nabla \phi + m^2 v\varphi]\,dx dy = \int_{\Gamma_R^0} h\varphi(\cdot, 0)\,dx$$

for every $\varphi \in C^1(\overline{B_R^+})$ such that $\varphi = 0$ on Γ_R^+.

Next we recall some regularity results whose proofs can be found in [181].

Proposition 4.2.8 ([181]) *Let $f, g \in L^q(\Gamma_1^0)$ for some $q > \frac{N}{2s}$ and $\eta, \psi \in L^r(B_1^+, y^{1-2s})$ for some $r > \frac{N+2-2s}{2}$. Let $v \in H^1(B_1^+, y^{1-2s})$ be a weak solution to*

$$\begin{cases} -\mathrm{div}(y^{1-2s}\nabla v) + y^{1-2s}\eta v = y^{1-2s}\psi & \text{in } B_1^+, \\ \frac{\partial v}{\partial \nu^{1-2s}} = f(x)v + g(x) & \text{on } \Gamma_1^0. \end{cases}$$

Then, $v \in C^{0,\alpha}(\overline{B_{1/2}^+})$ and, in addition,

$$\|v\|_{C^{0,\alpha}(\overline{B_{1/2}^+})} \le C\left(\|v\|_{L^2(B_1^+)} + \|g\|_{L^q(\Gamma_1^0)} + \|\eta\|_{L^r(B_1^+, y^{1-2s})}\right),$$

with $C, \alpha > 0$ depending only on N, s, $\|f\|_{L^q(\Gamma_1^0)}$, $\|\eta\|_{L^r(B_1^+, y^{1-2s})}$.

Proposition 4.2.9 ([181]) *Let $f, g \in C^k(\Gamma_1^0)$ and $\nabla_x^i \eta, \nabla_x^i \psi \in L^\infty(B_1^+)$, for some $k \ge 1$ and $i = 0, \ldots, k$. Let $v \in H^1(B_1^+, y^{1-2s})$ be a weak solution to*

$$\begin{cases} -\mathrm{div}(y^{1-2s}\nabla v) + y^{1-2s}\eta v = y^{1-2s}\psi & \text{in } B_1^+, \\ \frac{\partial v}{\partial \nu^{1-2s}} = f(x)v + g(x) & \text{on } \Gamma_1^0. \end{cases}$$

Then, for $i = 1, \ldots, k$, we have that $v \in C^{i,\alpha}(\overline{B_r^+})$ for some $r \in (0, 1)$ depending only on k, and, in addition,

$$\sum_{i=1}^{k} \|\nabla_x^i v\|_{C^{0,\alpha}(\overline{B_r^+})} \le C\left(\|v\|_{L^2(B_1^+, y^{1-2s})} + \|f\|_{C^k(\overline{\Gamma_r^0})} + \|g\|_{C^k(\overline{\Gamma_r^0})}\right.$$

$$\left. + \sum_{i=1}^{k} \|\nabla_x^i \eta, \nabla_x^i \psi\|_{L^\infty(B_1^+)}\right),$$

where $C, \alpha > 0$ depend only on N, s, k, r, $\|f\|_{L^\infty(\Gamma_{1/2}^0)}$, $\|\eta\|_{L^\infty(\Gamma_{1/2}^0)}$.

Lemma 4.2.10 ([181]) *Let* $g \in C^{0,\gamma}(\Gamma_1^0)$ *for some* $\gamma \in [0, 2 - 2s)$ *(when* $\gamma = 0$ *we mean* $g \in C^{0,\gamma} = L^{\infty}$*). Let* $v \in H_m^1(B_1^+, y^{1-2s})$ *be a weak solution to*

$$\begin{cases} -\mathrm{div}(y^{1-2s}\nabla v) + m^2 y^{1-2s} v = 0 & \text{in } B_1^+, \\ \frac{\partial v}{\partial v^{1-2s}} = g & \text{on } \Gamma_1^0. \end{cases}$$

Then, for every sufficiently small $t_0 > 0$ *there exist positive constants* C *and* $\alpha \geq 0$ *(with* $\alpha > 0$ *if* $\gamma > 0$*), depending only on* N, s, t_0, m, γ, *such that*

$$\left\| y^{1-2s} \frac{\partial v}{\partial y} \right\|_{C^{0,\alpha}(\Gamma_{1/8}^0 \times [0, t_0))} \leq C \left(\|v\|_{L^2(B_1^+, y^{1-2s})} + \|g\|_{C^{\gamma}(\Gamma_{1/2}^0)} \right).$$

We also have the following weak Harnack inequality.

Proposition 4.2.11 ([181]) *Let* $f, g \in L^q(\Gamma_1^0)$ *for some* $q > \frac{N}{2s}$ *and* $\eta, \psi \in L^r(B_1^+, y^{1-2s})$ *for some* $r > \frac{N+2-2s}{2}$*. Let* $v \in H^1(B_1^+, y^{1-2s})$ *be a nonnegative weak solution to*

$$\begin{cases} -\mathrm{div}(y^{1-2s}\nabla v) + y^{1-2s}\eta v \geq y^{1-2s}\psi & \text{in } B_1^+, \\ \frac{\partial v}{\partial v^{1-2s}} \geq f(x)v + g(x) & \text{on } \Gamma_1^0. \end{cases}$$

Then, for some $p_0 > 0$ *and any* $0 < r < r' < 1$,

$$\inf_{\bar{B}_r^+} v + \|g_-\|_{L^q(\Gamma_1^0)} + \|\psi_-\|_{L^r(B_1^+, y^{1-2s})} \geq C \|v\|_{L^{p_0}(B_{r'}^+, y^{1-2s})},$$

where $C > 0$ *depends only on* N, s, r, r', $\|f_-\|_{L^q(\Gamma_1^0)}$, $\|\eta^+\|_{L^r(B_1^+, y^{1-2s})}$.

In the spirit of [124], we prove some maximum principles:

Proposition 4.2.12 (Weak Maximum Principle) *Let* $v \in H_m^1(B_R^+, y^{1-2s})$ *be a weak solution to*

$$\begin{cases} -\mathrm{div}(y^{1-2s}\nabla v) + m^2 y^{1-2s} v \geq 0 & \text{in } B_R^+, \\ \frac{\partial v}{\partial v^{1-2s}} \geq 0 & \text{on } \Gamma_R^0, \\ v \geq 0 & \text{on } \Gamma_R^+. \end{cases}$$

Then, $v \geq 0$ *in* B_R^+.

Proof It is enough to use v_- as a test function in the weak formulation of the above problem.   □

Remark 4.2.13 In addition, the following strong maximum principle holds true: either $v \equiv 0$, or $v > 0$ in $B_R^+ \cup \Gamma_R^0$. In fact, v cannot vanish at an interior point, due to the classical strong maximum principle for strictly elliptic operators. Finally, the fact that v cannot vanish at a point in Γ_R^0 follows from the Hopf principle that we establish below.

Note that the same weak and strong maximum principles (and proofs) hold in other bounded domains of $\mathbb{R}_+^{N+1}$ different than B_R^+.

Proposition 4.2.14 (Hopf Principle) *Let $C_{R,1} = \Gamma_R^0 \times (0, 1)$ and $v \in H_m^1(C_{R,1}, y^{1-2s}) \cap C(\overline{C_{R,1}})$ be a weak solution to*

$$\begin{cases} -\operatorname{div}(y^{1-2s}\nabla v) + m^2 y^{1-2s} v \geq 0 & \text{in } C_{R,1}, \\ v > 0 & \text{in } C_{R,1}, \\ v(0, 0) = 0. \end{cases}$$

Then

$$\limsup_{y \to 0} -y^{1-2s} \frac{v(0, y)}{y} < 0.$$

In addition, if $y^{1-2s}\frac{\partial v}{\partial y} \in C(\overline{C_{R,1}})$, then

$$\frac{\partial v}{\partial v^{1-2s}}(0, 0) < 0.$$

Proof We modify in a convenient way the proof of Proposition 4.11 in [124]. Consider the function

$$w_A(x, y) = y^{2s-1}(y + Ay^2)\varphi(x),$$

where $A > 0$ is a constant that will be chosen later and $\varphi(x)$ is the first eigenfunction of $-\Delta_x + m^2$ in $\Gamma_{R/2}^0$ with Dirichlet boundary conditions, that is

$$\begin{cases} -\Delta_x \varphi + m^2\varphi = \lambda_1 \varphi & \text{in } \Gamma_{R/2}^0, \\ \varphi = 0 & \text{on } \partial\Gamma_{R/2}^0. \end{cases}$$

Note that $\lambda_1 > 0$ and that we can choose $\varphi > 0$ in $\Gamma_{R/2}^0$ so that $\|\varphi\|_{L^\infty(\Gamma_{R/2}^0)} = 1$. Then w_A satisfies the following problem:

$$\begin{cases} \operatorname{div}(y^{1-2s}\nabla w_A) - m^2 y^{1-2s} w_A = \varphi(x)[A(1+2s) - \lambda_1(y + Ay^2)] & \text{in } C_{R/2,1}, \\ w_A \geq 0 & \text{in } \overline{C_{R/2,1}}, \\ w_A = 0 & \text{on } \partial\Gamma_{R/2}^0 \times [0, 1). \end{cases}$$

Choosing A big enough, we get

$$\text{div}(y^{1-2s} \nabla w_A) - m^2 y^{1-2s} w_A \geq 0 \quad \text{in } \mathcal{C}_{R/2,1}.$$

Hence, for $\varepsilon > 0$,

$$\text{div}(y^{1-2s} \nabla (v - \varepsilon w_A)) - m^2 y^{1-2s} (v - \varepsilon w_A) \leq 0 \text{ in } \mathcal{C}_{R/2,1},$$

and $v - \varepsilon w_A = v \geq 0$ on $\partial \Gamma^0_{R/2} \times [0, 1)$. Moreover, taking $\varepsilon > 0$ small enough, we see that

$$v \geq \varepsilon w_A \quad \text{on } \Gamma^0_{R/2} \times \left\{ y = \frac{1}{2} \right\},$$

since v is continuous and positive on the closure of this set. Noting that $w_A = 0$ on $\Gamma^0_R \times \{y = 0\}$, we have

$$\begin{cases} \text{div}(y^{1-2s} \nabla (v - \varepsilon w_A)) - m^2 y^{1-2s} (v - \varepsilon w_A) \leq 0 & \text{in } \mathcal{C}_{R/2,1/2}, \\ v - \varepsilon w_A \geq 0 & \text{on } \partial \mathcal{C}_{R/2,1/2}. \end{cases}$$

Now using the weak maximum principle we deduce that

$$v - \varepsilon w_A \geq 0 \quad \text{in } \overline{\mathcal{C}_{R/2,1/2}},$$

which gives

$$\limsup_{y \to 0} -y^{1-2s} \frac{v(0, y)}{y} \leq \varepsilon \limsup_{y \to 0} -y^{1-2s} \frac{w_A(0, y)}{y} = -\varepsilon \varphi(0) < 0.$$

Now suppose that $y^{1-2s} \frac{\partial v}{\partial y} \in C(\overline{\mathcal{C}_{R,1}})$. Let $y_0 \leq \frac{1}{2}$. Since $v - \varepsilon w_A \geq 0$ in $[0, y_0]$ and $(v - \varepsilon w_A)(0, 0) = 0$, we have

$$\left(\frac{\partial v}{\partial y} - \varepsilon \frac{\partial w_A}{\partial y} \right) (0, y_1) \geq 0$$

for some $y_1 \in (0, y_0)$. Repeating this argument for a sequence of y_0's tending to 0, we see that $-y^{1-2s} \frac{\partial v}{\partial y} \leq -\varepsilon y^{1-2s} \frac{\partial w_A}{\partial y}$ at a sequence of points $(0, y_j) \to (0, 0)$ as $j \to \infty$. Using the continuity of $y^{1-2s} \frac{\partial v}{\partial y}$ up to $\{y = 0\}$ and the fact that $-\varepsilon y_j^{1-2s} \frac{\partial w_A}{\partial y}(0, y_j) \to -\varepsilon \varphi(0)$, we conclude that $\frac{\partial v}{\partial v^{1-2s}}(0, 0) < 0$. $\square$

Corollary 4.2.15 *Let d be a Hölder continuous function in Γ_R^0 and $v \in L^\infty(B_R^+) \cap H_m^1(B_R^+, y^{1-2s})$ be a weak solution to*

$$
\begin{cases}
-\mathrm{div}(y^{1-2s}\nabla v) + m^2 y^{1-2s} v = 0 & in\ B_R^+, \\
v \geq 0 & on\ B_R^+, \\
\frac{\partial v}{\partial v^{1-2s}} + d(x)v = 0 & on\ \Gamma_R^0.
\end{cases}
$$

Then, $v > 0$ in $B_R^+ \cup \Gamma_R^0$, unless $v \equiv 0$ in B_R^+.

Proof In view of Proposition 4.2.8 and Lemma 4.2.10, v and $y^{1-2s}\frac{\partial v}{\partial y}$ are $C^{0,\alpha}$ up to the boundary. Hence, the equation $\frac{\partial v}{\partial v^{1-2s}} + d(x)v = 0$ is satisfied pointwise on Γ_R^0. If v is not identically 0 in B_R^+, then $v > 0$ in B_R^+ by the strong maximum principle for the strictly elliptic operator operator

$$
-\mathrm{div}(y^{1-2s}\nabla v) + m^2 y^{1-2s} v.
$$

If $v(x_0, 0) = 0$ at some point $(x_0, 0) \in \Gamma_R^0$, then a rescaled version of Proposition 4.2.14 yields $\frac{\partial v}{\partial v^{1-2s}}(x_0, 0) < 0$. This gives a contradiction. $\qquad\square$

4.3   A Minimization Argument via the Extension Method

In this subsection we prove the existence of a ground state solution to (4.2.6).

Let us consider the functional

$$
\mathcal{I}_m(v) = \frac{1}{2}\|v\|^2_{H_m^1(\mathbb{R}_+^{N+1}, y^{1-2s})} - \frac{m^{2s}}{2}\|v(\cdot, 0)\|^2_{L^2(\mathbb{R}^N)}
$$

$$
+ \frac{\mu}{2}\|v(\cdot, 0)\|^2_{L^2(\mathbb{R}^N)} - \frac{1}{p}\|v(\cdot, 0)\|^p_{L^p(\mathbb{R}^N)}, \tag{4.3.1}
$$

defined for any $v \in H_m^1(\mathbb{R}_+^{N+1}, y^{1-2s})$. Firstly we note that

$$
\iint_{\mathbb{R}_+^{N+1}} y^{1-2s}(|\nabla v|^2 + m^2 v^2)\,dxdy + (\mu - m^{2s})\int_{\mathbb{R}^N} v^2(x, 0)\,dx
$$

is equivalent to the standard norm in $H_m^1(\mathbb{R}_+^{N+1}, y^{1-2s})$,

$$
\|v\|^2_{H_m^1(\mathbb{R}_+^{N+1}, y^{1-2s})} = \iint_{\mathbb{R}_+^{N+1}} y^{1-2s}(|\nabla v|^2 + m^2 v^2)\,dxdy.
$$

In fact, if $\mu \geq m^{2s}$, then

$$\iint_{\mathbb{R}^{N+1}_+} y^{1-2s}(|\nabla v|^2 + m^2 v^2)\,dx\,dy + (\mu - m^{2s})\int_{\mathbb{R}^N} v^2(x,0)\,dx \geq \|v\|^2_{H^1_m(\mathbb{R}^{N+1}_+,y^{1-2s})}$$

and using (4.2.2) we get

$$\iint_{\mathbb{R}^{N+1}_+} y^{1-2s}(|\nabla v|^2 + m^2 v^2)\,dx\,dy + (\mu - m^{2s})\int_{\mathbb{R}^N} v^2(x,0)\,dx \leq$$

$$\left(1 + \frac{\mu - m^{2s}}{m^{2s}}\right)\|v\|^2_{H^1_m(\mathbb{R}^{N+1}_+,y^{1-2s})}.$$

Now suppose that $\mu < m^{2s}$. Then

$$\iint_{\mathbb{R}^{N+1}_+} y^{1-2s}(|\nabla v|^2 + m^2 v^2)\,dx\,dy + (\mu - m^{2s})\int_{\mathbb{R}^N} v^2(x,0)\,dx \leq \|v\|^2_{H^1_m(\mathbb{R}^{N+1}_+,y^{1-2s})}$$

and by (4.2.2) it follows that

$$\iint_{\mathbb{R}^{N+1}_+} y^{1-2s}(|\nabla v|^2 + m^2 v^2)\,dx\,dy + (\mu - m^{2s})\int_{\mathbb{R}^N} v^2(x,0)\,dx$$

$$\geq \left(1 + \frac{\mu - m^{2s}}{m^{2s}}\right)\|v\|^2_{H^1_m(\mathbb{R}^{N+1}_+,y^{1-2s})}.$$

Thus

$$C_1(m,s,\mu)\|v\|^2_{H^1_m(\mathbb{R}^{N+1}_+,y^{1-2s})} \leq \|v\|^2_{H^1_m(\mathbb{R}^{N+1}_+,y^{1-2s})} + (\mu - m^{2s})\|v(\cdot,0)\|^2_{L^2(\mathbb{R}^N)}$$

$$\leq C_2(m,s,\mu)\|v\|^2_{H^1_m(\mathbb{R}^{N+1}_+,y^{1-2s})}. \tag{4.3.2}$$

Set

$$\|v\|^2_{e,m} = \iint_{\mathbb{R}^{N+1}_+} y^{1-2s}(|\nabla v|^2 + m^2 v^2)\,dx\,dy + (\mu - m^{2s})\int_{\mathbb{R}^N} v^2(x,0)\,dx.$$

In order to prove Theorem 4.1.1, we minimize $\mathcal{I}_m$ on the following Nehari manifold

$$\mathcal{N}_m = \{v \in H^1_m(\mathbb{R}^{N+1}_+, y^{1-2s}) \setminus \{0\} : \mathcal{J}_m(v) = 0\},$$

where $\mathcal{J}_m(v) = \langle \mathcal{I}'_m(v), v \rangle$, that is,

$$\mathcal{J}_m(v) = \|v\|^2_{H^1_m(\mathbb{R}^{N+1}_+, y^{1-2s})} - m^{2s}\|v(\cdot, 0)\|^2_{L^2(\mathbb{R}^N)} + \mu\|v(\cdot, 0)\|^2_{L^2(\mathbb{R}^N)} - \|v(\cdot, 0)\|^p_{L^p(\mathbb{R}^N)}$$

$$= \|v\|^2_{e,m} - \|v(\cdot, 0)\|^p_{L^p(\mathbb{R}^N)}.$$

Finally, we define

$$c_m = \min_{v \in \mathcal{N}_m} \mathcal{I}_m(v).$$

Proof of Theorem 4.1.1   We divide the argument into several steps.

Step 1   The set $\mathcal{N}_m$ is not empty.

Fix $v \in H^1_m(\mathbb{R}^{N+1}_+, y^{1-2s}) \setminus \{0\}$. Then

$$h(t) = \mathcal{I}_m(tv) = \frac{t^2}{2}\|v\|^2_{e,m} - \frac{t^p}{p}\|v(\cdot, 0)\|^p_{L^p(\mathbb{R}^N)}$$

achieves its maximum at some $\tau > 0$. Differentiating h with respect to t we have $\tau^{-1}\langle \mathcal{I}'_m(\tau v), \tau v \rangle = 0$, and so $\tau v \in \mathcal{N}_m$.

Step 2   Selection of an adequate minimizing sequence.

Let $(v_j) \subset \mathcal{N}_m$ be a minimizing sequence for $\mathcal{I}_m$ and set $u_j = v_j(\cdot, 0)$. Let $\tilde{u}_j$ be the symmetric-decreasing rearrangement of $|u_j|$. It is well known that (see [3, 249]) for all $w \in L^q(\mathbb{R}^N)$

$$\|w\|_{L^q(\mathbb{R}^N)} = \|\tilde{w}\|_{L^q(\mathbb{R}^N)} \quad \text{for any } q \in [1, \infty].$$

Now, we prove that for all $w \in H^s_m(\mathbb{R}^N)$, $|\tilde{w}|_{H^s_m(\mathbb{R}^N)} \leq |w|_{H^s_m(\mathbb{R}^N)}$, namely

$$\int_{\mathbb{R}^N} |(-\Delta + m^2)^{\frac{s}{2}}\tilde{w}(x)|^2\, dx \leq \int_{\mathbb{R}^N} |(-\Delta + m^2)^{\frac{s}{2}}w(x)|^2\, dx. \tag{4.3.3}$$

The case $m = 0$ is proved in [3] (see also [200]). When $m > 0$ we argue as in [64, 248, 249, 308]. Since $||w(x)| - |w(y)|| \leq |w(x) - w(y)|$ and by (4.2.1), we may assume that

w is non-negative. Recalling that $-y^{1-2s}\theta'(y) \to \kappa_s$ as $y \to 0$ (see Sect. 1.2.3), we note that

$$\kappa_s \int_{\mathbb{R}^N} |(-\Delta + m^2)^{\frac{s}{2}} w(x)|^2 \, dx = \kappa_s \int_{\mathbb{R}^N} (|\xi|^2 + m^2)^s |\mathcal{F}w(\xi)|^2 \, d\xi$$

$$= -\int_{\mathbb{R}^N} \lim_{t \to 0} t^{1-2s} \frac{\partial \theta(t\sqrt{|\xi|^2 + m^2})}{\partial t} |\mathcal{F}w(\xi)|^2 \, d\xi$$

or, equivalently,

$$-2s \int_{\mathbb{R}^N} \lim_{t \to 0} \frac{[\theta(t\sqrt{|\xi|^2 + m^2}) - 1]}{t^{2s}} |\mathcal{F}w(\xi)|^2 \, d\xi = \kappa_s \int_{\mathbb{R}^N} |(-\Delta + m^2)^{\frac{s}{2}} w(x)|^2 \, dx,$$

where θ is defined as in Remark 1.2.9. Since $r \in \mathbb{R}_+ \mapsto \frac{\theta(r)-1}{r^{2s}}$ is decreasing (because $K_s(r)$ is decreasing and positive [178, 336]), we can pass to the limit as $t \to 0$ under the integral sign to obtain that

$$2s \lim_{t \to 0} \int_{\mathbb{R}^N} \frac{[1 - \theta(t\sqrt{|\xi|^2 + m^2})]}{t^{2s}} |\mathcal{F}w(\xi)|^2 \, d\xi = \kappa_s \int_{\mathbb{R}^N} |(-\Delta + m^2)^{\frac{s}{2}} w(x)|^2 \, dx.$$

Let $t > 0$ and define

$$I_s^t(w) = \int_{\mathbb{R}^N} \frac{[1 - \theta(t\sqrt{|\xi|^2 + m^2})]}{t^{2s}} |\mathcal{F}w(\xi)|^2 \, d\xi.$$

By the Parseval identity,

$$I_s^t(w) = \frac{1}{t^{2s}} \left[\int_{\mathbb{R}^N} w^2(x) \, dx - \iint_{\mathbb{R}^{2N}} w(x) \widetilde{P}_m(x - y, t) w(y) \, dx dy \right], \tag{4.3.4}$$

where $\widetilde{P}_m$ is defined as in (4.2.5). Since the $L^2(\mathbb{R}^N)$ norm of w does not change under rearrangements and the second term on the right-hand side in (4.3.4) increases by Riesz's rearrangement inequality [249], we obtain $I_s^t(\tilde{w}) \le I_s^t(w)$ for all $t > 0$. Thus (4.3.3) follows by letting $t \to 0$.

Now, let $\tilde{v}_j$ be the unique solution to

$$\begin{cases} -\text{div}(y^{1-2s} \nabla \tilde{v}_j) + m^2 y^{1-2s} \tilde{v}_j = 0 & \text{in } \mathbb{R}_+^{N+1}, \\ \tilde{v}_j(\cdot, 0) = \tilde{u}_j & \text{on } \partial \mathbb{R}_+^{N+1}. \end{cases} \tag{4.3.5}$$

We recall that

$$\|\tilde{v}_j\|_{H_m^1(\mathbb{R}_+^{N+1}, y^{1-2s})} = |\tilde{u}_j|_{H_m^s(\mathbb{R}^N)}, \tag{4.3.6}$$

thanks to Theorem 4.2.1-(ii) (we recall that we are assuming $\kappa_s = 1$). Taking into account Theorem 4.2.1-(ii), (4.3.3) and (4.3.6), we get

$$\|\tilde{v}_j\|_{H^1_m(\mathbb{R}^{N+1}_+, y^{1-2s})} = |\tilde{u}_j|_{H^s_m(\mathbb{R}^N)} \leq |u_j|_{H^s_m(\mathbb{R}^N)} \leq \|v_j\|_{H^1_m(\mathbb{R}^{N+1}_+, y^{1-2s})},$$

so we deduce that

$$\mathcal{J}_m(\tilde{v}_j) \leq \mathcal{J}_m(v_j) = 0 \quad \text{and} \quad \mathcal{I}_m(\tilde{v}_j) \leq \mathcal{I}_m(v_j).$$

Proceeding as in the proof of Step 1, we can find $t_j > 0$ such that $\mathcal{J}_m(t_j \tilde{v}_j) = 0$. Since $\mathcal{J}_m(\tilde{v}_j) \leq 0$, we see that $t_j \leq 1$. Further, since $\mathcal{J}_m(v) = \langle \mathcal{I}'_m(v), v \rangle$, we have

$$0 = \mathcal{J}_m(t_j \tilde{v}_j) = \left[\mathcal{I}_m(t_j \tilde{v}_j) + \frac{1}{p} \int_{\mathbb{R}^N} |t \tilde{v}_j(x,0)|^p \, dx - \frac{1}{2} \int_{\mathbb{R}^N} |t_j \tilde{v}_j(x,0)|^p \, dx \right],$$

and because $0 < t_j \leq 1$ we obtain

$$
\begin{aligned}
\mathcal{I}_m(t_j \tilde{v}_j) &= \left(\frac{1}{2} - \frac{1}{p} \right) \int_{\mathbb{R}^N} |t_j \tilde{v}_j(x,0)|^p \, dx \\
&\leq \left(\frac{1}{2} - \frac{1}{p} \right) \int_{\mathbb{R}^N} |\tilde{v}_j(x,0)|^p \, dx \\
&= \left(\frac{1}{2} - \frac{1}{p} \right) \int_{\mathbb{R}^N} |v_j(x,0)|^p \, dx = \mathcal{I}(v_j).
\end{aligned}
$$

Then $(w_j) = (t_j \tilde{v}_j) \subset \mathcal{N}_m \cap X^m_{\text{rad}}$ is a minimizing sequence. Therefore, we can assume that for all $j \in \mathbb{N}$, w_j is non-negative and radially symmetric with respect to x.

Step 3   Passage to the limit.

By (4.3.2), we get

$$\left(\frac{1}{2} - \frac{1}{p} \right) C_1(m, s, \mu) \|w_j\|^2_{H^1_m(\mathbb{R}^{N+1}_+, y^{1-2s})} \leq \left(\frac{1}{2} - \frac{1}{p} \right) \|w_j\|^2_{e,m} = \mathcal{I}_m(w_j) \leq C.$$

Using Theorem 4.2.6 we may assume that

$$w_j \rightharpoonup w \text{ in } H^1_m(\mathbb{R}^{N+1}_+, y^{1-2s}), \tag{4.3.7}$$

$$w_j(\cdot, 0) \to w(\cdot, 0) \text{ in } L^q(\mathbb{R}^N) \quad \forall q \in (2, 2^*_s). \tag{4.3.8}$$

Thus (4.3.7) and (4.3.8) yield $\mathcal{I}_m(w) \leq c_m$ and $\mathcal{J}_m(w) \leq 0$. Now we show that w is not identically zero. Using the fact that $\mathcal{J}_m(w_j) = 0$ and (4.2.3), we can see that

$$
\int_{\mathbb{R}^N} |w_j(x,0)|^p \, dx = \iint_{\mathbb{R}^{N+1}_+} y^{1-2s}(|\nabla w_j|^2 + m^2 w_j^2) \, dx \, dy + (\mu - m^{2s}) \int_{\mathbb{R}^N} |w_j(x,0)|^2 \, dx
$$

$$
\geq C_1(m,s,\mu) \|w_j\|^2_{H^1_m(\mathbb{R}^{N+1}_+, y^a)}
$$

$$
\geq C(m,s,\mu,N,p) \left(\int_{\mathbb{R}^N} |w_j(x,0)|^p \, dx \right)^{\frac{2}{p}},
$$

that is, $\|w_j(\cdot,0)\|_{L^p(\mathbb{R}^N)} \geq C(m,s,\mu,N,p)^{\frac{1}{p-2}} > 0$. By (4.3.8) and $p \in (2, 2^*_s)$, we deduce that $\|w(\cdot,0)\|_{L^p(\mathbb{R}^N)} > 0$. Then, as above, we can find $\tau \in (0,1]$ such that $\mathcal{I}_m(\tau w) \leq c_m$ and $\mathcal{J}_m(\tau w) = 0$. Since $\mathcal{I}_m(\tau w) \geq c_m$, we have that $\mathcal{I}_m(\tau w) = c_m$.

Step 4   Conclusion.

Let v be the minimizer obtained above. Using the fact that $v \in \mathcal{N}_m$, we have

$$
\langle \mathcal{J}'_m(v), v \rangle = 2 \left(\iint_{\mathbb{R}^{N+1}_+} y^{1-2s}(|\nabla v|^2 + m^2 v^2) \, dx \, dy + (\mu - m^{2s}) \int_{\mathbb{R}^N} |v(x,0)|^2 \, dx \right)
$$

$$
- p \int_{\mathbb{R}^N} |v(x,0)|^p \, dx
$$

$$
= (2-p) \int_{\mathbb{R}^N} |v(x,0)|^p \, dx \neq 0.
$$

Consequently, we can find a Lagrange multiplier $\lambda \in \mathbb{R}$ such that

$$
\langle \mathcal{I}'_m(v), \varphi \rangle = \lambda \langle \mathcal{J}'_m(v), \varphi \rangle \tag{4.3.9}
$$

for any $\varphi \in X^m_{\mathrm{rad}}$. Taking $\varphi = v$ in (4.3.9) we deduce that $\lambda = 0$ and v is a nontrivial solution to (4.2.6). Therefore, $u = v(\cdot, 0)$ is a ground state solution to (4.1.1). Now Lemma 4.4.1 and Theorem 4.5.4 (see also Theorem 1.2.8) imply that $u \in C^{0,\alpha}(\mathbb{R}^N)$ for some $\alpha > 0$, and applying Proposition 4.2.11 we conclude that $u > 0$ in $\mathbb{R}^N$.   □

Remark 4.3.1 At the end of the proof of Theorem 4.1.1, we used the fact that, if $U \in H^1_m(\mathbb{R}^{N+1}_+, y^{1-2s})$ is a ground state solution to (4.2.6), then $U(\cdot, 0)$ is a ground state solution to (4.1.1). Conversely, if $u \in H^s_m(\mathbb{R}^N)$ is a ground state solution to (4.1.1), then the extension $\mathrm{Ext}_m(u)$ of u is a ground state solution of (4.2.6). The proofs of these facts are essentially based on the fact that the map $T : H^s_m(\mathbb{R}^N) \to H^1_m(\mathbb{R}^{N+1}_+, y^{1-2s})$ defined as $u \mapsto \mathrm{Ext}_m(u)$ is an isometry (if we insert the constant $\frac{1}{\sqrt{\kappa_s}}$ in the definition of the norm in $H^1_m(\mathbb{R}^{N+1}_+, y^{1-2s})$).

Remark 4.3.2 If we consider $(-\Delta + m^2)^s u = h(u)$ in $\mathbb{R}^N$, where h is a continuous function such that $h(t) = o(t)$ as $t \to 0$ and $h(t) = o(|t|^{2^*_s - 1})$ as $|t| \to \infty$, it is not hard to check that we can solve the constrained minimization problem *(here $H(t) = \int_0^t h(\tau)\, d\tau$)*

$$\inf \left\{ \int_{\mathbb{R}^N} |(-\Delta + m^2)^{\frac{s}{2}} u|^2 \, dx : \int_{\mathbb{R}^N} H(u)\, dx = 1 \right\}$$

in the framework of radially-symmetric functions, but getting rid of the associated Lagrange multiplier is a complicated task. This difficulty is again caused by the lack of scaling invariance for the fractional operator $(-\Delta + m^2)^s$. Recently, in [226], a mountain pass approach was used to study scalar field equations governed by (4.1.3) and involving general subcritical nonlinearities.

4.4   Regularity, Decay and Symmetry

In this section we analyze regularity, decay and symmetry properties of solutions to (4.1.1).

Lemma 4.4.1 *Let $v \in H^1_m(\mathbb{R}^{N+1}_+, y^{1-2s})$ be a weak solution to*

$$\begin{cases} -\mathrm{div}(y^{1-2s}\nabla v) + m^2 y^{1-2s} v = 0 & \text{in } \mathbb{R}^{N+1}_+, \\ \frac{\partial v}{\partial \nu^{1-2s}} = m^{2s} v + f(v) & \text{on } \partial \mathbb{R}^{N+1}_+, \end{cases} \tag{4.4.1}$$

where $f(v(x,0)) = -\mu v(x,0) + |v(x,0)|^{p-2}v$. Then, $v(\cdot, 0) \in L^q(\mathbb{R}^N)$ for all $q \in [2, \infty]$ and $v \in L^\infty(\mathbb{R}^{N+1}_+)$.

Proof We combine a Brezis–Kato argument [112] with a Moser iteration [278]. Since v is a weak solution to (4.4.1), we know that

$$\iint_{\mathbb{R}^{N+1}_+} y^{1-2s}(\nabla v \cdot \nabla \eta + m^2 v\eta)\, dxdy = \int_{\mathbb{R}^N} [m^{2s} v(x,0) + f(v(x,0))]\eta(x,0)\, dx \tag{4.4.2}$$

for all $\eta \in H^1_m(\mathbb{R}^{N+1}_+, y^{1-2s})$. For $K > 1$ and $\beta > 0$, let $w = vv_K^{2\beta} \in H^1_m(\mathbb{R}^{N+1}_+, y^{1-2s})$, where $v_K = \min\{|v|, K\}$. Taking $\eta = w$ in (4.4.2), we deduce that

$$\iint_{\mathbb{R}^{N+1}_+} y^{1-2s} v_K^{2\beta}(|\nabla v|^2 + m^2 v^2)\, dxdy + \iint_{D_K} 2\beta y^{1-2s} v_K^{2\beta}|\nabla v|^2 \, dxdy$$

$$= m^{2s} \int_{\mathbb{R}^N} v^2(x,0) v_K^{2\beta}(x,0)\, dx + \int_{\mathbb{R}^N} f(v(x,0))v(x,0)v_K^{2\beta}(x,0)\, dx \tag{4.4.3}$$

where $D_K = \{(x, y) \in \mathbb{R}^{N+1}_+ : |v(x, y)| \leq K\}$. It is easy to check that

$$\iint_{\mathbb{R}^{N+1}_+} y^{1-2s} |\nabla(v v_K^\beta)|^2 \, dx \, dy = \iint_{\mathbb{R}^{N+1}_+} y^{1-2s} v_K^{2\beta} |\nabla v|^2 \, dx \, dy$$

$$+ \iint_{D_K} (2\beta + \beta^2) y^{1-2s} v_K^{2\beta} |\nabla v|^2 \, dx \, dy. \qquad (4.4.4)$$

Then, combining (4.4.3) and (4.4.4), we get

$$\|v v_K^\beta\|^2_{H^1_m(\mathbb{R}^{N+1}_+, y^{1-2s})} = \iint_{\mathbb{R}^{N+1}_+} y^{1-2s} [|\nabla(v v_K^\beta)|^2 + m^2 v^2 v_K^{2\beta}] \, dx \, dy$$

$$= \iint_{\mathbb{R}^{N+1}_+} y^{1-2s} v_K^{2\beta} [|\nabla v|^2 + m^2 v^2] \, dx \, dy + \iint_{D_K} 2\beta \left(1 + \frac{\beta}{2}\right) y^{1-2s} v_K^{2\beta} |\nabla v|^2 \, dx \, dy$$

$$\leq c_\beta \left[\iint_{\mathbb{R}^{N+1}_+} y^{1-2s} v_K^{2\beta} [|\nabla v|^2 + m^2 v^2] \, dx \, dy + \iint_{D_K} 2\beta y^{1-2s} v_K^{2\beta} |\nabla v|^2 \, dx \, dy \right]$$

$$= c_\beta \int_{\mathbb{R}^N} m^{2s} v^2(x, 0) v_K^{2\beta}(x, 0) + f(v(x, 0)) v(x, 0) v_K^{2\beta}(x, 0) \, dx, \qquad (4.4.5)$$

where

$$c_\beta = 1 + \frac{\beta}{2}.$$

Now, we prove that there exist a constant $c > 0$ and a function $h \in L^{N/2s}(\mathbb{R}^N)$, $h \geq 0$ and independent of K and β, such that

$$m^{2s} v^2(\cdot, 0) v_K^{2\beta}(\cdot, 0) + f(v(\cdot, 0)) v(\cdot, 0) v_K^{2\beta}(\cdot, 0) \leq (c + h) v^2(\cdot, 0) v_K^{2\beta}(\cdot, 0) \quad \text{on } \mathbb{R}^N.$$
$$(4.4.6)$$

First, we note that

$$m^{2s} v^2(\cdot, 0) v_K^{2\beta}(\cdot, 0) + f(v(\cdot, 0)) v(\cdot, 0) v_K^{2\beta}(\cdot, 0)$$

$$\leq (m^{2s} + C) v^2(\cdot, 0) v_K^{2\beta}(\cdot, 0) + C |v(\cdot, 0)|^{p-2} v^2(\cdot, 0) v_K^{2\beta}(\cdot, 0) \quad \text{on } \mathbb{R}^N.$$

Moreover,

$$|v(\cdot, 0)|^{p-2} \leq 1 + h \quad \text{on } \mathbb{R}^N,$$

for some $h \in L^{N/2s}(\mathbb{R}^N)$. In fact, we can observe that

$$|v(\cdot, 0)|^{p-2} = \chi_{\{|v(\cdot,0)|\le 1\}}|v(\cdot, 0)|^{p-2} + \chi_{\{|v(\cdot,0)|>1\}}|v(\cdot, 0)|^{p-2}$$

$$\le 1 + \chi_{\{|v(\cdot,0)|>1\}}|v(\cdot, 0)|^{p-2} \quad \text{on } \mathbb{R}^N,$$

and that if $(p-2)\frac{N}{2s} < 2$, then

$$\int_{\mathbb{R}^N} \chi_{\{|v(\cdot,0)|>1\}}|v(x, 0)|^{\frac{N}{2s}(p-2)}\, dx \le \int_{\mathbb{R}^N} \chi_{\{|v(\cdot,0)|>1\}}|v(x, 0)|^2\, dx < \infty,$$

while if $2 \le (p-2)\frac{N}{2s}$ we deduce that $(p-2)\frac{N}{2s} \in [2, 2_s^*]$.

Taking into account (4.4.5) and (4.4.6) we obtain that

$$\|v v_K^\beta\|^2_{H_m^1(\mathbb{R}_+^{N+1}, y^{1-2s})} \le c_\beta \int_{\mathbb{R}^N} (c + h(x))v^2(x, 0)v_K^{2\beta}(x, 0)\, dx,$$

and by the monotone convergence theorem (v_K is nondecreasing with respect to K) we have, as $K \to \infty$,

$$\||v|^{\beta+1}\|^2_{H_m^1(\mathbb{R}_+^{N+1}, y^{1-2s})} \le c c_\beta \int_{\mathbb{R}^N} |v(x, 0)|^{2(\beta+1)}\, dx + c_\beta \int_{\mathbb{R}^N} h(x)|v(x, 0)|^{2(\beta+1)}\, dx. \tag{4.4.7}$$

Fix $M > 0$ and let $A_1 = \{h \le M\}$ and $A_2 = \{h > M\}$. Then,

$$\int_{\mathbb{R}^N} h(x)|v(x, 0)|^{2(\beta+1)}\, dx \le M\||v(\cdot, 0)|^{\beta+1}\|^2_{L^2(\mathbb{R}^N)} + \varepsilon(M)\||v(\cdot, 0)|^{\beta+1}\|^2_{L^{2_s^*}(\mathbb{R}^N)}, \tag{4.4.8}$$

where $\varepsilon(M) = \left(\int_{A_2} h^{N/2s}\, dx\right)^{\frac{2s}{N}} \to 0$ as $M \to \infty$. In view of (4.4.7) and (4.4.8),

$$\||v|^{\beta+1}\|^2_{H_m^1(\mathbb{R}_+^{N+1}, y^{1-2s})} \le c_\beta(c + M)\||v(\cdot, 0)|^{\beta+1}\|^2_{L^2(\mathbb{R}^N)} + c_\beta \varepsilon(M)\||v(\cdot, 0)|^{\beta+1}\|^2_{L^{2_s^*}(\mathbb{R}^N)}. \tag{4.4.9}$$

By (4.2.3), it follows that

$$\||v(\cdot, 0)|^{\beta+1}\|^2_{L^{2_s^*}(\mathbb{R}^N)} \le C_{2_s^*}^2 \||v|^{\beta+1}\|^2_{H_m^1(\mathbb{R}_+^{N+1}, y^{1-2s})}. \tag{4.4.10}$$

Then, choosing M large so that

$$\varepsilon(M)c_\beta C_{2_s^*}^2 < \frac{1}{2},$$

and using (4.4.9) and (4.4.10) we obtain

$$\||v(\cdot,0)|^{\beta+1}\|_{L^{2_s^*}(\mathbb{R}^N)}^2 \le 2C_{2_s^*}^2 c_\beta(c+M)\||v(\cdot,0)|^{\beta+1}\|_{L^2(\mathbb{R}^N)}^2. \tag{4.4.11}$$

Then we can start a bootstrap argument: since $v(\cdot,0) \in L^{2_s^*}(\mathbb{R}^N)$, we can apply (4.4.11) with $\beta_1+1 = \frac{N}{N-2s}$ to deduce that $v(\cdot,0) \in L^{\frac{(\beta_1+1)2N}{N-2s}}(\mathbb{R}^N) = L^{\frac{2N^2}{(N-2s)^2}}(\mathbb{R}^N)$. Applying again (4.4.11), after k iterations, we find $v(\cdot,0) \in L^{\frac{2N^k}{(N-2s)^k}}(\mathbb{R}^N)$, and so $v(\cdot,0) \in L^q(\mathbb{R}^N)$ for all $q \in [2,\infty)$.

Now we prove that $v(\cdot,0) \in L^\infty(\mathbb{R}^N)$. Since $v(\cdot,0) \in L^q(\mathbb{R}^N)$ for all $q \in [2,\infty)$, we have that $h \in L^{\frac{N}{s}}(\mathbb{R}^N)$. By the generalized Hölder inequality, we can see that for all $\lambda > 0$

$$\int_{\mathbb{R}^N} h(x)|v(x,0)|^{2(\beta+1)}\,dx \le \|h\|_{L^{\frac{N}{s}}(\mathbb{R}^N)}\||v(\cdot,0)|^{\beta+1}\|_{L^2(\mathbb{R}^N)}\||v(\cdot,0)|^{\beta+1}\|_{L^{2_s^*}(\mathbb{R}^N)}$$

$$\le \|h\|_{L^{\frac{N}{s}}(\mathbb{R}^N)}\left(\lambda\||v(\cdot,0)|^{\beta+1}\|_{L^2(\mathbb{R}^N)}^2 + \frac{1}{\lambda}\||v(\cdot,0)|^{\beta+1}\|_{L^{2_s^*}(\mathbb{R}^N)}^2\right).$$

Then, using (4.4.7) and (4.4.10), we deduce that

$$\||v(\cdot,0)|^{\beta+1}\|_{L^{2_s^*}(\mathbb{R}^N)}^2 \le C_{2_s^*}^2\||v|^{\beta+1}\|_{H_m^1(\mathbb{R}_+^{N+1},y^{1-2s})}^2$$

$$\le c_\beta C_{2_s^*}^2(c + \|h\|_{L^{\frac{N}{s}}(\mathbb{R}^N)}\lambda)\||v(\cdot,0)|^{\beta+1}\|_{L^2(\mathbb{R}^N)}^2 + C_{2_s^*}^2\frac{c_\beta\|h\|_{L^{\frac{N}{s}}(\mathbb{R}^N)}}{\lambda}\||v(\cdot,0)|^{\beta+1}\|_{L^{2_s^*}(\mathbb{R}^N)}^2. \tag{4.4.12}$$

Taking $\lambda > 0$ such that

$$\frac{c_\beta\|h\|_{L^{\frac{N}{s}}(\mathbb{R}^N)}}{\lambda}C_{2_s^*}^2 = \frac{1}{2},$$

we obtain that

$$\||v(\cdot,0)|^{\beta+1}\|_{L^{2_s^*}(\mathbb{R}^N)}^2 \le 2c_\beta(c + \|h\|_{L^{\frac{N}{s}}(\mathbb{R}^N)}\lambda)C_{2_s^*}^2\||v(\cdot,0)|^{\beta+1}\|_{L^2(\mathbb{R}^N)}^2$$

$$= M_\beta\||v(\cdot,0)|^{\beta+1}\|_{L^2(\mathbb{R}^N)}^2.$$

Now we can control the dependence on β of M_β as follows:

$$M_\beta \leq C c_\beta^2 \leq C(1+\beta)^2 \leq M_0^2 e^{2\sqrt{\beta+1}},$$

which implies that

$$\|v(\cdot,0)\|_{L^{2_s^*(\beta+1)}(\mathbb{R}^N)} \leq M_0^{\frac{1}{\beta+1}} e^{\frac{1}{\sqrt{\beta+1}}} \|v(\cdot,0)\|_{L^{2(\beta+1)}(\mathbb{R}^N)}.$$

Iterating this last inequality and choosing $\beta_0 = 0$ and $2(\beta_{n+1}+1) = 2_s^*(\beta_n+1)$, we deduce that

$$\|v(\cdot,0)\|_{L^{2_s^*(\beta_n+1)}(\mathbb{R}^N)} \leq M_0^{\sum_{i=0}^{n} \frac{1}{\beta_i+1}} e^{\sum_{i=0}^{n} \frac{1}{\sqrt{\beta_i+1}}} \|v(\cdot,0)\|_{L^{2(\beta_0+1)}(\mathbb{R}^N)}.$$

Since $1+\beta_n = (\frac{N}{N-2s})^n$, the series

$$\sum_{i=0}^{\infty} \frac{1}{\beta_i+1} \quad \text{and} \quad \sum_{i=0}^{\infty} \frac{1}{\sqrt{\beta_i+1}}$$

converge and we get

$$\|v(\cdot,0)\|_{L^\infty(\mathbb{R}^N)} = \lim_{n\to\infty} \|v(\cdot,0)\|_{L^{2_s^*(\beta_n+1)}(\mathbb{R}^N)} < \infty.$$

Now, using (4.4.12) with $\lambda = 1$ and the fact that $\|v(\cdot,0)\|_{L^q(\mathbb{R}^N)} \leq \tilde{C}$ for all $q \in [2,\infty]$, we have for all $\beta > 0$

$$\||v|^{\beta+1}\|^2_{H_m^1(\mathbb{R}^{N+1}_+, y^{1-2s})} \leq c_\beta (c + \|h\|_{L^{\frac{N}{s}}(\mathbb{R}^N)}) \tilde{C}^{2(\beta+1)} + c_\beta \|h\|_{L^{\frac{N}{s}}(\mathbb{R}^N)} \tilde{C}^{2(\beta+1)}.$$

Since Lemma 1.3.9 yields

$$\left(\iint_{\mathbb{R}^{N+1}_+} y^{1-2s} |v(x,y)|^{2\gamma(\beta+1)}\, dx dy \right)^{\frac{\beta+1}{2\gamma(\beta+1)}} = \left(\iint_{\mathbb{R}^{N+1}_+} y^{1-2s} ||v(x,y)|^{\beta+1}|^{2\gamma}\, dx dy \right)^{\frac{1}{2\gamma}}$$

$$\leq C' \||v|^{\beta+1}\|_{H_m^1(\mathbb{R}^{N+1}_+, y^{1-2s})}$$

where $\gamma = 1 + \frac{2}{N-2s}$, for some constant $C' > 0$ independent of β, it follows that

$$\left(\iint_{\mathbb{R}^{N+1}_+} y^{1-2s} |v(x,y)|^{2\gamma(\beta+1)}\, dx dy \right)^{\frac{2(\beta+1)}{2\gamma(\beta+1)}} \leq C'' c_\beta \tilde{C}^{2(\beta+1)}$$

for some constant $C'' > 0$ independent of β. Consequently,

$$\left(\iint_{\mathbb{R}^{N+1}_+} y^{1-2s}|v(x,y)|^{2\gamma(\beta+1)}\,dx\,dy\right)^{\frac{1}{2\gamma(\beta+1)}} \le (C''c_\beta)^{\frac{1}{2(\beta+1)}}\tilde{C}.$$

Note that $(C''c_\beta)^{\frac{1}{2(\beta+1)}}\tilde{C} \le \bar{C}$ for all $\beta > 0$, since

$$\frac{1}{\beta+1}\log\left(1+\frac{\beta}{2}\right) \le \frac{1}{2} \quad \text{for all } \beta > 0.$$

Fix $k > \bar{C}$ and define $A_k = \{(x,y) \in \mathbb{R}^{N+1}_+ : |v(x,y)| > k\}$. Then we obtain that

$$\bar{C} \ge \left(\iint_{\mathbb{R}^{N+1}_+} y^{1-2s}|v(x,y)|^{2\gamma(\beta+1)}\,dx\,dy\right)^{\frac{1}{2\gamma(\beta+1)}}$$

$$\ge k^{\frac{\gamma(\beta+1)-1}{\gamma(\beta+1)}}\left(\iint_{A_k} y^{1-2s}|v(x,y)|^2\,dx\,dy\right)^{\frac{1}{2\gamma(\beta+1)}},$$

whence

$$\left(\frac{\bar{C}}{k}\right)^{\beta+1-\frac{1}{\gamma}} \ge \frac{1}{\bar{C}^{\frac{1}{\gamma}}}\left(\iint_{A_k} y^{1-2s}|v(x,y)|^2\,dx\,dy\right)^{\frac{1}{2\gamma}}.$$

Letting $\beta \to \infty$ we have

$$\iint_{A_k} y^{1-2s}|v(x,y)|^2\,dx\,dy = 0 \implies |A_k| = 0,$$

i.e., $v \in L^\infty(\mathbb{R}^{N+1}_+)$. $\square$

In order to study the Hölder regularity of solutions of (4.1.1), we introduce the Hölder-Zygmund (or Lipschitz) spaces Λ_α; see [128, 241, 315, 326]. If $\alpha > 0$ and $\alpha \notin \mathbb{N}$ we set $\Lambda_\alpha = C^{[\alpha],\alpha-[\alpha]}(\mathbb{R}^N)$. If $\alpha = k \in \mathbb{N}$ we set $\Lambda_\alpha = \Lambda_k^*$, where

$$\Lambda_1^* = \left\{u \in L^\infty(\mathbb{R}^N) \cap C(\mathbb{R}^N) : \sup_{x,h\in\mathbb{R}^N,|h|>0} \frac{|u(x+h)+u(x-h)-2u(x)|}{|h|} < \infty\right\}$$

if $k = 1$, and

$$\Lambda_k^* = \left\{ u \in C^{k-1}(\mathbb{R}^N) : D^\gamma u \in \Lambda_1^* \text{ for all } |\gamma| \le k - 1 \right\},$$

if $k \in \mathbb{N}$, $k \ge 2$. Note that $C^{0,1}(\mathbb{R}^N) \ne \Lambda_1^*$ and $\Lambda_\beta \subset \Lambda_\alpha$ if $0 < \alpha < \beta$. Moreover, we have the following useful result:

Theorem 4.4.2 ([128,315,323]) *If $\alpha, \beta \ge 0$, then $(1 - \Delta)^{-\alpha}$ is an isomorphism from $\mathscr{L}_\beta^p$ to $\mathscr{L}_{\beta+2\alpha}^p$. If $\alpha \ge 0$ and $\beta > 0$, then $(1 - \Delta)^{-\alpha}$ is an isomorphism from Λ_β to $\Lambda_{\beta+2\alpha}$.*

Next we prove a stronger regularity result for positive solutions of (4.1.1).

Theorem 4.4.3 *Let $u \in H^s(\mathbb{R}^N)$ be a positive solution to (4.1.1). Then, $u \in C^{1,\alpha}(\mathbb{R}^N)$ and $u(x) \to 0$ as $|x| \to \infty$.*

Proof Let $g(u) = (m^{2s} - \mu)u + u^{p-1}$. By Lemma 4.4.1, $g(u) \in L^r(\mathbb{R}^N)$ for any $r \in [2, \infty]$. Then

$$u = G_{2s,m} * g(u) \in \mathscr{L}_{2s,m}^q(\mathbb{R}^N) = \{G_{2s,m} * h | h \in L^q(\mathbb{R}^N)\}$$

for any $q \in [2, \infty)$, where $G_{2s,m}$ is the Bessel kernel given by

$$\mathcal{F}G_{2s,m}(\xi) = (2\pi)^{-\frac{N}{2}}(|\xi|^2 + m^2)^{-s}.$$

Theorem 4.5.4 implies that $u \in C^{0,\beta}(\mathbb{R}^N)$ for all $\beta \in (0, 2s)$ if $2s \le 1$, and $u \in C^{1,\beta}(\mathbb{R}^N)$ for all $\beta \in (0, 2s - 1)$ if $2s > 1$. In particular, $u \in C^{0,\beta}(\mathbb{R}^N) \cap L^2(\mathbb{R}^N)$ implies that $|u(x)| \to 0$ as $|x| \to \infty$. Now, in view of Theorem 4.4.2 (see also Theorem 4.5.3), we can repeat the argument used in Theorem 3.2.11 for studying the regularity of solutions in the case of the operator $(-\Delta)^s$ to obtain the desired result. $\qquad\square$

Remark 4.4.4 We can also deduce the Hölder regularity of u arguing as follows. Let $U(x, y) = (\widetilde{P}_m(\cdot, y) * u)(x)$. Fix $x \in \mathbb{R}^N$. By Remark 4.2.5, the fact that $\theta(y) = \frac{2}{\Gamma(s)}(\frac{y}{2})^s K_s(y) \sim 2^{-s}$ as $y \to 0$, and Young's inequality, we get

$$\|U\|_{L^2(B_1(x)\times[0,1])} \le \|U\|_{L^2(\mathbb{R}^N \times[0,1])} = \|\widetilde{P}_m(\cdot, y) * u\|_{L^2(\mathbb{R}^N \times[0,1])}$$

$$\le \left(\int_0^1 \|\widetilde{P}_m(\cdot, y) * u\|_{L^2(\mathbb{R}^N)}^2 \, dy\right)^{\frac{1}{2}}$$

$$\le \left(\int_0^1 \|\widetilde{P}_m(\cdot, y)\|_{L^1(\mathbb{R}^N)}^2 \|u\|_{L^2(\mathbb{R}^N)}^2 \, dy\right)^{\frac{1}{2}}$$

$$= \|u\|_{L^2(\mathbb{R}^N)} \left(\int_0^1 \theta^2(my) \, dy \right)^{\frac{1}{2}}$$

$$\leq C_{m,s} \|u\|_{L^2(\mathbb{R}^N)}.$$

Then, using that $g(u) \in L^r(\mathbb{R}^N)$ for any $r \in [2, \infty]$, Proposition 4.2.8 and the translation invariance of (4.1.1) with respect to $x \in \mathbb{R}^N$, we deduce that $u \in C^{0,\alpha}(\mathbb{R}^N)$ for some $\alpha > 0$. The Hölder regularity of ∇u follows from Proposition 4.2.9.

Theorem 4.4.5 *Let v be a positive solution to* (4.2.6). *Then, $v \in C^{0,\alpha}(\overline{\mathbb{R}_+^{N+1}})$ and satisfies for all $y \geq 1$*

$$\sup_{x \in \mathbb{R}^N} |v(x, y)| \leq C \|v(\cdot, 0)\|_{L^2(\mathbb{R}^N)} y^{s - \frac{1}{2}} e^{-my}.$$

In particular, for any $\lambda \in (0, m)$, $|v(x, y)|e^{\lambda y} \to 0$ as $y \to \infty$.

Proof We argue as in [117, 152]. By Lemma 4.4.1 and Proposition 4.2.8, we deduce that $v \in C^{0,\alpha}(\overline{\mathbb{R}_+^{N+1}})$; see also [180]. To prove the decay at infinity, note that v is a solution to (4.2.4), and then use the Fourier transform with respect to the variable $x \in \mathbb{R}^N$ (see Remark 4.2.5) to obtain that

$$\mathcal{F}v(\xi, y) = \theta(y\sqrt{|\xi|^2 + m^2})\mathcal{F}u(\xi),$$

where $u(x) = v(x, 0) \in L^2(\mathbb{R}^N)$ and $\theta(y) \in H^1(\mathbb{R}_+, y^{1-2s})$ is defined as in Remark 1.2.9. Then,

$$v(x, y) = \left(\widetilde{P}_m(\cdot, y) * u \right)(x) = \int_{\mathbb{R}^N} \widetilde{P}_m(x - z, y)u(z) \, dz$$

where $\widetilde{P}_m(x, y)$ is given in (4.2.5). Therefore, by Parseval's identity,

$$|v(x, y)| \leq \|u\|_{L^2(\mathbb{R}^N)} \left(\int_{\mathbb{R}^N} |\theta(y\sqrt{|\xi|^2 + m^2})|^2 \, d\xi \right)^{\frac{1}{2}}.$$

Since $\theta(y)$ is continuous, for $y \geq 1$ we have

$$|\theta(y\sqrt{|\xi|^2 + m^2})| \leq C_1 (y\sqrt{|\xi|^2 + m^2})^{\frac{2s-1}{2}} e^{-y\sqrt{|\xi|^2 + m^2}},$$

which gives

$$|v(x, y)| \leq \|u\|_{L^2(\mathbb{R}^N)} C_1 \left(\int_{\mathbb{R}^N} (y\sqrt{|\xi|^2 + m^2})^{2s-1} e^{-2y\sqrt{|\xi|^2+m^2}} \, d\xi \right)^{\frac{1}{2}}. \qquad (4.4.13)$$

Now let us estimate the integral in the right-hand side of (4.4.13). We first assume that $s \in (0, \frac{1}{2}]$. Since $y\sqrt{|\xi|^2 + m^2} \geq my$ and $2s - 1 \leq 0$, we get

$$\int_{\mathbb{R}^N} (y\sqrt{|\xi|^2 + m^2})^{2s-1} e^{-2y\sqrt{|\xi|^2+m^2}} \, d\xi \leq (my)^{2s-1} \int_{\mathbb{R}^N} e^{-2y\sqrt{|\xi|^2+m^2}} \, d\xi. \qquad (4.4.14)$$

Next we study the integral on the right-hand side of (4.4.14). Take $R > 0$ such that $|\xi|^2 \geq 3m^2$ for all $\xi \in B_R^c$. Hence, $y\sqrt{|\xi|^2 + m^2} \geq 2my$ for all $\xi \in B_R^c$. On the other hand, $y\sqrt{|\xi|^2 + m^2} \geq |\xi|y$. Combining these two estimates, we find

$$-2y\sqrt{|\xi|^2 + m^2} \leq -(2my + |\xi|y) \quad \text{for all } \xi \in B_R^c.$$

Consequently, for all $y \geq 1$

$$
\begin{aligned}
\int_{\mathbb{R}^N} e^{-2y\sqrt{|\xi|^2+m^2}} \, dk &= \int_{B_R} e^{-2y\sqrt{|\xi|^2+m^2}} \, d\xi + \int_{B_R^c} e^{-2y\sqrt{|\xi|^2+m^2}} \, d\xi \\
&\leq \int_{B_R} e^{-2my} \, d\xi + \int_{B_R^c} e^{-(2my+|\xi|y)} \, d\xi \\
&\leq e^{-2my} |B_R| + e^{-2my} \int_{B_R^c} e^{-|\xi|} \, d\xi \\
&\leq C_2 e^{-2my}.
\end{aligned}
\qquad (4.4.15)
$$

Combining (4.4.13), (4.4.14) and (4.4.15), we see that

$$|v(x, y)| \leq C_1 \|u\|_{L^2(\mathbb{R}^N)} \left((my)^{2s-1} C_2 e^{-2my} \right)^{\frac{1}{2}}$$

$$\leq C_3 \|u\|_{L^2(\mathbb{R}^N)} y^{s-\frac{1}{2}} e^{-my}$$

for all $y \geq 1$. Now, let us treat the case $s \in (\frac{1}{2}, 1)$. Note that

$$\int_{\mathbb{R}^N} \left(y\sqrt{|\xi|^2 + m^2} \right)^{2s-1} e^{-2y\sqrt{|\xi|^2+m^2}} \, d\xi$$

$$= y^{2s-1} \int_{\mathbb{R}^N} \left(\sqrt{|\xi|^2 + m^2} \right)^{2s-1} e^{-2y\sqrt{|\xi|^2+m^2}} \, d\xi$$

$$\leq y^{2s-1} \int_{\mathbb{R}^N} \left(|\xi|^{2s-1} + m^{2s-1} \right) e^{-2y\sqrt{|\xi|^2+m^2}} \, d\xi$$

$$= y^{2s-1} \int_{\mathbb{R}^N} |\xi|^{2s-1} e^{-2y\sqrt{|\xi|^2+m^2}} \, d\xi + (my)^{2s-1} \int_{\mathbb{R}^N} e^{-2y\sqrt{|\xi|^2+m^2}} \, d\xi.$$

As before, choosing $R > 0$ such that $|\xi|^2 \geq 3m^2$ for all $\xi \in B_R^c$, we have

$$-2y\sqrt{|\xi|^2 + m^2} \leq -(2my + |\xi|y) \quad \text{for all } \xi \in B_R^c.$$

Hence, for $y \geq 1$, it holds that

$$\int_{\mathbb{R}^N} |\xi|^{2s-1} e^{-2y\sqrt{|\xi|^2+m^2}} \, d\xi = \int_{B_R} |\xi|^{2s-1} e^{-2y\sqrt{|\xi|^2+m^2}} \, d\xi + \int_{B_R^c} |\xi|^{2s-1} e^{-2y\sqrt{|\xi|^2+m^2}} \, d\xi$$

$$\leq e^{-2my} R^{2s-1} |B_R| + e^{-2my} \int_{\mathbb{R}^N} |\xi|^{2s-1} e^{-|\xi|} \, d\xi$$

$$\leq C_3 e^{-2my}.$$

This combined with (4.4.15) yields the required estimate in the case $s \in (\frac{1}{2}, 1)$. $\square$

Remark 4.4.6 If u is a positive solution to (4.1.1), then u has an exponential decay at infinity. Let $m = 1$ and fix $\delta \in (0, 1)$. Let $h \in L^2(\mathbb{R}^N)$ be such that $h \geq 0$ in $\mathbb{R}^N$, $h \not\equiv 0$ and h has compact support. For instance, assume that supp$(h) = B_r$ for some $r > 0$. Let $v \in H^s(\mathbb{R}^N)$ be the unique solution to $(-\Delta + 1)^s v - (1 - \delta)v = h$ in $\mathbb{R}^N$. Then $v = G_{2s} * ((1-\delta)+h)$, $v \in C^{0,\alpha}(\mathbb{R}^N)$ for some $\alpha > 0$, and $v > 0$ in $\mathbb{R}^N$ (remark that $G_{2s} > 0$). By [63,64], we know that the kernel $\mathcal{B}_{2s}(x) = \mathcal{F}^{-1}([(|\xi|^2 + 1)^s - (1 - \delta)]^{-1}) \leq C e^{-c|x|}$ for all $|x| \geq 2$. Set $R = \max\{r, 2\}$. Then, for all $|x| \geq 2R$, we get

$$v(x) = \int_{B_r} \mathcal{B}_{2s}(x - y)h(y) \, dy \leq C \|h\|_{L^\infty(\mathbb{R}^N)} \int_{B_r} e^{-c|x-y|} \, dy \leq C' e^{-c\frac{|x|}{R}},$$

where we used $2R \geq \frac{R}{R-1}$, since $R \geq 2$, and then

$$|x - y| \geq |x| - |y| \geq |x| - 1 \geq \frac{|x|}{R} \geq 2 \quad \text{for all } |x| \geq 2R.$$

From the continuity of v we deduce that $0 < v(x) \leq C_1 e^{-c_2|x|}$ for all $x \in \mathbb{R}^N$.

Next we use a comparison argument to deduce the desired estimate. Since $u \in C^{0,\alpha}(\mathbb{R}^N) \cap L^q(\mathbb{R}^N)$ for all $q \in [2, \infty]$, we infer that $u(x) \to 0$ as $|x| \to \infty$ and that the function $h(x) = (g(u(x)) - (1 - \delta_0)u(x))^+ > 0$ has compact support (note that $\lim_{t \to 0} g(t)/t = 1 - \mu < 1$), for some $\delta_0 \in (0, 1)$. Hence,

$$(-\Delta + 1)^s u - (1 - \delta_0)u = g(u) - (1 - \delta_0)u \leq h = (-\Delta + 1)^s v - (1 - \delta_0)v \quad \text{in } \mathbb{R}^N,$$

and this implies that $(-\Delta + 1)^s (u - v) - (1 - \delta_0)(u - v) \le 0$ in $\mathbb{R}^N$. Taking $(u - v)^+$ as a test function, and recalling the definition (4.1.3) and the elementary inequality

$$(x - y)(x^+ - y^+) \ge |x^+ - y^+|^2 \quad \text{for all } x, y \in \mathbb{R},$$

we obtain $u \le v$ in $\mathbb{R}^N$, that is, $u(x) \le Ce^{-c|x|}$ for all $x \in \mathbb{R}^N$.

Finally we use the method of moving planes to deduce the symmetry of positive solutions to (4.1.1).

Theorem 4.4.7 *Every positive solution* $u \in H^s(\mathbb{R}^N)$ *of* (4.1.1) *is radially symmetric and radially decreasing with respect to a point of* $\mathbb{R}^N$.

Proof Let v be the unique solution of (4.2.4) with boundary datum u. Let $\lambda > 0$ and consider the sets

$$\Sigma_\lambda = \{(x_1, \dots, x_N, y) : x_1 > \lambda, \ y \ge 0\}$$

and

$$T_\lambda = \{(x_1, \dots, x_N, y) : x_1 = \lambda, \ y \ge 0\}.$$

Let $v_\lambda(x, y) = v(2\lambda - x_1, \dots, x_N, y)$ and $w_\lambda = v_\lambda - v$. Then w_λ satisfies

$$\begin{cases} -\text{div}(y^{1-2s} \nabla w_\lambda) + m^2 y^{1-2s} w_\lambda = 0 & \text{in } \mathbb{R}^{N+1}_+, \\ \frac{\partial w_\lambda}{\partial v^{1-2s}} = (C_\lambda(x) + m^{2s} - \mu) w_\lambda & \text{on } \partial \mathbb{R}^{N+1}_+, \end{cases} \tag{4.4.16}$$

where

$$C_\lambda(x) = \begin{cases} \dfrac{v_\lambda^{p-1}(x,0) - v^{p-1}(x,0)}{v_\lambda(x,0) - v(x,0)}, & \text{if } v_\lambda(x, 0) \ne v(x, 0), \\ 0, & \text{if } v_\lambda(x, 0) = v(x, 0). \end{cases}$$

Let $w_\lambda^- = \min\{0, w_\lambda\}$. Note that as $\lambda \to \infty$, $C_\lambda(x) \to 0$ uniformly in $x \in \{x \in \mathbb{R}^N : x_1 > \lambda\}$, so that $w_\lambda^-(x, 0) \ne 0$ because $\lim_{|x| \to \infty} v(x, 0) = 0$ and $0 \le v_\lambda(x, 0) < v(x, 0)$ whenever $w_\lambda^-(x, 0) \ne 0$. Multiplying the weak formulation of (4.4.16) by w_λ^- and applying (4.2.2), we get

$$\iint_{\Sigma_\lambda} y^{1-2s} [|\nabla w_\lambda^-|^2 + m^2(w_\lambda^-)^2] \, dx \, dy = \int_{\{x_1 > \lambda\}} [C_\lambda(x) + m^{2s} - \mu](w_\lambda^-(x, 0))^2 \, dx$$

$$\le \int_{\{x_1 > \lambda\}} C_\lambda(x)(w_\lambda^-(x, 0))^2 \, dx + A(m, \mu, s) \iint_{\Sigma_\lambda} y^{1-2s} [|\nabla w_\lambda^-|^2 + m^2(w_\lambda^-)^2] \, dx \, dy$$

where

$$A(m, \mu, s) = \begin{cases} 1 - \frac{\mu}{m^{2s}}, & \text{if } m^{2s} - \mu > 0, \\ 0, & \text{if } m^{2s} - \mu \leq 0. \end{cases}$$

Choosing $\lambda > 0$ sufficiently large, we deduce that $w_\lambda^- \equiv 0$ on Σ_λ and consequently

$$w_\lambda(x, y) \geq 0 \quad \text{on } \Sigma_\lambda$$

for such $\lambda > 0$.

Now, we define

$$v = \inf\{\tau > 0 : w_\lambda \geq 0 \text{ on } \Sigma_\lambda \text{ for every } \lambda \geq \tau\}.$$

We distinguish two cases. First, we assume that $v > 0$ and prove that $w_v \equiv 0$. We argue by contradiction. By continuity, $w_v \geq 0$ on Σ_v, and by the strong maximum principle, $w_v > 0$ on the set

$$\Sigma_v' = \{(x_1, \ldots, x_N, y) : x_1 > v, \ y > 0\}.$$

We also have $w_v(x, 0) \geq 0$ on the set $\{x \in \mathbb{R}^N : x_1 \geq v\}$ by continuity. Furthermore, by the Hopf principle, $w_v(x, 0) > 0$ on the set $\{x \in \mathbb{R}^N : x_1 > v\}$. Indeed, if, by contradiction, there exists $\bar{x} \in \mathbb{R}^N$ such that $\bar{x}_1 > v$ and $w_v(\bar{x}, 0) = 0$, then by the Hopf principle we have $\frac{\partial w_v}{\partial v^{1-2s}}(\bar{x}, 0) < 0$, which is in contrast with

$$\frac{\partial w_v}{\partial v^{1-2s}}(\bar{x}, 0) = (C_\lambda(x) + m^{2s} - \mu)w_v(\bar{x}, 0) = 0.$$

Take $\lambda_j < v$ such that $\lambda_j \to v$ as $j \to \infty$. Let $r_0 > 0$ be such that $|C_v(x)| \leq \frac{\mu}{4}$ for every $|x| > r_0$. Note that, since $\|v_{\lambda_j}(\cdot, 0)\|_{C^1(\mathbb{R}^N)}$ is uniformly bounded, $D = \|C_{\lambda_j}\|_{L^\infty(\mathbb{R}^N)} < \infty$ and $|C_{\lambda_j}(x)| \leq \frac{\mu}{2}$ for every $|x| > r_0$ and $j \in \mathbb{N}$. Denote

$$B_{r_0}(p_j) = \{x \in \mathbb{R}^N : |x - p_j| < r_0\} \subset \mathbb{R}^N,$$

where $p_j = (\lambda_j, 0, \ldots, 0)$. As above, we obtain

$$\iint_{\Sigma_{\lambda_j}} y^{1-2s}[|\nabla w_{\lambda_j}^-|^2 + m^2(w_{\lambda_j}^-)^2] \, dx dy \leq \int_{\{x_1 > \lambda_j\}} (C_{\lambda_j} + m^{2s} - \mu)(w_{\lambda_j}^-(x, 0))^2 \, dx$$

$$\leq (D + m^{2s} + \mu) \int_{\{x_1 > \lambda_j\} \cap B_{r_0}(p_j)} (w_{\lambda_j}^-(x, 0))^2 \, dx +$$

$$(m^{2s} - \frac{\mu}{2}) \int_{\{x_1 > \lambda_j\} \setminus B_{r_0}(p_j)} (w_{\lambda_j}^-(x, 0))^2 \, dx$$

$$\leq (D + m^{2s} + \mu) \int_{\{x_1 > \lambda_j\} \cap B_{r_0}(p_j)} (w_{\lambda_j}^-(x, 0))^2 \, dx +$$

$$B(m, \mu, s) \int_{\{x_1 > \lambda_j\} \setminus B_{r_0}(p_j)} (w_{\lambda_j}^-(x, 0))^2 \, dx,$$

where

$$B(m, \mu, s) = \begin{cases} m^{2s} - \frac{\mu}{2}, & \text{if } m^{2s} - \frac{\mu}{2} > 0, \\ 0, & \text{if } m^{2s} - \frac{\mu}{2} \leq 0. \end{cases}$$

Since $w_\nu(x, 0) > 0$ on the set $\{x \in \mathbb{R}^N : x_1 > \nu\}$, the measure of the support E_j of $w_{\lambda_j}^-(\cdot, 0)$ on $B_{r_0}(p_j)$ goes to 0 as $j \to \infty$. Then, using the Hölder and Sobolev inequalities, we see that

$$\int_{\{x_1 > \lambda_j\} \cap B_{r_0}(p_j)} (w_{\lambda_j}^-(x, 0))^2 \, dx = \int_{\{x_1 > \lambda_j\}} \chi_{E_j}(x)(w_{\lambda_j}^-(x, 0))^2 \, dx$$

$$\leq \|\chi_{E_j}\|_{L^{N/2s}(\{x_1 > \lambda_j\})} \|w_{\lambda_j}^-(\cdot, 0)\|^2_{L^{2N/N-2s}(\{x_1 > \lambda_j\})}$$

$$\leq o(1) \int_{\Sigma_{\lambda_j}} y^{1-2s} |\nabla w_{\lambda_j}^-|^2 \, dx \, dy.$$

Therefore, if j is large, we have $w_{\lambda_j} \geq 0$ on Σ_{λ_j}. This gives a contradiction because of the minimality of ν. Hence, we conclude that $w_\nu \equiv 0$ on Σ_ν and get the symmetry in the x_1-direction.

Now assume $\nu = 0$. We repeat the above argument for $\lambda < 0$ and $w_\lambda = v_\lambda - v$ defined on

$$\Omega_\lambda = \{(x_1, \ldots, x_N, y) : x_1 < \lambda, \; y \geq 0\}.$$

Then $w_\lambda \geq 0$ for $|\lambda|$ sufficiently large. Let

$$\nu' = \sup\{\tau < 0 : w_\lambda \geq 0 \text{ on } \Omega_\lambda \text{ for every } \lambda \leq \tau\}.$$

If $\nu' < 0$, we get the symmetry as above. If $\nu' = 0$, then since $\nu = 0$, we have

$$v(-x_1, x_2, \ldots, x_N, y) \geq v(x_1, x_2, \ldots, x_N, y) \quad \text{on } \mathbb{R}^{N+1}_+.$$

Consequently, by replacing x_1 with $-x_1$, we deduce that

$$v(-x_1, x_2, \ldots, x_N, y) = v(x_1, x_2, \ldots, x_N, y) \quad \text{on } \mathbb{R}_+^{N+1}.$$

Using the same approach in any arbitrary direction x_i, we obtain the thesis. $\square$

We conclude this subsection by observing that every solution v of (4.4.1) satisfies the following Pohozaev-type identity:

$$\frac{N-2s}{2} \iint_{\mathbb{R}_+^{N+1}} y^{1-2s} |\nabla v|^2 \, dx dy + m^2 \frac{N+2-2s}{2} \iint_{\mathbb{R}_+^{N+1}} y^{1-2s} v^2 \, dx dy$$

$$= N\kappa_s \int_{\mathbb{R}^N} G(v(x,0)) \, dx, \tag{4.4.17}$$

where

$$G(v(x,0)) = m^{2s} \frac{v^2(x,0)}{2} + F(v(x,0)).$$

Indeed, arguing as in the proof of Theorem 3.5.1, we have the additional term

$$A_4(R, \delta) = -m^2 \iint_{D_{R,\delta}^+} y^{1-2s} v \left((x,y) \cdot \nabla v\right) dx dy = -m^2 \iint_{D_{R,\delta}^+} y^{1-2s} \left((x,y) \cdot \nabla \left(\frac{v^2}{2}\right)\right) dx dy$$

which in view of the divergence theorem can be written as

$$-m^2 \iint_{D_{R,\delta}^+} y^{1-2s} \left((x,y) \cdot \nabla \left(\frac{v^2}{2}\right)\right) dx dy = m^2 \frac{N+2-2s}{2} \iint_{D_{R,\delta}^+} y^{1-2s} v^2 \, dx dy$$

$$+ m^2 \int_{\partial D_{R,\delta}^1} y^{1-2s} y \frac{v^2}{2} \, d\sigma - m^2 \int_{\partial D_{R,\delta}^2} y^{1-2s} R \frac{v^2}{2} \, d\sigma.$$

Then one can verify that

$$\lim_{n\to\infty} \lim_{\delta\to 0} A_4(R, \delta) = m^2 \frac{N+2-2s}{2} \iint_{\mathbb{R}_+^{N+1}} y^{1-2s} v^2 \, dx dy$$

and this proves our claim.

On the other hand, as in [64, 116, 181, 184, 226, 308], we can obtain a Pohozaev identity for $(-\Delta + m^2)^s$. Indeed, noting that $\mathcal{F}v(\xi, y) = \mathcal{F}u(\xi)\theta(y\sqrt{|\xi|^2 + m^2})$, where $u =$

$v(\cdot, 0)$, we see that

$$\iint_{\mathbb{R}^{N+1}_+} y^{1-2s} v^2\, dxdy = \left(\int_{\mathbb{R}^N} |\mathcal{F}u(\xi)|^2 (|\xi|^2 + m^2)^{s-1}\, d\xi\right)\left(\int_0^\infty y^{1-2s}\theta^2(y)\, dy\right).$$

Recalling that θ is a solution to (1.2.10) and the asymptotic estimates (1.2.4)–(1.2.5), (1.2.11), an integration by parts yields (see [116])

$$\int_0^\infty y^{1-2s}\theta^2(y)\, dy = \int_0^\infty \left(\frac{y^{2-2s}}{2-2s}\right)' \theta^2(y)\, dy$$

$$= -\frac{1}{1-s}\int_0^\infty \theta(y)\theta'(y)y^{2-2s}\, dy$$

$$= -\frac{1}{1-s}\int_0^\infty \theta'(y)\left(\frac{1-2s}{y}\theta'(y) + \theta''(y)\right) y^{2-2s}\, dy$$

$$= -\frac{1-2s}{1-s}\int_0^\infty y^{1-2s}(\theta'(y))^2\, dy - \frac{1}{1-s}\int_0^\infty y^{2-2s}\theta'(y)\theta''(y)\, dy$$

$$= -\frac{1-2s}{1-s}\int_0^\infty y^{1-2s}(\theta'(y))^2\, dy + \int_0^\infty y^{1-2s}(\theta'(y))^2\, dy$$

$$= \frac{s}{1-s}\int_0^\infty y^{1-2s}(\theta'(y))^2\, dy$$

$$= \frac{s}{1-s}\kappa_s - \frac{s}{1-s}\int_0^\infty y^{1-2s}\theta^2(y)\, dy,$$

that is,

$$\int_0^\infty y^{1-2s}\theta^2(y)\, dy = s\kappa_s.$$

Consequently,

$$\iint_{\mathbb{R}^{N+1}_+} y^{1-2s} v^2\, dxdy = s\kappa_s \int_{\mathbb{R}^N} |\mathcal{F}u(\xi)|^2 (|\xi|^2 + m^2)^{s-1}\, d\xi.$$

Hence, by using Theorem 4.2.1 and (4.4.17), we discover that

$$\frac{N-2s}{2}\int_{\mathbb{R}^N} |\mathcal{F}u(\xi)|^2 (|\xi|^2 + m^2)^s\, d\xi + sm^2 \int_{\mathbb{R}^N} |\mathcal{F}u(\xi)|^2 (|\xi|^2 + m^2)^{s-1}\, d\xi$$

$$= N\int_{\mathbb{R}^N} G(u)\, dx.$$

4.5   Passage to the Limit as $m \to 0$

In this section we show that it is possible to pass to the limit in problem (4.2.6) and to find a nontrivial ground state to (4.1.4). To this end, we estimate c_m from above and below uniformly in $m > 0$, provided that m is sufficiently small.

Fix $0 < m < (\frac{\mu}{2})^{1/2s}$. Using the characterization of the infimum c_m on $\mathcal{N}_m$ we can verify that

$$c_m = \inf_{v \in \mathcal{N}_m} \mathcal{I}_m(v) = \inf_{v \in X_{m,\mathrm{rad}} \setminus \{0\}} \max_{t > 0} \mathcal{I}_m(tv).$$

Let us prove that there exist $\lambda > 0$ and $\delta > 0$ independent of m such that

$$\lambda \leq c_m \leq \delta. \tag{4.5.1}$$

Let

$$w(x, y) = v_0(x) \frac{1}{y + 1},$$

where v_0 is defined by setting

$$v_0(x) = \begin{cases} 1, & \text{if } |x| \leq 1, \\ 2 - |x|, & \text{if } 1 \leq |x| \leq 2, \\ 0, & \text{if } |x| \geq 2. \end{cases} \tag{4.5.2}$$

Then $w \in H_m^1(\mathbb{R}_+^{N+1}, y^{1-2s})$ and

$$\|w\|_{H_m^1(\mathbb{R}_+^{N+1}, y^{1-2s})}^2 = \left(\int_0^\infty y^{1-2s} \frac{1}{(y + 1)^2} \, dy \right) \left[\int_{\mathbb{R}^N} |\nabla_x v_0|^2 + m^2 v_0^2 \, dx \right]$$

$$+ \left(\int_0^\infty y^{1-2s} \frac{1}{(y + 1)^4} \, dy \right) \int_{\mathbb{R}^N} v_0^2 \, dx$$

$$\leq A \left[\int_{\mathbb{R}^N} |\nabla_x v_0|^2 + \left(\frac{\mu}{2} \right)^{1/s} v_0^2 \, dx \right] + B \int_{\mathbb{R}^N} v_0^2 \, dx = C.$$

Therefore,

$$\sup_{t>0} \mathcal{I}_m(tw) = \left(\frac{1}{2} - \frac{1}{p}\right) \frac{\|w\|_{e,m}^{\frac{p}{p-2}}}{\|w\|_{L^p(\mathbb{R}^N)}^{\frac{2}{p-2}}}$$

$$\leq \left(\frac{1}{2} - \frac{1}{p}\right) \frac{\left[\|w\|_{H_m^1(\mathbb{R}_+^{N+1}, y^{1-2s})}^2 + \mu\|v_0\|_{L^2(\mathbb{R}^N)}^2\right]^{\frac{p}{p-2}}}{\|v_0\|_{L^p(\mathbb{R}^N)}^{\frac{2}{p-2}}}$$

$$\leq \left(\frac{1}{2} - \frac{1}{p}\right) \frac{\left[C + \mu\|v_0\|_{L^2(\mathbb{R}^N)}^2\right]^{\frac{p}{p-2}}}{\|v_0\|_{L^p(\mathbb{R}^N)}^{\frac{2}{p-2}}} = \delta.$$

Since $\mathcal{I}_m(v_m) = c_m$ and $\mathcal{J}_m(v_m) = 0$,

$$c_m = \mathcal{I}_m(v_m) = \left(\frac{1}{2} - \frac{1}{p}\right) \int_{\mathbb{R}^N} |v_m(x,0)|^p \, dx.$$

Then, to deduce a lower bound for c_m, it is enough to estimate the $L^p(\mathbb{R}^N)$ norm of $v_m(\cdot, 0)$. Since $\mathcal{J}_m(v_m) = 0$, it follows from Remark 4.2.3 that

$$\|v_m(\cdot,0)\|_{L^p(\mathbb{R}^N)}^p = \iint_{\mathbb{R}_+^{N+1}} y^{1-2s}(|\nabla v_m|^2 + m^2 v_m^2)\, dx dy + (\mu - m^{2s}) \int_{\mathbb{R}^N} |v_m(x,0)|^2 \, dx$$

$$\geq \iint_{\mathbb{R}_+^{N+1}} y^{1-2s} |\nabla v_m|^2 \, dx dy + \frac{\mu}{2} \|v_m(\cdot,0)\|_{L^2(\mathbb{R}^N)}^2$$

$$\geq \kappa_s \frac{C(N,s)}{2} [v_m(\cdot,0)]_s^2 + \frac{\mu}{2} \|v_m(\cdot,0)\|_{L^2(\mathbb{R}^N)}^2$$

$$\geq C_{s,\mu} |v_m(\cdot,0)|_{H^s(\mathbb{R}^N)}^2 \geq C_{s,\mu,p} \|v_m(\cdot,0)\|_{L^p(\mathbb{R}^N)}^2,$$

that is,

$$\|v_m(\cdot,0)\|_{L^p(\mathbb{R}^N)} \geq (C_{s,\mu,p})^{\frac{1}{p-2}} \quad \text{and} \quad c_m \geq \left(\frac{1}{2} - \frac{1}{p}\right)(C_{s,\mu,p})^{\frac{p}{p-2}} = \lambda. \quad (4.5.3)$$

Now, using (4.5.1), we are able to prove the following result:

Theorem 4.5.1 *There exists $v \in H_{\mathrm{loc}}^1(\mathbb{R}_+^{N+1}, y^{1-2s})$ such that, as $m \to 0$, $v_m \rightharpoonup v$ in $L^2(\mathbb{R}^N \times [0, \varepsilon], y^{1-2s})$ for any $\varepsilon > 0$, $\nabla v_m \rightharpoonup \nabla v$ in $L^2(\mathbb{R}_+^{N+1}, y^{1-2s})$ and $v_m(\cdot,0) \to v(\cdot,0)$ in $L^q(\mathbb{R}^N)$ for any $q \in (2, 2_s^*)$. In particular $v(\cdot,0)$ is a nontrivial weak solution to (4.1.4).*

Proof Taking into account that $\mathcal{J}(v_m) = 0$, $c_m \leq \delta$, $0 < m < (\frac{\mu}{2})^{1/2s}$ and using Remark 4.2.3, we can see that

$$\delta^{1/p} \left(\frac{1}{2} - \frac{1}{p}\right)^{-1/p} \geq c_m^{1/p} \left(\frac{1}{2} - \frac{1}{p}\right)^{-1/p} = \|v_m(\cdot, 0)\|_{L^p(\mathbb{R}^N)}^p$$

$$= \iint_{\mathbb{R}^{N+1}_+} y^{1-2s}(|\nabla v_m|^2 + m^2 v_m^2)\,dx\,dy + (\mu - m^{2s}) \int_{\mathbb{R}^N} |v_m(x, 0)|^2\,dx$$

$$\geq \iint_{\mathbb{R}^{N+1}_+} y^{1-2s}|\nabla v_m|^2\,dx\,dy + \frac{\mu}{2}\|v_m(\cdot, 0)\|_{L^2(\mathbb{R}^N)}^2$$

$$\geq \kappa_s \frac{C(N, s)}{2}[v_m(\cdot, 0)]_s^2 + \frac{\mu}{2}\|v_m(\cdot, 0)\|_{L^2(\mathbb{R}^N)}^2$$

$$\geq C_{s,\mu}|v_m(\cdot, 0)|_{H^s(\mathbb{R}^N)}^2,$$

that is,

$$\iint_{\mathbb{R}^{N+1}_+} y^{1-2s}|\nabla v_m|^2\,dx\,dy \leq C_1$$

and

$$|v_m(\cdot, 0)|_{H^s(\mathbb{R}^N)}^2 \leq C_2.$$

Now, fix $\varepsilon > 0$ and let $v \in C_c^\infty(\overline{\mathbb{R}^{N+1}_+})$. For any $x \in \mathbb{R}^N$ and $y \in [0, \varepsilon]$, we have

$$v(x, y) = v(x, 0) + \int_0^y \frac{\partial v}{\partial y}(x, t)\,dt.$$

Since $(a + b)^2 \leq 2a^2 + 2b^2$ for all $a, b \geq 0$, we obtain

$$|v(x, y)|^2 \leq 2|v(x, 0)|^2 + 2\left(\int_0^y \left|\frac{\partial v}{\partial y}(x, t)\right| dt\right)^2,$$

and applying the Hölder inequality we deduce that

$$|v(x, y)|^2 \leq 2\left[|v(x, 0)|^2 + \left(\int_0^y t^{1-2s}\left|\frac{\partial v}{\partial y}(x, t)\right|^2 dt\right)\frac{y^{2s}}{2s}\right].$$

Multiplying both members by y^{1-2s}, we get

$$y^{1-2s}|v(x,y)|^2 \leq 2\left[y^{1-2s}|v(x,0)|^2 + \left(\int_0^y t^{1-2s}\left|\frac{\partial v}{\partial y}(x,t)\right|^2 dt\right)\frac{y}{2s}\right]. \qquad (4.5.4)$$

Integration of (4.5.4) over $\mathbb{R}^N \times [0,\varepsilon]$ yields

$$\|v\|^2_{L^2(\mathbb{R}^N \times [0,\varepsilon], y^{1-2s})} \leq \frac{\varepsilon^{2-2s}}{1-s}\|v(\cdot,0)\|^2_{L^2(\mathbb{R}^N)} + \frac{\varepsilon^2}{2s}\left\|\frac{\partial v}{\partial y}\right\|^2_{L^2(\mathbb{R}^{N+1}_+, y^{1-2s})}. \qquad (4.5.5)$$

By density, (4.5.5) holds for all $v \in H^1_m(\mathbb{R}^{N+1}_+, y^{1-2s})$. Then, replacing v by v_m, we can infer that

$$\|v_m\|^2_{L^2(\mathbb{R}^N \times [0,\varepsilon], y^{1-2s})} \leq C(\varepsilon, s)K(\delta, p)^2$$

for any $0 < m < (\frac{\mu}{2})^{1/2s}$. Hence, there exists $v \in H^1_{loc}(\mathbb{R}^{N+1}_+, y^{1-2s})$ such that

$$v_m \rightharpoonup v \quad \text{in } L^2(\mathbb{R}^N \times [0,\varepsilon], y^{1-2s}) \text{ for all } \varepsilon > 0, \qquad (4.5.6)$$

$$\nabla v_m \rightharpoonup \nabla v \quad \text{in } L^2(\mathbb{R}^{N+1}_+, y^{1-2s}), \qquad (4.5.7)$$

$$v_m(\cdot,0) \to v(\cdot,0) \quad \text{in } L^q(\mathbb{R}^N) \quad \forall q \in (2, 2^*_s), \qquad (4.5.8)$$

as $m \to 0$. Finally, we prove that $v(\cdot,0)$ is a nontrivial weak solution to (4.1.4). We proceed as in [31]. Fix $\eta \in C^\infty_c(\overline{\mathbb{R}^{N+1}_+})$ and let $\psi \in C^\infty([0,\infty))$ be defined by

$$\begin{cases} \psi = 1, & \text{if } 0 \leq y \leq 1, \\ 0 \leq \psi \leq 1, & \text{if } 1 \leq y \leq 2, \\ \psi = 0, & \text{if } y \geq 2. \end{cases} \qquad (4.5.9)$$

Let $\psi_R(y) = \psi(\frac{y}{R})$ for $R > 1$. Clearly, $\eta\psi_R \in H^1_m(\mathbb{R}^{N+1}_+, y^{1-2s})$. Then taking $\eta\psi_R$ in the weak formulation of (4.2.6) we have

$$\iint_{\mathbb{R}^{N+1}_+} y^{1-2s}[\nabla v_m \cdot \nabla(\eta\psi_R) + m^2 v_m \eta\psi_R] \, dx \, dy$$

$$+ (\mu - m^{2s})\int_{\mathbb{R}^N} v_m(x,0)\eta(x,0) \, dx = \int_{\mathbb{R}^N} |v_m(x,0)|^{p-2}v_m(x,0)\eta(x,0) \, dx.$$

Letting $m \to 0$ and using (4.5.6)–(4.5.8) we find

$$\iint_{\mathbb{R}_+^{N+1}} y^{1-2s} \nabla v \cdot \nabla(\eta \psi_R) \, dx \, dy + \mu \int_{\mathbb{R}^N} v(x,0) \eta(x,0) \, dx = \int_{\mathbb{R}^N} |v(x,0)|^{p-2} v(x,0) \eta(x,0) \, dx.$$

Letting $R \to \infty$ we deduce that

$$\iint_{\mathbb{R}_+^{N+1}} y^{1-2s} \nabla v \cdot \nabla \eta \, dx \, dy + \mu \int_{\mathbb{R}^N} v(x,0) \eta(x,0) \, dx = \int_{\mathbb{R}^N} |v(x,0)|^{p-2} v(x,0) \eta(x,0) \, dx$$

for all $\eta \in C_c^\infty(\overline{\mathbb{R}_+^{N+1}})$. Finally, $v(\cdot, 0)$ is not identically zero because of (4.5.3), (4.5.8) and $2 < p < 2_s^*$. Applying Proposition 1.3.11-(ii) we deduce that $v(\cdot, 0) > 0$ in $\mathbb{R}^N$. $\qquad\square$

4.5.1   Final Comments on $(-\Delta + m^2)^s$

Bearing in mind the asymptotic estimates (1.2.4) and (1.2.5) for K_ν, we can now prove the analogue of Lemma 1.2.1 for the operator $(-\Delta + m^2)^s$, with $m > 0$ and $s \in (0, 1)$.

Theorem 4.5.2 ([63]) *Let $s \in (0, 1)$ and $m > 0$. Then, for every $u \in \mathcal{S}(\mathbb{R}^N)$,*

$$(-\Delta + m^2)^s u(x) = m^{2s} u(x) + \frac{c_{N,s}}{2} m^{\frac{N+2s}{2}} \int_{\mathbb{R}^N} \frac{2u(x) - u(x+y) - u(x-y)}{|y|^{\frac{N+2s}{2}}} K_{\frac{N+2s}{2}}(m|y|) \, dy.$$

Proof Choosing the substitution $z = y - x$ in (4.1.3), we obtain

$$(-\Delta + m^2)^s u(x) = m^{2s} u(x) + c_{N,s} m^{\frac{N+2s}{2}} \text{P.V.} \int_{\mathbb{R}^N} \frac{u(x) - u(y)}{|x-y|^{\frac{N+2s}{2}}} K_{\frac{N+2s}{2}}(m|x-y|) \, dy$$

$$= m^{2s} u(x) + c_{N,s} m^{\frac{N+2s}{2}} \text{P.V.} \int_{\mathbb{R}^N} \frac{u(x) - u(x+z)}{|z|^{\frac{N+2s}{2}}} K_{\frac{N+2s}{2}}(m|z|) \, dz.$$
$$\tag{4.5.10}$$

Substituting $\tilde{z} = -z$ in the last term in (4.5.10), we get

$$\text{P.V.} \int_{\mathbb{R}^N} \frac{u(x+z) - u(x)}{|z|^{\frac{N+2s}{2}}} K_{\frac{N+2s}{2}}(m|z|) \, dz = \text{P.V.} \int_{\mathbb{R}^N} \frac{u(x-\tilde{z}) - u(x)}{|\tilde{z}|^{\frac{N+2s}{2}}} K_{\frac{N+2s}{2}}(m|\tilde{z}|) \, d\tilde{z},$$
$$\tag{4.5.11}$$

and so, after relabeling $\tilde{z}$ as z,

$$2\text{P.V.} \int_{\mathbb{R}^N} \frac{u(x+z) - u(x)}{|z|^{\frac{N+2s}{2}}} K_{\frac{N+2s}{2}}(m|z|)\, dz$$

$$= \text{P.V.} \int_{\mathbb{R}^N} \frac{u(x+z) - u(x)}{|z|^{\frac{N+2s}{2}}} K_{\frac{N+2s}{2}}(m|z|)\, dz$$

$$+ \text{P.V.} \int_{\mathbb{R}^N} \frac{u(x-z) - u(x)}{|z|^{\frac{N+2s}{2}}} K_{\frac{N+2s}{2}}(m|z|)\, dz$$

$$= \text{P.V.} \int_{\mathbb{R}^N} \frac{u(x+z) + u(x-z) - 2u(x)}{|z|^{\frac{N+2s}{2}}} K_{\frac{N+2s}{2}}(m|z|)\, dz. \qquad (4.5.12)$$

Hence, if we rename z as y in (4.5.10) and (4.5.12), we can write $(-\Delta + m^2)^s$ as

$$(-\Delta + m^2)^s u(x) = \frac{c_{N,s}}{2} m^{\frac{N+2s}{2}} \text{P.V.} \int_{\mathbb{R}^N} \frac{2u(x) - u(x+y) - u(x-y)}{|y|^{\frac{N+2s}{2}}} K_{\frac{N+2s}{2}}(m|y|)\, dy$$

$$+ m^{2s} u(x). \qquad (4.5.13)$$

Now, a second-order Taylor expansion shows that

$$\left| \frac{2u(x) - u(x+y) - u(x-y)}{|y|^{\frac{N+2s}{2}}} K_{\frac{N+2s}{2}}(m|y|) \right| \leq \frac{\|D^2 u\|_{L^\infty(\mathbb{R}^N)}}{|y|^{\frac{N+2s-4}{2}}} K_{\frac{N+2s}{2}}(m|y|).$$

From (1.2.4) we deduce that

$$\frac{|D^2 u|_\infty}{|y|^{\frac{N+2s-4}{2}}} K_{\frac{N+2s}{2}}(m|y|) \sim \frac{C}{|y|^{N+2s-2}} \qquad \text{as } |y| \to 0$$

which is integrable near 0. On the other hand, using (1.2.5), we get

$$\left| \frac{2u(x) - u(x+y) - u(x-y)}{|y|^{\frac{N+2s}{2}}} K_{\frac{N+2s}{2}}(m|y|) \right| \leq \frac{C\|u\|_{L^\infty(\mathbb{R}^N)}}{|y|^{\frac{N+2s}{2}}} K_{\frac{N+2s}{2}}(m|y|)$$

$$\sim \frac{C}{|y|^{\frac{N+2s+1}{2}}} e^{-m|y|} \qquad \text{as } |y| \to \infty$$

which is integrable near ∞. Therefore, we can remove the P.V. in (4.5.13). □

The next two theorems can be seen as the analogues of Propositions 1.3.1 and 1.3.2, respectively, for the operator $(-\Delta + m^2)^s$ with $m > 0$ and $s \in (0, 1)$ (more precisely, compare with Theorem 15 in [316] concerning the case $m = 0$). The first result is a direct consequence of Theorem 4.4.2 and provides Schauder-Zygmund estimates.

Theorem 4.5.3 *Let $s \in (0, 1)$, $m > 0$ and $\alpha \in (0, 1)$. Assume that $f \in C^{0,\alpha}(\mathbb{R}^N)$ and that $u \in L^\infty(\mathbb{R}^N)$ is a solution to $(-\Delta + m^2)^s u = f$ in $\mathbb{R}^N$.*

- *If $\alpha + 2s < 1$, then $u \in C^{0,\alpha+2s}(\mathbb{R}^N)$.*
- *If $1 < \alpha + 2s < 2$, then $u \in C^{1,\alpha+2s-1}(\mathbb{R}^N)$.*
- *If $2 < \alpha + 2s < 3$, then $u \in C^{2,\alpha+2s-2}(\mathbb{R}^N)$.*
- *If $\alpha + 2s = k \in \{1, 2\}$, then $u \in \Lambda_k^*$.*

The next result gives Schauder-Hölder-Zygmund estimates for bounded solutions to the equation $(-\Delta + m^2)^s u = f$ in $\mathbb{R}^N$ with $f \in L^\infty(\mathbb{R}^N)$. To prove it one can use the characterization of Lipschitz spaces in terms of the Poisson kernel $P_{\frac{1}{2}}(x, y)$ for $\mathbb{R}_+^{N+1}$ defined in Section 1.2 (see [241, 315] for more details). We argue as in [64]. Assume for simplicity $m = 1$. Define

$$U(x, y) = P_{\frac{1}{2}}(x, y) * u(x) = (P_{\frac{1}{2}}(x, y) * \mathcal{G}_{2s}(x)) * f(x) = \mathcal{G}_{2s}(x, y) * f(x).$$

By using the formula (59) with $l = 2$ and $\beta = 2s$ at pag. 149 in [315], we can find $C > 0$ such that

$$\left\| \frac{\partial^2 \mathcal{G}_{2s}}{\partial y^2}(x, y) \right\|_{L^1(\mathbb{R}^N)} \leq C y^{2s-2}, \quad y > 0.$$

Since $\frac{\partial^2 U}{\partial y^2}(x, y) = \frac{\partial^2 \mathcal{G}_{2s}}{\partial y^2}(x, y) * f(x)$ and $f \in L^\infty(\mathbb{R}^N)$, we use Young's inequality and Theorem 15.6 with $\alpha = 2s$ and $k = 2$ in [241], to deduce that $u \in \Lambda_{2s}$. Therefore, we have proved the following result:

Theorem 4.5.4 *Let $s \in (0, 1)$ and $m > 0$. Assume that $f \in L^\infty(\mathbb{R}^N)$ and that $u \in L^\infty(\mathbb{R}^N)$ is a solution to $(-\Delta + m^2)^s u = f$ in $\mathbb{R}^N$.*

- *If $2s < 1$, then $u \in C^{0,2s}(\mathbb{R}^N)$.*
- *If $2s = 1$, then $u \in \Lambda_1^*$.*
- *If $2s > 1$, then $u \in C^{1,2s-1}(\mathbb{R}^N)$.*

Ground States for a Superlinear Fractional Schrödinger Equation with Potentials

5.1 Nonlinearities with Subcritical Growth

5.1.1 Introduction

In this section we focus our attention on the study of the following fractional Schrödinger equation:

$$\begin{cases} (-\Delta)^s u + V(x)u = f(x, u) \text{ in } \mathbb{R}^N, \\ u \in H^s(\mathbb{R}^N), \end{cases} \tag{5.1.1}$$

where $s \in (0, 1)$, $N \geq 2$, the potential $V : \mathbb{R}^N \to \mathbb{R}$ satisfies the assumption

$(V1)$ $V \in C(\mathbb{R}^N, \mathbb{R})$ and $\alpha \leq V(x) \leq \beta$,

and the nonlinearity $f : \mathbb{R}^N \times \mathbb{R} \to \mathbb{R}$ fulfills the following conditions:

$(f1)$ $f \in C(\mathbb{R}^N \times \mathbb{R}, \mathbb{R})$ is 1-periodic in x and

$$\lim_{|t| \to \infty} \frac{f(x, t)}{|t|^{2_s^* - 1}} = 0, \quad \text{uniformly in } x \in \mathbb{R}^N;$$

$(f2)$ $f(x, t) = o(t)$ as $|t| \to 0$, uniformly in $x \in \mathbb{R}^N$.

© The Author(s), under exclusive license to Springer Nature Switzerland AG 2021
V. Ambrosio, *Nonlinear Fractional Schrödinger Equations in* $\mathbb{R}^N$,
Frontiers in Mathematics, https://doi.org/10.1007/978-3-030-60220-8_5

When $s = 1$, the equation in (5.1.1) formally reduces to the classical nonlinear Schrödinger equation

$$- \Delta u + V(x)u = f(x, u) \quad \text{in } \mathbb{R}^N, \tag{5.1.2}$$

which has been extensively studied in the last 20 years. Since we cannot review the huge bibliography, we just mention [28, 90, 165, 197, 299, 330, 340] and references therein, where several existence and multiplicity results are obtained under different assumptions on the potential V and the nonlinearity f. To deal with (5.1.2), many authors assume that the nonlinear term satisfied the following condition due to Ambrosetti and Rabinowitz [29]

$$\exists \mu > 2, \, R > 0: \, 0 < \mu F(x, t) \leq f(x, t)t \quad \forall |t| \geq R, \tag{AR}$$

where $F(x, t) = \int_0^t f(x, \tau) \, d\tau$.

Roughly speaking, the role of (AR) is to guarantee the boundedness of Palais–Smale sequences for the functional associated with the problem under consideration. However, although (AR) is a quite natural condition when we deal with superlinear elliptic problems, it is somewhat restrictive. In fact, by a direct integration of (AR), we can deduce that

$$(f3) \quad \lim_{|t| \to \infty} \frac{F(x, t)}{t^2} = \infty \text{ uniformly in } x \in \mathbb{R}^N.$$

Of course, also condition $(f3)$ characterizes the nonlinearity f to be superlinear at infinity. Anyway, if we consider the function $f(x, t) = t \log(1 + |t|)$, then it is easy to prove that f fulfills $(f3)$ but does not satisfy (AR). This means that there are functions that are superlinear at infinity and do not verify (AR). For this reason, in several works concerning superlinear problems, some authors tried to drop the condition (AR); see for instance [151, 231, 259, 271, 305] and references therein. For instance, Jeanjean in [231] introduced the following assumption:

$(f4)$ There exists $\lambda \geq 1$ such that

$$G(x, \theta t) \leq \lambda G(x, t) \quad \text{for } (x, t) \in \mathbb{R}^N \times \mathbb{R} \text{ and } \theta \in [0, 1],$$

where $G(x, t) = f(x, t)t - 2F(x, t)$.

In view of the above discussion and motivated by the fact that recently several authors [172, 183, 306] studied (5.1.1) assuming the well-known condition (AR), here we investigate solutions of (5.1.1) without requiring this condition. Our first main result can be stated as follows:

Theorem 5.1.1 ([32]) *Assume that f satisfies $(f1)$–$(f4)$ and V fulfills $(V1)$ and*

$(V2)$ $V(x)$ *is* 1-*periodic.*

Then there exists a nontrivial ground state solution $u \in H^s(\mathbb{R}^N)$ to (5.1.1).

In order to study our problem, we will look for the critical points for the functional

$$\mathcal{J}(u) = \frac{1}{2}\left([u]_s^2 + \int_{\mathbb{R}^N} V(x)u^2\,dx\right) - \int_{\mathbb{R}^N} F(x, u)\,dx.$$

Thanks to the assumptions on f, it is easy to see that $\mathcal{J}$ has a mountain pass geometry. Namely, setting

$$\Gamma = \{\gamma \in C([0, 1], H^s(\mathbb{R}^N)) : \gamma(0) = 0 \text{ and } \mathcal{J}(\gamma(1)) < 0\},$$

we have $\Gamma \neq \varnothing$ and

$$c = \inf_{\gamma \in \Gamma} \max_{t \in [0,1]} \mathcal{J}(\gamma(t))$$

is the mountain pass level for $\mathcal{J}$. The Ekeland variational principle [177] guarantees the existence of a Cerami sequence at the level c. Hence, by using some suitable variational arguments inspired by Jeanjean and Tanaka [233] and Liu [259] and the $\mathbb{Z}^N$-invariance of the problem (5.1.1), we will prove that every Cerami sequence for $\mathcal{J}$ is bounded and admits a subsequence which converges to a critical point for $\mathcal{J}$.

Finally, we will also consider the potential well case. We will assume that $V(x)$ satisfies, in addition to $(V1)$, the following condition:

$(V3)$ $V(x) < V_\infty = \lim_{|y|\to\infty} V(y) < \infty$ $\quad \forall x \in \mathbb{R}^N$.

We will also assume that $f(x, u) = b(x)f(u)$, where $b \in C(\mathbb{R}^N)$ and

$$0 < b_\infty = \lim_{|y|\to\infty} b(y) \leq b(x) \leq \bar{b} < \infty \tag{5.1.3}$$

for any $x \in \mathbb{R}^N$ and that f satisfies $(f1)$–$(f4)$. Therefore our problem becomes

$$\begin{cases} (-\Delta)^s u + V(x)u = b(x)f(u) \text{ in } \mathbb{R}^N, \\ u \in H^s(\mathbb{R}^N). \end{cases} \tag{5.1.4}$$

To study (5.1.4), we will use the energy comparison method presented in [233]. More precisely, by introducing the energy functional at infinity,

$$\mathcal{J}_\infty(u) = \frac{1}{2}\left([u]_s^2 + \int_{\mathbb{R}^N} V_\infty u^2\, dx\right) - \int_{\mathbb{R}^N} b_\infty F(u)\, dx,$$

we will show that, under the above assumptions on f and V, $\mathcal{J}$ has a nontrivial critical point provided that

$$c < m_\infty, \tag{5.1.5}$$

where

$$m_\infty = \inf\{\mathcal{J}_\infty(u) : u \neq 0 \text{ and } \mathcal{J}'_\infty(u) = 0\}.$$

In order to prove (5.1.5), we use the fact that our problem at infinity is autonomous:

$$(-\Delta)^s u = -V_\infty u + b_\infty f(u) \quad \text{in } \mathbb{R}^N,$$

so it admits a least one energy solution satisfying the Pohozaev identity. This information will be fundamental for deducing the existence of a path $\gamma \in \Gamma$ such that $\max_{t\in[0,1]} \mathcal{J}(\gamma(t)) < m_\infty$. Combining these facts, we will be able to prove our second result:

Theorem 5.1.2 ([32]) *Assume that V satisfies $(V1)$ and $(V3)$, and that f fulfills assumptions $(f1)$–$(f4)$. Then (5.1.4) has a ground state.*

5.1.2   Preliminaries

We start with the concept of weak solution for the equation

$$(-\Delta)^s u + V(x)u = g \quad \text{in } \mathbb{R}^N. \tag{5.1.6}$$

Definition 5.1.3 Given $g \in L^2(\mathbb{R}^N)$, we say that u is a weak solution to (5.1.6) if $u \in H^s(\mathbb{R}^N)$ and satisfies

$$\iint_{\mathbb{R}^{2N}} \frac{(u(x) - u(y))(v(x) - v(y))}{|x - y|^{N+2s}}\, dx\, dy + \int_{\mathbb{R}^N} V(x)uv\, dx = \int_{\mathbb{R}^N} gv\, dx$$

for all $v \in H^s(\mathbb{R}^N)$.

In order to deal with (5.1.1), we consider the functional on $H^s(\mathbb{R}^N)$ given by

$$\mathcal{J}(u) = \frac{1}{2}\left([u]_s^2 + \int_{\mathbb{R}^N} V(x)u^2 dx\right) - \int_{\mathbb{R}^N} F(x,u)\,dx.$$

By $(V1)$, it follows that

$$\|u\| = \left([u]_s^2 + \int_{\mathbb{R}^N} V(x)u^2\,dx\right)^{\frac{1}{2}}$$

is a norm that is equivalent to the standard one in $H^s(\mathbb{R}^N)$. For this reason, we will always write

$$\mathcal{J}(u) = \frac{1}{2}\|u\|^2 - \int_{\mathbb{R}^N} F(x,u)\,dx.$$

In particular, by the assumptions on f, we deduce that $\mathcal{J} \in C^1(H^s(\mathbb{R}^N), \mathbb{R})$. Moreover, $\mathcal{J}$ possesses a mountain pass geometry. More precisely, we have the following result whose simple proof is omitted.

Lemma 5.1.4 *Under assumptions $(f1)$–$(f4)$, there exist $r > 0$ and $v_0 \in H^s(\mathbb{R}^N)$ such that $\|v_0\| > r$ and*

$$b = \inf_{\|u\|=r} \mathcal{J}(u) > \mathcal{J}(0) = 0 \geq \mathcal{J}(v_0). \tag{5.1.7}$$

In particular,

$$\langle \mathcal{J}'(u), u \rangle = \|u\|^2 + o(\|u\|^2) \ \ as \ \|u\| \to 0,$$
$$\mathcal{J}(u) = \tfrac{1}{2}\|u\|^2 + o(\|u\|^2) \ \ \ \ as \ \|u\| \to 0,$$

and, consequently

(i) there exists $\eta > 0$ such that if v is a critical point for $\mathcal{J}$, then $\|v\| \geq \eta$;
(ii) for every $c > 0$ there exists $\eta_c > 0$ such that if $\mathcal{J}(v_n) \to c$, then $\|v_n\| \geq \eta_c$.

Therefore, by Lemma 5.1.4, it follows that

$$\Gamma = \{\gamma \in C([0,1], H^s(\mathbb{R}^N)) : \gamma(0) = 0 \text{ and } \mathcal{J}(\gamma(1)) < 0\} \neq \emptyset$$

and we can define the mountain pass level

$$c = \inf_{\gamma \in \Gamma} \max_{t \in [0,1]} \mathcal{J}(\gamma(t)). \tag{5.1.8}$$

Let us point out that, by (5.1.7), c is positive. Then, Theorem 2.2.15 ensures that there exists a Cerami sequence (v_n) at the level c for $\mathcal{J}$, that is

$$\mathcal{J}(v_n) \to c \quad \text{and} \quad (1 + \|v_n\|)\|\mathcal{J}'(v_n)\|_{H^{-s}(\mathbb{R}^N)} \to 0.$$

We conclude this subsection by proving that the primitive $F(x, t)$ of $f(x, t)$ is nonnegative.

Lemma 5.1.5 *Assume that f satisfies $(f1)$, $(f2)$, and $(f4)$. Then $F \geq 0$ in $\mathbb{R}^N \times \mathbb{R}$.*

Proof First we observe that by $(f4)$ we have

$$G(x, t) = f(x, t)t - 2F(x, t) \geq 0 \quad \text{for all } (x, t) \in \mathbb{R}^N \times \mathbb{R}.$$

Fix $t > 0$. For $x \in \mathbb{R}^N$, let us compute the derivative of $\frac{F(x,t)}{t^2}$ with respect to t:

$$\frac{\partial}{\partial t}\left(\frac{F(x, t)}{t^2}\right) = \frac{f(x, t)\, t^2 - 2t\, F(x, t)}{t^4} \geq 0. \tag{5.1.9}$$

Moreover, by $(f2)$, we get

$$\lim_{t \to 0} \frac{F(x, t)}{t^2} = 0. \tag{5.1.10}$$

Combining (5.1.9) and (5.1.10) we deduce that $F(x, t) \geq 0$ for all $(x, t) \in \mathbb{R}^N \times [0, \infty)$. Analogously, we obtain that $F(x, t) \geq 0$ for all $(x, t) \in \mathbb{R}^N \times (-\infty, 0]$.   □

5.1.3   Periodic Potentials

In this subsection we give the proof of the Theorem 5.1.1. We begin with the following result which guarantees the boundedness of Cerami sequences for the functional $\mathcal{J}$.

Lemma 5.1.6 *Assume that $(V1)$ and $(f1)$–$(f4)$ hold. Let $c \in \mathbb{R}$. Then every Cerami sequence for $\mathcal{J}$ is bounded.*

Proof Let (v_n) be a Cerami sequence for $\mathcal{J}$. Assume, by contradiction, that (v_n) is unbounded. Going if necessary to a subsequence, we may assume that

$$\mathcal{J}(v_n) \to c, \quad \|v_n\| \to \infty, \quad \|\mathcal{J}'(v_n)\|_{H^{-s}(\mathbb{R}^N)}\|v_n\| \to 0. \tag{5.1.11}$$

Now denote $w_n = \dfrac{v_n}{\|v_n\|}$. Clearly, (w_n) is bounded in $H^s(\mathbb{R}^N)$ and its elements have unit norm. We claim that (w_n) vanishes, i.e., that

$$\lim_{n \to \infty} \sup_{z \in \mathbb{R}^N} \int_{B_2(z)} w_n^2 \, dx = 0. \tag{5.1.12}$$

Indeed, if (5.1.12) does not hold, then there exists $\delta > 0$ such that

$$\sup_{z \in \mathbb{R}^N} \int_{B_2(z)} w_n^2 \, dx \geq \delta > 0.$$

Consequently, we can choose a sequence of points $(z_n) \subset \mathbb{R}^N$ such that

$$\int_{B_2(z_n)} w_n^2 \, dx \geq \frac{\delta}{2}.$$

Since the number of points in $\mathbb{Z}^N \cap B_2(z_n)$ is less than 4^N, there exists a point $\xi_n \in \mathbb{Z}^N \cap B_2(z_n)$ such that

$$\int_{B_2(\xi_n)} w_n^2 \, dx \geq K > 0, \tag{5.1.13}$$

where $K = \delta 2^{-(2N+1)}$. Set $\tilde{w}_n = w_n(\cdot + \xi_n)$. Using $(V1)$ and that w_n has unit norm, we deduce that

$$\begin{aligned}
\|\tilde{w}_n\|^2 &= [\tilde{w}_n]_s^2 + \int_{\mathbb{R}^N} V(x) \tilde{w}_n^2 \, dx \\
&\leq [\tilde{w}_n]_s^2 + \beta \int_{\mathbb{R}^N} V(x) \tilde{w}_n^2 \, dx \\
&= [w_n]_s^2 + \beta \int_{\mathbb{R}^N} w_n^2 \, dx \\
&\leq \frac{\beta}{\alpha} \left([w_n]_s^2 + \alpha \int_{\mathbb{R}^N} w_n^2 \, dx \right) \\
&\leq \frac{\beta}{\alpha} \left([w_n]_s^2 + \int_{\mathbb{R}^N} V(x) w_n^2 \, dx \right) = \frac{\beta}{\alpha},
\end{aligned}$$

that is, $(\tilde{w}_n)$ is bounded. By Lemma 1.1.8, we may assume, passing if necessary to a subsequence, that

$$\begin{aligned}
\tilde{w}_n &\to \tilde{w} \quad \text{in } L^2_{\text{loc}}(\mathbb{R}^N), \\
\tilde{w}_n(x) &\to \tilde{w}(x) \quad \text{for a.e. } x \in \mathbb{R}^N.
\end{aligned} \tag{5.1.14}$$

By (5.1.13) and (5.1.14),

$$\int_{B_2} \tilde{w}^2 \, dx = \lim_{n \to \infty} \int_{B_2} \tilde{w}_n^2 \, dx = \lim_{n \to \infty} \int_{B_2(\xi_n)} w_n^2 \, dx \geq K > 0, \qquad (5.1.15)$$

which implies that $\tilde{w} \neq 0$.

Let $\tilde{v}_n = \|v_n\| \tilde{w}_n$. Since $\tilde{w} \neq 0$, the set $A = \{x \in \mathbb{R}^N : \tilde{w} \neq 0\}$ has positive Lebesgue measure and $|\tilde{v}_n(x)| \to \infty$. In particular, by $(f3)$, we get

$$\frac{F(x, \tilde{v}_n(x))}{(\tilde{v}_n(x))^2} (\tilde{w}_n(x))^2 \to \infty. \qquad (5.1.16)$$

Let us observe that $f(x, t)$ is 1-periodic with respect to x, so

$$\int_{\mathbb{R}^N} F(x, v_n) \, dx = \int_{\mathbb{R}^N} F(x, \tilde{v}_n) \, dx. \qquad (5.1.17)$$

By (5.1.11), (5.1.16), (5.1.17) and Lemma 5.1.5, it follows that

$$\frac{1}{2} - \frac{c + o(1)}{\|v_n\|^2} = \int_{\mathbb{R}^N} \frac{F(x, v_n)}{\|v_n\|^2} \, dx = \int_{\mathbb{R}^N} \frac{F(x, \tilde{v}_n)}{\|v_n\|^2} \, dx \geq \int_A \frac{F(x, \tilde{v}_n)}{\tilde{v}_n^2} \tilde{w}_n^2 \, dx \to \infty, \qquad (5.1.18)$$

which gives a contradiction. Therefore (5.1.12) holds true. In particular, by Lemma 1.4.4, we get

$$w_n \to 0 \text{ in } L^q(\mathbb{R}^N) \quad \forall q \in (2, 2_s^*).$$

Now, let ρ be a positive real number. By $(f1)$–$(f3)$ and Lemma 5.1.5, we deduce that for every $\varepsilon > 0$ there exists $C_\varepsilon > 0$ such that

$$0 \leq F(x, \rho t) \leq \varepsilon(|t|^2 + |t|^{2_s^*}) + C_\varepsilon |t|^q. \qquad (5.1.19)$$

Since $\|w_n\| = 1$, by (1.1.1) there exists $\tilde{c} > 0$ such that

$$\|w_n\|_{L^2(\mathbb{R}^N)}^2 + \|w_n\|_{L^{2_s^*}(\mathbb{R}^N)}^{2_s^*} \leq \tilde{c}. \qquad (5.1.20)$$

Taking into account (5.1.19) and (5.1.20), we obtain

$$\limsup_{n \to \infty} \int_{\mathbb{R}^N} F(x, \rho w_n) \, dx \leq \limsup_{n \to \infty} \left[\varepsilon(\|w_n\|_{L^2(\mathbb{R}^N)}^2 + \|w_n\|_{L^{2_s^*}(\mathbb{R}^N)}^{2_s^*}) + C_\varepsilon(\|w_n\|_{L^q(\mathbb{R}^N)}^q) \right]$$

$$\leq \varepsilon \tilde{c},$$

and in view of the arbitrariness of ε we conclude that

$$\lim_{n \to \infty} \int_{\mathbb{R}^N} F(x, \rho w_n) \, dx = 0. \tag{5.1.21}$$

Now, let $(t_n) \subset [0, 1]$ be a sequence such that

$$\mathcal{J}(t_n v_n) = \max_{t \in [0,1]} \mathcal{J}(t v_n). \tag{5.1.22}$$

Using (5.1.11), we can verify that $2\sqrt{j} \|v_n\|^{-1} \in (0, 1)$ for n sufficiently large and $j \in \mathbb{N}$. Taking $\rho = 2\sqrt{j}$ in (5.1.21), we have

$$\mathcal{J}(t_n v_n) \geq \mathcal{J}(2\sqrt{j} \, w_n) = 2j - \int_{\mathbb{R}^N} F(x, 2\sqrt{j} \, w_n) \, dx \geq j$$

for n large enough and for all $j \in \mathbb{N}$. Then

$$\mathcal{J}(t_n v_n) \to \infty. \tag{5.1.23}$$

Since $\mathcal{J}(0) = 0$ and $\mathcal{J}(v_n) \to c$ we deduce that $t_n \in (0, 1)$. By (5.1.22), we obtain

$$\langle \mathcal{J}'(t_n v_n), t_n v_n \rangle = t_n \frac{d}{dt} \mathcal{J}(t v_n) \Big|_{t=t_n} = 0. \tag{5.1.24}$$

Indeed, (5.1.11), (5.1.24) and $(f4)$ show that

$$\begin{aligned}
\frac{2}{\lambda} \mathcal{J}(t_n v_n) &= \frac{1}{\lambda} \left(2\mathcal{J}(t_n v_n) - \langle \mathcal{J}'(t_n v_n), t_n v_n \rangle \right) \\
&= \frac{1}{\lambda} \int_{\mathbb{R}^N} (f(x, t_n v_n) t_n v_n - 2F(x, t_n v_n)) \, dx \\
&= \frac{1}{\lambda} \int_{\mathbb{R}^N} G(x, t_n v_n) \, dx \\
&\leq \int_{\mathbb{R}^N} G(x, t_n v_n) \, dx \\
&= \int_{\mathbb{R}^N} (f(x, v_n) v_n - 2F(x, v_n)) \, dx \\
&= 2\mathcal{J}(v_n) - \langle \mathcal{J}'(v_n), v_n \rangle \to 2c,
\end{aligned}$$

which is incompatible with (5.1.23). Thus (v_n) is bounded. $\qquad\square$

Remark 5.1.7 We stress that the conclusion of Lemma 5.1.6 holds true if we consider $f(x, t) = b(x)f(t)$ with $b \in C(\mathbb{R}^N, \mathbb{R})$ and $0 < b_0 \leq b(x) \leq b_1 < \infty$ for any $x \in \mathbb{R}^N$. In fact, in this case, the contradiction in (5.1.18) is obtained replacing (5.1.17) by

$$\int_{\mathbb{R}^N} b(x) F(v_n) \, dx \geq \frac{b_0}{b_1} \int_{\mathbb{R}^N} b(x) F(\tilde{v}_n) \, dx.$$

Now let us prove that, up to a subsequence, our bounded Cerami sequence (u_n) converges weakly to a non-trivial critical point for $\mathcal{J}$.

Proof of Theorem 5.1.1 Let c be the mountain pass level defined in (5.1.8). We know that $c > 0$ and that there exists a Cerami sequence (u_n) for $\mathcal{J}$, which is bounded in $H^s(\mathbb{R}^N)$ by Lemma 5.1.6. Denote

$$\delta = \lim_{n \to \infty} \sup_{z \in \mathbb{R}^N} \int_{B_2(z)} u_n^2 \, dx.$$

If $\delta = 0$, then Lemma 1.4.4 implies that $u_n \to 0$ in $L^q(\mathbb{R}^N)$ for all $q \in (2, 2_s^*)$. Analogously to (5.1.21), we can see that

$$\lim_{n \to \infty} \int_{\mathbb{R}^N} F(x, u_n) \, dx = 0,$$

$$\lim_{n \to \infty} \int_{\mathbb{R}^N} f(x, u_n) u_n \, dx = 0.$$

Consequently,

$$0 = \lim_{n \to \infty} \int_{\mathbb{R}^N} \left(\frac{1}{2} f(x, u_n) v_n - F(x, u_n) \right) dx = \lim_{n \to \infty} \left(\mathcal{J}(u_n) - \frac{1}{2} \langle \mathcal{J}'(u_n), u_n \rangle \right) = c$$

which is impossible because $c > 0$. Therefore, $\delta > 0$. As for (5.1.15), we can find a sequence $(\xi_n) \subset \mathbb{Z}^N$ and a positive constant K such that

$$\int_{B_2} w_n^2 \, dx = \int_{B_2(\xi_n)} u_n^2 \, dx > K, \tag{5.1.25}$$

where $w_n = u_n(\cdot + \xi_n)$. Note that $\|w_n\| = \|u_n\|$, so (w_n) is bounded in $H^s(\mathbb{R}^N)$. By Lemma 1.1.8, we can assume, up to a subsequence, that

$$w_n \rightharpoonup w \quad \text{in } H^s(\mathbb{R}^N),$$

$$w_n \to w \quad \text{in } L^2_{\text{loc}}(\mathbb{R}^N),$$

and by using (5.1.25) we have $w \neq 0$. Since (5.1.1) is $\mathbb{Z}^N$-invariant, (w_n) is a Cerami sequence for $\mathcal{J}$. Then,

$$\langle \mathcal{J}'(w), \phi \rangle = \lim_{n \to \infty} \langle \mathcal{J}'(w_n), \phi \rangle = 0$$

for all $\phi \in C_c^\infty(\mathbb{R}^N)$, that is, $\mathcal{J}'(w) = 0$ and w is a nontrivial solution to (5.1.1).

Now we want to prove that the problem (5.1.1) has a ground state. Let

$$m = \inf\{\mathcal{J}(v) : v \neq 0 \text{ and } \mathcal{J}'(v) = 0\}$$

and suppose that v is an arbitrary critical point for $\mathcal{J}$. By $(f4)$,

$$G(x, t) \geq 0 \quad \forall (x, t) \in \mathbb{R}^N \times \mathbb{R},$$

which implies that

$$\mathcal{J}(v) = \mathcal{J}(v) - \frac{1}{2}\langle \mathcal{J}'(v), v \rangle = \frac{1}{2}\int_{\mathbb{R}^N} G(x, v)\,dx \geq 0.$$

Hence, $m \geq 0$. Now, let (u_n) be a sequence of nontrivial critical points for $\mathcal{J}$ such that $\mathcal{J}(u_n) \to m$. By Lemma 5.1.4, there exists $\eta > 0$ such that

$$\|u_n\| \geq \eta. \tag{5.1.26}$$

Since u_n is a critical point for $\mathcal{J}$, we have that

$$(1 + \|u_n\|)\|\mathcal{J}'(u_n)\|_{H^{-s}(\mathbb{R}^N)} \to 0.$$

Therefore, (u_n) is a Cerami sequence at the level m and, by Lemma 5.1.6, (u_n) is bounded in $H^s(\mathbb{R}^N)$. Let

$$\delta = \lim_{n \to \infty} \sup_{z \in \mathbb{R}^N} \int_{B_2(z)} u_n^2\,dx.$$

As before, if $\delta = 0$, then

$$\lim_{n \to \infty} \int_{\mathbb{R}^N} f(x, u_n)u_n\,dx = 0,$$

whence

$$\|u_n\|^2 = \langle \mathcal{J}'(u_n), u_n \rangle + \int_{\mathbb{R}^N} f(x, u_n)u_n\,dx \to 0, \tag{5.1.27}$$

which is impossible because of (5.1.26). Thus $\delta > 0$. Arguing as before, we can find the translated sequence $w_n(x) = u_n(x + \xi_n)$ such that

$$\mathcal{J}'(w_n) = 0, \quad \mathcal{J}(w_n) = \mathcal{J}(u_n) \to m, \tag{5.1.28}$$

and (w_n) weakly converges to a nonzero critical point w for $\mathcal{J}$. Hence, by (5.1.28), the fact that $G \geq 0$ and Fatou's lemma,

$$
\begin{aligned}
\mathcal{J}(w) &= \mathcal{J}(w) - \frac{1}{2}\langle \mathcal{J}'(w), w \rangle \\
&= \frac{1}{2} \int_{\mathbb{R}^N} G(x, w) \, dx \\
&\leq \liminf_{n \to \infty} \frac{1}{2} \int_{\mathbb{R}^N} G(x, w_n) \, dx \\
&= \liminf_{n \to \infty} \left(\mathcal{J}(w_n) - \frac{1}{2}\langle \mathcal{J}'(w_n), w_n \rangle \right) = m.
\end{aligned}
\tag{5.1.29}
$$

Thus, w is a nontrivial critical point for $\mathcal{J}$ such that $\mathcal{J}(w) = m$.   □

5.1.4   Bounded Potentials

In this section we give the proof of the Theorem 5.1.2. The main ingredient of our proof is the following result which takes advantage of the Pohozaev identity for $(-\Delta)^s$.

Proposition 5.1.8 *Let $u \in H^s(\mathbb{R}^N)$ be a nontrivial critical point for*

$$\mathcal{I}(u) = \frac{1}{2}[u]_s^2 - \int_{\mathbb{R}^N} G(u) \, dx.$$

Then there exists $\gamma \in C([0, 1], H^s(\mathbb{R}^N))$ such that $\gamma(0) = 0$, $\mathcal{I}(\gamma(1)) < 0$, $u \in \gamma([0, 1])$ and

$$\max_{t \in [0,1]} \mathcal{I}(\gamma(t)) = \mathcal{I}(u).$$

Proof Let $u \in H^s(\mathbb{R}^N)$ be a nontrivial critical point for $\mathcal{I}$. We set for $t > 0$

$$u^t(x) = u\left(\frac{x}{t}\right).$$

By the Pohozaev identity,

$$\frac{N - 2s}{2}[u]_s^2 = N \int_{\mathbb{R}^N} G(u) \, dx,$$

which implies that

$$\mathcal{I}(u^t) = \frac{t^{N-2s}}{2}[u]_s^2 - t^N \int_{\mathbb{R}^N} G(u)\,dx = \left(\frac{1}{2}t^{N-2s} - \frac{N-2s}{2N}t^N\right)[u]_s^2.$$

Therefore, we deduce that $\max_{t>0} \mathcal{I}(u^t) = \mathcal{I}(u)$, $\mathcal{I}(u^t) \to -\infty$ as $t \to \infty$, and

$$\|u^t\|_{H^s(\mathbb{R}^N)}^2 = t^{N-2s}[u]_s^2 + t^N\|u\|_{L^2(\mathbb{R}^N)}^2 \to 0 \text{ as } t \to 0.$$

Choosing $\alpha > 1$ such that $\mathcal{I}(u^\alpha) < 0$ and setting

$$\gamma(t) = \begin{cases} u^{\alpha t}, & \text{for } t \in (0, 1], \\ 0, & \text{for } t = 0, \end{cases}$$

we get the conclusion. $\qquad\square$

Now we consider the following functionals:

$$\mathcal{J}(u) = \frac{1}{2}\|u\|^2 - \int_{\mathbb{R}^N} b(x)F(u)\,dx$$

and

$$\mathcal{J}_\infty(u) = \frac{1}{2}\left([u]_s^2 + \int_{\mathbb{R}^N} V_\infty u^2\,dx\right) - \int_{\mathbb{R}^N} b_\infty F(u)\,dx.$$

By $(V3)$, it follows that

$$\mathcal{J}(u) < \mathcal{J}_\infty(u) \quad \text{for any } u \in H^s(\mathbb{R}^N) \setminus \{0\}. \tag{5.1.30}$$

Taking into account Proposition 5.1.8, we prove the following result:

Lemma 5.1.9 *Assume that $V(x)$ satisfies $(V1)$ and $(V3)$ and f fulfills $(f1)$–$(f4)$. Then $\mathcal{J}$ has a nontrivial critical point.*

Proof Let c be the mountain pass level for $\mathcal{J}$. We know that $\mathcal{J}$ has a Cerami sequence (u_n) at the level c, which is bounded by Lemma 5.1.6. Then $u_n \rightharpoonup u$ in $H^s(\mathbb{R}^N)$ and $\mathcal{J}'(u) = 0$. We claim that $u \neq 0$. Assume, by contradiction, that $u = 0$. Thanks to $(V3)$, the fact that $u_n \to u$ in $L^2_{loc}(\mathbb{R}^N)$, Lemma 5.1.5 and (5.1.3),

$$|\mathcal{J}_\infty(u_n) - \mathcal{J}(u_n)| \leq \int_{\mathbb{R}^N} |V_\infty - V(x)||u_n|^2\,dx + \int_{\mathbb{R}^N} |b(x) - b_\infty|F(u_n)\,dx \to 0$$

and

$$\|\mathcal{J}_\infty'(u_n) - \mathcal{J}'(u_n)\|_{H^{-s}(\mathbb{R}^N)}$$

$$\leq \sup_{\substack{\phi \in H^s(\mathbb{R}^N) \\ \|\phi\|_{H^s(\mathbb{R}^N)} = 1}} \left\{ \left| \int_{\mathbb{R}^N} [V_\infty - V(x)] u_n \phi \, dx \right| + \left| \int_{\mathbb{R}^N} [b_\infty - b(x)] f(u_n) \phi \, dx \right| \right\} \to 0,$$

that is, (u_n) is a Palais–Smale sequence for $\mathcal{J}_\infty$ at the level c.

Now denote

$$\delta = \lim_{n \to \infty} \sup_{\xi \in \mathbb{R}^N} \int_{B_2(\xi)} u_n^2 \, dx. \tag{5.1.31}$$

If $\delta = 0$, then proceeding similarly to (5.1.27), we deduce that $\|u_n\|^2 \to 0$, which contradicts Lemma 5.1.4. Therefore, $\delta > 0$ and there exists $(\xi_n) \subset \mathbb{Z}^N$ such that

$$\int_{B_2(\xi_n)} u_n^2 \, dx \geq \frac{\delta}{2} > 0. \tag{5.1.32}$$

Let $v_n = u_n(x + \xi_n)$. Then

$$\|v_n\| = \|u_n\|,$$

$$\mathcal{J}_\infty(v_n) = \mathcal{J}_\infty(u_n),$$

$$\mathcal{J}_\infty'(v_n) = \mathcal{J}_\infty'(u_n).$$

Therefore, (v_n) is a bounded Palais–Smale sequence for $\mathcal{J}_\infty$. As in the proof of Theorem 5.1.1, we use (5.1.32) to deduce that $v_n \rightharpoonup v$ in $H^s(\mathbb{R}^N)$ and v is a nontrivial critical point for $\mathcal{J}_\infty$. Moreover, proceeding as in (5.1.29), we have

$$\mathcal{J}_\infty(v) \leq c.$$

Using Proposition 5.1.8 with $g(t) = b_\infty f(t) - V_\infty t$, we can find $\gamma_\infty \in C([0, 1], H^s(\mathbb{R}^N))$ such that $\gamma_\infty(0) = 0$, $\mathcal{J}_\infty(\gamma_\infty(1)) < 0$, $v \in \gamma_\infty([0, 1])$ and

$$\max_{t \in [0,1]} \mathcal{J}_\infty(\gamma_\infty(t)) = \mathcal{J}_\infty(v).$$

Since $0 \notin \gamma_\infty((0, 1])$, by (5.1.30), it follows that, for all $t \in (0, 1]$

$$\mathcal{J}(\gamma_\infty(t)) < \mathcal{J}_\infty(\gamma_\infty(t)). \tag{5.1.33}$$

In particular, $\mathcal{J}(\gamma_\infty(1)) \leq \mathcal{J}_\infty(\gamma_\infty(1)) < 0$, so $\gamma_\infty \in \Gamma$. Then, taking into account that $\mathcal{J}(0) = \mathcal{J}_\infty(0) = 0$, (5.1.33) and that $c > 0$, we deduce that

$$c \leq \max_{t\in[0,1]} \mathcal{J}(\gamma_\infty(t)) < \max_{t\in[0,1]} \mathcal{J}_\infty(\gamma_\infty(t)) = \mathcal{J}_\infty(v) \leq c,$$

which gives a contradiction. □

Remark 5.1.10 Since (u_n) is a Cerami sequence for $\mathcal{J}$ at the level c and $u_n \rightharpoonup u$ in $H^s(\mathbb{R}^N)$, a similar argument as in (5.1.29) shows that $\mathcal{J}(u) \leq c$.

Finally, we give the proof of Theorem 5.1.2.

Proof of Theorem 5.1.2 Let $m = \inf\{\mathcal{J}(u) : u \neq 0 \text{ and } \mathcal{J}'(u) = 0\}$ and let u denote the nontrivial critical point for $\mathcal{J}$ obtained in the previous lemma. Then (see Remark 5.1.10) one can verify that

$$0 \leq m \leq \mathcal{J}(u) \leq c. \tag{5.1.34}$$

Now, let (u_n) be a sequence of nontrivial critical points for $\mathcal{J}$ such that $\mathcal{J}(u_n) \to m$. As in the proof of Theorem 5.1.1, we have that (u_n) is a Cerami bounded sequence at the level m and $\delta > 0$, where δ is defined via (5.1.31). Extracting a subsequence, $u_n \rightharpoonup \tilde{u}$ in $H^s(\mathbb{R}^N)$, and $\tilde{u}$ is a critical point for $\mathcal{J}$ satisfying $\mathcal{J}(\tilde{u}) \leq m$ as in (5.1.29). Now, if $\tilde{u} = 0$, (u_n) is a bounded Palais–Smale sequence for $\mathcal{J}_\infty$ at the level m. Since $\delta > 0$, we deduce that (v_n), which is a suitable translation of (u_n), converges weakly to some critical point $v \neq 0$ for $\mathcal{J}_\infty$ and $\mathcal{J}_\infty(v) \leq m$.

Arguing as in the proof of Lemma 5.1.9, it follows from Proposition 5.1.8 that there exists $\gamma_\infty \in \Gamma_\infty \cap \Gamma$ such that

$$c \leq \max_{t\in[0,1]} \mathcal{J}(\gamma_\infty(t)) < \max_{t\in[0,1]} \mathcal{J}_\infty(\gamma_\infty(t)) = \mathcal{J}_\infty(v) \leq m,$$

which is a contradiction because of (5.1.34). Thus, $\tilde{u}$ is a nontrivial critical point for $\mathcal{J}$ such that $\mathcal{J}(\tilde{u}) = m$. □

5.2 Nonlinearities with Critical Growth

5.2.1 Introduction

This section is devoted to the existence of ground state solutions for the following nonlinear fractional elliptic problem:

$$\begin{cases} (-\Delta)^s u + V(x)u = f(u) & \text{in } \mathbb{R}^N, \\ u \in H^s(\mathbb{R}^N), \quad u > 0 \text{ in } \mathbb{R}^N, \end{cases} \tag{5.2.1}$$

with $s \in (0, 1)$ and $N \geq 2$. We assume that the potential $V : \mathbb{R}^N \to \mathbb{R}$ satisfies the following conditions:

($V1$) $V \in C^1(\mathbb{R}^N, \mathbb{R})$;
($V2$) there exists $V_0 > 0$ such that $\inf_{x \in \mathbb{R}^N} V(x) \geq V_0$;
($V3$) $V(x) \leq V_\infty = \lim_{|x| \to \infty} V(x) < \infty$ for all $x \in \mathbb{R}^N$;
($V4$) $\| \max\{x \cdot \nabla V(x), 0\} \|_{L^{\frac{N}{2s}}(\mathbb{R}^N)} < 2s S_*$.

Concerning the nonlinear term f, we assume that $f(t) = 0$ for $t \leq 0$ and

($f1$) $f \in C^1(\mathbb{R}_+, \mathbb{R})$;
($f2$) $\lim\limits_{t \to 0} \dfrac{f(t)}{t} = 0$;
($f3$) $\lim\limits_{t \to \infty} \dfrac{f(t)}{t^{2_s^* - 1}} = K > 0$;
($f4$) there exist $D > 0$ and $\max\left\{2, \dfrac{4s}{N - 2s}\right\} < q < 2_s^*$ such that

$$f(t) \geq K t^{2_s^* - 1} + D t^{q - 1} \text{ for all } t \geq 0.$$

We observe that assumptions ($f3$) and ($f4$) on the nonlinearity f enable us to consider the critical growth case. In the case $s = 1$, assumption ($f4$) was introduced in [345]. We point out that ($f4$) plays an important role in ensuring the existence of solutions for the problem (5.2.1). In fact, if we take $f(t) = (t^+)^{2_s^* - 1}$, then f satisfies ($f1$)–($f3$), and by using the Pohozaev identity for the fractional Laplacian, we can see that there are no nontrivial solutions to (5.2.1).

Our first main result concerns the existence of ground state solutions to (5.2.1) in the case of constant potentials.

Theorem 5.2.1 ([67]) *Let $s \in (0, 1)$ and $N \geq 2$. Assume that f satisfies ($f1$)–($f4$) and $V(x) \equiv V > 0$ is constant. Then (5.2.1) possesses a positive ground state solution $u \in H^s(\mathbb{R}^N)$.*

Let us sketch the proof of Theorem 5.2.1. In order to obtain the existence of a nontrivial solution to (5.2.1), we look for critical points of the Euler-Lagrange functional associated with (5.2.1), that is,

$$\mathcal{I}(u) = \frac{1}{2} \int_{\mathbb{R}^N} \left(|(-\Delta)^{\frac{s}{2}} u|^2 + V(x) u^2 \right) dx - \int_{\mathbb{R}^N} F(u) \, dx$$

for any $u \in H^s(\mathbb{R}^N)$, where $F(t) = \int_0^t f(\tau) \, d\tau$. In view of the assumptions on f, it is clear that $\mathcal{I}$ has a mountain pass geometry, but it is hard to verify the boundedness of

Palais–Smale sequences of $\mathcal{I}$. To overcome this difficulty, we use the idea in [231]. For $\lambda \in [\frac{1}{2}, 1]$, let us introduce the family of functionals

$$\mathcal{I}_\lambda(u) = \frac{1}{2} \int_{\mathbb{R}^N} \left(|(-\Delta)^{\frac{s}{2}} u|^2 + V(x) u^2 \right) dx - \lambda \int_{\mathbb{R}^N} F(u) \, dx.$$

As a first step, we prove that for every $\lambda \in [\frac{1}{2}, 1]$, $\mathcal{I}_\lambda$ has a mountain pass geometry and that $\mathcal{I}_\lambda$ admits a bounded Palais–Smale sequence (u_n) at the mountain pass level c_λ. More precisely, we use the following abstract version of Struwe's monotonicity trick [319] developed by Jeanjean [231]:

Theorem 5.2.2 ([231]) *Let $(X, \| \cdot \|)$ be a Banach space and $\Lambda \subset \mathbb{R}_+$ an interval. Let $(\mathcal{I}_\lambda)_{\lambda \in \Lambda}$ be a family of C^1-functionals on X of the form*

$$\mathcal{I}_\lambda(u) = A(u) - \lambda B(u), \quad \text{for } \lambda \in \Lambda,$$

where $B(u) \geq 0$ for all $u \in X$, and either $A(u) \to \infty$ or $B(u) \to \infty$ as $\|u\| \to \infty$. We assume that there exist $v_1, v_2 \in X$ such that

$$c_\lambda = \inf_{\gamma \in \Gamma} \max_{t \in [0,1]} \mathcal{I}_\lambda(\gamma(t)) > \max\{\mathcal{I}_\lambda(v_1), \mathcal{I}_\lambda(v_2)\}, \quad \text{for all } \lambda \in \Lambda,$$

where

$$\Gamma = \{\gamma \in C([0, 1], X) : \gamma(0) = v_1, \gamma(1) = v_2\}.$$

Then, for almost every $\lambda \in \Lambda$, there is a sequence $(u_n) \subset X$ such that

(i) (u_n) is bounded;
(ii) $\mathcal{I}_\lambda(u_n) \to c_\lambda$;
(iii) $\mathcal{I}'_\lambda(u_n) \to 0$ in X^*.

Moreover, the map $\lambda \mapsto c_\lambda$ is non-increasing and continuous from the left.

Since we are dealing with the critical case, we are able to prove that for any $\lambda \in [\frac{1}{2}, 1]$

$$0 < c_\lambda < \frac{s}{N} \frac{S_*^{\frac{N}{2s}}}{\lambda^{\frac{N-2s}{2s}}}.$$

Secondly, in the spirit of [236] (see also [260, 346]), we establish a global compactness result in the critical case, which gives a description of the bounded Palais–Smale sequences of $\mathcal{I}_\lambda$. Then, by using the fact that every solution of (5.2.1) satisfies the Pohozaev identity and the compactness lemma, we prove the existence of a bounded Palais–Smale sequence of $\mathcal{I}$ which converges to a positive solution to (5.2.1).

Now, we state our second main result which deals with the existence of ground state of (5.2.1) in the case in which V is not a constant.

Theorem 5.2.3 ([67]) *Let* $s \in (0, 1)$ *and* $N \geq 2$. *Assume that* f *fulfills* $(f1)$–$(f4)$ *and* V *satisfies* $(V1)$–$(V4)$, *and* $V(x) \not\equiv V_\infty$. *Then* (5.2.1) *admits a positive ground state solution* $u \in H^s(\mathbb{R}^N)$.

To deal with the non-autonomous case, we enlist some ideas developed in [236]. We consider the previous family of functionals $\mathcal{I}_\lambda$, and, since $\mathcal{I}_\lambda$ satisfies the assumptions of Theorem 5.2.2, we can deduce the existence of a Palais–Smale sequence (u_n^j) at the mountain pass level c_{λ_j}, where $\lambda_j \to 1$. Therefore, $u_n^j \rightharpoonup u_j$ in $H^s(\mathbb{R}^N)$, where u_j is a critical point of $\mathcal{I}_{\lambda_j}$. This time, the boundedness of the sequence (u_j) follows by the assumption $(V4)$. Moreover, we prove that (u_j) is a bounded Palais–Smale sequence of $\mathcal{I}$. To show that the bounded sequence (u_j) converges to a nontrivial weak solution of (5.2.1), we show that c_1 is strictly less than the least energy level m^∞ of the functional $\mathcal{I}^\infty$ associated to the "problem at infinity"

$$(-\Delta)^s u + V_\infty u = f(u) \quad \text{in } \mathbb{R}^N.$$

Together with an accurate description of the sequence as a sum of translated critical points, this allows us to infer that $u_j \rightharpoonup u$ in $H^s(\mathbb{R}^N)$, for some nontrivial critical point u of $\mathcal{I}$. Let us recall that when f is an odd function satisfying $(f1)$–$(f4)$ and V is constant, the existence of a radial positive ground state to (5.2.1) was proved in [36] via a minimization argument and by working in the space of radial functions $H^s_{\mathrm{rad}}(\mathbb{R}^N)$ which is compactly embedded into $L^p(\mathbb{R}^N)$ for all $p \in (2, 2^*_s)$. Here we present a different proof of this result (see Theorem 5.2.1) which is based on the global compactness lemma, which will be also useful for proving Theorem 5.2.3. In fact, we believe that the global compactness lemma is not only interesting for the aim of this chapter, but it can be also used to deal with other problems similar to (5.2.1). Let us also point out that by using the methods developed here, we are able to study (5.2.1) dealing with radial and non-radial potentials by a unified approach.

We conclude this introduction by mentioning that further interesting results for fractional elliptic problems with critical growth in $\mathbb{R}^N$ can be found in [170, 174, 216, 311, 344], and [41, 87, 310, 324] for problems in bounded domains.

5.2.2 Splitting Lemmas

In this subsection we collect some useful splitting results; see [1, 113, 154, 169, 340, 347] for some classical results.

Lemma 5.2.4 *Assume that $u_n \rightharpoonup u$ in $H^s(\mathbb{R}^N)$. Then we have*

$$\left| \int_{\mathbb{R}^N} \left[|u_n|^{2^*_s-2} u_n - |u|^{2^*_s-2} u - |u_n - u|^{2^*_s-2} (u_n - u) \right] w \, dx \right| = o_n(1) \|w\|_{H^s(\mathbb{R}^N)}$$

where $o_n(1) \to 0$ as $n \to \infty$, uniformly for any $w \in C_c^\infty(\mathbb{R}^N)$.

Proof We follow [262]. Using the mean value theorem, the boundedness of (u_n) in $H^s(\mathbb{R}^N)$, the generalized Hölder inequality, the fact that $u \in L^{2^*_s}(\mathbb{R}^N)$, and Theorem 1.1.8, we deduce that for every $\varepsilon > 0$ there exists $R = R(\varepsilon) > 0$ such that

$$\left| \int_{\mathbb{R}^N \setminus B_R} \left[|u_n|^{2^*_s-2} u_n - |u|^{2^*_s-2} u - |u_n - u|^{2^*_s-2} (u_n - u) \right] w \, dx \right|$$

$$\leq \left| \int_{\mathbb{R}^N \setminus B_R} \left[|u_n|^{2^*_s-2} u_n - |u_n - u|^{2^*_s-2} (u_n - u) \right] w \, dx \right| + \left| \int_{\mathbb{R}^N \setminus B_R} |u|^{2^*_s-2} u w \, dx \right|$$

$$\leq C \int_{\mathbb{R}^N \setminus B_R} \left(|u_n|^{2^*_s-2} + |u_n - u|^{2^*_s-2} \right) |u w| \, dx + \int_{\mathbb{R}^N \setminus B_R} |u|^{2^*_s-1} |w| \, dx$$

$$\leq C \|u_n\|_{L^{2^*_s}(\mathbb{R}^N)}^{2^*_s-2} \|u\|_{L^{2^*_s}(\mathbb{R}^N \setminus B_R)} \|w\|_{L^{2^*_s}(\mathbb{R}^N)} + C \|u\|_{L^{2^*_s}(\mathbb{R}^N \setminus B_R)}^{2^*_s-1} \|w\|_{L^{2^*_s}(\mathbb{R}^N)}$$

$$\leq C \varepsilon \|w\|_{H^s(\mathbb{R}^N)}. \tag{5.2.2}$$

On the other hand, for every $r > 0$, we have

$$\left| \int_{B_R} \left[|u_n|^{2^*_s-2} u_n - |u|^{2^*_s-2} u - |u_n - u|^{2^*_s-2} (u_n - u) \right] w \, dx \right|$$

$$\leq \int_{B_R \cap \{|u_n - u| \leq r\}} \left| |u_n|^{2^*_s-2} u_n - |u|^{2^*_s-2} u - |u_n - u|^{2^*_s-2} (u_n - u) \right| |w| \, dx$$

$$+ \int_{B_R \cap \{|u_n - u| \geq r\}} \left| |u_n|^{2^*_s-2} u_n - |u|^{2^*_s-2} u - |u_n - u|^{2^*_s-2} (u_n - u) \right| |w| \, dx$$

$$= I_1 + I_2. \tag{5.2.3}$$

Now, there exists $r = r(R)$ such that $r |B_R|^{\frac{1}{2^*_s}} \leq \varepsilon$. This fact together with the mean value theorem, the boundedness of (u_n) in $H^s(\mathbb{R}^N)$ and the generalized Hölder inequality implies that

$$I_1 \leq C \int_{B_R \cap \{|u_n - u| \leq r\}} \left(|u_n|^{2^*_s-2} + |u|^{2^*_s-2} + |u_n - u|^{2^*_s-2} \right) |(u_n - u) w| \, dx$$

$$\leq C \|u_n\|_{L^{2^*_s}(\mathbb{R}^N)}^{2^*_s-2} \|u_n - u\|_{L^{2^*_s}(B_R \cap \{|u_n - u| \leq r\})} \|w\|_{L^{2^*_s}(\mathbb{R}^N)}$$

$$+ C\|u\|_{L^{2_s^*}(\mathbb{R}^N)}^{2_s^*-2} \|u_n - u\|_{L^{2_s^*}(B_R \cap \{|u_n-u|\leq r\})} \|w\|_{L^{2_s^*}(\mathbb{R}^N)}$$

$$+ C\|u_n - u\|_{L^{2_s^*}(\mathbb{R}^N)}^{2_s^*-2} \|u_n - u\|_{L^{2_s^*}(B_R \cap \{|u_n-u|\leq r\})} \|w\|_{L^{2_s^*}(\mathbb{R}^N)}$$

$$\leq Cr|B_R|^{\frac{1}{2_s^*}} \|w\|_{H^s(\mathbb{R}^N)} \leq C\varepsilon \|w\|_{H^s(\mathbb{R}^N)}. \tag{5.2.4}$$

For such r, R fixed above, u_n strongly converges to u in $L^2(B_R)$ and then $|B_R \cap \{|u_n - u| \geq r\}| \to 0$ as $n \to \infty$. Therefore, by the dominated convergence theorem,

$$\int_{B_R \cap \{|u_n-u|\geq r\}} |u|^{2_s^*}\, dx \to 0 \quad \text{as } n \to \infty,$$

which implies that, for n large enough,

$$I_2 \leq C \int_{B_R \cap \{|u_n-u|\geq r\}} \left(|u_n|^{2_s^*-2} + |u_n - u|^{2_s^*-2} \right) |uw|\, dx + \int_{B_R \cap \{|u_n-u|\geq r\}} |u|^{2_s^*-1} |w|\, dx$$

$$\leq C\|u_n\|_{L^{2_s^*}(\mathbb{R}^N)}^{2_s^*-2} \|u\|_{L^{2_s^*}(B_R \cap \{|u_n-u|\geq r\})} \|w\|_{H^s(\mathbb{R}^N)}$$

$$+ C\|u_n - u\|_{L^{2_s^*}(\mathbb{R}^N)}^{2_s^*-2} \|u\|_{L^{2_s^*}(B_R \cap \{|u_n-u|\geq r\})} \|w\|_{H^s(\mathbb{R}^N)}$$

$$+ C\|u\|_{L^{2_s^*}(B_R \cap \{|u_n-u|\geq r\})}^{2_s^*-1} \|w\|_{H^s(\mathbb{R}^N)}$$

$$\leq C\|u\|_{L^{2_s^*}(B_R \cap \{|u_n-u|\geq r\})} \|w\|_{H^s(\mathbb{R}^N)} + C\|u\|_{L^{2_s^*}(B_R \cap \{|u_n-u|\geq r\})}^{2_s^*-1} \|w\|_{H^s(\mathbb{R}^N)}$$

$$\leq C\varepsilon \|w\|_{H^s(\mathbb{R}^N)}. \tag{5.2.5}$$

Combining (5.2.2), (5.2.3), (5.2.4), and (5.2.5) we get the thesis. □

Lemma 5.2.5 *Let* $f_1 : \mathbb{R} \to \mathbb{R}$ *be a continuous function such that*

$$\lim_{|t|\to\infty} \frac{f_1(t)}{|t|^{2_s^*-1}} = 0,$$

and

$$|f_1(t)| \leq C(|t| + |t|^{2_s^*-1}) \quad \text{for all } t \in \mathbb{R}.$$

If $u_n \rightharpoonup u$ *in* $H^s(\mathbb{R}^N)$ *and* $u_n \to u$ *a.e. in* $\mathbb{R}^N$, *then*

$$\lim_{n\to\infty} \left[\int_{\mathbb{R}^N} F_1(u_n)\, dx - \int_{\mathbb{R}^N} F_1(u)\, dx - \int_{\mathbb{R}^N} F_1(u_n - u)\, dx \right] = 0,$$

where $F_1(t) = \int_0^t f(\tau) \, d\tau$.

Proof The proof is standard. However, for the reader's convenience, we show it here. For $R > 0$, the mean value theorem shows that

$$\int_{\mathbb{R}^N} F_1(u_n) \, dx = \int_{B_R} F_1(u_n) \, dx + \int_{B_R^c} F_1(u_n) \, dx$$

$$= \int_{B_R} F_1(u_n) \, dx + \int_{B_R^c} F_1((u_n - u) + u) \, dx$$

$$= \int_{B_R} F_1(u_n) \, dx + \int_{B_R^c} F_1(u_n - u) \, dx + \int_{B_R^c} \left[\int_0^1 f_1(u_n - u + \theta u) u \, d\theta \right] dx.$$

We now write

$$\left| \int_{\mathbb{R}^N} F_1(u_n) \, dx - \int_{\mathbb{R}^N} F_1(u) \, dx - \int_{\mathbb{R}^N} F_1(u_n - u) \, dx \right|$$

$$\leq \left| \int_{B_R} (F_1(u_n) - F_1(u)) \, dx \right| + \left| \int_{B_R^c} F_1(u) \, dx \right|$$

$$+ \left| \int_{B_R} F_1(u_n - u) \, dx \right| + \left| \int_{B_R^c} \left[\int_0^1 f_1(u_n - u + \theta u) u \, d\theta \right] dx \right|.$$

From the growth assumption on f and applying the Hölder inequality we obtain that

$$\left| \int_{B_R^c} \left[\int_0^1 f_1(u_n - u + \theta u) u \, d\theta \right] dx \right|$$

$$\leq C \left[\int_{B_R^c} (|u_n| + |u|) |u| \, dx + \int_{B_R^c} (|u_n| + |u|)^{2_s^* - 1} |u| \, dx \right]$$

$$\leq C \left[\left(\int_{\mathbb{R}^N} (|u_n| + |u|)^2 \, dx \right)^{\frac{1}{2}} \left(\int_{B_R^c} |u|^2 \, dx \right)^{\frac{1}{2}} + \left(\int_{\mathbb{R}^N} (|u_n| + |u|)^{2_s^*} \, dx \right)^{\frac{2_s^* - 1}{2_s^*}} \left(\int_{B_R^c} |u|^{2_s^*} \, dx \right)^{\frac{1}{2_s^*}} \right]$$

$$\leq C \left[\left(\int_{B_R^c} |u|^2 \, dx \right)^{\frac{1}{2}} + \left(\int_{B_R^c} |u|^{2_s^*} \, dx \right)^{\frac{1}{2_s^*}} \right],$$

where in the last inequality we used the boundedness of (u_n) in $L^2(\mathbb{R}^N) \cap L^{2_s^*}(\mathbb{R}^N)$. Moreover,

$$\left| \int_{B_R^c} F_1(u) \, dx \right| \leq C \left[\int_{B_R^c} |u|^2 \, dx + \int_{B_R^c} |u|^{2_s^*} \, dx \right].$$

Hence, for every $\varepsilon > 0$ there exists $R = R_\varepsilon > 0$ such that

$$\left| \int_{B_R^c} F_1(u)\, dx \right| + \left| \int_{B_R^c} \left[\int_0^1 f_1(u_n - u + \theta u)u\, d\theta \right] dx \right| \leq \varepsilon.$$ (5.2.6)

On the other hand, it follows from Lemma 1.4.2 (with $P(t) = F(t)$ and $Q(t) = t^2 + |t|^{2_s^*}$) that

$$\lim_{n \to \infty} \int_{B_R} (F_1(u_n) - F_1(u))\, dx = 0$$ (5.2.7)

and

$$\lim_{n \to \infty} \int_{B_R} F_1(u_n - u)\, dx = 0.$$ (5.2.8)

Now the result follows from (5.2.6), (5.2.7), and (5.2.8).   □

Lemma 5.2.6 *Let $f_1 : \mathbb{R} \to \mathbb{R}$ be a continuous function such that*

$$\lim_{|t| \to 0} \frac{f_1(t)}{|t|} = 0 = \lim_{|t| \to \infty} \frac{f_1(t)}{|t|^{2_s^* - 1}}.$$

Let $(u_n) \subset H^s(\mathbb{R}^N)$ be a sequence such that $u_n \rightharpoonup u$ in $H^s(\mathbb{R}^N)$. Then we have

$$\left| \int_{\mathbb{R}^N} (f_1(u_n) - f_1(u) - f_1(u_n - u))\, w\, dx \right| = o_n(1)\|w\|_{H^s(\mathbb{R}^N)},$$

where $o_n(1) \to 0$ as $n \to \infty$, uniformly for any $w \in C_c^\infty(\mathbb{R}^N)$.

Proof We argue as in [7, 347]. For any fixed $\eta > 0$, since $f(t) = o(|t|)$ as $|t| \to 0$, we can choose $r_0 = r_0(\eta) \in (0, 1)$ such that

$$|f_1(t)| \leq \eta|t| \quad \text{for } |t| \leq 2r_0.$$ (5.2.9)

On the other hand, since $f(t) = o(|t|^{2_s^* - 1})$ as $|t| \to \infty$, we can pick $r_1 = r_1(\eta) > 2$ such that

$$|f_1(t)| \leq \eta|t|^{2_s^* - 1} \quad \text{for } |t| \geq r_1 - 1.$$ (5.2.10)

By the continuity of f, there exists $\delta = \delta(\eta) \in (0, r_0)$ such that

$$|f_1(t_1) - f_1(t_2)| \leq r_0 \eta \quad \text{for } |t_1 - t_2| \leq \delta,\ |t_1|, |t_2| \leq r_1 + 1.$$ (5.2.11)

Moreover, since $f(t) = o(|t|^{2_s^* - 1})$ as $|t| \to \infty$, there exists a positive constant $c = c(\eta)$ such that

$$|f_1(t)| \leq c(\eta)|t| + \eta|t|^{2_s^* - 1} \quad \text{for } t \in \mathbb{R}. \tag{5.2.12}$$

In what follows we estimate the integral:

$$\int_{\mathbb{R}^N \backslash B_R} |f_1(u_n - u) - f_1(u_n) - f_1(u)| |w| \, dx.$$

Using (5.2.12) and $u \in L^2(\mathbb{R}^N) \cap L^{2_s^*}(\mathbb{R}^N)$ we can find $R = R(\eta) > 0$ such that

$$\int_{\mathbb{R}^N \backslash B_R} |f_1(u)w| \, dx \leq c \left(\int_{\mathbb{R}^N \backslash B_R} |u|^{2_s^*} \, dx \right)^{\frac{2_s^* - 1}{2_s^*}} \|w\|_{L^{2_s^*}(\mathbb{R}^N)}$$

$$+ c \left(\int_{\mathbb{R}^N \backslash B_R} |u|^2 \, dx \right)^{\frac{1}{2}} \|w\|_{L^2(\mathbb{R}^N)}$$

$$\leq c\eta \|w\|_{H^s(\mathbb{R}^N)}.$$

Set $\mathcal{A}_n = \{x \in \mathbb{R}^N \backslash B_R : |u_n(x)| \leq r_0\}$. In view of (5.2.9), the Hölder inequality shows that

$$\int_{\mathcal{A}_n \cap \{|u| \leq \delta\}} |f_1(u_n) - f_1(u_n - u)| |w| \, dx \leq \eta (\|u_n\|_{L^2(\mathbb{R}^N)} + \|u_n - u\|_{L^2(\mathbb{R}^N)}) \|w\|_{L^2(\mathbb{R}^N)}$$

$$\leq c\eta \|w\|_{H^s(\mathbb{R}^N)}. \tag{5.2.13}$$

Let $\mathcal{B}_n = \{x \in \mathbb{R}^N \backslash B_R : |u_n(x)| \geq r_1\}$. Then (5.2.10) and Hölder's inequality yield

$$\int_{\mathcal{B}_n \cap \{|u| \leq \delta\}} |f_1(u_n) - f_1(u_n - u)| |w| \, dx \leq \eta (\|u_n\|_{L^{2_s^*}(\mathbb{R}^N)}^{2_s^* - 1} + \|u_n - u\|_{L^{2_s^*}(\mathbb{R}^N)}^{2_s^* - 1}) \|w\|_{L^{2_s^*}(\mathbb{R}^N)}$$

$$\leq c\eta \|w\|_{H^s(\mathbb{R}^N)}. \tag{5.2.14}$$

Finally, define $\mathcal{C}_n = \{x \in \mathbb{R}^N \backslash B_R : r_0 \leq |u_n(x)| \leq r_1\}$. Since $u_n \in H^s(\mathbb{R}^N)$, we know that $|\mathcal{C}_n| < \infty$. Then (5.2.11) gives

$$\int_{\mathcal{C}_n \cap \{|u| \leq \delta\}} |f_1(u_n) - f_1(u_n - u)| |w| \, dx \leq r_0 \eta \|w\|_{L^2(\mathbb{R}^N)} |\mathcal{C}_n|^{\frac{1}{2}}$$

$$\leq \eta \|u_n\|_{L^2(\mathbb{R}^N)} \|w\|_{L^2(\mathbb{R}^N)}$$

$$\leq c\eta \|w\|_{H^s(\mathbb{R}^N)}. \tag{5.2.15}$$

Combining (5.2.13), (5.2.14), and (5.2.15), we have that

$$\int_{(\mathbb{R}^N \setminus B_R) \cap \{|u| \le \delta\}} \left| f_1(u_n) - f_1(u_n - u) \right| |w| \, dx \le c\eta \|w\|_{H^s(\mathbb{R}^N)} \quad \text{for all } n \in \mathbb{N}.$$

$$(5.2.16)$$

Now, we note that (5.2.12) implies

$$|f_1(u_n) - f_1(u_n - u)| \le \eta(|u_n|^{2_s^* - 1} + |u_n - u|^{2_s^* - 1}) + c(\eta)(|u_n| + |u_n - u|),$$

and consequently

$$\int_{(\mathbb{R}^N \setminus B_R) \cap \{|u| \ge \delta\}} |f_1(u_n) - f_1(u_n - u)| |w| \, dx$$

$$\le \int_{(\mathbb{R}^N \setminus B_R) \cap \{|u| \ge \delta\}} \left[\eta(|u_n|^{2_s^* - 1} + |u_n - u|^{2_s^* - 1})|w| + c(\eta)(|u_n| + |u_n - u|)|w| \right] dx$$

$$\le c\eta \|w\|_{H^s(\mathbb{R}^N)} + \int_{(\mathbb{R}^N \setminus B_R) \cap \{|u| \ge \delta\}} c(\eta)(|u_n| + |u_n - u|)|w| \, dx.$$

Since $u \in H^s(\mathbb{R}^N)$, we get $|(\mathbb{R}^N \setminus B_R) \cap \{u \ge \delta\}| \to 0$ as $R \to \infty$. Then choosing $R = R(\eta)$ large enough we infer that

$$\int_{(\mathbb{R}^N \setminus B_R) \cap \{|u| \ge \delta\}} c(\eta)(|u_n| + |u_n - u|)|w| \, dx$$

$$\le c(\eta) \left(\|u_n\|_{L^{2_s^*}(\mathbb{R}^N)} + \|u_n - u\|_{L^{2_s^*}(\mathbb{R}^N)} \right) \|w\|_{L^{2_s^*}(\mathbb{R}^N)} |(\mathbb{R}^N \setminus B_R) \cap \{|u| \ge \delta\}|^{\frac{2_s^* - 2}{2_s^*}}$$

$$\le \eta \|w\|_{H^s(\mathbb{R}^N)},$$

where we used the generalized Hölder inequality. Therefore

$$\int_{(\mathbb{R}^N \setminus B_R) \cap \{|u| \ge \delta\}} \left| f_1(u_n) - f_1(u_n - u) \right| |w| \, dx \le c\eta \|w\|_{H^s(\mathbb{R}^N)} \quad \text{for all } n \in \mathbb{N},$$

which combined with (5.2.16) yields

$$\int_{\mathbb{R}^N \setminus B_R} \left| f_1(u_n) - f_1(u) - f_1(u_n - u) \right| |w| \, dx \le c\eta \|w\|_{H^s(\mathbb{R}^N)} \quad \text{for all } n \in \mathbb{N}.$$

$$(5.2.17)$$

Now, recalling that $u_n \rightharpoonup u$ in $H^s(\mathbb{R}^N)$ we may assume that, up to a subsequence, $u_n \to u$ strongly in $L^2(B_R)$ and there exists $h \in L^2(B_R)$ such that $|u_n(x)|, |u(x)| \le |h(x)|$ a.e.

$x \in B_R$. It is clear that

$$\int_{B_R} |f_1(u_n - u)||w| \, dx \le c\eta \|w\|_{H^s(\mathbb{R}^N)} \tag{5.2.18}$$

as long as n is large enough. Set $\mathcal{D}_n = \{x \in B_R : |u_n(x) - u(x)| \ge 1\}$. Thus

$$\int_{\mathcal{D}_n} |f_1(u_n) - f_1(u)||w| \, dx \le \int_{\mathcal{D}_n} \left(c(\eta)(|u| + |u_n|) + \eta(|u_n|^{2^*_s - 1} + |u|^{2^*_s - 1}) \right) |w| \, dx$$

$$\le c\eta \|w\|_{H^s(\mathbb{R}^N)} + 2c(\eta) \int_{\mathcal{D}_n} |h||w| \, dx$$

$$\le c\eta \|w\|_{H^s(\mathbb{R}^N)} + 2c(\eta) \left(\int_{\mathcal{D}_n} |h|^2 \, dx \right)^{\frac{1}{2}} \|w\|_{L^2(\mathbb{R}^N)}.$$

Observing that $|\mathcal{D}_n| \to 0$ as $n \to \infty$, we deduce that

$$\int_{\mathcal{D}_n} |f_1(u_n) - f_1(u)||w| \, dx \le c\eta \|w\|_{H^s(\mathbb{R}^N)}. \tag{5.2.19}$$

Since $u \in H^s(\mathbb{R}^N)$, we know that $|\{|u| \ge L\}| \to 0$ as $L \to \infty$, so there exists $L = L(\eta) > 0$ such that for all n

$$\int_{(B_R \setminus \mathcal{D}_n) \cap \{|u| \ge L\}} |f_1(u_n) - f_1(u)||w| \, dx$$

$$\le \int_{(B_R \setminus \mathcal{D}_n) \cap \{|u| \ge L\}} \left[\eta(|u_n|^{2^*_s - 1} + |u|^{2^*_s - 1})|w| + c(\eta)(|u_n| + |u|)|w| \right] dx$$

$$\le c\eta \|w\|_{H^s(\mathbb{R}^N)} + c(\eta)(\|u_n\|_{L^{2^*_s}(\mathbb{R}^N)} + \|u\|_{L^{2^*_s}(\mathbb{R}^N)}) \|w\|_{L^{2^*_s}(\mathbb{R}^N)} |(B_R \setminus \mathcal{D}_n) \cap \{|u| \ge L\}|^{\frac{2^*_s - 2}{2^*_s}}$$

$$\le c\eta \|w\|_{H^s(\mathbb{R}^N)}. \tag{5.2.20}$$

On the other hand, by the dominated convergence theorem,

$$\int_{(B_R \setminus \mathcal{D}_n) \cap \{|u| \le L\}} |f_1(u_n) - f_1(u)|^2 \, dx \to 0 \quad \text{as } n \to \infty.$$

Consequently,

$$\int_{(B_R \setminus \mathcal{D}_n) \cap \{|u| \le L\}} |f_1(u_n) - f_1(u)||w| \, dx \le c\eta \|w\|_{H^s(\mathbb{R}^N)} \tag{5.2.21}$$

for n large enough. Combining (5.2.19), (5.2.20), and (5.2.21) we have

$$\int_{B_R} |f_1(u_n) - f_1(u)||w| \, dx \leq c\eta \|w\|_{H^s(\mathbb{R}^N)}.$$

This and (5.2.18) yield

$$\int_{B_R} |f_1(u_n) - f_1(u) - f_1(u_n - u)||w| \, dx \leq c\eta \|w\|_{H^s(\mathbb{R}^N)}. \qquad (5.2.22)$$

Finally, (5.2.17) and (5.2.22) imply that for n large enough

$$\int_{\mathbb{R}^N} |f_1(u_n) - f_1(u) - f_1(u_n - u)||w| \, dx \leq c\eta \|w\|_{H^s(\mathbb{R}^N)}.$$

$\square$

5.2.3 A Compactness Lemma when the Potential is Constant

In this subsection we provide the proof of Theorem 5.2.1. For simplicity, we may assume that $K = 1$ in $(f3)$. Consider the space

$$H = \left\{ u \in H^s(\mathbb{R}^N) : \int_{\mathbb{R}^N} V(x) u^2 \, dx < \infty \right\},$$

endowed with the norm

$$\|u\| = \left([u]_s^2 + \int_{\mathbb{R}^N} V(x) u^2 dx \right)^{\frac{1}{2}}.$$

Clearly, H is a Hilbert space with respect to the inner product

$$\langle u, v \rangle_H = \iint_{\mathbb{R}^{2N}} \frac{(u(x) - u(y))(v(x) - v(y))}{|x - y|^{N+2s}} \, dx dy + \int_{\mathbb{R}^N} V(x) uv \, dx \quad \forall u, v \in H.$$

Using the assumptions $(V2)$ and $(V3)$, it is easy to prove that $\|\cdot\|$ is equivalent to the standard norm in $H^s(\mathbb{R}^N)$. In what follows, we use the symbol H^* to denote the topological dual of H. The functional $\mathcal{I} : H \to \mathbb{R}$ associated with (5.2.1) is defined as

$$\mathcal{I}(u) = \frac{1}{2} \|u\|^2 - \int_{\mathbb{R}^N} F(u) \, dx.$$

By Theorem 1.1.8 and the assumptions on f, it is clear that $\mathcal{I}$ is well defined, $\mathcal{I} \in C^1(H, \mathbb{R})$, and the differential of $\mathcal{I}$ is given by

$$\langle \mathcal{I}'_\lambda(u), \varphi \rangle = \langle u, \varphi \rangle_H - \int_{\mathbb{R}^N} f(u)\varphi \, dx$$

for all $u, \varphi \in H$. Moreover, the critical points of $\mathcal{I}$ are weak solutions to (5.2.1). For $\lambda \in [\frac{1}{2}, 1]$, we introduce the family of functionals $\mathcal{I}_\lambda : H \to \mathbb{R}$ defined by

$$\mathcal{I}_\lambda(u) = \frac{1}{2}\|u\|^2 - \lambda \int_{\mathbb{R}^N} F(u) \, dx.$$

Clearly, $\mathcal{I}_1 = \mathcal{I}$. Let us prove that $\mathcal{I}_\lambda$ satisfies the assumptions of Theorem 5.2.2.

Lemma 5.2.7 *Assume that $(V1)$–$(V2)$ and $(f1)$–$(f4)$ hold. Then, for almost every $\lambda \in [\frac{1}{2}, 1]$, there is a sequence $(u_n) \subset H$ such that*

(i) *(u_n) is bounded;*
(ii) *$\mathcal{I}_\lambda(u_n) \to c_\lambda = \inf_{\gamma \in \Gamma} \max_{t \in [0,1]} \mathcal{I}_\lambda(\gamma(t))$, where $\Gamma = \{\gamma \in C([0,1], H) : \gamma(0) = 0, \gamma(1) = v_2\}$ for some $v_2 \in H \setminus \{0\}$ such that $\mathcal{I}_\lambda(v_2) < 0$ for all $\lambda \in [\frac{1}{2}, 1]$;*
(iii) *$\mathcal{I}'_\lambda(u_n) \to 0$ on H^*.*

Moreover, if $V \in L^\infty(\mathbb{R}^N)$, then

$$c_\lambda < \frac{s}{N} \frac{S_*^{\frac{N}{2s}}}{\lambda^{\frac{N-2s}{2s}}}. \tag{5.2.23}$$

Proof We aim to apply Theorem 5.2.2 with $X = H$, $\Lambda = [\frac{1}{2}, 1]$, $A(u) = \frac{1}{2}\|u\|^2$, and $B(u) = \int_{\mathbb{R}^N} F(u) \, dx$. Clearly, $A(u) \to \infty$ as $\|u\| \to \infty$, and by the assumption $(f4)$, it follows that $B(u) \geq 0$ for any $u \in H$. Now, assumptions $(f1)$–$(f3)$ show that for every $\varepsilon > 0$ there exists $C_\varepsilon > 0$ such that

$$|F(t)| \leq \varepsilon t^2 + C_\varepsilon |t|^{2_s^*} \quad \text{for all } t \in \mathbb{R}.$$

Then, since $\lambda \in [\frac{1}{2}, 1]$, Theorem 1.1.8 and $(V2)$ imply that

$$\mathcal{I}_\lambda(u) \geq \frac{1}{2}\|u\|^2 - \lambda \left[\varepsilon \|u\|^2_{L^2(\mathbb{R}^N)} + C_\varepsilon \|u\|^{2_s^*}_{L^{2_s^*}(\mathbb{R}^N)} \right]$$

$$\geq \frac{1}{2}\|u\|^2 - \frac{\varepsilon}{V_0}\|u\|^2 - C_\varepsilon S_*^{-\frac{2_s^*}{2}}\|u\|^{2_s^*},$$

so there exist $\alpha > 0$ and $r > 0$, independent of λ, such that

$$\mathcal{I}_\lambda(u) \geq \alpha > 0 \quad \text{for any } u \quad \text{with } \|u\| = r.$$

Condition $(f4)$ and the fact that $\lambda \in [\frac{1}{2}, 1]$ imply that

$$\mathcal{I}_\lambda(u) \leq \frac{1}{2}\|u\|^2 - \frac{N-2s}{4N}\|u^+\|^{2_s^*}_{L^{2_s^*}(\mathbb{R}^N)} - \frac{D}{2q}\|u^+\|^q_{L^q(\mathbb{R}^N)} \tag{5.2.24}$$

so, taking $\varphi \in H$ such that $\varphi \geq 0$ and $\varphi \neq 0$, we can see that $\mathcal{I}_\lambda(t u) \to -\infty$ as $t \to \infty$. Hence, there exists $t_0 > 0$ such that $\|t_0\varphi\| > r$ and $\mathcal{I}_\lambda(t_0\varphi) < 0$ for all $\lambda \in [\frac{1}{2}, 1]$. Since $\mathcal{I}_\lambda(0) = 0$, we set $v_1 = 0$ and $v_2 = t_0\varphi$. Therefore, $\mathcal{I}_\lambda$ satisfies the assumptions of Theorem 5.2.2, and we can find a bounded Palais–Smale sequence for $\mathcal{I}_\lambda$ at the level c_λ.

Finally, we prove the estimate in (5.2.23). Let $\eta \in C_c^\infty(\mathbb{R}^N)$ be a cut-off function such that $0 \leq \eta \leq 1$, $\eta = 1$ on B_r and $\eta = 0$ on $\mathbb{R}^N \setminus B_{2r}$, where B_r denotes the ball in $\mathbb{R}^N$ of radius r centered at the origin. As in [310], for $\varepsilon > 0$, we define $u_\varepsilon(x) = \eta(x)U_\varepsilon(x)$, where

$$U_\varepsilon(x) = \frac{\kappa \varepsilon^{-\frac{N-2s}{2}}}{\left(\mu^2 + \left|\frac{x}{\varepsilon\, S_*^{\frac{1}{2s}}}\right|^2\right)^{\frac{N-2s}{2}}}$$

and $\kappa \in \mathbb{R}$ and $\mu > 0$ are fixed constants. We recall (see Remark 1.1.9) that the value S_* is achieved by U_ε. Now we set

$$v_\varepsilon = \frac{u_\varepsilon}{\|u_\varepsilon\|_{L^{2_s^*}(\mathbb{R}^N)}}.$$

As proved in [174, 310], v_ε satisfies the following useful estimates:

$$[v_\varepsilon]_s^2 \leq S_* + O(\varepsilon^{N-2s}), \tag{5.2.25}$$

$$\|v_\varepsilon\|^2_{L^2(\mathbb{R}^N)} = \begin{cases} O(\varepsilon^{2s}), & \text{if } N > 4s, \\ O(\varepsilon^{2s}|\log(\varepsilon)|), & \text{if } N = 4s, \\ O(\varepsilon^{N-2s}), & \text{if } N < 4s, \end{cases} \tag{5.2.26}$$

and

$$\|v_\varepsilon\|^q_{L^q(\mathbb{R}^N)} = \begin{cases} O(\varepsilon^{\frac{2N-(N-2s)q}{2}}), & \text{if } q > \frac{N}{N-2s}, \\ O(\varepsilon^{\frac{(N-2s)q}{2}}), & \text{if } q < \frac{N}{N-2s}. \end{cases} \tag{5.2.27}$$

By the definition of c_λ,

$$c_\lambda \leq \sup_{t \geq 0} \mathcal{I}_\lambda(tv_\varepsilon). \tag{5.2.28}$$

Next, we consider the following function for $t \geq 0$

$$k(t) = \frac{t^2}{2} \|v_\varepsilon\|^2 - \frac{t^{2_s^*}}{2_s^*} \lambda.$$

We observe that $k(t)$ attains its maximum at $t_0 = (\lambda^{-1}\|v_\varepsilon\|^2)^{\frac{1}{2_s^*-2}}$ and

$$k(t_0) = \frac{s}{N} \frac{1}{\lambda^{\frac{N-2s}{2s}}} \|v_\varepsilon\|^{\frac{N}{s}} = \frac{s}{N} \frac{1}{\lambda^{\frac{N-2s}{2s}}} \left([v_\varepsilon]_s^2 + \int_{\mathbb{R}^N} V(x)v_\varepsilon^2 \, dx \right)^{\frac{N}{2s}}. \tag{5.2.29}$$

Note that there exists $\tau \in (0, 1)$ such that for $\varepsilon < 1$

$$\sup_{t \in [0,\tau]} \mathcal{I}_\lambda(tv_\varepsilon) \leq \sup_{t \in [0,\tau]} \frac{t^2}{2} \|v_\varepsilon\|^2 < \frac{s}{N} \frac{S_*^{\frac{N}{2s}}}{\lambda^{\frac{N-2s}{2s}}}. \tag{5.2.30}$$

On the other hand, in view of $(f4)$, $\lambda \in [\frac{1}{2}, 1]$, (5.2.25) and (5.2.29), we get

$$\sup_{t \geq \tau} \mathcal{I}_\lambda(tv_\varepsilon) \leq \sup_{t \geq 0} k(t) - \lambda \frac{D}{q} \tau^q \|v_\varepsilon\|_{L^q(\mathbb{R}^N)}^q$$

$$\leq \frac{s}{N} \frac{1}{\lambda^{\frac{N-2s}{2s}}} \left([v_\varepsilon]_s^2 + \int_{\mathbb{R}^N} V(x)v_\varepsilon^2 \, dx \right)^{\frac{N}{2s}} - \frac{D}{2q} \tau^q \|v_\varepsilon\|_{L^q(\mathbb{R}^N)}^q$$

$$\leq \frac{s}{N} \frac{1}{\lambda^{\frac{N-2s}{2s}}} \left(S_* + O(\varepsilon^{N-2s}) + \int_{\mathbb{R}^N} V(x)v_\varepsilon^2 \, dx \right)^{\frac{N}{2s}} - C_0 \|v_\varepsilon\|_{L^q(\mathbb{R}^N)}^q.$$

Using the elementary inequality $(a+b)^p \leq a^p + p(a+b)^{p-1}b$ for all $a, b > 0$ and $p \geq 1$, and the assumption that $V \in L^\infty(\mathbb{R}^N)$, we have

$$\sup_{t \geq \tau} \mathcal{I}_\lambda(tv_\varepsilon) \leq \frac{s}{N} \frac{S_*^{\frac{N}{2s}}}{\lambda^{\frac{N-2s}{2s}}} + O(\varepsilon^{N-2s}) + C_1 \int_{\mathbb{R}^N} V(x)v_\varepsilon^2 \, dx - C_0 \|v_\varepsilon\|_{L^q(\mathbb{R}^N)}^q$$

$$\leq \frac{s}{N} \frac{S_*^{\frac{N}{2s}}}{\lambda^{\frac{N-2s}{2s}}} + O(\varepsilon^{N-2s}) + C_2 \|v_\varepsilon\|_{L^2(\mathbb{R}^N)}^2 - C_0 \|v_\varepsilon\|_{L^q(\mathbb{R}^N)}^q.$$

We distinguish the following cases:

If $N > 4s$, then $q \in (2, 2_s^*)$ and, in particular, $q > \frac{N}{N-2s}$. Hence, by using (5.2.26) and (5.2.27), we can see that

$$\sup_{t \geq \tau} \mathcal{I}_\lambda(tv_\varepsilon) \leq \frac{s}{N} \frac{S_*^{\frac{N}{2s}}}{\lambda^{\frac{N-2s}{2s}}} + O(\varepsilon^{N-2s}) + O(\varepsilon^{2s}) - O(\varepsilon^{\frac{2N-(N-2s)q}{2}}).$$

Since

$$\frac{2N - (N - 2s)q}{2} < 2s < N - 2s,$$

there exists $\varepsilon_0 > 0$ such that for any $\varepsilon \in (0, \varepsilon_0)$,

$$\sup_{t \geq \tau} \mathcal{I}_\lambda(tv_\varepsilon) < \frac{s}{N} \frac{S_*^{\frac{N}{2s}}}{\lambda^{\frac{N-2s}{2s}}}. \qquad (5.2.31)$$

If now $N = 4s$, then $q \in (2, 4)$ and, in particular, $q > \frac{N}{N-2s} = 2$, so from (5.2.26) and (5.2.27) we deduce that

$$\sup_{t \geq \tau} \mathcal{I}_\lambda(tv_\varepsilon) \leq \frac{s}{N} \frac{S_*^{\frac{N}{2s}}}{\lambda^{\frac{N-2s}{2s}}} + O(\varepsilon^{2s}) + O(\varepsilon^{2s}|\log(\varepsilon)|) - O(\varepsilon^{4s-sq}).$$

Since

$$\lim_{\varepsilon \to 0} \frac{\varepsilon^{4s-sq}}{\varepsilon^{2s}(1 + |\log(\varepsilon)|)} = \infty,$$

for any ε sufficiently small we have

$$\sup_{t \geq \tau} \mathcal{I}_\lambda(tv_\varepsilon) < \frac{s}{N} \frac{S_*^{\frac{N}{2s}}}{\lambda^{\frac{N-2s}{2s}}}. \qquad (5.2.32)$$

Finally, if $2s < N < 4s$, then $q \in (\frac{4s}{N-2s}, 2_s^*)$ and, in particular, $q > \frac{N}{N-2s}$. Hence, observing that $\frac{2N-(N-2s)q}{2} < N - 2s$, we get

$$\sup_{t \geq \tau} \mathcal{I}_\lambda(tv_\varepsilon) \leq \frac{s}{N} \frac{S_*^{\frac{N}{2s}}}{\lambda^{\frac{N-2s}{2s}}} + O(\varepsilon^{N-2s}) + O(\varepsilon^{N-2s}) - O(\varepsilon^{\frac{2N-(N-2s)q}{2}}) < \frac{s}{N} \frac{S_*^{\frac{N}{2s}}}{\lambda^{\frac{N-2s}{2s}}}.$$

$$(5.2.33)$$

for any $\varepsilon > 0$ small enough. Combining (5.2.28), (5.2.30) and (5.2.31)–(5.2.33), we can conclude that (5.2.23) holds.

$\square$

Remark 5.2.8 Let us note that for $\lambda \in [\frac{1}{2}, 1]$, if $(u_n) \subset H$ is such that

$$\|u_n\| \leq C, \quad \mathcal{I}_\lambda(u_n) \to c_\lambda, \quad \mathcal{I}'_\lambda(u_n) \to 0 \text{ in } H^*,$$

then we may assume that $u_n \geq 0$. In fact, using the equality $\langle \mathcal{I}'_\lambda(u_n), u_n^- \rangle = \langle \mu_n, u_n^- \rangle$, where $u^- = \min\{u, 0\}$ and $\mu_n \to 0$ in H^*, and the fact that $f(t) = 0$ if $t \leq 0$, we can infer that

$$\int_{\mathbb{R}^N} \left[(-\Delta)^{\frac{s}{2}} u_n (-\Delta)^{\frac{s}{2}} u_n^- + V(x)(u_n^-)^2 \right] dx = \langle \mu_n, u_n^- \rangle.$$

On the other hand, using the fact that

$$|x - y|(x^- - y^-) \geq |x^- - y^-|^2 \quad \forall x, y \in \mathbb{R},$$

we see that

$$\int_{\mathbb{R}^N} (-\Delta)^{\frac{s}{2}} u_n (-\Delta)^{\frac{s}{2}} u_n^- \, dx = \iint_{\mathbb{R}^{2N}} \frac{(u_n(x) - u_n(y))(u_n^-(x) - u_n^-(y))}{|x - y|^{N+2s}} \, dx dy$$

$$\geq \iint_{\mathbb{R}^{2N}} \frac{|u_n^-(x) - u_n^-(y)|^2}{|x - y|^{N+2s}} \, dx dy = [u_n^-]_s^2.$$

Then, $\|u_n^-\| = o_n(1)$, which also yields that $\|u_n^+\| \leq C$. Now, we prove that $\mathcal{I}_\lambda(u_n^+) \to c_\lambda$ and $\mathcal{I}'_\lambda(u_n^+) \to 0$ in H^* as $n \to \infty$. Clearly, $\|u_n\|^2 = \|u_n^+\|^2 + o_n(1)$. Using the conditions $(f2)$–$(f3)$, the mean value theorem, the decomposition $u_n = u_n^+ + u_n^-$, and Hölder's inequality, we deduce that

$$\left| \int_{\mathbb{R}^N} F(u_n) \, dx - \int_{\mathbb{R}^N} F(u_n^+) \, dx \right| \leq C \int_{\mathbb{R}^N} (|u_n| + |u_n|^{2_s^*-1})|u_n^-| \, dx$$

$$\leq C(\|u_n^-\|_{L^2(\mathbb{R}^N)} + \|u_n^-\|_{L^{2_s^*}(\mathbb{R}^N)}) = o_n(1)$$

and this shows that $\mathcal{I}_\lambda(u_n^+) \to c_\lambda$. We claim that $\mathcal{I}'_\lambda(u_n^+) \to 0$ in H^*. Fix $\varphi \in H$ such that $\|\varphi\| \leq 1$. Then, since $u_n^- = u_n - u_n^+$, we have

$$\langle \mathcal{I}'_\lambda(u_n), \varphi \rangle - \langle \mathcal{I}_\lambda(u_n^+), \varphi \rangle$$

$$= \int_{\mathbb{R}^N} (-\Delta)^{\frac{s}{2}} u_n (-\Delta)^{\frac{s}{2}} \varphi \, dx - \int_{\mathbb{R}^N} (-\Delta)^{\frac{s}{2}} u_n^+ (-\Delta)^{\frac{s}{2}} \varphi \, dx$$

$$+ \int_{\mathbb{R}^N} V(x) u_n \varphi \, dx - \int_{\mathbb{R}^N} V(x) u_n^+ \varphi \, dx - \lambda \int_{\mathbb{R}^N} (f(u_n) - f(u_n^+)) \varphi \, dx$$

$$= \int_{\mathbb{R}^N} (-\Delta)^{\frac{s}{2}} u_n^- (-\Delta)^{\frac{s}{2}} \varphi \, dx + \int_{\mathbb{R}^N} V(x) u_n^- \varphi \, dx - \lambda \int_{\mathbb{R}^N} f(u_n^-) \varphi \, dx + \langle \mu_n, \varphi \rangle,$$

for some $\mu_n \to 0$ in H^*. Therefore, by $(f2)$–$(f3)$, $(V2)$, $\|\varphi\| \leq 1$, and Hölder's inequality, we get

$$\left| \langle \mathcal{I}'_\lambda(u_n), \varphi \rangle - \langle \mathcal{I}_\lambda(u_n^+), \varphi \rangle \right|$$

$$\leq [u_n^-]_s [\varphi]_s + \|\sqrt{V} u_n^-\|_{L^2(\mathbb{R}^N)} \|\sqrt{V} \varphi\|_{L^2(\mathbb{R}^N)} + \lambda [C_1 \|u_n^-\|_{L^2(\mathbb{R}^N)} \|\varphi\|_{L^2(\mathbb{R}^N)}$$

$$+ C_2 \|u_n^-\|_{L^{2_s^*}(\mathbb{R}^N)}^{2_s^*-1} \|\varphi\|_{L^{2_s^*}(\mathbb{R}^N)}] + \|\mu_n\|_* \|\varphi\|$$

$$\leq C_3 \|u_n^-\| + C_4 \|u_n^-\|^{2_s^*-1} + \|\mu_n\|_* = o_n(1).$$

Since $\mathcal{I}'_\lambda(u_n) \to 0$ in H^*, we can deduce that $\mathcal{I}'_\lambda(u_n^+) \to 0$ in H^*.

Arguing as in Theorem 3.5.1 (see also [307]), we obtain the following fractional Pohozaev identity:

Lemma 5.2.9 *For $\lambda \in [\frac{1}{2}, 1]$, if u_λ is a critical point of $\mathcal{I}_\lambda$, then u_λ satisfies the following Pohozaev identity:*

$$\frac{N - 2s}{2} [u_\lambda]_s^2 + \frac{1}{2} \int_{\mathbb{R}^N} \nabla V(x) \cdot x \, u_\lambda^2 \, dx = N \int_{\mathbb{R}^N} \left[\lambda F(u) - \frac{1}{2} V(x) u_\lambda^2 \right] dx. \quad (5.2.34)$$

Remark 5.2.10 It is easy to check that if $(V1)$–$(V2)$ and $(f1)$–$(f3)$ hold, then there exists $\beta > 0$ independent of $\lambda \in [\frac{1}{2}, 1]$ such that any nontrivial critical point u_λ of $\mathcal{I}_\lambda$ satisfies $\|u_\lambda\| \geq \beta > 0$. In fact, by using $(f1)$–$(f3)$, we can see that for every $\varepsilon > 0$ there exists $C_\varepsilon > 0$ such that

$$|F(t)| \leq \varepsilon t^2 + C_\varepsilon |t|^{2_s^*} \quad \text{for all } t \in \mathbb{R}.$$

Taking into account that $\langle \mathcal{I}'_\lambda(u_\lambda), u_\lambda \rangle = 0$, $\lambda \leq 1$, $F \geq 0$, Theorem 1.1.8, $(V1)$–$(V2)$, we have

$$\|u_\lambda\|^2 = \lambda \int_{\mathbb{R}^N} F(u_\lambda) \, dx \leq \varepsilon C_1 \|u_\lambda\|^2 + C_\varepsilon C_2 \|u_\lambda\|^{2_s^*}$$

where $C_1, C_2 > 0$ depend only on V_0 and the best constant S_*. Choosing $\varepsilon > 0$ sufficiently small and by using that $u_\lambda \neq 0$, we deduce that there exists $\beta > 0$ such that $\|u_\lambda\| \geq \beta > 0$.

Now, we establish the following compactness lemma which will be useful in proving Theorem 5.2.1.

Lemma 5.2.11 *Assume that $V(x) \equiv V$ and f satisfies $(f1)$–$(f4)$. For $\lambda \in [\frac{1}{2}, 1]$, let $(u_n) \subset H$ be a bounded sequence in H such that $u_n \geq 0$, $\mathcal{I}_\lambda(u_n) \to c_\lambda$, $\mathcal{I}'_\lambda(u_n) \to 0$ in H^*. Moreover,*

$$c_\lambda < \frac{s}{N} \frac{S_*^{\frac{N}{2s}}}{\lambda^{\frac{N-2s}{2s}}}.$$

Then there exist a subsequence of (u_n), which we denote again by (u_n), and an integer $k \in \mathbb{N} \cup \{0\}$ and $w_\lambda^j \in H$ for $1 \leq j \leq k$, such that

(i) $u_n \rightharpoonup u_\lambda$ in H and $\mathcal{I}'_\lambda(u_\lambda) = 0$,
(ii) $w_\lambda^j \neq 0$ and $\mathcal{I}'_\lambda(w_\lambda^j) = 0$ for $1 \leq j \leq k$,
(iii) $c_\lambda = \mathcal{I}_\lambda(u_\lambda) + \sum_{j=1}^k \mathcal{I}_\lambda(w_\lambda^j)$,

where we agree that in the case $k = 0$, the above holds without w_λ^j.

Proof We divide the proof into five steps.

Step 1 Passing to a subsequence if necessary, we can assume that $u_n \rightharpoonup u_\lambda$ in H, where u_λ is a critical point of $\mathcal{I}_\lambda$.

Since (u_n) is bounded in H, up to a subsequence, we can suppose that $u_n \rightharpoonup u_\lambda$ in H, and, in view of Theorem 1.1.8, $u_n \to u_\lambda$ in $L_{loc}^r(\mathbb{R}^N)$ for all $r \in [1, 2_s^*)$. Then, for any $\varphi \in C_c^\infty(\mathbb{R}^N)$,

$$\langle \mathcal{I}'_\lambda(u_n), \varphi \rangle - \langle \mathcal{I}'_\lambda(u_\lambda), \varphi \rangle = \langle u_n - u_\lambda, \varphi \rangle_H - \lambda \int_{\mathbb{R}^N} [g(u_n) - g(u)]\varphi \, dx$$

$$- \lambda \int_{\mathbb{R}^N} (|u_n|^{2_s^*-2} u_n - |u_\lambda|^{2_s^*-2} u_\lambda)\varphi \, dx,$$

(5.2.35)

where $g(t) = f(t) - (t^+)^{2_s^*-1}$. Since $u_n \rightharpoonup u_\lambda$ in H, we get

$$\langle u_n - u_\lambda, \varphi \rangle_H \to 0. \tag{5.2.36}$$

Moreover, the sequence $(|u_n|^{2_s^*-2} u_n - |u_\lambda|^{2_s^*-2} u_\lambda)$ is bounded in $L^{\frac{2_s^*}{2_s^*-1}}(\mathbb{R}^N)$ and

$$|u_n|^{2_s^*-2} u_n \to |u_\lambda|^{2_s^*-2} u_\lambda \quad \text{a.e. in } \mathbb{R}^N,$$

so we conclude that

$$\int_{\mathbb{R}^N} (|u_n|^{2^*_s-2}u_n - |u_\lambda|^{2^*_s-2}u_\lambda)\varphi \, dx \to 0. \tag{5.2.37}$$

Using Lemma 1.4.2 and conditions $(f2)$ and $(f3)$, we infer that

$$\int_{\mathbb{R}^N} [g(u_n) - g(u)]\varphi \, dx \to 0. \tag{5.2.38}$$

Combining (5.2.35), (5.2.36), (5.2.37), (5.2.38) and the fact that $\mathcal{I}'_\lambda(u_n) \to 0$, we see that $\langle \mathcal{I}'_\lambda(u_\lambda), \varphi \rangle = 0$ for any $\varphi \in C_c^\infty(\mathbb{R}^N)$. Since $C_c^\infty(\mathbb{R}^N)$ is dense in $H^s(\mathbb{R}^N)$, we deduce that $\mathcal{I}'_\lambda(u_\lambda) = 0$, that is, (i) is satisfied. Now, we set $v_n^1 = u_n - u_\lambda$.

Step 2 If $\lim_{n\to\infty} \sup_{z\in\mathbb{R}^N} \int_{B_1(z)} |v_n^1|^2 dx = 0$, then $u_n \to u_\lambda$ in H and Lemma 5.2.11 holds with $k = 0$.

In view of Lemma 1.4.4,

$$v_n^1 \to 0 \text{ in } L^t(\mathbb{R}^N), \quad \forall t \in (2, 2^*_s). \tag{5.2.39}$$

Now, we observe that

$$\langle \mathcal{I}'_\lambda(u_n), v_n^1 \rangle = \langle u_n, v_n^1 \rangle_H - \lambda \int_{\mathbb{R}^N} f(u_n)v_n^1 \, dx = \|v_n^1\|^2 + \langle u_\lambda, v_n^1 \rangle_H - \lambda \int_{\mathbb{R}^N} f(u_n)v_n^1 \, dx,$$

that is,

$$\|v_n^1\|^2 = \langle \mathcal{I}'_\lambda(u_n), v_n^1 \rangle - \langle u_\lambda, v_n^1 \rangle_H + \lambda \int_{\mathbb{R}^N} f(u_n)v_n^1 \, dx.$$

By $\langle \mathcal{I}'_\lambda(u_\lambda), v_n^1 \rangle = 0$, $\langle \mathcal{I}'_\lambda(u_n), v_n^1 \rangle = o(1)$ and the definition of g, we have

$$\|v_n^1\|^2 = \langle \mathcal{I}'_\lambda(u_n), v_n^1 \rangle + \lambda \int_{\mathbb{R}^N} (f(u_n) - f(u_\lambda))v_n^1 dx$$

$$= \lambda \int_{\mathbb{R}^N} (g(u_n) - g(u_\lambda))v_n^1 dx + \lambda \int_{\mathbb{R}^N} (|u_n|^{2^*_s-2}u_n - |u_\lambda|^{2^*_s-2}u_\lambda)v_n^1 dx + o(1).$$

Now, by $(f1)$–$(f3)$, we know that for every $\varepsilon > 0$ there exists $C_\varepsilon > 0$ such that

$$|g(t)| \le \varepsilon(|t| + |t|^{2^*_s-1}) + C_\varepsilon |t|^{q-1} \quad \text{for all } t \in \mathbb{R}. \tag{5.2.40}$$

Therefore, taking into account (5.2.39) and (5.2.40), we see that

$$\|v_n^1\|^2 = \lambda \int_{\mathbb{R}^N} (|u_n|^{2_s^*-2}u_n - |u_\lambda|^{2_s^*-2}u_\lambda)v_n^1 \, dx + o(1).$$

By Lemma 5.2.4, it follows that

$$\left| \int_{\mathbb{R}^N} [|u_n|^{2_s^*-2}u_n - |u_\lambda|^{2_s^*-2}u_\lambda - |u_n - u_\lambda|^{2_s^*-2}(u_n - u_\lambda)]\varphi \, dx \right| = o(1)\|\varphi\|, \quad \forall \varphi \in H.$$
$$(5.2.41)$$

Taking $\varphi = v_n^1 = u_n - u_\lambda$ in (5.2.41), we deduce that

$$\|v_n^1\|^2 = \lambda \int_{\mathbb{R}^N} |v_n^1|^{2_s^*} \, dx + o(1). \tag{5.2.42}$$

Now Brezis-Lieb lemma [113] yields

$$\|v_n^1\|^2 = \|u_n\|^2 - \|u_\lambda\|^2 + o(1) \tag{5.2.43}$$

and

$$\int_{\mathbb{R}^N} |v_n^1|^{2_s^*} \, dx = \int_{\mathbb{R}^N} |u_n|^{2_s^*} \, dx - \int_{\mathbb{R}^N} |u_\lambda|^{2_s^*} \, dx + o(1), \tag{5.2.44}$$

and applying Lemma 5.2.5 we obtain

$$\int_{\mathbb{R}^N} G(v_n^1) \, dx = \int_{\mathbb{R}^N} G(u_n) \, dx - \int_{\mathbb{R}^N} G(u_\lambda) \, dx + o(1). \tag{5.2.45}$$

Next, combining (5.2.43)–(5.2.45), we have

$$c_\lambda - \mathcal{I}_\lambda(u_\lambda) \tag{5.2.46}$$
$$= \mathcal{I}_\lambda(u_n) - \mathcal{I}_\lambda(u_\lambda) + o(1)$$
$$= \left[\frac{1}{2}\|u_n\|^2 - \lambda \int_{\mathbb{R}^N} G(u_n) \, dx - \frac{\lambda}{2_s^*}\|u_n\|_{L^{2_s^*}(\mathbb{R}^N)}^{2_s^*} \right]$$
$$\quad - \left[\frac{1}{2}\|u_\lambda\|^2 - \lambda \int_{\mathbb{R}^N} G(u_\lambda) \, dx - \frac{\lambda}{2_s^*}\|u_\lambda\|_{L^{2_s^*}(\mathbb{R}^N)}^{2_s^*} \right] + o(1)$$
$$= \frac{1}{2}\|v_n^1\|^2 - \lambda \int_{\mathbb{R}^N} G(v_n^1) \, dx - \frac{\lambda}{2_s^*}\|v_n^1\|_{L^{2_s^*}(\mathbb{R}^N)}^{2_s^*} + o(1). \tag{5.2.47}$$

Using (5.2.39), (5.2.40), and (5.2.46) we infer that

$$c_\lambda - \mathcal{I}_\lambda(u_\lambda) = \frac{1}{2}\|v_n^1\|^2 - \frac{\lambda}{2_s^*}\int_{\mathbb{R}^N} |v_n^1|^{2_s^*}\, dx + o(1). \tag{5.2.48}$$

Since $\mathcal{I}_\lambda'(u_\lambda) = 0$, it follows from Lemma 5.2.9 that $\mathcal{I}_\lambda(u_\lambda) = \frac{s}{N}[u_\lambda]_s^2 \geq 0$. Then, in view of (5.2.48), we get

$$c_\lambda - \mathcal{I}_\lambda(u_\lambda) < \frac{s}{N}\frac{S_*^{\frac{N}{2s}}}{\lambda^{\frac{N-2s}{2s}}}.$$

Now, we may assume that $\|v_n^1\|^2 \to L \geq 0$. By (5.2.42), it follows that $\lambda\|v_n^1\|_{L^{2_s^*}(\mathbb{R}^N)}^{2_s^*} \to L$.

Suppose that $L > 0$. Then, by Theorem 1.1.8,

$$\|v_n^1\|_{L^{2_s^*}(\mathbb{R}^N)}^2 S_* \leq \|v_n^1\|^2,$$

so we can deduce that $L \geq \frac{S_*^{\frac{N}{2s}}}{\lambda^{\frac{N-2s}{2s}}}$. This fact and (5.2.48) yield

$$c_\lambda - \mathcal{I}_\lambda(u_\lambda) = \frac{s}{N}L \geq \frac{s}{N}\frac{S_*^{\frac{N}{2s}}}{\lambda^{\frac{N-2s}{2s}}}$$

which gives a contradiction. Hence, $\|v_n^1\| \to 0$ as $n \to \infty$.

Step 3 If there exists $(z_n) \subset \mathbb{R}^N$ such that $\int_{B_1(z_n)} |v_n^1|^2\, dx \to d > 0$, then, up to a subsequence, the following conditions hold:

(1) $|z_n| \to \infty$,
(2) $u_n(\cdot + z_n) \rightharpoonup w_\lambda \neq 0$ in H,
(3) $\mathcal{I}_\lambda'(w_\lambda) = 0$.

We may assume that there exists $(z_n) \subset \mathbb{R}^N$ such that

$$\int_{B_1(z_n)} |v_n^1|^2\, dx \geq \frac{d}{2} > 0.$$

Set $\tilde{v}_n^1(x) = v_n^1(x+z_n)$. Then $\tilde{v}_n^1$ is bounded in H and we may assume that $\tilde{v}_n^1 \rightharpoonup \tilde{v}^1$ in H.

Since

$$\int_{B_1} |\tilde{v}_n^1|^2\, dx \geq \frac{d}{2},$$

we get

$$\int_{B_1} |\tilde{v}^1|^2 \, dx \geq \frac{d}{2},$$

that is, $\tilde{v}^1 \neq 0$. Using the fact that $v_n^1 \rightharpoonup 0$ in H, we deduce that (z_n) is unbounded, so we may assume that $|z_n| \to \infty$. Now, we set $\tilde{u}_n(x) = u_n(x + z_n) \rightharpoonup w_\lambda \neq 0$. As in Step 1, we can see that $\langle \mathcal{I}_\lambda'(\tilde{u}_n), \varphi \rangle - \langle \mathcal{I}_\lambda'(w_\lambda), \varphi \rangle \to 0$, for all $\varphi \in C_c^\infty(\mathbb{R}^N)$. On the other hand, since $|z_n| \to \infty$, we have

$$\langle \mathcal{I}_\lambda'(\tilde{u}_n), \varphi \rangle = \langle \mathcal{I}_\lambda'(u_n), \varphi(\cdot - z_n) \rangle \to 0$$

for all $\varphi \in C_c^\infty(\mathbb{R}^N)$, so we conclude that $\langle \mathcal{I}_\lambda'(w_\lambda), \varphi \rangle = 0$, for all $\varphi \in C_c^\infty(\mathbb{R}^N)$.

Step 4 If there exist $m \geq 1$, $(y_n^k) \subset \mathbb{R}^N$, $w_\lambda^k \in H$ for $1 \leq k \leq m$ such that
   (i) $|y_n^k| \to \infty$, $|y_n^k - y_n^h| \to \infty$ if $k \neq h$,
   (ii) $u_n(\cdot + y_n^k) \rightharpoonup w_\lambda^k \neq 0$ in H, for any $1 \leq k \leq m$,
   (iii) $w_\lambda^k \geq 0$ and $\mathcal{I}_\lambda'(w_\lambda^k) = 0$ for any $1 \leq k \leq m$,
then one of the following conclusions must hold:
(1) If $\sup_{z \in \mathbb{R}^N} \int_{B_1(z)} \left| u_n - u_0 - \sum_{k=1}^m w_\lambda^k(\cdot - y_n^k) \right|^2 dx \to 0$, then

$$\left\| u_n - u_0 - \sum_{k=1}^m w_\lambda^k(\cdot - y_n^k) \right\| \to 0.$$

(2) If there exists $(z_n) \subset \mathbb{R}^N$ such that

$$\int_{B_1(z_n)} \left| u_n - u_0 - \sum_{k=1}^m w_\lambda^k(\cdot - y_n^k) \right|^2 dx \to d > 0,$$

then up to a subsequence, the following conditions hold:
   (i) $|z_n| \to \infty$, $|z_n - y_n^k| \to \infty$ for any $1 \leq k \leq m$,
   (ii) $u_n(\cdot + z_n) \rightharpoonup w_\lambda^{m+1} \neq 0$ in H,
   (iii) $w_\lambda^{m+1} \geq 0$ and $\mathcal{I}_\lambda'(w_\lambda^{m+1}) = 0$.

Assume that (1) holds. Set $\xi_n = u_n - u_0 - \sum_{k=1}^m w_\lambda^k(\cdot - y_n^k)$. Then, using Lemma 1.4.4, we see that

$$\xi_n \to 0 \quad \text{in } L^t(\mathbb{R}^N) \text{ for all } t \in (2, 2_s^*). \tag{5.2.49}$$

By the definition of ξ_n and the fact that $\langle \mathcal{I}_\lambda'(u_\lambda), \xi_n \rangle = 0 = \langle \mathcal{I}_\lambda'(w_\lambda^k), \xi_n \rangle$,

$$\|\xi_n\|^2 = \langle \mathcal{I}_\lambda'(u_n), \xi_n \rangle + \lambda \int_{\mathbb{R}^N} \left(f(u_n) - f(u_\lambda) \right) \xi_n \, dx - \lambda \sum_{k=1}^m \int_{\mathbb{R}^N} f(w_\lambda^k) \xi_n(\cdot + y_n^k) \, dx.$$

In view of (5.2.40) and (5.2.49), we deduce that

$$\|\xi_n\|^2 = \lambda \int_{\mathbb{R}^N} \left(|u_n|^{2_s^*-2} u_n - |u_\lambda|^{2_s^*-2} u_\lambda \right) \xi_n \, dx - \lambda \sum_{k=1}^{m} \int_{\mathbb{R}^N} |w_\lambda^k|^{2_s^*-2} w_\lambda^k \xi_n(\cdot + y_n^k) \, dx + o(1).$$

Recalling (5.2.41), we observe that

$$\|\xi_n\|^2 = \lambda \int_{\mathbb{R}^N} |u_n - u_\lambda|^{2_s^*-2}(u_n - u_\lambda)\xi_n \, dx - \lambda \int_{\mathbb{R}^N} |w_\lambda^1|^{2_s^*-2} w_\lambda^1 \xi_n(\cdot + y_n^1) \, dx$$

$$- \lambda \sum_{k=2}^{m} \int_{\mathbb{R}^N} |w_\lambda^k|^{2_s^*-2} w_\lambda^k \xi_n(\cdot + y_n^k) \, dx + o(1)$$

$$= \lambda \int_{\mathbb{R}^N} \left| u_n(\cdot + y_n^1) - u_\lambda(\cdot + y_n^1) \right|^{2_s^*-2} \left(u_n(\cdot + y_n^1) - u_\lambda(\cdot + y_n^1) \right) \xi_n(\cdot + y_n^1) \, dx$$

$$- \lambda \int_{\mathbb{R}^N} |w_\lambda^1|^{2_s^*-2} w_\lambda^1 \xi_n(\cdot + y_n^1) \, dx - \lambda \sum_{k=2}^{m} \int_{\mathbb{R}^N} |w_\lambda^k|^{2_s^*-2} w_\lambda^k \xi_n(\cdot + y_n^k) \, dx + o(1).$$

Since $|y_n^1| \to \infty$, and $u_n(\cdot + y_n^1) \rightharpoonup w_\lambda^1$ in H, we deduce that $u_n(\cdot + y_n^1) - u_\lambda(\cdot + y_n^1) \rightharpoonup w_\lambda^1$ in H. Consequently,

$$\|\xi_n\|^2 = \lambda \int_{\mathbb{R}^N} \left| u_n(\cdot + y_n^1) - u_\lambda(\cdot + y_n^1) - w_\lambda^1 \right|^{2_s^*-2} \left(u_n(\cdot + y_n^1) - u_\lambda(\cdot + y_n^1) - w_\lambda^1 \right) \xi_n(\cdot + y_n^1) \, dx$$

$$- \lambda \sum_{k=2}^{m} \int_{\mathbb{R}^N} |w_\lambda^k|^{2_s^*-2} w_\lambda^k \xi_n(\cdot + y_n^k) \, dx + o(1).$$

Iterating this procedure, we conclude that

$$\|\xi_n\|^2 = \lambda \int_{\mathbb{R}^N} |\xi_n|^{2_s^*} \, dx + o(1). \qquad (5.2.50)$$

Now, since $u_n(\cdot + y_n^1) - u_\lambda(\cdot + y_n^1) \rightharpoonup w_\lambda^1$ in H, we can argue as in Step 2 to see that

$$c_\lambda - \mathcal{I}_\lambda(u_\lambda) = \frac{1}{2}\|u_n - u_\lambda\|^2 - \lambda \int_{\mathbb{R}^N} G(u_n - u_\lambda) \, dx - \frac{\lambda}{2_s^*} \int_{\mathbb{R}^N} |u_n - u_\lambda|^{2_s^*} \, dx + o(1)$$

$$= \frac{1}{2}\|u_n(\cdot + y_n^1) - u_\lambda(\cdot + y_n^1) - w_\lambda^1\|^2 - \lambda \int_{\mathbb{R}^N} G(u_n(\cdot + y_n^1) - u_\lambda(\cdot + y_n^1) - w_\lambda^1) \, dx$$

$$- \frac{\lambda}{2_s^*} \int_{\mathbb{R}^N} \left| u_n(\cdot + y_n^1) - u_\lambda(\cdot + y_n^1) - w_\lambda^1 \right|^{2_s^*} \, dx + \mathcal{I}_\lambda(w_\lambda^1) + o(1).$$

Continuing this process, we obtain that

$$c_\lambda - \mathcal{I}_\lambda(u_\lambda) - \sum_{k=1}^m \mathcal{I}_\lambda(w_\lambda^k) = \frac{1}{2}\|\xi_n\|^2 - \lambda \int_{\mathbb{R}^N} G(\xi_n)\,dx - \frac{\lambda}{2_s^*}\int_{\mathbb{R}^N} |\xi_n|^{2_s^*}\,dx + o(1),$$

$$(5.2.51)$$

which together with (5.2.49) yields

$$c_\lambda - \mathcal{I}_\lambda(u_\lambda) - \sum_{k=1}^m \mathcal{I}_\lambda(w_\lambda^k) = \frac{1}{2}\|\xi_n\|^2 - \frac{\lambda}{2_s^*}\int_{\mathbb{R}^N} |\xi_n|^{2_s^*}\,dx + o(1). \qquad (5.2.52)$$

Then, taking into account (5.2.50) and (5.2.52), we can argue as in Step 2 to infer that

$$\left\| u_n - u_0 - \sum_{k=1}^m w_\lambda^k(\cdot - y_n^k) \right\| = \|\xi_n\| \to 0 \quad \text{as } n \to \infty.$$

Now, we assume that (2) holds. The proof of this is standard (see [236]), so we skip the details here.

Step 5 Conclusion.

Using Step 1, we can verify that Lemma 5.2.11 (i) holds. If the assumption of Step 2 holds, then Lemma 5.2.11 holds with $k = 0$. Otherwise, the assumption of Step 3 holds. We set $(y_n^1) = (z_n)$ and $w_\lambda^1 = w_\lambda$. Now, if (1) of Step 4 holds with $m = 1$, from (5.2.52), we obtain the conclusion of Lemma 5.2.11. Otherwise, item (2) of Step 4 must hold, and by setting $(y_n^2) = (z_n)$, $w_\lambda^2 = w_\lambda^2$, we iterate Step 4. Then, to conclude the proof, we have to show that item (1) of Step 4 must occur after a finite number of iterations. Note that, for all $m \geq 1$, we have

$$\lim_{n\to\infty} \left(\|u_n\|^2 - \|u_\lambda\|^2 - \sum_{k=1}^m \|w_\lambda^k\|^2 \right) = \lim_{n\to\infty} \left\| u_n - u_\lambda - \sum_{k=1}^m w_\lambda^k(\cdot - y_n^k) \right\|^2 \geq 0.$$

In fact, by using items (i) and (ii) of Step 4 and the fact that $u_n \rightharpoonup u_\lambda$ in H, we see that

$$\left\| u_n - u_\lambda - \sum_{k=1}^m w_\lambda^k(\cdot - y_n^k) \right\|^2$$

$$= \|u_n\|^2 + \|u_\lambda\|^2 + \sum_{k=1}^m \|w_\lambda^k\|^2 - 2\langle u_n, u_\lambda \rangle_H - 2\sum_{k=1}^m \langle u_n, w_\lambda^k(\cdot - y_n^k) \rangle_H$$

$$+ 2 \sum_{k=1}^{m} \langle u_\lambda, w_\lambda^k (\cdot - y_n^k) \rangle_H + 2 \sum_{h,k} \langle w_\lambda^h (\cdot - y_n^h), w_\lambda^k (\cdot - y_n^k) \rangle_H$$

$$= \|u_n\|^2 - \|u_\lambda\|^2 - \sum_{k=1}^{m} \|w_\lambda^k\|^2 + o(1). \tag{5.2.53}$$

On the other hand, by Remark 5.2.10, $\|w_\lambda^k\| \geq \beta$ for some $\beta > 0$ that does not depend on λ. Thus, using (5.2.53) and the fact that (u_n) is bounded in H, we deduce that (1) in Step 4 must occur after a finite number of iterations. This together with (5.2.52) shows that Lemma 5.2.11 holds.

□

Before giving the proof of the main result of this subsection, we prove the following lemma.

Lemma 5.2.12 *Under the assumptions of Theorem 5.2.1, for almost every $\lambda \in [\frac{1}{2}, 1]$, $\mathcal{I}_\lambda$ has a positive critical point.*

Proof By Lemma 5.2.7, for almost every $\lambda \in [\frac{1}{2}, 1]$ there exists a bounded sequence $(u_n) \subset H$ such that

$$\mathcal{I}_\lambda(u_n) \to c_\lambda, \quad \mathcal{I}_\lambda'(u_n) \to 0 \quad \text{in } H^*. \tag{5.2.54}$$

In the light of Remark 5.2.8, we may assume that $u_n \geq 0$ in H. In addition,

$$c_\lambda \in \left(0, \frac{s}{N} \frac{S_*^{\frac{N}{2s}}}{\lambda^{\frac{N-2s}{2s}}} \right).$$

Then, up to a subsequence, we may assume that $u_n \rightharpoonup u_\lambda$ in H. If $u_\lambda \neq 0$, then we are finished. Otherwise, we may assume that $u_n \rightharpoonup 0$ in H. Now, we aim to show that there exists $\delta > 0$ such that

$$\lim_{n \to \infty} \sup_{y \in \mathbb{R}^N} \int_{B_1(y)} |u_n|^2 \, dx \geq \delta > 0. \tag{5.2.55}$$

Suppose (5.2.55) fails. Then, by Lemma 1.4.4,

$$u_n \to 0 \text{ in } L^t(\mathbb{R}^N), \ \forall t \in (2, 2_s^*). \tag{5.2.56}$$

Using (5.2.40) and (5.2.56), we see that $\int_{\mathbb{R}^N} G(u_n)\,dx = o(1)$ and $\int_{\mathbb{R}^N} g(u_n)u_n\,dx = o(1)$. This and (5.2.54) yield

$$\frac{1}{2}\|u_n\|^2 - \frac{\lambda}{2_s^*}\int_{\mathbb{R}^N}|u_n|^{2_s^*}dx = c_\lambda + o(1) \tag{5.2.57}$$

and

$$\|u_n\|^2 - \lambda\int_{\mathbb{R}^N}|u_n|^{2_s^*}dx = o(1). \tag{5.2.58}$$

Since $c_\lambda > 0$, we may assume that $\|u_n\|^2 \to L$ for some $L > 0$. By the Sobolev embedding, we can infer that

$$L \geq \frac{S_*^{\frac{N}{2s}}}{\lambda^{\frac{N-2s}{2s}}},$$

which in conjunction with (5.2.57) and (5.2.58) implies that

$$c_\lambda \geq \frac{s}{N}\frac{S_*^{\frac{N}{2s}}}{\lambda^{\frac{N-2s}{2s}}};$$

we reached a contradiction. Thus, (5.2.55) holds true, and we can find a sequence $(y_n) \subset \mathbb{R}^N$ such that $|y_n| \to \infty$ and $\int_{B_1(y_n)}|u_n|^2\,dx \geq \frac{\delta}{2} > 0$.

Set $v_n = u_n(\cdot + y_n)$. Using (5.2.54), we derive that $\mathcal{I}_\lambda(v_n) \to c_\lambda$ and $\mathcal{I}'_\lambda(v_n) \to 0$. In view of (5.2.55), we can deduce that $v_n \rightharpoonup v_\lambda \neq 0$ in H and $\mathcal{I}'_\lambda(v_\lambda) = 0$. Since $\langle \mathcal{I}'_\lambda(v_\lambda), v_\lambda^-\rangle = 0$, it is easy to check that $v_\lambda \geq 0$ in $\mathbb{R}^N$. By Lemma 3.2.14 and combining Proposition 1.3.1 with Proposition 1.3.2, we deduce that $v_\lambda \in C^{1,\alpha}(\mathbb{R}^N)$. Finally, by Theorem 1.3.5, we obtain that $v_\lambda > 0$ in $\mathbb{R}^N$. □

Now, we are ready to prove the existence of a positive ground state to (5.2.1) when V is constant.

Proof of Theorem 5.2.1 Using Lemma 5.2.12, for almost every $\lambda \in [\frac{1}{2}, 1]$, there exists $(u_n) \subset H$ such that $u_n \geq 0$ in H, $\mathcal{I}_\lambda(u_n) \to c_\lambda \in \left(0, \frac{s}{N}\frac{S_*^{\frac{N}{2s}}}{\lambda^{\frac{N-2s}{2s}}}\right)$, $\mathcal{I}'_\lambda(u_n) \to 0$ in H^*, and $u_n \rightharpoonup u_\lambda > 0$ in H. In view of Lemma 5.2.11, we can see that

$$c_\lambda = \mathcal{I}_\lambda(u_\lambda) + \sum_{j=1}^{k}\mathcal{I}_\lambda(w_\lambda^j),$$

$\mathcal{I}'_\lambda(u_\lambda) = 0$ and $\mathcal{I}'_\lambda(w^j_\lambda) = 0$ for $1 \leq j \leq k$. Then Lemma 5.2.9 implies that $\mathcal{I}_\lambda(u_\lambda) > 0$ and $\mathcal{I}_\lambda(w^j_\lambda) \geq 0$ for $1 \leq j \leq k$, so we have $c_\lambda \geq \mathcal{I}_\lambda(u_\lambda) > 0$. Thus, there exists $(\lambda_n) \subset [\frac{1}{2}, 1]$ such that $\lambda_n \to 1$, $u_{\lambda_n} \in H$, $u_{\lambda_n} > 0$, $\mathcal{I}'_{\lambda_n}(u_{\lambda_n}) = 0$, $c_{\lambda_n} \geq \mathcal{I}_{\lambda_n}(u_{\lambda_n}) > 0$ and $c_{\lambda_n} \in \left(0, \frac{s}{N} \frac{S_*^{\frac{N}{2s}}}{\lambda_n^{\frac{N-2s}{2s}}}\right)$. Using the fact that $\mathcal{I}'_{\lambda_n}(u_{\lambda_n}) = 0$ and Lemma 5.2.9, we infer that

$$c_{\lambda_n} \geq \mathcal{I}_{\lambda_n}(u_{\lambda_n}) = \frac{s}{N}[u_{\lambda_n}]_s^2 > 0.$$

Moreover, in view of (1.1.1) we have $\|u_{\lambda_n}\|_{L^{2^*_s}(\mathbb{R}^N)} \leq C$ for all $n \in \mathbb{N}$. Putting together $(f1)$–$(f3)$ and Lemma 5.2.9, we see that for every $\varepsilon > 0$ there exists $C_\varepsilon > 0$ such that

$$\frac{N-2s}{2N}[u_{\lambda_n}]_s^2 + \frac{1}{2}\int_{\mathbb{R}^N} V u_{\lambda_n}^2 \, dx = \lambda_n \int_{\mathbb{R}^N} F(u_{\lambda_n}) \, dx \leq \varepsilon\|u_{\lambda_n}\|_{L^2(\mathbb{R}^N)}^2 + C_\varepsilon\|u_{\lambda_n}\|_{L^{2^*_s}(\mathbb{R}^N)}^{2^*_s},$$

which implies that

$$\frac{V}{2}\int_{\mathbb{R}^N} u_{\lambda_n}^2 \, dx \leq \varepsilon\|u_{\lambda_n}\|_{L^2(\mathbb{R}^N)}^2 + C_\varepsilon C.$$

Therefore, choosing $\varepsilon \in (0, \frac{V}{2})$, we deduce that (u_{λ_n}) is bounded in H. Now, we may assume that $\lim_{n\to\infty} \mathcal{I}_{\lambda_n}(u_{\lambda_n})$ exists. Since the map $\lambda \mapsto c_\lambda$ is left continuous (see Theorem 5.2.2), we have

$$0 \leq \lim_{n\to\infty} \mathcal{I}_{\lambda_n}(u_{\lambda_n}) \leq c_1 < \frac{s}{N}S_*^{\frac{N}{2s}}.$$

Then, using the fact that

$$\mathcal{I}(u_{\lambda_n}) = \mathcal{I}_{\lambda_n}(u_{\lambda_n}) + (\lambda_n - 1)\int_{\mathbb{R}^N} F(u_{\lambda_n}) \, dx$$

and $\|u_{\lambda_n}\| \leq C$, we infer that

$$0 \leq \lim_{n\to\infty} \mathcal{I}(u_{\lambda_n}) \leq c_1 < \frac{s}{N}S_*^{\frac{N}{2s}} \tag{5.2.59}$$

and

$$\lim_{n\to\infty} \mathcal{I}'(u_{\lambda_n}) = 0. \tag{5.2.60}$$

In view of Remark 5.2.10, there exists $\beta > 0$ independent of λ_n such that $\|u_{\lambda_n}\| \geq \beta$. Moreover, we know that (u_{λ_n}) is bounded in H, so we can argue as in the proof of Lemma 5.2.12 to obtain the existence of a positive solution u_0 to (5.2.1).

By Lemma 5.2.11, we can also see that

$$\mathcal{I}(u_0) \leq \lim_{n \to \infty} \mathcal{I}(u_{\lambda_n}) \leq c_1 < \frac{s}{N} S_*^{\frac{N}{2s}}.$$

Let us define

$$m = \inf\{\mathcal{I}(u) : u \in H, u \neq 0, \mathcal{I}'(u) = 0\}.$$

Since $\mathcal{I}'(u_0) = 0$, we have that $m \leq \mathcal{I}(u_0) < \frac{s}{N} S_*^{\frac{N}{2s}}$, and using Lemma 5.2.9, we get that $0 \leq m < \frac{s}{N} S_*^{\frac{N}{2s}}$. By the definition of m, we can find $(u_n) \subset H$ such that $\mathcal{I}(u_n) \to m$ and $\mathcal{I}'(u_n) = 0$. Taking into account Remark 5.2.10, we deduce that $\|u_n\| \geq \beta > 0$ for some β independent of n. Moreover, it is easy to see that (u_n) is bounded in H. In virtue of Remark 5.2.8, we may assume that $u_n \geq 0$ in H. Then, bearing in mind that $\|u_n\| \geq \beta > 0$, we can proceed as in the proof of Lemma 5.2.12 to show that there exists a sequence $(v_n) \subset H$ such that $v_n \geq 0$ in H, $v_n \rightharpoonup v_0 \neq 0$ in H, $\mathcal{I}(v_n) \to m$ and $\mathcal{I}'(v_n) = 0$. Using Lemma 5.2.11, we infer that $\mathcal{I}(v_0) \leq m$ and $\mathcal{I}'(v_0) = 0$. Since $\mathcal{I}'(v_0) = 0$, we also have $\mathcal{I}(v_0) \geq m$. Consequently, $v_0 \geq 0$ in $\mathbb{R}^N$, $v_0 \neq 0$, and satisfies $\mathcal{I}(v_0) = m$ and $\mathcal{I}'(v_0) = 0$. By Lemma 3.2.14 and combining Proposition 1.3.1 with Proposition 1.3.2, we deduce that $v_0 \in C^{1,\alpha}(\mathbb{R}^N)$. Finally, by using Theorem 1.3.5, we obtain that $v_0 > 0$ in $\mathbb{R}^N$. $\qquad \square$

5.2.4 Ground States when the Potential is Not Constant

In this subsection we establish the existence of a ground state to (5.2.1) under the assumption that V is a non-constant potential. Thus, we will assume that $V(x) \not\equiv V_\infty$. For $\lambda \in [\frac{1}{2}, 1]$, we introduce the following family of functionals defined for $u \in H$:

$$\mathcal{I}_\lambda^\infty(u) = \frac{1}{2} \left([u]_s^2 + V_\infty \|u\|_{L^2(\mathbb{R}^N)}^2\right) - \lambda \int_{\mathbb{R}^N} F(u) \, dx.$$

Arguing as in the proof of Theorem 3.2.13 we can derive the following result.

Lemma 5.2.13 *For $\lambda \in [\frac{1}{2}, 1]$, if $w_\lambda \in H$ is a nontrivial critical point of $\mathcal{I}_\lambda^\infty$, then there exists $\gamma_\lambda \in C([0, 1], H)$ such that $\gamma_\lambda(0) = 0$, $\mathcal{I}_{\frac{1}{2}}^\infty(\gamma_\lambda(1)) < 0$, $w_\lambda \in \gamma_\lambda([0, 1])$, $0 \notin \gamma_\lambda((0, 1])$ and $\max_{t \in [0,1]} \mathcal{I}_\lambda^\infty(\gamma_\lambda(t)) = \mathcal{I}_\lambda^\infty(w_\lambda)$.*

Proof We only give a sketch of the proof. Let us define

$$\gamma_\lambda(t)(x) = \begin{cases} w_\lambda(\frac{x}{t}), & \text{for } t > 0, \\ 0, & \text{for } t = 0. \end{cases}$$

Then,

$$\|\gamma_\lambda(t)\|^2 = t^{N-2s}[w_\lambda]_s^2 + t^N V_\infty \|w_\lambda\|_{L^2(\mathbb{R}^N)}^2,$$

$$\mathcal{I}_\lambda^\infty(\gamma_\lambda(t)) = \frac{t^{N-2s}}{2}[w_\lambda]_s^2 + \frac{t^N}{2} V_\infty \|w_\lambda\|_{L^2(\mathbb{R}^N)}^2 - t^N \lambda \int_{\mathbb{R}^N} F(w_\lambda) \, dx. \quad (5.2.61)$$

Hence, $\gamma_\lambda \in C([0, \infty), H)$. By Lemma 5.2.9, we know that

$$\lambda \int_{\mathbb{R}^N} F(w_\lambda) \, dx = \frac{N-2s}{2N}[w_\lambda]_s^2 + \frac{1}{2} V_\infty \|w_\lambda\|_{L^2(\mathbb{R}^N)}^2, \quad (5.2.62)$$

which together with (5.2.61) and (5.2.62) yields

$$\mathcal{I}_\lambda^\infty(\gamma_\lambda(t)) = \left(\frac{t^{N-2s}}{2} - t^N \frac{N-2s}{2N}\right)[w_\lambda]_s^2.$$

Then the proof follows along the lines of Theorem 3.2.13.   □

Remark 5.2.14 By Theorem 5.2.1, we know that if $(f1)$–$(f4)$ hold, then for $\lambda \in [\frac{1}{2}, 1]$, $\mathcal{I}_\lambda^\infty$ has a ground state.

Lemma 5.2.15 *Under the same assumptions of Theorem 5.2.3, for almost every $\lambda \in [\frac{1}{2}, 1]$, the functional $\mathcal{I}_\lambda$ has a positive critical point.*

Proof In view of Lemma 5.2.7 and Remark 5.2.8, we may assume that for almost every $\lambda \in [\frac{1}{2}, 1]$, there exists a sequence $(u_n) \subset H$ such that $u_n \geq 0$ in H, $u_n \rightharpoonup u_\lambda$ in H, $\mathcal{I}_\lambda(u_n) \to c_\lambda \in (0, \frac{s}{N} \frac{S_*^{\frac{N}{2s}}}{\lambda^{\frac{N-2s}{2s}}})$ and $\mathcal{I}_\lambda'(u_n) \to 0$ in H^*.

Now, let us prove that $u_\lambda \neq 0$. We argue by contradiction and suppose that $u_\lambda = 0$. As in the proof of Lemma 5.2.12, we can find a sequence $(y_n) \subset \mathbb{R}^N$ such that $|y_n| \to \infty$ and $v_n = u_n(\cdot + y_n) \rightharpoonup v_\lambda \neq 0$ in H. Furthermore, using the fact that $u_n \rightharpoonup 0$ in H, we can see that $\mathcal{I}_\lambda^\infty(u_n) \to c_\lambda$ and $(\mathcal{I}_\lambda^\infty)'(u_n) \to 0$. Thus, $\mathcal{I}_\lambda^\infty(v_n) \to c_\lambda$ and $(\mathcal{I}_\lambda^\infty)'(v_n) \to 0$. Since $v_n \rightharpoonup v_\lambda \neq 0$ in H, we have that $(\mathcal{I}_\lambda^\infty)'(v_\lambda) = 0$. In view of Lemma 5.2.11, we get $c_\lambda \geq \mathcal{I}_\lambda^\infty(v_\lambda)$. It follows from Remark 5.2.14 that $\mathcal{I}_\lambda^\infty$ has a ground state w_λ. Thus, $c_\lambda \geq \mathcal{I}_\lambda^\infty(w_\lambda)$. By Lemma 5.2.13, we can find a path $\gamma_\lambda \in C([0, 1], H)$ such that $\gamma_\lambda(0) = 0$, $\mathcal{I}_{\frac{1}{2}}^\infty(\gamma_\lambda(1)) < 0$, $w_\lambda \in \gamma_\lambda([0, 1])$, $0 \notin \gamma_\lambda((0, 1])$ and $\max_{t \in [0,1]} \mathcal{I}_\lambda^\infty(\gamma_\lambda(t)) = \mathcal{I}_\lambda^\infty(w_\lambda)$. Consequently,

$$c_\lambda \geq \mathcal{I}_\lambda^\infty(w_\lambda) = \max_{t \in [0,1]} \mathcal{I}_\lambda^\infty(\gamma_\lambda(t)). \quad (5.2.63)$$

Taking into account $(V3)$, $V \not\equiv V_\infty$ and $0 \notin \gamma_\lambda((0, 1])$, we see that $\mathcal{I}_\lambda(\gamma_\lambda(t)) < \mathcal{I}_\lambda^\infty(\gamma_\lambda(t))$ for all $t \in (0, 1]$. Now, we take $v_1 = 0$ and $v_2 = \gamma_\lambda(1)$ in Theorem 5.2.2. Then, using the definition of c_λ and (5.2.63), we see that

$$c_\lambda \leq \max_{t \in [0,1]} \mathcal{I}_\lambda(\gamma_\lambda(t)) < \max_{t \in [0,1]} \mathcal{I}_\lambda^\infty(\gamma_\lambda(t)) \leq c_\lambda, \tag{5.2.64}$$

which gives a contradiction. Consequently, $u_\lambda \neq 0$ and, in particular, $u_\lambda \geq 0$ in $\mathbb{R}^N$. Now, by $(V2)$, we see that v_λ is a weak subsolution of $(-\Delta)^s u_\lambda + V_0 u_\lambda = \lambda f(u_\lambda)$ in $\mathbb{R}^N$, and using Lemma 3.2.14 we deduce that $u_\lambda \in L^\infty(\mathbb{R}^N)$. Propositions 1.3.1 and 1.3.2 imply that $u_\lambda \in C^{1,\alpha}(\mathbb{R}^N)$, and by applying Theorem 1.3.5 we get $u_\lambda > 0$ in $\mathbb{R}^N$. □

At this point we establish the following lemma which will play a fundamental role in the proof of Theorem 5.2.3.

Lemma 5.2.16 *Assume that $(V1)$–$(V3)$ and $(f1)$–$(f4)$ are satisfied. For $\lambda \in [\frac{1}{2}, 1]$, let $(u_n) \subset H$ be a bounded sequence in H such that $u_n \geq 0$ in H, $\mathcal{I}_\lambda(u_n) \to c_\lambda \in \left(0, \frac{s}{N} \frac{S_*^{\frac{N}{2s}}}{\lambda^{\frac{N-2s}{2s}}}\right)$ and $\mathcal{I}_\lambda'(u_n) \to 0$ in H^*.*
   Then there exists a subsequence of (u_n), which we denote again by (u_n), such that

(i) $u_n \rightharpoonup u$ in H and $\mathcal{I}_\lambda'(u_\lambda) = 0$,
(ii) $\mathcal{I}_\lambda(u_\lambda) \leq c_\lambda$.

Proof Since (u_n) is bounded in H, up to a subsequence, we may suppose that $u_n \rightharpoonup u_\lambda$ in H. Then, proceeding as in the proof of Step 1 in Lemma 5.2.11, and using the assumption $(V3)$, we can see that $\mathcal{I}_\lambda'(u_\lambda) = 0$, that is, (i) is satisfied.
   Set $w_n^1 = u_n - u_\lambda$. Similarly to the proof of Lemma 5.2.11, we deduce that

$$c_\lambda - \mathcal{I}_\lambda(u_\lambda) = \frac{1}{2}\|w_n^1\|^2 - \lambda \int_{\mathbb{R}^N} G(w_n^1)\,dx - \frac{\lambda}{2_s^*}\|w_n^1\|_{L^{2_s^*}(\mathbb{R}^N)}^{2_s^*} + o(1). \tag{5.2.65}$$

Using Lemma 5.2.6, we see that for any $\varphi \in H$

$$\left| \int_{\mathbb{R}^N} \left(g(u_n) - g(u_\lambda) - g(w_n^1) \right) \varphi\,dx \right| = o(1)\|\varphi\|. \tag{5.2.66}$$

Next, for $\lambda \in [\frac{1}{2}, 1]$, we introduce the following functionals on H:

$$H_\lambda(u) = \frac{1}{2}\|u\|^2 - \lambda \int_{\mathbb{R}^N} G(u)\,dx - \frac{\lambda}{2_s^*} \int_{\mathbb{R}^N} |u|^{2_s^*}\,dx,$$

$$H_\lambda^\infty(u) = \frac{1}{2}\int_{\mathbb{R}^N}\left(|(-\Delta)^{\frac{s}{2}}u|^2 + V_\infty u^2\right)dx - \lambda\int_{\mathbb{R}^N}G(u)\,dx - \frac{\lambda}{2_s^*}\int_{\mathbb{R}^N}|u|^{2_s^*}\,dx,$$

$$J_\lambda^\infty(u) = \frac{1}{2}\int_{\mathbb{R}^N}\left(|(-\Delta)^{\frac{s}{2}}u|^2 + V_\infty u^2\right)dx - \frac{\lambda}{2_s^*}\int_{\mathbb{R}^N}|u|^{2_s^*}\,dx.$$

Thus (5.2.65) becomes

$$c_\lambda - \mathcal{I}_\lambda(u_\lambda) = H_\lambda(w_n^1) + o(1). \tag{5.2.67}$$

Using (5.2.41) and (5.2.66) we have, for any $\varphi \in H$,

$$|\langle \mathcal{I}_\lambda'(u_n) - \mathcal{I}_\lambda'(u_\lambda), \varphi\rangle - \langle H_\lambda'(w_n^1), \varphi\rangle| = o(1)\|\varphi\|, \tag{5.2.68}$$

which gives

$$H_\lambda'(w_n^1) = o(1). \tag{5.2.69}$$

Taking into account (5.2.67), (5.2.69), $(V3)$ and the fact that $w_n^1 \rightharpoonup 0$ in H, we see that

$$c_\lambda - \mathcal{I}_\lambda(u_\lambda) = H_\lambda^\infty(w_n^1) + o(1) \tag{5.2.70}$$

and

$$(H_\lambda^\infty)'(w_n^1) = o(1). \tag{5.2.71}$$

Now we distinguish two cases.

Case 1: $\lim_{n\to\infty}\sup_{y\in\mathbb{R}^N}\int_{B_1(y)}|w_n^1|^2\,dx = 0$.
    By Lemma 1.4.4,

$$w_n^1 \to 0 \quad \text{in } L^t(\mathbb{R}^N), \ \forall t \in (2, 2_s^*). \tag{5.2.72}$$

Combining (5.2.40) and (5.2.70)–(5.2.72), we deduce that

$$c_\lambda - \mathcal{I}_\lambda(u_\lambda) = J_\lambda^\infty(w_n^1) + o(1) \quad \text{and} \quad J_\lambda'(w_n^1) = o(1)$$

which gives

$$c_\lambda - \mathcal{I}_\lambda(u_\lambda) = \frac{\lambda s}{N}\|w_n^1\|_{L^{2_s^*}(\mathbb{R}^N)}^{2_s^*} + o(1),$$

and then $c_\lambda \geq \mathcal{I}_\lambda(u_\lambda)$.

Case 2: $\lim_{n\to\infty} \sup_{y\in\mathbb{R}^N} \int_{B_1(y)} |w_n^1|^2 \, dx \geq \delta_1$ for some $\delta_1 > 0$.

Thus, there exists $y_n^1 \in \mathbb{R}^N$, $|y_n^1| \to \infty$, such that $\int_{B_1(y_n^1)} |w_n^1|^2 \, dx \geq \frac{\delta_1}{2}$. Then $w_n^1(\cdot + y_n^1) \rightharpoonup w_\lambda^1 \neq 0$ in H,

$$c_\lambda - \mathcal{I}_\lambda(u_\lambda) = H_\lambda^\infty(w_n^1(\cdot + y_n^1)) + o(1) \tag{5.2.73}$$

and

$$(H_\lambda^\infty)'(w_n^1(\cdot + y_n^1)) = o(1). \tag{5.2.74}$$

By (5.2.74) we have $(H_\lambda^\infty)'(w_n^1) = 0$. Now, if $c_\lambda - \mathcal{I}_\lambda(u_\lambda) < \frac{s}{N} \frac{S_*^{\frac{N}{2s}}}{\lambda^{\frac{N-2s}{2s}}}$, then we can proceed as in the proof of Lemma 5.2.11 to obtain the desired conclusion.

Otherwise, we set $w_n^2 = w_n^1(\cdot + y_n^1) - w_\lambda^1$, and repeating the same arguments of (5.2.67) and (5.2.69), we see that

$$c_\lambda - \mathcal{I}_\lambda(u_\lambda) - H_\lambda^\infty(w_\lambda^1) + o(1) = H_\lambda^\infty(w_n^2) \tag{5.2.75}$$

and

$$(H_\lambda^\infty)'(w_n^2) = o(1). \tag{5.2.76}$$

Then either

$$\lim_{n\to\infty} \sup_{y\in\mathbb{R}^N} \int_{B_1(y)} |w_n^2|^2 \, dx = 0, \tag{5.2.77}$$

or there exists $\delta_2 > 0$ such that

$$\lim_{n\to\infty} \sup_{y\in\mathbb{R}^N} \int_{B_1(y)} |w_n^2|^2 \, dx \geq \delta_2 > 0. \tag{5.2.78}$$

Suppose (5.2.77) holds. Then, by Case 1, we deduce that

$$c_\lambda - \mathcal{I}_\lambda(u_\lambda) - H_\lambda^\infty(w_\lambda^1) \geq 0,$$

and by Lemma 5.2.9 we get $H_\lambda^\infty(w_\lambda^1) \geq 0$. This two facts give $c_\lambda - \mathcal{I}_\lambda(u_\lambda) \geq 0$.

Now, suppose (5.2.78) holds. Repeating this procedure, we can find $w_n^i \in H$, $y_n^i \in \mathbb{R}^N$, $|y_n^i| \to \infty$, $i \in \mathbb{N}$, such that $w_n^i(\cdot + y_n^i) \rightharpoonup w_\lambda^i \neq 0$ in H, $(H_\lambda^\infty)'(w_\lambda^i) = 0$,

$$c_\lambda - \mathcal{I}_\lambda(u_\lambda) - \sum_{i=1}^{j} H_\lambda^\infty(w_\lambda^i) + o(1) = H_\lambda^\infty(w_n^{j+1}) \tag{5.2.79}$$

and

$$(H_\lambda^\infty)'(w_n^{j+1}) = o(1), \tag{5.2.80}$$

where

$$w_n^{j+1} = w_n^j(\cdot + y_n^j) - w_\lambda^j, \quad j \in \mathbb{N}.$$

Since $(H_\lambda^\infty)'(w_\lambda^i) = 0$, Lemma 5.2.9 implies that

$$H_\lambda^\infty(w_\lambda^i) = \frac{s}{N}[w_\lambda^i]_s^2. \tag{5.2.81}$$

Now we show that there exists $\alpha > 0$ independent of i such that

$$[w_\lambda^i]_s \geq \alpha. \tag{5.2.82}$$

In fact, using that $(H_\lambda^\infty)'(w_\lambda^i) = 0$, $\lambda \in [\frac{1}{2}, 1]$, and conditions $(f1)$–$(f3)$, we can see that for any $\varepsilon > 0$ there exists $C_\varepsilon > 0$ such that

$$\int_{\mathbb{R}^N} \left(|(-\Delta)^{\frac{s}{2}} w_\lambda^i|^2 + V_\infty |w_\lambda^i|^2 \right) dx \leq \varepsilon \int_{\mathbb{R}^N} |w_\lambda^i|^2 \, dx + C_\varepsilon \int_{\mathbb{R}^N} |w_\lambda^i|^{2_s^*} \, dx$$

$$\leq \frac{\varepsilon}{V_\infty} \int_{\mathbb{R}^N} V_\infty |w_\lambda^i|^2 \, dx + C_\varepsilon \int_{\mathbb{R}^N} |w_\lambda^i|^{2_s^*} \, dx.$$

Choosing $\varepsilon \in (0, V_\infty)$, we can infer that

$$[w_\lambda^i]_s^2 \leq C \|w_\lambda^i\|_{L^{2_s^*}(\mathbb{R}^N)}^{2_s^*},$$

which together with (1.1.1) gives (5.2.82). Then, combining (5.2.81) and (5.2.82), for some $j = k$, we obtain that

$$c_\lambda - \mathcal{I}_\lambda(u_\lambda) - \sum_{i=1}^{j} H_\lambda^\infty(w_\lambda^i) < \frac{s}{N} \frac{S_*^{\frac{N}{2s}}}{\lambda^{\frac{N-2s}{2s}}}.$$

The conclusion follows by applying Lemma 5.2.11. $\square$

We end this subsection by giving the

Proof of Theorem 5.2.3 In view of Lemma 5.2.15, for almost every $\lambda \in [\frac{1}{2}, 1]$, there exists a sequence $(u_n) \subset H$ such that $\mathcal{I}_\lambda(u_n) \to c_\lambda \in \left(0, \frac{S}{N} \frac{S_*^{\frac{N}{2s}}}{\lambda^{\frac{N-2s}{2s}}}\right)$, $\mathcal{I}'_\lambda(u_n) \to 0$ in H^*, $u_n \rightharpoonup u_\lambda \neq 0$ in H.

By Lemma 5.2.16, we deduce that $\mathcal{I}_\lambda(u_\lambda) \leq c_\lambda$ and $\mathcal{I}'_\lambda(u_\lambda) = 0$. Hence, we can find a sequence $(\lambda_n) \subset [\frac{1}{2}, 1]$ such that $\lambda_n \to 1$, $c_{\lambda_n} \in (0, \frac{S}{N} \frac{S_*^{\frac{N}{2s}}}{\lambda^{\frac{N-2s}{2s}}})$, $u_{\lambda_n} \in H$ such that $\mathcal{I}'_{\lambda_n}(u_{\lambda_n}) = 0$, $\mathcal{I}_{\lambda_n}(u_{\lambda_n}) \leq c_{\lambda_n}$. Let us show that there exists a positive constant C such that

$$\|u_{\lambda_n}\| \leq C \quad \text{for all } n \in \mathbb{N}. \tag{5.2.83}$$

By $(V4)$, there exists $\theta \in (0, 2s)$ such that

$$\| \max\{x \cdot \nabla V, 0\}\|_{L^{\frac{N}{2s}}(\mathbb{R}^N)} \leq \theta S_*. \tag{5.2.84}$$

Since $\mathcal{I}_{\lambda_n}(u_{\lambda_n}) \leq c_{\frac{1}{2}}$ and $\mathcal{I}'_{\lambda_n}(u_{\lambda_n}) = 0$, Lemma 5.2.9, the Hölder inequality, Theorem 1.1.8 and (5.2.84) imply that

$$s[u_{\lambda_n}]_s^2 = \frac{N}{2}[u_{\lambda_n}]_s^2 + \frac{N}{2}\int_{\mathbb{R}^N} V(x)u_{\lambda_n}^2 \, dx + \frac{1}{2}\int_{\mathbb{R}^N} x \cdot \nabla V(x)u_{\lambda_n}^2 \, dx - \lambda_n N \int_{\mathbb{R}^N} F(u_{\lambda_n}) \, dx$$

$$= N\mathcal{I}_{\lambda_n}(u_{\lambda_n}) + \frac{1}{2}\int_{\mathbb{R}^N} x \cdot \nabla V(x)u_{\lambda_n}^2 \, dx \leq Nc_{\frac{1}{2}} + \frac{\theta}{2}[u_{\lambda_n}]_s^2, \tag{5.2.85}$$

and consequently $[u_{\lambda_n}]_s \leq C$ for any $n \in \mathbb{N}$. Next, combining $\mathcal{I}'_{\lambda_n}(u_{\lambda_n}) = 0$, $\lambda_n \in [\frac{1}{2}, 1]$, $(f1)$–$(f3)$ and (1.1.1), we see that for any $\varepsilon \in (0, V_0)$

$$V_0\|u_{\lambda_n}\|^2_{L^2(\mathbb{R}^N)} \leq \int_{\mathbb{R}^N} \left(|(-\Delta)^{\frac{s}{2}}u_{\lambda_n}|^2 + V(x)u_{\lambda_n}^2\right) dx$$

$$\leq \int_{\mathbb{R}^N} f(u_{\lambda_n})u_{\lambda_n} \, dx$$

$$\leq \varepsilon\|u_{\lambda_n}\|^2_{L^2(\mathbb{R}^N)} + C_\varepsilon\|u_{\lambda_n}\|^{2^*_s}_{L^{2^*_s}(\mathbb{R}^N)}$$

$$\leq \varepsilon\|u_{\lambda_n}\|^2_{L^2(\mathbb{R}^N)} + C'_\varepsilon[u_{\lambda_n}]_s^{2^*_s}. \tag{5.2.86}$$

Using $[u_{\lambda_n}]_s \leq C$, the estimate (5.2.86) yields $\|u_{\lambda_n}\|_{L^2(\mathbb{R}^N)} \leq C$ for all $n \in \mathbb{N}$. In view of conditions $(V2)$ and $(V3)$, we deduce that

$$0 \leq \int_{\mathbb{R}^N} V(x)u_{\lambda_n}^2 \, dx \leq V_\infty \|u_{\lambda_n}\|_{L^2(\mathbb{R}^N)}^2 \leq V_\infty C^2,$$

which completes the proof of (5.2.83).

Now, we note that

$$\mathcal{I}(u_{\lambda_n}) = \mathcal{I}_{\lambda_n}(u_{\lambda_n}) + (\lambda_n - 1) \int_{\mathbb{R}^N} F(u_{\lambda_n}) \, dx,$$

so we can infer that

$$\lim_{n \to \infty} \mathcal{I}(u_{\lambda_n}) \leq c_1 < \frac{s}{N} S_*^{\frac{N}{2s}},$$

and

$$\lim_{n \to \infty} \mathcal{I}'(u_{\lambda_n}) = 0.$$

By Remark 5.2.10, there exists $\beta > 0$ independent of λ_n such that $\|u_{\lambda_n}\| \geq \beta$. Since $\|u_{\lambda_n}\| \leq C$ for all $n \in \mathbb{N}$, we can proceed as in the proof of Lemma 5.2.15 to show that $u_{\lambda_n} \rightharpoonup u_0 \neq 0$ in H. Then, by using Lemma 5.2.16, we see that

$$\mathcal{I}(u_0) \leq \lim_{n \to \infty} \mathcal{I}(u_{\lambda_n}) \leq c_1 < \frac{s}{N} S_*^{\frac{N}{2s}} \quad \text{and} \quad \mathcal{I}'(u_0) = 0.$$

Now set

$$m = \inf\{\mathcal{I}(u) : u \in H, u \neq 0, \mathcal{I}'(u) = 0\}.$$

Since $\mathcal{I}'(u_0) = 0$, it is clear that $m \leq \mathcal{I}(u_0) < \frac{s}{N} S_*^{\frac{N}{2s}}$. Now, by using the definition of m, we can find a sequence $(v_n) \subset H$ such that $v_n \neq 0$, $\mathcal{I}(v_n) \to m$ and $\mathcal{I}'(v_n) = 0$. Arguing as in (5.2.85) and (5.2.86), we can show that (v_n) is bounded in H, and that there exists $\beta > 0$ independent of n such that $\|v_n\| \geq \beta$. This means that $m > -\infty$. Proceeding in much the same way as in the proof of Lemma 5.2.15, we can see that $v_n \rightharpoonup v_0 \neq 0$ in H. Then, by using Lemma 5.2.16, we deduce that $\mathcal{I}'(v_0) = 0$ and $\mathcal{I}(v_0) \leq m$. Since $\mathcal{I}'(v_0) = 0$, we also have that $\mathcal{I}(v_0) \geq m$. Therefore, we have proved that $v_0 \neq 0$ is such that $\mathcal{I}(v_0) = m$ and $\mathcal{I}'(v_0) = 0$, that is, v_0 is a ground state of (5.2.1). Standard arguments show that $v_0 > 0$ in $\mathbb{R}^N$. $\qquad\square$

Fractional Schrödinger Equations with Rabinowitz Condition

6.1 Introduction

In the first part of this chapter we focus our attention on the existence, multiplicity and concentration of positive solutions for the fractional elliptic problem

$$\begin{cases} \varepsilon^{2s}(-\Delta)^s u + V(x)u = f(u) & \text{in } \mathbb{R}^N, \\ u \in H^s(\mathbb{R}^N), \quad u > 0 \text{ in } \mathbb{R}^N, \end{cases} \tag{6.1.1}$$

where $\varepsilon > 0$ is a small parameter, $s \in (0, 1)$, $N > 2s$. We require that $V : \mathbb{R}^N \to \mathbb{R}$ is a continuous function satisfying the following condition introduced by Rabinowitz [299]:

$$V_\infty = \liminf_{|x| \to \infty} V(x) > V_0 = \inf_{x \in \mathbb{R}^N} V(x) > 0, \tag{V}$$

and we consider both cases $V_\infty < \infty$ and $V_\infty = \infty$. Concerning the nonlinearity $f : \mathbb{R} \to \mathbb{R}$ we assume that

(f_1) $f \in C(\mathbb{R}, \mathbb{R})$ and $f(t) = 0$ for all $t < 0$;

(f_2) $\lim_{t \to 0} \dfrac{f(t)}{t} = 0$;

(f_3) there exists $q \in (2, 2_s^*)$ such that $\lim_{|t| \to \infty} \dfrac{f(t)}{t^{q-1}} = 0$;

(f_4) there exists $\vartheta \in (2, q)$ such that

$$0 < \vartheta F(t) = \vartheta \int_0^t f(\tau) \, d\tau \leq t f(t) \quad \text{for all } t > 0;$$

(f_5) the map $t \mapsto \dfrac{f(t)}{t}$ is increasing in $(0, \infty)$.

© The Author(s), under exclusive license to Springer Nature Switzerland AG 2021
V. Ambrosio, *Nonlinear Fractional Schrödinger Equations in $\mathbb{R}^N$*,
Frontiers in Mathematics, https://doi.org/10.1007/978-3-030-60220-8_6

When $s = 1$, equation (6.1.1) boils down to the classical nonlinear Schrödinger equation

$$-\varepsilon^2 \Delta u + V(x)u = f(u) \text{ in } \mathbb{R}^N, \tag{6.1.2}$$

for which the existence and the multiplicity of solutions have been extensively studied in the last 30 years by many authors; see [13, 28, 90, 100, 146, 197, 284, 299, 318, 330].

For instance, Rabinowitz in [299] investigated the existence of positive solutions to (6.1.2) for $\varepsilon > 0$ small enough, under the assumption that f satisfies (f_4) and the potential $V(x)$ fulfills (V). Wang [330] showed that these solutions concentrate at global minimum points of $V(x)$. We recall that a solution u_ε of (6.1.2) is said to concentrate at $x_0 \in \mathbb{R}^N$ as $\varepsilon \to 0$ if

$$\forall \delta > 0, \quad \exists \varepsilon_0, R > 0 : u_\varepsilon(x) \leq \delta, \quad \forall |x - x_0| \geq \varepsilon R, \ \varepsilon < \varepsilon_0.$$

Using a local mountain pass approach, del Pino and Felmer in [165] proved the existence of a single-spike solution to (6.1.2) which concentrates around a local minimum of V, by assuming that there exists a bounded open set Λ in $\mathbb{R}^N$ such that

$$\inf_{x \in \Lambda} V(x) < \min_{x \in \partial \Lambda} V(x),$$

and considering nonlinearities f satisfying $(f4)$ and the monotonicity assumption on $t \mapsto \frac{f(t)}{t}$. Subsequently, many authors introduced some new variational methods to extend the results obtained in [165] to a wider class of nonlinearities. For more details we refer to [121, 123, 235, 341].

In the nonlocal setting, there are only few results concerning the existence and the concentration phenomena of solutions for the fractional equation (6.1.1), maybe because many important techniques developed in the local framework cannot be adapted so easily to the fractional case.

Next, we recall some recent results related to the concentration phenomenon of solutions for the nonlinear fractional Schrödinger equation (6.1.1).

Chen and Zheng [141] studied, via the Lyapunov–Schmidt reduction method, the concentration phenomenon for solutions of (6.1.1) with $f(t) = |t|^\alpha t$, and under suitable limitations on the dimension N of the space and the fractional powers s, Davila et al. [159] showed via Lyapunov–Schmidt reduction that if the potential V satisfies

$$V \in C^{1,\alpha}(\mathbb{R}^N) \cap L^\infty(\mathbb{R}^N) \text{ and } \inf_{x \in \mathbb{R}^N} V(x) > 0,$$

then (6.1.1) has multi-peak solutions. Fall et al. [182] established necessary and sufficient conditions for the smooth potential V to produce concentration of solutions of (6.1.1) when the parameter ε converges to zero. In particular, when V is coercive and has a unique global minimum, then ground-states concentrate at this point. The multiplicity of positive solutions to (6.1.1) under condition (V) and involving subcritical and critical nonlinearities was considered in [6, 193] and [311], respectively. Alves and Miyagaki [19] investigated the existence and the concentration of positive solutions to (6.1.1), via a penalization approach and extension method [127]. He and Zou [216] used variational methods and the Lusternik–Schnirelman theory to study (6.1.1) when $f(t) = g(t) + t^{2^*_s - 1}$ and g satisfies the conditions (f_1), (f'_2), (f_3) and (f_4). In [35] the author extended the results in [19] and [216], obtaining the existence and the multiplicity of solutions to (6.1.1) when f has subcritical, critical or supercritical growth. Finally, we would like also to mention the paper [158] in which the concentration phenomenon for a nonlocal Schrödinger equation with Dirichlet datum is considered.

Let us denote

$$M = \{x \in \mathbb{R}^N : V(x) = V_0\} \quad \text{and} \quad M_\delta = \{x \in \mathbb{R}^N : \text{dist}(x, M) \le \delta\}, \text{ for } \delta > 0.$$

Our first main result can be stated as follows:

Theorem 6.1.1 ([74]) *Let $N > 2s$, and suppose that V satisfies (V) and f fulfills (f_1)–(f_5). Then, for every $\delta > 0$, there exists $\varepsilon_\delta > 0$ such that, for any $\varepsilon \in (0, \varepsilon_\delta)$, problem (6.1.1) has at least* $\text{cat}_{M_\delta}(M)$ *positive solutions. Moreover, if u_ε denotes one of these solutions and $x_\varepsilon \in \mathbb{R}^N$ is a global maximum point of u_ε, then*

$$\lim_{\varepsilon \to 0} V(x_\varepsilon) = V_0.$$

Exploiting the variational nature of problem (6.1.1), we look for critical points of the functional

$$\mathcal{J}_\varepsilon(u) = \frac{1}{2} \iint_{\mathbb{R}^{2N}} \frac{|u(x) - u(y)|^2}{|x - y|^{N+2s}} \, dx dy + \frac{1}{2} \int_{\mathbb{R}^N} V(\varepsilon x) u^2 \, dx - \int_{\mathbb{R}^N} F(u) \, dx,$$

defined on a suitable subspace of $H^s(\mathbb{R}^N)$. Since f is only continuous, we cannot apply standard Nehari manifold arguments for C^1 functionals; see [27, 280, 340]. Indeed, we cannot proceed as in [14], where the authors considered the corresponding local problem to (6.1.1) (with $p = 2$) under the assumptions $f \in C^1$ and

(f_6) there exist $C > 0$ and $\sigma \in (2, 2^*_s)$ such that

$$f'(t)t^2 - f(t)t \ge Ct^\sigma \quad \text{for all } t \ge 0.$$

To overcome the indicated difficulty, we use some variants of critical point theorems due to Szulkin and Weth [321, 322]; see Sect. 2.4. As usual, the presence of the fractional Laplacian operator makes our analysis more delicate and intriguing. In order to obtain multiple critical points, we employ a technique introduced by Benci and Cerami [95], which consists in making precise comparisons between the category of some sublevel sets of $\mathcal{J}_\varepsilon$ and the category of the set M. Then, after proving that the levels of compactness are strongly related to the behavior of the potential $V(x)$ at infinity (see Proposition 6.3.14), we can apply Lusternik–Schnirelman theory to deduce the existence of multiple positive solutions u_ε's of (6.1.1). Finally, we show that each u_ε concentrates around a global minimum point of V as $\varepsilon \to 0$. To do this, we first adapt the Moser iteration technique [278] in the fractional setting to obtain L^∞-estimates (independent of ε) for the translated sequence $v_\varepsilon = u_\varepsilon(\cdot + \tilde{y}_\varepsilon)$ of u_ε, with $\varepsilon \tilde{y}_\varepsilon \to y \in M$. Then, taking into account some elliptic regularity estimates presented in Chap. 1, we infer that $v_\varepsilon(x) \to 0$ as $|x| \to \infty$ uniformly in ε. This step will be fundamental in deducing the desired concentration result.

Secondly, we consider the following fractional problem involving the critical Sobolev exponent:

$$\begin{cases} \varepsilon^{2s}(-\Delta)^s u + V(x)u = f(u) + |u|^{2^*_s - 2}u & \text{in } \mathbb{R}^N, \\ u \in H^s(\mathbb{R}^N), \quad u > 0 \text{ in } \mathbb{R}^N. \end{cases} \tag{6.1.3}$$

In order to deal with the critical growth of the nonlinearity, we assume that f satisfies (f_1)–(f_5), and the following technical condition:

(f'_6) there exist $q_1 \in (2, 2^*_s)$ and $\lambda > 0$ such that $f(t) \geq \lambda t^{q_1 - 1}$ for any $t > 0$, where λ is such that

- $\lambda > 0$ if either $N \geq 4s$, or $2s < N < 4s$ and $2^*_s - 2 < q_1 < 2^*_s$,
- λ is sufficiently large if $2s < N < 4s$ and $2 < q_1 \leq 2^*_s - 2$.

Then we are able to obtain our second result:

Theorem 6.1.2 ([74]) *Let $N > 2s$, and suppose that (V) and (f_1)–(f_5) and (f'_6) hold. Then, for every $\delta > 0$, there exists $\varepsilon_\delta > 0$ such that, for any $\varepsilon \in (0, \varepsilon_\delta)$, problem (6.1.3) has at least $\text{cat}_{M_\delta}(M)$ positive solutions. Moreover, if u_ε denotes one of these solutions and $x_\varepsilon \in \mathbb{R}^N$ is a global maximum point of u_ε, then*

$$\lim_{\varepsilon \to 0} V(x_\varepsilon) = V_0.$$

We note that Theorem 6.1.2 improves and extends, in the fractional setting, Theorem 1.1 in [186], where the author assumed that $f \in C^1$. The approach developed in this case follows the arguments used to analyze the subcritical case. Anyway, this new problem presents an extra difficulty due to the fact that the level of non-compactness is affected by the critical growth of the nonlinearity. To overcome this hitch we adapt some calculations performed in [310] and we prove that the functional associated with (6.1.3) satisfies the Palais–Smale condition at every level

$$0 < c < \frac{s}{N} S_*^{\frac{N}{2s}},$$

where

$$S_* = \inf_{u \in \mathcal{D}^{s,2}(\mathbb{R}^N)} \frac{[u]_s^2}{\|u\|_{L^{2_s^*}(\mathbb{R}^N)}^2}.$$

We point out that the results presented here improve the ones obtained in [193, 311] in which it is assumed that $f \in C^1$. Here, we follow the approach in [74], where the multiplicity of positive solutions for a fractional Schrödinger equation involving the fractional p-Laplacian operator is obtained.

6.2   Preliminaries

In this preliminary section we introduce some notations and prove some technical lemmas that will be used later.

We fix $\varepsilon > 0$ and introduce the fractional Sobolev space

$$\mathcal{H}_\varepsilon = \left\{ u \in H^s(\mathbb{R}^N) : \int_{\mathbb{R}^N} V(\varepsilon x) u^2 dx < \infty \right\}$$

endowed with the norm

$$\|u\|_\varepsilon = \left([u]_s^2 + \int_{\mathbb{R}^N} V(\varepsilon x) u^2 \, dx \right)^{\frac{1}{2}}.$$

Clearly, $\mathcal{H}_\varepsilon$ is a Hilbert space with the inner product

$$\langle u, v \rangle_\varepsilon = \langle u, v \rangle_{\mathcal{D}^{s,2}(\mathbb{R}^N)} + \int_{\mathbb{R}^N} V(\varepsilon x) uv \, dx \quad \forall u, v \in \mathcal{H}_\varepsilon.$$

In view of assumption (V), Theorems 1.1.8 and 1.1.7, it is easy to check that the following results hold.

Lemma 6.2.1 *The space $\mathcal{H}_\varepsilon$ is continuously embedded in $H^s(\mathbb{R}^N)$. Therefore, $\mathcal{H}_\varepsilon$ is continuously embedded in $L^r(\mathbb{R}^N)$ for any $r \in [2, 2_s^*]$ and compactly embedded in $L^r_{\text{loc}}(\mathbb{R}^N)$ for any $r \in [1, 2_s^*)$.*

Lemma 6.2.2 $C_c^\infty(\mathbb{R}^N)$ *is dense in $\mathcal{H}_\varepsilon$.*

Moreover, when V is coercive, we get the following compactness lemma.

Lemma 6.2.3 *Let $V_\infty = \infty$. Then $\mathcal{H}_\varepsilon$ is compactly embedded in $L^r(\mathbb{R}^N)$ for any $r \in [2, 2_s^*)$.*

Proof For simplicity, we assume that $\varepsilon = 1$. Let $r = 2$. From Lemma 6.2.1 we know that $\mathcal{H}_1 \subset L^2(\mathbb{R}^N)$. Let (u_n) be a sequence such that $u_n \rightharpoonup 0$ in $\mathcal{H}_1$. Then, $u_n \rightharpoonup 0$ in $H^s(\mathbb{R}^N)$. Let us define

$$M = \sup_{n \in \mathbb{N}} \|u_n\|_1 < \infty. \tag{6.2.1}$$

Since V is coercive, for every $\eta > 0$ there exists $R = R_\eta > 0$ such that

$$\frac{1}{V(x)} \leq \eta, \quad \text{for any } |x| \geq R. \tag{6.2.2}$$

Since $u_n \to 0$ in $L^2(B_R)$, it follows that there exists $n_0 > 0$ such that

$$\int_{B_R} u_n^2 \, dx \leq \eta \quad \text{for any } n \geq n_0. \tag{6.2.3}$$

Hence, for any $n \geq n_0$, by (6.2.1)–(6.2.3), we have

$$\int_{\mathbb{R}^N} u_n^2 \, dx = \int_{B_R} u_n^2 \, dx + \int_{\mathbb{R}^N \setminus B_R} u_n^2 \, dx$$

$$\leq \eta + \eta \int_{\mathbb{R}^N \setminus B_R} V(x) u_n^2 \, dx \leq \eta(1 + M^2).$$

Therefore, $u_n \to 0$ in $L^2(\mathbb{R}^N)$.

For $r \in (2, 2_s^*)$, the interpolation inequality and Theorem 6.2.1 yield

$$\|u_n\|_{L^r(\mathbb{R}^N)} \leq C[u_n]_s^\alpha \|u_n\|_{L^2(\mathbb{R}^N)}^{1-\alpha},$$

where

$$\frac{1}{r} = \frac{\alpha}{2} + \frac{1-\alpha}{2_s^*},$$

and using the conclusion with $r = 2$ we get the required result.   □

Finally, we prove a splitting lemma for the functions f and F. This result is consequence of Lemmas 5.2.5 and 5.2.6 proved in Sect. 5.2.2, but here we prefer to give an alternative proof which can also be used to obtain splitting results in other situations.

Lemma 6.2.4 *Let* $(u_n) \subset \mathcal{H}_\varepsilon$ *be a sequence such that* $u_n \rightharpoonup u$ *in* $\mathcal{H}_\varepsilon$*, and* $v_n = u_n - u$*. Then we have*

$$\int_{\mathbb{R}^N} \big(F(v_n) - F(u_n) + F(u) \big) \, dx = o_n(1) \tag{6.2.4}$$

and

$$\sup_{\|w\|_\varepsilon \le 1} \int_{\mathbb{R}^N} \big| (f(v_n) - f(u_n) + f(u)) w \big| \, dx = o_n(1). \tag{6.2.5}$$

Proof We follow [74] (see also [113]). We begin by proving (6.2.4). Let us note that

$$F(v_n) - F(u_n) = \int_0^1 \frac{d}{dt} F(u_n - tu) \, dt = -\int_0^1 u f(u_n - tu) \, dt.$$

In view of (f_2) and (f_3), for any $\delta > 0$ there exists $c_\delta > 0$ such that

$$|f(t)| \le 2\delta |t| + c_\delta |t|^{2_s^* - 1} \quad \text{for all } t \in \mathbb{R}, \tag{6.2.6}$$

$$|F(t)| \le \delta |t|^2 + c_\delta' |t|^{2_s^*} \quad \text{for all } t \in \mathbb{R}. \tag{6.2.7}$$

Using (6.2.6) with $\delta = 1$ and $(|a| + |b|)^r \le C(r)(|a|^r + |b|^r)$ for any $a, b \in \mathbb{R}$ and $r \ge 1$, we can see that

$$|F(v_n) - F(u_n)| \le C|u_n||u| + C|u|^2 + C|u_n|^{2_s^* - 1}|u| + C|u|^{2_s^*}. \tag{6.2.8}$$

Fix $\eta > 0$. Applying the Young inequality $ab \le \eta a^r + C(\eta) b^{r'}$ for all $a, b > 0$, with $r, r' \in (1, \infty)$ such that $\frac{1}{r} + \frac{1}{r'} = 1$, to the first and third term on the right-hand side of (6.2.8), we deduce that

$$|F(v_n) - F(u_n)| \le \eta(|u_n|^2 + |u_n|^{2_s^*}) + C_\eta(|u|^2 + |u|^{2_s^*})$$

which together with (6.2.7) with $\delta = \eta$ implies that

$$|F(v_n) - F(u_n) + F(u)| \leq \eta(|u_n|^2 + |u_n|^{2_s^*}) + C_\eta'(|u|^2 + |u|^{2_s^*}).$$

Let

$$G_{\eta,n}(x) = \max\left\{|F(v_n) - F(u_n) + F(u)| - \eta(|u_n|^2 + |u_n|^{2_s^*}), 0\right\}.$$

Then $G_{\eta,n} \to 0$ a.e. in $\mathbb{R}^N$ as $n \to \infty$, and $0 \leq G_{\eta,n} \leq C_\eta'(|u|^2 + |u|^{2_s^*}) \in L^1(\mathbb{R}^N)$. By the dominated convergence theorem,

$$\int_{\mathbb{R}^N} G_{\eta,n}(x)\, dx \to 0 \quad \text{as } n \to \infty.$$

On the other hand, the definition of $G_{\eta,n}$ implies that

$$|F(v_n) - F(u_n) + F(u)| \leq \eta(|u_n|^2 + |u_n|^{2_s^*}) + G_{\eta,n}$$

which together with the boundedness of (u_n) in $L^2(\mathbb{R}^N) \cap L^{2_s^*}(\mathbb{R}^N)$ yields

$$\limsup_{n \to \infty} \int_{\mathbb{R}^N} |F(v_n) - F(u_n) + F(u)|\, dx \leq C\eta.$$

By the arbitrariness of $\eta > 0$ we deduce that (6.2.4) holds.

Now we prove (6.2.5). Here we follow [1, 74, 169]. We claim that there is a subsequence (u_{n_j}) of (u_n) such that, for all $\eta > 0$ there exists $r_\eta > 0$ satisfying

$$\limsup_{j \to \infty} \int_{B_j \setminus B_r} |u_{n_j}|^\tau\, dx \leq \eta \tag{6.2.9}$$

for all $r \geq r_\eta$, where $\tau \in [2, 2_s^*)$ is fixed. To verify (6.2.9) note that, for each $j \in \mathbb{N}$, $\int_{B_j} |u_n|^\tau\, dx \to \int_{B_j} |u|^\tau\, dx$ as $n \to \infty$, so there exists $\hat{n}_j \in \mathbb{N}$ such that

$$\int_{B_j} (|u_n|^\tau - |u|^\tau)\, dx < \frac{1}{j} \quad \text{for all } n = \hat{n}_j + i, \ i = 1, 2, 3, \ldots.$$

Without loss of generality we can assume that $\hat{n}_{j+1} \geq \hat{n}_j$. In particular, for $n_j = \hat{n}_j + j$ we have

$$\int_{B_j} (|u_{n_j}|^\tau - |u|^\tau)\, dx < \frac{1}{j}.$$

Observe that there is $r_\eta > 0$ such that

$$\int_{\mathbb{R}^N \setminus B_r} |u|^\tau \, dx < \eta \qquad (6.2.10)$$

for all $r \geq r_\eta$. Since

$$\int_{B_j \setminus B_r} |u_{n_j}|^\tau \, dx = \int_{B_j} (|u_{n_j}|^\tau - |u|^\tau) \, dx + \int_{B_j \setminus B_r} |u|^\tau \, dx + \int_{B_r} (|u|^\tau - |u_{n_j}|^\tau) \, dx$$

$$\leq \frac{1}{j} + \int_{\mathbb{R}^N \setminus B_r} |u|^\tau \, dx + \int_{B_r} (|u|^\tau - |u_{n_j}|^\tau) \, dx$$

$$\leq \frac{1}{j} + \eta + \int_{B_r} (|u|^\tau - |u_{n_j}|^\tau) \, dx,$$

(6.2.9) now follows.

Let first (u_{n_j}) be a subsequence of (u_n) such that (6.2.9) holds for $\tau = 2$. Repeating the argument, we can then find a subsequence $(u_{n_{j_h}})$ of (u_{n_j}) such that (6.2.9) holds for $\tau = q$. Therefore, for notational convenience, we assume in what follows that (6.2.9) holds for both $\tau = 2$ and $\tau = q$ with the same subsequence.

Let $\phi : [0, \infty) \to [0, 1]$ be a smooth function such that $\phi(t) = 1$ if $t \in [0, 1]$ and $\phi(t) = 0$ if $t \in [2, \infty)$, and define $\tilde{u}_j(x) = \phi\left(\frac{2|x|}{j}\right) u(x)$. Applying Lemma 1.4.8 and the dominated convergence theorem, we can see that

$$\tilde{u}_j \to u \quad \text{in } H^s(\mathbb{R}^N). \qquad (6.2.11)$$

Set $h_j = u - \tilde{u}_j$. Observe that for all $w \in \mathcal{H}_\varepsilon$ one has that

$$\int_{\mathbb{R}^N} [f(u_{n_j}) - f(v_{n_j}) - f(u)]w \, dx =$$

$$= \int_{\mathbb{R}^N} [f(u_{n_j}) - f(u_{n_j} - \tilde{u}_j) - f(\tilde{u}_j)]w \, dx$$

$$+ \int_{\mathbb{R}^N} [f(v_{n_j} + h_j) - f(v_{n_j})]w \, dx + \int_{\mathbb{R}^N} [f(\tilde{u}_j) - f(u)]w \, dx$$

$$= I_j + II_j + III_j. \qquad (6.2.12)$$

Clearly, (6.2.11) ensures that

$$\lim_{j \to \infty} \sup_{\|w\|_\varepsilon \leq 1} |III_j| = 0. \qquad (6.2.13)$$

Let us show that

$$\lim_{j \to \infty} \sup_{\|w\|_\varepsilon \le 1} |I_j| = 0. \tag{6.2.14}$$

Invoking Lemma 6.2.1 and conditions (f_2)–(f_3) we have, for all $r > 0$,

$$\lim_{j \to \infty} \sup_{\|w\|_\varepsilon \le 1} \left| \int_{B_r} [f(u_{n_j}) - f(u_{n_j} - \tilde{u}_j) - f(\tilde{u}_j)] w \, dx \right| = 0. \tag{6.2.15}$$

For any $\eta > 0$ let $r_\eta > 0$ be so large that (6.2.9) and (6.2.10) hold. Then, it follows from (6.2.10) and (6.2.11) that

$$\limsup_{j \to \infty} \int_{B_j \setminus B_r} |\tilde{u}_j|^\tau \, dx \le \int_{\mathbb{R}^N \setminus B_r} |u|^\tau \, dx \le \eta \tag{6.2.16}$$

for all $r \ge r_\eta$. Hence, since (f_2)–(f_3) and $\tilde{u}_j = 0$ on $\mathbb{R}^N \setminus B_j$ for all $j \in \mathbb{N}$, the Hölder inequality, Lemma 6.2.1, (6.2.9), and (6.2.16) imply that

$$\limsup_{j \to \infty} \left| \int_{\mathbb{R}^N \setminus B_r} [f(u_{n_j}) - f(u_{n_j} - \tilde{u}_j) - f(\tilde{u}_j)] w \, dx \right|$$

$$= \limsup_{j \to \infty} \left| \int_{B_j \setminus B_r} [f(u_{n_j}) - f(u_{n_j} - \tilde{u}_j) - f(\tilde{u}_j)] w \, dx \right|$$

$$\le C \limsup_{j \to \infty} \left[\|u_{n_j}\|_{L^2(B_j \setminus B_r)} + \|\tilde{u}_j\|_{L^2(B_j \setminus B_r)} \right] \|w\|_\varepsilon$$

$$+ C \limsup_{j \to \infty} \left[\|u_{n_j}\|_{L^q(B_j \setminus B_r)}^{q-1} + \|\tilde{u}_j\|_{L^q(B_j \setminus B_r)}^{q-1} \right] \|w\|_\varepsilon$$

$$\le C \eta^{\frac{1}{2}} + C \eta^{\frac{q-1}{q}}. \tag{6.2.17}$$

Combining (6.2.15) and (6.2.17), we obtain (6.2.14) holds true. Finally we verify that

$$\lim_{j \to \infty} \sup_{\|w\|_\varepsilon \le 1} |II_j| = 0. \tag{6.2.18}$$

Set

$$g(t) = \begin{cases} \frac{f(t)}{|t|}, & \text{if } t \ne 0, \\ 0, & \text{if } t = 0. \end{cases}$$

In the light of (f_1)–(f_3), we can see that $g \in C(\mathbb{R})$ and $|g(t)| \le C(1 + |t|^{q-2})$ for all $t \in \mathbb{R}$. For any $a > 0$ and any $j \in \mathbb{N}$, we set $C_j^a = \{x \in \mathbb{R}^N : |v_{n_j}(x)| \le a\}$ and $D_j^a = \mathbb{R}^N \setminus C_j^a$. Since (v_{n_j}) is bounded in $L^2(\mathbb{R}^N)$, we can see that

$$|D_j^a| \le \frac{1}{a^2} \int_{D_j^a} |v_{n_j}|^2 \, dx \le \frac{C}{a^2} \to 0 \text{ as } a \to \infty.$$

Then, in view of (f_2)–(f_3), the Hölder inequality and the boundedness of (h_j) we have that

$$\left| \int_{D_j^a} [f(v_{n_j} + h_j) - f(v_{n_j})] w \, dx \right|$$

$$\le C \int_{D_j^a} [|v_{n_j}| + |v_{n_j}|^{q-1} + |h_j| + |h_j|^{q-1}] |w| \, dx$$

$$\le C(|D_j^a|^{\frac{2_s^* - 2}{2_s^*}} \|v_{n_j}\|_{L^{2_s^*}(\mathbb{R}^N)} \|w\|_{L^{2_s^*}(\mathbb{R}^N)} + |D_j^a|^{\frac{2_s^* - q}{2_s^*}} \|v_{n_j}\|_{L^{2_s^*}(\mathbb{R}^N)}^{q-1} \|w\|_{L^{2_s^*}(\mathbb{R}^N)})$$

$$+ C(|D_j^a|^{\frac{2_s^* - 2}{2_s^*}} \|h_j\|_{L^{2_s^*}(\mathbb{R}^N)} \|w\|_{L^{2_s^*}(\mathbb{R}^N)} + |D_j^a|^{\frac{2_s^* - q}{2_s^*}} \|h_j\|_{L^{2_s^*}(\mathbb{R}^N)}^{q-1} \|w\|_{L^{2_s^*}(\mathbb{R}^N)})$$

$$\le C(|D_j^a|^{\frac{2_s^* - 2}{2_s^*}} + |D_j^a|^{\frac{2_s^* - q}{2_s^*}}) \|w\|_\varepsilon,$$

which implies that there exists $\tilde{a} = \tilde{a}_\eta > 0$ such that

$$\left| \int_{D_j^{\tilde{a}}} [f(v_{n_j} + h_j) - f(v_{n_j})] w \, dx \right| \le \eta \tag{6.2.19}$$

uniformly in $\|w\|_\varepsilon \le 1$. Since g is uniformly continuous in the interval $[-\tilde{a}, \tilde{a}]$, there exists $\delta > 0$ such that

$$|g(t + h) - g(t)| \le \eta \quad \text{for all } t \in [-\tilde{a}, \tilde{a}] \text{ and } |h| \le \delta. \tag{6.2.20}$$

Let $V_j^\delta = \{x \in \mathbb{R}^N : |h_j(x)| \le \delta\}$ and $W_j^\delta = \mathbb{R}^N \setminus V_j^\delta$. Noting that

$$|C_j^{\tilde{a}} \cap W_j^\delta| \le |W_j^\delta| \le \frac{1}{\delta^2} \int_{W_j^\delta} |h_j|^2 \, dx \le \frac{1}{\delta^2} \|h_j\|_{L^2(\mathbb{R}^N)}^2 \to 0 \quad \text{as } j \to \infty,$$

we can argue as before to infer that there exists $j_0 \in \mathbb{N}$ such that

$$\left| \int_{C_j^{\tilde{a}} \cap W_j^{\delta}} [f(v_{n_j} + h_j) - f(v_{n_j})] w \, dx \right| \leq \eta \quad \text{for all } j \geq j_0, \tag{6.2.21}$$

uniformly in $\|w\|_{\varepsilon} \leq 1$. Now observe that

$$[f(v_{n_j} + h_j) - f(v_{n_j})]w = g(v_{n_j} + h_j)[|v_{n_j} + h_j| - |v_{n_j}|]w + [g(v_{n_j} + h_j) - g(v_{n_j})]|v_{n_j}|w.$$

In view of (6.2.11), we can find $j_1 \in \mathbb{N}$ such that $j_1 \geq j_0$ and

$$\|h_j\|_{L^{\tau}(\mathbb{R}^N)} < \eta \quad \text{for all } j \geq j_1. \tag{6.2.22}$$

Taking into account (6.2.20), (6.2.22) and the boundedness of (v_{n_j}), we can see that for all $j \geq j_1$

$$\left| \int_{C_j^{\tilde{a}} \cap V_j^{\delta}} [g(v_{n_j} + h_j) - g(v_{n_j})]|v_{n_j}|w \, dx \right|$$

$$\leq \eta \int_{C_j^{\tilde{a}} \cap V_j^{\delta}} |v_{n_j}||w| \, dx$$

$$\leq \eta \|v_{n_j}\|_{L^2(\mathbb{R}^N)} \|w\|_{L^2(\mathbb{R}^N)}$$

$$\leq C\eta,$$

uniformly in $\|w\|_{\varepsilon} \leq 1$. On the other hand, using the fact that $|g(t)| \leq C(1 + |t|^{q-2})$, the Hölder inequality and (6.2.22), we have

$$\left| \int_{C_j^{\tilde{a}} \cap V_j^{\delta}} g(v_{n_j} + h_j)[|v_{n_j} + h_j| - |v_{n_j}|]w \, dx \right|$$

$$\leq C \int_{\mathbb{R}^N} (1 + |v_{n_j} + h_j|^{q-2})|h_j||w| \, dx$$

$$\leq C[\|h_j\|_{L^2(\mathbb{R}^N)} \|w\|_{L^2(\mathbb{R}^N)} + \|v_{n_j}\|_{L^q(\mathbb{R}^N)}^{q-2} \|h_j\|_{L^q(\mathbb{R}^N)} \|w\|_{L^q(\mathbb{R}^N)} + \|h_j\|_{L^q(\mathbb{R}^N)}^{q-1} \|w\|_{L^q(\mathbb{R}^N)}]$$

$$\leq C[\eta + \eta^{q-1}].$$

This estimates in conjunction with (6.2.21) and the equality $C_j^{\tilde{a}} = (C_j^{\tilde{a}} \cap V_j^{\delta}) \cup (C_j^{\tilde{a}} \cap W_j^{\delta})$ shows that

$$\left| \int_{C_j^{\tilde{a}}} [f(v_{n_j} + h_j) - f(v_{n_j})]w \, dx \right| \leq C(\eta + \eta^{q-1}) \quad \text{for all } j \geq j_1,$$

uniformly in $\|w\|_{\varepsilon} \leq 1$, which together with (6.2.19) yields (6.2.18). $\qquad \square$

6.3 The Subcritical Case

6.3.1 The Nehari Method for (6.1.1)

Making the change of variable $x \mapsto \varepsilon x$, we are led to consider the problem

$$\begin{cases} (-\Delta)^s u + V(\varepsilon x)u = f(u) & \text{in } \mathbb{R}^N, \\ u \in H^s(\mathbb{R}^N), \quad u > 0 \text{ in } \mathbb{R}^N. \end{cases} \tag{P_ε}$$

To find weak solutions of (P_ε), we look for critical points of the functional

$$\mathcal{J}_\varepsilon(u) = \frac{1}{2}\|u\|_\varepsilon^2 - \int_{\mathbb{R}^N} F(u)\,dx,$$

which is well defined on $\mathcal{H}_\varepsilon$. It is standard to check that (f_2)–(f_3) ensure that given $\xi > 0$ there exists $C_\xi > 0$ such that

$$|f(t)| \le \xi|t| + C_\xi|t|^{q-1} \quad \forall t \in \mathbb{R}, \tag{6.3.1}$$

$$|F(t)| \le \frac{\xi}{2}t^2 + \frac{C_\xi}{q}|t|^q \quad \forall t \in \mathbb{R}. \tag{6.3.2}$$

On the other hand, hypothesis (f_5) implies that

$$t \mapsto \frac{1}{2}f(t)t - F(t) \quad \text{is increasing for any } t > 0. \tag{6.3.3}$$

By Lemma 6.2.1, it is readily seen that $\mathcal{J}_\varepsilon \in C^1(\mathcal{H}_\varepsilon, \mathbb{R})$ and its differential $\mathcal{J}_\varepsilon'$ is given by

$$\langle \mathcal{J}_\varepsilon'(u), \varphi \rangle = \langle u, v \rangle_\varepsilon - \int_{\mathbb{R}^N} f(u)\varphi\,dx,$$

for any $u, \varphi \in \mathcal{H}_\varepsilon$.

Now, we introduce the Nehari manifold associated with $\mathcal{J}_\varepsilon$, that is,

$$\mathcal{N}_\varepsilon = \left\{ u \in \mathcal{H}_\varepsilon \setminus \{0\} : \langle \mathcal{J}_\varepsilon'(u), u \rangle = 0 \right\}.$$

Then

$$\|u\|_\varepsilon^2 = \int_{\mathbb{R}^N} f(u)u\,dx$$

for all $u \in \mathcal{N}_\varepsilon$, which together with (f_4) implies that

$$\mathcal{J}_\varepsilon(u) = \mathcal{J}_\varepsilon(u) - \frac{1}{2}\langle \mathcal{J}_\varepsilon'(u), u\rangle = \int_{\mathbb{R}^N}\left[\frac{1}{2}f(u)u - F(u)\right]dx \geq 0 \qquad (6.3.4)$$

for all $u \in \mathcal{N}_\varepsilon$

Since f is merely continuous, the next results are very important for overcoming the non-differentiability of $\mathcal{N}_\varepsilon$. We begin by proving some properties for the functional $\mathcal{J}_\varepsilon$.

Lemma 6.3.1 *Under assumptions (V) and (f_1)–(f_5), for $\varepsilon > 0$ we have:*

(i) *$\mathcal{J}_\varepsilon'$ maps bounded sets in $\mathcal{H}_\varepsilon$ into bounded sets in $\mathcal{H}_\varepsilon$.*
(ii) *$\mathcal{J}_\varepsilon'$ is weakly sequentially continuous in $\mathcal{H}_\varepsilon$.*
(iii) *$\mathcal{J}_\varepsilon(t_n u_n) \to -\infty$ as $t_n \to \infty$, where $u_n \in K$ and $K \subset \mathcal{H}_\varepsilon \setminus \{0\}$ is a compact subset.*

Proof

(i) Let (u_n) be a bounded sequence in $\mathcal{H}_\varepsilon$. Then, for all $v \in \mathcal{H}_\varepsilon$, we deduce from (6.3.1) and Lemma 6.2.1 that

$$\langle \mathcal{J}_\varepsilon'(u_n), v\rangle \leq C\|u_n\|_\varepsilon\|v\|_\varepsilon + C\|u_n\|_\varepsilon^{q-1}\|v\|_\varepsilon \leq C.$$

(ii) Let $u_n \rightharpoonup u$ in $\mathcal{H}_\varepsilon$. By Lemma 6.2.1, $u_n \to u$ in $L^r_{loc}(\mathbb{R}^N)$ for all $r \in [1, 2_s^*)$ and $u_n \to u$ a.e. in $\mathbb{R}^N$. Then, for all $v \in C_c^\infty(\mathbb{R}^N)$, it follows from (6.3.1) and the dominated convergence theorem that

$$\langle \mathcal{J}_\varepsilon'(u_n), v\rangle \to \langle \mathcal{J}_\varepsilon'(u), v\rangle. \qquad (6.3.5)$$

In view of Lemma 6.2.2, for any $v \in \mathcal{H}_\varepsilon$ we can take $(v_j) \subset C_c^\infty(\mathbb{R}^N)$ such that $\|v_j - v\|_\varepsilon \to 0$ as $j \to \infty$. Note that (6.3.1) and Lemma 6.2.1 yield

$$|\langle \mathcal{J}_\varepsilon'(u_n), v\rangle - \langle \mathcal{J}_\varepsilon'(u), v\rangle|$$
$$\leq |\langle \mathcal{J}_\varepsilon'(u_n) - \mathcal{J}_\varepsilon'(u), v_j\rangle| + |\langle \mathcal{J}_\varepsilon'(u_n) - \mathcal{J}_\varepsilon'(u), v - v_j\rangle|$$
$$\leq |\langle \mathcal{J}_\varepsilon'(u_n) - \mathcal{J}_\varepsilon'(u), v_j\rangle| + C\int_{\mathbb{R}^N}(|u_n| + |u| + |u_n|^{q-1} + |u|^{q-1})|v - v_j|\,dx$$
$$\leq |\langle \mathcal{J}_\varepsilon'(u_n) - \mathcal{J}_\varepsilon'(u), v_j\rangle| + C\|v_j - v\|_\varepsilon.$$

For any $\xi > 0$, fix $j \in \mathbb{N}$ such that $\|v_j - v\|_\varepsilon < \frac{\xi}{2C}$. By (6.3.5), there is $n_0 \in \mathbb{N}$ such that

$$|\langle \mathcal{J}_\varepsilon'(u_n) - \mathcal{J}_\varepsilon'(u), v_j \rangle| < \frac{\xi}{2} \qquad \text{for all } n \geq n_0.$$

Thus

$$|\langle \mathcal{J}_\varepsilon'(u_n), v \rangle - \langle \mathcal{J}_\varepsilon'(u), v \rangle| < \xi \text{ for all } n \geq n_0$$

and this shows that $\mathcal{J}_\varepsilon'$ is weakly sequentially continuous in $\mathcal{H}_\varepsilon$.

(iii) Without loss of generality we may assume that $\|u\|_\varepsilon = 1$ for each $u \in K$. For $u_n \in K$, after passing to a subsequence, we obtain that $u_n \to u \in \mathbb{S}_\varepsilon$. Then, using (f_4), letting $t_n \to \infty$ and applying Fatou's lemma, we see that

$$\mathcal{J}_\varepsilon(t_n u_n) = \frac{t_n^2}{2} \|u_n\|_\varepsilon^2 - \int_{\mathbb{R}^N} F(t_n u_n)\, dx$$

$$\leq t_n^\vartheta \left(\frac{\|u_n\|_\varepsilon^2}{t_n^{\vartheta-2}} - \int_{\mathbb{R}^N} \frac{F(t_n u_n)}{t_n^\vartheta}\, dx \right) \to -\infty \qquad \text{as } n \to \infty.$$

$\square$

Lemma 6.3.2 *Under the assumptions of Lemma 6.3.1, for $\varepsilon > 0$ we have:*

(i) *for every $u \in \mathbb{S}_\varepsilon$, there exists a unique $t_u > 0$ such that $t_u u \in \mathcal{N}_\varepsilon$. Moreover, $m_\varepsilon(u) = t_u u$ is the unique maximum of $\mathcal{J}_\varepsilon$ on $\mathcal{H}_\varepsilon$, where $\mathbb{S}_\varepsilon = \{u \in \mathcal{H}_\varepsilon : \|u\|_\varepsilon = 1\}$.*

(ii) *The set $\mathcal{N}_\varepsilon$ is bounded away from 0. Furthermore, $\mathcal{N}_\varepsilon$ is closed in $\mathcal{H}_\varepsilon$.*

(iii) *There exists $\alpha > 0$ such that $t_u \geq \alpha$ for each $u \in \mathbb{S}_\varepsilon$ and, for each compact subset $W \subset \mathbb{S}_\varepsilon$, there exists a constant $C_W > 0$ such that $t_u \leq C_W$ for all $u \in W$.*

(iv) *$\mathcal{N}_\varepsilon$ is a regular manifold diffeomorphic to the unit sphere $\mathbb{S}_\varepsilon$.*

(v) *$c_\varepsilon = \inf_{\mathcal{N}_\varepsilon} \mathcal{J}_\varepsilon \geq \rho > 0$ and $\mathcal{J}_\varepsilon$ is bounded below on $\mathcal{N}_\varepsilon$, where ρ is independent of ε.*

Proof

(i) For each $u \in \mathbb{S}_\varepsilon$ and $t > 0$, set $h(t) = \mathcal{J}_\varepsilon(tu)$. It is clear that $h(0) = 0$ and that $h(t) < 0$ for t large. Let us prove that $h(t) > 0$ for $t > 0$ small. Indeed, by (6.3.2) and Lemma 6.2.1, it follows that

$$h(t) = \frac{t^2}{2} \|u\|_\varepsilon^2 - \int_{\mathbb{R}^N} F(tu)\, dx$$

$$\geq \frac{t^2}{2}\|u\|_\varepsilon^2 - \frac{\xi t^2}{2}\|u\|_{L^2(\mathbb{R}^N)}^2 - C_\xi \frac{t^q}{q}\|u\|_{L^q(\mathbb{R}^N)}^q$$

$$\geq \frac{t^2}{2}\|u\|_\varepsilon^2 - \frac{\xi t^2}{2V_0}\|u\|_\varepsilon^2 - C_\xi \frac{t^q}{q}\|u\|_\varepsilon^q.$$

Choosing $\xi \in (0, V_0)$ we deduce that $h(t) > 0$ for $t > 0$ sufficiently small. Therefore, $\max_{t\geq 0} h(t)$ is achieved at some $t_u > 0$ satisfying $h'(t_u) = 0$ and $t_u u \in \mathcal{N}_\varepsilon$. Next, we prove the uniqueness of such a t_u. Suppose, by contradiction, that there exist $t'_u > t_u > 0$ such that $t_u u, t'_u u \in \mathcal{N}_\varepsilon$. Then

$$t_u^2 \|u\|_\varepsilon^2 = \int_{\mathbb{R}^N} f(t_u u) t_u u \, dx \quad \text{and} \quad (t'_u)^2 \|u\|_\varepsilon^2 = \int_{\mathbb{R}^N} f(t'_u u) t'_u u \, dx,$$

whence

$$\left(\frac{1}{t'_u} - \frac{1}{t_u}\right) \|u\|_\varepsilon^2 = \int_{\mathbb{R}^N} \left(\frac{f(t'_u u)}{t'_u u} - \frac{f(t_u u)}{t_u u}\right) u^2 \, dx.$$

Using (f_5) and $t'_u > t_u > 0$, we deduce that this identity makes no sense.

(ii) By virtue of (6.3.1) and Lemma 6.2.1, we can see that for any $u \in \mathcal{N}_\varepsilon$

$$\|u\|_\varepsilon^p = \int_{\mathbb{R}^N} f(u) u \, dx \leq C\xi \|u\|_\varepsilon^2 + C_\xi \|u\|_\varepsilon^q,$$

which implies that $\|u\|_\varepsilon \geq \kappa$ for some $\kappa > 0$.

Now we prove that the set $\mathcal{N}_\varepsilon$ is closed in $\mathcal{H}_\varepsilon$. Let $(u_n) \subset \mathcal{N}_\varepsilon$ be such that $u_n \to u$ in $\mathcal{H}_\varepsilon$. In view of Lemma 6.3.1, $\mathcal{J}'_\varepsilon(u_n)$ is bounded, and we deduce that

$$\langle \mathcal{J}'_\varepsilon(u_n), u_n\rangle - \langle \mathcal{J}'_\varepsilon(u), u\rangle = \langle \mathcal{J}'_\varepsilon(u_n), u_n - u\rangle + \langle \mathcal{J}'_\varepsilon(u_n) - \mathcal{J}'_\varepsilon(u), u\rangle \to 0,$$

that is, $\langle \mathcal{J}'_\varepsilon(u), u\rangle = 0$. This fact combined with $\|u\|_\varepsilon \geq \kappa$ implies that

$$\|u\|_\varepsilon = \lim_{n\to\infty} \|u_n\|_\varepsilon \geq \kappa > 0,$$

and then $u \in \mathcal{N}_\varepsilon$.

(iii) For $(u_n) \subset \mathbb{S}_\varepsilon$ there exists $t_{u_n} > 0$ such that $t_{u_n} u_n \in \mathcal{N}_\varepsilon$. The proof of (ii) shows that

$$t_{u_n} = \|t_{u_n} u_n\|_\varepsilon \geq \kappa,$$

which implies that $t_{u_n} \nrightarrow 0$ as $n \to \infty$. Now let us prove that $t_u \leq C_W$ for all $u \in W \subset \mathbb{S}_\varepsilon$. Assume, by contradiction, that there exists a sequence $(u_n) \subset W \subset \mathbb{S}_\varepsilon$ such that $t_{u_n} \to \infty$. Since W is compact, we can find $u \in W$ such that $u_n \to u$ in $\mathcal{H}_\varepsilon$

and $u_n \to u$ a.e. in $\mathbb{R}^N$. Using Lemma 6.3.1-(iii), we deduce that $\mathcal{J}_\varepsilon(t_{u_n} u_n) \to -\infty$ as $n \to \infty$, which gives a contradiction because of (6.3.4).

(iv) Define the maps $\hat{m}_\varepsilon : \mathcal{H}_\varepsilon \setminus \{0\} \to \mathcal{N}_\varepsilon$ and $m_\varepsilon : \mathbb{S}_\varepsilon \to \mathcal{N}_\varepsilon$ by setting

$$\hat{m}_\varepsilon(u) = t_u u \quad \text{and} \quad m_\varepsilon = \hat{m}_\varepsilon|_{\mathbb{S}_\varepsilon}. \tag{6.3.6}$$

In view of (i)–(iii) and Proposition 2.4.2, we deduce that m_ε is a homeomorphism between $\mathbb{S}_\varepsilon$ and $\mathcal{N}_\varepsilon$ and the inverse of m_ε is given by $m_\varepsilon^{-1}(u) = \frac{u}{\|u\|_\varepsilon}$. Therefore, $\mathcal{N}_\varepsilon$ is a regular manifold diffeomorphic to $\mathbb{S}_\varepsilon$.

(v) For $\varepsilon > 0$, $t > 0$ and $u \in \mathcal{H}_\varepsilon \setminus \{0\}$, we see that (6.3.2) yields

$$\mathcal{J}_\varepsilon(tu) \geq \frac{t^2}{2}\left(1 - \frac{\xi}{V_0}\right)\|u\|_\varepsilon^2 - t^q C_\xi \|u\|_\varepsilon^q.$$

Choosing $\xi > 0$ small enough, we can find $\rho > 0$ such that $\mathcal{J}_\varepsilon(tu) \geq \rho > 0$ for $t > 0$ small enough. On the other hand, we deduce from (i)–(iii) that (see (2.4.1))

$$c_\varepsilon = \inf_{u \in \mathcal{N}_\varepsilon} \mathcal{J}_\varepsilon(u) = \inf_{u \in \mathcal{H}_\varepsilon \setminus \{0\}} \max_{t > 0} \mathcal{J}_\varepsilon(tu) = \inf_{u \in \mathbb{S}_\varepsilon} \max_{t > 0} \mathcal{J}_\varepsilon(tu) \tag{6.3.7}$$

which implies that $c_\varepsilon \geq \rho$ and $\mathcal{J}_\varepsilon|_{\mathcal{N}_\varepsilon} \geq \rho$.

$\square$

Now we introduce the functionals $\hat{\Psi}_\varepsilon : \mathcal{H}_\varepsilon \setminus \{0\} \to \mathbb{R}$ and $\Psi_\varepsilon : \mathbb{S}_\varepsilon \to \mathbb{R}$ defined by

$$\hat{\Psi}_\varepsilon(u) = \mathcal{J}_\varepsilon(\hat{m}_\varepsilon(u)) \quad \text{and} \quad \Psi_\varepsilon = \hat{\Psi}_\varepsilon|_{\mathbb{S}_\varepsilon},$$

where $\hat{m}_\varepsilon(u) = t_u u$ is given in (6.3.6). As in [322] (see Proposition 2.4.3 and Corollary 2.4.4) we have the following result:

Lemma 6.3.3 *Under the assumptions of Lemma 6.3.1, for $\varepsilon > 0$ we have:*

(i) $\Psi_\varepsilon \in C^1(\mathbb{S}_\varepsilon, \mathbb{R})$, *and*

$$\langle \Psi_\varepsilon'(w), v \rangle = \|m_\varepsilon(w)\|_\varepsilon \langle \mathcal{J}_\varepsilon'(m_\varepsilon(w)), v \rangle \quad \text{for all } v \in T_w \mathbb{S}_\varepsilon.$$

(ii) (w_n) *is a Palais–Smale sequence for Ψ_ε if and only if $(m_\varepsilon(w_n))$ is a Palais–Smale sequence for $\mathcal{J}_\varepsilon$. If $(u_n) \subset \mathcal{N}_\varepsilon$ is a bounded Palais–Smale sequence for $\mathcal{J}_\varepsilon$, then $(m_\varepsilon^{-1}(u_n))$ is a Palais–Smale sequence for Ψ_ε.*

(iii) $u \in \mathbb{S}_\varepsilon$ *is a critical point of Ψ_ε if and only if $m_\varepsilon(u)$ is a critical point of $\mathcal{J}_\varepsilon$. Moreover, the corresponding critical values coincide and*

$$\inf_{\mathbb{S}_\varepsilon} \Psi_\varepsilon = \inf_{\mathcal{N}_\varepsilon} \mathcal{J}_\varepsilon = c_\varepsilon.$$

Finally, let us show that $\mathcal{J}_\varepsilon$ possesses a mountain pass geometry [29].

Lemma 6.3.4 *The functional $\mathcal{J}_\varepsilon$ satisfies the following conditions:*

(i) *There exist $\alpha, \rho > 0$ such that $\mathcal{J}_\varepsilon(u) \geq \alpha$, with $\|u\|_\varepsilon = \rho$.*
(ii) *There exists $e \in \mathcal{H}_\varepsilon$ such that $\|e\|_\varepsilon > \rho$ and $\mathcal{J}_\varepsilon(e) < 0$.*

Proof

(i) Using (6.3.2), the inequality $V_0 \leq V(\varepsilon x)$, and Lemma 6.2.1, we get

$$\mathcal{J}_\varepsilon(u) \geq \frac{1}{2}\|u\|_\varepsilon^2 - \frac{\xi}{2V_0}\int_{\mathbb{R}^N} V(\varepsilon x)u^2\,dx - \frac{C_\xi}{q}\int_{\mathbb{R}^N} |u|^q\,dx$$

$$\geq \left(\frac{1}{2} - \frac{\xi}{2V_0}\right)\|u\|_\varepsilon^2 - \frac{C_\xi C_q}{q}\|u\|_\varepsilon^q.$$

Choosing $\xi \in (0, V_0)$, there exist $\alpha, \rho > 0$ such that

$$\mathcal{J}_\varepsilon(u) \geq \alpha > 0, \quad \text{with } \|u\|_\varepsilon = \rho.$$

(ii) By (f_4),

$$F(t) \geq C_1 t^\vartheta - C_2 \quad \text{for any } t \geq 0,$$

for some $C_1, C_2 > 0$. Taking $\varphi \in C_c^\infty(\mathbb{R}^N)$ such that $\varphi \geq 0$ and $\varphi \not\equiv 0$, we have

$$\mathcal{J}_\varepsilon(t\varphi) \leq \frac{t^2}{2}\|\varphi\|_\varepsilon^2 - t^\vartheta C_1 \int_{\text{supp}(\varphi)} \varphi^\vartheta\,dx + C_2|\text{supp}(\varphi)| \to -\infty \quad \text{as } t \to \infty.$$

□

By Lemma 6.3.4 and using a variant of the mountain pass theorem without the Palais–Smale condition (see Remark 2.2.10), we see that there exists a Palais–Smale sequence $(u_n) \subset \mathcal{H}_\varepsilon$ for $\mathcal{J}_\varepsilon$ at the level c_ε', that is,

$$\mathcal{J}_\varepsilon(u_n) \to c_\varepsilon' \quad \text{and} \quad \mathcal{J}_\varepsilon'(u_n) \to 0 \text{ in } \mathcal{H}_\varepsilon^*,$$

where

$$c_\varepsilon' = \inf_{\gamma \in \Gamma_\varepsilon} \max_{t \in [0,1]} \mathcal{J}_\varepsilon(\gamma(t)).$$

and

$$\Gamma_\varepsilon = \{\gamma \in C([0, 1], \mathcal{H}_\varepsilon) : \mathcal{J}_\varepsilon(0) = 0, \mathcal{J}_\varepsilon(\gamma(1)) < 0\}.$$

Motivated by Rabinowitz [299], we use an equivalent characterization of c'_ε that is more adequate for our purpose, namely

$$c'_\varepsilon = \inf_{u \in \mathcal{H}_\varepsilon \backslash \{0\}} \max_{t > 0} \mathcal{J}_\varepsilon(tu) = c_\varepsilon,$$

where in the last equality we used (6.3.7).

We note that, by using (f_4), any Palais–Smale sequence (u_n) of $\mathcal{J}_\varepsilon$ is bounded in $\mathcal{H}_\varepsilon$. Indeed, for any $n \in \mathbb{N}$, we have

$$
\begin{aligned}
C(1 + \|u_n\|_\varepsilon) &\geq \mathcal{J}_\varepsilon(u_n) - \frac{1}{\vartheta}\langle \mathcal{J}'_\varepsilon(u_n), u_n \rangle \\
&= \left(\frac{1}{2} - \frac{1}{\vartheta}\right)\|u_n\|_\varepsilon^2 + \frac{1}{\vartheta}\int_{\mathbb{R}^N}(f(u_n)u_n - \vartheta F(u_n))\,dx \\
&\geq \left(\frac{1}{2} - \frac{1}{\vartheta}\right)\|u_n\|_\varepsilon^2,
\end{aligned}
$$

which gives the desired result.

Remark 6.3.5 Arguing as in Remark 5.2.8, we may always suppose that $u_n \geq 0$ in $\mathbb{R}^N$ for all $n \in \mathbb{N}$.

6.3.2   The Autonomous Subcritical Problem

Let us consider the family of autonomous problems related to (P_ε), that is, for $\mu > 0$

$$
\begin{cases}
(-\Delta)^s u + \mu u = f(u) & \text{in } \mathbb{R}^N, \\
u \in H^s(\mathbb{R}^N), \quad u > 0 \text{ in } \mathbb{R}^N.
\end{cases}
\tag{P_μ}
$$

The functional energy corresponding to (P_μ) is given by

$$\mathcal{I}_\mu(u) = \frac{1}{2}\|u\|_\mu^2 - \int_{\mathbb{R}^N} F(u)\,dx$$

which is well defined on the space $\mathbb{X}_\mu = H^s(\mathbb{R}^N)$ endowed with the norm

$$\|u\|_\mu = \left([u]_s^2 + \mu\|u\|_{L^2(\mathbb{R}^N)}^2\right)^{\frac{1}{2}}.$$

Clearly, $\mathcal{I}_\mu \in C^1(\mathbb{X}_\mu, \mathbb{R})$ and its differential $\mathcal{I}_\mu'$ is given by

$$\langle \mathcal{I}_\mu'(u), \varphi \rangle = \langle u, v \rangle_{\mathcal{D}^{s,2}(\mathbb{R}^N)} + \mu \int_{\mathbb{R}^N} u\varphi \, dx - \int_{\mathbb{R}^N} f(u)\varphi \, dx$$

for any $u, \varphi \in \mathbb{X}_\mu$. Let us introduce the Nehari manifold associated with $\mathcal{I}_\mu$, that is,

$$\mathcal{M}_\mu = \left\{ u \in \mathbb{X}_\mu \setminus \{0\} : \langle \mathcal{I}_\mu'(u), u \rangle = 0 \right\}.$$

Note that assumption (f_4) implies that

$$\mathcal{I}_\mu(u) = \int_{\mathbb{R}^N} \left(\frac{1}{2} f(u)u - F(u) \right) dx \geq \left(\frac{1}{2} - \frac{1}{\vartheta} \right) \|u\|_\mu^2 \quad \text{for all } u \in \mathcal{M}_\mu. \quad (6.3.8)$$

Arguing as in the previous section and using (6.3.8), it is easy to prove the following lemma.

Lemma 6.3.6 *Under the assumptions of Lemma 6.3.1, for $\mu > 0$ we have:*

 (i) *for all $u \in \mathbb{S}_\mu$, there exists a unique $t_u > 0$ such that $t_u u \in \mathcal{M}_\mu$. Moreover, $m_\mu(u) = t_u u$ is the unique maximum of $\mathcal{I}_\mu$ on $\mathbb{X}_\mu$, where $\mathbb{S}_\mu = \{u \in \mathbb{X}_\mu : \|u\|_\mu = 1\}$.*
 (ii) *The set $\mathcal{M}_\mu$ is bounded away from 0. Furthermore, $\mathcal{M}_\mu$ is closed in $\mathbb{X}_\mu$.*
(iii) *There exists $\alpha > 0$ such that $t_u \geq \alpha$ for each $u \in \mathbb{S}_\mu$ and, for each compact subset $W \subset \mathbb{S}_\mu$, there exists $C_W > 0$ such that $t_u \leq C_W$ for all $u \in W$.*
(iv) *$\mathcal{M}_\mu$ is a regular manifold diffeomorphic to the unit sphere $\mathbb{S}_\mu$.*
 (v) *$d_\mu = \inf_{\mathcal{M}_\mu} \mathcal{I}_\mu > 0$ and $\mathcal{I}_\mu$ is bounded below on $\mathcal{M}_\mu$ by some positive constant.*

Now we define the following functionals $\hat{\Psi}_\mu : \mathbb{X}_\mu \setminus \{0\} \to \mathbb{R}$ and $\Psi_\mu : \mathbb{S}_\mu \to \mathbb{R}$ by setting

$$\hat{\Psi}_\mu(u) = \mathcal{I}_\mu(\hat{m}_\mu(u)) \quad \text{and} \quad \Psi_\mu = \hat{\Psi}_\mu|_{\mathbb{S}_\mu}.$$

The inverse of the mapping m_μ to $\mathbb{S}_\mu$ is given by

$$m_\mu^{-1} : \mathcal{M}_\mu \to \mathbb{S}_\mu, \quad m_\mu^{-1}(u) = \frac{u}{\|u\|_\mu}.$$

We have the next result:

Lemma 6.3.7 *Under the assumptions of Lemma 6.3.1, for $\mu > 0$ it holds that:*

(i) $\Psi_\mu \in C^1(\mathbb{S}_\mu, \mathbb{R})$, *and*

$$\langle \Psi'_\mu(w), v \rangle = \|m_\mu(w)\|_\mu \langle \mathcal{I}'_\mu(m_\mu(w)), v \rangle \quad \text{for all } v \in T_w \mathbb{S}_\mu.$$

(ii) (w_n) *is a Palais–Smale sequence for* Ψ_μ *if and only if* $(m_\mu(w_n))$ *is a Palais–Smale sequence for* $\mathcal{I}_\mu$. *If* $(u_n) \subset \mathcal{M}_\mu$ *is a bounded Palais–Smale sequence for* $\mathcal{I}_\mu$, *then* $(m_\mu^{-1}(u_n))$ *is a Palais–Smale sequence for* Ψ_μ.

(iii) $u \in \mathbb{S}_\mu$ *is a critical point of* Ψ_μ *if and only if* $m_\mu(u)$ *is a critical point of* $\mathcal{I}_\mu$. *Moreover, the corresponding critical values coincide and*

$$\inf_{\mathbb{S}_\mu} \Psi_\mu = \inf_{\mathcal{M}_\mu} \mathcal{I}_\mu = d_\mu.$$

Remark 6.3.8 As in (6.3.7), items *(i)–(iii)* of Lemma 6.3.2 imply that c_μ admits the following minimax characterization:

$$d_\mu = \inf_{u \in \mathcal{M}_\mu} \mathcal{I}_\mu(u) = \inf_{u \in \mathbb{X}_\mu \setminus \{0\}} \max_{t > 0} \mathcal{I}_\mu(tu) = \inf_{u \in \mathbb{S}_\mu} \max_{t > 0} \mathcal{I}_\mu(tu). \qquad (6.3.9)$$

Arguing as in the proof of Lemma 6.3.4, it is easy to verify that $\mathcal{I}_\mu$ has a mountain pass geometry. Thus we can use a variant of the mountain pass theorem without the Palais–Smale condition (see Remark 2.2.10) to find a sequence (u_n) in $\mathbb{X}_\mu$ such that

$$\mathcal{I}_\mu(u_n) \to d'_\mu = \inf_{\gamma \in \Gamma_\mu} \max_{t \in [0,1]} \mathcal{I}_\mu(\gamma(t)) \quad \text{and} \quad \mathcal{I}'_\mu(u_n) \to 0 \text{ in } \mathbb{X}^*_\mu,$$

where $\Gamma_\mu = \{\gamma \in C([0, 1], \mathbb{X}_\mu) : \mathcal{I}_\mu(0) = 0, \mathcal{I}_\mu(\gamma(1)) < 0\}$. By using (f_4), we can see that (u_n) is bounded in $\mathbb{X}_\mu$. In what follows, we use the equivalent characterization of d'_μ given by

$$d'_\mu = \inf_{u \in \mathbb{X}_\mu \setminus \{0\}} \max_{t > 0} \mathcal{I}_\mu(tu) = d_\mu,$$

where in the last equality we used (6.3.9).

We now study the minimizing sequences for $\mathcal{I}_\mu$.

Lemma 6.3.9 *Let* $(u_n) \subset \mathcal{M}_\mu$ *be such that* $\mathcal{I}_\mu(u_n) \to d_\mu$. *Then,* (u_n) *is bounded in* $\mathbb{X}_\mu$. *Moreover, there exist a sequence* $(y_n) \subset \mathbb{R}^N$ *and constants* $R, \beta > 0$ *such that*

$$\liminf_{n \to \infty} \int_{B_R(y_n)} u_n^2 \, dx \geq \beta > 0.$$

Proof By using (f_4), we have

$$d_\mu = \mathcal{I}_\mu(u_n) - \frac{1}{\vartheta}\langle \mathcal{I}'_\mu(u_n), u_n\rangle$$

$$= \left(\frac{1}{2} - \frac{1}{\vartheta}\right)\|u_n\|_\mu^2 + \frac{1}{\vartheta}\int_{\mathbb{R}^N}(f(u_n)u_n - \vartheta F(u_n))\,dx$$

$$\geq \left(\frac{1}{2} - \frac{1}{\vartheta}\right)\|u_n\|_\mu^2,$$

that is, (u_n) is bounded in $\mathbb{X}_\mu$. Now, to prove the second conclusion of the lemma, we argue by contradiction. Assume that for any $R > 0$ it holds that

$$\lim_{n\to\infty}\sup_{y\in\mathbb{R}^N}\int_{B_R(y)}u_n^2\,dx = 0.$$

Since (u_n) is bounded in $\mathbb{X}_\mu$, it follows from Lemma 1.4.4 that

$$u_n \to 0 \text{ in } L^t(\mathbb{R}^N) \quad \text{for any } t \in (2, 2_s^*). \tag{6.3.10}$$

Fix $\xi \in (0, \mu)$. Since $\langle \mathcal{I}'_\mu(u_n), u_n\rangle = 0$, (6.3.1) and the boundedness of (u_n) in $\mathbb{X}_\mu$ imply that

$$0 \leq \|u_n\|_\mu^2 \leq \xi\int_{\mathbb{R}^N}u_n^2\,dx + C_\xi\int_{\mathbb{R}^N}|u_n|^q\,dx \leq \frac{\xi}{\mu}\|u_n\|_\mu^2 + C_\xi\|u_n\|_{L^q(\mathbb{R}^N)}^q,$$

whence

$$\left(1 - \frac{\xi}{\mu}\right)\|u_n\|_\mu^2 \leq C_\xi\|u_n\|_{L^q(\mathbb{R}^N)}^q.$$

In the light of (6.3.10), we obtain that $u_n \to 0$ in $\mathbb{X}_\mu$ and then $\mathcal{I}_\mu(u_n) \to 0$, which is a contradiction because $d_\mu > 0$.   □

Next we prove the main result for the autonomous problem (P_μ).

Lemma 6.3.10 *The problem (P_μ) has at least one positive ground state solution.*

Proof By item (v) of Lemma 6.3.6, $d_\mu > 0$ for each $\mu > 0$. Moreover, if $u \in \mathcal{M}_\mu$ satisfies $\mathcal{I}_\mu(u) = d_\mu$, then $m_\mu^{-1}(u)$ is a minimizer of Ψ_μ and therefore a critical point of Ψ_μ. In view of Lemma 6.3.7, we see that u is a critical point of $\mathcal{I}_\mu$. It remains to show that there exists a minimizer of $\mathcal{I}_\mu|_{\mathcal{M}_\mu}$. Theorem 2.2.1 yields a sequence $(v_n) \subset \mathbb{S}_\mu$ such that $\Psi_\mu(v_n) \to d_\mu$ and $\Psi'_\mu(v_n) \to 0$ as $n \to \infty$. Let $u_n = m_\mu(v_n) \in \mathcal{M}_\mu$. Then,

$\mathcal{I}_\mu(u_n) \to d_\mu$ and $\mathcal{I}'_\mu(u_n) \to 0$ in $\mathbb{X}^*_\mu$ as $n \to \infty$. It is easy to see that (u_n) is bounded in $\mathbb{X}_\mu$ and $u_n \rightharpoonup u$ in $H^s(\mathbb{R}^N)$. Clearly, $\mathcal{I}'_\mu(u) = 0$.

Assume that $u \neq 0$. Then $u \in \mathcal{M}_\mu$. If we can show that

$$\|u_n\|_\mu \to \|u\|_\mu, \tag{6.3.11}$$

then we can use the fact that $\mathbb{X}_\mu$ is a Hilbert space to deduce that $u_n \to u$ in $\mathbb{X}_\mu$. Consequently, $\mathcal{I}_\mu(u) = d_\mu$.

Let us verify that (6.3.11) holds. By Fatou's lemma,

$$\|u\|^2_\mu \leq \liminf_{n\to\infty} \|u_n\|^2_\mu. \tag{6.3.12}$$

Suppose, by contradiction, that

$$\|u\|^2_\mu < \limsup_{n\to\infty} \|u_n\|^2_\mu. \tag{6.3.13}$$

Let us note that

$$\begin{aligned}
d_\mu + o_n(1) &= \mathcal{I}_\mu(u_n) - \frac{1}{\vartheta}\langle \mathcal{I}'_\mu(u_n), u_n \rangle \\
&= \left(\frac{1}{2} - \frac{1}{\vartheta}\right)\|u_n\|^2_\mu + \int_{\mathbb{R}^N}\left[\frac{1}{\vartheta}f(u_n)u_n - F(u_n)\right]dx.
\end{aligned} \tag{6.3.14}$$

Then, recalling that

$$\limsup_{n\to\infty}(a_n + b_n) \geq \limsup_{n\to\infty} a_n + \liminf_{n\to\infty} b_n$$

and $\vartheta > 2$, Fatou's lemma, (6.3.13), (6.3.14), (f_4) and $u \in \mathcal{M}_\mu$ imply that

$$\begin{aligned}
d_\mu &\geq \left(\frac{1}{2} - \frac{1}{\vartheta}\right)\limsup_{n\to\infty}\|u_n\|^2_\mu + \liminf_{n\to\infty}\int_{\mathbb{R}^N}\left[\frac{1}{\vartheta}f(u_n)u_n - F(u_n)\right]dx \\
&> \left(\frac{1}{2} - \frac{1}{\vartheta}\right)\|u\|^2_\mu + \int_{\mathbb{R}^N}\left[\frac{1}{\vartheta}f(u)u - F(u)\right]dx \\
&= \mathcal{I}_\mu(u) - \frac{1}{\vartheta}\langle \mathcal{I}'_\mu(u), u \rangle = \mathcal{I}_\mu(u) \geq d_\mu,
\end{aligned}$$

which gives a contradiction. Consequently, by (6.3.12),

$$\|u\|^2_\mu \leq \liminf_{n\to\infty}\|u_n\|^2_\mu \leq \limsup_{n\to\infty}\|u_n\|^2_\mu \leq \|u\|^2_\mu$$

and this implies that (6.3.11) holds true.

Finally, we consider the case $u = 0$. Arguing as in the proof of Lemma 6.3.9, we can find a sequence $(y_n) \subset \mathbb{R}^N$ and constants $R, \beta > 0$ such that

$$\liminf_{n \to \infty} \int_{B_R(y_n)} u_n^2 \, dx \geq \beta > 0.$$

Set $v_n = u_n(\cdot + y_n)$. Then, using that $\mathcal{M}_\mu$ and $\mathcal{I}_\mu$ are invariant under translations, we infer that (v_n) is a Palais–Smale sequence for $\mathcal{I}_\mu$ at the level d_μ, $(v_n) \subset \mathcal{M}_\mu$ and $v_n \rightharpoonup v \neq 0$ in $H^s(\mathbb{R}^N)$. Thus we can proceed as above to deduce that $v_n \to v$ strongly in $\mathbb{X}_\mu$ and thus $v \in \mathcal{M}_\mu$ and $\mathcal{I}_\mu(v) = d_\mu$.

Let u be the ground state solution of (6.1.1) obtained above. Then u is positive. Indeed, since $\langle \mathcal{I}'_\mu(u), u^- \rangle = 0$, $f(t) = 0$ for $t \leq 0$ and $(x - y)(x^- - y^-) \geq |x^- - y^-|^2$ for $x, y \in \mathbb{R}$, where $x^- = \min\{x, 0\}$, we see that

$$\|u^-\|_\mu^2 \leq \langle u, u^- \rangle_{\mathcal{D}^{s,2}(\mathbb{R}^N)} + \int_{\mathbb{R}^N} \mu u u^- \, dx = \int_{\mathbb{R}^N} f(u) u^- \, dx = 0$$

which implies that $u^- = 0$, that is, $u \geq 0$. By Proposition 3.2.14, $u \in L^\infty(\mathbb{R}^N)$, and using Proposition 1.3.2 we can see that $u \in C^{0,\alpha}(\mathbb{R}^N)$. Applying Proposition 1.3.11-(ii) (or Theorem 1.3.5), we conclude that $u > 0$ in $\mathbb{R}^N$.   □

Following the above arguments, we obtain the next compactness result:

Theorem 6.3.11 *Let $(u_n) \subset \mathcal{M}_{V_0}$ be a sequence such that $\mathcal{I}_{V_0}(u_n) \to d_{V_0}$. Then we have either*

(i) *(u_n) has a subsequence that converges strongly in $H^s(\mathbb{R}^N)$, or*
(ii) *there exists a sequence $(\tilde{y}_n) \subset \mathbb{R}^N$ such that, up to a subsequence, $v_n = u_n(\cdot + \tilde{y}_n)$ converges strongly in $H^s(\mathbb{R}^N)$.*

In particular, there exists a minimizer for d_{V_0}.

Proof By Lemma 6.3.9, we know that (u_n) is bounded in $\mathbb{X}_{V_0}$. From Lemma 6.3.7, $v_n = m_\varepsilon^{-1}(u_n)$ is a minimizing sequence of Ψ_{V_0}. By Theorem 2.2.1, we may assume that $\Psi_{V_0}(v_n) \to d_{V_0}$ and $\Psi'_{V_0}(v_n) \to 0$. Then $\mathcal{I}_{V_0}(u_n) \to d_{V_0}$, $\mathcal{I}'_{V_0}(u_n) \to 0$ and $\langle \mathcal{I}'_{V_0}(u_n), u_n \rangle = 0$ where $u_n = m_{V_0}(v_n)$. Hence, up to a subsequence, we may assume that there exists $u \in H^s(\mathbb{R}^N)$ such that $u_n \rightharpoonup u$ in $H^s(\mathbb{R}^N)$. At this point we can argue as in the proof of Lemma 6.3.10 by considering the cases $u \neq 0$ and $u = 0$.   □

6.3.3 An Existence Result for (6.1.1)

In this section we focus on the existence of solutions to (6.1.1) under the assumption that ε is sufficiently small. We begin with the next useful lemma.

Lemma 6.3.12 *Let $(u_n) \subset \mathcal{N}_\varepsilon$ be a sequence such that $\mathcal{J}_\varepsilon(u_n) \to c$ and $u_n \rightharpoonup 0$ in $\mathcal{H}_\varepsilon$. Then either*

(a) $u_n \to 0$ in $\mathcal{H}_\varepsilon$, or
(b) *there are a sequence $(y_n) \subset \mathbb{R}^N$ and constants $R, \beta > 0$ such that*

$$\liminf_{n \to \infty} \int_{B_R(y_n)} u_n^2 \, dx \geq \beta > 0.$$

Proof Assume that (b) does not hold true. Then, for any $R > 0$,

$$\lim_{n \to \infty} \sup_{y \in \mathbb{R}^N} \int_{B_R(y)} u_n^2 \, dx = 0.$$

Since (u_n) is bounded in $\mathcal{H}_\varepsilon$, Lemma 1.4.4 shows that

$$u_n \to 0 \text{ in } L^t(\mathbb{R}^N) \quad \text{for any } t \in (2, 2_s^*). \tag{6.3.15}$$

Now, we can argue as in the proof of Lemma 6.3.9 to deduce that $\|u_n\|_\varepsilon \to 0$ as $n \to \infty$. $\qquad \square$

To obtain a compactness result for $\mathcal{J}_\varepsilon$, we need the following auxiliary lemma.

Lemma 6.3.13 *Assume that $V_\infty < \infty$. Let $(v_n) \subset \mathcal{N}_\varepsilon$ be a sequence such that $\mathcal{J}_\varepsilon(v_n) \to d$ with $v_n \rightharpoonup 0$ in $\mathcal{H}_\varepsilon$. If $v_n \not\to 0$ in $\mathcal{H}_\varepsilon$, then $d \geq d_{V_\infty}$, where d_{V_∞} is the infimum of $\mathcal{I}_{V_\infty}$ over $\mathcal{M}_{V_\infty}$.*

Proof Let $(t_n) \subset (0, \infty)$ be such that $(t_n v_n) \subset \mathcal{M}_{V_\infty}$.

Claim 1 *We have*

$$\limsup_{n \to \infty} t_n \leq 1.$$

Suppose, by contradiction, that there exist $\delta > 0$ and a subsequence, still denoted by (t_n), such that

$$t_n \geq 1 + \delta \quad \text{for all } n \in \mathbb{N}. \tag{6.3.16}$$

Since $\langle \mathcal{J}'_\varepsilon(v_n), v_n \rangle = 0$, we deduce that

$$[v_n]_s^2 + \int_{\mathbb{R}^N} V(\varepsilon x) v_n^2 \, dx = \int_{\mathbb{R}^N} f(v_n) v_n \, dx. \tag{6.3.17}$$

Further, since $t_n v_n \in \mathcal{M}_{V_\infty}$, we also have

$$t_n^2 [v_n]_s^2 + t_n^2 V_\infty \int_{\mathbb{R}^N} v_n^2 \, dx = \int_{\mathbb{R}^N} f(t_n v_n) t_n v_n \, dx. \tag{6.3.18}$$

Combining (6.3.17) and (6.3.18) we obtain

$$\int_{\mathbb{R}^N} \left(\frac{f(t_n v_n)}{t_n v_n} - \frac{f(v_n)}{v_n} \right) v_n^2 \, dx = \int_{\mathbb{R}^N} (V_\infty - V(\varepsilon x)) \, v_n^2 \, dx.$$

Hypothesis (V) ensures that, given $\zeta > 0$ there exists $R = R(\zeta) > 0$ such that

$$V(\varepsilon x) \geq V_\infty - \zeta \quad \text{for any } |x| \geq R. \tag{6.3.19}$$

Now, since $v_n \to 0$ in $L^2(B_R)$ and since the sequence (v_n) is bounded in $\mathcal{H}_\varepsilon$, we infer that

$$\int_{\mathbb{R}^N} (V_\infty - V(\varepsilon x)) \, v_n^2 \, dx$$

$$= \int_{B_R} (V_\infty - V(\varepsilon x)) \, v_n^2 \, dx + \int_{B_R^c} (V_\infty - V(\varepsilon x)) \, v_n^2 \, dx$$

$$\leq V_\infty \int_{B_R} v_n^2 \, dx + \zeta \int_{B_R^c} v_n^2 \, dx$$

$$\leq o_n(1) + \frac{\zeta}{V_0} \|v_n\|_\varepsilon^2 \leq o_n(1) + \zeta C.$$

Thus

$$\int_{\mathbb{R}^N} \left(\frac{f(t_n v_n)}{t_n v_n} - \frac{f(v_n)}{v_n} \right) v_n^2 \, dx \leq \zeta C + o_n(1). \tag{6.3.20}$$

Since $v_n \nrightarrow 0$ in $\mathcal{H}_\varepsilon$, we apply Lemma 6.3.12 to deduce the existence of a sequence $(y_n) \subset \mathbb{R}^N$ and two positive numbers $\bar{R}$ and β such that

$$\int_{B_{\bar{R}}(y_n)} v_n^2 \, dx \geq \beta > 0. \tag{6.3.21}$$

Define $\bar{v}_n = v_n(x + y_n)$. By condition (V) and the boundedness of (v_n) in $\mathcal{H}_\varepsilon$,

$$\|\bar{v}_n\|_{V_0}^2 = \|v_n\|_{V_0}^2 \leq [v_n]_s^2 + \int_{\mathbb{R}^N} V(\varepsilon x) v_n^2 \, dx = \|v_n\|_\varepsilon^2 \leq C,$$

therefore $(\bar{v}_n)$ is bounded in $H^s(\mathbb{R}^N)$. Taking into account that $H^s(\mathbb{R}^N)$ is a reflexive Banach space, we may assume that $\bar{v}_n \rightharpoonup \bar{v}$ in $H^s(\mathbb{R}^N)$. By (6.3.21), there exists $\Omega \subset \mathbb{R}^N$ with positive measure and such that $\bar{v} > 0$ in Ω. Using (6.3.16), assumption (f_5) and (6.3.20), we can infer that

$$0 < \int_\Omega \left(\frac{f((1+\delta)\bar{v}_n)}{((1+\delta)\bar{v}_n)} - \frac{f(\bar{v}_n)}{\bar{v}_n} \right) \bar{v}_n^2 \, dx \leq \zeta C + o_n(1).$$

Letting $n \to \infty$ and applying Fatou's lemma we obtain

$$0 < \int_\Omega \left(\frac{f((1+\delta)\bar{v})}{((1+\delta)\bar{v})} - \frac{f(\bar{v})}{\bar{v}} \right) \bar{v}^2 \, dx \leq \zeta C$$

for any $\zeta > 0$, which is a contradiction.

Now, we distinguish the following cases:

Case 1 Assume that $\limsup_{n\to\infty} t_n = 1$. Thus, there exists (t_n) such that $t_n \to 1$. Recalling that $\mathcal{J}_\varepsilon(v_n) \to d$, we have

$$
\begin{aligned}
d + o_n(1) &= \mathcal{J}_\varepsilon(v_n) \\
&= \mathcal{J}_\varepsilon(v_n) - \mathcal{I}_{V_\infty}(t_n v_n) + \mathcal{I}_{V_\infty}(t_n v_n) \\
&\geq \mathcal{J}_\varepsilon(v_n) - \mathcal{I}_{V_\infty}(t_n v_n) + d_{V_\infty}.
\end{aligned}
\tag{6.3.22}
$$

Let us compute the difference $\mathcal{J}_\varepsilon(v_n) - \mathcal{I}_{V_\infty}(t_n v_n)$:

$$
\begin{aligned}
&\mathcal{J}_\varepsilon(v_n) - \mathcal{I}_{V_\infty}(t_n v_n) \\
&= \frac{(1 - t_n^2)}{2}[v_n]_s^2 + \frac{1}{2}\int_{\mathbb{R}^N}(V(\varepsilon x) - t_n^2 V_\infty) v_n^2 \, dx + \int_{\mathbb{R}^N}(F(t_n v_n) - F(v_n)) \, dx.
\end{aligned}
\tag{6.3.23}
$$

Using condition (V), $v_n \to 0$ in $L^2(B_R)$, $t_n \to 1$, (6.3.19), and

$$V(\varepsilon x) - t_n^2 V_\infty = (V(\varepsilon x) - V_\infty) + (1 - t_n^2)V_\infty \geq -\zeta + (1 - t_n^2)V_\infty \quad \text{for } |x| \geq R,$$

we get

$$\int_{\mathbb{R}^N} \left(V(\varepsilon x) - t_n^2 V_\infty \right) v_n^2 \, dx$$

$$= \int_{B_R} \left(V(\varepsilon x) - t_n^2 V_\infty \right) v_n^2 \, dx + \int_{B_R^c} \left(V(\varepsilon x) - t_n^2 V_\infty \right) v_n^2 \, dx$$

$$\geq (V_0 - t_n^2 V_\infty) \int_{B_R} v_n^2 \, dx - \zeta \int_{B_R^c} v_n^2 \, dx + V_\infty (1 - t_n^2) \int_{B_R^c} v_n^2 \, dx$$

$$\geq o_n(1) - \zeta C. \tag{6.3.24}$$

On the other hand, since (v_n) is bounded in $\mathcal{H}_\varepsilon$,

$$\frac{(1 - t_n^2)}{2} [v_n]_s^2 = o_n(1). \tag{6.3.25}$$

Putting together (6.3.23), (6.3.24), and (6.3.25), we obtain

$$\mathcal{J}_\varepsilon(v_n) - \mathcal{I}_{V_\infty}(t_n v_n) = \int_{\mathbb{R}^N} (F(t_n v_n) - F(v_n)) \, dx + o_n(1) - \zeta C. \tag{6.3.26}$$

Now, we claim that

$$\int_{\mathbb{R}^N} (F(t_n v_n) - F(v_n)) \, dx = o_n(1). \tag{6.3.27}$$

Indeed, applying the mean value theorem and (6.3.1) we have

$$\int_{\mathbb{R}^N} |F(t_n v_n) - F(v_n)| \, dx \leq C|t_n - 1| \left(\|v_n\|_{L^2(\mathbb{R}^N)}^2 + \|v_n\|_{L^q(\mathbb{R}^N)}^q \right),$$

so thanks to the boundedness of (v_n) in $\mathcal{H}_\varepsilon$ the claim is proved. Now (6.3.22), (6.3.26), and (6.3.27) imply that

$$d + o_n(1) \geq o_n(1) - \zeta C + d_{V_\infty},$$

and passing to the limit as $\zeta \to 0$ we get $d \geq d_{V_\infty}$.

Case 2 Assume that $\limsup_{n \to \infty} t_n = t_0 < 1$. Then there is a subsequence, still denoted by (t_n), such that $t_n \to t_0$ and $t_n < 1$ for any $n \in \mathbb{N}$. Let us observe that

$$d + o_n(1) = \mathcal{J}_\varepsilon(v_n) - \frac{1}{2}\langle \mathcal{J}_\varepsilon'(v_n), v_n \rangle = \int_{\mathbb{R}^N} \left(\frac{1}{2} f(v_n) v_n - F(v_n) \right) dx. \tag{6.3.28}$$

Using the facts that $t_n v_n \in \mathcal{M}_{V_\infty}$, (6.3.3) and (6.3.28), we obtain

$$
d_{V_\infty} \leq \mathcal{I}_{V_\infty}(t_n v_n) = \mathcal{I}_{V_\infty}(t_n v_n) - \frac{1}{2}\langle \mathcal{I}'_{V_\infty}(t_n v_n), t_n v_n \rangle
$$

$$
= \int_{\mathbb{R}^N} \left(\frac{1}{2} f(t_n v_n) t_n v_n - F(t_n v_n) \right) dx
$$

$$
\leq \int_{\mathbb{R}^N} \left(\frac{1}{2} f(v_n) v_n - F(v_n) \right) dx = d + o_n(1).
$$

Taking the limit as $n \to \infty$ we get $d \geq d_{V_\infty}$.

$\square$

At this point we are able to prove the following compactness result.

Proposition 6.3.14 *Let $(u_n) \subset \mathcal{N}_\varepsilon$ be such that $\mathcal{J}_\varepsilon(u_n) \to c$, where $c < d_{V_\infty}$ if $V_\infty < \infty$, and $c \in \mathbb{R}$ if $V_\infty = \infty$. Then (u_n) has a subsequence that converges strongly in $\mathcal{H}_\varepsilon$.*

Proof It is readily seen that (u_n) is bounded in $\mathcal{H}_\varepsilon$. Then, up to a subsequence, we may assume that

$$
u_n \rightharpoonup u, \ \text{in } \mathcal{H}_\varepsilon,
$$

$$
u_n \to u, \ \text{in } L^q_{\text{loc}}(\mathbb{R}^N) \quad \forall q \in [1, 2^*_s), \tag{6.3.29}
$$

$$
u_n \to u, \ \text{a.e. in } \mathbb{R}^N.
$$

Using assumptions (f_2)–(f_3), (6.3.29) and Lemma 6.2.2, it is routine to check that $\mathcal{J}'_\varepsilon(u) = 0$. Now, let $v_n = u_n - u$. By the Brezis-Lieb lemma [113] and Lemma 6.2.4,

$$
\mathcal{J}_\varepsilon(v_n) = \frac{\|u_n\|^2_\varepsilon}{2} - \frac{\|u\|^2_\varepsilon}{2} - \int_{\mathbb{R}^N} F(u_n)\, dx + \int_{\mathbb{R}^N} F(u)\, dx + o_n(1)
$$

$$
= \mathcal{J}_\varepsilon(u_n) - \mathcal{J}_\varepsilon(u) + o_n(1)
$$

$$
= c - \mathcal{J}_\varepsilon(u) + o_n(1) = d + o_n(1), \tag{6.3.30}
$$

and $\mathcal{J}'_\varepsilon(v_n) = o_n(1)$. Assume that $V_\infty < \infty$. It follows from (6.3.30) and (6.3.4) that

$$
d \leq c < d_{V_\infty}
$$

which together with Lemma 6.3.13 gives $v_n \to 0$ in $\mathcal{H}_\varepsilon$, that is, $u_n \to u$ in $\mathcal{H}_\varepsilon$.

Let us consider the case $V_\infty = \infty$. Then we use Lemma 6.2.3 to deduce that $v_n \to 0$ in $L^r(\mathbb{R}^N)$ for all $r \in [2, 2_s^*)$. This fact combined with assumptions (f_2) and (f_3) implies that

$$\int_{\mathbb{R}^N} f(v_n)v_n \, dx = o_n(1). \tag{6.3.31}$$

Since $\langle \mathcal{J}_\varepsilon'(v_n), v_n \rangle = 0$ and applying (6.3.31) we infer that

$$\|v_n\|_\varepsilon^2 = o_n(1),$$

which yields $u_n \to u$ in $\mathcal{H}_\varepsilon$. □

We end this section by establishing the existence of a positive solution to (P_ε) when $\varepsilon > 0$ is small enough.

Theorem 6.3.15 *Assume that (V) and (f_1)–(f_5) hold. Then there exists $\varepsilon_0 > 0$ such that, for any $\varepsilon \in (0, \varepsilon_0)$, problem (P_ε) admits a positive ground state solution.*

Proof From item (v) of Lemma 6.3.2, we know that $c_\varepsilon \geq \rho > 0$ for each $\varepsilon > 0$. Moreover, if $u \in \mathcal{N}_\varepsilon$ satisfies $\mathcal{J}_\varepsilon(u) = c_\varepsilon$, then $m_\varepsilon^{-1}(u)$ is a minimizer of Ψ_ε and it is a critical point of Ψ_ε. In view of Lemma 6.3.3, we see that u is a critical point of $\mathcal{J}_\varepsilon$.

Let us show that there exists a minimizer of $\mathcal{J}_\varepsilon|_{\mathcal{N}_\varepsilon}$. Theorem 2.2.1 yields a sequence $(v_n) \subset \mathbb{S}_\varepsilon$ such that $\Psi_\varepsilon(v_n) \to c_\varepsilon$ and $\Psi_\varepsilon'(v_n) \to 0$ as $n \to \infty$. Let $u_n = m_\varepsilon(v_n) \in \mathcal{N}_\varepsilon$. Then, by Lemma 6.3.3, $\mathcal{J}_\varepsilon(u_n) \to c_\varepsilon$, $\langle \mathcal{J}_\varepsilon'(u_n), u_n \rangle = 0$ and $\mathcal{J}_\varepsilon'(u_n) \to 0$ as $n \to \infty$. Therefore, (u_n) is a Palais–Smale sequence for $\mathcal{J}_\varepsilon$ at level c_ε. Consequently, (u_n) is bounded in $\mathcal{H}_\varepsilon$, and we denote by u its weak limit. It is easy to verify that $\mathcal{J}_\varepsilon'(u) = 0$.

When $V_\infty = \infty$, we use Lemma 6.2.3 to deduce that $\mathcal{J}_\varepsilon(u) = c_\varepsilon$ and $\mathcal{J}_\varepsilon'(u) = 0$.

Now, we deal with the case $V_\infty < \infty$. By virtue of Proposition 6.3.14, it is enough to show that $c_\varepsilon < d_{V_\infty}$ for small $\varepsilon > 0$. Without loss of generality, we may suppose that

$$V(0) = V_0 = \inf_{x \in \mathbb{R}^N} V(x).$$

Let $\mu \in \mathbb{R}$ be such that $\mu \in (V_0, V_\infty)$. Clearly, $d_{V_0} < d_\mu < d_{V_\infty}$. By Lemma 6.3.10, it follows that the autonomous problem (P_{V_0}) admits a positive ground state $w \in H^s(\mathbb{R}^N)$. Let $\eta_r \in C_c^\infty(\mathbb{R}^N)$ be a cut-off function such that $\eta_r = 1$ in B_r and $\eta_r = 0$ in B_{2r}^c. Let us define $w_r(x) = \eta_r(x)w(x)$, and take $t_r > 0$ such that

$$\mathcal{I}_\mu(t_r w_r) = \max_{t \geq 0} \mathcal{I}_\mu(t w_r).$$

We claim that there exists r sufficiently large for which $\mathcal{I}_\mu(t_r w_r) < d_{V_\infty}$.

Assume, by contradiction, that $\mathcal{I}_\mu(t_r w_r) \geq d_{V_\infty}$ for any $r > 0$. Since $w_r \to w$ in $H^s(\mathbb{R}^N)$ as $r \to \infty$ thanks to Lemma 1.4.8, $t_r w_r$ and w belong to $\mathcal{M}_\mu$ and using (f_5), we have that $t_r \to 1$. Therefore,

$$d_{V_\infty} \leq \liminf_{r\to\infty} \mathcal{I}_\mu(t_r w_r) = \mathcal{I}_\mu(w) = d_\mu,$$

which leads to a contradiction because $d_{V_\infty} > d_\mu$. Hence, there exists $r > 0$ such that

$$\mathcal{I}_\mu(t_r w_r) = \max_{\tau\geq 0} \mathcal{I}_\mu(\tau(t_r w_r)) \quad \text{and} \quad \mathcal{I}_\mu(t_r w_r) < d_{V_\infty}. \tag{6.3.32}$$

Condition (V) implies that there exists $\varepsilon_0 > 0$ such that

$$V(\varepsilon x) \leq \mu \quad \text{for all } x \in \text{supp}(w_r), \quad \varepsilon \in (0, \varepsilon_0). \tag{6.3.33}$$

Therefore, using (6.3.32) and (6.3.33), we deduce that for all $\varepsilon \in (0, \varepsilon_0)$

$$c_\varepsilon \leq \max_{\tau\geq 0} \mathcal{J}_\varepsilon(\tau(t_r w_r)) \leq \max_{\tau\geq 0} \mathcal{I}_\mu(\tau(t_r w_r)) = \mathcal{I}_\mu(t_r w_r) < d_{V_\infty}$$

which implies that $c_\varepsilon < d_{V_\infty}$ for any $\varepsilon > 0$ sufficiently small.

Now, if u is a nonnegative ground state of (P_ε), then we can use a Moser iteration argument (see Lemma 6.3.23 below) to deduce that $u \in L^\infty(\mathbb{R}^N) \cap C_{loc}^{0,\alpha}(\mathbb{R}^N)$ and that $u(x) \to 0$ as $|x| \to \infty$. By Theorem 1.3.5, $u > 0$ in $\mathbb{R}^N$, as needed. $\qquad\square$

6.3.4 A Multiplicity Result for (6.1.1)

This section deals with the multiplicity of solutions to (6.1.1). We begin by proving the following result which will be needed to implement the barycenter machinery.

Proposition 6.3.16 *Let $\varepsilon_n \to 0$ and $(u_n) = (u_{\varepsilon_n}) \subset \mathcal{N}_{\varepsilon_n}$ be such that $\mathcal{J}_{\varepsilon_n}(u_n) \to d_{V_0}$. Then there exists $(\tilde{y}_n) = (\tilde{y}_{\varepsilon_n}) \subset \mathbb{R}^N$ such that $v_n(x) = u_n(x + \tilde{y}_n)$ has a subsequence that converges in $H^s(\mathbb{R}^N)$. Moreover, up to a subsequence, $y_n \to y \in M$, where $y_n = \varepsilon_n \tilde{y}_n$.*

Proof Since $\langle \mathcal{J}'_{\varepsilon_n}(u_n), u_n \rangle = 0$ and $\mathcal{J}_{\varepsilon_n}(u_n) \to d_{V_0}$, we deduce that (u_n) is bounded in $\mathcal{H}_\varepsilon$. From $d_{V_0} > 0$, we can infer that $\|u_n\|_{\varepsilon_n} \nrightarrow 0$. Therefore, as in the proof of Lemma 6.3.12, we can find a sequence $(\tilde{y}_n) \subset \mathbb{R}^N$ and constants $R, \beta > 0$ such that

$$\liminf_{n\to\infty} \int_{B_R(\tilde{y}_n)} u_n^2 \, dx \geq \beta. \tag{6.3.34}$$

Let us define

$$v_n(x) = u_n(x + \tilde{y}_n).$$

Thanks to the boundedness of (u_n) and (6.3.34), we may assume that $v_n \rightharpoonup v$ in $H^s(\mathbb{R}^N)$ for some $v \neq 0$. Let $(t_n) \subset (0, \infty)$ be such that $w_n = t_n v_n \in \mathcal{M}_{V_0}$, and we set $y_n = \varepsilon_n \tilde{y}_n$.

Thus, using the change of variables $z \mapsto x + \tilde{y}_n$, the fact that $V(x) \geq V_0$ and the translation invariance, we can see that

$$d_{V_0} \leq \mathcal{I}_{V_0}(w_n) \leq \mathcal{J}_{\varepsilon_n}(t_n v_n) \leq \mathcal{J}_{\varepsilon_n}(u_n) = d_{V_0} + o_n(1).$$

Hence, $\mathcal{I}_{V_0}(w_n) \to d_{V_0}$. This fact and $(w_n) \subset \mathcal{M}_{V_0}$ imply that there exists $K > 0$ such that $\|w_n\|_{V_0} \leq K$ for all $n \in \mathbb{N}$. Moreover, we can prove that the sequence (t_n) is bounded in $\mathbb{R}$. In fact, $v_n \nrightarrow 0$ in $H^s(\mathbb{R}^N)$, so there exists $\alpha > 0$ such that $\|v_n\|_{V_0} \geq \alpha$. Consequently,

$$|t_n|\alpha \leq \|t_n v_n\|_{V_0} = \|w_n\|_{V_0} \leq K,$$

for all $n \in \mathbb{N}$, and so $|t_n| \leq \frac{K}{\alpha}$ for all $n \in \mathbb{N}$. Therefore, up to a subsequence, we may suppose that $t_n \to t_0 \geq 0$. Let us show that $t_0 > 0$. Otherwise, if $t_0 = 0$, from the boundedness of (v_n), we get $w_n = t_n v_n \to 0$ in $H^s(\mathbb{R}^N)$, that is $\mathcal{I}_{V_0}(w_n) \to 0$, in contrast with $d_{V_0} > 0$. Thus $t_0 > 0$. Let w be the weak limit of w_n in $H^s(\mathbb{R}^N)$. Since $t_n \to t_0 > 0$ and $v_n \rightharpoonup v \neq 0$ in $H^s(\mathbb{R}^N)$, by the uniqueness of the weak limit we have that $w_n \rightharpoonup w = t_0 v \neq 0$ in $H^s(\mathbb{R}^N)$. Hence, we have proved that

$$\mathcal{I}_{V_0}(w_n) \to d_{V_0} \quad \text{and} \quad w_n \rightharpoonup w \neq 0 \text{ in } H^s(\mathbb{R}^N).$$

Now Theorem 6.3.11 implies that $w_n \to w$ in $H^s(\mathbb{R}^N)$, and then $w \in \mathcal{M}_{V_0}$ and $v_n \to v \neq 0$ in $H^s(\mathbb{R}^N)$.

Let us show that (y_n) has a subsequence such that $y_n \to y \in M$. First, we prove that (y_n) is bounded in $\mathbb{R}^N$. Assume, by contradiction, that (y_n) is not bounded, that is, there exists a subsequence, still denoted by (y_n), such that $|y_n| \to \infty$.

First, we deal with the case $V_\infty = \infty$. Since $(u_n) \subset \mathcal{N}_{\varepsilon_n}$, a change of variable shows that

$$\int_{\mathbb{R}^N} V(\varepsilon_n x + y_n) v_n^2 \, dx \leq [v_n]_s^2 + \int_{\mathbb{R}^N} V(\varepsilon_n x + y_n) v_n^2 \, dx$$

$$= \|u_n\|_{\varepsilon_n}^2 = \int_{\mathbb{R}^N} f(u_n) u_n \, dx = \int_{\mathbb{R}^N} f(v_n) v_n \, dx.$$

Then Fatou's lemma and the fact that $v_n \to v$ in $H^s(\mathbb{R}^N)$ imply that

$$\infty = \liminf_{n\to\infty} \int_{\mathbb{R}^N} V(\varepsilon_n x + y_n) v_n^2 \, dx \leq \liminf_{n\to\infty} \int_{\mathbb{R}^N} f(v_n) v_n \, dx = \int_{\mathbb{R}^N} f(v) v \, dx < \infty,$$

which gives a contradiction.

Let us consider the case $V_\infty < \infty$. Taking into account that $w \in \mathcal{M}_{V_0}$, that $w_n \to w$ strongly in $H^s(\mathbb{R}^N)$, condition (V), Fatou's lemma and using the translation invariance, we have

$$d_{V_0} = \mathcal{I}_{V_0}(w) < \mathcal{I}_{V_\infty}(w)$$

$$\leq \liminf_{n\to\infty} \left[\frac{1}{2}[w_n]_s^2 + \frac{1}{2}\int_{\mathbb{R}^N} V(\varepsilon_n x + y_n) w_n^2 \, dx - \int_{\mathbb{R}^N} F(w_n)\, dx \right]$$

$$= \liminf_{n\to\infty} \left[\frac{t_n^2}{2}[u_n]_s^2 + \frac{t_n^2}{2}\int_{\mathbb{R}^N} V(\varepsilon_n x) u_n^2 \, dx - \int_{\mathbb{R}^N} F(t_n u_n)\, dx \right]$$

$$= \liminf_{n\to\infty} \mathcal{J}_{\varepsilon_n}(t_n u_n) \leq \liminf_{n\to\infty} \mathcal{J}_{\varepsilon_n}(u_n) = d_{V_0}, \tag{6.3.35}$$

which is impossible. Thus (y_n) is bounded in $\mathbb{R}^N$ and, up to a subsequence, we may assume that $y_n \to y$. If $y \notin M$, then $V_0 < V(y)$ and we can argue as above to get a contradiction. Therefore, we conclude that $y \in M$. □

Let $\delta > 0$ be fixed. Let ψ be a smooth nonincreasing cut-off function defined in $[0, \infty)$ such that $\psi = 1$ in $[0, \frac{\delta}{2}]$, $\psi = 0$ in $[\delta, \infty)$, $0 \leq \psi \leq 1$ and $|\psi'| \leq c$ for some $c > 0$. For any $y \in M$, we define

$$\Upsilon_{\varepsilon,y}(x) = \psi(|\varepsilon x - y|)\omega\left(\frac{\varepsilon x - y}{\varepsilon}\right),$$

where $\omega \in H^s(\mathbb{R}^N)$ is a positive ground state solution to (P_{V_0}), the existence of which is guaranteed by Lemma 6.3.10. Let $t_\varepsilon > 0$ be the unique positive number such that

$$\mathcal{J}_\varepsilon(t_\varepsilon \Upsilon_{\varepsilon,y}) = \max_{t\geq 0} \mathcal{J}_\varepsilon(t \Upsilon_{\varepsilon,y})$$

and define the map $\Phi_\varepsilon : M \to \mathcal{N}_\varepsilon$ by setting $\Phi_\varepsilon(y) = t_\varepsilon \Upsilon_{\varepsilon,y}$. By construction, Φ_ε has compact support for any $y \in M$.

Lemma 6.3.17 *The functional Φ_ε has the property that*

$$\lim_{\varepsilon\to 0} \mathcal{J}_\varepsilon(\Phi_\varepsilon(y)) = d_{V_0}, \quad \text{uniformly in } y \in M. \tag{6.3.36}$$

Proof Suppose, by contradiction, that there exist $\delta_0 > 0$, $(y_n) \subset M$ and $\varepsilon_n \to 0$ such that

$$|\mathcal{J}_{\varepsilon_n}(\Phi_{\varepsilon_n}(y_n)) - d_{V_0}| \geq \delta_0. \tag{6.3.37}$$

Let us note that Lemma 1.4.8 and the dominated convergence theorem imply that

$$\lim_{n \to \infty} \|\Upsilon_{\varepsilon_n, y_n}\|_{\varepsilon_n}^2 = \|\omega\|_{V_0}^2 \in (0, \infty). \tag{6.3.38}$$

Since $\langle \mathcal{J}'_{\varepsilon_n}(t_{\varepsilon_n} \Upsilon_{\varepsilon_n, y_n}), t_{\varepsilon_n} \Upsilon_{\varepsilon_n, y_n} \rangle = 0$, we can use the change of variable $z = \frac{\varepsilon_n x - y_n}{\varepsilon_n}$ to see that

$$\|t_{\varepsilon_n} \Upsilon_{\varepsilon_n, y_n}\|_{\varepsilon_n}^2 = \int_{\mathbb{R}^N} f(t_{\varepsilon_n} \Upsilon_{\varepsilon_n}) t_{\varepsilon_n} \Upsilon_{\varepsilon_n} \, dx$$

$$= \int_{\mathbb{R}^N} f(t_{\varepsilon_n} \psi(|\varepsilon_n z|) \omega(z)) t_{\varepsilon_n} \psi(|\varepsilon_n z|) \omega(z) \, dz. \tag{6.3.39}$$

Now, let us prove that $t_{\varepsilon_n} \to 1$. First we show that $t_{\varepsilon_n} \to t_0 < \infty$. Again by contradiction, suppose that $|t_{\varepsilon_n}| \to \infty$. Since $\psi(|x|) = 1$ for $x \in B_{\frac{\delta}{2}}$ and $B_{\frac{\delta}{2}} \subset B_{\frac{\delta}{2\varepsilon_n}}$ for n sufficiently large, (6.3.39) and (f_5) give

$$\|\Upsilon_{\varepsilon_n, y_n}\|_{\varepsilon_n}^2 \geq \int_{B_{\frac{\delta}{2}}} \frac{f(t_{\varepsilon_n} \omega(z))}{t_{\varepsilon_n} \omega(z)} \omega^2(z) \, dz \geq \frac{f(t_{\varepsilon_n} \omega(\bar{z}))}{t_{\varepsilon_n} \omega(\bar{z})} \int_{B_{\frac{\delta}{2}}} \omega^2(z) \, dz, \tag{6.3.40}$$

where $\bar{z} \in \mathbb{R}^N$ is such that $\omega(\bar{z}) = \min\{\omega(z) : |z| \leq \frac{\delta}{2}\} > 0$. From (f_4) and $t_{\varepsilon_n} \to \infty$, we see that (6.3.40) implies that $\|\Upsilon_{\varepsilon_n, y_n}\|_{\varepsilon_n}^2 \to \infty$, which contradicts (6.3.38). Therefore, up to a subsequence, we may assume that $t_{\varepsilon_n} \to t_0 \geq 0$. If $t_0 = 0$, we can use the growth assumptions on f, (6.3.38) and (6.3.39) to get a contradiction. Hence, $t_0 > 0$. Let us show that $t_0 = 1$. Indeed, taking the limit as $n \to \infty$ in (6.3.39), we obtain that

$$\|\omega\|_{V_0}^2 = \int_{\mathbb{R}^N} \frac{f(t_0 \omega)}{t_0} \omega \, dx,$$

so recalling that $\omega \in \mathcal{M}_{V_0}$ and using (f_5), we get $t_0 = 1$. This fact and the dominated convergence theorem yield

$$\lim_{n \to \infty} \int_{\mathbb{R}^N} F(t_{\varepsilon_n} \Upsilon_{\varepsilon_n, y_n}) \, dx = \int_{\mathbb{R}^N} F(\omega) \, dx. \tag{6.3.41}$$

Passing to the limit as $n \to \infty$ in

$$\mathcal{J}_\varepsilon(\Phi_{\varepsilon_n}(y_n)) = \frac{t_{\varepsilon_n}^2}{2} \|\Upsilon_{\varepsilon_n, y_n}\|_{\varepsilon_n}^2 - \int_{\mathbb{R}^N} F(t_{\varepsilon_n} \Upsilon_{\varepsilon_n, y_n}) \, dx,$$

and using (6.3.38) and (6.3.41), we infer that

$$\lim_{n \to \infty} \mathcal{J}_{\varepsilon_n}(\Phi_{\varepsilon_n}(y_n)) = \mathcal{I}_{V_0}(\omega) = d_{V_0}$$

which is impossible in view of (6.3.37).   □

Now, we are in the position to introduce the barycenter map. For any $\delta > 0$, let $\rho = \rho(\delta) > 0$ be such that $M_\delta \subset B_\rho$, and we consider $\chi : \mathbb{R}^N \to \mathbb{R}^N$ given by

$$\chi(x) = \begin{cases} x, & \text{if } |x| < \rho, \\ \frac{\rho x}{|x|}, & \text{if } |x| \geq \rho. \end{cases}$$

We define the barycenter map $\beta_\varepsilon : \mathcal{N}_\varepsilon \to \mathbb{R}^N$ by the formula

$$\beta_\varepsilon(u) = \frac{\int_{\mathbb{R}^N} \chi(\varepsilon x) u^2(x)\, dx}{\int_{\mathbb{R}^N} u^2(x)\, dx}.$$

Lemma 6.3.18 *The functional Φ_ε has the property that*

$$\lim_{\varepsilon \to 0} \beta_\varepsilon(\Phi_\varepsilon(y)) = y, \quad \text{uniformly in } y \in M. \tag{6.3.42}$$

Proof Suppose, by contradiction, that there exist $\delta_0 > 0$, $(y_n) \subset M$ and $\varepsilon_n \to 0$ such that

$$|\beta_{\varepsilon_n}(\Phi_{\varepsilon_n}(y_n)) - y_n| \geq \delta_0. \tag{6.3.43}$$

Using the definitions of $\Phi_{\varepsilon_n}(y_n)$, β_{ε_n}, ψ and the change of variable $z = \frac{\varepsilon_n x - y_n}{\varepsilon_n}$, we see that

$$\beta_{\varepsilon_n}(\Phi_{\varepsilon_n}(y_n)) = y_n + \frac{\int_{\mathbb{R}^N} [\chi(\varepsilon_n z + y_n) - y_n](\psi(|\varepsilon_n z|)\omega(z))^2\, dz}{\int_{\mathbb{R}^N} (\psi(|\varepsilon_n z|)\omega(z))^2\, dz}.$$

Since $(y_n) \subset M \subset B_\rho$, the dominated convergence theorem shows that

$$|\beta_{\varepsilon_n}(\Phi_{\varepsilon_n}(y_n)) - y_n| = o_n(1),$$

which contradicts (6.3.43).   □

At this point, we introduce the following subset $\widetilde{\mathcal{N}}_\varepsilon$ of $\mathcal{N}_\varepsilon$:

$$\widetilde{\mathcal{N}}_\varepsilon = \{u \in \mathcal{N}_\varepsilon : \mathcal{J}_\varepsilon(u) \leq d_{V_0} + h(\varepsilon)\},$$

where $h(\varepsilon) = \sup_{y \in M} |\mathcal{J}_\varepsilon(\Phi_\varepsilon(y)) - d_{V_0}|$. From Lemma 6.3.17, we know that $h(\varepsilon) \to 0$ as $\varepsilon \to 0$. By the definition of $h(\varepsilon)$, we deduce that, for all $y \in M$ and $\varepsilon > 0$, $\Phi_\varepsilon(y) \in \tilde{\mathcal{N}}_\varepsilon$ and $\tilde{\mathcal{N}}_\varepsilon \neq \emptyset$.

Lemma 6.3.19 *For any $\delta > 0$,*

$$\lim_{\varepsilon \to 0} \sup_{u \in \tilde{\mathcal{N}}_\varepsilon} \operatorname{dist}(\beta_\varepsilon(u), M_\delta) = 0.$$

Proof Let $\varepsilon_n \to 0$ as $n \to \infty$. By definition, there exists $(u_n) \subset \tilde{\mathcal{N}}_{\varepsilon_n}$ such that

$$\sup_{u \in \tilde{\mathcal{N}}_{\varepsilon_n}} \inf_{y \in M_\delta} |\beta_{\varepsilon_n}(u) - y| = \inf_{y \in M_\delta} |\beta_{\varepsilon_n}(u_n) - y| + o_n(1).$$

Therefore, it suffices to prove that there exists $(y_n) \subset M_\delta$ such that

$$\lim_{n \to \infty} |\beta_{\varepsilon_n}(u_n) - y_n| = 0. \tag{6.3.44}$$

Recalling that $(u_n) \subset \tilde{\mathcal{N}}_{\varepsilon_n} \subset \mathcal{N}_{\varepsilon_n}$, we deduce that

$$d_{V_0} \le c_{\varepsilon_n} \le \mathcal{J}_{\varepsilon_n}(u_n) \le d_{V_0} + h(\varepsilon_n),$$

which implies that $\mathcal{J}_{\varepsilon_n}(u_n) \to d_{V_0}$. Using Proposition 6.3.16, there exists $(\tilde{y}_n) \subset \mathbb{R}^N$ such that $y_n = \varepsilon_n \tilde{y}_n \in M_\delta$ for n sufficiently large. Thus

$$\beta_{\varepsilon_n}(u_n) = y_n + \frac{\int_{\mathbb{R}^N} [\chi(\varepsilon_n z + y_n) - y_n](u_n(z + \tilde{y}_n))^2 \, dz}{\int_{\mathbb{R}^N} (u_n(z + \tilde{y}_n))^2 \, dz}.$$

Since $u_n(\cdot + \tilde{y}_n)$ converges strongly in $H^s(\mathbb{R}^N)$ and $\varepsilon_n z + y_n \to y \in M_\delta$, we infer that $\beta_{\varepsilon_n}(u_n) = y_n + o_n(1)$, that is, (6.3.44) holds. $\qquad\square$

Now we show that (P_ε) admits at least $\operatorname{cat}_{M_\delta}(M)$ positive solutions. To this end, we recall the following result for critical points involving Lusternik–Schnirelman category. For more details one can consult [147, 267].

Theorem 6.3.20 *Let U be a $C^{1,1}$ complete Riemannian manifold (modeled on a Hilbert space). Assume that $h \in C^1(U, \mathbb{R})$ is bounded from below and satisfies $-\infty < \inf_U h < d < k < \infty$. Moreover, assume that h satisfies the Palais–Smale condition on the sublevel $\{u \in U : h(u) \le k\}$ and that d is not a critical level for h. Then*

$$\operatorname{card}\{u \in h^d : \nabla h(u) = 0\} \ge \operatorname{cat}_{h^d}(h^d),$$

where $h^d = \{u \in U : h(u) \le d\}$.

With a view to apply Theorem 6.3.20, the following abstract lemma provides a very useful tool in that it relates the topology of some sublevel of a functional to the topology of some subset of the space $\mathbb{R}^N$. For the proof, an easy application of the definitions of category and of homotopy equivalence between maps, we refer to [95, 147].

Lemma 6.3.21 *Let I, I_1 and I_2 be closed sets with $I_1 \subset I_2$, and let $\pi : I \to I_2$ and $\psi : I_1 \to I$ be two continuous maps such that $\pi \circ \psi$ is homotopy equivalent to the embedding $j : I_1 \to I_2$. Then $\mathrm{cat}_I(I) \geq \mathrm{cat}_{I_2}(I_1)$.*

Since $\mathcal{N}_\varepsilon$ is not a C^1 submanifold of $\mathcal{H}_\varepsilon$, we cannot apply directly Theorem 6.3.20. Fortunately, by Lemma 6.3.2, we know that the mapping m_ε is a homeomorphism between $\mathcal{N}_\varepsilon$ and $\mathbb{S}_\varepsilon$, and $\mathbb{S}_\varepsilon$ is a C^1 submanifold of $\mathcal{H}_\varepsilon$. So we can apply Theorem 6.3.20 to $\Psi_\varepsilon(u) = \mathcal{J}_\varepsilon(\hat{m}_\varepsilon(u))|_{\mathbb{S}_\varepsilon} = \mathcal{J}_\varepsilon(m_\varepsilon(u))$, where Ψ_ε is given in Lemma 6.3.3.

Theorem 6.3.22 *Assume that (V) and (f_1)–(f_5) hold. Then, for every $\delta > 0$ there exists $\bar{\varepsilon}_\delta > 0$ such that, for any $\varepsilon \in (0, \bar{\varepsilon}_\delta)$, problem (P_ε) has at least $\mathrm{cat}_{M_\delta}(M)$ positive solutions.*

Proof For any $\varepsilon > 0$, we define $\alpha_\varepsilon : M \to \mathbb{S}_\varepsilon$ by setting $\alpha_\varepsilon(y) = m_\varepsilon^{-1}(\Phi_\varepsilon(y))$. Using Lemma 6.3.17 and the definition of Ψ_ε, we see that

$$\lim_{\varepsilon \to 0} \Psi_\varepsilon(\alpha_\varepsilon(y)) = \lim_{\varepsilon \to 0} \mathcal{J}_\varepsilon(\Phi_\varepsilon(y)) = d_{V_0}, \quad \text{uniformly in } y \in M.$$

Set $\widetilde{\mathbb{S}}_\varepsilon = \{w \in \mathbb{S}_\varepsilon : \Psi_\varepsilon(w) \leq d_{V_0} + h(\varepsilon)\}$, where $h(\varepsilon) = \sup_{y \in M} |\Psi_\varepsilon(\alpha_\varepsilon(y)) - d_{V_0}| \to 0$ as $\varepsilon \to 0$. Thus, $\alpha_\varepsilon(y) \in \widetilde{\mathbb{S}}_\varepsilon$ for all $y \in M$, and this yields $\widetilde{\mathbb{S}}_\varepsilon \neq \emptyset$ for all $\varepsilon > 0$.

By Lemmas 6.3.2, 6.3.3, 6.3.17, and 6.3.19, we can find $\bar{\varepsilon} = \bar{\varepsilon}_\delta > 0$ such that the following diagram

$$M \xrightarrow{\Phi_\varepsilon} \widetilde{\mathcal{N}}_\varepsilon \xrightarrow{m_\varepsilon^{-1}} \widetilde{\mathbb{S}}_\varepsilon \xrightarrow{m_\varepsilon} \widetilde{\mathcal{N}}_\varepsilon \xrightarrow{\beta_\varepsilon} M_\delta$$

is well defined for any $\varepsilon \in (0, \bar{\varepsilon})$. By Lemma 6.3.18, for $\varepsilon > 0$ small enough, we can write $\beta_\varepsilon(\Phi_\varepsilon(y)) = y + \theta(\varepsilon, y)$ for $y \in M$, where $|\theta(\varepsilon, y)| < \frac{\delta}{2}$ uniformly for $y \in M$. Let $H(t, y) = y + (1 - t)\theta(\varepsilon, y)$. Then we see that $H : [0, 1] \times M \to M_\delta$ is continuous, $H(0, y) = \beta_\varepsilon(\Phi_\varepsilon(y))$ and $H(1, y) = y$ for all $y \in M$. Hence, $H(t, y)$ is a homotopy between $\beta_\varepsilon \circ \Phi_\varepsilon = (\beta_\varepsilon \circ m_\varepsilon) \circ \alpha_\varepsilon$ and the inclusion map $id : M \to M_\delta$. This fact and Lemma 6.3.21 imply that $\mathrm{cat}_{\widetilde{\mathbb{S}}_\varepsilon}(\widetilde{\mathbb{S}}_\varepsilon) \geq \mathrm{cat}_{M_\delta}(M)$. On the other hand, let us choose a function $h(\varepsilon) > 0$ such that $h(\varepsilon) \to 0$ as $\varepsilon \to 0$ and $d_{V_0} + h(\varepsilon)$ is not a critical level for $\mathcal{J}_\varepsilon$. For $\varepsilon > 0$ small enough, we deduce from Theorem 6.3.15 that $\mathcal{J}_\varepsilon$ satisfies the Palais–Smale condition in $\widetilde{\mathcal{N}}_\varepsilon$. Then, by item (ii) of Lemma 6.3.3, we infer that Ψ_ε satisfies the Palais–Smale condition in $\widetilde{\mathbb{S}}_\varepsilon$. Applying Theorem 6.3.20 we see that Ψ_ε has at least $\mathrm{cat}_{\widetilde{\mathbb{S}}_\varepsilon}(\widetilde{\mathbb{S}}_\varepsilon)$ critical points on $\widetilde{\mathbb{S}}_\varepsilon$. In view of item (iii) of Lemma 6.3.3, we conclude that $\mathcal{J}_\varepsilon$ admits at least $\mathrm{cat}_{M_\delta}(M)$ critical points. $\qquad\square$

6.3.5 Concentration Phenomenon for (6.1.1)

Let us prove the following result which plays a fundamental role in the study of the behavior of maximum points of solutions to (6.1.1).

Lemma 6.3.23 *Let v_n be a solution of the following problem:*

$$\begin{cases} (-\Delta)^s v_n + V_n(x)v_n = f(v_n) & in\ \mathbb{R}^N, \\ v_n \in H^s(\mathbb{R}^N), \quad v_n > 0\ in\ \mathbb{R}^N, \end{cases} \tag{6.3.45}$$

where $V_n(x) \geq V_0$ for all $x \in \mathbb{R}^N$. Assume that $v_n \to v$ in $H^s(\mathbb{R}^N)$ for some $v \not\equiv 0$. Then, $v_n \in L^\infty(\mathbb{R}^N)$ and there exists $C > 0$ such that $\|v_n\|_{L^\infty(\mathbb{R}^N)} \leq C$ for all $n \in \mathbb{N}$. In addition, $v_n(x) \to 0$ as $|x| \to \infty$, uniformly in $n \in \mathbb{N}$.

Proof We use a Moser iteration argument [278]. For any $L > 0$ and $\beta > 1$, let $\gamma(t) = t t_L^{2(\beta-1)}$ for $t \geq 0$, where $t_L = \min\{t, L\}$, and $\Gamma(t) = \int_0^t (\gamma'(\tau))^{\frac{1}{2}} d\tau$. Thus $\gamma(v_n) = v_n v_{L,n}^{2(\beta-1)} \in \mathcal{H}_\varepsilon$, where $v_{L,n} = \min\{v_n, L\}$. Moreover, arguing as in the proof of Proposition 3.2.14, we have the following estimates:

$$|\Gamma(v_n(x)) - \Gamma(v_n(y))|^2 \leq (v_n(x) - v_n(y))((v_n v_{L,n}^{2(\beta-1)})(x) - (v_n v_{L,n}^{2(\beta-1)})(y)),$$

and

$$[\Gamma(v_n)]_s^2 \geq S_* \|\Gamma(v_n)\|_{L^{2_s^*}(\mathbb{R}^N)}^2 \geq \left(\frac{1}{\beta}\right)^2 S_* \|v_n v_{L,n}^{\beta-1}\|_{L^{2_s^*}(\mathbb{R}^N)}^2.$$

Choosing $\gamma(v_n) = v_n v_{L,n}^{2(\beta-1)}$ as test function in (6.3.45) we deduce that

$$\left(\frac{1}{\beta}\right)^2 S_* \|v_n v_{L,n}^{\beta-1}\|_{L^{2_s^*}(\mathbb{R}^N)}^2 + \int_{\mathbb{R}^N} V_n(x) v_n^2 v_{L,n}^{2(\beta-1)}\, dx$$

$$\leq \iint_{\mathbb{R}^{2N}} \frac{(v_n(x) - v_n(y))}{|x-y|^{N+2s}} ((v_n v_{L,n}^{2(\beta-1)})(x) - (v_n v_{L,n}^{2(\beta-1)})(y))\, dx\, dy + \int_{\mathbb{R}^N} V_n(x) v_n^2 v_{L,n}^{2(\beta-1)}\, dx$$

$$= \int_{\mathbb{R}^N} f(v_n) v_n v_{L,n}^{2(\beta-1)}\, dx. \tag{6.3.46}$$

On the other hand, by (f_2)–(f_3), we know that for any $\xi > 0$ there exists $C_\xi > 0$ such that

$$|f(t)| \leq \xi|t| + C_\xi|t|^{2^*_s-1} \qquad \text{for all } t \in \mathbb{R}. \tag{6.3.47}$$

Taking $\xi \in (0, V_0)$, and combining (V), (6.3.46) and (6.3.47), we see that

$$\|v_n v_{L,n}^{\beta-1}\|^2_{L^{2^*_s}(\mathbb{R}^N)} \leq C\beta^2 \int_{\mathbb{R}^N} v_n^{2^*_s} v_{L,n}^{2(\beta-1)} \, dx. \tag{6.3.48}$$

Now, we take $\beta = \frac{2^*_s}{2}$ and fix $R > 0$. Observing that $0 \leq v_{L,n} \leq v_n$, we deduce that

$$\int_{\mathbb{R}^N} v_n^{2^*_s} v_{L,n}^{2(\beta-1)} \, dx = \int_{\mathbb{R}^N} v_n^{2^*_s-2} v_n^2 v_{L,n}^{2^*_s-2} \, dx$$

$$= \int_{\mathbb{R}^N} v_n^{2^*_s-2} (v_n v_{L,n}^{\frac{2^*_s-2}{2}})^2 \, dx$$

$$\leq \int_{\{v_n<R\}} R^{2^*_s-2} v_n^{2^*_s} \, dx + \int_{\{v_n>R\}} v_n^{2^*_s-2} (v_n v_{L,n}^{\frac{2^*_s-2}{2}})^2 \, dx$$

$$\leq \int_{\{v_n<R\}} R^{2^*_s-2} v_n^{2^*_s} \, dx + \left(\int_{\{v_n>R\}} v_n^{2^*_s} \, dx\right)^{\frac{2^*_s-2}{2^*_s}} \left(\int_{\mathbb{R}^N} (v_n v_{L,n}^{\frac{2^*_s-2}{2}})^{2^*_s} \, dx\right)^{\frac{2}{2^*_s}}. \tag{6.3.49}$$

Since $v_n \to v$ in $H^s(\mathbb{R}^N)$, for $R > 0$ sufficiently large, independent of n, we have that

$$\left(\int_{\{v_n>R\}} v_n^{2^*_s} \, dx\right)^{\frac{2^*_s-2}{2^*_s}} \leq \frac{1}{2C\beta^2}. \tag{6.3.50}$$

Putting together (6.3.48), (6.3.49), and (6.3.50) we get

$$\left(\int_{\mathbb{R}^N} (v_n v_{L,n}^{\frac{2^*_s-2}{2}})^{2^*_s} \, dx\right)^{\frac{2}{2^*_s}} \leq C\beta^2 \int_{\mathbb{R}^N} R^{2^*_s-2} v_n^{2^*_s} \, dx < \infty$$

and taking the limit as $L \to \infty$, we conclude that $v_n \in L^{\frac{(2^*_s)^2}{2}}(\mathbb{R}^N)$.

Next, since $0 \leq v_{L,n} \leq v_n$, passing to the limit as $L \to \infty$ in (6.3.48), we have

$$\left(\int_{\mathbb{R}^N} v_n^{2^*_s\beta} \, dx\right)^{\frac{2}{2^*_s}} \leq C\beta^2 \int_{\mathbb{R}^N} v_n^{2^*_s+2(\beta-1)} \, dx,$$

which implies that

$$\left(\int_{\mathbb{R}^N} v_n^{2_s^* \beta}\, dx\right)^{\frac{1}{2_s^*(\beta-1)}} \leq (C\beta)^{\frac{1}{\beta-1}}\left(\int_{\mathbb{R}^N} v_n^{2_s^*+2(\beta-1)}\, dx\right)^{\frac{1}{2(\beta-1)}}. \tag{6.3.51}$$

For $m \geq 1$ we define β_{m+1} inductively so that

$$2_s^* + 2(\beta_{m+1} - 1) = 2_s^* \beta_m \text{ and } \beta_1 = \frac{2_s^*}{2}.$$

Therefore,

$$\beta_{m+1} = 1 + \left(\frac{2_s^*}{2}\right)^m (\beta_1 - 1),$$

which together with (6.3.51) yields

$$\left(\int_{\mathbb{R}^N} v_n^{2_s^* \beta_{m+1}}\, dx\right)^{\frac{1}{2_s^*(\beta_{m+1}-1)}} \leq (C\beta_{m+1})^{\frac{1}{\beta_{m+1}-1}}\left(\int_{\mathbb{R}^N} v_n^{2_s^* \beta_m}\, dx\right)^{\frac{1}{2_s^*(\beta_m-1)}}.$$

Let us define

$$D_m = \left(\int_{\mathbb{R}^N} v_n^{2_s^* \beta_m}\, dx\right)^{\frac{1}{2_s^*(\beta_m-1)}}.$$

Using an iteration argument, we can find $C_0 > 0$ independent of m such that

$$D_{m+1} \leq \prod_{k=1}^{m} (C\beta_{k+1})^{\frac{1}{\beta_{k+1}-1}} D_1 \leq C_0 D_1.$$

Taking the limit as $m \to \infty$ we get $\|v_n\|_{L^\infty(\mathbb{R}^N)} \leq K$ for all $n \in \mathbb{N}$. In particular, by using interpolation in L^p spaces,

$$\|v_n\|_{L^r(\mathbb{R}^N)} \leq C \text{ for all } n \in \mathbb{N} \text{ and } r \in [2, \infty]. \tag{6.3.52}$$

From $v_n \to v$ in $H^s(\mathbb{R}^N)$ and (6.3.52), we see that $v_n \to v$ in $L^r(\mathbb{R}^N)$ for all $r \in [2, \infty)$.

Let $w_n(x, y) = \text{Ext}(v_n) = P_s(x, y) * v_n(x)$ be the s-harmonic extension of v_n. Note that w_n solves (see Sect. 1.2.3 in Chap. 1)

$$\begin{cases} -\operatorname{div}(y^{1-2s}\nabla w_n) = 0 & \text{in } \mathbb{R}_+^{N+1}, \\ w_n(\cdot, 0) = v_n & \text{on } \partial\mathbb{R}_+^{N+1}, \\ \frac{\partial w}{\partial \nu^{1-2s}} = -V_n(x)v_n + f(v_n) & \text{on } \partial\mathbb{R}_+^{N+1}. \end{cases}$$

Since $\int_{\mathbb{R}^N} P_s(z, y) \, dz = 1$ for all $y > 0$, inequality (6.3.52) implies that

$$|w_n(x, y)| \leq \|v_n\|_{L^\infty(\mathbb{R}^N)} \int_{\mathbb{R}^N} P_s(z, y) \, dz \leq K \text{ for all } (x, y) \in \mathbb{R}^{N+1}_+, n \in \mathbb{N}.$$

Using the above estimate and (6.3.52), it follows from Proposition 1.3.11-(iii) that $v_n \in C^{0,\alpha}_{\text{loc}}(\mathbb{R}^N)$ (alternatively, one can combine (6.3.52) with Corollary 5.5 in [222] to deduce the Hölder continuity of v_n). Now, recalling that

$$\sqrt{\frac{\kappa_s C(N, s)}{2}} [\omega]_s = \|\text{Ext}(\omega)\|_{X^s_0(\mathbb{R}^{N+1}_+)} \quad \text{for all } \omega \in H^s(\mathbb{R}^N),$$

the convergence $v_n \to v$ in $H^s(\mathbb{R}^N)$ and Lemma 1.3.9-(i) show that $w_n \to \text{Ext}(v)$ in $L^{2\gamma}(\mathbb{R}^{N+1}_+, y^{1-2s})$, where $\gamma = 1 + \frac{2}{N-2s}$. Fix $x_0 \in \mathbb{R}^N$ and $\delta \in (0, V_0)$. Since $V_n \geq V_0 > 0$ in $\mathbb{R}^N$ and $|f(v_n)| \leq \delta v_n + C_\delta v_n^{2^*_s - 1}$, we deduce that w_n is a weak subsolution to (1.3.3) in $Q_1(x_0, 0) = B_1(x_0) \times (0, 1)$, with $a(x) = 0$ and $b(x) = C_\delta v_n^{2^*_s - 1}(x)$. By Proposition 1.3.11-(i) (with $\nu = 2\gamma$),

$$v_n(x_0) \leq C(\|w_n\|_{L^{2\gamma}(Q_1(x_0,0), y^{1-2s})} + |v_n^{2^*_s - 1}|_{L^q(B_1(x_0))}) \quad \text{for all } n \in \mathbb{N},$$

where $q > \frac{N}{2s}$ is fixed and $C > 0$ is a constant depending only on N, s, q, γ and not on $n \in \mathbb{N}$ and x_0. Note that $q(2^*_s - 1) \in (2, \infty)$ because $N > 2s$ and $q > \frac{N}{2s}$. Using the strong convergence of (w_n) in $L^{2\gamma}(\mathbb{R}^{N+1}_+)$ and (v_n) in $L^{q(2^*_s-1)}(\mathbb{R}^N)$, respectively, we conclude that $v_n(x_0) \to 0$ as $|x_0| \to \infty$ uniformly in $n \in \mathbb{N}$. ☐

Lemma 6.3.24 *Under the assumptions of Lemma 6.3.23, there exists $\delta > 0$ such that $\|v_n\|_{L^\infty(\mathbb{R}^N)} \geq \delta$ for all $n \in \mathbb{N}$.*

Proof Suppose, by contradiction, that $\|v_n\|_{L^\infty(\mathbb{R}^N)} \to 0$ as $n \to \infty$. Using (f_2), there exists $n_0 \in \mathbb{N}$ such that $\frac{f(\|v_n\|_{L^\infty(\mathbb{R}^N)})}{\|v_n\|_{L^\infty(\mathbb{R}^N)}} < \frac{V_0}{2}$ for all $n \geq n_0$. Therefore, in view of (f_5), we see that

$$[v_n]^2_s + V_0 \|v_n\|^2_{L^2(\mathbb{R}^N)} \leq \int_{\mathbb{R}^N} \frac{f(\|v_n\|_{L^\infty(\mathbb{R}^N)})}{\|v_n\|_{L^\infty(\mathbb{R}^N)}} |v_n|^2 \, dx \leq \frac{V_0}{2} \|v_n\|^2_{L^2(\mathbb{R}^N)},$$

for all $n \geq n_0$, which implies that $\|v_n\|_{V_0} = 0$ for all $n \geq n_0$: a contradiction. ☐

We end this section by establishing Theorem 6.1.1.

Proof of Theorem 6.1.1 In the light of Theorem 6.3.22, we know that for any $\delta > 0$ there exists $\bar{\varepsilon}_\delta > 0$ such that, for any $\varepsilon \in (0, \bar{\varepsilon}_\delta)$, problem (P_ε) has at least $\mathrm{cat}_{M_\delta}(M)$ positive solutions. Let u_{ε_n} be a solution to (P_{ε_n}). Then $v_n(x) = u_{\varepsilon_n}(x + \tilde{y}_n)$ is a solution to (6.3.45) with $V_n(x) = V(\varepsilon_n x + \varepsilon_n \tilde{y}_n)$, where $(\tilde{y}_n)$ is given in Proposition 6.3.16. Moreover, up to a subsequence, it follows from Proposition 6.3.16 that $v_n \to v \neq 0$ in $H^s(\mathbb{R}^N)$ and $y_n = \varepsilon_n \tilde{y}_n \to y \in M$. Let p_n be a global maximum point of v_n. Using Lemmas 6.3.23 and 6.3.24, we see that $p_n \in B_R$ for some $R > 0$. Consequently, $z_{\varepsilon_n} = p_n + \tilde{y}_n$ is a global maximum point of u_{ε_n} and then $\varepsilon_n z_{\varepsilon_n} = \varepsilon_n p_n + \varepsilon_n \tilde{y}_n \to y \in M$ because (p_n) is bounded. This fact and the continuity of V imply that

$$\lim_{n \to \infty} V(\varepsilon_n z_{\varepsilon_n}) = V(y) = V_0.$$

Now, if u_ε is a positive solution to (P_ε), then $w_\varepsilon(x) = u_\varepsilon(\frac{x}{\varepsilon})$ is a positive solution to (6.1.1). Thus, the maximum points η_ε and z_ε of w_ε and u_ε, respectively, satisfy the equality $\eta_\varepsilon = \varepsilon z_\varepsilon$, and

$$\lim_{\varepsilon \to 0} V(\eta_\varepsilon) = V_0.$$

This ends the proof of theorem.   □

6.4   The Critical Case

6.4.1   The Nehari Method for (6.1.3)

In this section we deal with the critical problem (6.1.3). Since many calculations are adaptations to those presented earlier, we will emphasize only the differences between the subcritical and the critical cases.

Using a change of variable, we can consider the following problem

$$\begin{cases} (-\Delta)^s u + V(\varepsilon x)u = f(u) + |u|^{2_s^*-2}u \ \text{ in } \mathbb{R}^N, \\ u \in H^s(\mathbb{R}^N), \quad u > 0 \ \text{ in } \mathbb{R}^N. \end{cases} \tag{P_ε^*}$$

We consider the energy functional $\mathcal{J}_\varepsilon : \mathcal{H}_\varepsilon \to \mathbb{R}$ given by

$$\mathcal{J}_\varepsilon(u) = \frac{1}{2}\|u\|_\varepsilon^2 - \int_{\mathbb{R}^N} F(u)\,dx - \frac{1}{2_s^*}\|u\|_{L^{2_s^*}(\mathbb{R}^N)}^{2_s^*}$$

and introduce the Nehari manifold associated with $\mathcal{J}_\varepsilon$, that is

$$\mathcal{N}_\varepsilon = \left\{ u \in \mathcal{H}_\varepsilon \setminus \{0\} : \langle \mathcal{J}_\varepsilon'(u), u \rangle = 0 \right\}.$$

Arguing as in the previous section we can prove that the next lemmas hold true.

Lemma 6.4.1 *The functional $\mathcal{J}_\varepsilon$ satisfies the following conditions:*

(i) *There exist $\alpha, \rho > 0$ such that $\mathcal{J}_\varepsilon(u) \geq \alpha$ with $\|u\|_\varepsilon = \rho$.*
(ii) *There exists $e \in \mathcal{H}_\varepsilon$ such that $\|e\|_\varepsilon > \rho$ and $\mathcal{J}_\varepsilon(e) < 0$.*

Lemma 6.4.2 *Under assumptions (V) and (f_1)–(f_5) and (f_6'), for $\varepsilon > 0$ we have:*

(i) $\mathcal{J}_\varepsilon'$ *maps bounded sets in $\mathcal{H}_\varepsilon$ into bounded sets in $\mathcal{H}_\varepsilon$.*
(ii) $\mathcal{J}_\varepsilon'$ *is weakly sequentially continuous in $\mathcal{H}_\varepsilon$.*
(iii) $\mathcal{J}_\varepsilon(t_n u_n) \to -\infty$ *as $t_n \to \infty$, where $u_n \in K$ and $K \subset \mathcal{H}_\varepsilon \setminus \{0\}$ is a compact subset.*

Lemma 6.4.3 *Under the assumptions of Lemma 6.4.2, for $\varepsilon > 0$ we have:*

(i) *For every $u \in \mathbb{S}_\varepsilon$, there exists a unique $t_u > 0$ such that $t_u u \in \mathcal{N}_\varepsilon$. Moreover, $m_\varepsilon(u) = t_u u$ is the unique maximum of $\mathcal{J}_\varepsilon$ on $\mathcal{H}_\varepsilon$, where $\mathbb{S}_\varepsilon = \{u \in \mathcal{H}_\varepsilon : \|u\|_\varepsilon = 1\}$.*
(ii) *The set $\mathcal{N}_\varepsilon$ is bounded away from 0. Furthermore, $\mathcal{N}_\varepsilon$ is closed in $\mathcal{H}_\varepsilon$.*
(iii) *There exists $\alpha > 0$ such that $t_u \geq \alpha$ for each $u \in \mathbb{S}_\varepsilon$ and, for each compact subset $W \subset \mathbb{S}_\varepsilon$, there exists $C_W > 0$ such that $t_u \leq C_W$ for all $u \in W$.*
(iv) $\mathcal{N}_\varepsilon$ *is a regular manifold diffeomorphic to the unit sphere $\mathbb{S}_\varepsilon$.*
(v) $c_\varepsilon = \inf_{\mathcal{N}_\varepsilon} \mathcal{J}_\varepsilon \geq \rho > 0$ *and $\mathcal{J}_\varepsilon$ is bounded below on $\mathcal{N}_\varepsilon$, where ρ is independent of ε.*

Now, we introduce the functionals $\hat{\Psi}_\varepsilon : \mathcal{H}_\varepsilon \setminus \{0\} \to \mathbb{R}$ and $\Psi_\varepsilon : \mathbb{S}_\varepsilon \to \mathbb{R}$ defined by

$$\hat{\Psi}_\varepsilon(u) = \mathcal{J}_\varepsilon(\hat{m}_\varepsilon(u)) \quad \text{and} \quad \Psi_\varepsilon = \hat{\Psi}_\varepsilon|_{\mathbb{S}_\varepsilon}.$$

Since the inverse of the mapping m_ε to $\mathbb{S}_\varepsilon$ is given by

$$m_\varepsilon^{-1} : \mathcal{N}_\varepsilon \to \mathbb{S}_\varepsilon, \quad m_\varepsilon^{-1}(u) = \frac{u}{\|u\|_\varepsilon},$$

the following result holds true:

Lemma 6.4.4 *Under the assumptions of Lemma 6.4.2, for $\varepsilon > 0$, we have:*

(i) $\Psi_\varepsilon \in C^1(\mathbb{S}_\varepsilon, \mathbb{R})$, *and*

$$\langle \Psi_\varepsilon'(w), v \rangle = \|m_\varepsilon(w)\|_\varepsilon \langle \mathcal{J}_\varepsilon'(m_\varepsilon(w)), v \rangle \quad \text{for } v \in T_w \mathbb{S}_\varepsilon.$$

(ii) (w_n) is a Palais–Smale sequence for Ψ_ε if and only if $(m_\varepsilon(w_n))$ is a Palais–Smale sequence for $\mathcal{J}_\varepsilon$. If $(u_n) \subset \mathcal{N}_\varepsilon$ is a bounded Palais–Smale sequence for $\mathcal{J}_\varepsilon$, then $(m_\varepsilon^{-1}(u_n))$ is a Palais–Smale sequence for Ψ_ε.

(iii) $u \in \mathbb{S}_\varepsilon$ is a critical point of Ψ_ε if and only if $m_\varepsilon(u)$ is a critical point of $\mathcal{J}_\varepsilon$. Moreover, the corresponding critical values coincide and

$$\inf_{\mathbb{S}_\varepsilon} \Psi_\varepsilon = \inf_{\mathcal{N}_\varepsilon} \mathcal{J}_\varepsilon = c_\varepsilon.$$

6.4.2 The Autonomous Critical Problem

Let us consider the following family of autonomous critical problems, with $\mu > 0$,

$$\begin{cases} (-\Delta)^s u + \mu u = f(u) + |u|^{2_s^*-2}u & \text{in } \mathbb{R}^N, \\ u \in H^s(\mathbb{R}^N), \quad u > 0 \text{ in } \mathbb{R}^N. \end{cases} \tag{P_μ^*}$$

Let us introduce the energy functional $\mathcal{I}_\mu : \mathbb{X}_\mu \to \mathbb{R}$ defined as

$$\mathcal{I}_\mu(u) = \frac{1}{2}\|u\|_\mu^2 - \int_{\mathbb{R}^N} F(u)\,dx - \frac{1}{2_s^*}\|u\|_{L^{2_s^*}(\mathbb{R}^N)}^{2_s^*},$$

and the Nehari manifold associated with $\mathcal{I}_\mu$, given by

$$\mathcal{M}_\mu = \left\{ u \in \mathbb{X}_\mu \setminus \{0\} : \langle \mathcal{I}_\mu'(u), u \rangle = 0 \right\}.$$

Similarly to the autonomous subcritical case, we have the following result:

Lemma 6.4.5 *Under the assumptions of Lemma 6.4.2, for $\mu > 0$ we have:*

(i) *for every $u \in \mathbb{S}_\mu$, there exists a unique $t_u > 0$ such that $t_u u \in \mathcal{M}_\mu$. Moreover, $m_\mu(u) = t_u u$ is the unique maximum of $\mathcal{I}_\mu$ on $\mathcal{W}_\varepsilon$, where $\mathbb{S}_\mu = \{u \in \mathbb{X}_\mu : \|u\|_\mu = 1\}$.*

(ii) *The set $\mathcal{M}_\mu$ is bounded away from 0. Furthermore, $\mathcal{M}_\mu$ is closed in $\mathbb{X}_\mu$.*

(iii) *There exists $\alpha > 0$ such that $t_u \geq \alpha$ for each $u \in \mathbb{S}_\mu$ and, for each compact subset $W \subset \mathbb{S}_\mu$, there exists $C_W > 0$ such that $t_u \leq C_W$ for all $u \in W$.*

(iv) *$\mathcal{M}_\mu$ is a regular manifold diffeomorphic to the unit sphere $\mathbb{S}_\mu$.*

(v) *$d_\mu = \inf_{\mathcal{M}_\mu} \mathcal{I}_\mu > 0$ and $\mathcal{I}_\mu$ is bounded below on $\mathcal{M}_\mu$ by some positive constant.*

Let us define the functionals $\hat{\Psi}_\mu : \mathbb{X}_\mu \setminus \{0\} \to \mathbb{R}$ and $\Psi_\mu : \mathbb{S}_\mu \to \mathbb{R}$ by setting

$$\hat{\Psi}_\mu(u) = \mathcal{I}_\mu(\hat{m}_\mu(u)) \quad \text{and} \quad \Psi_\mu = \hat{\Psi}_\mu|_{\mathbb{S}_\mu}.$$

Then we obtain the following result:

Lemma 6.4.6 *Under the assumptions of Lemma 6.4.2, for $\mu > 0$ we have:*

(i) $\Psi_\mu \in C^1(\mathbb{S}_\mu, \mathbb{R})$, *and*

$$\langle \Psi'_\mu(w), v \rangle = \|m_\mu(w)\|_\mu \langle \mathcal{I}'_\mu(m_\mu(w)), v \rangle \quad \text{for all } v \in T_w \mathbb{S}_\mu.$$

(ii) (w_n) *is a Palais–Smale sequence for* Ψ_μ *if and only if* $(m_\mu(w_n))$ *is a Palais–Smale sequence for* $\mathcal{I}_\mu$. *If* $(u_n) \subset \mathcal{M}_\mu$ *is a bounded Palais–Smale sequence for* $\mathcal{I}_\mu$, *then* $(m_\mu^{-1}(u_n))$ *is a Palais–Smale sequence for* Ψ_μ.

(iii) $u \in \mathbb{S}_\mu$ *is a critical point of* Ψ_μ *if and only if* $m_\mu(u)$ *is a critical point of* $\mathcal{I}_\mu$. *Moreover, the corresponding critical values coincide and*

$$\inf_{\mathbb{S}_\mu} \Psi_\mu = \inf_{\mathcal{M}_\mu} \mathcal{I}_\mu = d_\mu.$$

Remark 6.4.7 As in the previous section, we have the following variational characterization of the infimum of $\mathcal{I}_\mu$ over $\mathcal{M}_\mu$:

$$d_\mu = \inf_{u \in \mathcal{M}_\mu} \mathcal{I}_\mu(u) = \inf_{u \in \mathbb{X}_\mu \setminus \{0\}} \max_{t > 0} \mathcal{I}_\mu(tu) = \inf_{u \in \mathbb{S}_\mu} \max_{t > 0} \mathcal{I}_\mu(tu).$$

It is easy to check that $\mathcal{I}_\mu$ has a mountain pass geometry. Hence, we can define the mountain pass level

$$d'_\mu = \inf_{\gamma \in \Gamma_\mu} \max_{t \in [0,1]} \mathcal{I}_\mu(\gamma(t)),$$

where $\Gamma_\mu = \{\gamma \in C([0, 1], \mathbb{X}_\mu) : \mathcal{I}_\mu(0) = 0, \mathcal{I}_\mu(\gamma(1)) < 0\}$. In what follows we use the following equivalent characterization of d'_μ which is more appropriate for our purpose:

$$d'_\mu = \inf_{u \in \mathbb{X}_\mu \setminus \{0\}} \max_{t > 0} \mathcal{I}_\mu(tu) = d_\mu,$$

where in the last equality we used (6.3.9). Standard arguments show that any Palais–Smale sequence of $\mathcal{I}_\mu$ is bounded in $\mathbb{X}_\mu$.

In order to obtain the existence of a nontrivial solution to the autonomous critical problem, we need to prove the next fundamental result.

Lemma 6.4.8 *For any $\mu > 0$, there exists $v \in \mathbb{X}_\mu \setminus \{0\}$ such that*

$$\max_{t \geq 0} \mathcal{I}_\mu(tv) < \frac{s}{N} S_*^{\frac{N}{2s}}.$$

In particular, $0 < d_\mu < \frac{s}{N} S_*^{\frac{N}{2s}}$.

Proof We know (see Remark 1.1.9), for any $\varepsilon > 0$, S_* is achieved by

$$U_\varepsilon(x) = \frac{\kappa \, \varepsilon^{-\frac{N-2s}{2}}}{\left(\mu^2 + \left|\frac{x}{\varepsilon \, S_*^{\frac{1}{2s}}}\right|^2\right)^{\frac{N-2s}{2}}},$$

where $\kappa \in \mathbb{R}$ and $\mu > 0$ are fixed constants. Let $\eta \in C_c^\infty(\mathbb{R}^N)$ be a cut-off function such that $\eta = 1$ in B_1, supp$(\eta) \subset B_2$ and $0 \le \eta \le 1$. Set $v_\varepsilon(x) = \eta(x)U_\varepsilon(x)$. It follows from [310] that

$$[v_\varepsilon]_s^2 = S_*^{\frac{N}{2s}} + O(\varepsilon^{N-2s}) \quad \text{and} \quad \|v_\varepsilon\|_{L^{2_s^*}(\mathbb{R}^N)}^{2_s^*} = S_*^{\frac{N}{2s}} + O(\varepsilon^N).$$

Define

$$u_\varepsilon = \frac{v_\varepsilon}{\|v_\varepsilon\|_{L^{2_s^*}(\mathbb{R}^N)}}. \tag{6.4.1}$$

Then we have the following estimates (see [174, 310]):

$$[u_\varepsilon]_s^2 \le S_* + O(\varepsilon^{N-2s}), \tag{6.4.2}$$

$$\|u_\varepsilon\|_{L^2(\mathbb{R}^N)}^2 = \begin{cases} O(\varepsilon^{2s}), & \text{if } N > 4s, \\ O(\varepsilon^{2s} \log\left(\frac{1}{\varepsilon}\right)), & \text{if } N = 4s, \\ O(\varepsilon^{N-2s}), & \text{if } N < 4s, \end{cases} \tag{6.4.3}$$

and

$$\|u_\varepsilon\|_{L^q(\mathbb{R}^N)}^q \ge \begin{cases} O(\varepsilon^{N-\frac{(N-2s)q}{2}}), & \text{if } q > \frac{N}{N-2s}, \\ O(\varepsilon^{\frac{N}{2}}|\log(\varepsilon)|), & \text{if } q = \frac{N}{N-2s}, \\ O(\varepsilon^{\frac{(N-2s)q}{2}}), & \text{if } q < \frac{N}{N-2s}. \end{cases} \tag{6.4.4}$$

Now, using (f_6'), we have that

$$\mathcal{I}_\mu(tu_\varepsilon) = \frac{t^2}{2}\|u_\varepsilon\|_\mu^2 - \int_{\mathbb{R}^N} F(tu_\varepsilon)\, dx - \frac{t^{2_s^*}}{2_s^*}$$

$$\le \frac{t^2}{2}\|u_\varepsilon\|_\mu^2 - \lambda t^{q_1}\|u_\varepsilon\|_{L^{q_1}(\mathbb{R}^N)}^{q_1} - \frac{t^{2_s^*}}{2_s^*}. \tag{6.4.5}$$

Let $y(t) = \frac{t^2}{2}\|u_\varepsilon\|_\mu^2 - \frac{t^{2_s^*}}{2_s^*}$. Clearly, $y(t)$ attains its maximum at $t_\varepsilon = \|u_\varepsilon\|_\mu^{\frac{2}{2_s^*-2}}$ and $y(t_\varepsilon) = \frac{s}{N}\|u_\varepsilon\|_\mu^{\frac{N}{s}}$. Observe that there exists $t_1 \in (0, 1)$ such that for $\varepsilon < 1$,

$$\max_{t \in [0,t_1]} \mathcal{I}_\mu(tu_\varepsilon) \le \max_{t \in [0,t_1]} \frac{t^2}{2}\|u_\varepsilon\|_\mu^2 < \frac{s}{N}S_*^{\frac{N}{2s}}. \tag{6.4.6}$$

On the other hand, by (6.4.5) and using (6.4.2), (6.4.3), (6.4.4), we can find $\varepsilon_0 > 0$ such that $\lim_{t \to \infty} \mathcal{I}_\mu(tu_\varepsilon) = -\infty$ uniformly for $\varepsilon \in (0, \varepsilon_0)$. Then, there exists $t_2 > 0$ such that for $\varepsilon \in (0, \varepsilon_0)$,

$$\max_{t \in [t_2,\infty)} \mathcal{I}_\mu(tu_\varepsilon) < \frac{s}{N}S_*^{\frac{N}{2s}}. \tag{6.4.7}$$

Now, we note that

$$\max_{t \in [t_1,t_2]} \mathcal{I}_\mu(tu_\varepsilon) \le \max_{t \in [t_1,t_2]} \left[y(t) - \lambda t^{q_1}\|u_\varepsilon\|_{L^{q_1}(\mathbb{R}^N)}^{q_1} \right]$$

$$\le y(t_\varepsilon) - \lambda t_1^{q_1}\|u_\varepsilon\|_{L^{q_1}(\mathbb{R}^N)}^{q_1}$$

$$\le \frac{s}{N}([u_\varepsilon]_s^2 + \mu\|u_\varepsilon\|_{L^2(\mathbb{R}^N)}^2)^{\frac{N}{2s}} - \lambda t_1^{q_1}\|u_\varepsilon\|_{L^{q_1}(\mathbb{R}^N)}^{q_1}.$$

Recalling the following elementary inequality

$$(a + b)^r \le a^r + r(a + b)^{r-1}b \quad \text{for all } a, b > 0, r \ge 1,$$

and gathering the estimates (6.4.2), (6.4.3), (6.4.4), we get

$$\max_{t \in [t_1,t_2]} \mathcal{I}_\mu(tu_\varepsilon) \le \begin{cases} \frac{s}{N}S_*^{\frac{N}{2s}} + O(\varepsilon^{N-2s}) + O(\varepsilon^{2s}) - \lambda C\|u_\varepsilon\|_{q_1}^{q_1}, & \text{if } N > 4s, \\ \frac{s}{N}S_*^{\frac{N}{2s}} + (\varepsilon^{2s}(1 + |\log\varepsilon|)) - \lambda C\|u_\varepsilon\|_{q_1}^{q_1}, & \text{if } N = 4s, \\ \frac{s}{N}S_*^{\frac{N}{2s}} + O(\varepsilon^{N-2s}) - \lambda C\|u_\varepsilon\|_{q_1}^{q_1}, & \text{if } N < 4s. \end{cases}$$

In what follows, we show that, for $\varepsilon > 0$ small enough,

$$\max_{t \in [t_1,t_2]} \mathcal{I}_\mu(tu_\varepsilon) < \frac{s}{N}S_*^{\frac{N}{2s}}. \tag{6.4.8}$$

Assume that $N > 4s$. Then $q_1 > 2 > \frac{N}{N-2s}$ and using (6.4.4), we have

$$\max_{t \in [t_1,t_2]} \mathcal{I}_\mu(tu_\varepsilon) \le \frac{s}{N}S_*^{\frac{N}{2s}} + O(\varepsilon^{N-2s}) + O(\varepsilon^{2s}) - O(\varepsilon^{N-\frac{(N-2s)}{2}q_1}).$$

Since

$$N - \frac{(N-2s)}{2}q_1 < 2s < N - 2s,$$

we infer that

$$\max_{t \in [t_1,t_2]} \mathcal{I}_\mu(tu_\varepsilon) < \frac{s}{N}S_*^{\frac{N}{2s}},$$

as long as $\varepsilon > 0$ is sufficiently small.

If $N = 4s$, then $q_1 > 2 = \frac{N}{N-2s}$, and in view of (6.4.4), we obtain that

$$\max_{t \in [t_1,t_2]} \mathcal{I}_\mu(tu_\varepsilon) \leq \frac{s}{N}S_*^{\frac{N}{2s}} + O(\varepsilon^{2s}(1+|\log \varepsilon|)) - O(\varepsilon^{4s-sq_1}).$$

Observing that $q_1 > 2$ yields

$$\lim_{\varepsilon \to 0} \frac{\varepsilon^{4s-sq_1}}{\varepsilon^{2s}(1+|\log \varepsilon|)} = \infty,$$

we get the conclusion for $\varepsilon > 0$ small enough.

Now let us consider the case $N < 4s$. First, we assume that $2_s^* - 2 < q_1 < 2_s^*$. Then,

$$q_1 > 2_s^* - 2 > \frac{N}{N-2s}$$

which combined with (6.4.4) gives

$$\max_{t \in [t_1,t_2]} \mathcal{I}_\mu(tu_\varepsilon) \leq \frac{s}{N}S_*^{\frac{N}{2s}} + O(\varepsilon^{N-2s}) - O(\varepsilon^{N-\frac{(N-2s)}{2}q_1}).$$

Since

$$N - \frac{(N-2s)}{2}q_1 < N - 2s < 2s,$$

we have for $\varepsilon > 0$ small enough

$$\max_{t \in [t_1,t_2]} \mathcal{I}_\mu(tu_\varepsilon) < \frac{s}{N}S_*^{\frac{N}{2s}}.$$

Second, we deal with the case $2 < q_1 \leq 2_s^* - 2$. We distinguish the following subcases:

$$2 < q_1 < \frac{N}{N-2s}, \quad q_1 = \frac{N}{N-2s} \quad \text{and} \quad \frac{N}{N-2s} < q_1 \leq 2_s^* - 2.$$

If $2 < q_1 < \frac{N}{N-2s}$ then, by (6.4.4), it holds

$$\max_{t \in [t_1, t_2]} \mathcal{I}_\mu(tu_\varepsilon) \leq \frac{s}{N} S_*^{\frac{N}{2s}} + O(\varepsilon^{N-2s}) - \lambda O(\varepsilon^{\frac{(N-2s)}{2}q_1})$$

and noting that

$$N - 2s < \frac{(N - 2s)}{2} q_1,$$

we can take $\lambda = \varepsilon^{-\mu}$, with $\mu > \frac{(N-2s)(q_1-2)}{2}$, to get the desired conclusion.
If $q_1 = \frac{N}{N-2s}$, then by (6.4.4) we have

$$\max_{t \in [t_1, t_2]} \mathcal{I}_\mu(tu_\varepsilon) \leq \frac{s}{N} S_*^{\frac{N}{2s}} + O(\varepsilon^{N-2s}) - \lambda O(\varepsilon^{\frac{N}{2}} |\log \varepsilon|),$$

and taking $\lambda = \varepsilon^{-\mu}$, with $\mu > 2s - \frac{N}{2}$, we deduce the assertion for $\varepsilon > 0$ small enough.
Finally, when $\frac{N}{N-2s} < q_1 \leq 2_s^* - 2$, it follows from (6.4.4) that

$$\max_{t \in [t_1, t_2]} \mathcal{I}_\mu(tu_\varepsilon) \leq \frac{s}{N} S_*^{\frac{N}{2s}} + O(\varepsilon^{N-2s}) - \lambda O(\varepsilon^{N - \frac{(N-2s)}{2}q_1}),$$

and choosing $\lambda = \varepsilon^{-\mu}$, with $\mu > 2s - \frac{(N-2s)}{2} q_1$, we have the thesis for $\varepsilon > 0$ sufficiently small.

Putting together (6.4.6), (6.4.7), and (6.4.8), we conclude that

$$\max_{t \geq 0} \mathcal{I}_\mu(tu_\varepsilon) < \frac{s}{N} S_*^{\frac{N}{2s}}.$$

Since $d_\mu \leq \max_{t \geq 0} \mathcal{I}_\mu(tu_\varepsilon)$, we obtain the desired estimate. $\qquad\square$

Now, we prove the following lemma.

Lemma 6.4.9 *Let the sequence* $(u_n) \subset \mathcal{M}_\mu$ *be such that* $\mathcal{I}_\mu(u_n) \to d_\mu$. *Then,* (u_n) *is bounded in* $\mathbb{X}_\mu$, *and there are a sequence* $(y_n) \subset \mathbb{R}^N$ *and constants* $R, \beta > 0$ *such that*

$$\liminf_{n \to \infty} \int_{B_R(y_n)} u_n^2 \, dx \geq \beta > 0.$$

Proof It is easy to check that (u_n) is bounded in $\mathbb{X}_\mu$. Now, we assume that for any $R > 0$,

$$\lim_{n \to \infty} \sup_{y \in \mathbb{R}^N} \int_{B_R(y)} u_n^2 \, dx = 0.$$

The boundedness of (u_n) and Lemma 1.4.4, imply that

$$u_n \to 0 \text{ in } L^r(\mathbb{R}^N) \quad \text{for any } r \in (2, 2_s^*). \tag{6.4.9}$$

Using (6.3.1), (6.3.2), and (6.4.9), we deduce that

$$\left| \int_{\mathbb{R}^N} f(u_n) u_n \, dx \right| \leq \xi \|u_n\|^2_{L^2(\mathbb{R}^N)} + o_n(1) \tag{6.4.10}$$

and

$$\left| \int_{\mathbb{R}^N} F(u_n) \, dx \right| \leq C\xi \|u_n\|^2_{L^2(\mathbb{R}^N)} + o_n(1). \tag{6.4.11}$$

Letting $\xi \to 0$ in (6.4.10) and (6.4.11), we see that

$$\int_{\mathbb{R}^N} f(u_n) u_n \, dx = o_n(1) \quad \text{and} \quad \int_{\mathbb{R}^N} F(u_n) \, dx = o_n(1). \tag{6.4.12}$$

Taking into account that $\langle \mathcal{J}_\varepsilon'(u_n), u_n \rangle = 0$ and using (6.4.12), we obtain

$$\|u_n\|^2_\mu - \|u_n\|^{2_s^*}_{L^{2_s^*}(\mathbb{R}^N)} = o_n(1).$$

Since (u_n) is bounded in $\mathbb{X}_\mu$, we may assume that

$$\|u_n\|^2_\mu \to \ell \geq 0 \quad \text{and} \quad \|u_n\|^{2_s^*}_{L^{2_s^*}(\mathbb{R}^N)} \to \ell \geq 0. \tag{6.4.13}$$

Suppose, by contradiction, that $\ell > 0$. By virtue of (6.4.12) and (6.4.13),

$$d_\mu = \mathcal{I}_\mu(u_n) + o_n(1) = \frac{1}{2}\|u_n\|^2_\mu - \int_{\mathbb{R}^N} F(u_n) \, dx - \frac{1}{2_s^*}\|u_n\|^{2_s^*}_{L^{2_s^*}(\mathbb{R}^N)} + o_n(1)$$

$$= \frac{\ell}{2} - \frac{\ell}{2_s^*} + o_n(1) = \frac{s}{N}\ell + o_n(1),$$

that is, $\ell = \frac{N}{s} d_\mu$. On the other hand, by Theorem 1.1.8,

$$S_* \|u_n\|^2_{L^{2_s^*}(\mathbb{R}^N)} \leq [u_n]^2_s + \mu \|u_n\|^2_{L^2(\mathbb{R}^N)} = \|u_n\|^2_\mu$$

and taking the limit as $n \to \infty$ we obtain that

$$S_* \ell^{\frac{2}{2_s^*}} \leq \ell.$$

Since $\ell = \frac{N}{s}d_\mu$, we get $d_\mu \geq \frac{s}{N}S_*^{\frac{N}{2s}}$, which is impossible in view of Lemma 6.4.8. $\square$

Now we prove the main result for the autonomous critical case.

Lemma 6.4.10 *The problem* (P_μ^*) *has at least one positive ground state solution.*

Proof The proof follows the arguments used in Lemma 6.3.10. We only need to replace (6.3.14) by

$$d_\mu + o_n(1) = \mathcal{I}_\mu(u_n) - \frac{1}{\vartheta}\langle \mathcal{I}_\mu'(u_n), u_n \rangle$$

$$= \left(\frac{1}{2} - \frac{1}{\vartheta}\right)\|u_n\|_\mu^2 + \int_{\mathbb{R}^N}\left(\frac{1}{\vartheta}f(u_n)u_n - F(u_n)\right)dx + \left(\frac{1}{\vartheta} - \frac{1}{2_s^*}\right)\|u_n\|_{L^{2_s^*}(\mathbb{R}^N)}^{2_s^*},$$

and then recalling that

$$\limsup_{n\to\infty}(a_n + b_n + c_n) \geq \limsup_{n\to\infty}a_n + \liminf_{n\to\infty}(b_n + c_n)$$

$$\geq \limsup_{n\to\infty}a_n + \liminf_{n\to\infty}b_n + \liminf_{n\to\infty}c_n,$$

we deduce that

$$d_\mu \geq \left(\frac{1}{2} - \frac{1}{\vartheta}\right)\limsup_{n\to\infty}\|u_n\|_\mu^2 + \liminf_{n\to\infty}\int_{\mathbb{R}^N}\left(\frac{1}{\vartheta}f(u_n)u_n - F(u_n)\right)dx$$

$$+ \left(\frac{1}{\vartheta} - \frac{1}{2_s^*}\right)\liminf_{n\to\infty}\|u_n\|_{L^{2_s^*}(\mathbb{R}^N)}^{2_s^*}.$$

Moreover, we use Lemma 6.4.9 instead of Lemma 6.3.9. Thus, we obtain the existence of a ground state solution to (P_μ^*). Let us note that all the calculations above can be repeated word by word, replacing $\mathcal{I}_\mu$ by

$$\mathcal{I}_\mu^+(u) = \frac{1}{2}\|u\|_\mu^2 - \int_{\mathbb{R}^N}F(u^+)\,dx - \frac{1}{2_s^*}\|u^+\|_{L^{2_s^*}(\mathbb{R}^N)}^{2_s^*}.$$

In this way we find a ground state solution $u \in \mathbb{X}_\mu$ to the equation

$$(-\Delta)^s u + \mu u = f(u^+) + (u^+)^{2_s^*-1} \quad \text{in } \mathbb{R}^N,$$

and standard arguments show that $u > 0$ in $\mathbb{R}^N$. $\square$

Finally, similarly to Theorem 6.3.11, we have the following compactness result for the critical autonomous case:

Theorem 6.4.11 *Let $(u_n) \subset \mathcal{M}_{V_0}$ be a sequence such that $\mathcal{I}_{V_0}(u_n) \to d_{V_0}$. Then either*

(i) *(u_n) has a subsequence that converges strongly in $H^s(\mathbb{R}^N)$, or*
(ii) *there exists a sequence $(\tilde{y}_n) \subset \mathbb{R}^N$ such that, up to a subsequence, $v_n = u_n(\cdot + \tilde{y}_n)$ converges strongly in $H^s(\mathbb{R}^N)$.*

In particular, there exists a minimizer for d_{V_0}.

6.4.3   An Existence Result for the Critical Case

Arguing as in the proof of Lemma 6.4.9, we can prove the "critical" version of Lemma 6.3.12.

Lemma 6.4.12 *Let $0 < d < \frac{s}{N}S_*^{\frac{N}{2s}}$ and let $(u_n) \subset \mathcal{N}_\varepsilon$ be a sequence such that $\mathcal{J}_\varepsilon(u_n) \to d$ and $u_n \rightharpoonup 0$ in $\mathcal{H}_\varepsilon$. Then either*

(a) *$u_n \to 0$ in $\mathcal{H}_\varepsilon$, or*
(b) *there are a sequence $(y_n) \subset \mathbb{R}^N$ and constants $R, \beta > 0$ such that*

$$\liminf_{n \to \infty} \int_{B_R(y_n)} u_n^2 \, dx \geq \beta > 0.$$

The next result is obtained following the lines of the proof of Lemma 6.3.13.

Lemma 6.4.13 *Assume that $V_\infty < \infty$ and let $(v_n) \subset \mathcal{N}_\varepsilon$ be a sequence such that $\mathcal{J}_\varepsilon(v_n) \to d$ with $0 < d < \frac{s}{N}S_*^{\frac{N}{2s}}$ and $v_n \rightharpoonup 0$ in $\mathcal{H}_\varepsilon$. If $v_n \nrightarrow 0$ in $\mathcal{H}_\varepsilon$, then $d \geq d_{V_\infty}$, where d_{V_∞} is the infimum of $\mathcal{I}_{V_\infty}$ over $\mathcal{M}_{V_\infty}$.*

Now, we establish a compactness result in the critical case.

Proposition 6.4.14 *Let $(u_n) \subset \mathcal{N}_\varepsilon$ be a sequence such that $\mathcal{J}_\varepsilon(u_n) \to c$, where $c < d_{V_\infty}$ if $V_\infty < \infty$ and $c < \frac{s}{N}S_*^{\frac{N}{2s}}$ if $V_\infty = \infty$. Then (u_n) admits a convergent subsequence in $\mathcal{H}_\varepsilon$.*

Proof Since $\mathcal{J}_\varepsilon(u_n) \to c$ and $\mathcal{J}_\varepsilon'(u_n) = 0$, we can see that (u_n) is bounded in $\mathcal{H}_\varepsilon$ and, up to a subsequence, we may assume that $u_n \rightharpoonup u$ in $\mathcal{H}_\varepsilon$. Clearly, $\mathcal{J}_\varepsilon'(u) = 0$.

Let $v_n = u_n - u$. Using the Brezis-Lieb lemma [113] and Lemma 3.3 in [269], we know that

$$\|v_n\|_{L^{2^*_s}(\mathbb{R}^N)}^{2^*_s} = \|u_n\|_{L^{2^*_s}(\mathbb{R}^N)}^{2^*_s} - \|u\|_{L^{2^*_s}(\mathbb{R}^N)}^{2^*_s} + o_n(1)$$

and

$$\int_{\mathbb{R}^N} \left| |v_n|^{2^*_s-2}v_n - |u_n|^{2^*_s-2}u_n + |u|^{2^*_s-2}u \right|^{\frac{2^*_s}{2^*_s-1}} dx = o_n(1).$$

Then, arguing as in Proposition 6.3.14, we see that

$$\mathcal{J}_\varepsilon(v_n) = \mathcal{J}_\varepsilon(u_n) - \mathcal{J}_\varepsilon(u) + o_n(1) = c - \mathcal{J}_\varepsilon(u) + o_n(1) = d + o_n(1)$$

and

$$\mathcal{J}_\varepsilon'(v_n) = o_n(1).$$

Note that, by (f_4), it holds that

$$\mathcal{J}_\varepsilon(u) = \mathcal{J}_\varepsilon(u) - \frac{1}{2}\langle \mathcal{J}_\varepsilon'(u), u \rangle = \int_{\mathbb{R}^N} \left[\frac{1}{2} f(u)u - F(u) \right] dx + \left(\frac{1}{2} - \frac{1}{2^*_s} \right) \|u\|_{L^{2^*_s}(\mathbb{R}^N)}^{2^*_s} \geq 0. \tag{6.4.14}$$

If $V_\infty < \infty$, then from (6.4.14) we deduce that $d \leq c < d_{V_\infty}$. In view of Lemma 6.4.13 we have that $v_n \to 0$ in $\mathcal{H}_\varepsilon$, that is $u_n \to u$ in $\mathcal{H}_\varepsilon$.

Let us consider the case $V_\infty = \infty$. Then, we can use Lemma 6.2.3 to deduce that $v_n \to 0$ in $L^r(\mathbb{R}^N)$ for all $r \in [2, 2^*_s)$, which in conjunction with (f_2) and (f_3) implies that

$$\int_{\mathbb{R}^N} f(v_n)v_n \, dx = o_n(1) \quad \text{and} \quad \int_{\mathbb{R}^N} F(v_n) \, dx = o_n(1). \tag{6.4.15}$$

Putting together (6.4.15) and the fact that $\langle \mathcal{J}_\varepsilon'(v_n), v_n \rangle = o_n(1)$, we deduce that

$$\|v_n\|_\varepsilon^2 = \|v_n\|_{L^{2^*_s}(\mathbb{R}^N)}^{2^*_s} + o_n(1).$$

Since (v_n) is bounded in $\mathcal{H}_\varepsilon$, we may assume that $\|v_n\|_\varepsilon^2 \to \ell$ and $\|v_n\|_{L^{2^*_s}(\mathbb{R}^N)}^{2^*_s} \to \ell$, for some $\ell \geq 0$. Let us show that $\ell = 0$. If, by contradiction, $\ell > 0$, then since $\mathcal{J}_\varepsilon(v_n) = d + o_n(1)$, we get

$$\frac{1}{2}\|v_n\|_\varepsilon^2 - \frac{1}{2^*_s}\|v_n\|_{L^{2^*_s}(\mathbb{R}^N)}^{2^*_s} = d + o_n(1),$$

whence

$$\frac{s}{N}\|v_n\|_\varepsilon^2 = d + o_n(1).$$

Taking the limit as $n \to \infty$ we see that $\frac{s}{N}\ell = d$, that is, $\ell = d\frac{N}{s}$. Therefore, by Theorem 1.1.8,

$$\|v_n\|_\varepsilon^2 \geq S_*\|v_n\|_{L^{2_s^*}(\mathbb{R}^N)}^2 = S_* \left(\|v_n\|_{L^{2_s^*}(\mathbb{R}^N)}^{2_s^*} \right)^{\frac{2}{2_s^*}},$$

and letting $n \to \infty$ we get $\ell \geq S_*^{\frac{N}{2s}}$. Thus we have $d \geq \frac{s}{N}S_*^{\frac{N}{2s}}$. Since $d \leq c < \frac{s}{N}S_*^{\frac{N}{2s}}$ we get a contradiction. Hence, $\ell = 0$ and $u_n \to u$ in $\mathcal{H}_\varepsilon$. □

Finally we can state the following existence result to (6.1.3) for $\varepsilon > 0$ small enough.

Theorem 6.4.15 *Assume that (V) and (f_1)–(f_5) and (f_6') hold. Then there exists $\varepsilon_0 > 0$ such that, for any $\varepsilon \in (0, \varepsilon_0)$, problem (P_ε^*) admits a ground state solution.*

Proof It is enough to proceed as in the proof of Theorem 6.3.15 once Lemmas 6.3.2, 6.3.3, Proposition 6.3.14 and Lemma 6.3.10 are replaced by Lemmas 6.4.3, 6.4.4, Proposition 6.4.14 and Lemma 6.4.10, respectively. □

6.4.4   A Multiplicity Result for (6.1.3)

In this section we study the multiplicity of solutions to (6.1.3). Arguing as in the proof of Proposition 6.3.16 and using Lemma 6.4.11 instead of Lemma 6.3.11, we can deduce the following result.

Proposition 6.4.16 *Let $\varepsilon_n \to 0$ and $(u_n) = (u_{\varepsilon_n}) \subset \mathcal{N}_{\varepsilon_n}$ be such that $\mathcal{J}_{\varepsilon_n}(u_n) \to d_{V_0}$. Then there exists $(\tilde{y}_n) = (\tilde{y}_{\varepsilon_n}) \subset \mathbb{R}^N$ such that $v_n(x) = u_n(x + \tilde{y}_n)$ has a convergent subsequence in $H^s(\mathbb{R}^N)$. Moreover, up to a subsequence, $y_n \to y \in M$, where $y_n = \varepsilon_n \tilde{y}_n$.*

Now, fix $\delta > 0$ and let $\omega \in H^s(\mathbb{R}^N)$ be a positive ground state solution to problem $(P_{V_0}^*)$, the existence of which is guaranteed by Lemma 6.4.10. Let ψ be a smooth nonincreasing cut-off function defined in $[0, \infty)$, such that $\psi = 1$ in $[0, \frac{\delta}{2}]$, $\psi = 0$ in $[\delta, \infty)$, $0 \leq \psi \leq 1$ and $|\psi'| \leq c$ for some $c > 0$. For any $y \in M$, we define

$$\Upsilon_{\varepsilon,y}(x) = \psi(|\varepsilon x - y|)\omega \left(\frac{\varepsilon x - y}{\varepsilon} \right),$$

and then take $t_\varepsilon > 0$ such that

$$\mathcal{J}_\varepsilon(t_\varepsilon \Upsilon_{\varepsilon,y}) = \max_{t \geq 0} \mathcal{J}_\varepsilon(t_\varepsilon \Upsilon_{\varepsilon,y}).$$

Also, define $\Phi_\varepsilon : M \to \mathcal{N}_\varepsilon$ by $\Phi_\varepsilon(y) = t_\varepsilon \Upsilon_{\varepsilon,y}$.

Lemma 6.4.17 *The functional Φ_ε has the property that*

$$\lim_{\varepsilon \to 0} \mathcal{J}_\varepsilon(\Phi_\varepsilon(y)) = d_{V_0}, \quad \text{uniformly in } y \in M. \tag{6.4.16}$$

Proof Suppose, by contradiction, that there exist $\delta_0 > 0$, $(y_n) \subset M$ and $\varepsilon_n \to 0$ such that

$$|\mathcal{J}_{\varepsilon_n}(\Phi_{\varepsilon_n}(y_n)) - d_{V_0}| \geq \delta_0. \tag{6.4.17}$$

Using Lemma 1.4.8 we know that

$$\lim_{n \to \infty} \|\Upsilon_{\varepsilon_n,y_n}\|_{\varepsilon_n}^2 = \|\omega\|_{V_0}^2. \tag{6.4.18}$$

On the other hand, by the definition of t_ε, we see that $\langle \mathcal{J}'_{\varepsilon_n}(t_{\varepsilon_n} \Upsilon_{\varepsilon_n,y_n}), t_{\varepsilon_n} \Upsilon_{\varepsilon_n,y_n} \rangle = 0$ which implies

$$\|t_{\varepsilon_n} \Upsilon_{\varepsilon_n,y_n}\|_{\varepsilon_n}^2 = \int_{\mathbb{R}^N} f(t_{\varepsilon_n} \Upsilon_{\varepsilon_n}) t_{\varepsilon_n} \Upsilon_{\varepsilon_n} \, dx + \int_{\mathbb{R}^N} (t_{\varepsilon_n} \Upsilon_{\varepsilon_n})^{2_s^*} dx$$

$$= \int_{\mathbb{R}^N} f(t_{\varepsilon_n} \psi(|\varepsilon_n z|) \omega(z)) t_{\varepsilon_n} \psi(|\varepsilon_n z|) \omega(z) \, dz$$

$$+ t_{\varepsilon_n}^{2_s^*} \int_{\mathbb{R}^N} (\psi(|\varepsilon_n z|) \omega(z))^{2_s^*} \, dz. \tag{6.4.19}$$

Let us prove that $t_{\varepsilon_n} \to 1$. First, we show that $t_{\varepsilon_n} \to t_0 < \infty$. If, by contradiction, $|t_{\varepsilon_n}| \to \infty$, then since $\psi(|x|) = 1$ for $x \in B_{\frac{\delta}{2}}$ and $B_{\frac{\delta}{2}} \subset B_{\frac{\delta}{2\varepsilon_n}}$ for n sufficiently large, we see that (6.4.19) yields

$$\|\Upsilon_{\varepsilon_n,y_n}\|_{\varepsilon_n}^2 \geq t_{\varepsilon_n}^{2_s^*-2} \int_{B_{\frac{\delta}{2}}} (\omega(z))^{2_s^*} dz$$

$$\geq t_{\varepsilon_n}^{2_s^*-2} |B_{\frac{\delta}{2}}| \left(\min_{|z| \leq \frac{\delta}{2}} \omega(z) \right)^{2_s^*} \to \infty, \tag{6.4.20}$$

which is impossible in view of (6.4.18). Hence, we can suppose that $t_{\varepsilon_n} \to t_0 \geq 0$. From the growth assumptions on f and (6.4.18), we deduce that $t_0 > 0$.

Let us prove that $t_0 = 1$. Taking the limit as $n \to \infty$ in (6.4.19), we can see that

$$\|\omega\|_{V_0}^2 = \int_{\mathbb{R}^N} \frac{f(t_0\omega)}{t_0}\omega\, dx + t_0^{2_s^*-2} \int_{\mathbb{R}^N} \omega^{2_s^*}\, dx,$$

and using that $\omega \in \mathcal{M}_{V_0}$ and condition (f_5), we deduce that $t_0 = 1$. This and the dominated convergence theorem show that

$$\lim_{n \to \infty} \int_{\mathbb{R}^N} F(t_{\varepsilon_n} \Upsilon_{\varepsilon_n, y_n})\, dx = \int_{\mathbb{R}^N} F(\omega)\, dx$$

and

$$\lim_{n \to \infty} \int_{\mathbb{R}^N} (t_{\varepsilon_n} \Upsilon_{\varepsilon_n, y_n})^{2_s^*}\, dx = \int_{\mathbb{R}^N} \omega^{2_s^*}\, dx.$$

Consequently,

$$\lim_{n \to \infty} \mathcal{J}_{\varepsilon_n}(\Phi_{\varepsilon_n}(y_n)) = \mathcal{I}_{V_0}(\omega) = d_{V_0}$$

which leads to a contradiction because of (6.4.17). □

For any $\delta > 0$, let $\rho = \rho(\delta) > 0$ be such that $M_\delta \subset B_\rho$. Define $\chi : \mathbb{R}^N \to \mathbb{R}^N$ by

$$\chi(x) = \begin{cases} x, & \text{if } |x| < \rho, \\ \frac{\rho x}{|x|}, & \text{if } |x| \geq \rho. \end{cases}$$

Consider the barycenter map $\beta_\varepsilon : \mathcal{N}_\varepsilon \to \mathbb{R}^N$ given by

$$\beta_\varepsilon(u) = \frac{\int_{\mathbb{R}^N} \chi(\varepsilon x) u^2(x)\, dx}{\int_{\mathbb{R}^N} u^2(x)\, dx}.$$

Arguing as in the proof of Lemma 6.3.18 we obtain the next result.

Lemma 6.4.18 *The functional Φ_ε has the property that*

$$\lim_{\varepsilon \to 0} \beta_\varepsilon(\Phi_\varepsilon(y)) = y, \quad \text{uniformly in } y \in M. \tag{6.4.21}$$

Define

$$\tilde{\mathcal{N}}_\varepsilon = \{u \in \mathcal{N}_\varepsilon : \mathcal{J}_\varepsilon(u) \leq d_{V_0} + h(\varepsilon)\},$$

where $h(\varepsilon) = \sup_{y \in M} |\mathcal{J}_\varepsilon(\Phi_\varepsilon(y)) - d_{V_0}|$. From Lemma 6.4.17 we deduce that $h(\varepsilon) \to 0$ as $\varepsilon \to 0$. By the definition of $h(\varepsilon)$, we know that, for every $y \in M$ and $\varepsilon > 0$, $\Phi_\varepsilon(y) \in \tilde{\mathcal{N}}_\varepsilon$ and $\tilde{\mathcal{N}}_\varepsilon \neq \emptyset$. In addition, proceeding as in the proof of Lemma 6.3.19, we get the following lemma:

Lemma 6.4.19 *For any $\delta > 0$, there holds that*

$$\lim_{\varepsilon \to 0} \sup_{u \in \tilde{\mathcal{N}}_\varepsilon} \operatorname{dist}(\beta_\varepsilon(u), M_\delta) = 0.$$

Finally, we can prove the main result related to (6.1.3).

Proof of Theorem 6.1.2 Given $\delta > 0$ and taking into account Lemmas 6.4.3, 6.4.4, 6.4.17, 6.4.18, 6.4.19, Proposition 6.4.14, Theorem 6.4.15 and recalling that $0 < d_{V_0} < \frac{s}{N} S_*^{\frac{N}{2s}}$ (see Lemma 6.4.8), we can argue as in the proof of Theorem 6.3.22 to deduce that (6.1.3) admits at least $\operatorname{cat}_{M_\delta}(M)$ positive solutions for all $\varepsilon > 0$ sufficiently small. The concentration of solutions is obtained by means of the arguments used in the proof of Theorem 6.1.1. Indeed, the proof of Lemma 6.3.23 works also in the critical case and the proof of Lemma 6.3.24 can be easily adapted to this case. $\square$

6.5 A Remark on the Ambrosetti-Rabinowitz Condition

As discussed in Chap. 5, the Ambrosetti-Rabinowitz condition plays a fundamental role in the verification of the boundedness of Palais–Smale sequences of the energy functional associated with the problem under consideration. Anyway, as proved in [75] (see also [6]), we can see that Theorems 6.1.1 and 6.1.2 are still true if we replace (f_4) by the condition

(f_4') $\frac{F(t)}{t^2} \to \infty$ as $t \to \infty$.

In what follows, we only prove some technical results which are used to implement the arguments used in the previous subsections. First, we need the following variant of Lions lemma inspired by Ramos et al. [300].

Lemma 6.5.1 *Let (u_n) be a bounded sequence in $H^s(\mathbb{R}^N)$ such that*

$$\lim_{n \to \infty} \sup_{y \in \mathbb{R}^N} \int_{B_R(y)} |u_n|^{2_s^*} \, dx = 0,$$

for some $R > 0$. Then, $u_n \to 0$ in $L^t(\mathbb{R}^N)$ for all $t \in (2, 2_s^*]$.

Proof Applying the Hölder inequality we have

$$\int_{B_R(y)} |u_n|^{2_s^*} \, dx \leq \left(\int_{B_R(y)} |u_n|^{2_s^*} \, dx \right)^{\frac{2_s^* - 2}{2_s^*}} \left(\int_{B_R(y)} |u_n|^{2_s^*} \, dx \right)^{\frac{2}{2_s^*}}.$$

Now, covering $\mathbb{R}^N$ by balls of radius R in such a way that each point of $\mathbb{R}^N$ is contained in at most $N + 1$ balls and using the fact that (u_n) is bounded in $H^s(\mathbb{R}^N)$, we obtain that

$$\|u_n\|_{L^{2_s^*}(\mathbb{R}^N)}^{2_s^*} \leq C \sup_{y \in \mathbb{R}^N} \left(\int_{B_R(y)} |u_n|^{2_s^*} \, dx \right)^{\frac{2_s^* - 2}{2_s^*}} \to 0 \quad \text{as } n \to \infty.$$

An interpolation argument concludes the proof of lemma. $\square$

In order to apply the Lusternik–Schnirelman category theory, we have to prove that $\mathcal{J}_\varepsilon$ satisfies the Palais–Smale condition on $\mathcal{N}_\varepsilon$. Since we do not assume (AR), we prove the following auxiliary result to deduce the boundedness of Palais–Smale sequences.

Lemma 6.5.2 *Let $(u_n) \subset \mathcal{N}_\varepsilon$ be a sequence such that $\mathcal{J}_\varepsilon(u_n) \to c$. Then, (u_n) is bounded in $\mathcal{H}_\varepsilon$.*

Proof Since we study both subcritical and critical cases, we consider the nonlinearity $f(t) + \gamma |t|^{2_s^* - 2} t$, where $\gamma \in \{0, 1\}$. Assume, by contradiction, that, up to subsequences, $\|u_n\|_\varepsilon \to \infty$ as $n \to \infty$. Set $v_n = u_n \|u_n\|_\varepsilon^{-1}$. Note that (v_n) is bounded in $\mathcal{H}_\varepsilon$ (since $\|v_n\|_\varepsilon = 1$ for all $n \in \mathbb{N}$). Let us prove that

$$\lim_{n \to \infty} \sup_{y \in \mathbb{R}^N} \int_{B_R(y)} |v_n|^{2_s^*} \, dx = 0$$

for all $R > 0$. Suppose this is not the case. Then there are $R, \delta > 0$, $(y_n) \subset \mathbb{R}^N$ such that

$$\liminf_{n \to \infty} \int_{B_R(y_n)} |v_n|^{2_s^*} \, dx \geq \delta > 0.$$

Set $\tilde{v}_n = v_n(\cdot + \tilde{y}_n)$. Then, $(\tilde{v}_n)$ is bounded in $\mathcal{H}_\varepsilon$ and $\tilde{v}_n \rightharpoonup \tilde{v}$ in $\mathcal{H}_\varepsilon$ and strongly in $L^t_{\text{loc}}(\mathbb{R}^N)$ for all $t \in [1, 2_s^*)$, for some $\tilde{v} \neq 0$. Define $\tilde{u}_n = \|u_n\|_\varepsilon \tilde{v}_n$. By (f_4') we deduce that

$$\frac{F(\tilde{u}_n)}{\tilde{u}_n^2} \tilde{v}_n^2 \to \infty \quad \text{a.e.} \quad \text{in } A = \{x \in \mathbb{R}^N : \tilde{v} \neq 0\}.$$

Then using Fatou's lemma we have that

$$\frac{1}{2} - \frac{c + o_n(1)}{\|u_n\|_\varepsilon^2} = \int_{\mathbb{R}^N} \frac{F(u_n) + \gamma |u_n|^{2_s^*}}{\|u_n\|_\varepsilon^2} \, dx \geq \int_A \frac{F(\tilde{u}_n)}{\tilde{u}_n^2} \tilde{v}_n^2 \, dx \to \infty,$$

which gives a contradiction. Therefore, by Lemma 6.5.1, we infer that $v_n \to 0$ in $L^t(\mathbb{R}^N)$ for all $t \in (2, 2_s^*]$. Now, using conditions (f_2)–(f_3), we see that for every $r > 1$ and for every $\tau > 0$ there exists a constant $C_\tau > 0$ such that

$$|F(rt)| \leq \tau |rt|^2 + C_\tau |rt|^{2_s^*} \quad \text{for all } t \in \mathbb{R}.$$

Therefore,

$$\limsup_{n \to \infty} \left| \int_{\mathbb{R}^N} F(rv_n) + \gamma \frac{|rv_n|^{2_s^*}}{2_s^*} \, dx \right| \leq \tau r^2 \quad \text{for all } \tau > 0,$$

which implies that

$$\limsup_{n \to \infty} \int_{\mathbb{R}^N} F(rv_n) + \gamma \frac{|rv_n|^{2_s^*}}{2_s^*} \, dx = 0. \tag{6.5.1}$$

On the other hand, $\|u_n\|_\varepsilon^{-1} r \in (0, 1)$ for all n sufficiently large, and we have

$$\mathcal{J}_\varepsilon(u_n) = \max_{t \geq 0} \mathcal{J}_\varepsilon(tu_n) \geq \mathcal{J}_\varepsilon(r\|u_n\|_\varepsilon^{-1} u_n) = \mathcal{J}_\varepsilon(rv_n) = \frac{r^2}{2} - \int_{\mathbb{R}^N} \left[F(rv_n) + \gamma \frac{|rv_n|^{2_s^*}}{2_s^*} \right] dx,$$

which together with (6.5.1) yields

$$\liminf_{n \to \infty} \mathcal{J}_\varepsilon(u_n) \geq r^2 \quad \text{for all } r > 1.$$

Letting $r \to \infty$ we deduce that

$$\infty > c = \liminf_{n \to \infty} \mathcal{J}_\varepsilon(u_n) = \infty$$

a contradiction. □

6.6 Further Generalizations: The Fractional p-Laplacian Operator

One can show (see [74, 75]) that the existence and multiplicity results obtained in the previous subsections can be extended to the following fractional p-Laplacian type problem:

$$\begin{cases} \varepsilon^{sp}(-\Delta)_p^s u + V(x)|u|^{p-2}u = f(u) + \gamma|u|^{p_s^*-2}u \ \text{ in } \mathbb{R}^N, \\ u \in W^{s,p}(\mathbb{R}^N), \ u > 0 \text{ in } \mathbb{R}^N, \end{cases} \tag{6.6.1}$$

where $\varepsilon > 0$ is a parameter, $\gamma \in \{0, 1\}$, $s \in (0, 1)$, $1 < p < \infty$, $N > sp$, and f is a continuous function with subcritical growth at infinity. Here, $(-\Delta)_p^s$ is the fractional p-Laplacian operator which may be defined, up to a normalization constant depending on N, p and s, by setting

$$(-\Delta)_p^s u(x) = 2 \lim_{r \to 0} \int_{\mathbb{R}^N \setminus B_r(x)} \frac{|u(x) - u(y)|^{p-2}(u(x) - u(y))}{|x - y|^{N+sp}} \, dy \quad (x \in \mathbb{R}^N)$$

for any $u \in C_c^\infty(\mathbb{R}^N)$. The above operator can be regarded as the fractional analogue of the p-Laplacian operator $\Delta_p u = \text{div}(|\nabla u|^{p-2}\nabla u)$.

We recall that the recent years have seen a surge of interest in nonlocal and fractional problems involving the fractional p-Laplacian operator because of the presence of two features: the nonlinearity of the operator and its nonlocal character. For this reason, several existence and regularity results have been established by many authors. Franzina and Palatucci [201] discussed some basic properties of the eigenfunctions of a class of nonlocal operators whose model is $(-\Delta)_p^s$ (see also [252]). Mosconi et al. [277] used an abstract linking theorem based on the cohomological index to obtain nontrivial solutions to the Brezis–Nirenberg problem [114] for the fractional p-Laplacian operator. Di Castro et al. [167] established interior Hölder regularity results for fractional p-minimizers (see also [222]). In [33] the author obtained the existence of infinitely many solutions for a superlinear fractional p-Laplacian equation with sign-changing potential. We also mention [56, 110, 163, 164, 329, 335] for other interesting contributions.

We stress that the operator $(-\Delta)_p^s$ is not linear when $p \neq 2$, so more technical difficulties arise in the study of (6.6.1). For instance, we cannot make use of the s-harmonic extension by Caffarelli and Silvestre [127] to apply well-known variational techniques in the study of local degenerate elliptic problems. Clearly, the arguments developed in the case $p = 2$ are not trivially adaptable and some technical lemmas are needed to overcome the non-Hilbertian structure of the fractional Sobolev spaces $W^{s,p}(\mathbb{R}^N)$ when $p \neq 2$. In particular, when $\gamma = 1$, more refined estimates coming from [110, 277] are used to obtain the corresponding version of Lemma 6.4.8. For more details we refer the interested reader to [74].

Fractional Schrödinger Equations with del Pino-Felmer Assumptions

7

7.1 Introduction

This chapter is concerned with the existence and concentration of positive solutions for the following fractional Laplacian problem

$$\begin{cases} \varepsilon^{2s}(-\Delta)^s u + V(x)u = f(u) + \gamma |u|^{2^*_s-2}u & \text{in } \mathbb{R}^N, \\ u \in H^s(\mathbb{R}^N), \quad u > 0 \text{ in } \mathbb{R}^N, \end{cases} \tag{7.1.1}$$

where $\varepsilon > 0$ is a small parameter, $s \in (0, 1)$, $N > 2s$, $2^*_s = \frac{2N}{N-2s}$ is the fractional critical Sobolev exponent, $\gamma \in \{0, 1\}$. Throughout the chapter we will assume that $V : \mathbb{R}^N \to \mathbb{R}$ is a continuous potential satisfying the following assumptions due to del Pino and Felmer [165]:

(V_1) there exists $V_1 > 0$ such that $V_1 = \inf_{x \in \mathbb{R}^N} V(x)$;
(V_2) there exists a bounded open set $\Lambda \subset \mathbb{R}^N$ such that

$$0 < V_0 = \inf_{x \in \Lambda} V(x) < \min_{x \in \partial\Lambda} V(x).$$

The nonlinearity $f : \mathbb{R} \to \mathbb{R}$ is a continuous function such that $f(t) = 0$ for $t \leq 0$ and fulfills the following conditions if $\gamma = 0$:

(f_1) $f(t) = o(t)$ as $t \to 0$;
(f_2) there exists $q \in (2, 2^*_s)$ such that

$$\lim_{t \to \infty} \frac{f(t)}{t^{q-1}} = 0;$$

© The Author(s), under exclusive license to Springer Nature Switzerland AG 2021
V. Ambrosio, *Nonlinear Fractional Schrödinger Equations in* $\mathbb{R}^N$,
Frontiers in Mathematics, https://doi.org/10.1007/978-3-030-60220-8_7

(f_3) there exists $\vartheta \in (2, q)$ such that $0 < \vartheta F(t) = \vartheta \int_0^t f(\tau)\,d\tau \le tf(t)$ for all $t > 0$;

(f_4) the function $t \mapsto \dfrac{f(t)}{t}$ is increasing in $(0, \infty)$.

When $\gamma = 1$ we require that f satisfies (f_1), (f_3), (f_4) and the following technical condition:

(f_2') there exist $q, \sigma \in (2, 2_s^*)$ and $\lambda > 0$ such that

$$f(t) \ge \lambda t^{q-1} \quad \forall t > 0, \quad \lim_{t \to \infty} \frac{f(t)}{t^{\sigma-1}} = 0,$$

where λ is such that

- $\lambda > 0$ if either $N > 4s$, or $2s < N < 4s$ and $2_s^* - 2 < q < 2_s^*$,
- λ is sufficiently large if $2s < N < 4s$ and $2 < q \le 2_s^* - 2$.

Differently from Chap. 6, here we focus on the existence and concentration of positive solutions to (7.1.1) under the local conditions (V_1)-(V_2) on the potential V. We observe that no restriction on the global behavior of V is imposed other than (V_1). In particular, V is not required to be bounded or to belong to a Kato class. Moreover, we consider the cases when f has subcritical, critical or supercritical growth.

Our first main result is the following:

Theorem 7.1.1 ([19, 35, 68]) *Assume that (V_1)-(V_2) and (f_1)–(f_4) hold. Then, there exists $\varepsilon_0 > 0$ such that, for all $\varepsilon \in (0, \varepsilon_0)$, problem (7.1.1) admits a positive solution. Moreover, if $x_\varepsilon \in \mathbb{R}^N$ denotes a global maximum point of u_ε, then*

$$\lim_{\varepsilon \to 0} V(x_\varepsilon) = V_0,$$

and there exists $C > 0$ such that

$$0 < u_\varepsilon(x) \le \frac{C\,\varepsilon^{N+2s}}{\varepsilon^{N+2s} + |x - x_\varepsilon|^{N+2s}} \quad \text{for all } x \in \mathbb{R}^N.$$

The proof of Theorem 7.1.1 is obtained by using some variational techniques inspired by [19, 35, 68, 165]. Since we don't have any information about the behavior of the potential V at infinity, we adapt the penalization method introduced by del Pino and Felmer in [165]. It consists in making a suitable modification on f, solving a modified problem and then checking that, for $\varepsilon > 0$ small enough, the solutions of the new problem are indeed solutions of the original one. This last step is rather difficult and it will be obtained by combining a Moser iteration argument [278] and some useful regularity elliptic estimates.

Our second result concerns the critical case.

Theorem 7.1.2 ([68, 216]) *Assume that* (V_1)-(V_2) *and* (f_1), (f_2'), (f_3) *with* $\vartheta \in (2, \sigma)$, (f_4) *hold. Then, there exists* $\varepsilon_0 > 0$ *such that, for all* $\varepsilon \in (0, \varepsilon_0)$, *problem* (7.1.1) *admits a positive solution. Moreover, if* $x_\varepsilon \in \mathbb{R}^N$ *denotes a global maximum point of* u_ε, *then*

$$\lim_{\varepsilon \to 0} V(x_\varepsilon) = V_0.$$

The proof of Theorem 7.1.2 follows some arguments in [12, 35, 68, 190, 216]. In this case an additional difficulty appears in the study of (7.1.1), due to the presence of the critical exponent. Anyway, we will see that the approach used in the subcritical case works, after a more careful analysis, in the critical case. Moreover, in order to prove some compactness properties for the energy modified functional, we make use of the concentration-compactness lemma established in Section 1.5.

Finally, we consider a supercritical version of the problem under consideration, namely

$$\varepsilon^{2s}(-\Delta)^s u + V(x)u = |u|^{q-2}u + \lambda|u|^{2_s^*-2}u \quad \text{in } \mathbb{R}^N, \tag{7.1.2}$$

where $2 < q < 2_s^* \le r$ and $\lambda > 0$. In this case, our main result can be stated as follows:

Theorem 7.1.3 ([35]) *Assume that* (V_1)-(V_2) *hold. Then there exists* $\lambda_0 > 0$ *with the following property: for any* $\lambda \in (0, \lambda_0)$ *there exists* $\varepsilon_\lambda > 0$ *such that, for all* $\varepsilon \in (0, \varepsilon_\lambda)$, *problem* (7.1.2) *admits a positive solution. Moreover, if* $x_\varepsilon \in \mathbb{R}^N$ *denotes a global maximum point of* u_ε, *then*

$$\lim_{\varepsilon \to 0} V(x_\varepsilon) = V_0.$$

The proof of this result is based on some arguments developed in [136, 188, 297]. Since $r > 2_s^*$, we cannot use directly variational techniques because the corresponding functional is not well defined on the Sobolev space $H^s(\mathbb{R}^N)$. To overcome this difficulty, we first truncate suitably the nonlinearity involving the supercritical exponent, in order to deal with a new subcritical problem. Taking into account Theorem 7.1.1, we know that an existence result for this truncated problem holds true. Then, we deduce a priori bounds for these solutions, and by using a Moser iteration technique [278], we are able to show that for $\lambda > 0$ sufficiently small the solutions of the truncated problem also satisfy the original problem (7.1.1).

7.2　The Subcritical Case

7.2.1　The Modified Problem

Using the change of variable $x \mapsto \varepsilon x$, we reduce the study of (7.1.1) to the investigation of the problem

$$
\begin{cases}
(-\Delta)^s u + V(\varepsilon x) u = f(u) & \text{in } \mathbb{R}^N, \\
u \in H^s(\mathbb{R}^N), \quad u > 0 \text{ in } \mathbb{R}^N.
\end{cases} \tag{7.2.1}
$$

Now, we introduce a penalized function in the spirit of [165]. First of all, without loss of generality, we may assume that

$$
0 \in \Lambda \quad \text{and} \quad V(0) = V_0 = \inf_\Lambda V.
$$

Take $K > \frac{\vartheta}{\vartheta - 2}$ and $a > 0$ such that $f(a) = \frac{V_1}{K} a$, and set

$$
\tilde{f}(t) =
\begin{cases}
f(t), & \text{if } t \le a, \\
\frac{V_1}{K} t, & \text{if } t > a,
\end{cases}
$$

and

$$
g(x, t) =
\begin{cases}
\chi_\Lambda(x) f(t) + (1 - \chi_\Lambda(x)) \tilde{f}(t), & \text{if } t \ge 0, \\
0, & \text{if } t < 0.
\end{cases}
$$

It is easy to check that $g : \mathbb{R}^N \times \mathbb{R} \to \mathbb{R}$ is a Carathéodory function satisfying the following properties:

(g_1)　$\displaystyle \lim_{t \to 0} \frac{g(x, t)}{t} = 0$ uniformly with respect to $x \in \mathbb{R}^N$;

(g_2)　$g(x, t) \le f(t)$ for all $x \in \mathbb{R}^N, t > 0$;

(g_3)　(i) $0 < \vartheta G(x, t) = \vartheta \displaystyle\int_0^t g(x, \tau)\, d\tau \le g(x, t) t$ for all $x \in \Lambda$ and $t > 0$,

　　(ii) $0 \le 2G(x, t) \le g(x, t) t \le \dfrac{V_1}{K} t^2$ for all $x \in \mathbb{R}^N \setminus \Lambda$ and $t > 0$;

(g_4)　for each $x \in \Lambda$ the function $t \mapsto \dfrac{g(x, t)}{t}$ is increasing in $(0, \infty)$, and for each $x \in$ $\mathbb{R}^N \setminus \Lambda$ the function $t \mapsto \dfrac{g(x, t)}{t}$ is increasing in $(0, a)$.

Then, we consider the following modified problem:

$$\begin{cases} (-\Delta)^s u + V(\varepsilon x)u = g(\varepsilon x, u) & \text{in } \mathbb{R}^N, \\ u \in H^s(\mathbb{R}^N), \quad u > 0 \text{ in } \mathbb{R}^N, \end{cases} \tag{7.2.2}$$

In view of the definition of g, we will look for weak solutions to (7.2.2) having the property

$$u(x) \le a \quad \text{for any } x \in \mathbb{R}^N \setminus \Lambda_\varepsilon,$$

where $\Lambda_\varepsilon = \Lambda / \varepsilon$. In order to study the problem (7.2.2), we seek the critical points of the functional

$$\mathcal{J}_\varepsilon(u) = \frac{1}{2}\|u\|_\varepsilon^2 - \int_{\mathbb{R}^N} G(\varepsilon x, u)\, dx,$$

which is well defined for all $u : \mathbb{R}^N \to \mathbb{R}$ belonging to the fractional space

$$\mathcal{H}_\varepsilon = \left\{ u \in H^s(\mathbb{R}^N) : \int_{\mathbb{R}^N} V(\varepsilon x)u^2\, dx < \infty \right\},$$

endowed with the norm

$$\|u\|_\varepsilon = \left([u]_s^2 + \int_{\mathbb{R}^N} V(\varepsilon x)u^2\, dx \right)^{\frac{1}{2}}.$$

Clearly, $\mathcal{H}_\varepsilon$ is a Hilbert space with the inner product

$$\langle u, v \rangle_\varepsilon = \langle u, v \rangle_{\mathcal{D}^{s,2}(\mathbb{R}^N)} + \int_{\mathbb{R}^N} V(\varepsilon x)uv\, dx.$$

Standard arguments show that the functional $\mathcal{J}_\varepsilon$ belongs to $C^1(\mathcal{H}_\varepsilon, \mathbb{R})$ and that its differential is given by

$$\langle \mathcal{J}_\varepsilon'(u), v \rangle = \langle u, v \rangle_\varepsilon - \int_{\mathbb{R}^N} g(\varepsilon x, u)v\, dx$$

for any $u, v \in \mathcal{H}_\varepsilon$. Now let us show that $\mathcal{J}_\varepsilon$ possesses a mountain pass geometry [29]:

Lemma 7.2.1 *The functional $\mathcal{J}_\varepsilon$ has a mountain pass geometry:*

(a) *there exist $\alpha, \rho > 0$ such that $\mathcal{J}_\varepsilon(u) \ge \alpha$ with $\|u\|_\varepsilon = \rho$;*
(b) *there exists $e \in \mathcal{H}_\varepsilon$ such that $\|e\|_\varepsilon > \rho$ and $\mathcal{J}_\varepsilon(e) < 0$.*

Proof (a) By (g_1), (g_2) and (f_2), we see that for every $\xi > 0$ there exists $C_\xi > 0$ such that

$$|g(x,t)| \leq \xi|t| + C_\xi|t|^{2_s^*-1} \quad \text{for any } (x,t) \in \mathbb{R}^N \times \mathbb{R}.$$

Therefore,

$$\mathcal{J}_\varepsilon(u) \geq \frac{1}{2}\|u\|_\varepsilon^2 - \int_{\mathbb{R}^N} G(\varepsilon x, u)\,dx \geq \frac{1}{2}\|u\|_\varepsilon^2 - \xi C\|u\|_\varepsilon^2 - C_\xi C\|u\|_\varepsilon^{2_s^*},$$

and we can find $\alpha, \rho > 0$ such that $\mathcal{J}_\varepsilon(u) \geq \alpha$ with $\|u\|_\varepsilon = \rho$.

(b) Using (g_3)-(i), we deduce that for any $u \in C_c^\infty(\mathbb{R}^N)$ such that $u \geq 0$, $u \not\equiv 0$ and $\operatorname{supp}(u) \subset \Lambda_\varepsilon$

$$\mathcal{J}_\varepsilon(\tau u) \leq \frac{\tau^2}{2}\|u\|_\varepsilon^2 - \int_{\Lambda_\varepsilon} G(\varepsilon x, \tau u)\,dx$$

$$\leq \frac{\tau^2}{2}\|u\|_\varepsilon^2 - C_1\tau^\vartheta \int_{\Lambda_\varepsilon} u^\vartheta\,dx + C_2 \quad \text{for any } \tau > 0,$$

for some positive constants C_1 and C_2. Since $\vartheta \in (2, 2_s^*)$, we see that $\mathcal{J}_\varepsilon(\tau u) \to -\infty$ as $\tau \to \infty$.

$\square$

Invoking a variant of the mountain pass theorem without the Palais-Smale condition (see Remark 2.2.10), we can find a sequence $(u_n) \subset \mathcal{H}_\varepsilon$ such that

$$\mathcal{J}_\varepsilon(u_n) \to c_\varepsilon \quad \text{and} \quad \mathcal{J}_\varepsilon'(u_n) \to 0 \text{ in } \mathcal{H}_\varepsilon^*,$$

where

$$c_\varepsilon = \inf_{\gamma \in \Gamma_\varepsilon} \max_{t \in [0,1]} \mathcal{J}_\varepsilon(\gamma(t)) \quad \text{and} \quad \Gamma_\varepsilon = \{v \in \mathcal{H}_\varepsilon : \mathcal{J}_\varepsilon(0) = 0, \mathcal{J}_\varepsilon(\gamma(1)) < 0\}.$$

As in [299], we can use the following equivalent characterization of c_ε that is more appropriate for our purposes:

$$c_\varepsilon = \inf_{u \in \mathcal{H}_\varepsilon \setminus \{0\}} \max_{t \geq 0} \mathcal{J}_\varepsilon(tu).$$

Moreover, in view of the monotonicity of g, it is easy to check that for any non-negative $u \in \mathcal{H}_\varepsilon \setminus \{0\}$ there exists a unique $t_0 = t_0(u) > 0$ such that

$$\mathcal{J}_\varepsilon(t_0 u) = \max_{t \geq 0} \mathcal{J}_\varepsilon(tu).$$

In the next lemma we prove that every Palais-Smale sequence of $\mathcal{J}_\varepsilon$ is bounded.

Lemma 7.2.2 *Let $(u_n) \subset \mathcal{H}_\varepsilon$ be a Palais-Smale sequence at the level c for $\mathcal{J}_\varepsilon$. Then (u_n) is bounded in $\mathcal{H}_\varepsilon$.*

Proof Since (u_n) is a Palais-Smale sequence at the level c, we have

$$\mathcal{J}_\varepsilon(u_n) \to c \quad \text{and} \quad \mathcal{J}'_\varepsilon(u_n) \to 0 \text{ in } \mathcal{H}^*_\varepsilon.$$

By (g_3) we deduce that for all $n \in \mathbb{N}$

$$
\begin{aligned}
C(1 + \|u_n\|_\varepsilon) &\geq \mathcal{J}_\varepsilon(u_n) - \frac{1}{\vartheta}\langle \mathcal{J}'_\varepsilon(u_n), u_n \rangle \\
&= \left(\frac{\vartheta - 2}{2\vartheta}\right)\|u_n\|^2_\varepsilon + \frac{1}{\vartheta}\int_{\mathbb{R}^N \setminus \Lambda_\varepsilon}[g(\varepsilon x, u_n)u_n - \vartheta G(\varepsilon x, u_n)]\,dx \\
&\quad + \frac{1}{\vartheta}\int_{\Lambda_\varepsilon}[g(\varepsilon x, u_n)u_n - \vartheta G(\varepsilon x, u_n)]\,dx \\
&\geq \left(\frac{\vartheta - 2}{2\vartheta}\right)\|u_n\|^2_\varepsilon + \frac{1}{\vartheta}\int_{\mathbb{R}^N \setminus \Lambda_\varepsilon}[g(\varepsilon x, u_n)u_n - \vartheta G(\varepsilon x, u_n)]\,dx \\
&\geq \left(\frac{\vartheta - 2}{2\vartheta}\right)\|u_n\|^2_\varepsilon - \left(\frac{\vartheta - 2}{2\vartheta}\right)\frac{1}{K}\int_{\mathbb{R}^N \setminus \Lambda_\varepsilon}V(\varepsilon x)u_n^2\,dx \\
&\geq \left(\frac{\vartheta - 2}{2\vartheta}\right)\left(1 - \frac{1}{K}\right)\|u_n\|^2_\varepsilon.
\end{aligned}
$$

Since $\vartheta > 2$ and $K > 1$, we conclude that (u_n) is bounded in $\mathcal{H}_\varepsilon$. $\qquad\square$

Remark 7.2.3 Arguing as in Remark 5.2.8, we may always assume that the Palais-Smale sequence (u_n) is non-negative in $\mathbb{R}^N$.

Lemma 7.2.4 *$\mathcal{J}_\varepsilon$ satisfies the $(PS)_c$ condition at any level $c \in \mathbb{R}$.*

Proof Let $(u_n) \subset \mathcal{H}_\varepsilon$ be a Palais-Smale sequence for $\mathcal{J}_\varepsilon$ at the level c. Our aim is to prove that for any $\xi > 0$ there exists $R = R_\xi > 0$ such that

$$\limsup_{n \to \infty} \int_{\mathbb{R}^N \setminus B_R}\int_{\mathbb{R}^N}\frac{|u_n(x) - u_n(y)|^2}{|x - y|^{N+2s}}\,dxdy + \int_{\mathbb{R}^N \setminus B_R}V(\varepsilon x)u_n^2\,dx < \xi. \tag{7.2.3}$$

Assume that the above claim is true and we show how it can be used to finish the proof of lemma. By Lemma 7.2.2, we may assume that $u_n \rightharpoonup u$ in $\mathcal{H}_\varepsilon$. Since $\mathcal{H}_\varepsilon$ is compactly embedded in $L^p(\mathcal{K})$ for all $p \in [1, 2^*_s)$ and compact sets $\mathcal{K} \subset \mathbb{R}^N$, g has subcritical growth, and $C_c^\infty(\mathbb{R}^N)$ is dense in $\mathcal{H}_\varepsilon$, it is easy to check that $\mathcal{J}_\varepsilon'(u) = 0$. In particular,

$$\|u\|_\varepsilon^2 = \int_{\mathbb{R}^N} g(\varepsilon x, u)u \, dx.$$

Recalling that $\langle \mathcal{J}_\varepsilon'(u_n), u_n \rangle = o_n(1)$, we have that

$$\|u_n\|_\varepsilon^2 = \int_{\mathbb{R}^N} g(\varepsilon x, u_n)u_n \, dx + o_n(1).$$

Using (7.2.3) and (V_1), we can see that for all $\xi > 0$ there exists $R = R_\xi > 0$ such that

$$\limsup_{n\to\infty} \int_{\mathbb{R}^N \setminus B_R} u_n^2 \, dx \leq \frac{1}{V_1} \limsup_{n\to\infty} \int_{\mathbb{R}^N \setminus B_R} V(\varepsilon x)u_n^2 \, dx < \frac{\xi}{V_1}.$$

Since $u \in L^2(\mathbb{R}^N)$, we may assume, eventually taking $R > 0$ larger, that

$$\int_{\mathbb{R}^N \setminus B_R} u^2 \, dx < \xi.$$

These last two inequalities combined with the fact that $\mathcal{H}_\varepsilon$ is compactly embedded in $L^2(B_R)$ imply that

$$\limsup_{n\to\infty} \|u_n - u\|_{L^2(\mathbb{R}^N)}^2 = \limsup_{n\to\infty} \left[\|u_n - u\|_{L^2(B_R)}^2 + \|u_n - u\|_{L^2(\mathbb{R}^N \setminus B_R)}^2 \right]$$

$$= \lim_{n\to\infty} \|u_n - u\|_{L^2(B_R)}^2 + \limsup_{n\to\infty} \|u_n - u\|_{L^2(\mathbb{R}^N \setminus B_R)}^2$$

$$\leq C\xi,$$

and letting $\xi \to 0$ we see that $u_n \to u$ in $L^2(\mathbb{R}^N)$. By interpolation on the L^p-spaces and the boundedness of (u_n) in $L^{2_s^*}(\mathbb{R}^N)$, we conclude that $u_n \to u$ in $L^p(\mathbb{R}^N)$ for all $p \in [2, 2_s^*)$. Consequently, by (g_1), (g_2), and the dominated convergence theorem,

$$\lim_{n\to\infty} \int_{\mathbb{R}^N} g(\varepsilon x, u_n)u_n \, dx = \int_{\mathbb{R}^N} g(\varepsilon x, u)u \, dx.$$

Therefore,

$$\lim_{n\to\infty} \|u_n\|_\varepsilon^2 = \|u\|_\varepsilon^2,$$

and using the fact that $\mathcal{H}_\varepsilon$ is a Hilbert space we deduce that $u_n \to u$ in $\mathcal{H}_\varepsilon$ as $n \to \infty$.

Now we prove that (7.2.3) holds. Let $\eta_R \in C^\infty(\mathbb{R}^N)$ be such that $0 \le \eta_R \le 1$, $\eta_R = 0$ in $B_{\frac{R}{2}}$, $\eta_R = 1$ in $\mathbb{R}^N \setminus B_R$ and $\|\nabla \eta_R\|_{L^\infty(\mathbb{R}^N)} \le \frac{C}{R}$ for some $C > 0$ independent of R. Since $\langle \mathcal{J}'_\varepsilon(u_n), \eta_R u_n \rangle = o_n(1)$, we have

$$\langle u_n, u_n \eta_R \rangle_{\mathcal{D}^{s,2}(\mathbb{R}^N)} + \int_{\mathbb{R}^N} V(\varepsilon x) \eta_R u_n^2 \, dx = \int_{\mathbb{R}^N} g(\varepsilon x, u_n) u_n \eta_R \, dx + o_n(1).$$

Fix $R > 0$ such that $\Lambda_\varepsilon \subset B_{R/2}$. Using (g_3)-(ii), we have

$$\iint_{\mathbb{R}^{2N}} \eta_R(x) \frac{|u_n(x) - u_n(y)|^2}{|x - y|^{N+2s}} \, dx dy + \int_{\mathbb{R}^N} V(\varepsilon x) u_n^2 \eta_R \, dx$$

$$\le -\left(\iint_{\mathbb{R}^{2N}} u_n(y) \frac{(u_n(x) - u_n(y))(\eta_R(x) - \eta_R(y))}{|x - y|^{N+2s}} \, dx dy \right)$$

$$+ \frac{1}{K} \int_{\mathbb{R}^N} V(\varepsilon x) u_n^2 \eta_R \, dx + o_n(1). \qquad (7.2.4)$$

By Hölder's inequality and the boundedness of (u_n) in $\mathcal{H}_\varepsilon$ it follows that

$$\left| \left(\iint_{\mathbb{R}^{2N}} u_n(y) \frac{(u_n(x) - u_n(y))(\eta_R(x) - \eta_R(y))}{|x - y|^{N+2s}} \, dx dy \right) \right|$$

$$\le \left(\iint_{\mathbb{R}^{2N}} \frac{|u_n(x) - u_n(y)|^2}{|x - y|^{N+2s}} \, dx dy \right)^{\frac{1}{2}} \left(\iint_{\mathbb{R}^{2N}} |u_n(y)|^2 \frac{|\eta_R(x) - \eta_R(y)|^2}{|x - y|^{N+2s}} \, dx dy \right)^{\frac{1}{2}}$$

$$\le C \left(\iint_{\mathbb{R}^{2N}} |u_n(y)|^2 \frac{|\eta_R(x) - \eta_R(y)|^2}{|x - y|^{N+2s}} \, dx dy \right)^{\frac{1}{2}}. \qquad (7.2.5)$$

Further, by Lemma 1.4.5,

$$\lim_{R \to \infty} \limsup_{n \to \infty} \iint_{\mathbb{R}^{2N}} |u_n(y)|^2 \frac{|\eta_R(x) - \eta_R(y)|^2}{|x - y|^{N+2s}} \, dx dy = 0. \qquad (7.2.6)$$

Filially, (7.2.4), (7.2.5) and (7.2.6), imply that

$$\lim_{R \to \infty} \limsup_{n \to \infty} \left[\int_{\mathbb{R}^N \setminus B_R} \int_{\mathbb{R}^N} \frac{|u_n(x) - u_n(y)|^2}{|x - y|^{N+2s}} \, dx dy + \left(1 - \frac{1}{K}\right) \int_{\mathbb{R}^N \setminus B_R} V(\varepsilon x) u_n^2 \, dx \right] = 0$$

which establishes (7.2.3).   □

Now, we are ready to provide an existence result for (7.2.2).

Theorem 7.2.5 *Assume that* (V_1)-(V_2) *and* (f_1)–(f_4) *hold. Then, for all* $\varepsilon > 0$, *problem* (7.2.2) *admits a positive mountain pass solution.*

Proof Taking into account Lemma 7.2.1, Lemma 7.2.4 and applying Theorem 2.2.9, we can see that there exists $u \in \mathcal{H}_\varepsilon$ such that $\mathcal{J}_\varepsilon(u) = c_\varepsilon$ and $\mathcal{J}'_\varepsilon(u) = 0$. Since $\langle \mathcal{J}'_\varepsilon(u), u^- \rangle = 0$, where $u^- = \min\{u, 0\}$, it is easy to check that $u \geq 0$ in $\mathbb{R}^N$. Indeed, since $g(x, t) = 0$ for $t \leq 0$ and $(x - y)(x^- - y^-) \geq |x^- - y^-|^2$, where $x^- = \min\{x, 0\}$, we get

$$\|u^-\|_\varepsilon^2 \leq \langle u, u^- \rangle_{\mathcal{D}^{s,2}(\mathbb{R}^N)} + \int_{\mathbb{R}^N} V(\varepsilon x)uu^-\, dx = \int_{\mathbb{R}^N} g(\varepsilon x, u)u^-\, dx = 0,$$

which implies that $u^- = 0$, that is, $u \geq 0$. Moreover, proceeding as in the proof of Lemma 7.2.9 below, we see that $u \in L^\infty(\mathbb{R}^N) \cap C^{0,\alpha}_{loc}(\mathbb{R}^N)$, and applying Theorem 1.3.5, we can conclude that $u > 0$ in $\mathbb{R}^N$.   □

Now, we deal with the following family of autonomous problems, with $\mu > 0$:

$$\begin{cases} (-\Delta)^s u + \mu u = f(u) & \text{in } \mathbb{R}^N, \\ u \in H^s(\mathbb{R}^N), & u > 0 \text{ in } \mathbb{R}^N. \end{cases} \tag{7.2.7}$$

It is clear that the Euler-Lagrange functional associated with (7.2.7) is given by

$$\mathcal{I}_\mu(u) = \frac{1}{2}\left([u]_s^2 + \mu \int_{\mathbb{R}^N} u^2\, dx\right) - \int_{\mathbb{R}^N} F(u)\, dx.$$

Let us denote by $\mathbb{X}_\mu$ the fractional Sobolev space $H^s(\mathbb{R}^N)$, endowed with the norm

$$\|u\|_\mu^2 = [u]_s^2 + \mu\|u\|_{L^2(\mathbb{R}^N)}^2.$$

The Nehari manifold associated with $\mathcal{I}_\mu$ is given by

$$\mathcal{M}_\mu = \{u \in \mathbb{X}_\mu \setminus \{0\} : \langle \mathcal{I}'_\mu(u), u \rangle = 0\}.$$

As in Chap. 6, it is easy to check that $\mathcal{I}_\mu$ possesses a mountain pass geometry, and denoting by m_μ the corresponding mountain pass level, we have that

$$d_\mu = \inf_{\mathcal{M}_\mu} \mathcal{I}_\mu = \inf_{u \in \mathbb{X}_\mu \setminus \{0\}} \max_{t \geq 0} \mathcal{I}_\mu(tu).$$

In view of Lemma 6.3.10, we can state

Theorem 7.2.6 *For all* $\mu > 0$, *problem* (7.2.7) *admits a positive ground state solution.*

In what follows, we establish a very useful relation between c_ε and d_{V_0}:

Lemma 7.2.7 *The following relation holds:*

$$\limsup_{\varepsilon \to 0} c_\varepsilon \leq d_{V_0}.$$

Proof Let $\omega_\varepsilon(x) = \psi_\varepsilon(x)\omega(x)$, where ω is the positive ground state of (7.2.7) whose existence is established by Theorem 7.2.6 with $\mu = V_0$, and $\psi_\varepsilon(x) = \psi(\varepsilon x)$, with $\psi \in C_c^\infty(\mathbb{R}^N)$ such that, $0 \leq \psi \leq 1$, $\psi(x) = 1$ if $|x| \leq \frac{1}{2}$ and $\psi(x) = 0$ if $|x| \geq 1$. For simplicity, we assume that $\mathrm{supp}(\psi) \subset B_1 \subset \Lambda$. By Lemma 1.4.8 and the dominated convergence theorem,

$$\omega_\varepsilon \to \omega \quad \text{in } H^s(\mathbb{R}^N) \quad \text{and} \quad \mathcal{I}_{V_0}(\omega_\varepsilon) \to \mathcal{I}_{V_0}(\omega) = d_{V_0} \tag{7.2.8}$$

as $\varepsilon \to 0$. Now, for each $\varepsilon > 0$ there exists $t_\varepsilon > 0$ such that

$$\mathcal{J}_\varepsilon(t_\varepsilon \omega_\varepsilon) = \max_{t \geq 0} \mathcal{J}_\varepsilon(t \omega_\varepsilon).$$

Then, $\langle \mathcal{J}'_\varepsilon(t_\varepsilon \omega_\varepsilon), \omega_\varepsilon \rangle = 0$ and this implies that

$$[\omega_\varepsilon]_s^2 + \int_{\mathbb{R}^N} V(\varepsilon x)\omega_\varepsilon^2 \, dx = \int_{\mathbb{R}^N} \frac{f(t_\varepsilon \omega_\varepsilon)}{t_\varepsilon \omega_\varepsilon} \omega_\varepsilon^2 \, dx. \tag{7.2.9}$$

From (7.2.8), (7.2.9) and the growth assumptions on f, we obtain that $t_\varepsilon \to t_0 \in (0, \infty)$. Taking the limit as $\varepsilon \to 0$ in (7.2.9) we get

$$[\omega]_s^2 + \int_{\mathbb{R}^N} V_0 \omega^2 \, dx = \int_{\mathbb{R}^N} \frac{f(t_0 \omega)}{t_0 \omega} \omega^2 \, dx \tag{7.2.10}$$

which together with (f_4), $\omega \in \mathcal{M}_{V_0}$ and (7.2.10) implies that $t_0 = 1$.
On the other hand,

$$c_\varepsilon \leq \max_{t \geq 0} \mathcal{J}_\varepsilon(t \omega_\varepsilon) = \mathcal{J}_\varepsilon(t_\varepsilon \omega_\varepsilon) = \mathcal{I}_{V_0}(t_\varepsilon \omega_\varepsilon) + \frac{t_\varepsilon^2}{2} \int_{\mathbb{R}^N} (V_\varepsilon(x) - V_0)\omega_\varepsilon^2 \, dx.$$

Since $V(\varepsilon \cdot)$ is bounded on the support of ω_ε, we use the dominated convergence theorem, (7.2.8) and the above inequality to obtain the desired result. $\qquad\square$

We conclude this section by proving the following compactness result which will be fundamental for showing that, for $\varepsilon > 0$ small enough, the solutions of the modified problem are also solutions of the original one.

Lemma 7.2.8 *Let $\varepsilon_n \to 0$ and $(u_n) = (u_{\varepsilon_n}) \subset \mathcal{H}_{\varepsilon_n}$ be such that $\mathcal{J}_{\varepsilon_n}(u_n) = c_{\varepsilon_n}$ and $\mathcal{J}'_{\varepsilon_n}(u_n) = 0$. Then there exists $(\tilde{y}_n) = (\tilde{y}_{\varepsilon_n}) \subset \mathbb{R}^N$ such that $\tilde{u}_n(x) = u_n(x + \tilde{y}_n)$ has a convergent subsequence in $H^s(\mathbb{R}^N)$. Moreover, up to a subsequence, $y_n = \varepsilon_n \tilde{y}_n \to y_0$ for some $y_0 \in \Lambda$ such that $V(y_0) = V_0$.*

Proof Using $\langle \mathcal{J}'_{\varepsilon_n}(u_n), u_n \rangle = 0$ and (g_1), (g_2), it is easy to see that there is $\kappa > 0$ such that

$$\|u_n\|_{\varepsilon_n} \geq \kappa > 0 \quad \text{for all } n \in \mathbb{N}.$$

Taking into account that $\mathcal{J}_{\varepsilon_n}(u_n) = c_{\varepsilon_n}$, $\langle \mathcal{J}'_{\varepsilon_n}(u_n), u_n \rangle = 0$ and Lemma 7.2.7, we can argue as in the proof of Lemma 7.2.2 to deduce that (u_n) is bounded in $\mathcal{H}_{\varepsilon_n}$. Moreover, we can find a sequence $(\tilde{y}_n) \subset \mathbb{R}^N$ and constants $R, \alpha > 0$ such that

$$\liminf_{n \to \infty} \int_{B_R(\tilde{y}_n)} u_n^2 \, dx \geq \alpha. \tag{7.2.11}$$

Indeed, if this limit inequality does not hold, we can use Lemma 1.4.4 to deduce that $u_n \to 0$ in $L^p(\mathbb{R}^N)$ for all $p \in (2, 2_s^*)$. In view of $\langle \mathcal{J}'_{\varepsilon_n}(u_n), u_n \rangle = 0$ and the growth assumptions on g, it follows that $\|u_n\|_{\varepsilon_n} \to 0$ as $n \to \infty$, which gives a contradiction.

Set $\tilde{u}_n(x) = u_n(x + \tilde{y}_n)$. Then, $(\tilde{u}_n)$ is bounded in $H^s(\mathbb{R}^N)$, and we may assume that

$$\tilde{u}_n \rightharpoonup \tilde{u} \quad \text{weakly in } H^s(\mathbb{R}^N). \tag{7.2.12}$$

Moreover, $\tilde{u} \neq 0$ because

$$\int_{B_R} \tilde{u}^2 \, dx \geq \alpha. \tag{7.2.13}$$

Now, we define $y_n = \varepsilon_n \tilde{y}_n$. Let us begin by proving that (y_n) is bounded in $\mathbb{R}^N$. To this end, it is enough to show the following claim:

Claim 1 $\lim_{n \to \infty} \text{dist}(y_n, \overline{\Lambda}) = 0$.

Indeed, assuming that this is not the case, there exist a $\delta > 0$ and a subsequence of (y_n), still denoted (y_n), such that

$$\text{dist}(y_n, \overline{\Lambda}) \geq \delta \quad \text{for all } n \in \mathbb{N}.$$

Then we can find $r > 0$ such that $B_r(y_n) \subset \Lambda^c$ for all $n \in \mathbb{N}$. Since $\tilde{u} \geq 0$ and $C_c^\infty(\mathbb{R}^N)$ is dense in $H^s(\mathbb{R}^N)$, we can find a sequence $(\psi_j) \subset C_c^\infty(\mathbb{R}^N)$ such that $\psi_j \geq 0$ and $\psi_j \to \tilde{u}$

in $H^s(\mathbb{R}^N)$. Fixing $j \in \mathbb{N}$ and using $\psi = \psi_j$ as test function in $\langle \mathcal{J}'_{\varepsilon_n}(u_n), \psi \rangle = 0$ we get

$$\langle \tilde{u}_n, \psi_j \rangle_{\mathcal{D}^{s,2}(\mathbb{R}^N)} + \int_{\mathbb{R}^N} V(\varepsilon_n x + \varepsilon_n \tilde{y}_n) \tilde{u}_n \psi_j \, dx = \int_{\mathbb{R}^N} g(\varepsilon_n x + \varepsilon_n \tilde{y}_n, \tilde{u}_n) \psi_j \, dx.$$

(7.2.14)

Recalling that $u_n, \psi_j \geq 0$ and the definition of g, we have

$$\int_{\mathbb{R}^N} g(\varepsilon_n x + \varepsilon_n \tilde{y}_n, \tilde{u}_n) \psi_j \, dx$$

$$= \int_{B_{r/\varepsilon_n}} g(\varepsilon_n x + \varepsilon_n \tilde{y}_n, \tilde{u}_n) \psi_j \, dx + \int_{\mathbb{R}^N \setminus B_{r/\varepsilon_n}} g(\varepsilon_n x + \varepsilon_n \tilde{y}_n, \tilde{u}_n) \psi_j \, dx$$

$$\leq \frac{V_1}{K} \int_{B_{r/\varepsilon_n}} \tilde{u}_n \psi_j \, dx + \int_{\mathbb{R}^N \setminus B_{r/\varepsilon_n}} f(\tilde{u}_n) \psi_j \, dx,$$

which together with (7.2.14) implies that

$$\langle \tilde{u}_n, \psi_j \rangle_{\mathcal{D}^{s,2}(\mathbb{R}^N)} + A \int_{\mathbb{R}^N} \tilde{u}_n \psi_j \, dx \leq \int_{\mathbb{R}^N \setminus B_{r/\varepsilon_n}} f(\tilde{u}_n) \psi_j \, dx,$$

(7.2.15)

where we set $A = V_1(1 - \frac{1}{K})$. By (7.2.12), the fact that ψ_j has compact support in $\mathbb{R}^N$ and since $\varepsilon_n \to 0$, we deduce that as $n \to \infty$

$$\langle \tilde{u}_n, \psi_j \rangle_{\mathcal{D}^{s,2}(\mathbb{R}^N)} \to \langle \tilde{u}, \psi_j \rangle_{\mathcal{D}^{s,2}(\mathbb{R}^N)}$$

and

$$\int_{\mathbb{R}^N \setminus B_{r/\varepsilon_n}} f(\tilde{u}_n) \psi_j \, dx \to 0.$$

The above limits and (7.2.15) show that

$$\langle \tilde{u}, \psi_j \rangle_{\mathcal{D}^{s,2}(\mathbb{R}^N)} + A \int_{\mathbb{R}^N} \tilde{u} \psi_j \, dx \leq 0,$$

and taking the limit as $j \to \infty$ we obtain

$$\|\tilde{u}\|_A^2 = [\tilde{u}]_s^2 + A \|\tilde{u}\|_{L^2(\mathbb{R}^N)}^2 \leq 0$$

which contradicts (7.2.13). Hence, there exists a subsequence of (y_n) such that $y_n \to y_0 \in \overline{\Lambda}$.

Claim 2 $y_0 \in \Lambda$.

From (g_2) and (7.2.14) we can see that

$$\langle \tilde{u}_n, \psi_j \rangle_{\mathcal{D}^{s,2}(\mathbb{R}^N)} + \int_{\mathbb{R}^N} V(\varepsilon_n x + \varepsilon_n \tilde{y}_n) \tilde{u}_n \psi_j \, dx \leq \int_{\mathbb{R}^N} f(\tilde{u}_n) \psi_j \, dx.$$

Letting $n \to \infty$ yields

$$\langle \tilde{u}, \psi_j \rangle_{\mathcal{D}^{s,2}(\mathbb{R}^N)} + \int_{\mathbb{R}^N} V(y_0) \tilde{u} \psi_j \, dx \leq \int_{\mathbb{R}^N} f(\tilde{u}) \psi_j \, dx,$$

and passing to the limit as $j \to \infty$ we have

$$[\tilde{u}]_s^2 + \int_{\mathbb{R}^N} V(y_0) \tilde{u}^2 \, dx \leq \int_{\mathbb{R}^N} f(\tilde{u}) \tilde{u} \, dx.$$

Then there exists $\tau \in (0, 1)$ such that $\tau \tilde{u} \in \mathcal{M}_{V(y_0)}$. Therefore, denoting by $d_{V(y_0)}$ the mountain pass level associated with $\mathcal{I}_{V(y_0)}$, we have

$$d_{V(y_0)} \leq \mathcal{I}_{V(y_0)}(\tau \tilde{u}) \leq \liminf_{n \to \infty} \mathcal{J}_{\varepsilon_n}(u_n) = \liminf_{n \to \infty} c_{\varepsilon_n} \leq d_{V_0},$$

from which we deduce that $V(y_0) \leq V(0) = V_0$. Since $V_0 = \inf_{\bar{\Lambda}} V$, we can infer that $V(y_0) = V_0$. Using (V_2), we obtain that $y_0 \notin \partial \Lambda$, that is $y_0 \in \Lambda$.

Claim 3 $\tilde{u}_n \to \tilde{u}$ in $H^s(\mathbb{R}^N)$ as $n \to \infty$.

Let us define

$$\tilde{\Lambda}_n = \frac{\Lambda - \varepsilon_n \tilde{y}_n}{\varepsilon_n}$$

and consider

$$\tilde{\chi}_n^1(x) = \begin{cases} 1, & \text{if } x \in \tilde{\Lambda}_n, \\ 0, & \text{if } x \in \mathbb{R}^N \setminus \tilde{\Lambda}_n, \end{cases}$$

$$\tilde{\chi}_n^2(x) = 1 - \tilde{\chi}_n^1(x).$$

Now, we introduce the following functions for all $x \in \mathbb{R}^N$

$$h_n^1(x) = \left(\frac{1}{2} - \frac{1}{\vartheta} \right) V(\varepsilon_n x + \varepsilon_n \tilde{y}_n) \tilde{u}_n^2(x) \tilde{\chi}_n^1(x),$$

$$h^1(x) = \left(\frac{1}{2} - \frac{1}{\vartheta} \right) V(y_0) \tilde{u}^2(x),$$

$$h_n^2(x) = \left[\left(\frac{1}{2} - \frac{1}{\vartheta} \right) V(\varepsilon_n x + \varepsilon_n \tilde{y}_n) \tilde{u}_n^2(x) + \frac{1}{\vartheta} g(\varepsilon_n x + \varepsilon_n \tilde{y}_n, \tilde{u}_n(x)) \tilde{u}_n(x) \right.$$

$$- G(\varepsilon_n\, x + \varepsilon_n\, \tilde{y}_n, \tilde{u}_n(x)) \Big] \tilde{\chi}_n^2(x)$$

$$\geq \left(\left(\frac{1}{2} - \frac{1}{\vartheta} \right) - \frac{1}{2K} \right) V(\varepsilon_n\, x + \varepsilon_n\, \tilde{y}_n) \tilde{u}_n^2(x) \tilde{\chi}_n^2(x),$$

$$h_n^3(x) = \left(\frac{1}{\vartheta} g(\varepsilon_n\, x + \varepsilon_n\, \tilde{y}_n, \tilde{u}_n(x)) \tilde{u}_n(x) - G(\varepsilon_n\, x + \varepsilon_n\, \tilde{y}_n, \tilde{u}_n(x)) \right) \tilde{\chi}_n^1(x)$$

$$= \left[\frac{1}{\vartheta} f(\tilde{u}_n(x)) \tilde{u}_n(x) - F(\tilde{u}_n(x)) \right] \tilde{\chi}_n^1(x),$$

$$h^3(x) = \frac{1}{\vartheta} f(\tilde{u}(x)) \tilde{u}(x) - F(\tilde{u}(x)).$$

In view of (f_3) and (g_3), the above functions are non-negative. Moreover, by (7.2.12) and Claim 2,

$$\tilde{u}_n(x) \to \tilde{u}(x) \quad \text{a.e. } x \in \mathbb{R}^N,$$

$$y_n = \varepsilon_n\, \tilde{y}_n \to y_0 \in \Lambda,$$

which implies that

$$\tilde{\chi}_n^1(x) \to 1, \ h_n^1(x) \to h^1(x), \ h_n^2(x) \to 0 \text{ and } h_n^3(x) \to h^3(x) \text{ a.e. } x \in \mathbb{R}^N.$$

Then, using Lemma 7.2.7, Fatou's lemma and a change of variable, we obtain that

$$d_{V_0} \geq \limsup_{n\to\infty} c_{\varepsilon_n} = \limsup_{n\to\infty} \left(\mathcal{J}_{\varepsilon_n}(u_n) - \frac{1}{\vartheta} \langle \mathcal{J}_{\varepsilon_n}'(u_n), u_n \rangle \right)$$

$$\geq \limsup_{n\to\infty} \left[\left(\frac{1}{2} - \frac{1}{\vartheta} \right) [\tilde{u}_n]_s^2 + \int_{\mathbb{R}^N} (h_n^1 + h_n^2 + h_n^3)\, dx \right]$$

$$\geq \liminf_{n\to\infty} \left[\left(\frac{1}{2} - \frac{1}{\vartheta} \right) [\tilde{u}_n]_s^2 + \int_{\mathbb{R}^N} (h_n^1 + h_n^2 + h_n^3)\, dx \right]$$

$$\geq \left(\frac{1}{2} - \frac{1}{\vartheta} \right) [\tilde{u}]_s^2 + \int_{\mathbb{R}^N} (h^1 + h^3)\, dx \geq d_{V_0}.$$

Accordingly,

$$\lim_{n\to\infty} [\tilde{u}_n]_s^2 = [\tilde{u}]_s^2 \tag{7.2.16}$$

and

$$h_n^1 \to h^1, \ h_n^2 \to 0 \quad \text{and} \quad h_n^3 \to h^3 \quad \text{in } L^1(\mathbb{R}^N).$$

Hence,

$$\lim_{n \to \infty} \int_{\mathbb{R}^N} V(\varepsilon_n x + \varepsilon_n \tilde{y}_n) \tilde{u}_n^2 \, dx = \int_{\mathbb{R}^N} V(y_0) \tilde{u}^2 \, dx,$$

from which we deduce that

$$\lim_{n \to \infty} \|\tilde{u}_n\|_{L^2(\mathbb{R}^N)}^2 = \|\tilde{u}\|_{L^2(\mathbb{R}^N)}^2. \tag{7.2.17}$$

Finally, (7.2.16) and (7.2.17) and the fact that $H^s(\mathbb{R}^N)$ is a Hilbert space imply that

$$\|\tilde{u}_n - \tilde{u}\|_{V_0}^2 = \|\tilde{u}_n\|_{V_0}^2 - \|\tilde{u}\|_{V_0}^2 + o_n(1) = o_n(1),$$

which ends the proof of lemma. □

7.2.2 Proof of Theorem 7.1.1

This last section is devoted to the proof of Theorem 7.1.1. First, we use a Moser iteration argument [278] to prove the following useful L^∞-estimate for the solutions of the modified problem (7.2.2).

Lemma 7.2.9 *Let $(\tilde{u}_n)$ be the sequence given in Lemma 7.2.8. Then, $\tilde{u}_n \in L^\infty(\mathbb{R}^N)$ and there exists $C > 0$ such that*

$$\|\tilde{u}_n\|_{L^\infty(\mathbb{R}^N)} \leq C \quad \text{for all } n \in \mathbb{N}.$$

Moreover,

$$|\tilde{u}_n(x)| \to 0 \text{ as } |x| \to \infty, \text{ uniformly in } n \in \mathbb{N}.$$

Proof Arguing as in the first part of the proof of Lemma 6.3.23 we see that

$$\left(\frac{1}{\beta}\right)^2 S_* \|\tilde{u}_n \tilde{u}_{L,n}^{\beta-1}\|_{L^{2_s^*}(\mathbb{R}^N)}^2 + \int_{\mathbb{R}^N} V_n(x) \tilde{u}_n^2 \tilde{u}_{L,n}^{2(\beta-1)} \, dx$$

$$\leq \iint_{\mathbb{R}^{2N}} \frac{(\tilde{u}_n(x) - \tilde{u}_n(y))}{|x-y|^{N+2s}} ((\tilde{u}_n \tilde{u}_{L,n}^{2(\beta-1)})(x) - (\tilde{u}_n \tilde{u}_{L,n}^{2(\beta-1)})(y)) \, dx dy$$

$$+ \int_{\mathbb{R}^N} V_n(x) \tilde{u}_n^2 \tilde{u}_{L,n}^{2(\beta-1)} \, dx$$

$$= \int_{\mathbb{R}^N} g_n(x, \tilde{u}_n) \tilde{u}_n \tilde{u}_{L,n}^{2(\beta-1)} \, dx, \tag{7.2.18}$$

where we used the notations $V_n(x) = V(\varepsilon_n x + \varepsilon_n \tilde{y}_n)$ and $g_n(x, \tilde{u}_n) = g(\varepsilon_n x + \varepsilon_n \tilde{y}_n, \tilde{u}_n)$. On the other hand, from (g_1) and (g_2), we know that for any $\xi > 0$ there exists $C_\xi > 0$ such that

$$|g_n(x, \tilde{u}_n)| \leq \xi |\tilde{u}_n| + C_\xi |\tilde{u}_n|^{2_s^* - 1}. \tag{7.2.19}$$

Taking $\xi \in (0, V_1)$, and using (V_1) and (7.2.19), we see that (7.2.18) implies that

$$\|\tilde{u}_n \tilde{u}_{L,n}^{\beta-1}\|_{L^{2_s^*}(\mathbb{R}^N)}^2 \leq C\beta^2 \int_{\mathbb{R}^N} |\tilde{u}_n|^{2_s^*} \tilde{u}_{L,n}^{2(\beta-1)} \, dx.$$

Now we can exploit the arguments in Lemma 6.3.23 to finish the proof. □

Remark 7.2.10 Next we prove in a different way that $\lim_{|x| \to \infty} \tilde{u}_n(x) = 0$ uniformly in $n \in \mathbb{N}$. From (V_1), we know that $\tilde{u}_n$ is a subsolution to the equation

$$(-\Delta)^s \tilde{u}_n + V_1 \tilde{u}_n = h_n \text{ in } \mathbb{R}^N,$$

where $h_n(x) = g_n(x, \tilde{u}_n)$. Since $\tilde{u}_n \to \tilde{u} \neq 0$ in $H^s(\mathbb{R}^N)$ (by Lemma 7.2.8) and $\|\tilde{u}_n\|_{L^\infty(\mathbb{R}^N)} \leq C$ for all $n \in \mathbb{N}$ (by Lemma 7.2.9), interpolation on the L^p spaces shows that $h_n \to h = f(\tilde{u})$ in $L^q(\mathbb{R}^N)$ for any $q \in [2, \infty)$, and there exists a constant $c_1 > 0$ such that $\|h_n\|_{L^\infty(\mathbb{R}^N)} \leq c_1$ for all $n \in \mathbb{N}$. Now, let z_n be the unique solution of

$$(-\Delta)^s z_n + V_1 z_n = h_n \text{ in } \mathbb{R}^N.$$

Then we can write

$$z_n(x) = (\mathcal{K} * h_n)(x) = \int_{\mathbb{R}^N} \mathcal{K}(x - y) h_n(y) \, dy, \tag{7.2.20}$$

where the kernel $\mathcal{K}(x) = \mathcal{F}^{-1}((|\xi|^{2s} + V_1)^{-1})$ satisfies the following properties (see *[183, 199]* and Remark 3.2.18):

(b_1) $\mathcal{K}$ is positive, radially symmetric and smooth in $\mathbb{R}^N \setminus \{0\}$;
(b_2) there exists a positive constant $C > 0$ such that $\mathcal{K}(x) \leq \frac{K_1}{|x|^{N+2s}}$ for all $x \in \mathbb{R}^N \setminus \{0\}$;
(b_3) $\mathcal{K} \in L^q(\mathbb{R}^N)$ for any $q \in [1, \frac{N}{N-2s})$.

At this point we borrow some arguments used in *[19]* to show that $z_n(x) \to 0$ as $|x| \to \infty$ uniformly in $n \in \mathbb{N}$. Note that, for any $\delta > 0$, it holds

$$0 \le z_n(x) = (\mathcal{K} * h_n)(x) = \int_{\mathcal{A}_\delta} \mathcal{K}(x-y)h_n(y)\,dy + \int_{\mathcal{B}_\delta} \mathcal{K}(x-y)h_n(y)\,dy \tag{7.2.21}$$

where

$$\mathcal{A}_\delta = \left\{ y \in \mathbb{R}^N : |y-x| \ge \frac{1}{\delta} \right\} \quad \text{and} \quad \mathcal{B}_\delta = \left\{ y \in \mathbb{R}^N : |y-x| < \frac{1}{\delta} \right\}.$$

From (b_2) we deduce that

$$\int_{\mathcal{A}_\delta} \mathcal{K}(x-y)h_n(y)\,dy \le K_1 \|h_n\|_{L^\infty(\mathbb{R}^N)} \int_{\mathcal{A}_\delta} \frac{dy}{|x-y|^{N+2s}}$$

$$\le c_1 \delta^{2s} K_1 \int_{|\xi| \ge 1} \frac{d\xi}{|\xi|^{N+2s}} = C_1 \delta^{2s}. \tag{7.2.22}$$

On the other hand,

$$\int_{\mathcal{B}_\delta} \mathcal{K}(x-y)|h_n(y)|\,dy \le \int_{\mathcal{B}_\delta} \mathcal{K}(x-y)|h_n(y) - h(y)|\,dy + \int_{\mathcal{B}_\delta} \mathcal{K}(x-y)|h(y)|\,dy.$$

Fix $q > 1$ with $q \approx 1$ and $q' > 2$ such that $\frac{1}{q} + \frac{1}{q'} = 1$. From (b_3) and Hölder's inequality we have that

$$\int_{\mathcal{B}_\delta} \mathcal{K}(x-y)|h_n(y)|\,dy \le \|\mathcal{K}\|_{L^q(\mathbb{R}^N)} \|h_n - h\|_{L^{q'}(\mathbb{R}^N)} + \|\mathcal{K}\|_{L^q(\mathbb{R}^N)} \|h\|_{L^{q'}(\mathcal{B}_\delta)}.$$

Since $\|h_n - h\|_{L^{q'}(\mathbb{R}^N)} \to 0$ as $n \to \infty$ and $\|h\|_{L^{q'}(\mathcal{B}_\delta)} \to 0$ as $|x| \to \infty$, we deduce that there exist $R > 0$ and $n_0 \in \mathbb{N}$ such that

$$\int_{\mathcal{B}_\delta} \mathcal{K}(x-y)|h_n(y)|\,dy \le \delta \tag{7.2.23}$$

for all $n \ge n_0$ and $|x| \ge R$. Putting together *(7.2.22)* and *(7.2.23)* we obtain that

$$\int_{\mathbb{R}^N} \mathcal{K}(x-y)|h_n(y)|\,dy \le C_1 \delta^{2s} + \delta. \tag{7.2.24}$$

for all $n \ge n_0$ and $|x| \ge R$.

The same approach can be used to prove that for each $n \in \{1, \ldots, n_0 - 1\}$, there is $R_n > 0$ such that

$$\int_{\mathbb{R}^N} \mathcal{K}(x - y)|h_n(y)|\, dy \leq C_1 \delta^{2s} + \delta$$

for all $|x| \geq R_n$. Hence, increasing R if necessary, we must have

$$\int_{\mathbb{R}^N} \mathcal{K}(x - y)|h_n(y)|\, dy \leq C_1 \delta^{2s} + \delta$$

for $|x| \geq R$, uniformly in $n \in \mathbb{N}$. Letting $\delta \to 0$ we get the desired result for z_n. Since by comparison we see that $0 \leq \tilde{u}_n \leq z_n$ in $\mathbb{R}^N$, we obtain the assertion. $\blacksquare$

Now we are ready to give the proof of the main result of this section.

Proof of Theorem 7.1.1 We begin by showing that there exists $\varepsilon_0 > 0$ such that for any $\varepsilon \in (0, \varepsilon_0)$ and any mountain pass solution $u_\varepsilon \in \mathcal{H}_\varepsilon$ of (7.2.2), it holds

$$\|u_\varepsilon\|_{L^\infty(\mathbb{R}^N \setminus \Lambda_\varepsilon)} < a. \tag{7.2.25}$$

Assume, by contradiction, that for some subsequence (ε_n) such that $\varepsilon_n \to 0$, we can find $u_n = u_{\varepsilon_n} \in \mathcal{H}_{\varepsilon_n}$ such that $\mathcal{J}_{\varepsilon_n}(u_n) = c_{\varepsilon_n}$, $\mathcal{J}'_{\varepsilon_n}(u_n) = 0$ and

$$\|u_n\|_{L^\infty(\mathbb{R}^N \setminus \Lambda_{\varepsilon_n})} \geq a. \tag{7.2.26}$$

In view of Lemma 7.2.8, we can find $(\tilde{y}_n) \subset \mathbb{R}^N$ such that $\tilde{u}_n = u_n(\cdot + \tilde{y}_n) \to \tilde{u}$ in $H^s(\mathbb{R}^N)$ and $\varepsilon_n \tilde{y}_n \to y_0$ for some $y_0 \in \Lambda$ such that $V(y_0) = V_0$.

Now, if we choose $r > 0$ so that $B_r(y_0) \subset B_{2r}(y_0) \subset \Lambda$, we can see that $B_{\frac{r}{\varepsilon_n}}(\frac{y_0}{\varepsilon_n}) \subset \Lambda_{\varepsilon_n}$. Then, for any $y \in B_{\frac{r}{\varepsilon_n}}(\tilde{y}_n)$,

$$\left| y - \frac{y_0}{\varepsilon_n} \right| \leq |y - \tilde{y}_n| + \left| \tilde{y}_n - \frac{y_0}{\varepsilon_n} \right| < \frac{1}{\varepsilon_n}(r + o_n(1)) < \frac{2r}{\varepsilon_n} \quad \text{for } n \text{ sufficiently large.}$$

Therefore,

$$\mathbb{R}^N \setminus \Lambda_{\varepsilon_n} \subset \mathbb{R}^N \setminus B_{\frac{r}{\varepsilon_n}}(\tilde{y}_n) \tag{7.2.27}$$

for any n big enough. Using Lemma 7.2.9 we see that

$$\tilde{u}_n(x) \to 0 \quad \text{as } |x| \to \infty \tag{7.2.28}$$

uniformly in $n \in \mathbb{N}$. Therefore, there exists $R > 0$ such that

$$\tilde{u}_n(x) < a \quad \text{for any } |x| \geq R, n \in \mathbb{N}.$$

Hence, $u_n(x) < a$ for any $x \in \mathbb{R}^N \setminus B_R(\tilde{y}_n)$ and $n \in \mathbb{N}$. On the other hand, by (7.2.27), there exists $\nu \in \mathbb{N}$ such that for any $n \geq \nu$ and $r/\varepsilon_n > R$,

$$\mathbb{R}^N \setminus \Lambda_{\varepsilon_n} \subset \mathbb{R}^N \setminus B_{\frac{r}{\varepsilon_n}}(\tilde{y}_n) \subset \mathbb{R}^N \setminus B_R(\tilde{y}_n),$$

which implies that $u_n(x) < a$ for any $x \in \mathbb{R}^N \setminus \Lambda_{\varepsilon_n}$ and $n \geq \nu$. This is impossible in view of (7.2.26). Since $u_\varepsilon \in \mathcal{H}_\varepsilon$ satisfies (7.2.25), by the definition of g it follows that u_ε is a solution of (7.2.1) for all $\varepsilon \in (0, \varepsilon_0)$. Consequently, $\hat{u}_\varepsilon(x) = u_\varepsilon(x/\varepsilon)$ is a solution to (7.1.1) for $\varepsilon \in (0, \varepsilon_0)$. We also notice that $\hat{u}_\varepsilon \in L^\infty(\mathbb{R}^N) \cap C_{\text{loc}}^{0,\alpha}(\mathbb{R}^N)$.

In what follows, we study the behavior of the maximum points of solutions to problem (7.1.1). Take $\varepsilon_n \to 0$ and consider a sequence $(u_n) \subset \mathcal{H}_{\varepsilon_n}$ of solutions to (7.2.1) as above. Let us observe that (g_1) implies that there exists $\omega \in (0, a)$ such that

$$g(\varepsilon x, t)t = f(t)t \leq \frac{V_1}{K} t^2 \quad \text{for any } x \in \mathbb{R}^N, 0 \leq t \leq \omega. \tag{7.2.29}$$

Arguing as before, we can find $R > 0$ such that

$$\|u_n\|_{L^\infty(\mathbb{R}^N \setminus B_R(\tilde{y}_n))} < \omega. \tag{7.2.30}$$

Moreover, up to a subsequence, we may assume that

$$\|u_n\|_{L^\infty(B_R(\tilde{y}_n))} \geq \omega. \tag{7.2.31}$$

Indeed, if (7.2.31) does not hold, then in view of (7.2.30), $\|u_n\|_{L^\infty(\mathbb{R}^N)} < \omega$. Then, using that $\langle \mathcal{J}'_{\varepsilon_n}(u_n), u_n \rangle = 0$ and (7.2.29) we infer

$$\|u_n\|_{\varepsilon_n}^2 = \int_{\mathbb{R}^N} g(\varepsilon_n x, u_n)u_n \, dx \leq \frac{V_1}{K} \int_{\mathbb{R}^N} u_n^2 \, dx$$

which yields $\|u_n\|_{\varepsilon_n} = 0$, which is impossible. Hence, (7.2.31) holds true.

Let $p_n \in \mathbb{R}^N$ be a global maximum point of u_n. Taking into account (7.2.30) and (7.2.31), we deduce that $p_n \in B_R(\tilde{y}_n)$. Therefore, $p_n = \tilde{y}_n + q_n$ for some $q_n \in B_R$. Consequently, $\eta_n = \varepsilon_n \tilde{y}_n + \varepsilon_n q_n$ is a global maximum point of $\hat{u}_n(x) = u_n(x/\varepsilon_n)$. Since $|q_n| < R$ for all $n \in \mathbb{N}$ and $\varepsilon_n \tilde{y}_n \to y_0$, the continuity of V implies that

$$\lim_{n \to \infty} V(\eta_n) = V(y_0) = V_0.$$

Finally, we prove a decay estimate for $\hat{u}_n$. By Lemma 3.2.17 and scaling, there exists a positive continuous function w such that

$$0 < w(x) \le \frac{C}{1 + |x|^{N+2s}} \quad \text{for all } x \in \mathbb{R}^N, \tag{7.2.32}$$

and w satisfies in the classical sense

$$(-\Delta)^s w + \frac{V_1}{2} w = 0 \quad \text{in } \mathbb{R}^N \setminus \overline{B}_{R_1}, \tag{7.2.33}$$

for a suitable $R_1 > 0$. Using (g_1) and (7.2.28), we can find $R_2 > 0$ sufficiently large such that

$$(-\Delta)^s \tilde{u}_n + \frac{V_1}{2} \tilde{u}_n = g_n(x, \tilde{u}_n) - \left(V_n - \frac{V_1}{2} \right) \tilde{u}_n$$

$$\le g_n(x, \tilde{u}_n) - \frac{V_1}{2} \tilde{u}_n \le 0 \text{ in } \mathbb{R}^N \setminus \overline{B}_{R_2}. \tag{7.2.34}$$

Let $R_3 = \max\{R_1, R_2\} > 0$ and set

$$c = \min_{\overline{B}_{R_3}} w > 0 \text{ and } \tilde{w}_n = (d+1)w - c\tilde{u}_n, \tag{7.2.35}$$

where $d = \sup_{n \in \mathbb{N}} \|\tilde{u}_n\|_{L^\infty(\mathbb{R}^N)} < \infty$. Let us show that

$$\tilde{w}_n \ge 0 \text{ in } \mathbb{R}^N. \tag{7.2.36}$$

First, we observe that (7.2.33), (7.2.34) and (7.2.35) yield

$$\tilde{w}_n \ge cd + w - cd > 0 \text{ in } \overline{B}_{R_3}, \tag{7.2.37}$$

$$(-\Delta)^s \tilde{w}_n + \frac{V_1}{2} \tilde{w}_n \ge 0 \text{ in } \mathbb{R}^N \setminus \overline{B}_{R_3}. \tag{7.2.38}$$

Then we can use Lemma 1.3.8 with $\Omega = \mathbb{R}^N \setminus \overline{B}_{R_3}$ to verify that (7.2.36) holds. In view of (7.2.32) and (7.2.36),

$$0 < \tilde{u}_n(x) \le \left(\frac{d+1}{c} \right) w(x) \le \frac{\tilde{C}}{1 + |x|^{N+2s}} \quad \text{for all } x \in \mathbb{R}^N, n \in \mathbb{N}, \tag{7.2.39}$$

for some constant $\tilde{C} > 0$. Since $\hat{u}_n(x) = u_n(\frac{x}{\varepsilon_n}) = \tilde{u}_n(\frac{x}{\varepsilon_n} - \tilde{y}_n)$ and $\eta_n = \varepsilon_n \tilde{y}_n + \varepsilon_n q_n$, we can use (7.2.39) to deduce that

$$
\begin{aligned}
0 < \hat{u}_n(x) = u_n \left(\frac{x}{\varepsilon_n} \right) &= \tilde{u}_n \left(\frac{x}{\varepsilon_n} - \tilde{y}_n \right) \\
&\leq \frac{\tilde{C}}{1 + |\frac{x}{\varepsilon_n} - \tilde{y}_n|^{N+2s}} \\
&= \frac{\tilde{C}\,\varepsilon_n^{N+2s}}{\varepsilon_n^{N+2s} + |x - \varepsilon_n \tilde{y}_n|^{N+2s}} \\
&\leq \frac{\tilde{C}\,\varepsilon_n^{N+2s}}{\varepsilon_n^{N+2s} + |x - \eta_n|^{N+2s}} \qquad \text{for all } x \in \mathbb{R}^N.
\end{aligned}
$$

This ends the proof of Theorem 7.1.2.   □

7.3 The Critical Case

7.3.1 The Modified Critical Problem and the Local (PS) Condition

To establish the existence of a nontrivial solution to (7.1.1) in the case $\gamma = 1$, we first modify the nonlinearity in a suitable way. Using the change of variable $x \mapsto \varepsilon x$, we consider the following problem:

$$
\begin{cases}
(-\Delta)^s u + V(\varepsilon x)u = f(u) + |u|^{2_s^* - 2}u & \text{in } \mathbb{R}^N, \\
u \in H^s(\mathbb{R}^N), \quad u > 0 \text{ in } \mathbb{R}^N.
\end{cases}
$$

Without loss of generality we may assume that

$$
0 \in \Lambda \quad \text{and} \quad V(0) = V_0 = \inf_\Lambda V.
$$

Take $K > \frac{\vartheta}{\vartheta - 2}$ and $a > 0$ such that $f(a) + a^{2_s^* - 1} = \frac{V_1}{K}a$, and define

$$
\tilde{f}(t) =
\begin{cases}
f(t) + (t^+)^{2_s^* - 1}, & \text{if } t \leq a, \\
\frac{V_1}{K}t, & \text{if } t > a,
\end{cases}
$$

and

$$
g(x, t) =
\begin{cases}
\chi_\Lambda(x)(f(t) + t^{2_s^* - 1}) + (1 - \chi_\Lambda(x))\tilde{f}(t), & \text{if } t \geq 0, \\
0, & \text{if } t < 0.
\end{cases}
$$

It is easy to check that g satisfies the following properties:

(k_1) $\lim\limits_{t \to 0} \dfrac{g(x,t)}{t} = 0$ uniformly with respect to $x \in \mathbb{R}^N$;

(k_2) $g(x,t) \le f(t) + t^{2^*_s - 1}$ for all $x \in \mathbb{R}^N$, $t > 0$;

(k_3) (i) $0 < \vartheta G(x,t) = \vartheta \displaystyle\int_0^t g(x,\tau)\, d\tau \le g(x,t)t$ for all $x \in \Lambda$ and $t > 0$,

(ii) $0 \le 2G(x,t) \le g(x,t)t \le \dfrac{V_1}{K} t^2$ for all $x \in \mathbb{R}^N \setminus \Lambda$ and $t > 0$;

(k_4) for each $x \in \Lambda$ the function $t \mapsto \dfrac{g(x,t)}{t}$ is increasing in $(0, \infty)$, and for each $x \in$ $\mathbb{R}^N \setminus \Lambda$ the function $t \mapsto \dfrac{g(x,t)}{t}$ is increasing in $(0, a)$.

Then, we consider the following modified problem:

$$\begin{cases} (-\Delta)^s u + V(\varepsilon x)u = g(\varepsilon x, u) & \text{in } \mathbb{R}^N, \\ u \in H^s(\mathbb{R}^N), \quad u > 0 \text{ in } \mathbb{R}^N. \end{cases} \tag{7.3.1}$$

Next we look for weak solutions of (7.3.1) having the property

$$u(x) \le a \quad \text{for any } x \in \mathbb{R}^N \setminus \Lambda_\varepsilon.$$

Let us introduce the functional $\mathcal{J}_\varepsilon : \mathcal{H}_\varepsilon \to \mathbb{R}$ defined as

$$\mathcal{J}_\varepsilon(u) = \frac{1}{2}\|u\|_\varepsilon^2 - \int_{\mathbb{R}^N} G(\varepsilon x, u)\, dx,$$

As in the proof of Lemma 7.2.1, we can see that $\mathcal{J}_\varepsilon$ has a mountain pass geometry; we denote by c_ε the corresponding mountain pass level. In order to study the compactness properties of $\mathcal{J}_\varepsilon$, we need to prove some technical results.

Lemma 7.3.1 *It holds,* $0 < c_\varepsilon < \frac{s}{N} S_*^{\frac{N}{2s}}.$

Proof It is enough to argue as in the proof of Lemma 6.4.8 by assuming that the support of the cut-off function is contained in Λ_ε. $\qquad\square$

Lemma 7.3.2 *Let* $c \in \mathbb{R}$ *be such that* $0 < c < \frac{s}{N} S_*^{\frac{N}{2s}}$. *Then* $\mathcal{J}_\varepsilon$ *satisfies the Palais-Smale condition at the level* c.

Proof Let $(u_n) \subset \mathcal{H}_\varepsilon$ be a Palais-Smale sequence for $\mathcal{J}_\varepsilon$ at the level c. We note that (u_n) is bounded in $\mathcal{H}_\varepsilon$ because thanks to (k_3) we have

$$
C(1 + \|u_n\|_\varepsilon) \geq \mathcal{J}_\varepsilon(u_n) - \frac{1}{\vartheta} \langle \mathcal{J}_\varepsilon'(u_n), u_n \rangle
$$

$$
\geq \left(\frac{1}{2} - \frac{1}{\vartheta} \right) \|u_n\|_\varepsilon^2 + \frac{1}{\vartheta} \int_{\mathbb{R}^N \setminus \Lambda_\varepsilon} [g(\varepsilon x, u_n)u_n - \vartheta G(\varepsilon x, u_n)] \, dx
$$

$$
\geq \left(\frac{\vartheta - 2}{2\vartheta} \right) \left(1 - \frac{1}{K} \right) \|u_n\|_\varepsilon^2,
$$

and recalling that $K > 1$ and $\vartheta > 2$ we get the boundedness. Then we may assume that $u_n \rightharpoonup u$ in $\mathcal{H}_\varepsilon$. Since $\langle \mathcal{J}_\varepsilon'(u_n), u_n \rangle = o_n(1)$, we can see that

$$
\|u_n\|_\varepsilon^2 = \int_{\mathbb{R}^N} g(\varepsilon x, u_n)u_n \, dx + o_n(1). \tag{7.3.2}
$$

On the other hand, standard calculations show that u is a critical point of $\mathcal{J}_\varepsilon$ and thus

$$
\|u\|_\varepsilon^2 + \int_{\mathbb{R}^N} V(\varepsilon x)u^2 \, dx = \int_{\mathbb{R}^N} g(\varepsilon x, u)u \, dx. \tag{7.3.3}
$$

Let us show that (u_n) strongly converges to u in $\mathcal{H}_\varepsilon$. To this end, it is enough to show that $\|u_n\|_\varepsilon \to \|u\|_\varepsilon$, which in view of (7.3.2) and (7.3.3) means to verify that

$$
\lim_{n \to \infty} \int_{\mathbb{R}^N} g(\varepsilon x, u_n)u_n \, dx = \int_{\mathbb{R}^N} g(\varepsilon x, u)u \, dx. \tag{7.3.4}
$$

We begin by showing that for each $\xi > 0$ there exists $R = R_\xi > 0$ such that

$$
\limsup_{n \to \infty} \int_{\mathbb{R}^N \setminus B_R} \int_{\mathbb{R}^N} \frac{|u_n(x) - u_n(y)|^2}{|x - y|^{N+2s}} \, dx dy + \int_{\mathbb{R}^N \setminus B_R} V(\varepsilon x)u_n^2 \, dx < \xi. \tag{7.3.5}
$$

We may assume that R is chosen so that $\Lambda_\varepsilon \subset B_{\frac{R}{2}}$. Let η_R be a cut-off function such that $\eta_R = 0$ on $B_{\frac{R}{2}}$, $\eta_R = 1$ on $\mathbb{R}^N \setminus B_R$, $0 \leq \eta \leq 1$ and $\|\nabla \eta_R\|_{L^\infty(\mathbb{R}^N)} \leq \frac{C}{R}$. Since (u_n) is a bounded Palais-Smale sequence, we have

$$
\langle \mathcal{J}_\varepsilon'(u_n), \eta_R u_n \rangle = o_n(1).
$$

From (k_3)-(ii), we get

$$\iint_{\mathbb{R}^{2N}} \eta_R(x) \frac{|u_n(x) - u_n(y)|^2}{|x-y|^{N+2s}}\, dxdy + \left(\iint_{\mathbb{R}^{2N}} \frac{(\eta_R(x) - \eta_R(y))(u_n(x) - u_n(y))}{|x-y|^{N+2s}} u_n(y)\, dxdy \right)$$

$$+ \int_{\mathbb{R}^N} V(\varepsilon x) u_n^2 \eta_R\, dx = \int_{\mathbb{R}^N} g(\varepsilon x, u_n) \eta_R u_n\, dx + o_n(1) \leq \frac{1}{K} \int_{\mathbb{R}^N} V(\varepsilon x) u_n^2 \eta_R\, dx + o_n(1).$$

Then, using the definition of η_R, the fact that $K > 1$, the Hölder inequality, the boundedness of (u_n) in $\mathcal{H}_\varepsilon$, and Remark 1.4.6, we have that

$$\int_{\mathbb{R}^N \setminus B_R} \int_{\mathbb{R}^N} \frac{|u_n(x) - u_n(y)|^2}{|x-y|^{N+2s}}\, dxdy + \left(1 - \frac{1}{K}\right) \int_{\mathbb{R}^N \setminus B_R} V(\varepsilon x) u_n^2\, dx$$

$$\leq \iint_{\mathbb{R}^{2N}} \frac{|u_n(x) - u_n(y)|^2}{|x-y|^{N+2s}} \eta_R(x)\, dxdy + \left(1 - \frac{1}{K}\right) \int_{\mathbb{R}^N} V(\varepsilon x) u_n^2 \eta_R\, dx$$

$$\leq - \left(\iint_{\mathbb{R}^{2N}} \frac{(\eta_R(x) - \eta_R(y))(u_n(x) - u_n(y))}{|x-y|^{N+2s}} u_n(y)\, dxdy \right) + o_n(1) \qquad (7.3.6)$$

$$\leq C \left(\iint_{\mathbb{R}^{2N}} \frac{|\eta_R(x) - \eta_R(y)|^2}{|x-y|^{N+2s}} |u_n(y)|^2\, dxdy \right)^{\frac{1}{2}} + o_n(1)$$

$$\leq \frac{C}{R^s} + o_n(1),$$

so we deduce that

$$\lim_{R\to\infty} \limsup_{n\to\infty} \iint_{\mathbb{R}^{2N}} \frac{|\eta_R(x) - \eta_R(y)|^2}{|x-y|^{N+2s}} |u_n(y)|^2\, dxdy = 0. \qquad (7.3.7)$$

Putting together (7.3.6) and (7.3.7), we infer that (7.3.5) holds. Now, note that the fractional Sobolev inequality (1.1.1) in Theorem 1.1.8, shows that

$$\left(\int_{\mathbb{R}^N \setminus B_R} |u_n|^{2_s^*}\, dx \right)^{\frac{2}{2_s^*}} \leq \left(\int_{\mathbb{R}^N} |u_n \eta_R|^{2_s^*}\, dx \right)^{\frac{2}{2_s^*}} \leq C[u_n \eta_R]_s^2.$$

From $0 \leq \eta_R \leq 1$, (7.3.6) and Remark 1.4.6, we see that

$$[u_n \eta_R]_s^2 = \iint_{\mathbb{R}^{2N}} \frac{|(u_n(x) - u_n(y))\eta_R(x) + (\eta_R(x) - \eta_R(y))u_n(y)|^2}{|x-y|^{N+2s}}\, dxdy$$

$$\leq C \left[\iint_{\mathbb{R}^{2N}} \frac{|u_n(x) - u_n(y)|^2}{|x-y|^{N+2s}} \eta_R^2(x)\, dxdy + \iint_{\mathbb{R}^{2N}} \frac{|\eta_R(x) - \eta_R(y)|^2}{|x-y|^{N+2s}} |u_n(y)|^2\, dxdy \right]$$

$$\leq C \left[\iint_{\mathbb{R}^{2N}} \frac{|u_n(x) - u_n(y)|^2}{|x - y|^{N+2s}} \eta_R(x) \, dx \, dy + \frac{C}{R^{2s}} \right]$$

$$\leq \frac{C}{R^s} + o_n(1) + \frac{C}{R^{2s}},$$

and thus

$$\lim_{R \to \infty} \limsup_{n \to \infty} \int_{\mathbb{R}^N \setminus B_R} |u_n|^{2_s^*} \, dx = 0. \tag{7.3.8}$$

Now, we note that (7.3.5) and (V_1) yield

$$\lim_{R \to \infty} \limsup_{n \to \infty} \int_{\mathbb{R}^N \setminus B_R} |u_n|^2 \, dx = 0, \tag{7.3.9}$$

and using interpolation on L^p-spaces and the boundedness of (u_n) in $L^{2_s^*}(\mathbb{R}^N)$, we also deduce that for all $p \in (2, 2_s^*)$

$$\lim_{R \to \infty} \limsup_{n \to \infty} \int_{\mathbb{R}^N \setminus B_R} |u_n|^p \, dx = 0. \tag{7.3.10}$$

Consequently, using (k_2), (f_1), (f_2'), (7.3.8), (7.3.9) and (7.3.10), we see that for every $\xi > 0$ there exists $R = R_\xi > 0$ such that

$$\limsup_{n \to \infty} \int_{\mathbb{R}^N \setminus B_R} g(\varepsilon x, u_n) u_n \, dx \leq C \limsup_{n \to \infty} \int_{\mathbb{R}^N \setminus B_R} (|u_n|^2 + |u_n|^\sigma + |u_n|^{2_s^*}) \, dx$$

$$\leq C\xi. \tag{7.3.11}$$

On the other hand, choosing $R > 0$ large enough, we may assume that

$$\int_{\mathbb{R}^N \setminus B_R} g(\varepsilon x, u) u \, dx < \xi. \tag{7.3.12}$$

Then, (7.3.11) and (7.3.12) yield

$$\limsup_{n \to \infty} \left| \int_{\mathbb{R}^N \setminus B_R} g(\varepsilon x, u_n) u_n \, dx - \int_{\mathbb{R}^N \setminus B_R} g(\varepsilon x, u) u \, dx \right| < C\xi \quad \text{for all } \xi > 0,$$

which implies that

$$\lim_{n \to \infty} \int_{\mathbb{R}^N \setminus B_R} g(\varepsilon x, u_n) u_n \, dx = \int_{\mathbb{R}^N \setminus B_R} g(\varepsilon x, u) u \, dx. \tag{7.3.13}$$

Using the definition of g, it follows that

$$g(\varepsilon x, u_n)u_n \leq f(u_n)u_n + a^{2_s^*} + \frac{V_1}{K}u_n^2 \text{ for any } x \in \mathbb{R}^N \setminus \Lambda_\varepsilon.$$

Since $B_R \cap (\mathbb{R}^N \setminus \Lambda_\varepsilon)$ is bounded, we can use the above estimate, (f_1), (f_2'), Theorem 1.1.8, and the dominated convergence theorem to infer that, as $n \to \infty$,

$$\lim_{n \to \infty} \int_{B_R \cap (\mathbb{R}^N \setminus \Lambda_\varepsilon)} g(\varepsilon x, u_n)u_n \, dx = \int_{B_R \cap (\mathbb{R}^N \setminus \Lambda_\varepsilon)} g(\varepsilon x, u)u \, dx. \tag{7.3.14}$$

At this point, we aim to show that

$$\lim_{n \to \infty} \int_{\Lambda_\varepsilon} (u_n^+)^{2_s^*} \, dx = \int_{\Lambda_\varepsilon} (u^+)^{2_s^*} \, dx. \tag{7.3.15}$$

Indeed, if we assume that (7.3.15) is true, then from (k_2), (f_1), (f_2'), Theorem 1.1.8, and the dominated convergence theorem we deduce that

$$\lim_{n \to \infty} \int_{\Lambda_\varepsilon \cap B_R} g(\varepsilon x, u_n)u_n \, dx = \int_{\Lambda_\varepsilon \cap B_R} g(\varepsilon x, u)u \, dx. \tag{7.3.16}$$

Now combining (7.3.13), (7.3.14) and (7.3.16), we conclude that (7.3.4) holds.

So let us prove that (7.3.15) is satisfied. Since (u_n^+) is bounded in $\mathcal{H}_\varepsilon$, we may assume that $|(-\Delta)^{\frac{s}{2}}u_n^+|^2 \rightharpoonup \mu$ and $(u_n^+)^{2_s^*} \rightharpoonup \nu$, where μ and ν are two bounded non-negative measures on $\mathbb{R}^N$. Applying Lemma 1.5.1 we can find an at most countable index set I and sequences $(x_i)_{i \in I} \subset \mathbb{R}^N$, $(\mu_i)_{i \in I}$, $(\nu_i)_{i \in I} \subset (0, \infty)$ such that

$$\mu \geq |(-\Delta)^{\frac{s}{2}}u^+|^2 + \sum_{i \in I} \mu_i \delta_{x_i},$$

$$\nu = |u^+|^{2_s^*} + \sum_{i \in I} \nu_i \delta_{x_i} \quad \text{and} \quad S_* \nu_i^{\frac{2}{2_s^*}} \leq \mu_i \quad \forall i \in I, \tag{7.3.17}$$

where δ_{x_i} is the Dirac mass at the point x_i. Let us show that $(x_i)_{i \in I} \cap \Lambda_\varepsilon = \emptyset$. Assume, by contradiction, that $x_i \in \Lambda_\varepsilon$ for some $i \in I$. For any $\rho > 0$, we define $\psi_\rho(x) = \psi(\frac{x - x_i}{\rho})$ where $\psi \in C_c^\infty(\mathbb{R}^N)$ is such that $\psi = 1$ in B_1, $\psi = 0$ in $\mathbb{R}^N \setminus B_2$, $0 \leq \psi \leq 1$ and $\|\nabla \psi\|_{L^\infty(\mathbb{R}^N)} \leq 2$. We suppose that $\rho > 0$ is such that $\text{supp}(\psi_\rho) \subset \Lambda_\varepsilon$. Since the sequence $(\psi_\rho u_n^+)$ is bounded in $\mathcal{H}_\varepsilon$, we see that $\langle \mathcal{J}_\varepsilon'(u_n), \psi_\rho u_n^+ \rangle = o_n(1)$, and so

$$
\iint_{\mathbb{R}^{2N}} \psi_\rho(y) \frac{|u_n^+(x) - u_n^+(y)|^2}{|x - y|^{N+2s}} \, dx \, dy
$$

$$
\leq \iint_{\mathbb{R}^{2N}} \psi_\rho(y) \frac{(u_n(x) - u_n(y))(u_n^+(x) - u_n^+(y))}{|x - y|^{N+2s}} \, dx \, dy
$$

$$
\leq - \left(\iint_{\mathbb{R}^{2N}} \frac{(\psi_\rho(x) - \psi_\rho(y))(u_n(x) - u_n(y))}{|x - y|^{N+2s}} u_n^+(x) \, dx \, dy \right)
$$

$$
+ \int_{\mathbb{R}^N} \psi_\rho f(u_n) u_n^+ \, dx + \int_{\mathbb{R}^N} \psi_\rho (u_n^+)^{2_s^*} \, dx + o_n(1), \tag{7.3.18}
$$

where we used that $(x - y)(x^+ - y^+) \geq |x^+ - y^+|^2$ for all $x, y \in \mathbb{R}$, and (V_1). Since f has subcritical growth and ψ_ρ has compact support, we have

$$
\lim_{\rho \to 0} \lim_{n \to \infty} \int_{\mathbb{R}^N} \psi_\rho f(u_n) u_n^+ \, dx = \lim_{\rho \to 0} \int_{\mathbb{R}^N} \psi_\rho f(u) u^+ \, dx = 0. \tag{7.3.19}
$$

Now we show that

$$
\lim_{\rho \to 0} \lim_{n \to \infty} \sup \left(\iint_{\mathbb{R}^{2N}} \frac{(\psi_\rho(x) - \psi_\rho(y))(u_n(x) - u_n(y))}{|x - y|^{N+2s}} u_n^+(x) \, dx \, dy \right) = 0. \tag{7.3.20}
$$

Using the Hölder inequality and the fact that (u_n) is bounded in $\mathcal{H}_\varepsilon$, we obtain that

$$
\left| \iint_{\mathbb{R}^{2N}} \frac{(\psi_\rho(x) - \psi_\rho(y))(u_n(x) - u_n(y))}{|x - y|^{N+2s}} u_n^+(x) \, dx \, dy \right|
$$

$$
\leq C \left(\iint_{\mathbb{R}^{2N}} |u_n^+(x)|^2 \frac{|\psi_\rho(x) - \psi_\rho(y)|^2}{|x - y|^{N+2s}} \, dx \, dy \right)^{\frac{1}{2}}.
$$

By Lemma 1.4.7, we deduce that

$$
\lim_{\rho \to 0} \lim_{n \to \infty} \sup \iint_{\mathbb{R}^{2N}} |u_n^+(x)|^2 \frac{|\psi_\rho(x) - \psi_\rho(y)|^2}{|x - y|^{N+2s}} \, dx \, dy = 0 \tag{7.3.21}
$$

and this implies that (7.3.20) holds.

Therefore, using (7.3.17) and taking the limit as $n \to \infty$ and $\rho \to 0$ in (7.3.18), we deduce that (7.3.19) and (7.3.20) yield $\nu_i \geq \mu_i$. From the last statement in (7.3.17) it follows that $v_i \geq S_*^{\frac{2}{2_s^*}}$, and using (f_4) and (k_3) we get

$$c = \mathcal{J}_\varepsilon(u_n) - \frac{1}{2}\langle \mathcal{J}_\varepsilon'(u_n), u_n \rangle + o_n(1)$$

$$= \int_{\mathbb{R}^N \setminus \Lambda_\varepsilon} \left[\frac{1}{2} g(\varepsilon x, u_n)u_n - G(\varepsilon x, u_n) \right] dx + \int_{\Lambda_\varepsilon} \left[\frac{1}{2} f(u_n)u_n - F(u_n) \right] dx$$

$$+ \frac{s}{N} \int_{\Lambda_\varepsilon} (u_n^+)^{2_s^*} dx + o_n(1)$$

$$\geq \frac{s}{N} \int_{\Lambda_\varepsilon} (u_n^+)^{2_s^*} dx + o_n(1)$$

$$\geq \frac{s}{N} \int_{\Lambda_\varepsilon} \psi_\rho (u_n^+)^{2_s^*} dx + o_n(1).$$

Then, by (7.3.17) and letting $n \to \infty$, we see that

$$c \geq \frac{s}{N} \sum_{\{i \in I : x_i \in \Lambda_\varepsilon\}} \psi_\rho(x_i)v_i = \frac{s}{N} \sum_{\{i \in I : x_i \in \Lambda_\varepsilon\}} v_i \geq \frac{s}{N} S_*^{\frac{N}{2s}},$$

which gives a contradiction. This ends the proof of (7.3.15). $\qquad\qquad \square$

Remark 7.3.3 Arguing as in Remark 5.2.8, we may always suppose that the Palais-Smale sequence (u_n) is non-negative in $\mathbb{R}^N$.

Now Lemma 7.3.1, Lemma 7.3.2 and Theorem 2.2.9, yield the following result:

Theorem 7.3.4 *Assume that (V_1)-(V_2) and (f_1), (f_2'), (f_3), (f_4) hold. Then, for all $\varepsilon > 0$, problem (7.3.1) admits a positive mountain pass solution.*

7.3.2   Proof of Theorem 7.1.2

Consider the following family of autonomous problems, with $\mu > 0$:

$$\begin{cases} (-\Delta)^s u + \mu u = f(u) + |u|^{2_s^* - 2}u & \text{in } \mathbb{R}^N, \\ u \in H^s(\mathbb{R}^N), \quad u > 0 \text{ in } \mathbb{R}^N. \end{cases} \tag{7.3.22}$$

By Lemma 6.4.10 we can state the following result.

Theorem 7.3.5 *Problem (7.3.22) admits a positive ground state solution.*

Arguing as in the proof of Lemma 7.2.7 (and recalling Lemma 6.4.8), we have the following relation between c_ε and d_{V_0}:

Lemma 7.3.6 *The following relation holds:*

$$\limsup_{\varepsilon \to 0} c_\varepsilon \leq d_{V_0} < \frac{s}{N} S_*^{\frac{N}{2s}}.$$

As in the previous section, we need to prove the following compactness result which will be fundamental for showing that, for $\varepsilon > 0$ small enough, the solutions of the modified problem are also solutions of the original problem.

Lemma 7.3.7 *Let $\varepsilon_n \to 0$ and $(u_n) = (u_{\varepsilon_n}) \subset \mathcal{H}_{\varepsilon_n}$ be such that $\mathcal{J}_{\varepsilon_n}(u_n) = c_{\varepsilon_n}$ and $\mathcal{J}'_{\varepsilon_n}(u_n) = 0$. Then there exists a sequence $(\tilde{y}_n) = (\tilde{y}_{\varepsilon_n}) \subset \mathbb{R}^N$ such that $\tilde{u}_n(x) = u_n(x + \tilde{y}_n)$ has a convergent subsequence in $H^s(\mathbb{R}^N)$. Moreover, up to a subsequence, $y_n = \varepsilon_n \tilde{y}_n \to y_0$ for some $y_0 \in \Lambda$ such that $V(y_0) = V_0$.*

Proof The proof of this result can be done essentially as in Lemma 7.2.8. However, due to the presence of the fractional critical exponent, some modifications are needed. More precisely, to find a sequence $(\tilde{y}_n) \subset \mathbb{R}^N$ and constants $R, \alpha > 0$ such that

$$\liminf_{n \to \infty} \int_{B_R(\tilde{y}_n)} u_n^2 \, dx \geq \alpha,$$

we proceed as follows. Assume, by contradiction, that the above limit inequality does not hold. By Lemma 1.4.4, we see that $u_n \to 0$ in $L^r(\mathbb{R}^N)$ for all $r \in (2, 2_s^*)$. By (f_1) and (f_2'), it follows that

$$\int_{\mathbb{R}^N} F(u_n) \, dx = \int_{\mathbb{R}^N} f(u_n) u_n \, dx = o_n(1).$$

This implies that

$$\int_{\mathbb{R}^N} G(\varepsilon_n x, u_n) \, dx \leq \frac{1}{2_s^*} \int_{\Lambda_{\varepsilon_n} \cup \{u_n \leq a\}} (u_n^+)^{2_s^*} \, dx + \frac{V_1}{2K} \int_{\Lambda_{\varepsilon_n}^c \cap \{u_n > a\}} u_n^2 \, dx + o_n(1)$$

$$\tag{7.3.23}$$

and

$$\int_{\mathbb{R}^N} g(\varepsilon_n x, u_n) u_n \, dx = \int_{\Lambda_{\varepsilon_n} \cup \{u_n \leq a\}} (u_n^+)^{2_s^*} \, dx + \frac{V_1}{K} \int_{\Lambda_{\varepsilon_n}^c \cap \{u_n > a\}} u_n^2 \, dx + o_n(1),$$

$$\tag{7.3.24}$$

where $\Lambda_{\varepsilon_n}^c = \mathbb{R}^N \setminus \Lambda_{\varepsilon_n}$. Taking into account that $\langle \mathcal{J}'_{\varepsilon_n}(u_n), u_n \rangle = 0$ and (7.3.24), we deduce that

$$\|u_n\|_{\varepsilon_n}^2 - \frac{V_1}{K} \int_{\Lambda_{\varepsilon_n}^c \cap \{u_n > a\}} u_n^2 \, dx = \int_{\Lambda_{\varepsilon_n} \cup \{u_n \le a\}} (u_n^+)^{2_s^*} \, dx + o_n(1). \tag{7.3.25}$$

Let $\ell \ge 0$ be such that

$$\|u_n\|_{\varepsilon_n}^2 - \frac{V_1}{K} \int_{\Lambda_{\varepsilon_n}^c \cap \{u_n > a\}} u_n^2 \, dx \to \ell.$$

It is easy to see that $\ell > 0$: otherwise, $\|u_n\|_{\varepsilon_n} \to 0$ and this is impossible because $\|u_n\|_{\varepsilon_n} \ge \kappa > 0$. It follows from (7.3.25) that

$$\int_{\Lambda_{\varepsilon_n} \cup \{u_n \le a\}} (u_n^+)^{2_s^*} dx \to \ell.$$

Using $\mathcal{J}_{\varepsilon_n}(u_n) - \frac{1}{2_s^*} \langle \mathcal{J}'_{\varepsilon}(u_n), u_n \rangle = c_{\varepsilon_n}$, (7.3.23) and (7.3.24) we see that

$$\frac{s}{N} \ell \le \liminf_{n \to \infty} c_{\varepsilon_n}. \tag{7.3.26}$$

On the other hand, by the definition of S_*, we have that

$$\|u_n\|_{\varepsilon_n}^2 - \frac{V_1}{K} \int_{\Lambda_{\varepsilon_n}^c \cap \{u_n > a\}} u_n^2 \, dx \ge S_* \left(\int_{\Lambda_{\varepsilon_n} \cup \{u_n \le a\}} (u_n^+)^{2_s^*} dx \right)^{\frac{2}{2_s^*}},$$

and taking the limit as $n \to \infty$ we infer that

$$\ell \ge S_* \ell^{\frac{2}{2_s^*}}. \tag{7.3.27}$$

Then, by $\ell > 0$, (7.3.26) and (7.3.27), we deduce that

$$\liminf_{n \to \infty} c_{\varepsilon_n} \ge \frac{s}{N} S_*^{\frac{N}{2s}},$$

which contradicts Lemma 7.3.6.

Finally, instead of the functions h_n^3 and h^3 introduced in Lemma 7.2.8, we need to consider the following functions:

$$\tilde{h}_n^3(x) = \left(\frac{1}{\vartheta} g(\varepsilon_n x + \varepsilon_n \tilde{y}_n, \tilde{u}_n(x))\tilde{u}_n(x) - G(\varepsilon_n x + \varepsilon_n \tilde{y}_n, \tilde{u}_n(x))\right) \tilde{\chi}_n^1(x)$$

$$= \left[\frac{1}{\vartheta} \left(f(\tilde{u}_n(x))\tilde{u}_n(x) + (\tilde{u}_n(x))^{2_s^*}\right) - \left(F(\tilde{u}_n(x)) + \frac{1}{2_s^*}(\tilde{u}_n(x))^{2_s^*}\right)\right] \tilde{\chi}_n^1(x)$$

$$\tilde{h}^3(x) = \frac{1}{\vartheta} \left(f(\tilde{u}(x))\tilde{u}(x) + (\tilde{u}(x))^{2_s^*}\right) - \left(F(\tilde{u}(x)) + \frac{1}{2_s^*}(\tilde{u}(x))^{2_s^*}\right).$$

$\square$

Proof of Theorem 7.1.2 In the light of Lemma 7.3.7 and Lemma 7.2.9, we can argue as in the proof of Theorem 7.1.1 to obtain the desired result. $\square$

7.4   A Supercritical Fractional Schrödinger Equation

7.4.1   The Truncated Problem

In this last section we study the existence of positive solutions for the following supercritical fractional problem:

$$(-\Delta)^s u + V(\varepsilon x)u = |u|^{q-2}u + \lambda|u|^{r-2}u \quad \text{in } \mathbb{R}^N. \tag{7.4.1}$$

Motivated by Chabrowski and Yang [136] and Rabinowitz [297], we truncate the nonlinearity $f(u) = |u|^{q-2}u + \lambda|u|^{r-2}u$ as follows. Let $K > 0$ be a real number, whose value will be fixed later, and set

$$f_\lambda(t) = \begin{cases} 0, & \text{if } t \leq 0, \\ t^{q-1} + \lambda t^{r-1}, & \text{if } 0 < t < K, \\ (1 + \lambda K^{r-q})t^{q-1}, & \text{if } t \geq K. \end{cases}$$

Then it is readily verified that f_λ satisfies the assumptions (f_1)–(f_4) $((f_3)$ holds with $\vartheta = q > 2)$. In particular,

$$f_\lambda(t) \leq (1 + \lambda K^{r-q})t^{q-1} \text{ for all } t \geq 0. \tag{7.4.2}$$

Now consider the following truncated problem:

$$(-\Delta)^s u + V(\varepsilon x)u = f_\lambda(u) \quad \text{in } \mathbb{R}^N, \tag{7.4.3}$$

and the corresponding functional $\mathcal{J}_{\varepsilon,\lambda} : \mathcal{H}_\varepsilon \to \mathbb{R}$, defined as

$$\mathcal{J}_{\varepsilon,\lambda}(u) = \frac{1}{2}\|u\|_\varepsilon^2 - \int_{\mathbb{R}^N} F_\lambda(u)\, dx.$$

We also introduce the autonomous functional $\mathcal{I}_{V_0,\lambda} : H^s(\mathbb{R}^N) \to \mathbb{R}$, given by

$$\mathcal{I}_{V_0,\lambda}(u) = \frac{1}{2}\|u\|_{V_0}^2 - \int_{\mathbb{R}^N} F_\lambda(u)\, dx.$$

Using Theorem 7.1.1, we know that for every $\lambda \geq 0$ there exists $\bar{\varepsilon}(\lambda) > 0$ such that, for every $\varepsilon \in (0, \bar{\varepsilon}(\lambda))$, problem (7.4.3) admits a positive solution $u_{\varepsilon,\lambda}$.

Next we prove an auxiliary result which shows that the $\mathcal{H}_\varepsilon$-norm of $u_{\varepsilon,\lambda}$ can be estimated from above by a constant independent of λ.

Lemma 7.4.1 *There exists $\bar{C} > 0$ such that, for any $\varepsilon > 0$ sufficiently small, $\|u_{\varepsilon,\lambda}\|_\varepsilon \leq \bar{C}$.*

Proof From the proof of Theorem 7.1.1, we know that $u_{\varepsilon,\lambda}$ satisfies the inequality

$$\mathcal{J}_{\varepsilon,\lambda}(u_{\varepsilon,\lambda}) \leq d_{V_0,\lambda} + h_\lambda(\varepsilon),$$

where $d_{V_0,\lambda}$ is the mountain pass level related to the functional $\mathcal{I}_{V_0,\lambda}$ and $h_\lambda(\varepsilon) \to 0$ as $\varepsilon \to 0$. Then, decreasing $\bar{\varepsilon}(\lambda)$ if necessary, we may assume that

$$\mathcal{J}_{\varepsilon,\lambda}(u_{\varepsilon,\lambda}) \leq d_{V_0,\lambda} + 1 \tag{7.4.4}$$

for any $\varepsilon \in (0, \bar{\varepsilon}(\lambda))$. Since $d_{V_0,\lambda} \leq d_{V_0,0}$ for any $\lambda \geq 0$, we infer that

$$\mathcal{J}_{\varepsilon,\lambda}(u_{\varepsilon,\lambda}) \leq d_{V_0,0} + 1 \tag{7.4.5}$$

for any $\varepsilon \in (0, \bar{\varepsilon}(\lambda))$. Moreover, using (f_3), we see that

$$\mathcal{J}_{\varepsilon,\lambda}(u_{\varepsilon,\lambda}) = \mathcal{J}_{\varepsilon,\lambda}(u_{\varepsilon,\lambda}) - \frac{1}{q}\langle \mathcal{J}'_{\varepsilon,\lambda}(u_{\varepsilon,\lambda}), u_{\varepsilon,\lambda}\rangle$$

$$= \left(\frac{1}{2} - \frac{1}{q}\right)\|u_{\varepsilon,\lambda}\|_\varepsilon^2 + \frac{1}{q}\int_{\mathbb{R}^N} f_\lambda(u_{\varepsilon,\lambda})u_{\varepsilon,\lambda} - q\,F_\lambda(u_{\varepsilon,\lambda})\, dx$$

$$\geq \left(\frac{1}{2} - \frac{1}{q}\right)\|u_{\varepsilon,\lambda}\|_\varepsilon^2. \tag{7.4.6}$$

Finally, (7.4.5) and (7.4.6) show that

$$\|u_{\varepsilon,\lambda}\|_\varepsilon \le \left[\left(\frac{2q}{q-2}\right)(d_{V_0,0}+1)\right]^{\frac{1}{2}} = \bar{C} \quad \forall \varepsilon \in (0, \bar{\varepsilon}(\lambda)).$$

□

7.4.2　A Moser Type Iteration Argument

Let $\varepsilon \in (0, \bar{\varepsilon}(\lambda))$ and assume that $\bar{\varepsilon}(\lambda)$ is small in such a way that the estimate in Lemma 7.4.1 holds true. We aim to prove that $u_{\varepsilon,\lambda}$ is a positive solution of the original problem (7.1.2) for small λ. To this end, we will show that we can find $K_0 > 0$ such that for any $K \ge K_0$, there exists $\lambda_0 = \lambda_0(K) > 0$ such that

$$\|u_{\varepsilon,\lambda}\|_{L^\infty(\mathbb{R}^N)} \le K \text{ for all } \lambda \in [0, \lambda_0].$$

In what follows we use a Moser iteration argument [278] (see also [136, 188, 297]). For simplicity we will write u instead of $u_{\varepsilon,\lambda}$.

Proof of Theorem 7.1.3　For any $L > 0$, let $u_L = \min\{u, L\} \ge 0$ and $w_L = uu_L^{\beta-1}$, where $\beta > 1$ will be chosen later. Taking $u_L^{2(\beta-1)}u$ in (7.4.3) we see that

$$\left(\iint_{\mathbb{R}^{2N}} \frac{(u(x)-u(y))}{|x-y|^{N+2s}}(u(x)u_L^{2(\beta-1)}(x) - u(y)u_L^{2(\beta-1)}(y))\,dxdy\right)$$
$$= \int_{\mathbb{R}^N} f_\lambda(u)uu_L^{2(\beta-1)}\,dx - \int_{\mathbb{R}^N} V(\varepsilon x)u^2 u_L^{2(\beta-1)}\,dx. \tag{7.4.7}$$

Putting together (7.4.7), (7.4.2) and (V_1) we get

$$\left(\iint_{\mathbb{R}^{2N}} \frac{(u(x)-u(y))}{|x-y|^{N+2s}}(u(x)u_L^{2(\beta-1)}(x) - u(y)u_L^{2(\beta-1)}(y))\,dxdy\right)$$
$$\le C_{\lambda,K}\int_{\mathbb{R}^N} u^q u_L^{2(\beta-1)}\,dx, \tag{7.4.8}$$

where $C_{\lambda,K} = 1 + \lambda K^{r-q}$. Arguing as in the first part of Lemma 3.2.14, we can see that

$$\|w_L\|_{L^{2^*_s}(\mathbb{R}^N)}^2 \le C_0\beta^2 \iint_{\mathbb{R}^{2N}} \frac{(u(x)-u(y))}{|x-y|^{N+2s}}(u(x)u_L^{2(\beta-1)}(x) - u(y)u_L^{2(\beta-1)}(y))\,dxdy.$$

$$\tag{7.4.9}$$

Since $u^q u_L^{2(\beta-1)} = u^{q-2} w_L^2$, we can use (7.4.8), (7.4.9), and the Hölder inequality to deduce that

$$\|w_L\|^2_{L^{2^*_s}(\mathbb{R}^N)} \leq C_1 \beta^2 C_{\lambda,K} \left(\int_{\mathbb{R}^N} u^{2^*_s} dx \right)^{\frac{q-2}{2^*_s}} \left(\int_{\mathbb{R}^N} w_L^{\alpha^*_s} dx \right)^{\frac{2}{\alpha^*_s}}, \qquad (7.4.10)$$

where

$$\alpha^*_s = \frac{2 \cdot 2^*_s}{2^*_s - (q-2)} \in (2, 2^*_s)$$

and $C_1 > 0$ is independent of ε and λ. Then, by $\mathcal{D}^{s,2}(\mathbb{R}^N) \subset L^{2^*_s}(\mathbb{R}^N)$ and Lemma 7.4.1, we obtain

$$\|w_L\|^2_{L^{2^*_s}(\mathbb{R}^N)} \leq C_2 \beta^2 C_{\lambda,K} \bar{C}^{q-2} \|w_L\|^2_{L^{\alpha^*_s}(\mathbb{R}^N)} \qquad (7.4.11)$$

for some $C_2 > 0$ that is independent of ε and λ. Note that, if $|u|^\beta \in L^{\alpha^*_s}(\mathbb{R}^N)$, the definition of w_L, the fact that $u_L \leq u$ and (7.4.11) imply that

$$\|w_L\|^2_{L^{2^*_s}(\mathbb{R}^N)} \leq C_2 \beta^2 C_{\lambda,K} \bar{C}^{q-2} \left(\int_{\mathbb{R}^N} u^{\beta \alpha^*_s} dx \right)^{\frac{2}{\alpha^*_s}} < \infty. \qquad (7.4.12)$$

Taking the limit as $L \to \infty$ in (7.4.12) and using Fatou's lemma, we have

$$\|u\|_{L^{\beta 2^*_s}(\mathbb{R}^N)} \leq (C_3 C_{\lambda,K})^{\frac{1}{2\beta}} \beta^{\frac{1}{\beta}} \|u\|_{L^{\beta \alpha^*_s}(\mathbb{R}^N)} \qquad (7.4.13)$$

whenever $|u|^{\beta \alpha^*_s} \in L^1(\mathbb{R}^N)$, where $C_3 = C_2 \bar{C}^{q-2}$.

Set $\beta = \frac{2^*_s}{\alpha^*_s} > 1$ and we note that, since $u \in L^{2^*_s}(\mathbb{R}^N)$, the above inequality holds for this choice of β. Then, observing that $\beta^2 \alpha^*_s = \beta 2^*_s$, it follows that (7.4.13) holds with β replaced by β^2, so we have

$$\|u\|_{L^{\beta^2 2^*_s}(\mathbb{R}^N)} \leq (C_3 C_{\lambda,K})^{\frac{1}{2\beta^2}} \beta^{\frac{2}{\beta^2}} \|u\|_{L^{\beta^2 \alpha^*_s}(\mathbb{R}^N)} \leq (C_3 C_{\lambda,K})^{\frac{1}{2}\left(\frac{1}{\beta}+\frac{1}{\beta^2}\right)} \beta^{\frac{1}{\beta}+\frac{2}{\beta^2}} \|u\|_{L^{\beta \alpha^*_s}(\mathbb{R}^N)}.$$

Iterating this process and using the fact that $\beta \alpha^*_s = 2^*_s$ we deduce that for every $m \in \mathbb{N}$

$$\|u\|_{L^{\beta^m 2^*_s}(\mathbb{R}^N)} \leq (C_3 C_{\lambda,K})^{\sum_{j=1}^m \frac{1}{2\beta^j}} \beta^{\sum_{j=1}^m j \beta^{-j}} \|u\|_{L^{2^*_s}(\mathbb{R}^N)}. \qquad (7.4.14)$$

Letting $m \to \infty$ in (7.4.14) and using the embedding $\mathcal{D}^{s,2}(\mathbb{R}^N) \subset L^{2^*_s}(\mathbb{R}^N)$ and

Lemma 7.4.1, we obtain

$$\|u\|_{L^\infty(\mathbb{R}^N)} \le (C_3 C_{\lambda,K})^{\gamma_1} \beta^{\gamma_2} C_4, \tag{7.4.15}$$

where $C_4 = S_*^{-\frac{1}{2}} \bar{C}$ and

$$\gamma_1 = \frac{1}{2} \sum_{j=1}^\infty \frac{1}{\beta^j} < \infty \quad \text{and} \quad \gamma_2 = \sum_{j=1}^\infty \frac{j}{\beta^j} < \infty.$$

Next, we will find suitable values of K and λ such that the following inequality holds:

$$(C_3 C_{\lambda,K})^{\gamma_1} \beta^{\gamma_2} C_4 \le K,$$

or, equivalently,

$$1 + \lambda K^{r-q} \le C_3^{-1} \beta^{-\frac{\gamma_2}{\gamma_1}} (K C_4^{-1})^{\frac{1}{\gamma_1}}.$$

Take $K > 0$ such that

$$\frac{(K C_4^{-1})^{\frac{1}{\gamma_1}}}{C_3 \beta^{\frac{\gamma_2}{\gamma_1}}} - 1 > 0$$

and fix $\lambda_0 > 0$ such that

$$\lambda \le \lambda_0 \le \left[\frac{(K C_4^{-1})^{\frac{1}{\gamma_1}}}{C_3 \beta^{\frac{\gamma_2}{\gamma_1}}} - 1 \right] \frac{1}{K^{r-q}}.$$

Then, using (7.4.15), we conclude that

$$\|u\|_{L^\infty(\mathbb{R}^N)} \le K \text{ for all } \lambda \in [0, \lambda_0].$$

$\square$

7.5   Some Extensions to Fractional Schrödinger Systems

It is also possible consider the nonlocal counterpart of the following elliptic system of Schrödinger equations (see for instance [4, 16, 17, 20, 80, 118, 189]):

$$\begin{cases} -\varepsilon^2 \Delta u + V(x)u = G_u(u,v) & \text{in } \mathbb{R}^N, \\ -\varepsilon^2 \Delta v + W(x)v = G_v(u,v) & \text{in } \mathbb{R}^N, \\ u, v > 0 & \text{in } \mathbb{R}^N. \end{cases}$$

More precisely, we can study the existence, multiplicity and concentration phenomenon of positive solutions for the following nonlinear fractional Schrödinger system

$$\begin{cases} \varepsilon^{2s}(-\Delta)^s u + V(x)u = Q_u(u, v) + \frac{\gamma}{2_s^*}K_u(u, v) & \text{in } \mathbb{R}^N, \\ \varepsilon^{2s}(-\Delta)^s v + W(x)v = Q_v(u, v) + \frac{\gamma}{2_s^*}K_v(u, v) & \text{in } \mathbb{R}^N, \\ u, v > 0 & \text{in } \mathbb{R}^N, \end{cases} \tag{7.5.1}$$

where $\varepsilon > 0$ is a small parameter, $s \in (0, 1)$, $N > 2s$, $\gamma \in \{0, 1\}$, $V : \mathbb{R}^N \to \mathbb{R}$ and $W : \mathbb{R}^N \to \mathbb{R}$ are Hölder continuous potentials, Q and K are homogeneous functions, with K having critical growth. We assume that there exist a bounded open set $\Lambda \subset \mathbb{R}^N$, $x_0 \in \mathbb{R}^N$ and $\rho_0 > 0$ such that:

($H1$) $V(x), W(x) \geq \rho_0$ for any $x \in \partial\Lambda$;
($H2$) $V(x_0), W(x_0) < \rho_0$;
($H3$) $V(x) \geq V(x_0) > 0, W(x) \geq W(x_0) > 0$ for any $x \in \mathbb{R}^N$.

Concerning the function $Q \in C^2(\mathbb{R}_+^2, \mathbb{R})$, where $\mathbb{R}_+^2 = [0, \infty) \times [0, \infty)$, we suppose that it satisfies the following conditions:

($Q1$) there exists $p \in (2, 2_s^*)$ such that $Q(tu, tv) = t^p Q(u, v)$ for any $t > 0$, $(u, v) \in \mathbb{R}_+^2$;
($Q2$) there exists $C > 0$ such that $|Q_u(u, v)| + |Q_v(u, v)| \leq C(u^{p-1} + v^{p-1})$ for any $(u, v) \in \mathbb{R}_+^2$;
($Q3$) $Q_u(0, 1) = 0 = Q_v(1, 0)$;
($Q4$) $Q_u(1, 0) = 0 = Q_v(0, 1)$;
($Q5$) $Q(u, v) > 0$ for any $u, v > 0$;
($Q6$) $Q_u(u, v), Q_v(u, v) \geq 0$ for any $(u, v) \in \mathbb{R}_+^2$.

Further, we assume that $K \in C^2(\mathbb{R}_+^2, \mathbb{R})$ fulfills the following hypotheses:

($K1$) $K(tu, tv) = t^{2_s^*}K(u, v)$ for all $t > 0$, $(u, v) \in \mathbb{R}_+^2$;
($K2$) the 1-homogeneous function $G : \mathbb{R}_+^2 \to \mathbb{R}$ given by $G(u^{2_s^*}, v^{2_s^*}) = K(u, v)$ is concave;
($K3$) there exists $c > 0$ such that $|K_u(u, v)| + |K_v(u, v)| \leq c(u^{2_s^*-1} + v^{2_s^*-1})$ for all $(u, v) \in \mathbb{R}_+^2$;
($K4$) $K_u(0, 1) = 0 = K_v(1, 0)$;
($K5$) $K_u(1, 0) = 0 = K_v(0, 1)$;
($K6$) $K(u, v) > 0$ for any $u, v > 0$;
($K7$) $K_u(u, v), K_v(u, v) \geq 0$ for all $(u, v) \in \mathbb{R}_+^2$.

Since we are interested in positive solutions of (7.5.1), we extend the functions Q and K to the whole of $\mathbb{R}^2$ by setting $Q(u, v) = K(u, v) = 0$ if $u \le 0$ or $v \le 0$. We note that the p-homogeneity of Q implies that the following identity holds:

$$pQ(u, v) = uQ_u(u, v) + vQ_v(u, v) \quad \text{for any } (u, v) \in \mathbb{R}^2,$$

and

$$p(p - 1)Q(u, v) = u^2 Q_{uu}(u, v) + 2uv Q_{uv}(u, v) + v^2 Q_{vv}(u, v) \quad \text{for any } (u, v) \in \mathbb{R}^2.$$

As a model for Q, we present the following example given in [161]. Let $q \ge 1$ and

$$\mathcal{P}_q(u, v) = \sum_{\alpha_i + \beta_i = q} a_i u^{\alpha_i} v^{\beta_i},$$

where $i \in \{1, \dots, k\}$, $\alpha_i, \beta_i \ge 1$ and $a_i \in \mathbb{R}$. The following functions and their possible combinations, with an appropriate choice of the coefficients a_i, satisfy assumptions $(Q1)$-$(Q5)$ on Q

$$Q_1(u, v) = \mathcal{P}_p(u, v), \quad Q_2(u, v) = \sqrt[r]{\mathcal{P}_\ell(u, v)} \quad \text{and} \quad Q_3(u, v) = \frac{\mathcal{P}_{\ell_1}(u, v)}{\mathcal{P}_{\ell_2}(u, v)},$$

with $r = \ell p$ and $\ell_1 - \ell_2 = p$.

In order to state precisely the main results, we need to introduce some notations. Fix $\xi \in \mathbb{R}^N$, and consider the following autonomous system:

$$\begin{cases} (-\Delta)^s u + V(\xi)u = Q_u(u, v) + \frac{\gamma}{2_s^*} K_u(u, v) & \text{in } \mathbb{R}^N, \\ (-\Delta)^s v + W(\xi)v = Q_v(u, v) + \frac{\gamma}{2_s^*} K_v(u, v) & \text{in } \mathbb{R}^N, \\ u, v > 0 & \text{in } \mathbb{R}^N. \end{cases}$$

Let $\mathcal{J}_\xi : H^s(\mathbb{R}^N) \times H^s(\mathbb{R}^N) \to \mathbb{R}$ be the Euler-Lagrange functional associated with the above problem, namely

$$\mathcal{J}_\xi(u, v) = \frac{1}{2} \|(u, v)\|_\xi^2 - \int_{\mathbb{R}^N} Q(u, v) \, dx - \frac{\gamma}{2_s^*} \int_{\mathbb{R}^N} K(u, v) \, dx,$$

where

$$\|(u, v)\|_\xi^2 = \int_{\mathbb{R}^N} \left(|(-\Delta)^{\frac{s}{2}} u|^2 + |(-\Delta)^{\frac{s}{2}} v|^2 \right) dx + \int_{\mathbb{R}^N} (V(\xi)u^2 + W(\xi)v^2) \, dx.$$

By the assumptions on Q and K, it follows that $\mathcal{J}_\xi$ possesses a mountain pass geometry (see [52]), so we can consider the mountain pass value

$$C(\xi) = \inf_{\gamma \in \Gamma} \max_{t \in [0,1]} \mathcal{J}_\xi(\gamma(t)),$$

where

$$\Gamma = \left\{ \gamma \in C([0,1], H^s(\mathbb{R}^N) \times H^s(\mathbb{R}^N)) : \gamma(0) = 0, \mathcal{J}_\xi(\gamma(1)) < 0 \right\}.$$

Moreover, the function $\xi \mapsto C(\xi)$ is continuous; in addition, $C(\xi)$ can be also characterized as

$$C(\xi) = \inf_{(u,v) \in \mathcal{N}_\xi} \mathcal{J}_\xi(u, v),$$

where $\mathcal{N}_\xi$ is the Nehari manifold associated with $\mathcal{J}_\xi$. Then, for any fixed $\xi \in \mathbb{R}^N$, $C(\xi)$ is achieved, and in view of condition $(H3)$ we deduce that $C(x_0) \leq C(\xi)$ for any $\xi \in \mathbb{R}^N$, which implies that

$$M = \left\{ x \in \mathbb{R}^N : C(x) = \inf_{\xi \in \mathbb{R}^N} C(\xi) \right\} \neq \emptyset.$$

Furthermore, we can prove that

$$C^* = C(x_0) = \inf_{\xi \in \Lambda} C(\xi) < \min_{\xi \in \partial\Lambda} C(\xi).$$

Then, we are able to state the main results for (7.5.1) whose proofs can be found in [54] when $\gamma = 0$, and [40] when $\gamma = 1$:

Theorem 7.5.1 ([40, 54]) *Assume that, if $\gamma = 0$, $(H1)$–$(H3)$ and $(Q1)$–$(Q6)$ hold. When $\gamma = 1$, we suppose that $(H1)$–$(H3)$, $(Q1)$–$(Q6)$ and $(K1)$–$(K7)$ are satisfied. In addition, we make the following technical assumption on Q:*

$(Q7)$ $Q(u, v) \geq \lambda u^{\tilde{\alpha}} v^{\tilde{\beta}}$ for any $(u, v) \in \mathbb{R}^2_+$ with $1 < \tilde{\alpha}, \tilde{\beta} < 2^*_s, \tilde{\alpha} + \tilde{\beta} = q_1 \in (2, 2^*_s)$, and λ verifying

- *$\lambda > 0$ if either $N \geq 4s$, or $2s < N < 4s$ and $2^*_s - 2 < q_1 < 2^*_s$;*
- *λ is sufficiently large if $2s < N < 4s$ and $2 < q_1 \leq 2^*_s - 2$.*

Then, for every $\delta > 0$ satisfying

$$M_\delta = \{x \in \mathbb{R}^N : \operatorname{dist}(x, M) \leq \delta\} \subset \Lambda,$$

there exists $\varepsilon_\delta > 0$ such that, for any $\varepsilon \in (0, \varepsilon_\delta)$, system (7.5.1) admits at least $\mathrm{cat}_{M_\delta}(M)$ solutions. Moreover, if $(u_\varepsilon, v_\varepsilon)$ is a solution to (7.5.1) and P_ε and Q_ε are global maximum points of u_ε and v_ε, respectively, then $C(P_\varepsilon), C(Q_\varepsilon) \to C(x_0)$ as $\varepsilon \to 0$, and the following estimates hold

$$u_\varepsilon(x) \le \frac{C\,\varepsilon^{N+2s}}{\varepsilon^{N+2s} + |x - P_\varepsilon|^{N+2s}} \quad and \quad v_\varepsilon(x) \le \frac{C\,\varepsilon^{N+2s}}{\varepsilon^{N+2s} + |x - Q_\varepsilon|^{N+2s}} \quad \forall x \in \mathbb{R}^N.$$

The above theorem is obtained by combining a variant of the penalization approach [165] considered in [4] (see also [16, 17]), Lusternik-Schnirelman theory and some ideas used in [19, 35, 216] to deal with fractional Schrödinger equations. Further results for fractional elliptic systems can be found in [52, 210, 261, 332].

Fractional Schrödinger Equations with Superlinear or Asymptotically Linear Nonlinearities

8.1 Introduction

In this paper we investigate the existence and the concentration phenomenon of positive solutions for the following fractional problem:

$$\begin{cases} \varepsilon^{2s}(-\Delta)^s u + V(x)u = f(u) & \text{in } \mathbb{R}^N, \\ u \in H^s(\mathbb{R}^N), \ u > 0 & \text{in } \mathbb{R}^N, \end{cases} \tag{8.1.1}$$

where $\varepsilon > 0$ is a small parameter, $s \in (0, 1)$ and $N \geq 2$. The external potential $V : \mathbb{R}^N \to \mathbb{R}$ is a Hölder continuous function and bounded below away from zero, that is, there exists $V_0 > 0$ such that

$$V(x) \geq V_0 > 0 \quad \text{for all } x \in \mathbb{R}^N. \tag{8.1.2}$$

Concerning the nonlinearity $f : \mathbb{R} \to \mathbb{R}$, we assume that it satisfies the following basic assumptions:

$(f1)$ $f \in C^1(\mathbb{R}, \mathbb{R})$;
$(f2)$ $\lim_{t \to 0} \frac{f(t)}{t} = 0$;
$(f3)$ there exists $p \in (1, 2_s^* - 1)$ such that $\lim_{t \to \infty} \frac{f(t)}{t^p} = 0$.

In this chapter we aim to study the existence of positive solutions to (8.1.1) concentrating around local minima of the potential $V(x)$, under the assumptions that the nonlinearity f is asymptotically linear or superlinear at infinity, and without assuming the monotonicity of $f(t)/t$. We also consider superlinear nonlinearities which do not satisfy the Ambrosetti–

© The Author(s), under exclusive license to Springer Nature Switzerland AG 2021
V. Ambrosio, *Nonlinear Fractional Schrödinger Equations in* $\mathbb{R}^N$,
Frontiers in Mathematics, https://doi.org/10.1007/978-3-030-60220-8_8

Rabinowitz condition [29]; see also [6,37,42,52,75] for some related works. We recall that the condition (AR) and the assumption $f(t)/t$ is increasing for $t > 0$ play a fundamental role when we have to verify the boundedness of Palais–Smale sequences and apply Nehari manifold arguments, respectively.

Now, we state our main result:

Theorem 8.1.1 ([47]) *Let us assume that $f(t)$ satisfies $(f1)$–$(f3)$ and either*

($f4$) there exists $\mu > 2$ such that $0 < \mu F(t) \leq f(t)t$ for any $t > 0$,

where $F(t) = \displaystyle\int_0^t f(\tau)\,d\tau$, or the following condition ($f5$):

(i) *There exists $a \in (0, \infty]$ such that $\lim_{t \to \infty} \frac{f(t)}{t} = a$.*
(ii) *There exists a constant $D \geq 1$ such that*

$$\widehat{F}(t) \leq D\widehat{F}(\bar{t}) \quad 0 \leq t \leq \bar{t}, \tag{8.1.3}$$

where $\widehat{F}(t) = \frac{1}{2}f(t)t - F(t)$.

Let $\Lambda \subset \mathbb{R}^N$ be a bounded open set such that

$$\inf_{\Lambda} V < \min_{\partial \Lambda} V \tag{8.1.4}$$

and, when $a < \infty$ in ($f5$),

$$\inf_{\Lambda} V < a. \tag{8.1.5}$$

Then, there exists $\varepsilon_0 > 0$ such that, for any $\varepsilon \in (0, \varepsilon_0]$, problem (8.1.1) admits a positive solution $u_\varepsilon(x)$. Moreover, if x_ε denotes a global maximum point of u_ε, then we have

(1) *$V(x_\varepsilon) \to \inf_{x \in \Lambda} V(x)$;*
(2) *there exists $C > 0$ such that*

$$u_\varepsilon(x) \leq \frac{C\varepsilon^{N+2s}}{\varepsilon^{N+2s} + |x - x_\varepsilon|^{N+2s}} \quad \text{for all } x \in \mathbb{R}^N.$$

We would like to note that Theorem 8.1.1 extends and improves Theorem 7.1.1, because we do not require any monotonicity assumption on $f(t)/t$, and we are able to deal with a more general class of nonlinearities, including the asymptotically linear case (see condition ($f5$)). Moreover, our result is in clear agreement with that for the classical local counterpart, that is Theorem 1.1 in [235]. Clearly, a more involved and accurate analysis with respect to the case $s = 1$ will be needed due to the nonlocal character of $(-\Delta)^s$.

Now, we give the main ideas for the proof of Theorem 8.1.1. After rescaling equation (8.1.1) with the change of variable $v(x) = u(\varepsilon x)$, we introduce a modified functional J_ε and we prove that it satisfies a mountain pass geometry [29]. Then, we investigate the boundedness of Cerami sequences for J_ε, and we give two types of boundedness results: one when ε is fixed, the other one to deduce uniform boundedness when $\varepsilon \to 0$. Through a careful study of the behavior of bounded Cerami sequences (v_ε), as $\varepsilon \to 0$, we prove that there exists a subsequence (v_{ε_j}) which converges, in a suitable sense, to a sum of translated critical points of certain autonomous functionals. This concentration-compactness type result will be useful for showing that an appropriate translated sequence $v_{\varepsilon_j}(\cdot + y_{\varepsilon_j})$ converges to a least energy solution ω^1. Finally, we prove L^∞-estimates (uniformly in $j \in \mathbb{N}$) and deduce some information about the behavior at infinity of the translated sequence, which allows one to obtain a positive solution of the rescaled problem.

8.2 Modification of the Nonlinearity

Since we are looking for positive solutions of (8.1.1), we can assume that $f(t) = 0$ for any $t \leq 0$. Arguing as in [235], we can prove the following useful properties of the function f:

Lemma 8.2.1 *Assume that* $(f1)$–$(f3)$ *hold. Then we have:*

(i) *For every* $\delta > 0$ *there exists* $C_\delta > 0$ *such that*

$$|f(t)| \leq \delta|t| + C_\delta|t|^p \quad \text{for all } t \in \mathbb{R}. \tag{8.2.1}$$

(ii) *If* $(f4)$ *holds, then* $f(t) \geq 0$ *for all* $t \geq 0$.
(iii) *If* $(f5)$ *holds, then* $f(t) \geq 0$, $\widehat{F}(t) \geq 0$, *and* $\frac{d}{dt}\left(\frac{F(t)}{t^2}\right) \geq 0$ *for all* $t \geq 0$.
(iv) *If* $t \mapsto \frac{f(t)}{t}$ *is increasing for* $t \in (0, \infty)$, *then* $(f5)$ *is satisfied with* $D = 1$.

Now, assume that $f(t)$ satisfies $(f1)$–$(f3)$ and that

$$V_0 < a = \lim_{\xi \to \infty} \frac{f(t)}{t} \in (0, \infty].$$

Take $v \in (0, \frac{V_0}{2})$ and we define

$$\underline{f}(t) = \begin{cases} \min\{f(t), vt\}, & \text{if } t \geq 0, \\ 0, & \text{if } t < 0. \end{cases}$$

Using $(f2)$ we can find $r_\nu > 0$ such that

$$\underline{f}(t) = f(t) \quad \text{for all } |t| \le r_\nu.$$

Moreover, it holds that

$$\underline{f}(t) = \begin{cases} \nu t, & \text{for large } t \ge 0, \\ 0, & \text{for } t \le 0. \end{cases}$$

For technical reasons, it is convenient to choose ν as follows:

If $(f4)$ holds, then we take $\nu > 0$ such that

$$\frac{\nu}{2V_0} < \frac{1}{2} - \frac{1}{\mu}. \tag{8.2.2}$$

When $(f5)$ is satisfied, we choose $\nu \in (0, \frac{V_0}{2})$ such that ν is a regular value of $t \in (0, \infty) \mapsto \frac{f(t)}{t}$. Since $\lim_{t \to 0} \frac{f(t)}{t} = 0$ and $\lim_{t \to \infty} \frac{f(t)}{t} = a > V_0 > \nu$, if ν is a regular value of $\frac{f(t)}{t}$ we deduce that

$$k_\nu = \text{card}\{t \in (0, \infty) : f(t) = \nu t\} < \infty. \tag{8.2.3}$$

Now, let $\Lambda \subset \mathbb{R}^N$ be a bounded open set such that $\partial\Lambda \in C^\infty$, and we assume that Λ satisfies (8.1.4). Take an open set $\Lambda' \subset \Lambda$ with smooth boundary $\partial\Lambda'$ and define a function $\chi \in C^\infty(\mathbb{R}^N, \mathbb{R})$ such that

$$\inf_{\Lambda \setminus \Lambda'} V > \inf_\Lambda V,$$

$$\min_{\partial\Lambda'} V > \inf_{\Lambda'} V = \inf_\Lambda V,$$

$$\chi(x) = 1 \quad \text{for } x \in \Lambda',$$

$$\chi(x) \in (0, 1) \quad \text{for } x \in \Lambda \setminus \overline{\Lambda'},$$

$$\chi(x) = 0 \quad \text{for } x \in \mathbb{R}^N \setminus \Lambda.$$

Without loss of generality, we suppose that $0 \in \Lambda'$ and $V(0) = \inf_\Lambda V$.

Finally, we introduce the penalty function

$$g(x, t) = \chi(x)f(t) + (1 - \chi(x))\underline{f}(t) \quad \text{for } (x, t) \in \mathbb{R}^N \times \mathbb{R},$$

and set

$$F(t) = \int_0^t \underline{f}(\tau) \, d\tau,$$

$$G(x, t) = \int_0^t g(x, \tau) \, d\tau = \chi(x) F(t) + (1 - \chi(x)) \underline{F}(t).$$

As in [235], it is easy to check that the following properties concerning $\underline{f}(t)$ and $g(x, t)$ hold.

Lemma 8.2.2

(i) $\underline{f}(t) = 0$, $\underline{F}(t) = 0$ for all $t \le 0$.
(ii) $\underline{f}(t) \le vt$, $\underline{F}(t) \le F(t)$ for all $t \ge 0$.
(iii) $\underline{f}(t) \le f(t)$ for all $t \ge 0$.
(iv) If $f(t)$ satisfies either $(f4)$ or $(f5)$, then $\underline{f}(t) \ge 0$ for all $t \in \mathbb{R}$.
(v) If $f(t)$ satisfies $(f5)$, then $\underline{f}(t)$ also satisfies $(f5)$. Moreover, $\widehat{\underline{F}}(t) \ge 0$ for all $t \ge 0$.

Corollary 8.2.3

(i) $g(x, t) \le f(t)$, $G(x, t) \le F(t)$ for all $(x, t) \in \mathbb{R}^N \times \mathbb{R}$.
(ii) $g(x, t) = f(t)$ if $|t| < r_v$.
(iii) For every $\delta > 0$ there exists a $C_\delta > 0$ such that

$$|g(x, t)| \le \delta |t| + C_\delta |t|^p \quad \text{for all } (x, t) \in \mathbb{R}^N \times \mathbb{R}.$$

(iv) if $f(t)$ satisfies $(f5)$–(ii), then $g(x, t)$ satisfies

$$\widehat{G}(x, t) \le D^{k_v} \widehat{G}(x, \bar{t}) \quad \text{for all } 0 \le t \le \bar{t},$$

where $\widehat{G}(x, t) = \frac{1}{2} g(x, t) t - G(x, t)$, $D \ge 1$ is given in $(f5)$–(ii) and k_v is given in (8.2.3).

In what follows, we investigate the existence of positive solutions u_ε of the modified problem

$$\begin{cases} \varepsilon^{2s} (-\Delta)^s u + V(x) u = g(x, u) & \text{in } \mathbb{R}^N, \\ u \in H^s(\mathbb{R}^N), \, u > 0 & \text{in } \mathbb{R}^N, \end{cases} \tag{8.2.4}$$

with the property

$$u_\varepsilon(x) \le r_v \quad \text{for } x \in \mathbb{R}^N \setminus \Lambda'.$$

In view of the definition of g, these functions u_ε are also solutions of (8.1.1).

8.3   Mountain Pass Argument

Using the change of variable $v(x) = u(\varepsilon x)$, one can show that (8.2.4) is equivalent to the following problem:

$$\begin{cases} (-\Delta)^s v + V(\varepsilon x)v = g(\varepsilon x, v) & \text{in } \mathbb{R}^N, \\ u \in H^s(\mathbb{R}^N), \, u > 0 & \text{in } \mathbb{R}^N. \end{cases} \tag{8.3.1}$$

The energy functional associated with (8.3.1) is given by

$$J_\varepsilon(v) = \frac{1}{2} \int_{\mathbb{R}^N} \left(|(-\Delta)^{\frac{s}{2}} v|^2 + V(\varepsilon x)v^2 \right) dx - \int_{\mathbb{R}^N} G(\varepsilon x, v) \, dx \quad \forall v \in \mathcal{H}_\varepsilon$$

where the fractional space

$$\mathcal{H}_\varepsilon = \left\{ v \in H^s(\mathbb{R}^N) : \int_{\mathbb{R}^N} V(\varepsilon x)v^2 \, dx < \infty \right\}$$

is endowed with the norm

$$\|v\|_\varepsilon = \left(\int_{\mathbb{R}^N} \left(|(-\Delta)^{\frac{s}{2}} v|^2 + V(\varepsilon x)v^2 \right) dx \right)^{\frac{1}{2}}.$$

Since $V_0 > 0$, we can equip $H^s(\mathbb{R}^N)$ with the equivalent norm

$$\|v\|_0 = \left(\int_{\mathbb{R}^N} \left(|(-\Delta)^{\frac{s}{2}} v|^2 + V_0 v^2 \right) dx \right)^{\frac{1}{2}}.$$

Clearly,

$$\|v\|_0 \leq \|v\|_\varepsilon, \tag{8.3.2}$$

so $\mathcal{H}_\varepsilon \subset H^s(\mathbb{R}^N)$ and $\mathcal{H}_\varepsilon$ is continuously embedded in $L^r(\mathbb{R}^N)$ for any $2 \leq r \leq 2_s^*$, and there exists $C_r' > 0$ such that

$$\|v\|_{L^r(\mathbb{R}^N)} \leq C_r' \|v\|_0. \tag{8.3.3}$$

In what follows, we denote by $\mathcal{H}_\varepsilon^*$ the dual of $\mathcal{H}_\varepsilon$. Let us prove that J_ε possesses a mountain pass geometry [29] that is uniform with respect to ε.

Lemma 8.3.1 $J_\varepsilon \in C^1(\mathcal{H}_\varepsilon, \mathbb{R})$ *and satisfies the following properties:*

(i) $J_\varepsilon(0) = 0$;
(ii) *there exist $\rho_0 > 0$ and $\delta_0 > 0$, independent of $\varepsilon \in (0, 1]$, such that*

$$J_\varepsilon(v) \geq \delta_0 \quad \text{for all } \|v\|_0 = \rho_0,$$
$$J_\varepsilon(v) > 0 \quad \text{for all } 0 < \|v\|_0 \leq \rho_0;$$

(iii) *there exist $v_0 \in C_c^\infty(\mathbb{R}^N)$ and $\varepsilon_0 > 0$ such that $J_\varepsilon(v_0) < 0$ for all $\varepsilon \in (0, \varepsilon_0]$.*

Proof Obviously, $J_\varepsilon \in C^1(\mathcal{H}_\varepsilon, \mathbb{R})$ and $J_\varepsilon(0) = 0$. Since $\underline{F} \leq F$, taking $\delta = \frac{V_0}{2}$ in (8.2.1) we get

$$J_\varepsilon(v) = \frac{1}{2}\|v\|_\varepsilon^2 - \int_{\mathbb{R}^N} \left(\chi(\varepsilon x)F(v) + (1 - \chi(\varepsilon x))\underline{F}(v)\right) dx$$

$$\geq \frac{1}{2}\|v\|_\varepsilon^2 - \int_{\mathbb{R}^N} F(v)\, dx$$

$$\geq \frac{1}{2}\|v\|_0^2 - \frac{V_0}{4}\|v\|_{L^2(\mathbb{R}^N)}^2 - C_{\frac{V_0}{2}}\|v\|_{L^{p+1}(\mathbb{R}^N)}^{p+1}$$

$$\geq \frac{\|v\|_0^2}{4} - \tilde{C}_{p+1}C_{\frac{V_0}{2}}\|v\|_0^{p+1},$$

where we used (8.3.2) and (8.3.3) with $r = p + 1$. Thus (ii) is satisfied.

To verify that (iii) holds, we take $v_0 \in C_c^\infty(\mathbb{R}^N)$ such that

$$\frac{1}{2}\int_{\mathbb{R}^N} \left(|(-\Delta)^{\frac{s}{2}}v_0|^2 + V(0)v_0^2\right) dx - \int_{\mathbb{R}^N} F(v_0)\, dx < 0.$$

This choice is lawful due to the fact that $V(0) < \lim_{z \to \infty} \frac{f(z)}{z}$, so the existence of a such v_0 is guaranteed by Theorem 3.1.1 (see Lemma 8.4.2), where is proved that

$$v \mapsto \frac{1}{2}\int_{\mathbb{R}^N} \left(|(-\Delta)^{\frac{s}{2}}v|^2 + V(0)v^2\right) dx - \int_{\mathbb{R}^N} F(v)\, dx$$

has a mountain pass geometry. Since $0 \in \Lambda'$, we observe that

$$J_\varepsilon(v_0) \to \frac{1}{2}\int_{\mathbb{R}^N} \left(|(-\Delta)^{\frac{s}{2}}v_0|^2 + V(0)v_0^2\right) dx - \int_{\mathbb{R}^N} F(v_0)\, dx < 0 \text{ as } \varepsilon \to 0,$$

i.e., (iii) is satisfied for ε sufficiently small. $\qquad\square$

Since J_ε has a mountain pass geometry, for each $\varepsilon \in (0, \varepsilon_0]$, we define the mountain pass value

$$c_\varepsilon = \inf_{\gamma \in \Gamma_\varepsilon} \max_{t \in [0,1]} J_\varepsilon(\gamma(t)) \tag{8.3.4}$$

where

$$\Gamma_\varepsilon = \{\gamma \in C([0, 1], \mathcal{H}_\varepsilon) : \gamma(0) = 0 \text{ and } J_\varepsilon(\gamma(1)) < 0\}. \tag{8.3.5}$$

Using Lemma 8.3.1, we are able to give the following estimate for c_ε.

Corollary 8.3.2 *There exist $m_1, m_2 > 0$ such that for any $\varepsilon \in (0, \varepsilon_0]$*

$$m_1 \leq c_\varepsilon \leq m_2.$$

Proof For any $\gamma \in \Gamma_\varepsilon$ we have

$$\gamma([0, 1]) \cap \{v \in \mathcal{H}_\varepsilon : \|v\|_0 = \rho\} \neq \emptyset.$$

Hence, by using Lemma 8.3.1, we deduce that

$$\max_{t \in [0,1]} J_\varepsilon(\gamma(t)) \geq \inf_{\|v\|_0 = \rho_0} J_\varepsilon(v) \geq \delta_0.$$

Set $\gamma_0(t) = t v_0$, where $v_0 \in C_c^\infty(\mathbb{R}^N)$ is supplied by Lemma 8.3.1. Then we see that

$$c_\varepsilon = \inf_{\gamma \in \Gamma_\varepsilon} \left(\max_{t \in [0,1]} J_\varepsilon(\gamma(t)) \right) \leq \max_{t \in [0,1]} J_\varepsilon(\gamma_0(t)) \leq \sup_{\varepsilon \in (0, \varepsilon_0]} \left(\max_{t \in [0,1]} J_\varepsilon(\gamma_0(t)) \right).$$

Putting $m_1 = \delta_0$ and $m_2 = \sup_{\varepsilon \in (0, \varepsilon_0]} \left(\max_{t \in [0,1]} J_\varepsilon(\gamma_0(t)) \right)$ we get the desired result. $\quad\square$

Using Lemma 8.3.1 and Theorem 2.2.15, we deduce that for all $\varepsilon \in (0, \varepsilon_0]$ there exists a Cerami sequence $(v_j) \subset \mathcal{H}_\varepsilon$ such that

$$J_\varepsilon(v_j) \to b_\varepsilon$$

$$(1 + \|v_j\|_\varepsilon) \|J_\varepsilon'(v_j)\|_{\mathcal{H}_\varepsilon^*} \to 0 \quad \text{as } j \to \infty.$$

We will show that (v_j) is bounded in $\mathcal{H}_\varepsilon$ and has a convergent subsequence. Thus J_ε has a critical point v_ε satisfying $J_\varepsilon(v_\varepsilon) = b_\varepsilon$ and $J_\varepsilon'(v_\varepsilon) = 0$. We also show that (v_ε) is bounded in the sense that

$$\limsup_{\varepsilon \to 0} \|v_\varepsilon\|_\varepsilon < \infty. \tag{8.3.6}$$

We start by establishing this second type of boundedness.

Lemma 8.3.3 *Assume that f satisfies $(f1)$–$(f3)$ and either $(f4)$, or $(f5)$. Suppose that there exists a sequence $(v_\varepsilon)_{\varepsilon \in (0,\varepsilon_1]}$, with $\varepsilon_1 \in (0, \varepsilon_0]$, such that*

$$v_\varepsilon \in \mathcal{H}_\varepsilon,$$

$$J_\varepsilon(v_\varepsilon) \in [m_1, m_2] \quad \forall \varepsilon \in (0, \varepsilon_1], \tag{8.3.7}$$

$$(1 + \|v_\varepsilon\|_\varepsilon)\|J_\varepsilon'(v_\varepsilon)\|_{\mathcal{H}_\varepsilon^*} \to 0 \quad as \ \varepsilon \to 0 \tag{8.3.8}$$

with $0 < m_1 < m_2$. Then (8.3.6) holds.

Proof First, assume that $(f4)$ holds. Let (v_ε) be a sequence satisfying (8.3.7) and (8.3.8). Then one can see that (8.3.7) yields

$$J_\varepsilon(v_\varepsilon) = \frac{1}{2}\|v_\varepsilon\|_\varepsilon^2 - \int_{\mathbb{R}^N} \left((1 - \chi(\varepsilon x))\underline{F}(v_\varepsilon) + \chi(\varepsilon x)F(v_\varepsilon)\right) dx \le m_2. \tag{8.3.9}$$

Moreover, by (8.3.8), for any ε sufficiently small we have

$$|\langle J_\varepsilon'(v_\varepsilon), v_\varepsilon \rangle| \le \|J_\varepsilon'(v_\varepsilon)\|_{\mathcal{H}_\varepsilon^*}\|v_\varepsilon\|_\varepsilon \le \|J_\varepsilon'(v_\varepsilon)\|_{\mathcal{H}_\varepsilon^*}(1 + \|v_\varepsilon\|_\varepsilon) \le 1,$$

that is

$$\left|\|v_\varepsilon\|_\varepsilon^2 - \int_{\mathbb{R}^N} \left((1 - \chi(\varepsilon x))\underline{f}(v_\varepsilon)v_\varepsilon + \chi(\varepsilon x)f(v_\varepsilon)v_\varepsilon\right) dx\right| \le 1. \tag{8.3.10}$$

Taking into account (8.3.9), (8.3.10) and $(f4)$ we get

$$\left(\frac{1}{2} - \frac{1}{\mu}\right)\|v_\varepsilon\|_\varepsilon^2 \le \int_{\mathbb{R}^N} (1 - \chi(\varepsilon x))\left(\underline{F}(v_\varepsilon) - \frac{1}{\mu}\underline{f}(v_\varepsilon)v_\varepsilon\right) dx + m_2 + \frac{1}{\mu}.$$

Using items (i) and (iv) of Lemma 8.2.2, we know that $t\underline{f}(t) \ge 0$ for all $t \in \mathbb{R}$, so we obtain

$$\left(\frac{1}{2} - \frac{1}{\mu}\right)\|v_\varepsilon\|_\varepsilon^2 \le \int_{\mathbb{R}^N} (1 - \chi(\varepsilon x))\underline{F}(v_\varepsilon)\, dx + m_2 + \frac{1}{\mu}. \tag{8.3.11}$$

On the other hand, by item (ii) of Lemma 8.2.2,

$$\underline{F}(t) \leq \frac{\nu t^2}{2} \quad \text{for all } t \in \mathbb{R}.$$

Then

$$\int_{\mathbb{R}^N} (1 - \chi(\varepsilon x))\underline{F}(v_\varepsilon)\, dx \leq \frac{1}{2}\nu \|v_\varepsilon\|^2_{L^2(\mathbb{R}^N)} \leq \frac{\nu}{2V_0}\|v_\varepsilon\|^2_\varepsilon,$$

which together with (8.3.11) shows that

$$\left(\frac{1}{2} - \frac{1}{\mu}\right) \|v_\varepsilon\|^2_\varepsilon \leq \frac{\nu}{2V_0}\|v_\varepsilon\|^2_\varepsilon + m_2 + \frac{1}{\mu}.$$

In view of (8.2.2) we get

$$\cdot \|v_\varepsilon\|^2_\varepsilon \leq \frac{m_2 + \frac{1}{\mu}}{\left[\left(\frac{1}{2} - \frac{1}{\mu}\right) - \frac{\nu}{2V_0}\right]},$$

which implies that $\|v_\varepsilon\|_\varepsilon$ is bounded if ε is small enough.

Now, let us suppose that $(f5)$ holds. Arguing by contradiction, we assume that

$$\limsup_{\varepsilon \to 0} \|v_\varepsilon\|_\varepsilon = \infty.$$

Let $\varepsilon_j \to 0$ be a subsequence such that $\|v_{\varepsilon_j}\|_{\varepsilon_j} \to \infty$. For simplicity, we denote ε_j still by ε. Set $w_\varepsilon = \frac{v_\varepsilon}{\|v_\varepsilon\|_{H^s_\varepsilon}}$. Clearly, $\|w_\varepsilon\|_0 = \frac{\|v_\varepsilon\|_0}{\|v_\varepsilon\|_\varepsilon} \leq \frac{\|v_\varepsilon\|_\varepsilon}{\|v_\varepsilon\|_\varepsilon} = 1$. Moreover, we can see that there exists $C_1 > 0$ independent of ε such that

$$\|\chi_\varepsilon w_\varepsilon\|_0 \leq C_1, \tag{8.3.12}$$

where $\chi_\varepsilon(x) = \chi(\varepsilon x)$. Indeed, since $0 \leq \chi \leq 1$, $(|a| + |b|)^2 \leq 2(|a|^2 + |b|^2)$, $\varepsilon \in (0, \varepsilon_1]$ and $s \in (0, 1)$, we obtain that

$$\iint_{\mathbb{R}^{2N}} \frac{|\chi(\varepsilon x)w_\varepsilon(x) - \chi(\varepsilon y)w_\varepsilon(y)|^2}{|x - y|^{N+2s}}\, dxdy + \int_{\mathbb{R}^N} V_0(\chi_\varepsilon w_\varepsilon)^2\, dx$$

$$\leq 2\iint_{\mathbb{R}^{2N}} \frac{|\chi(\varepsilon x) - \chi(\varepsilon y)|^2}{|x - y|^{N+2s}} w^2_\varepsilon(x)\, dxdy + 2\iint_{\mathbb{R}^{2N}} \frac{|w_\varepsilon(x) - w_\varepsilon(y)|^2}{|x - y|^{N+2s}}\, dxdy$$

$$+ \int_{\mathbb{R}^N} V_0 w^2_\varepsilon\, dx$$

$$\leq 2\,\varepsilon^2\,\|\nabla\chi\|^2_{L^\infty(\mathbb{R}^N)}\int_{\mathbb{R}^N}w_\varepsilon^2(x)\,dx\int_{|z|\leq 1}\frac{1}{|z|^{N+2s-2}}\,dz$$

$$+8\int_{\mathbb{R}^N}w_\varepsilon^2(x)\,dx\int_{|z|>1}\frac{1}{|z|^{N+2s}}\,dz+2\iint_{\mathbb{R}^{2N}}\frac{|w_\varepsilon(x)-w_\varepsilon(y)|^2}{|x-y|^{N+2s}}\,dxdy$$

$$+\int_{\mathbb{R}^N}V_0 w_\varepsilon^2\,dx$$

$$\leq\left((1-s)^{-1}\varepsilon_1^2\,\|\nabla\chi\|^2_{L^\infty(\mathbb{R}^N)}\omega_{N-1}+4s^{-1}\omega_{N-1}+V_0\right)\|w_\varepsilon\|^2_{L^2(\mathbb{R}^N)}+2[w_\varepsilon]_s^2$$

$$\leq C_1\|w_\varepsilon\|^2_0\leq C_1.$$

Now, (8.3.8) implies that $\langle J'_\varepsilon(v_\varepsilon),\varphi\rangle=o(1)$ for any $\varphi\in\mathcal{H}_\varepsilon$, that is

$$\int_{\mathbb{R}^N}[(-\Delta)^{\frac{s}{2}}v_\varepsilon(-\Delta)^{\frac{s}{2}}\varphi+V(\varepsilon x)v_\varepsilon\varphi]\,dx$$

$$=\int_{\mathbb{R}^N}[\chi_\varepsilon f(v_\varepsilon)+(1-\chi_\varepsilon)\underline{f}(v_\varepsilon)]\varphi\,dx+o(1),$$

or equivalently

$$\int_{\mathbb{R}^N}\left[(-\Delta)^{\frac{s}{2}}w_\varepsilon(-\Delta)^{\frac{s}{2}}\varphi+V(\varepsilon x)w_\varepsilon\varphi\right]dx$$

$$=\int_{\mathbb{R}^N}\left[\chi_\varepsilon\frac{f(v_\varepsilon)}{v_\varepsilon}w_\varepsilon+(1-\chi_\varepsilon)\frac{\underline{f}(v_\varepsilon)}{v_\varepsilon}w_\varepsilon\right]\varphi\,dx+o(1). \tag{8.3.13}$$

Taking $\varphi=w_\varepsilon^-=\min\{w_\varepsilon,0\}$ in (8.3.13) and recalling that

$$(x-y)(x^--y^-)\geq|x^--y^-|^2\quad\text{for all }x,y\in\mathbb{R},$$

and that $f(t)=\underline{f}(t)=0$ for all $t\leq 0$, we have

$$\int_{\mathbb{R}^N}\left[|(-\Delta)^{\frac{s}{2}}w_\varepsilon^-|^s+V(\varepsilon x)(w_\varepsilon^-)^2\right]dx$$

$$\leq\int_{\mathbb{R}^N}\left[\chi_\varepsilon\frac{f(v_\varepsilon)}{v_\varepsilon}w_\varepsilon+(1-\chi_\varepsilon)\frac{\underline{f}(v_\varepsilon)}{v_\varepsilon}w_\varepsilon\right]w_\varepsilon^-\,dx+o(1)=o(1),$$

so we see that

$$\|w_\varepsilon^-\|_\varepsilon^2\to 0\quad\text{as }\varepsilon\to 0. \tag{8.3.14}$$

Now, we observe that one of the following two cases must occur.

Case 1: $\limsup_{\varepsilon \to 0} \left(\sup_{z \in \mathbb{R}^N} \int_{B_1(z)} |\chi_\varepsilon(x) w_\varepsilon|^2 dx \right) > 0$;

Case 2: $\limsup_{\varepsilon \to 0} \left(\sup_{z \in \mathbb{R}^N} \int_{B_1(z)} |\chi_\varepsilon(x) w_\varepsilon|^2 dx \right) = 0$.

Step 1   Case 1 cannot occur under assumption $(f5)$ with $a = \infty$.

Suppose, on the contrary, that Case 1 occurs. Then, up to a subsequence, there exist $(x_\varepsilon) \subset \mathbb{R}^N$, $d > 0$ and $x_0 \in \overline{\Lambda}$ such that

$$\int_{B_1(x_\varepsilon)} |\chi_\varepsilon w_\varepsilon|^2 \, dx \to d > 0, \tag{8.3.15}$$

$$\varepsilon x_\varepsilon \to x_0 \in \overline{\Lambda}. \tag{8.3.16}$$

Indeed, the existence of (y_ε) satisfying (8.3.15) is clear. Moreover, (8.3.15) implies that $B_1(x_\varepsilon) \cap \mathrm{supp}(\chi_\varepsilon) \neq \emptyset$, so there exists $z_\varepsilon \in \mathrm{supp}(\chi_\varepsilon)$ such that $\chi(\varepsilon z_\varepsilon) \neq 0$ and $|z_\varepsilon - x_\varepsilon| < 1$. Hence $|\varepsilon x_\varepsilon - \varepsilon z_\varepsilon| < \varepsilon$ yields $\varepsilon x_\varepsilon \in N_\varepsilon(\Lambda) = \{z \in \mathbb{R}^N : \mathrm{dist}(z, \Lambda) < \varepsilon\}$, and we may assume that (8.3.16) holds. Since $\|w_\varepsilon\|_0 \leq 1$, we may assume that

$$w_\varepsilon(\cdot + x_\varepsilon) \rightharpoonup w_0 \quad \text{in } H^s(\mathbb{R}^N). \tag{8.3.17}$$

In view of (8.3.16) and (8.3.17) we have

$$(\chi_\varepsilon w_\varepsilon)(\cdot + x_\varepsilon) \rightharpoonup \chi(x_0) w_0 \quad \text{in } H^s(\mathbb{R}^N).$$

To prove this, fix $\varphi \in H^s(\mathbb{R}^N)$ and note that

$$\iint_{\mathbb{R}^{2N}} \frac{(\chi_\varepsilon w_\varepsilon)(x + x_\varepsilon) - (\chi_\varepsilon w_\varepsilon)(y + x_\varepsilon)}{|x - y|^{N+2s}} (\varphi(x) - \varphi(y)) \, dx dy$$

$$= \iint_{\mathbb{R}^{2N}} \frac{(h_\varepsilon w_\varepsilon)(x + x_\varepsilon) - (h_\varepsilon w_\varepsilon)(y + x_\varepsilon)}{|x - y|^{N+2s}} (\varphi(x) - \varphi(y)) \, dx dy$$

$$+ \iint_{\mathbb{R}^{2N}} \frac{(w_\varepsilon(x + x_\varepsilon) - w_\varepsilon(y + x_\varepsilon))}{|x - y|^{N+2s}} (\varphi(x) - \varphi(y)) \chi(x_0) \, dx dy$$

$$= A_\varepsilon + B_\varepsilon,$$

where $h_\varepsilon(x) = \chi_\varepsilon(x) - \chi(x_0)$. In view of (8.3.17) we know that

$$B_\varepsilon \to \iint_{\mathbb{R}^{2N}} \frac{(w_0(x) - w_0(y))}{|x - y|^{N+2s}} (\varphi(x) - \varphi(y)) \chi(x_0) \, dx dy.$$

Now, we observe that

$$A_\varepsilon = \iint_{\mathbb{R}^{2N}} \frac{(h_\varepsilon(x + x_\varepsilon) - h_\varepsilon(y + x_\varepsilon))}{|x - y|^{N+2s}} (\varphi(x) - \varphi(y)) w_\varepsilon(x + x_\varepsilon) \, dx dy$$

$$+ \iint_{\mathbb{R}^{2N}} \frac{(w_\varepsilon(x + x_\varepsilon) - w_\varepsilon(y + x_\varepsilon))}{|x - y|^{N+2s}} (\varphi(x) - \varphi(y)) h_\varepsilon(y + x_\varepsilon) \, dx dy$$

$$= A_\varepsilon^1 + A_\varepsilon^2.$$

Using Hölder's inequality, (8.3.16), (8.3.17) and the dominated convergence theorem, we see that

$$|A_\varepsilon^2| \leq C \left(\iint_{\mathbb{R}^{2N}} \frac{|\varphi(x) - \varphi(y)|^2}{|x - y|^{N+2s}} |h_\varepsilon(y + x_\varepsilon)|^2 \, dx dy \right)^{\frac{1}{2}} \to 0.$$

On the other hand

$$|A_\varepsilon^1| \leq [\varphi]_s \left(\iint_{\mathbb{R}^{2N}} \frac{|h_\varepsilon(x + x_\varepsilon) - h_\varepsilon(y + x_\varepsilon)|^2}{|x - y|^{N+2s}} |w_\varepsilon(x + x_\varepsilon)|^2 \, dx dy \right)^{\frac{1}{2}} \to 0$$

because

$$\iint_{\mathbb{R}^{2N}} \frac{|h_\varepsilon(x + x_\varepsilon) - h_\varepsilon(y + x_\varepsilon)|^2}{|x - y|^{N+2s}} |w_\varepsilon(x + x_\varepsilon)|^2 \, dx dy$$

$$\leq \int_{\mathbb{R}^N} |w_\varepsilon(x + x_\varepsilon)|^2 \, dx \left[\int_{|y-x|>\frac{1}{\varepsilon}} \frac{4 dy}{|x - y|^{N+2s}} + \int_{|y-x|<\frac{1}{\varepsilon}} \frac{\varepsilon^2 \|\nabla \chi\|_{L^\infty(\mathbb{R}^N)}^2 dy}{|x - y|^{N+2s-2}} \right]$$

$$\leq C \varepsilon^{2s} \int_{\mathbb{R}^N} |w_\varepsilon(x + x_\varepsilon)|^2 \, dx \leq C \varepsilon^{2s} \to 0.$$

Now, let us show that $\chi(x_0) \neq 0$ and $w_0 \geq 0$ ($\neq 0$). If, by contradiction, $\chi(x_0) = 0$, then the dominated convergence theorem, (8.3.15), (8.3.17) and Theorem 1.1.8 imply that

$$0 < d = \lim_{\varepsilon \to 0} \int_{B_1(x_\varepsilon)} |\chi_\varepsilon w_\varepsilon|^2 \, dx$$

$$= \lim_{\varepsilon \to 0} \int_{B_1} |\chi_\varepsilon w_\varepsilon|^2 (x + x_\varepsilon) \, dx$$

$$= \int_{B_1} |\chi(x_0) w_0(x)|^2 \, dx = 0,$$

which is impossible. For the same reason $w_0 \neq 0$. Using (8.3.14) and (8.3.17) we can see that $w_0 \geq 0$ in $\mathbb{R}^N$. Thus, there exists a set $K \subset \mathbb{R}^N$ such that

$$|K| > 0 \tag{8.3.18}$$

$$w_\varepsilon(x + x_\varepsilon) \to w_0(x) > 0 \quad \text{for } x \in K. \tag{8.3.19}$$

Taking $\varphi = w_\varepsilon$ in (8.3.13), we get

$$1 = \|w_\varepsilon\|_{H_\varepsilon^s}^2 = \int_{\mathbb{R}^N} \left[\chi_\varepsilon \frac{f(v_\varepsilon)}{v_\varepsilon} w_\varepsilon^2 + (1 - \chi_\varepsilon) \frac{f(v_\varepsilon)}{v_\varepsilon} w_\varepsilon^2 \right] dx + o(1),$$

and using item (iv) of Lemma 8.2.2, we deduce that

$$\limsup_{\varepsilon \to 0} \int_{\mathbb{R}^N} \chi_\varepsilon \frac{f(v_\varepsilon)}{v_\varepsilon} w_\varepsilon^2 \, dx \leq 1, \tag{8.3.20}$$

that is,

$$\limsup_{\varepsilon \to 0} \int_{\mathbb{R}^N} \chi(\varepsilon x + \varepsilon x_\varepsilon) \frac{f(v_\varepsilon(x + x_\varepsilon))}{v_\varepsilon(x + x_\varepsilon)} w_\varepsilon^2(x + x_\varepsilon) \, dx \leq 1.$$

In view of (8.3.18), (8.3.19) and the definition of w_ε, we obtain

$$v_\varepsilon(x + x_\varepsilon) = w_\varepsilon(x + x_\varepsilon) \|v_\varepsilon\|_\varepsilon \to w_0(x) \cdot (\infty) = \infty \quad \forall x \in K.$$

This, together with $\lim_{\xi \to \infty} \frac{f(\xi)}{\xi} = a = \infty$ and Fatou's lemma yields

$$\liminf_{\varepsilon \to 0} \int_{\mathbb{R}^N} \chi_\varepsilon(x + x_\varepsilon) \frac{f(v_\varepsilon(x + x_\varepsilon))}{w_\varepsilon(x + x_\varepsilon)} w_\varepsilon^2(x + x_\varepsilon) \, dx$$

$$\geq \liminf_{\varepsilon \to 0} \int_K \chi_\varepsilon(x + x_\varepsilon) \frac{f(v_\varepsilon(x + x_\varepsilon))}{v_\varepsilon(x + x_\varepsilon)} w_\varepsilon^2(x + x_\varepsilon) \, dx = \infty,$$

which contradicts (8.3.20).

Step 2 Case 1 cannot take place under assumption $(f5)$ with $a < \infty$.

As in Step 1, we extract a subsequence and assume that (8.3.15), (8.3.16) and (8.3.17) hold with $\chi(x_0) \neq 0$ and $w_0 \geq 0$ ($\not\equiv 0$). We aim to prove that w_0 is a weak solution to

$$(-\Delta)^s w_0 + V(x_0)w_0 = (\chi(x_0)a + (1 - \chi(x_0))\nu)w_0 \quad \text{in } \mathbb{R}^N. \tag{8.3.21}$$

This leads to a contradiction, since $(-\Delta)^s$ has no eigenvalues in $H^s(\mathbb{R}^N)$ (this fact can be seen by using the Pohozaev identity for the fractional Laplacian).

Fix $\varphi \in C_c^\infty(\mathbb{R}^N)$. Taking into account (8.3.16), (8.3.17) and the continuity of V, we see that

$$\int_{\mathbb{R}^N} \left[(-\Delta)^{\frac{s}{2}} w_\varepsilon(x + x_\varepsilon)(-\Delta)^{\frac{s}{2}} \varphi(x) + V(\varepsilon x + \varepsilon x_\varepsilon) w_\varepsilon \varphi \right] dx$$

$$\to \int_{\mathbb{R}^N} \left[(-\Delta)^{\frac{s}{2}} w_0 (-\Delta)^{\frac{s}{2}} \varphi + V(x_0) w_0 \varphi \right] dx. \tag{8.3.22}$$

Let us show that

$$\int_{\mathbb{R}^N} \frac{g(\varepsilon x + \varepsilon x_\varepsilon, v_\varepsilon(x + x_\varepsilon))}{v_\varepsilon(x + x_\varepsilon)} w_\varepsilon \varphi \, dx \to (\chi(x_0)a + (1 - \chi(x_0))v) \int_{\mathbb{R}^N} w_0 \varphi \, dx. \tag{8.3.23}$$

Take $R > 1$ such that $\mathrm{supp}\,(\varphi) \subset B_R$. Then, using the fact that $H^s(\mathbb{R}^N)$ is compactly embedded in $L^2_{\mathrm{loc}}(\mathbb{R}^N)$, we see that $\|w_\varepsilon - w_0\|^2_{L^2(B_R)} \to 0$. Hence, there exists $h \in L^2(B_R)$ such that

$$|w_\varepsilon| \leq h \text{ a.e. in } B_R.$$

Since $a < \infty$, there exists $C > 0$ such that $|\frac{g(x,t)}{t}| \leq C$ for all $t > 0$. Recall that

$$\frac{g(x, t)}{t} \to \chi(x)a + (1 - \chi(x))v < \infty \quad \text{as } t \to \infty.$$

Then

$$\left| \frac{g(\varepsilon x + \varepsilon x_\varepsilon, v_\varepsilon(x + x_\varepsilon))}{v_\varepsilon(x + x_\varepsilon)} w_\varepsilon \varphi \right| \leq C \|\varphi\|_{L^\infty(\mathbb{R}^N)} |w_\varepsilon|$$

$$\leq C \|\varphi\|_{L^\infty(\mathbb{R}^N)} h \in L^1(B_R), \tag{8.3.24}$$

and

$$\frac{g(\varepsilon x + \varepsilon x_\varepsilon, v_\varepsilon(x + x_\varepsilon))}{v_\varepsilon(x + x_\varepsilon)} w_\varepsilon(x) \to [\chi(x_0)a + (1 - \chi(x_0))v] w_0(x) \quad \text{a.e. in } B_R. \tag{8.3.25}$$

In fact, if $w_0(x) = 0$, by $\left| \frac{g(x,t)}{t} \right| \leq C$ for all $t > 0$ and $w_\varepsilon \to w_0 = 0$ a.e. in B_R, we get

$$\left| \frac{g(\varepsilon x + \varepsilon x_\varepsilon, v_\varepsilon(x + x_\varepsilon))}{v_\varepsilon(x + x_\varepsilon)} w_\varepsilon \right| \leq C |w_\varepsilon| \to 0 \quad \text{a.e. in } B_R.$$

If $w_0(x) \neq 0$, then $v_\varepsilon(x + x_\varepsilon) = w_\varepsilon(x + x_\varepsilon)\|v_\varepsilon\|_{H^s_\varepsilon} \to \infty$, and since $w_\varepsilon \to w_0$ a.e. in B_R, we have

$$\frac{g(\varepsilon x + \varepsilon x_\varepsilon, v_\varepsilon(x + x_\varepsilon))}{v_\varepsilon(x + x_\varepsilon)} w_\varepsilon \to [\chi(x_0)a + (1 - \chi(x_0))v]w_0 \quad \text{a.e. in } B_R.$$

Then (8.3.25) holds. Taking into account (8.3.24) and (8.3.25), we infer that (8.3.23) is true thanks to the dominated convergence theorem. Putting together $\langle J'_\varepsilon(v_\varepsilon), \varphi \rangle = o(1)$, (8.3.22) and (8.3.23) we obtain (8.3.21).

Step 3 Case 2 cannot take place.

Assume the contrary. Since (8.3.12) holds and

$$\lim_{\varepsilon \to 0} \sup_{z \in \mathbb{R}^N} \int_{B_1(z)} |\chi_\varepsilon w_\varepsilon|^2 \, dx = 0,$$

Lemma 1.4.4 shows that $\|\chi_\varepsilon w_\varepsilon\|_{L^{p+1}(\mathbb{R}^N)} \to 0$. Now, for any $L > 1$ we can see that

$$J_\varepsilon(Lw_\varepsilon) = \frac{1}{2}L^2 - \int_{\mathbb{R}^N} \left[\chi_\varepsilon F(Lw_\varepsilon) + (1 - \chi_\varepsilon)\underline{F}(Lw_\varepsilon) \right] dx.$$

By item (ii) of Lemma 8.2.2 and since $v \in (0, \frac{V_0}{2})$, we have

$$\int_{\mathbb{R}^N} (1 - \chi_\varepsilon)\underline{F}(Lw_\varepsilon) \, dx \leq \int_{\mathbb{R}^N} \frac{1}{2}vL^2|w_\varepsilon|^2 \, dx$$

$$\leq \int_{\mathbb{R}^N} \frac{V_0}{4}L^2|w_\varepsilon|^2 \, dx$$

$$\leq \frac{L^2}{4}\|w_\varepsilon\|_0 \leq \frac{L^2}{4}.$$

Consequently,

$$J_\varepsilon(Lw_\varepsilon) \geq \frac{1}{4}L^2 - \int_{\mathbb{R}^N} \chi_\varepsilon F(Lw_\varepsilon) \, dx. \tag{8.3.26}$$

Using (8.2.1), Hölder's inequality and the fact that $\|\chi_\varepsilon w_\varepsilon\|_{L^{p+1}(\mathbb{R}^N)} \to 0$, we see that

$$\int_{\mathbb{R}^N} \chi_\varepsilon F(Lw_\varepsilon) \, dx \leq \int_{\mathbb{R}^N} \left[\frac{\delta}{2}L^2|w_\varepsilon|^2 + C_\delta \frac{|Lw_\varepsilon|^{p+1}}{p+1}\chi_\varepsilon(x) \right] dx$$

$$\leq \delta L^2 \|w_\varepsilon\|^2_{L^2(\mathbb{R}^N)} + C_\delta L^{p+1} \|w_\varepsilon\|^p_{L^{p+1}(\mathbb{R}^N)} \|\chi_\varepsilon w_\varepsilon\|_{L^{p+1}(\mathbb{R}^N)}$$

$$\leq \frac{\delta L^2}{V_0^2}\|w_\varepsilon\|^2_\varepsilon + o(1). \tag{8.3.27}$$

Putting together (8.3.26) and (8.3.27) we have

$$J_\varepsilon(Lw_\varepsilon) \geq \frac{1}{4}L^2 - \frac{\delta L^2}{V_0^2}\|w_\varepsilon\|_\varepsilon^2 + o(1) \quad \forall \delta > 0,$$

and thanks to the arbitrariness of $\delta > 0$, we get

$$\liminf_{\varepsilon \to 0} J_\varepsilon(Lw_\varepsilon) \geq \frac{1}{4}L^2.$$

Since $\|v_\varepsilon\|_\varepsilon \to \infty$, we can see that $\frac{L}{\|v_\varepsilon\|_\varepsilon} \in (0, 1)$ for ε sufficiently small, and so

$$\max_{t\in[0,1]} J_\varepsilon(tv_\varepsilon) \geq J_\varepsilon\left(\frac{L}{\|v_\varepsilon\|_\varepsilon}v_\varepsilon\right) \geq \frac{1}{4}L^2.$$

Take $L > 0$ sufficiently large so that $m_2 < \frac{1}{4}L^2$ and recall that, by (8.3.7), $J_\varepsilon(v_\varepsilon) \leq m_2$. Then we can find $t_\varepsilon \in (0, 1)$ such that

$$J_\varepsilon(t_\varepsilon v_\varepsilon) = \max_{t\in[0,1]} J_\varepsilon(tv_\varepsilon).$$

Hence

$$J_\varepsilon(t_\varepsilon v_\varepsilon) = \max_{t\in[0,1]} J_\varepsilon(tv_\varepsilon) \geq \frac{1}{4}L^2 \to \infty \quad \text{as } L \to \infty,$$

that is

$$J_\varepsilon(t_\varepsilon v_\varepsilon) \to \infty \quad \text{as } \varepsilon \to 0. \tag{8.3.28}$$

Now, since $\langle J_\varepsilon'(t_\varepsilon v_\varepsilon), (t_\varepsilon v_\varepsilon)\rangle = 0$, (8.3.8) and Corollary 8.2.3-(iv) imply that

$$J_\varepsilon(t_\varepsilon v_\varepsilon) = J_\varepsilon(t_\varepsilon v_\varepsilon) - \frac{1}{2}\langle J_\varepsilon'(t_\varepsilon v_\varepsilon), (t_\varepsilon v_\varepsilon)\rangle$$

$$= \int_{\mathbb{R}^N} \widehat{G}(\varepsilon x, t_\varepsilon v_\varepsilon)\,dx$$

$$\leq D^{k_v} \int_{\mathbb{R}^N} \widehat{G}(\varepsilon x, v_\varepsilon)\,dx$$

$$= D^{k_v}\left(J_\varepsilon(v_\varepsilon) - \frac{1}{2}\langle J_\varepsilon'(v_\varepsilon), v_\varepsilon\rangle\right)$$

$$\leq D^{k_v} m_2 + o(1) \tag{8.3.29}$$

which contradicts (8.3.28). Then the Case 2 can not take place.

Step 4   Conclusion.

Steps 1, 2 and 3, show that $\|v_\varepsilon\|_\varepsilon$ is bounded as $\varepsilon \to 0$.   □

In the next lemma we prove that every Cerami sequence $(v_j) \subset \mathcal{H}_\varepsilon$ at level c_ε is bounded and admits a convergent subsequence in H_ε^s.

Lemma 8.3.4 *Assume that f satisfies $(f1)$–$(f3)$ and either $(f4)$ or $(f5)$. Then there exists $\varepsilon_1 \in (0, \varepsilon_0]$ such that for every $\varepsilon \in (0, \varepsilon_1]$ and every sequence $(v_j) \subset \mathcal{H}_\varepsilon$ satisfying*

$$J_\varepsilon(v_j) \to c > 0, \tag{8.3.30}$$

$$(1 + \|v_j\|_\varepsilon)\|J_\varepsilon'(v_j)\|_{\mathcal{H}_\varepsilon^*} \to 0 \text{ as } j \to \infty, \tag{8.3.31}$$

for some $c > 0$, one has that

(i) *$\|v_j\|_\varepsilon$ is bounded as $j \to \infty$;*
(ii) *there exist (j_k) and $v_0 \in \mathcal{H}_\varepsilon$ such that $v_{j_k} \to v_0$ strongly in $\mathcal{H}_\varepsilon$.*

Proof The proof of (i) can be carried out in much the same way as the one of Lemma 8.3.3, with suitable modifications. More precisely, in Step 1 of Lemma 8.3.3, for a given sequence (v_j), there exists $(x_j) \subset \mathbb{R}^N$ such that $\int_{B_1(x_j)} |\chi_\varepsilon w_j|^2 dx \to d > 0$. The sequence (x_j) satisfies $\varepsilon x_j \in N_\varepsilon(\Lambda)$, and we may assume that $\varepsilon x_j \to x_0 \in \overline{N_\varepsilon(\Lambda)}$, where x_0 is such that $\chi(\varepsilon x + x_0) \neq 0$ in B_1.

In Step 2 we replace (8.3.21) by

$$(-\Delta)^s w_0 + V(\varepsilon x + x_0)w_0 = (\chi(\varepsilon x + x_0)a + (1 - \chi(\varepsilon x + x_0))v)w_0 \quad \text{in } \mathbb{R}^N \tag{8.3.32}$$

where $w_0 \in H^s(\mathbb{R}^N)$ is non-negative and not identically zero. Indeed, by the maximum principle, $w_0 > 0$ in $\mathbb{R}^N$. Set $\tilde{w}(x) = w_0(\frac{x - x_0}{\varepsilon})$. Then $\tilde{w}$ satisfies

$$\varepsilon^{2s}(-\Delta)^s \tilde{w} + V(x)\tilde{w} = (\chi(x)a + (1 - \chi(x))v)\tilde{w} \quad \text{in } \mathbb{R}^N. \tag{8.3.33}$$

We aim to prove that this is impossible for $\varepsilon > 0$ sufficiently small. Using the Caffarelli-Silvestre extension technique, we know that $\tilde{W} = \text{Ext}(\tilde{w})$ is a solution to the following problem

$$\begin{cases} -\varepsilon^{2s}\text{div}(y^{1-2s}\nabla \tilde{W}) = 0 & \text{in } \mathbb{R}_+^{N+1}, \\ \tilde{W}(\cdot, 0) = \tilde{w} & \text{on } \partial\mathbb{R}_+^{N+1}, \\ \frac{\partial \tilde{W}}{\partial \nu^{1-2s}} = -V(x)\tilde{w} + (\chi(x)a + (1 - \chi(x))v)\tilde{w} & \text{on } \partial\mathbb{R}_+^{N+1}. \end{cases}$$

Take $r > 0$ sufficiently small such that

$$\chi(x) = 1 \text{ and } V(x) < a \quad \text{for } x \in B_r.$$

Let us recall the following notations:

$$B_r^+ = \{(x, y) \in \mathbb{R}_+^{N+1} : y > 0, |(x, y)| < r\},$$

$$\Gamma_r^+ = \{(x, y) \in \mathbb{R}_+^{N+1} : y \geq 0, |(x, y)| = r\},$$

$$\Gamma_r^0 = \{(x, 0) \in \partial \mathbb{R}_+^{N+1} : |x| < r\},$$

and define

$$H^1_{0, \Gamma_r^+}(B_r^+) = \{V \in H^1(B_r^+, y^{1-2s}) : V \equiv 0 \text{ on } \Gamma_r^+\}.$$

Let

$$\mu_r = \inf \left\{ \iint_{B_r^+} y^{1-2s} |\nabla U|^2 \, dx dy : U \in H^1_{0, \Gamma_r^+}(B_r^+), \int_{\Gamma_r^0} u^2 \, dx = 1 \right\}.$$

By the compactness of the embedding $H^1_{0, \Gamma_r^+}(B_r^+) \Subset L^2(\Gamma_r^0)$, it is not difficult to see that the infimum is achieved by a function $U_r \in H^1_{\Gamma_r^+}(B_r^+) \setminus \{0\}$. Moreover, we may assume that $U_r \geq 0$, since $|U|$ is a minimizer whenever U is a minimizer. Then U_r is a solution, not identically zero, of

$$\begin{cases} -\text{div}(y^{1-2s} \nabla U_r) = 0 & \text{in } B_r^+, \\ \dfrac{\partial U_r}{\partial \nu^{1-2s}} = \mu_r u_r & \text{on } \Gamma_r^0, \\ U_r = 0 & \text{on } \Gamma_r^+. \end{cases} \qquad (8.3.34)$$

By the strong maximum principle, $U_r > 0$ on $B_r^+ \cup \Gamma_r^0$. Note $\mu_r \geq 0$ and μ_r is a nonincreasing function of r. Indeed, μ_r is decreasing in r. In fact, if by contradiction we assume that $r_1 < r_2$ and $\mu_{r_1} = \mu_{r_2}$, we can multiply the equation $\text{div}(y^{1-2s} \nabla U_{r_1}) = 0$ by U_{r_2}, and after an integration by parts, we can use the equalities satisfied by U_{r_1} and U_{r_2}, and the assumption $\mu_{r_1} = \mu_{r_2}$, to deduce that

$$\int_{\Gamma_{r_1}^+} \frac{\partial U_{r_1}}{\partial \nu^{1-2s}} U_{r_2} \, d\sigma = 0.$$

This gives a contradiction, because of $U_{r_2} > 0$ and $\dfrac{\partial U_{r_1}}{\partial \nu^{1-2s}} < 0$ on $\Gamma_{r_1}^+$.

Next, extend U_r by setting $U_r = 0$ in $\mathbb{R}^{N+1}_+ \setminus B_r^+$, so that $U_r \in H^1(\mathbb{R}^{N+1}_+, y^{1-2s})$. Therefore,

$$\varepsilon^{2s} \mu_r \int_{\Gamma_r^0} u_r \tilde{w} \, dx = \iint_{B_r^+} y^{1-2s} \varepsilon^{2s} \nabla \tilde{W} \cdot \nabla U_r \, dx dy$$

$$= -\int_{\Gamma_r^0} (V(x) - a) \tilde{w} u_r \, dx$$

that is

$$\int_{\Gamma_r^0} (V(x) - a + \varepsilon^{2s} \mu_r) \tilde{w} u_r \, dx = 0. \tag{8.3.35}$$

But this is impossible because of $V(x) - a + \mu_r \varepsilon^{2s} < 0$ in Γ_r^0 for $\varepsilon > 0$ small and $u_r \tilde{w} > 0$ in Γ_r^0.

In order to verify (ii), fix $\varepsilon \in (0, \varepsilon_1]$ and (v_j) satisfying (8.3.30) and (8.3.31). Using (i), we see that (v_j) is bounded in $\mathcal{H}_\varepsilon$. Up to a subsequence, we may assume that

$$v_j \rightharpoonup v_0 \text{ in } \mathcal{H}_\varepsilon.$$

To show that this convergence is actually strong, we follow Lemma 7.2.4 in which we observed that it suffices to show that

$$\lim_{R \to \infty} \limsup_{j \to \infty} \int_{|x| \geq R} \left(|(-\Delta)^{\frac{s}{2}} v_j|^2 + V(\varepsilon x) v_j^2 \right) dx = 0. \tag{8.3.36}$$

Let $\eta_R \in C^\infty(\mathbb{R}^N)$ be a cut-off function such that

$$\begin{cases} \eta_R(x) = 0, & \text{for } |x| \leq \frac{R}{2}, \\ \eta_R(x) = 1, & \text{for } |x| \geq R, \\ 0 \leq \eta_R(x) \leq 1, & \text{for } x \in \mathbb{R}^N, \\ |\nabla \eta_R(x)| \leq \frac{C}{R}, & \text{for } x \in \mathbb{R}^N. \end{cases}$$

Take $R > 0$ such that $\Lambda_\varepsilon \subset B_{\frac{R}{2}}$. Since $(v_j \eta_R)$ is bounded in $\mathcal{H}_\varepsilon$, we see that $\langle J_\varepsilon'(v_j), \eta_R v_j \rangle = o_j(1)$. Hence,

$$\int_{\mathbb{R}^N} (-\Delta)^{\frac{s}{2}} v_j (-\Delta)^{\frac{s}{2}} (v_j \eta_R) \, dx + \int_{\mathbb{R}^N} V(\varepsilon x) v_j^2 \eta_R \, dx$$

$$= \int_{\mathbb{R}^N} f(v_j) v_j \eta_R \, dx + o_j(1)$$

$$\leq \nu \int_{\mathbb{R}^N} v_j^2 \eta_R \, dx + o_j(1).$$

By our choice of v, we can find $\alpha \in (0, 1)$ such that

$$\int_{\mathbb{R}^N} (-\Delta)^{\frac{s}{2}} v_j (-\Delta)^{\frac{s}{2}} (v_j \eta_R) \, dx + \alpha \int_{\mathbb{R}^N} V(\varepsilon x) v_j^2 \eta_R \, dx \le o_j(1). \qquad (8.3.37)$$

Now we observe that

$$\int_{\mathbb{R}^N} (-\Delta)^{\frac{s}{2}} v_j (-\Delta)^{\frac{s}{2}} (v_j \eta_R) \, dx$$

$$= \iint_{\mathbb{R}^{2N}} \frac{(v_j(x) - v_j(y))(v_j(x)\eta_R(x) - v_j(y)\eta_R(y))}{|x - y|^{N+2s}} \, dx dy$$

$$= \iint_{\mathbb{R}^{2N}} \eta_R(x) \frac{|v_j(x) - v_j(y)|^2}{|x - y|^{N+2s}} \, dx dy$$

$$+ \iint_{\mathbb{R}^{2N}} \frac{(v_j(x) - v_j(y))(\eta_R(x) - \eta_R(y))}{|x - y|^{N+2s}} v_j(y) \, dx dy$$

$$= A_{R,j} + B_{R,j}. \qquad (8.3.38)$$

Clearly,

$$A_{R,j} \ge \int_{|x| \ge R} \int_{\mathbb{R}^N} \frac{|v_j(x) - v_j(y)|^2}{|x - y|^{N+2s}} \, dx dy. \qquad (8.3.39)$$

Using Hölder's inequality, Lemma 1.4.5 and the fact that (v_j) is bounded in $H^s(\mathbb{R}^N)$, we see that

$$\limsup_{R \to \infty} \limsup_{j \to \infty} |B_{R,j}|$$

$$\le \limsup_{R \to \infty} \limsup_{j \to \infty} \left(\iint_{\mathbb{R}^{2N}} \frac{|v_j(x) - v_j(y)|^2}{|x - y|^{N+2s}} \, dx dy \right)^{\frac{1}{2}} \left(\iint_{\mathbb{R}^{2N}} \frac{|\eta_R(x) - \eta_R(y)|^2}{|x - y|^{N+2s}} |v_j(y)|^2 \, dx dy \right)^{\frac{1}{2}}$$

$$\le C \limsup_{R \to \infty} \limsup_{j \to \infty} \left(\iint_{\mathbb{R}^{2N}} \frac{|\eta_R(x) - \eta_R(y)|^2}{|x - y|^{N+2s}} |v_j(y)|^2 \, dx dy \right)^{\frac{1}{2}} = 0. \qquad (8.3.40)$$

Combining (8.3.37)–(8.3.40) and using the fact that

$$\int_{|x| \ge R} V(\varepsilon x) v_j^2 \, dx \le \int_{\mathbb{R}^N} V(\varepsilon x) v_j^2 \eta_R \, dx,$$

we infer that (8.3.36) holds.   □

Taking into account Lemma 8.3.3 and Lemma 8.3.4 we deduce the following result:

Corollary 8.3.5 *There exists $\varepsilon_1 \in (0, \varepsilon_0]$ such that for every $\varepsilon \in (0, \varepsilon_1]$ there exists a critical point $v_\varepsilon \in \mathcal{H}_\varepsilon$ of $J_\varepsilon(v)$ satisfying $J_\varepsilon(v_\varepsilon) = c_\varepsilon$, where $c_\varepsilon \in [m_1, m_2]$ is defined as in (8.3.4)–(8.3.5). Moreover there exists a constant $M > 0$ independent of $\varepsilon \in (0, \varepsilon_1]$ such that $\|v_\varepsilon\|_\varepsilon \leq M$ for any $\varepsilon \in (0, \varepsilon_1]$.*

8.4   Limit Equation

In the next section we will see that the sequence of critical points obtained in Corollary 8.3.5 converges, in some sense, to a sum of translated critical points of certain autonomous functionals. As proved in [39], least energy solutions for autonomous nonlinear scalar field equations admit a mountain pass characterization. This property will play a fundamental role in the proof of Theorem 8.1.1. For this reason, in this section we collect some important results on autonomous functionals associated with "limit equations".

Firstly, we introduce some notations and definitions that will be useful later. For $x_0 \in \mathbb{R}^N$ we define the autonomous functional $\Phi_{x_0} : H^s(\mathbb{R}^N) \to \mathbb{R}$ by setting

$$\Phi_{x_0}(v) = \frac{1}{2} \int_{\mathbb{R}^N} \left(|(-\Delta)^{\frac{s}{2}} v|^2 + V(x_0)v^2 \right) dx - \int_{\mathbb{R}^N} G(x_0, v)\, dx.$$

It is routine to check that $\Phi_{x_0} \in C^1(H^s(\mathbb{R}^N), \mathbb{R})$ and critical points of Φ_{x_0} are weak solutions to the equation

$$(-\Delta)^s u + V(x_0)u = g(x_0, u) \text{ in } \mathbb{R}^N. \tag{8.4.1}$$

We note that, if u is a solution to (8.2.4), then $v(x) = u(\varepsilon x + x_0)$ satisfies

$$(-\Delta)^s v + V(\varepsilon x + x_0)v = g(\varepsilon x + x_0, v) \quad \text{in } \mathbb{R}^N, \tag{8.4.2}$$

that is, (8.4.1) can be seen as the limit equation of (8.4.2) as $\varepsilon \to 0$.

For any $x_0 \in \mathbb{R}^N$ and $u, v \in H^s(\mathbb{R}^N)$ we denote

$$\langle u, v \rangle_{\mathcal{H}_\varepsilon} = \int_{\mathbb{R}^N} (-\Delta)^{\frac{s}{2}} u (-\Delta)^{\frac{s}{2}} v + V(\varepsilon x)uv\, dx,$$

$$\langle u, v \rangle_{x_0} = \int_{\mathbb{R}^N} (-\Delta)^{\frac{s}{2}} u (-\Delta)^{\frac{s}{2}} v + V(x_0)uv\, dx,$$

$$|v|_{x_0}^2 = \int_{\mathbb{R}^N} |(-\Delta)^{\frac{s}{2}} v|^2 + V(x_0)v^2\, dx.$$

Finally, we define

$$H(x, t) = -\frac{1}{2}V(x)t^2 + G(x, t)$$

and

$$\Omega = \left\{ x \in \mathbb{R}^N : \sup_{t>0} H(x, t) > 0 \right\}.$$

Remark 8.4.1

(i) $\Omega \subset \Lambda$ and $0 \in \{x \in \Lambda' : V(x) = \inf_{y \in \Lambda} V(y)\} \subset \Omega$.
(ii) If $(f3)$ or $(f5)$ with $a = \infty$ holds, then $\Omega = \Lambda$.

Now, we state the following Jeanjean-Tanaka type result [234] proved in Chap. 3 (see Theorem 3.1.1) in connection with the study of nonlinear scalar field equation with fractional diffusion

$$(-\Delta)^s u = h(u) \text{ in } \mathbb{R}^N, \quad u \in H^s(\mathbb{R}^N), \tag{8.4.3}$$

where $h \in C^1(\mathbb{R}, \mathbb{R})$ is an odd function satisfying the following Berestycki–Lions type assumptions [100]:

$(h1)$ $-\infty < \liminf_{t \to 0} h(t)/t \le \limsup_{t \to 0} h(t)/t < 0$;
$(h2)$ $\lim_{|t| \to \infty} \frac{h(t)}{|t|^{2_s^*-1}} = 0$;
$(h3)$ there exists $\bar{t} > 0$ such that $H(\bar{t}) > 0$.

We recall that the existence of a solution to (8.4.3) has been established in [39, 139].

Lemma 8.4.2 *[39] Assume that* $h \in C^1(\mathbb{R}, \mathbb{R})$ *is an odd function satisfying the Berestycki–Lions type assumptions* $(h1)$–$(h3)$*. Let* $\tilde{I} : H^s(\mathbb{R}^N) \to \mathbb{R}$ *be the functional defined by*

$$\tilde{I}(u) = \int_{\mathbb{R}^N} \left(\frac{1}{2}|(-\Delta)^{\frac{s}{2}}u|^2 - H(u) \right) dx.$$

Then $\tilde{I}$ *has a mountain pass geometry and* $c = m$*, where* m *is defined as*

$$m = \inf\{\tilde{I}(u) : u \in H^s(\mathbb{R}^N) \setminus \{0\} \text{ is a solution to } (8.4.3)\}, \tag{8.4.4}$$

and

$$c = \inf_{\gamma \in \Gamma} \max_{t \in [0,1]} \tilde{I}(\gamma(t)),$$

where $\Gamma = \{\gamma \in C([0,1], H^s(\mathbb{R}^N)) : \gamma(0) = 0, \tilde{I}(\gamma(1)) < 0\}$.

Moreover, for any least energy solution $\omega(x)$ *of (8.4.3) there exists a path* $\gamma \in \Gamma$ *such that*

$$\tilde{I}(\gamma(t)) \leq m = \tilde{I}(\omega) \quad \text{for all } t \in [0, 1], \tag{8.4.5}$$

$$\omega \in \gamma([0, 1]). \tag{8.4.6}$$

At this point, we give the proof of the following lemma which we will use in the next section to obtain a concentration-compactness type result.

Lemma 8.4.3 *Assume that* f *satisfies* $(f1)$–$(f3)$. *Then we have*

(i) $\Phi_{x_0}(v)$ *has non-zero critical points if and only if* $x_0 \in \Omega$.
(ii) *There exists* $\delta_1 > 0$, *independent of* $x_0 \in \mathbb{R}^N$, *such that* $\left| v \right|_{x_0} \geq \delta_1$ *for any non-zero critical point* v *of* Φ_{x_0}.

Proof Firstly, we extend $f(t)$ to an odd function on $\mathbb{R}$. Let us consider the function

$$h(t) = -V(x_0)t + g(x_0, t),$$

that is $h(t) = H'(x_0, t)$. Clearly h is odd. Now we show that h satisfies assumptions $(h1)$–$(h3)$. By $(f2)$ and $(f3)$, it follows that $(h1)$ and $(h2)$ hold.

Since $\Omega = \{x \in \mathbb{R}^N : \sup_{t>0} H(x,t) > 0\}$, we see that $(h3)$ holds if and only if $x_0 \in \Omega$. Then (i) follows by Theorem 1 in [35] (see also Theorem 1.1 in [139]).

Now let v be a non-zero critical point of Φ_{x_0}. Then

$$\langle \Phi'_{x_0}(v), v \rangle = 0 \Longrightarrow \int_{\mathbb{R}^N} \left(|(-\Delta)^{\frac{s}{2}} v|^2 + V(x_0) v^2 \right) dx - \int_{\mathbb{R}^N} g(x_0, v) v \, dx = 0.$$

Using item (i) of Corollary 8.2.3, we get

$$\|v\|_0^2 - \int_{\mathbb{R}^N} f(v) v \, dx \leq 0,$$

so by (8.2.1) it follows that for any $\delta \in (0, V_0)$,

$$\|v\|_0^2 \le \delta \|v\|_{L^2(\mathbb{R}^N)}^2 + C_\delta \|v\|_{L^{p+1}(\mathbb{R}^N)}^{p+1}$$

$$\le \frac{\delta}{V_0} \|v\|_0^2 + C_\delta C'_{p+1} \|v\|_0^{p+1}.$$

Then

$$\left(1 - \frac{\delta}{V_0}\right) \|v\|_0^2 \le C_\delta C'_{p+1} \|v\|_0^{p+1},$$

and we can find $\delta_1 > 0$ such that $\|v\|_0 \ge \delta_1$ for any $x_0 \in \mathbb{R}^N$ and for any non-zero critical point v. Since $\left| v \right|_{x_0} \ge \|v\|_0$, we conclude that (ii) is verified. □

For $x \in \mathbb{R}^N$, set

$$m(x) = \begin{cases} \text{least energy level of } \Phi_x(v), & \text{if } x \in \Omega, \\ \infty, & \text{if } x \in \mathbb{R}^N \setminus \Omega. \end{cases}$$

By Lemma 8.4.2, we can see that $m(x)$ is equal to the mountain pass value for $\Phi_x(v)$ if $x \in \Omega$, that is

$$m(x) = \inf_{\gamma \in \Gamma} \left(\max_{t \in [0,1]} \Phi_x(\gamma(t)) \right),$$

where $\Gamma = \{ \gamma \in C([0, 1], H^s(\mathbb{R}^N)) : \gamma(0) = 0 \text{ and } \Phi_x(\gamma(1)) < 0 \}$.

Lemma 8.4.4

$$m(x_0) = \inf_{x \in \mathbb{R}^N} m(x) \text{ if and only if } x_0 \in \Lambda \text{ and } V(x_0) = \inf_{x \in \Lambda} V(x).$$

In particular, $m(0) = \inf_{x \in \mathbb{R}^N} m(x)$.

Proof Fix $x_0 \in \Lambda$ such that $V(x_0) = \inf_{x \in \Lambda} V(x)$. We note that $x_0 \in \Lambda'$. Otherwise, if $x_0 \in \Lambda \setminus \Lambda'$, then

$$V(x_0) \ge \inf_{x \in \Lambda \setminus \Lambda'} V(x) > \inf_{x \in \Lambda} V(x),$$

which is impossible. Hence, $x_0 \in \Lambda'$ and $\chi(x_0) = 1$. Moreover, $x_0 \in \Omega$ by Remark 8.4.1. Now, using the fact that $V(x) \ge V(x_0)$ in Λ and $G(x, t) \le F(t)$ for any $(x, t) \in \mathbb{R}^N \times \mathbb{R}$, we see that for any $\bar{x} \in \Omega$

$$\Phi_{\bar{x}}(v) = \frac{1}{2}\|(-\Delta)^{\frac{s}{2}}v\|^2_{L^2(\mathbb{R}^N)} + \frac{1}{2}V(\bar{x})\|v\|^2_{L^2(\mathbb{R}^N)} - \int_{\mathbb{R}^N} G(\bar{x}, v)\, dx$$

$$\geq \frac{1}{2}\|(-\Delta)^{\frac{s}{2}}v\|^2_{L^2(\mathbb{R}^N)} + \frac{1}{2}V(x_0)\|v\|^2_{L^2(\mathbb{R}^N)} - \int_{\mathbb{R}^N} F(v)\, dx$$

$$= \Phi_{x_0}(v) \text{ for any } v \in H^s(\mathbb{R}^N).$$

This implies that $m(x_0) \leq m(x)$ for all $x \in \mathbb{R}^N$, so $m(x_0) \leq \inf_{x \in \mathbb{R}^N} m(x) \leq m(x_0)$ that is $m(x_0) = \inf_{x \in \mathbb{R}^N} m(x)$.

Now fix $x' \in \Lambda$ such that $V(x') > V(x_0)$. Take $\gamma \in \Gamma$ such that (8.4.5) and (8.4.6) hold with $\tilde{I}(v) = \Phi_{x'}(v)$. Then we deduce that

$$m(x_0) \leq \max_{t \in [0,1]} \Phi_{x_0}(\gamma(t)) < \max_{t \in [0,1]} \Phi_{x'}(\gamma(t)) = m(x').$$

$\square$

Finally, we note that the function $m(x)$ is continuous (see [47]).

Proposition 8.4.5 *The function $m(x) : \mathbb{R}^N \mapsto (-\infty, \infty]$ is continuous in the following sense:*

$$m(x_j) \to m(x_0) \quad \text{when } x_j \to x_0 \in \Omega,$$

$$m(x_j) \to \infty \quad \text{when } x_j \to x_0 \in \mathbb{R}^N \setminus \Omega.$$

8.5 ε-Dependent Concentration-Compactness Result

This section is devoted to the study of the behavior as $\varepsilon \to 0$ of critical points (v_ε) obtained in Corollary 8.3.5. More generally, we consider (v_ε) such that

$$v_\varepsilon \in \mathcal{H}_\varepsilon, \tag{8.5.1}$$

$$J_\varepsilon(v_\varepsilon) \to c \in \mathbb{R}, \tag{8.5.2}$$

$$(1 + \|v_\varepsilon\|_\varepsilon)\|J'_\varepsilon(v_\varepsilon)\|_{\mathcal{H}^*_\varepsilon} \to 0, \tag{8.5.3}$$

$$\|v_\varepsilon\|_\varepsilon \leq m, \tag{8.5.4}$$

where c and m are independent of ε.

We begin by proving the following concentration-compactness type result.

Lemma 8.5.1 *Assume that f satisfies $(f1)$–$(f3)$ and $(v_\varepsilon)_{\varepsilon \in (0,\varepsilon_1]}$ satisfies the conditions (8.5.1)–(8.5.4). Then there exist a subsequence $\varepsilon_j \to 0$, $l \in \mathbb{N} \cup \{0\}$, sequences $(y_{\varepsilon_j}^k) \subset \mathbb{R}^N$, $x^k \in \Omega$, $\omega^k \in H^s(\mathbb{R}^N) \setminus \{0\}$ $(k = 1, \ldots, l)$ such that*

$$|y_{\varepsilon_j}^k - y_{\varepsilon_j}^{k'}| \to \infty \quad \text{as } j \to \infty, \text{ for } k \neq k', \tag{8.5.5}$$

$$\varepsilon_j y_{\varepsilon_j}^k \to x^k \in \Omega \quad \text{as } j \to \infty, \tag{8.5.6}$$

$$\omega^k \not\equiv 0 \text{ and } \Phi'_{x^k}(\omega^k) = 0, \tag{8.5.7}$$

$$\left\| v_{\varepsilon_j} - \psi_{\varepsilon_j} \left(\sum_{k=1}^{l} \omega^k(\cdot - y_{\varepsilon_j}^k) \right) \right\|_{\varepsilon_j} \to 0 \quad \text{as } j \to \infty, \tag{8.5.8}$$

$$J_{\varepsilon_j}(v_{\varepsilon_j}) \to \sum_{k=1}^{l} \Phi_{x^k}(\omega^k). \tag{8.5.9}$$

Here $\psi_\varepsilon(x) = \psi(\varepsilon x)$, and $\psi \in C_c^\infty(\mathbb{R}^N)$ is such that $\psi(x) = 1$ for $x \in \Lambda$ and $0 \leq \psi \leq 1$ on $\mathbb{R}^N$. When $l = 0$, we have $\|v_{\varepsilon_j}\|_{\varepsilon_j} \to 0$ and $J_{\varepsilon_j}(v_{\varepsilon_j}) \to 0$.

Remark 8.5.2 Let us note that $\sup \psi(\varepsilon x) V(\varepsilon x) < \infty$. Moreover, for all $w \in H^s(\mathbb{R}^N)$, $\psi_\varepsilon w \in H^s_\varepsilon$ and there exists a constant $C > 0$, independent of ε, such that

$$\|\psi_\varepsilon w\|_\varepsilon \leq C\|w\|_0. \tag{8.5.10}$$

Remark 8.5.3 For any $w \in H^s(\mathbb{R}^N)$ and for any sequence $(y_\varepsilon) \subset \mathbb{R}^N$ such that $\varepsilon y_\varepsilon \to x_0 \in \Lambda$, we have

$$\|\psi_\varepsilon w(\cdot - y_\varepsilon)\|_\varepsilon^2$$

$$= \int_{\mathbb{R}^N} |(-\Delta)^{\frac{s}{2}} (\psi(\varepsilon x + \varepsilon y_\varepsilon) w(x))|^2 + V(\varepsilon x + \varepsilon y_\varepsilon)(\psi(\varepsilon x + \varepsilon y_\varepsilon) w(x))^2 \, dx$$

$$\to \int_{\mathbb{R}^N} |(-\Delta)^{\frac{s}{2}} w|^2 + V(x_0) w^2 \, dx = \left| w \right|_{x_0}^2 \quad \text{as } \varepsilon \to 0. \tag{8.5.11}$$

We first prove that

$$\int_{\mathbb{R}^N} |(-\Delta)^{\frac{s}{2}} (\psi(\varepsilon x + \varepsilon y_\varepsilon) w(x))|^2 \, dx \to \int_{\mathbb{R}^N} |(-\Delta)^{\frac{s}{2}} w(x)|^2 \, dx. \tag{8.5.12}$$

Then,

$$\iint_{\mathbb{R}^{2N}} \frac{|\psi(\varepsilon x + \varepsilon y_\varepsilon)w(x) - \psi(\varepsilon y + \varepsilon y_\varepsilon)w(y)|^2}{|x-y|^{N+2s}} \, dx \, dy$$

$$= \iint_{\mathbb{R}^{2N}} \frac{|\psi(\varepsilon x + \varepsilon y_\varepsilon) - \psi(\varepsilon y + \varepsilon y_\varepsilon)|^2}{|x-y|^{N+2s}} |w(x)|^2 \, dx \, dy$$

$$+ \iint_{\mathbb{R}^{2N}} \frac{|w(x) - w(y)|^2}{|x-y|^{N+2s}} (\psi(\varepsilon y + \varepsilon y_\varepsilon))^2 \, dx \, dy$$

$$+ 2 \iint_{\mathbb{R}^{2N}} \frac{(\psi(\varepsilon x + \varepsilon y_\varepsilon) - \psi(\varepsilon y + \varepsilon y_\varepsilon))(w(x) - w(y))}{|x-y|^{N+2s}} w(x)\psi(\varepsilon y + \varepsilon y_\varepsilon) \, dx \, dy$$

$$= A_\varepsilon + B_\varepsilon + 2C_\varepsilon.$$

Now, by the dominated convergence theorem and the fact that $\psi(\varepsilon \cdot + \varepsilon y_\varepsilon) \to 1$, we see that $B_\varepsilon \to [w]_s^2$. On the other hand

$$A_\varepsilon = \int_{\mathbb{R}^N} dx \int_{|x-y| \le \frac{1}{\varepsilon}} \frac{|\psi(\varepsilon x + \varepsilon y_\varepsilon) - \psi(\varepsilon y + \varepsilon y_\varepsilon)|^2}{|x-y|^{N+2s}} |w(x)|^2 \, dy$$

$$+ \int_{\mathbb{R}^N} dx \int_{|x-y| > \frac{1}{\varepsilon}} \frac{|\psi(\varepsilon x + \varepsilon y_\varepsilon) - \psi(\varepsilon y + \varepsilon y_\varepsilon)|^2}{|x-y|^{N+2s}} |w(x)|^2 \, dy$$

$$\le \varepsilon^2 \|\nabla \psi\|_{L^\infty(\mathbb{R}^N)}^2 \omega_{N-1} \int_{\mathbb{R}^N} |w(x)|^2 \, dx \int_0^{\frac{1}{\varepsilon}} \frac{1}{z^{2s-1}} \, dz$$

$$+ 4\omega_{N-1} \int_{\mathbb{R}^N} |w(x)|^2 \, dx \int_{\frac{1}{\varepsilon}}^\infty \frac{1}{z^{2s+1}} \, dz$$

$$= \varepsilon^{2s} \omega_{N-1} \left(\frac{\|\nabla \psi\|_{L^\infty(\mathbb{R}^N)}^2}{2 - 2s} + \frac{2}{s} \right) \int_{\mathbb{R}^N} |w(x)|^2 \, dx \to 0 \quad \text{as } \varepsilon \to 0, \qquad (8.5.13)$$

and since

$$|C_\varepsilon| \le [w]_s \sqrt{A_\varepsilon} \to 0,$$

we infer that (8.5.12) holds. Since

$$\int_{\mathbb{R}^N} V(\varepsilon x + \varepsilon y_\varepsilon)|\psi(\varepsilon x + \varepsilon y_\varepsilon)w(x)|^2 \, dx \to \int_{\mathbb{R}^N} V(x_0)|w(x)|^2 \, dx, \qquad (8.5.14)$$

it follows from (8.5.12) and (8.5.14) that (8.5.11).

Proof We divide the proof into several steps. In what follows, we write ε instead of ε_j.

Step 1 Up to a subsequence, $v_\varepsilon \rightharpoonup v_0$ in $H^s(\mathbb{R}^N)$ and v_0 is a critical point of $\Phi_0(v)$.

Using (8.5.4) and (8.3.2), we see that $\|v_\varepsilon\|_0 \le m$. Thus (v_ε) is bounded in $H^s(\mathbb{R}^N)$ and we can assume that $v_\varepsilon \rightharpoonup v_0$ in $H^s(\mathbb{R}^N)$.

Let us show that v_0 is a critical point of $\Phi_0(v)$, that is, $\langle \Phi_0'(v_0), \varphi \rangle = 0$ for all $\varphi \in H^s(\mathbb{R}^N)$. Since $C_c^\infty(\mathbb{R}^N)$ is dense in $H^s(\mathbb{R}^N)$, it is enough to prove it for all $\varphi \in C_c^\infty(\mathbb{R}^N)$. Fix $\varphi \in C_c^\infty(\mathbb{R}^N)$. It follows from (8.5.3) that

$$\int_{\mathbb{R}^N} \left[(-\Delta)^{\frac{s}{2}} v_\varepsilon (-\Delta)^{\frac{s}{2}} \varphi + V(\varepsilon x) v_\varepsilon \varphi - g(\varepsilon x, v_\varepsilon)\varphi \right] dx \to 0.$$

Now we show that

$$\langle J_\varepsilon'(v_\varepsilon), \varphi \rangle = \langle v_\varepsilon, \varphi \rangle_{\mathcal{H}_\varepsilon} - \int_{\mathbb{R}^N} g(\varepsilon x, v_\varepsilon)\varphi \, dx \to \langle v_0, \varphi \rangle_0 - \int_{\mathbb{R}^N} g(0, v_0)\varphi \, dx.$$

Note that

$$\langle v_\varepsilon, \varphi \rangle_{\mathcal{H}_\varepsilon} - \langle v_0, \varphi \rangle_0$$

$$= \int_{\mathbb{R}^N} (-\Delta)^{\frac{s}{2}} (v_\varepsilon - v_0)(-\Delta)^{\frac{s}{2}} \varphi \, dx + \int_{\mathbb{R}^N} [V(\varepsilon x) - V(0)] v_\varepsilon \varphi \, dx$$

$$+ V(0) \int_{\mathbb{R}^N} (v_\varepsilon - v_0)\varphi \, dx$$

$$= A_\varepsilon^1 + A_\varepsilon^2 + A_\varepsilon^3.$$

Then $A_\varepsilon^1, A_\varepsilon^3 \to 0$ because $v_\varepsilon \rightharpoonup v_0$ in $H^s(\mathbb{R}^N)$, and

$$|A_\varepsilon^2| \le C \|V(\varepsilon \cdot) - V(0)\|_{L^\infty(\text{supp}(\varphi))} \|v_\varepsilon\|_{H^s} \|\varphi\|_{L^2(\mathbb{R}^N)}$$

$$\le C' \|V(\varepsilon \cdot) - V(0)\|_{L^\infty(\text{supp}(\varphi))} \to 0.$$

On the other hand, using item (iii) of Corollary 8.2.3 and the fact that $H^s(\mathbb{R}^N) \Subset L_{\text{loc}}^q(\mathbb{R}^N)$ for any $q \in [1, 2_s^*)$, we have

$$\int_{\mathbb{R}^N} g(\varepsilon x, v_\varepsilon)\varphi \, dx \to \int_{\mathbb{R}^N} g(0, v_0)\varphi \, dx.$$

Hence

$$\langle \Phi_0'(v_0), \varphi \rangle = \int_{\mathbb{R}^N} \left[(-\Delta)^{\frac{s}{2}} v_0 (-\Delta)^{\frac{s}{2}} \varphi + V(0) v_0 \varphi - g(0, v_0)\varphi \right] dx = 0.$$

If $v_0 \ne 0$, we set $y_\varepsilon^1 = 0$ and $\omega^1 = v_0$.

Step 2 Suppose that there exist $n \in \mathbb{N} \cup \{0\}$, $(y_\varepsilon^k) \subset \mathbb{R}^N$, $x^k \in \Omega$, $\omega^k \in H^s(\mathbb{R}^N)$ $(k = 1, \ldots, n)$ such that (8.5.5), (8.5.6), (8.5.7) of Lemma 8.5.1 hold for $k = 1, \ldots, n$ and

$$v_\varepsilon(\cdot + y_\varepsilon^k) \rightharpoonup \omega^k \quad \text{in } H^s(\mathbb{R}^N) \text{ for } k = 1, \ldots, n. \tag{8.5.15}$$

Moreover, we assume that

$$\sup_{y \in \mathbb{R}^N} \int_{B_1(y)} \left| v_\varepsilon - \psi_\varepsilon \sum_{k=1}^n \omega^k(x - y_\varepsilon^k) \right|^2 dx \to 0. \tag{8.5.16}$$

Then

$$\left\| v_\varepsilon - \psi_\varepsilon \sum_{k=1}^n \omega^k(\cdot - y_\varepsilon^k) \right\|_\varepsilon^2 \to 0. \tag{8.5.17}$$

Set

$$\xi_\varepsilon(x) = v_\varepsilon(x) - \psi_\varepsilon(x) \sum_{k=1}^n \omega^k(x - y_\varepsilon^k).$$

Inequality (8.5.10) implies that

$$\|\xi_\varepsilon\|_\varepsilon \leq \|v_\varepsilon\|_\varepsilon + \left\| \psi_\varepsilon \sum_{k=1}^n \omega^k(\cdot - y_\varepsilon^k) \right\|_\varepsilon$$

$$\leq m + C \sum_{k=1}^n \|\omega^k\|_0,$$

and using the fact that $\|\xi_\varepsilon\|_0 \leq \|\xi_\varepsilon\|_\varepsilon$, we deduce that (ξ_ε) is bounded in $H^s(\mathbb{R}^N)$.

By (8.5.16) and Lemma 1.4.4, $\|\xi_\varepsilon\|_{L^{p+1}(\mathbb{R}^N)} \to 0$ as $\varepsilon \to 0$. A direct calculation shows that

$$\|\xi_\varepsilon\|_\varepsilon^2 = \left\langle v_\varepsilon - \psi_\varepsilon \sum_{k=1}^n \omega^k(\cdot - y_\varepsilon^k), \xi_\varepsilon \right\rangle_{\mathcal{H}_\varepsilon}$$

$$= \langle v_\varepsilon, \xi_\varepsilon \rangle_{\mathcal{H}_\varepsilon} - \sum_{k=1}^n \langle \psi_\varepsilon \omega^k(\cdot - y_\varepsilon^k), \xi_\varepsilon \rangle_{\mathcal{H}_\varepsilon}. \tag{8.5.18}$$

We claim that

$$\langle \psi_\varepsilon \omega^k(\cdot - y_\varepsilon^k), \xi_\varepsilon \rangle_{\mathcal{H}_\varepsilon} = \langle \omega^k(\cdot - y_\varepsilon^k), \psi_\varepsilon \xi_\varepsilon \rangle_{x^k} + o(1) \tag{8.5.19}$$

for all $k = 1, \ldots, n$. Indeed

$$\langle \psi_\varepsilon \omega^k(\cdot - y_\varepsilon^k), \xi_\varepsilon \rangle_{H_\varepsilon^s} - \langle \omega^k(\cdot - y_\varepsilon^k), \psi_\varepsilon \xi_\varepsilon \rangle_{x^k}$$

$$= \left[\iint_{\mathbb{R}^{2N}} \frac{(\psi_\varepsilon(x) - \psi_\varepsilon(y))(\xi_\varepsilon(x) - \xi_\varepsilon(y))\omega^k(x - y_\varepsilon^k)}{|x - y|^{N+2s}} \, dx dy \right.$$

$$\left. - \iint_{\mathbb{R}^{2N}} \frac{(\psi_\varepsilon(x) - \psi_\varepsilon(y))(\omega^k(x - y_\varepsilon^k) - \omega^k(y - y_\varepsilon^k))\xi_\varepsilon(x)}{|x - y|^{N+2s}} \, dx dy \right]$$

$$+ \int_{\mathbb{R}^N} (V(\varepsilon x + \varepsilon y_\varepsilon^k) - V(x^k))\psi(\varepsilon x + \varepsilon y_\varepsilon^k)\omega^k(x)\xi_\varepsilon(x + y_\varepsilon^k) \, dx$$

$$= B_\varepsilon^1 + B_\varepsilon^2.$$

Note that

$$\left| \iint_{\mathbb{R}^{2N}} \frac{(\psi_\varepsilon(x) - \psi_\varepsilon(y))(\xi_\varepsilon(x) - \xi_\varepsilon(y))\omega^k(x - y_\varepsilon^k)}{|x - y|^{N+2s}} \, dx dy \right|$$

$$\leq \left(\iint_{\mathbb{R}^{2N}} \frac{|\xi_\varepsilon(x) - \xi_\varepsilon(y)|^2}{|x - y|^{N+2s}} \, dx dy \right)^{\frac{1}{2}} \left(\iint_{\mathbb{R}^{2N}} \frac{|\psi_\varepsilon(x) - \psi_\varepsilon(y)|^2(\omega^k(x - y_\varepsilon^k))^2}{|x - y|^{N+2s}} \, dx dy \right)^{\frac{1}{2}},$$

and

$$\left| \iint_{\mathbb{R}^{2N}} \frac{(\psi_\varepsilon(x) - \psi_\varepsilon(y))(\omega^k(x - y_\varepsilon^k) - \omega^k(y - y_\varepsilon^k))\xi_\varepsilon(x)}{|x - y|^{N+2s}} \, dx dy \right|$$

$$\leq \left(\iint_{\mathbb{R}^{2N}} \frac{|\omega^k(x - y_\varepsilon^k) - \omega^k(y - y_\varepsilon^k)|^2}{|x - y|^{N+2s}} \, dx dy \right)^{\frac{1}{2}} \left(\iint_{\mathbb{R}^{2N}} |\xi_\varepsilon(x)|^2 \frac{|\psi_\varepsilon(x) - \psi_\varepsilon(y)|^2}{|x - y|^{N+2s}} \, dx dy \right)^{\frac{1}{2}}.$$

Hence, since $\|\xi_\varepsilon\|_0 \leq \bar{C}_1$ and $\|\omega^k\|_0 \leq \bar{C}_2$ for some $\bar{C}_1, \bar{C}_2 > 0$, we can argue as in the proof of (8.5.13) to see that $B_\varepsilon^1 \to 0$ as $\varepsilon \to 0$. We note that the quantity $(V(\varepsilon x + \varepsilon y_\varepsilon^k) - V(x^k))\psi(\varepsilon x + \varepsilon y_\varepsilon^k)$ is bounded in $L^\infty(\mathbb{R}^N)$. By (8.5.5) and (8.5.15) we deduce that

$$\xi_\varepsilon(\cdot + y_\varepsilon^k) \rightharpoonup 0 \quad \text{in } H^s(\mathbb{R}^N)$$

$$\xi_\varepsilon(\cdot + y_\varepsilon^k) \to 0 \quad \text{in } L^2_{loc}(\mathbb{R}^N). \tag{8.5.20}$$

Then $B_\varepsilon^2 \to 0$ and we conclude that (8.5.19) holds.

Putting together (8.5.18) and (8.5.19) we see that

$$\|\xi_\varepsilon\|_\varepsilon^2 = \langle v_\varepsilon, \xi_\varepsilon \rangle_{\mathcal{H}_\varepsilon} - \sum_{k=1}^{n} \langle \omega^k(\cdot - y_\varepsilon^k), \psi_\varepsilon \xi_\varepsilon \rangle_{x^k} + o(1)$$

$$= \langle J_\varepsilon'(v_\varepsilon), \xi_\varepsilon \rangle + \int_{\mathbb{R}^N} g(\varepsilon x, v_\varepsilon) \xi_\varepsilon \, dx - \sum_{k=1}^{n} \left(\langle \Phi_{x^k}'(\omega^k(\cdot - y_\varepsilon^k)), \psi_\varepsilon \xi_\varepsilon \rangle \right.$$

$$\left. + \int_{\mathbb{R}^N} g(x^k, \omega^k(x - y_\varepsilon^k)) \psi_\varepsilon \xi_\varepsilon \, dx \right) + o(1)$$

$$= \int_{\mathbb{R}^N} g(\varepsilon x, v_\varepsilon) \xi_\varepsilon \, dx - \sum_{k=1}^{n} \int_{\mathbb{R}^N} g(x^k, \omega^k(x - y_\varepsilon^k)) \psi_\varepsilon \xi_\varepsilon \, dx + o(1)$$

$$= C_\varepsilon^1 - \sum_{k=1}^{n} C_\varepsilon^2 + o(1).$$

By Corollary 8.2.3-(iii),

$$|C_\varepsilon^1| \leq \delta \int_{\mathbb{R}^N} |v_\varepsilon \xi_\varepsilon| \, dx + C_\delta \int_{\mathbb{R}^N} |v_\varepsilon|^p |\xi_\varepsilon| \, dx$$

$$\leq \delta \|v_\varepsilon\|_{L^2(\mathbb{R}^N)} \|\xi_\varepsilon\|_{L^2(\mathbb{R}^N)} + C_\delta \|v_\varepsilon\|_{L^{p+1}(\mathbb{R}^N)}^p \|\xi_\varepsilon\|_{L^{p+1}(\mathbb{R}^N)},$$

and using that $\|\xi_\varepsilon\|_{L^{p+1}(\mathbb{R}^N)} \to 0$ as $\varepsilon \to 0$, the boundedness of $\|v_\varepsilon\|_{L^2(\mathbb{R}^N)}$ and $\|\xi_\varepsilon\|_{L^2(\mathbb{R}^N)}$, and the arbitrariness of δ, we get $C_\varepsilon^1 \to 0$. In view of (8.5.20), we see that $C_\varepsilon^2 \to 0$. Hence, $\|\xi_\varepsilon\|_\varepsilon \to 0$ and (8.5.17) holds.

Step 3 Suppose that there exist $n \in \mathbb{N} \cup \{0\}$, $(y_\varepsilon^k) \subset \mathbb{R}^N$, $x^k \in \Omega$, $\omega^k \in H^s(\mathbb{R}^N) \setminus \{0\}$ $(k = 1, \ldots, n)$ such that (8.5.5), (8.5.6), (8.5.7) and (8.5.15) hold. We also assume that there exists $z_\varepsilon \in \mathbb{R}^N$ such that

$$\int_{B_1(z_\varepsilon)} \left| v_\varepsilon - \psi_\varepsilon \sum_{k=1}^{n} \omega^k(x - y_\varepsilon^k) \right|^2 dx \to c > 0. \tag{8.5.21}$$

Then there exist $x^{k+1} \in \Omega$ and $\omega^{k+1} \in H^s(\mathbb{R}^N) \setminus \{0\}$ such that

$$|z_\varepsilon - y_\varepsilon^k| \to \infty \quad \text{for all } k = 1, \ldots, n, \tag{8.5.22}$$

$$\varepsilon z_\varepsilon \to x^{k+1} \in \Omega, \tag{8.5.23}$$

$$v_\varepsilon(\cdot + z_\varepsilon) \rightharpoonup \omega^{k+1} \not\equiv 0 \quad \text{in } H^s(\mathbb{R}^N), \tag{8.5.24}$$

$$\Phi_{x^{k+1}}'(\omega^{k+1}) = 0. \tag{8.5.25}$$

It is routine to prove that z_ε satisfies (8.5.22) and that there exists $\omega^{k+1} \in H^s(\mathbb{R}^N) \setminus \{0\}$ satisfying (8.5.24).

Let us show that (8.5.23) holds. First, we prove that $\limsup_{\varepsilon \to 0} |\varepsilon z_\varepsilon| < \infty$. Assume, by contradiction, that $|\varepsilon z_\varepsilon| \to \infty$. Let $\varphi \in C_c^\infty(\mathbb{R}^N)$ be a cut-off function such that $\varphi \geq 0$, $\varphi(0) = 1$ and let $\varphi_R(x) = \varphi(x/R)$. Since $(\varphi_R(\cdot - z_\varepsilon)v_\varepsilon)$ is bounded in $\mathcal{H}_\varepsilon$, we obtain

$$\langle J_\varepsilon'(v_\varepsilon), \varphi_R(\cdot - z_\varepsilon)v_\varepsilon \rangle \to 0 \quad \text{as } \varepsilon \to 0,$$

that is

$$\int_{\mathbb{R}^N} (-\Delta)^{\frac{s}{2}} v_\varepsilon(x + z_\varepsilon)(-\Delta)^{\frac{s}{2}} (\varphi_R(x)v_\varepsilon(x + z_\varepsilon)) + V(\varepsilon x + \varepsilon z_\varepsilon)v_\varepsilon^2(x + z_\varepsilon)\varphi_R(x)\,dx$$

$$- \int_{\mathbb{R}^N} g(\varepsilon x + \varepsilon z_\varepsilon, v_\varepsilon(x + z_\varepsilon))v_\varepsilon(x + z_\varepsilon)\varphi_R(x)\,dx \to 0. \tag{8.5.26}$$

Since $|\varepsilon z_\varepsilon| \to \infty$,

$$g(\varepsilon x + \varepsilon z_\varepsilon, v_\varepsilon(x + z_\varepsilon)) = \underline{f}(v_\varepsilon(x + z_\varepsilon)) \text{ on supp}(\varphi_R)$$

for any ε small enough. Moreover, $\varphi_R(x) \to 1$ as $R \to \infty$ and

$$|\underline{f}(\omega^{k+1})\omega^{k+1}\varphi_R| \leq C_1|\omega^{k+1}|^2 + C_2|\omega^{k+1}|^{p+1} \in L^1(\mathbb{R}^N).$$

in view of Lemma 8.2.2-(iii) and Lemma 8.2.1-(i). Hence, by invoking the dominated convergence theorem, we infer that

$$\lim_{R \to \infty} \lim_{\varepsilon \to 0} \int_{\mathbb{R}^N} g(\varepsilon x + \varepsilon z_\varepsilon, v_\varepsilon(x + z_\varepsilon))v_\varepsilon(x + z_\varepsilon)\varphi_R(x)\,dx$$

$$= \lim_{R \to \infty} \int_{\mathbb{R}^N} \underline{f}(\omega^{k+1})\omega^{k+1}\varphi_R\,dx$$

$$= \int_{\mathbb{R}^N} \underline{f}(\omega^{k+1})\omega^{k+1}\,dx. \tag{8.5.27}$$

On the other hand, using (8.5.24), Hölder's inequality and Lemma 1.4.5 (with $\eta_R = 1 - \varphi_R$), we can see that

$$\lim_{R \to \infty} \limsup_{\varepsilon \to 0} \iint_{\mathbb{R}^{2N}} \frac{(v_\varepsilon(x + z_\varepsilon) - v_\varepsilon(y + z_\varepsilon))(\varphi_R(x) - \varphi_R(y))}{|x - y|^{N+2s}} v_\varepsilon(y + z_\varepsilon)\,dxdy = 0,$$

$$\tag{8.5.28}$$

and applying Fatou's lemma and (8.5.24) we get

$$\lim_{R\to\infty} \liminf_{\varepsilon\to 0} \iint_{\mathbb{R}^{2N}} \frac{|v_\varepsilon(x+z_\varepsilon) - v_\varepsilon(y+z_\varepsilon)|^2}{|x-y|^{N+2s}} \varphi_R(x)\,dx\,dy$$

$$\geq \iint_{\mathbb{R}^{2N}} \frac{|\omega^{k+1}(x) - \omega^{k+1}(y)|^2}{|x-y|^{N+2s}}\,dx\,dy. \qquad (8.5.29)$$

Taking into account (8.5.26), (8.5.27), (8.5.28) and (8.5.29), we deduce that

$$\iint_{\mathbb{R}^{2N}} \frac{|\omega^{k+1}(x) - \omega^{k+1}(y)|^2}{|x-y|^{N+2s}}\,dx\,dy + \int_{\mathbb{R}^N} V_0(\omega^{k+1})^2 - \underline{f}(\omega^{k+1})\omega^{k+1}\,dx \leq 0.$$
$$(8.5.30)$$

By Lemma 8.2.2 (i)–(ii) and (8.5.30),

$$\iint_{\mathbb{R}^{2N}} \frac{|\omega^{k+1}(x) - \omega^{k+1}(y)|^2}{|x-y|^{N+2s}}\,dx\,dy + \int_{\mathbb{R}^N} (V_0 - v)(\omega^{k+1})^2\,dx \leq 0.$$

Since $V_0 > v$, we infer that $\omega^{k+1} \equiv 0$, which contradicts (8.5.24).

Then, $\limsup_{\varepsilon\to 0} |\varepsilon z_\varepsilon| < \infty$ and there exists $x^{k+1} \in \mathbb{R}^N$ such that $\varepsilon z_\varepsilon \to x^{k+1}$. This and the fact that $\langle J_\varepsilon'(v_\varepsilon), \varphi(\cdot - z_\varepsilon) \rangle \to 0$ for any $\varphi \in C_c^\infty(\mathbb{R}^N)$ show that $\Phi'_{x^{k+1}}(\omega^{k+1}) = 0$. Since $\omega^{k+1} \not\equiv 0$, it follows that $x^{k+1} \in \Omega$ by Lemma 8.4.3 (i).

Step 4 Conclusion.

Suppose first that $v_0 \neq 0$. Set $y_\varepsilon^1 = 0$, $x^1 = 0$, $\omega^1 = v_0$. If $\|v_\varepsilon - \psi_\varepsilon \omega^1\|_\varepsilon \to 0$, then (8.5.5)–(8.5.8) are satisfied by $0 \in \Omega$, $v_0 \neq 0$ and $\Phi_0'(v_0) = 0$. If $\|v_\varepsilon - \psi_\varepsilon \omega^1\|_\varepsilon \not\to 0$, then (8.5.16) in Step 2 does not occur, and there exists (z_ε) satisfying (8.5.21) in Step 3. In view of Step 3, there exist x^2, ω^2 satisfying (8.5.22)–(8.5.25). Then set $y_\varepsilon^2 = z_\varepsilon$. If $\|v_\varepsilon - \psi_\varepsilon(\omega^1 + \omega^2(\cdot - y_\varepsilon^2))\|_\varepsilon \to 0$, then (8.5.5)–(8.5.8) hold because $|y_\varepsilon^2 - y_\varepsilon^1| = |z_\varepsilon| \to \infty$, $\varepsilon y_\varepsilon^2 \to x^2 \in \Omega$ and $\Phi'_{x^2}(\omega^2) = 0$. Otherwise, we can use Steps 2 and 3 to continue this procedure. Now we assume that $v_0 \equiv 0$. If $\|v_\varepsilon\|_\varepsilon \to 0$, we are done. Otherwise, condition (8.5.16) in Step 2 does not occur, and we can find (z_ε) satisfying (8.5.21) in Step 3. Applying Step 3, there exist x^1 and ω^1 satisfying (8.5.22)–(8.5.25). Thus, we set $y_\varepsilon^1 = z_\varepsilon$. If $\|v_\varepsilon - \psi_\varepsilon(\omega^1(\cdot - y_\varepsilon^1))\|_\varepsilon \to 0$, we are done. Otherwise, we use Steps 2 and 3 and we continue this procedure. At this point, we aim to show that the procedure stops after a finite number of steps.

First we show that, under assumptions (8.5.5)–(8.5.7) and (8.5.15),

$$\lim_{\varepsilon\to 0} \left\| v_\varepsilon - \psi_\varepsilon \sum_{k=1}^n \omega^k(\cdot - y_\varepsilon^k) \right\|_\varepsilon^2 = \lim_{\varepsilon\to 0} \|v_\varepsilon\|_\varepsilon^2 - \sum_{k=1}^n |\omega^k|_{x^k}^2. \qquad (8.5.31)$$

Note that

$$
\left\| v_\varepsilon - \psi_\varepsilon \sum_{k=1}^n \omega^k(\cdot - y_\varepsilon^k) \right\|_\varepsilon^2
$$

$$
= \|v_\varepsilon\|_\varepsilon^2 - 2\sum_{k=1}^n \langle v_\varepsilon, \psi_\varepsilon \omega^k(\cdot - y_\varepsilon^k)\rangle_{\mathcal{H}_\varepsilon} + \sum_{k,k'} \langle \psi_\varepsilon \omega^k(\cdot - y_\varepsilon^k), \psi_\varepsilon \omega^{k'}(\cdot - y_\varepsilon^{k'})\rangle_{\mathcal{H}_\varepsilon}.
$$

(8.5.32)

Let us verify that

$$
\langle v_\varepsilon, \psi_\varepsilon \omega^k(\cdot - y_\varepsilon^k)\rangle_{\mathcal{H}_\varepsilon} \to \int_{\mathbb{R}^N} |(-\Delta)^{\frac{s}{2}}\omega^k|^2 + V(x^k)(\omega^k)^2\, dx = |\omega^k|_{x^k}^2.
$$

(8.5.33)

Indeed,

$$
\langle v_\varepsilon, \psi_\varepsilon \omega^k(\cdot - y_\varepsilon^k)\rangle_{\mathcal{H}_\varepsilon}
$$

$$
= \iint_{\mathbb{R}^{2N}} \frac{(v_\varepsilon(x + y_\varepsilon^k) - v_\varepsilon(y + y_\varepsilon^k))(\psi_\varepsilon(x + y_\varepsilon^k) - \psi_\varepsilon(y + y_\varepsilon^k))}{|x - y|^{N+2s}}\omega^k(x)\, dx\, dy
$$

$$
+ \iint_{\mathbb{R}^{2N}} \frac{(v_\varepsilon(x + y_\varepsilon^k) - v_\varepsilon(y + y_\varepsilon^k))(\omega^k(x) - \omega^k(y))}{|x - y|^{N+2s}}\psi_\varepsilon(y + y_\varepsilon^k)\, dx\, dy
$$

$$
+ \int_{\mathbb{R}^N} V(\varepsilon x + \varepsilon y_\varepsilon^k)\psi_\varepsilon(x + y_\varepsilon^k)v_\varepsilon(x + y_\varepsilon^k)\omega^k(x)\, dx
$$

$$
= D_\varepsilon^1 + D_\varepsilon^2 + D_\varepsilon^3.
$$

Using Hölder's inequality and the boundedness of $v_\varepsilon(\cdot + y_\varepsilon^k)$ we can argue as in the proof of (8.5.13) to see that $D_\varepsilon^1 \to 0$. Concerning D_ε^2, we observe that

$$
D_\varepsilon^2 = \iint_{\mathbb{R}^{2N}} \frac{[(v_\varepsilon(x + y_\varepsilon^k) - v_\varepsilon(y + y_\varepsilon^k))(\omega^k(x) - \omega^k(y))]}{|x - y|^{N+2s}}\, dx\, dy
$$

$$
+ \iint_{\mathbb{R}^{2N}} \frac{(\psi_\varepsilon(y + y_\varepsilon^k) - 1)(v_\varepsilon(x + y_\varepsilon^k) - v_\varepsilon(y + y_\varepsilon^k))(\omega^k(x) - \omega^k(y))}{|x - y|^{N+2s}}\, dx\, dy
$$

$$
= D_\varepsilon^{2,1} + D_\varepsilon^{2,2}.
$$

Since $v_\varepsilon(\cdot + y_\varepsilon^k) \rightharpoonup \omega^k$ in $H^s(\mathbb{R}^N)$, we obtain that $D_\varepsilon^{2,1} \to [\omega^k]_s^2$. On the other hand, using Hölder's inequality and the fact that $v_\varepsilon(\cdot + y_\varepsilon^k)$ is bounded in $H^s(\mathbb{R}^N)$, the dominated convergence theorem implies that

$$|D_\varepsilon^{2,2}| \le C \left(\iint_{\mathbb{R}^{2N}} \frac{|(\psi_\varepsilon(x + y_\varepsilon^k) - 1)(\omega^k(x) - \omega^k(y))|^2}{|x - y|^{N+2s}} \, dx dy \right)^{\frac{1}{2}} \to 0.$$

Since $D_\varepsilon^3 \to \int_{\mathbb{R}^N} V(x^k)(\omega^k)^2 \, dx$, we deduce that (8.5.33) holds. In a similar fashion, we can obtain

$$\langle \psi_\varepsilon \omega^k(\cdot - y_\varepsilon^k), \psi_\varepsilon \omega^{k'}(\cdot - y_\varepsilon^{k'}) \rangle_{\mathcal{H}_\varepsilon} \to \begin{cases} 0, & \text{if } k \ne k' \\ \left| \omega^k \right|_{x^k}^2, & \text{if } k = k'. \end{cases} \tag{8.5.34}$$

Combining (8.5.32), (8.5.33) and (8.5.34), we infer that (8.5.31) holds. Now, (8.5.31) yields that

$$\sum_{k=1}^n \left| \omega^k \right|_{x^k}^2 \le \lim_{\varepsilon \to 0} \|v_\varepsilon\|_\varepsilon^2,$$

and using Lemma 8.4.3-(ii) and (8.5.4) we get

$$\delta_1^2 n \le \lim_{\varepsilon \to 0} \|v_\varepsilon\|_\varepsilon^2 \le m^2.$$

Therefore, the procedure to find (y_ε^k), x^k, ω^k can not be iterated infinitely many times, so there exist $l \in \mathbb{N} \cup \{0\}$, (y_ε^k), x^k, ω^k such that (8.5.5)–(8.5.8) hold. Clearly, (8.5.9) follows in a standard way by (8.5.5)–(8.5.8). $\qquad \square$

In the next lemma we investigate the behavior of c_ε as $\varepsilon \to 0$.

Lemma 8.5.4 *Let $(c_\varepsilon)_{\varepsilon \in (0, \varepsilon_1]}$ be the mountain pass value of J_ε defined in (8.3.4)–(8.3.5). Then*

$$c_\varepsilon \to m(0) = \inf_{x \in \mathbb{R}^N} m(x) \quad \text{as } \varepsilon \to 0.$$

Proof By Lemma 8.4.2, we can find a path $\gamma \in C([0, 1], H^s(\mathbb{R}^N))$ such that $\gamma(0) = 0$, $\Phi_0(\gamma(1)) < 0$, $\Phi_0(\gamma(t)) \le m(0)$ for all $t \in [0, 1]$, and

$$\max_{t \in [0,1]} \Phi_0(\gamma(t)) = m(0).$$

Take $\varphi \in C_c^\infty(\mathbb{R}^N)$ such that $\varphi(0) = 1$ and $\varphi \geq 0$, and set

$$\gamma_R(t)(x) = \varphi\left(\frac{x}{R}\right)\gamma(t)(x).$$

Then, it is easy to check that $\gamma_R(t) \in C([0, 1], \mathcal{H}_\varepsilon)$, $\gamma_R(0) = 0$ and $\Phi_0(\gamma_R(1)) < 0$ for any $R > 1$ sufficiently large. Therefore, $\gamma_R(t) \in \Gamma_\varepsilon$. Now, fixed $R > 0$, we can see that $\max_{t\in[0,1]} |J_\varepsilon(\gamma_R(t)) - \Phi_0(\gamma_R(t))| \to 0$ as $\varepsilon \to 0$. Hence, for any $R > 1$ large enough, we get

$$c_\varepsilon \leq \max_{t\in[0,1]} J_\varepsilon(\gamma_R(t)) \to \max_{t\in[0,1]} \Phi_0(\gamma_R(t)) \quad \text{as } \varepsilon \to 0.$$

On the other hand

$$\max_{t\in[0,1]} \Phi_0(\gamma_R(t)) \to m(0) \quad \text{as } R \to \infty,$$

so we deduce that $\limsup_{\varepsilon\to 0} c_\varepsilon \leq m(0)$.

To complete the proof, we show that $\liminf_{\varepsilon\to 0} c_\varepsilon \geq m(0)$. Let $v_\varepsilon \in \mathcal{H}_\varepsilon$ be a critical point of $J_\varepsilon(v)$ associated with c_ε. By Lemma 8.5.1, there exist $\varepsilon_j \to 0, l \in \mathbb{N} \cup \{0\}$, $(y_{\varepsilon_j}^k) \subset \mathbb{R}^N$, $x^k \in \Omega$, $\omega^k \in H^s(\mathbb{R}^N) \setminus \{0\}$ $(k = 1, \ldots, l)$ satisfying (8.5.5)–(8.5.9). If, by contradiction, $l = 0$, then (8.5.9) yields $c_{\varepsilon_j} = J_{\varepsilon_j}(v_{\varepsilon_j}) \to 0$, which contradicts Corollary 8.3.2. Consequently, $l \geq 1$ and using (8.5.9) and Lemma 8.4.4, we have

$$\liminf_{j\to\infty} c_{\varepsilon_j} = \sum_{k=1}^l \Phi_{x^k}(\omega^k) \geq \sum_{k=1}^l m(x^k) \geq lm(0) \geq m(0).$$

$\square$

In view of Lemma 8.5.4, we deduce the following result.

Lemma 8.5.5 *For any $\varepsilon \in (0, \varepsilon_1]$, let v_ε denote a critical point of J_ε corresponding to c_ε. Then for any sequence $\varepsilon_j \to 0$ we can find a subsequence, still denoted by ε_j, and $y_{\varepsilon_j}, x^1, \omega^1$ such that*

$$\varepsilon_j y_{\varepsilon_j} \to x^1, \tag{8.5.35}$$

$$x^1 \in \Lambda' : V(x^1) = \inf_{x\in\Lambda} V(x), \tag{8.5.36}$$

$$\omega^1(x) \text{ is a least energy solution of } \Phi'_{x^1}(v) = 0, \tag{8.5.37}$$

$$\|v_{\varepsilon_j} - \psi_{\varepsilon_j} w^1(\cdot - y_{\varepsilon_j})\|_{\varepsilon_j} \to 0, \tag{8.5.38}$$

$$J_{\varepsilon_j}(v_{\varepsilon_j}) \to m(x^1) = m(0). \tag{8.5.39}$$

8.6   Proof of Theorem 8.1.1

In this last section we provide the proof of Theorem 8.1.1. Corollary 8.3.5 shows that there exists $\varepsilon_1 \in (0, \varepsilon_0]$ such that for any $\varepsilon \in (0, \varepsilon_1]$, there exists a critical point $v_\varepsilon \in \mathcal{H}_\varepsilon$ of J_ε satisfying $J_\varepsilon(v_\varepsilon) = c_\varepsilon$. Then, by Lemma 8.5.5, we know that for any sequence $\varepsilon_j \to 0$, there exists a subsequence ε_j and $(y_{\varepsilon_j}) \subset \mathbb{R}^N$, $x^1 \in \Lambda'$, $\omega^1 \in H^s(\mathbb{R}^N) \setminus \{0\}$ satisfying (8.5.35)–(8.5.39). Note that $v_{\varepsilon_j} \in L^\infty(\mathbb{R}^N) \cap C_{loc}^{0,\alpha}(\mathbb{R}^N)$ (see Lemma 7.2.9), and by Theorem 1.3.5 we have $v_{\varepsilon_j} > 0$ in $\mathbb{R}^N$. In view of (8.3.2) and (8.5.38) we obtain

$$\|v_{\varepsilon_j} - \psi_{\varepsilon_j}\omega^1(\cdot - y_{\varepsilon_j})\|_{H^s(\mathbb{R}^N)} \to 0. \tag{8.6.1}$$

We also note that (8.5.31) and (8.6.1) yield

$$\lim_{j \to \infty} \|v_{\varepsilon_j}\|_{H^s_{\varepsilon_j}}^2 = |\omega^1|_{x^1}^2 \neq 0. \tag{8.6.2}$$

Let $\tilde{v}_{\varepsilon_j}(x) = v_{\varepsilon_j}(x + y_{\varepsilon_j})$. Arguing as in the proof of (8.5.13), and using that $\psi(x^1) = 1$, (8.5.35), and the dominated convergence theorem, we see that

$$[\psi_{\varepsilon_j}(\cdot + y_{\varepsilon_j})\omega^1 - \omega^1]_s^2$$
$$\leq 2 \iint_{\mathbb{R}^{2N}} \frac{|\psi_{\varepsilon_j}(x + y_{\varepsilon_j}) - \psi_{\varepsilon_j}(y + y_{\varepsilon_j})|^2}{|x - y|^{N+2s}} (\omega^1(x))^2 \, dx\,dy$$
$$+ 2 \iint_{\mathbb{R}^{2N}} \frac{|\psi_{\varepsilon_j}(y + y_{\varepsilon_j}) - 1|^2}{|x - y|^{N+2s}} |\omega^1(x) - \omega^1(y)|^2 \, dx\,dy \to 0.$$

Clearly,

$$\int_{\mathbb{R}^N} |\psi_{\varepsilon_j}(x + y_{\varepsilon_j})\omega^1 - \omega^1|^2 \, dx \to 0.$$

These two facts together with (8.6.1) imply that

$$\|\tilde{v}_{\varepsilon_j} - \omega^1\|_{H^s(\mathbb{R}^N)} \to 0. \tag{8.6.3}$$

Arguing as in the proof of Lemma 7.2.9, we can find $K > 0$ such that

$$\|\tilde{v}_{\varepsilon_j}\|_{L^\infty(\mathbb{R}^N)} \leq K \quad \text{for all } j \in \mathbb{N}, \tag{8.6.4}$$

and (see also Remark 7.2.10)

$$\tilde{v}_{\varepsilon_j}(x) \to 0 \quad \text{as } |x| \to \infty \tag{8.6.5}$$

uniformly in $j \in \mathbb{N}$. Moreover, using interpolation in L^q spaces, we see that

$$\tilde{v}_{\varepsilon_j} \to \omega^1 \text{ in } L^q(\mathbb{R}^N), \text{ for any } q \in [2, \infty),$$
$$h_j(x) = g(\varepsilon_j x + \varepsilon_j y_{\varepsilon_j}, \tilde{v}_{\varepsilon_j}) \to f(\omega^1) \text{ in } L^q(\mathbb{R}^N), \text{ for any } q \in [2, \infty). \tag{8.6.6}$$

Now let us prove that $\tilde{v}_{\varepsilon_j}$ is a solution to (8.1.1) for small $\varepsilon_j > 0$. Since $\varepsilon_j y_{\varepsilon_j} \to x^1 \in \Lambda'$, there exists $r > 0$ such that for some subsequence, still denoted by itself, we have

$$B_r(\varepsilon_j y_{\varepsilon_j}) \subset \Lambda' \quad \text{for all } j \in \mathbb{N}.$$

By setting $\Lambda'_\varepsilon = \frac{\Lambda'}{\varepsilon}$, we can see that

$$B_{\frac{r}{\varepsilon_j}}(y_{\varepsilon_j}) \subset \Lambda'_{\varepsilon_j} \quad \text{for all } j \in \mathbb{N}$$

which yields

$$\mathbb{R}^N \setminus \Lambda'_{\varepsilon_j} \subset \mathbb{R}^N \setminus B_{\frac{r}{\varepsilon_j}}(y_{\varepsilon_j}) \quad \text{for all } j \in \mathbb{N}.$$

In view of (8.6.5), there exists $R > 0$ such that

$$\tilde{v}_{\varepsilon_j}(x) < r_v \quad \text{for all } |x| \geq R, j \in \mathbb{N}$$

so that

$$v_{\varepsilon_j}(x) = \tilde{v}_{\varepsilon_j}(x - y_{\varepsilon_j}) < r_v \quad \text{for all } x \in \mathbb{R}^N \setminus B_R(y_{\varepsilon_j}), j \in \mathbb{N}.$$

On the other hand, there exists $j_0 \in \mathbb{N}$ such that

$$\mathbb{R}^N \setminus \Lambda'_{\varepsilon_j} \subset \mathbb{R}^N \setminus B_{\frac{r}{\varepsilon_j}}(y_{\varepsilon_j}) \subset \mathbb{R}^N \setminus B_R(y_{\varepsilon_j}) \quad \text{for all } j \geq j_0.$$

Hence,

$$v_{\varepsilon_j}(x) < r_v \quad \text{for all } x \in \mathbb{R}^N \setminus \Lambda'_{\varepsilon_j}, j \geq j_0. \tag{8.6.7}$$

Now, up to a subsequence, we may assume that

$$\|v_{\varepsilon_j}\|_{L^\infty(B_R(y_{\varepsilon_j}))} \geq r_\nu \quad \text{for all } j \geq j_0. \tag{8.6.8}$$

Otherwise, if this is not the case, we have $\|v_{\varepsilon_j}\|_{L^\infty(\mathbb{R}^N)} < r_\nu$, and taking into account the definition of g and our choice of r_ν, we get

$$g(\varepsilon_j x, v_{\varepsilon_j})v_{\varepsilon_j} = f(v_{\varepsilon_j})v_{\varepsilon_j} \leq \nu v_{\varepsilon_j}^2 \leq \frac{V_0}{2} v_{\varepsilon_j}^2.$$

Then, since $\langle J'_{\varepsilon_j}(v_{\varepsilon_j}), v_{\varepsilon_j}\rangle = 0$, we deduce that

$$\|v_{\varepsilon_j}\|_{\varepsilon_j}^2 = \int_{\mathbb{R}^N} f(v_{\varepsilon_j})v_{\varepsilon_j}\, dx \leq \frac{V_0}{2}\int_{\mathbb{R}^N} v_{\varepsilon_j}^2\, dx$$

which implies that $\lim_{j\to\infty}\|v_{\varepsilon_j}\|_{\varepsilon_j}^2 = 0$, and this is a contradiction in view of (8.6.2). Therefore, combining (8.6.7) and (8.6.8), we deduce that if $z_{\varepsilon_j} \in \mathbb{R}^N$ is a global maximum point of v_{ε_j} then z_{ε_j} belongs to $B_R(y_{\varepsilon_j})$. Hence, $z_{\varepsilon_j} = y_{\varepsilon_j} + \bar{z}_{\varepsilon_j}$ for some $\bar{z}_{\varepsilon_j} \in B_R$. Recalling that the associated solution of our problem (8.1.1) is of the form $u_{\varepsilon_j}(x) = v_{\varepsilon_j}(\frac{x}{\varepsilon_j})$, we conclude that $x_{\varepsilon_j} = \varepsilon_j y_{\varepsilon_j} + \varepsilon_j \bar{z}_{\varepsilon_j}$ is a global maximum point of u_{ε_j}. Since $(\bar{z}_{\varepsilon_j}) \subset B_R$ is bounded and $\varepsilon_j y_{\varepsilon_j} \to x^1 \in \Lambda'$, we obtain

$$\lim_{j\to\infty} V(x_{\varepsilon_j}) = V(x^1) = \inf_{x\in\Lambda} V(x).$$

Therefore, we have proved that there exists $\varepsilon_0 > 0$ such that for any $\varepsilon \in (0, \varepsilon_0]$, problem (8.1.1) admits a positive solution $u_\varepsilon(x) = v_\varepsilon(\frac{x}{\varepsilon})$ satisfying (1) of Theorem 8.1.1. We note that $u_\varepsilon \in L^\infty(\mathbb{R}^N) \cap C^{0,\alpha}_{loc}(\mathbb{R}^N)$. Finally, one can argue as at the end of the proof of Theorem 7.1.1 to deduce (2).   □

Multiplicity and Concentration Results for a Fractional Choquard Equation

9

9.1   Introduction

This chapter deals with the following class of nonlinear fractional Choquard problems:

$$
\begin{cases}
\varepsilon^{2s}(-\Delta)^s u + V(x)u = \varepsilon^{\mu-N} \left(\frac{1}{|x|^\mu} * F(u) \right) f(u) & \text{in } \mathbb{R}^N, \\
u \in H^s(\mathbb{R}^N), \quad u > 0 \text{ in } \mathbb{R}^N,
\end{cases}
\tag{9.1.1}
$$

where $\varepsilon > 0$ is a small parameter, $s \in (0, 1)$, $N > 2s$ and $0 < \mu < 2s$. The potential $V : \mathbb{R}^N \to \mathbb{R}$ is a continuous function satisfying the following del Pino-Felmer type hypotheses [165]:

(V_1) $\inf_{x \in \mathbb{R}^N} V(x) = V_0 > 0$;
(V_2) there exists a bounded open set $\Lambda \subset \mathbb{R}^N$ such that

$$
V_0 < \min_{x \in \partial \Lambda} V(x) \text{ and } M = \{x \in \Lambda : V(x) = V_0\} \neq \emptyset.
$$

Without loss of generality, we may assume that $0 \in M$. Concerning the nonlinearity $f : \mathbb{R} \to \mathbb{R}$, we assume that f is a continuous function such that $f(t) = 0$ for $t < 0$, and satisfies the following conditions:

(f_1) $\lim_{t \to 0} \dfrac{f(t)}{t} = 0$;

(f_2) there exists $q \in (2, \frac{2^*_s}{2}(2 - \frac{\mu}{N}))$ such that $\lim_{t \to \infty} \dfrac{f(t)}{t^{q-1}} = 0$;

© The Author(s), under exclusive license to Springer Nature Switzerland AG 2021
V. Ambrosio, *Nonlinear Fractional Schrödinger Equations in $\mathbb{R}^N$*,
Frontiers in Mathematics, https://doi.org/10.1007/978-3-030-60220-8_9

(f_3) the following Ambrosetti–Rabinowitz type condition [29] holds:

$$0 < 4F(t) = 4 \int_0^t f(\tau)\, d\tau \le 2f(t)t \ \text{ for all } t > 0;$$

(f_4) The map $t \mapsto \dfrac{f(t)}{t}$ is increasing for every $t > 0$.

We recall that the problem (9.1.1) is motivated by the search of standing wave solutions for the fractional Schrödinger equation

$$\iota \frac{\partial \psi}{\partial t} = (-\Delta)^s \psi + V(x)\psi - \left(I_s * |\Psi|^q\right) |\Psi|^{q-2}\Psi \quad (t, x) \in \mathbb{R} \times \mathbb{R}^N,$$

where $I_s(x) = \dfrac{\Gamma\left(\frac{N-s}{2}\right)}{\pi^{\frac{N}{2}} 2^s \Gamma\left(\frac{s}{2}\right)} |x|^{s-N}$ is the Riesz kernel defined in Chap. 1.

When $s = 1$, $V(x) \equiv 1$, $\varepsilon = 1$, $\mu = 1$ and $F(u) = \frac{|u|^2}{2}$, equation in (9.1.1) becomes the so-called Choquard equation

$$-\Delta u + u = \left(I_2 * |u|^2\right) u \text{ in } \mathbb{R}^3, \tag{9.1.2}$$

introduced at least as early as 1954, in a work by Pekar [289] describing the quantum mechanics of a polaron at rest. In 1976 Choquard used (9.1.2) to describe an electron trapped in its own hole, in a certain approximation to Hartree–Fock Theory of one-component plasma [248]. More recently, Penrose [290] used it as a model of self-gravitating matter. In this context (9.1.2) is usually called the nonlinear Schrödinger-Newton equation. From a mathematical point of view, equation (9.1.2) and its generalizations have been widely investigated. The early existence and symmetry results are due to Lieb [248] and Lions [254]. Later, Ma and Zhao [266] classified all positive solutions to (9.1.2) and proved that they must be radially symmetric and monotonically decreasing about some fixed point. Moroz and Van Schaftingen [274] investigated existence, qualitative properties and decay asymptotics of positive ground state solutions for a generalized Choquard equation. Alves and Yang [25] studied multiplicity and concentration of positive solutions for a quasilinear Choquard equation. Further results on Choquard equations can be found in [1, 145, 275, 276, 339] and references therein.

In the case $s \in (0, 1)$, only few recent papers considered fractional Choquard equations like (9.1.1). In [156] d'Avenia et al. considered the fractional Choquard equation

$$(-\Delta)^s u + u = \left(\frac{1}{|x|^\mu} * |u|^p\right) |u|^{p-2} u \text{ in } \mathbb{R}^N,$$

and proved regularity, existence and non-existence, symmetry and decay properties of solutions. Coti Zelati and Nolasco [153] established the existence of ground state solutions for a pseudo-relativistic Hartree-equation via critical point theory. Shen et al. [312] investigated the existence of ground state solutions for a fractional Choquard equation involving a nonlinearity that satisfies Berestycki–Lions type assumptions. Belchior et al. [93] dealt with existence, regularity and polynomial decay for a fractional Choquard equation involving the fractional p-Laplacian.

In this chapter, we focus our attention on the multiplicity and the concentration of positive solutions of (9.1.1), involving a potential and a continuous nonlinearity satisfying the assumptions (V_1)–(V_2) and (f_1)–(f_4), respectively. In particular, we are interested in relating the number of positive solutions of (9.1.1) with the topology of the set $M = \{x \in \Lambda : V(x) = V_0\}$.

Our main result can be stated as follows:

Theorem 9.1.1 ([46]) *Suppose that V fulfills (V_1)–(V_2), $0 < \mu < 2s$ and f satisfies (f_1)–(f_4) with $2 < q < \frac{2(N-\mu)}{N-2s}$. Then, for every $\delta > 0$ such that $M_\delta = \{x \in \mathbb{R}^N : \text{dist}(x, M) \leq \delta\} \subset \Lambda$, there exists $\varepsilon_\delta > 0$ such that, for any $\varepsilon \in (0, \varepsilon_\delta)$, problem (9.1.1) has at least $\text{cat}_{M_\delta}(M)$ positive solutions. Moreover, if u_ε denotes one of these positive solutions and $x_\varepsilon \in \mathbb{R}^N$ is a global maximum point of u_ε, then*

$$\lim_{\varepsilon \to 0} V(x_\varepsilon) = V_0.$$

Firstly, we note that the restriction on the exponent q is justified by the Hardy-Littlewood-Sobolev inequality (see Theorem 9.2.1 in Chap. 2). Indeed, if $F(u) = |u|^q$, then the term

$$\int_{\mathbb{R}^N} \left(\frac{1}{|x|^\mu} * F(u) \right) F(u) \, dx$$

is well defined if $F(u) \in L^t(\mathbb{R}^N)$ for $t > 1$ such that $\frac{2}{t} + \frac{\mu}{N} = 2$. Hence, recalling that $H^s(\mathbb{R}^N)$ is continuously embedded in $L^r(\mathbb{R}^N)$ for any $r \in [2, 2_s^*]$, we need to require that $tq \in [2, 2_s^*]$ and this leads to assuming that

$$2 - \frac{\mu}{N} \leq q \leq \frac{2_s^*}{2} \left(2 - \frac{\mu}{N} \right).$$

Here we only consider the case $q > 2$. On the other hand, inspired by [25], we adapt the del Pino-Felmer penalization technique [165] by imposing further restrictions on the exponent q. Indeed, we treat the convolution $\frac{1}{|x|^\mu} * F(u)$ as a bounded term and introduce the monotonicity condition on f at the same time. In this way, the assumptions $0 < \mu < 2s$ and $2 < q < \frac{2(N-\mu)}{N-2s}$ allow us to apply the penalization method.

Differently from the case $s = 1$ considered in [25], in our setting a more accurate investigation is needed due to the presence of two nonlocal terms. Moreover, the nonlinearity appearing in (9.1.1) is only continuous (while $f \in C^1$ in [25]), so to overcome the non-differentiability of the associated Nehari manifold, we will use the abstract critical point results in Sect. 2.4. Concerning the multiplicity result for the modified problem, our approach resembles some ideas in [95] based on the comparison between the category of some sublevel sets of the modified functional and the category of the set M. Finally, in order to prove that the solutions of the modified problem are solutions of the problem (9.1.1), we use a Moser iteration argument [278] and the boundedness of the convolution term to obtain the desired result. In Sect. 9.2 we introduce the functional setting and the modified problem. Section 9.3 is devoted to the existence of positive solutions to the autonomous problem associated to (9.1.1). In Sect. 9.4, we obtain a multiplicity result by using Lusternik–Schnirelman theory. Finally, by means of a Moser iteration scheme, we are able to prove that, for ε small enough, the solutions of the modified problem are indeed solutions of (9.1.1). We conclude this introduction by mentioning that a multiplicity result for a fractional p-Choquard equation with a global condition on the potential was established in [56].

9.2   Variational Framework

First, we recall the Hardy–Littlewood–Sobolev inequality, which will be frequently used along this section:

Theorem 9.2.1 ([249]) *Let $r, t > 1$ and $0 < \mu < N$ such that $\frac{1}{r} + \frac{\mu}{N} + \frac{1}{t} = 2$. Let $f \in L^r(\mathbb{R}^N)$ and $h \in L^t(\mathbb{R}^N)$. Then there exists a sharp constant $C(r, N, \mu, t) > 0$, independent of f and h, such that*

$$\int_{\mathbb{R}^N} \int_{\mathbb{R}^N} \frac{f(x)h(y)}{|x - y|^\mu} \, dx\, dy \leq C(r, N, \mu, t) \|f\|_{L^r(\mathbb{R}^N)} \|h\|_{L^t(\mathbb{R}^N)}.$$

For any $\varepsilon > 0$, we consider the fractional Sobolev space

$$\mathcal{H}_\varepsilon = \left\{ u \in H^s(\mathbb{R}^N) : \int_{\mathbb{R}^N} V(\varepsilon x) u^2 dx < \infty \right\},$$

which is a Hilbert space with the inner product

$$\langle u, v \rangle_\varepsilon = \langle u, v \rangle_{\mathcal{D}^{s,2}(\mathbb{R}^N)} + \int_{\mathbb{R}^N} V(\varepsilon x) uv \, dx \quad \forall u, v \in \mathcal{H}_\varepsilon$$

and the associated norm $\|u\|_\varepsilon = \sqrt{\langle u, u \rangle_\varepsilon}$, $u \in \mathcal{H}_\varepsilon$.

Using the change of variable $u(x) \mapsto u(\varepsilon x)$, we see that (9.1.1) is equivalent to the following problem:

$$\begin{cases} (-\Delta)^s u + V(\varepsilon x)u = \left(\frac{1}{|x|^\mu} * F(u) \right) f(u) & \text{in } \mathbb{R}^N, \\ u \in H^s(\mathbb{R}^N), \quad u > 0 \text{ in } \mathbb{R}^N. \end{cases} \tag{9.2.1}$$

Choose $\ell_0 > 0$, to be determined later, and take $a > 0$ such that $\frac{f(a)}{a} = \frac{V_0}{\ell_0}$. We introduce the functions

$$\tilde{f}(t) = \begin{cases} f(t), & \text{if } t \le a, \\ \frac{V_0}{\ell_0} t, & \text{if } t > a, \end{cases}$$

and

$$g(x, t) = \chi_\Lambda(x) f(t) + (1 - \chi_\Lambda(x)) \tilde{f}(t),$$

where χ_Λ is the characteristic function on Λ, and we write $G(x, t) = \int_0^t g(x, \tau)\, d\tau$.

By assumptions (f_1)–(f_4), the function g satisfies the following properties:

(g_1) $\displaystyle \lim_{t \to 0} \frac{g(x, t)}{t} = 0$ uniformly in $x \in \mathbb{R}^N$;

(g_2) $\displaystyle \lim_{t \to \infty} \frac{g(x, t)}{t^{q-1}} = 0$ uniformly in $x \in \mathbb{R}^N$;

(g_3) $0 < 4G(x, t) \le 2g(x, t)t$ for any $x \in \Lambda$ and $t > 0$, and
$0 \le 2G(x, t) \le g(x, t)t \le \frac{V_0}{\ell_0} t^2$ for any $x \in \mathbb{R}^N \setminus \Lambda$ and $t > 0$;

(g_4) $t \mapsto g(x, t)$ and $t \mapsto \frac{G(x,t)}{t}$ are increasing for all $x \in \mathbb{R}^N$ and $t > 0$.

Thus we consider the auxiliary problem

$$\begin{cases} (-\Delta)^s u + V(\varepsilon x)u = \left(\frac{1}{|x|^\mu} * G(\varepsilon x, u) \right) g(\varepsilon x, u) & \text{in } \mathbb{R}^N, \\ u \in H^s(\mathbb{R}^N), \quad u > 0 \text{ in } \mathbb{R}^N, \end{cases} \tag{9.2.2}$$

and we note that if u is a solution of (9.2.2) such that

$$u(x) \le a \text{ for all } x \in \mathbb{R}^N \setminus \Lambda_\varepsilon, \tag{9.2.3}$$

where $\Lambda_\varepsilon = \{x \in \mathbb{R}^N : \varepsilon x \in \Lambda\}$, then u solves (9.2.1).

To study weak solutions to (9.2.2), we look for critical points of the C^1-functional $J_\varepsilon : \mathcal{H}_\varepsilon \to \mathbb{R}$ defined by

$$J_\varepsilon(u) = \frac{1}{2}\|u\|_\varepsilon^2 - \Sigma_\varepsilon(u),$$

where

$$\Sigma_\varepsilon(u) = \frac{1}{2}\int_{\mathbb{R}^N} \left(\frac{1}{|x|^\mu} * G(\varepsilon x, u)\right) G(\varepsilon x, u)\, dx.$$

Lemma 9.2.2 *J_ε has a mountain pass geometry, that is*

(i) *there exist $\alpha, \rho > 0$ such that $J_\varepsilon(u) \geq \alpha$ for any $u \in H_\varepsilon^s$ such that $\|u\|_\varepsilon = \rho$;*

(ii) *there exists $e \in \mathcal{H}_\varepsilon$ such that $\|e\|_\varepsilon > \rho$ and $J_\varepsilon(e) < 0$.*

Proof From (g_1) and (g_2) we deduce that that for every $\eta > 0$ there exists $C_\eta > 0$ such that

$$|g(\varepsilon x, t)| \leq \eta|t| + C_\eta|t|^{q-1} \quad \text{for } (x, t) \in \mathbb{R}^N \times \mathbb{R}. \tag{9.2.4}$$

Using Theorem 9.2.1 and (9.2.4), we get

$$\left|\int_{\mathbb{R}^N} \left(\frac{1}{|x|^\mu} * G(\varepsilon x, u)\right) G(\varepsilon x, u)\, dx\right| \leq C\|G(\varepsilon x, u)\|_{L^t(\mathbb{R}^N)}\|G(\varepsilon x, u)\|_{L^t(\mathbb{R}^N)}$$

$$\leq C\left(\int_{\mathbb{R}^N}(|u|^2 + |u|^q)^t\, dx\right)^{\frac{2}{t}}, \tag{9.2.5}$$

where $t = \frac{2N}{2N-\mu}$. Since $2 < q < \frac{2(N-\mu)}{N-2s}$, we see that $tq \in (2, 2_s^*)$, and by Theorem 1.1.8 we have

$$\left(\int_{\mathbb{R}^N}(|u|^2 + |u|^q)^t\, dx\right)^{\frac{2}{t}} \leq C(\|u\|_\varepsilon^2 + \|u\|_\varepsilon^q)^2. \tag{9.2.6}$$

Inequalities (9.2.5) and (9.2.6) imply that

$$\left|\int_{\mathbb{R}^N} \left(\frac{1}{|x|^\mu} * G(\varepsilon x, u)\right) G(\varepsilon x, u)\, dx\right| \leq C(\|u\|_\varepsilon^2 + \|u\|_\varepsilon^q)^2 \leq C(\|u\|_\varepsilon^4 + \|u\|_\varepsilon^{2q}).$$

Consequently,

$$J(u) \geq \frac{1}{2}\|u\|_\varepsilon^2 - C(\|u\|_\varepsilon^4 + \|u\|_\varepsilon^{2q}),$$

and thanks to $q > 2$ we conclude that (i) holds. Now fix a non-negative function $u_0 \in H^s(\mathbb{R}^N)$ such that $u \not\equiv 0$ and $\mathrm{supp}(u_0) \subset \Lambda_\varepsilon$, and set

$$h(t) = \Sigma_\varepsilon \left(\frac{tu_0}{\|u_0\|_\varepsilon} \right) \quad \text{for } t > 0.$$

Since $G(\varepsilon x, u_0) = F(u_0)$, using (f_3) we deduce that

$$
\begin{aligned}
h'(t) &= \Sigma'_\varepsilon \left(\frac{tu_0}{\|u_0\|_\varepsilon} \right) \frac{u_0}{\|u_0\|_\varepsilon} \\
&= \int_{\mathbb{R}^N} \left(\frac{1}{|x|^\mu} * F \left(\frac{tu_0}{\|u_0\|_\varepsilon} \right) \right) f \left(\frac{tu_0}{\|u_0\|_\varepsilon} \right) \frac{u_0}{\|u_0\|_\varepsilon} \, dx \\
&= \frac{4}{t} \int_{\mathbb{R}^N} \frac{1}{2} \left(\frac{1}{|x|^\mu} * F \left(\frac{tu_0}{\|u_0\|_\varepsilon} \right) \right) \frac{1}{2} f \left(\frac{tu_0}{\|u_0\|_\varepsilon} \right) \frac{tu_0}{\|u_0\|_\varepsilon} \, dx \\
&> \frac{4}{t} h(t).
\end{aligned}
$$

$$(9.2.7)$$

Integrating (9.2.7) on $[1, t\|u_0\|_\varepsilon]$ with $t > \frac{1}{\|u_0\|_\varepsilon}$, we obtain

$$h(t\|u_0\|_\varepsilon) \geq h(1)(t\|u_0\|_\varepsilon)^4,$$

which gives

$$\Sigma_\varepsilon(tu_0) \geq \Sigma_\varepsilon \left(\frac{u_0}{\|u_0\|_\varepsilon} \right) \|u_0\|_\varepsilon^4 t^4.$$

Therefore,

$$J_\varepsilon(tu_0) = \frac{t^2}{2} \|u_0\|_\varepsilon^2 - \Sigma_\varepsilon(tu_0) \leq C_1 t^2 - C_2 t^4 \quad \text{for } t > \frac{1}{\|u_0\|_\varepsilon}.$$

Taking $e = tu_0$ with t sufficiently large, we conclude that (ii) holds. $\qquad \square$

Lemma 9.2.2 and a variant of the mountain pass theorem without the (PS) condition establish the existence of a Palais–Smale sequence $(u_n) \subset \mathcal{H}_\varepsilon$ at the mountain pass level c'_ε that can be also characterized as

$$c'_\varepsilon = \inf_{u \in \mathcal{H}_\varepsilon \setminus \{0\}} \max_{t > 0} J_\varepsilon(tu).$$

Since $\text{supp}(u_0) \subset \Lambda_\varepsilon$, there exists $\kappa > 0$ independent of ε, ℓ, a such that $c'_\varepsilon < \kappa$ for small $\varepsilon > 0$. We now define

$$\mathcal{B} = \{u \in H^s(\mathbb{R}^N) : \|u\|_\varepsilon^2 \leq 4(\kappa + 1)\}$$

and set

$$\tilde{K}_\varepsilon(u)(x) = \frac{1}{|x|^\mu} * G(\varepsilon x, u).$$

With the above notations, we are able to show the following estimate.

Lemma 9.2.3 *Assume that (f_1)–(f_3) hold and $2 < q < \frac{2(N-\mu)}{N-2s}$. Then there exists $\ell_0 > 0$ such that*

$$\frac{\sup_{u \in \mathcal{B}} \|\tilde{K}_\varepsilon(u)\|_{L^\infty(\mathbb{R}^N)}}{\ell_0} \leq \frac{1}{2} \quad \text{for all } \varepsilon > 0.$$

Proof We first show that there exists $C_0 > 0$ such that

$$\sup_{u \in \mathcal{B}} \|\tilde{K}_\varepsilon(u)\|_{L^\infty(\mathbb{R}^N)} \leq C_0 \quad \text{for all } \varepsilon > 0. \tag{9.2.8}$$

We observe that

$$|G(\varepsilon x, u)| \leq C(|u|^2 + |u|^q) \quad \text{for all } \varepsilon > 0. \tag{9.2.9}$$

Consequently,

$$\begin{aligned}
|\tilde{K}_\varepsilon(u)(x)| &= \left| \int_{\mathbb{R}^N} \frac{G(\varepsilon x, u)}{|x - y|^\mu} \, dy \right| \\
&\leq \left| \int_{|x-y| \leq 1} \frac{G(\varepsilon x, u)}{|x - y|^\mu} \, dy \right| + \left| \int_{|x-y| > 1} \frac{G(\varepsilon x, u)}{|x - y|^\mu} \, dy \right| \\
&\leq C \int_{|x-y| \leq 1} \frac{|u(y)|^2 + |u(y)|^q}{|x - y|^\mu} \, dy + C \int_{\mathbb{R}^N} (|u|^2 + |u|^q) \, dy \\
&\leq C \int_{|x-y| \leq 1} \frac{|u(y)|^2 + |u(y)|^q}{|x - y|^\mu} \, dy + C, \tag{9.2.10}
\end{aligned}$$

where in the last line we used Theorem 1.1.8 and $\|u\|_\varepsilon^2 \leq 4(\kappa + 1)$. Now, we take

$$t \in \left(\frac{N}{N - \mu}, \frac{N}{N - 2s} \right] \quad \text{and} \quad r \in \left(\frac{N}{N - \mu}, \frac{2N}{q(N - 2s)} \right].$$

Hölder's inequality, Theorem 1.1.8 and the fact that $\|u\|_\varepsilon^2 \le 4(\kappa + 1)$ imply that

$$\int_{|x-y|\le 1} \frac{|u(y)|^2}{|x - y|^\mu}\, dy \le \left(\int_{|x-y|\le 1} |u|^{2t}\, dy\right)^{\frac{1}{t}} \left(\int_{|x-y|\le 1} \frac{1}{|x - y|^{\frac{t\mu}{t-1}}}\, dy\right)^{\frac{t-1}{t}}$$

$$\le C_*(4(\kappa + 1))^2 \left(\int_0^1 \rho^{N-1-\frac{t\mu}{t-1}}\, d\rho\right)^{\frac{t-1}{t}} < \infty, \qquad (9.2.11)$$

because $N - 1 - \frac{t\mu}{t-1} > -1$. Similarly, we get

$$\int_{|x-y|\le 1} \frac{|u(y)|^q}{|x - y|^\mu}\, dy \le \left(\int_{|x-y|\le 1} |u|^{rq}\, dy\right)^{\frac{1}{r}} \left(\int_{|x-y|\le 1} \frac{1}{|x - y|^{\frac{r\mu}{r-1}}}\, dy\right)^{\frac{r-1}{r}}$$

$$\le C_*(4(\kappa + 1))^q \left(\int_0^1 \rho^{N-1-\frac{r\mu}{r-1}}\, d\rho\right)^{\frac{r-1}{r}} < \infty, \qquad (9.2.12)$$

because $N - 1 - \frac{r\mu}{r-1} > -1$. Combining (9.2.11) and (9.2.12) we have that

$$\int_{|x-y|\le 1} \frac{|u(y)|^2 + |u(y)|^q}{|x - y|^\mu}\, dy \le C \quad \text{for all } x \in \mathbb{R}^N,$$

which in view of (9.2.10) yields (9.2.8). Then there exists $\ell_0 > 0$ such that

$$\frac{\sup_{u\in\mathcal{B}} \|\tilde{K}_\varepsilon(u)\|_{L^\infty(\mathbb{R}^N)}}{\ell_0} \le \frac{C_0}{\ell_0} \le \frac{1}{2} \quad \text{for all } \varepsilon > 0.$$

$\square$

Let $a > 0$ be the unique number such that

$$\frac{f(a)}{a} = \frac{V_0}{\ell_0},$$

where ℓ_0 is given in Lemma 9.2.3, and consider the penalized problem (9.2.2) with these choices.

Let us introduce the Nehari manifold associated with (9.2.2), that is

$$\mathcal{N}_\varepsilon = \{u \in \mathcal{H}_\varepsilon \setminus \{0\} : \langle J_\varepsilon'(u), u \rangle = 0\}.$$

Using Theorem 9.2.1 and (g_1)–(g_2), we see that

$$\|u\|_{\varepsilon}^2 \leq C(\|u\|_{\varepsilon}^4 + \|u\|_{\varepsilon}^{2q}),$$

for all $u \in \mathcal{N}_{\varepsilon}$, so there exists $r > 0$, independent of $\varepsilon > 0$, such that

$$\|u\|_{\varepsilon} \geq r \quad \text{for all } u \in \mathcal{N}_{\varepsilon}. \tag{9.2.13}$$

Let $\mathbb{S}_{\varepsilon}$ denote the unit sphere in $\mathcal{H}_{\varepsilon}$. Since f is only continuous, the next two results will play a fundamental role to overcome the non-differentiability of $\mathcal{N}_{\varepsilon}$.

Lemma 9.2.4 *Suppose that (V_1)–(V_2) and (f_1)–(f_4) are satisfied. Then, the following facts hold true:*

(a) *For $u \in \mathcal{H}_{\varepsilon} \setminus \{0\}$, let $h_u : \mathbb{R}_+ \to \mathbb{R}$ be defined by $h_u(t) = J_{\varepsilon}(tu)$. Then, there is a unique $t_u > 0$ such that $h_u'(t) > 0$ in $(0, t_u)$ and $h_u'(t) < 0$ in (t_u, ∞).*
(b) *There is a $\tau > 0$, independent of u, such that $t_u \geq \tau$ for every $u \in \mathbb{S}_{\varepsilon}$. Moreover, for each compact set $W \subset \mathbb{S}_{\varepsilon}$, there is a $C_W > 0$ such that $t_u \leq C_W$ for every $u \in W$.*
(c) *The map $\hat{m}_{\varepsilon} : \mathcal{H}_{\varepsilon} \setminus \{0\} \to \mathcal{N}_{\varepsilon}$ given by $\hat{m}_{\varepsilon}(u) = t_u u$ is continuous and $m_{\varepsilon} = \hat{m}|_{\mathbb{S}_{\varepsilon}}$ is a homeomorphism between $\mathbb{S}_{\varepsilon}$ and $\mathcal{N}_{\varepsilon}$. Moreover, $m_{\varepsilon}^{-1}(u) = \frac{u}{\|u\|_{\varepsilon}}$.*

Proof

(a) From the proof of Lemma 9.2.2, we deduce that $h_u(0) = 0$, $h_u(t) > 0$ for $t > 0$ small and $h_u(t) < 0$ for t large. Then, by the continuity of h_u, it is easy to see that there exists $t_u > 0$ such that $\max_{t \geq 0} h_u(t) = h_u(t_u)$, $t_u u \in \mathcal{N}_{\varepsilon}$ and $h_u v(t_u) = 0$. Noting that

$$tu \in \mathcal{N}_{\varepsilon} \Longleftrightarrow \|u\|_{\varepsilon}^2 = \int_{\mathbb{R}^N} \left(\frac{1}{|x|^{\mu}} * \frac{G(\varepsilon x, tu)}{t} \right) g(\varepsilon x, tu)u \, dx,$$

and using (g_4), we get the uniqueness of a such t_u.
(b) Let $u \in \mathbb{S}_{\varepsilon}$. Recalling that $h_u'(t_u) = 0$, and using (g_1), (g_2), Theorem 9.2.1 (see estimates in Lemma 9.2.2), and Theorem 1.1.8, we get that for any small $\xi > 0$,

$$t_u^2 = \int_{\mathbb{R}^N} \tilde{K}_{\varepsilon}(t_u u) g(\varepsilon x, t_u u) t_u u \, dx \leq \xi C_1 t_u^4 + C_2 C_{\xi} t_u^{2q}.$$

Since $q > 2$, there exists a $\tau > 0$, independent of u, such that $t_u \geq \tau$. Now, by (g_3),

$$J_{\varepsilon}(v) = J_{\varepsilon}(v) - \frac{1}{4}\langle J_{\varepsilon}'(v), v \rangle$$

$$= \frac{1}{4}\|v\|_\varepsilon^2 - \frac{1}{4}\int_{\mathbb{R}^N} \tilde{K}_\varepsilon(u)[2G(\varepsilon x, u) - g(\varepsilon x, u)u]\,dx$$

$$\geq \frac{1}{4}\|v\|_\varepsilon^2 \quad \text{for all } v \in \mathcal{N}_\varepsilon. \tag{9.2.14}$$

Hence, if $W \subset \mathbb{S}_\varepsilon$ is a compact set and $(u_n) \subset W$ is such that $t_{u_n} \to \infty$, it follows that $u_n \to u$ in H_ε^s and $J_\varepsilon(t_{u_n}u_n) \to -\infty$. Taking $v_n = t_{u_n}u_n \in \mathcal{N}_\varepsilon$ in (9.2.14), we get

$$0 < \frac{1}{4} \leq \frac{J_\varepsilon(t_{u_n}u_n)}{t_{u_n}^2} \leq 0 \quad \text{as } n \to \infty$$

which gives a contradiction.

(c) Since (a) and (b) hold, we can apply Proposition 2.4.2 to deduce the thesis. $\qquad\square$

Remark 9.2.5 From the estimates in (b), we deduce that J_ε is coercive on $\mathcal{N}_\varepsilon$. Indeed, for all $u \in \mathcal{N}_\varepsilon$

$$J_\varepsilon(u) \geq \frac{1}{4}\|u\|_\varepsilon^2 \to \infty \quad \text{as } \|u\|_\varepsilon \to \infty.$$

This estimate in conjunction with (9.2.13) implies that $J_\varepsilon|_{\mathcal{N}_\varepsilon}$ is bounded below by some positive constant independent of ε.

Let us define the maps $\hat{\psi}_\varepsilon : \mathcal{H}_\varepsilon \setminus \{0\} \to \mathbb{R}$ by $\hat{\psi}_\varepsilon(u) = J_\varepsilon(\hat{m}_\varepsilon(u))$, and $\psi_\varepsilon = \hat{\psi}|_{\mathbb{S}_\varepsilon}$. The next result is a consequence of Lemma 9.2.4, Proposition 2.4.3 and Corollary 2.4.4.

Proposition 9.2.6 *Suppose that (V_1)–(V_2) and (f_1)–(f_4) hold. Then we have:*

(a) $\hat{\psi}_\varepsilon \in C^1(\mathcal{H}_\varepsilon \setminus \{0\}, \mathbb{R})$ *and*

$$\langle \hat{\psi}_\varepsilon'(u), v \rangle = \frac{\|\hat{m}_\varepsilon(u)\|_\varepsilon}{\|u\|_\varepsilon} \langle J_\varepsilon'(\hat{m}_\varepsilon(u)), v \rangle,$$

for every $u \in \mathcal{H}_\varepsilon \setminus \{0\}$ and $v \in \mathcal{H}_\varepsilon$.

(b) $\psi_\varepsilon \in C^1(\mathbb{S}_\varepsilon, \mathbb{R})$ *and* $\langle \psi_\varepsilon'(u), v \rangle = \|m_\varepsilon(u)\|_\varepsilon \langle J_\varepsilon'(m_\varepsilon(u)), v \rangle$, *for every $v \in T_u\mathbb{S}_\varepsilon$.*

(c) *If (u_n) is a Palais–Smale sequence for ψ_ε, then $(m_\varepsilon(u_n))$ is a Palais–Smale sequence for J_ε. Moreover, if $(u_n) \subset \mathcal{N}_\varepsilon$ is a bounded Palais–Smale sequence for J_ε, then $(m_\varepsilon^{-1}(u_n))$ is a Palais–Smale sequence for ψ_ε.*

(d) u is a critical point of ψ_ε if and only if $m_\varepsilon(u)$ is a nontrivial critical point for J_ε. Moreover, the corresponding critical values coincide and

$$\inf_{u \in \mathbb{S}_\varepsilon} \psi_\varepsilon(u) = \inf_{u \in \mathcal{N}_\varepsilon} J_\varepsilon(u).$$

Remark 9.2.7 As in [322], we have the following characterization of the infimum of J_ε on $\mathcal{N}_\varepsilon$:

$$c_\varepsilon = \inf_{u \in \mathcal{N}_\varepsilon} J_\varepsilon(u) = \inf_{u \in \mathcal{H}_\varepsilon \setminus \{0\}} \max_{t>0} J_\varepsilon(tu) = \inf_{u \in \mathbb{S}_\varepsilon} \max_{t>0} J_\varepsilon(tu).$$

In the next result we show that J_ε verifies a local compactness condition.

Lemma 9.2.8 *J_ε satisfies the $(PS)_c$ condition for all $c \in [c_\varepsilon, \kappa]$.*

Proof Let $(u_n) \subset \mathcal{H}_\varepsilon$ be a Palais–Smale sequence at the level c, that is $J_\varepsilon(u_n) \to c$ and $J_\varepsilon'(u_n) \to 0$. We divide the proof in two main steps.

Step 1   For every $\eta > 0$ there exists $R = R_\eta > 0$ such that

$$\limsup_{n \to \infty} \int_{\mathbb{R}^N \setminus B_R} \left(|(-\Delta)^{\frac{s}{2}} u_n|^2 + V(\varepsilon x) u_n^2 \right) dx < \eta. \tag{9.2.15}$$

Condition (g_3) implies that

$$
\begin{aligned}
J_\varepsilon(u_n) - \frac{1}{4} \langle J_\varepsilon'(u_n), u_n \rangle &= \left(\frac{1}{2} - \frac{1}{4} \right) \|u_n\|_{H_\varepsilon^s}^2 + \frac{1}{4} \int_{\mathbb{R}^N} \tilde{K}_\varepsilon(u_n) g(\varepsilon x, u_n) u_n \, dx \\
&\quad - \frac{1}{2} \int_{\mathbb{R}^N} \tilde{K}_\varepsilon(u_n) G(\varepsilon x, u_n) \, dx \\
&\geq \frac{1}{4} \|u_n\|_\varepsilon^2,
\end{aligned}
$$

which shows that (u_n) is bounded in $\mathcal{H}_\varepsilon$. Moreover, there exists $n_0 \in \mathbb{N}$ such that

$$\|u_n\|_\varepsilon^2 \leq 4(\kappa + 1) \quad \text{for all } n \geq n_0.$$

Then we may assume that $u_n \rightharpoonup u$ in $\mathcal{H}_\varepsilon$ and $u_n \to u$ in $L_{\text{loc}}^r(\mathbb{R}^N)$ for any $r \in [1, 2_s^*)$. Moreover, by Lemma 9.2.3, we deduce that

$$\frac{\sup_{n \geq n_0} \|\tilde{K}_\varepsilon(u_n)\|_{L^\infty(\mathbb{R}^N)}}{\ell_0} \leq \frac{1}{2}. \tag{9.2.16}$$

Fix $R > 0$ such that $\Lambda_\varepsilon \subset B_{R/2}$ and take $\psi_R \in C^\infty(\mathbb{R}^N)$ such that $\psi_R = 0$ in $B_{R/2}$, $\psi_R = 1$ in B_R^c, $0 \le \psi_R \le 1$ and $\|\nabla\psi_R\|_{L^\infty(\mathbb{R}^N)} \le C/R$ for some $C > 0$ independent of R. Since the sequence $(u_n\psi_R)$ is bounded in $\mathcal{H}_\varepsilon$, condition (g_3) implies that

$$\int_{\mathbb{R}^N} \left((-\Delta)^{\frac{s}{2}} u_n (-\Delta)^{\frac{s}{2}} (u_n\psi_R) + V(\varepsilon x)\psi_R u_n^2 \right) dx$$

$$= \langle J'_\varepsilon(u_n), u_n\psi_R \rangle + \int_{\mathbb{R}^N} \tilde{K}_\varepsilon(u_n) g(\varepsilon x, u_n) u_n\psi_R \, dx$$

$$\le o_n(1) + \int_{\mathbb{R}^N} \frac{\tilde{K}_\varepsilon(u_n)}{\ell_0} V(\varepsilon x) u_n^2 \psi_R \, dx,$$

whence

$$\int_{\mathbb{R}^N} \left(|(-\Delta)^{\frac{s}{2}} u_n|^2 \psi_R + V(\varepsilon x) u_n^2 \psi_R \right) dx \le \int_{\mathbb{R}^N} \frac{\tilde{K}_\varepsilon(u_n)}{\ell_0} V(\varepsilon x) u_n^2 \psi_R \, dx + o_n(1)$$

$$- \iint_{\mathbb{R}^{2N}} \frac{(u_n(x) - u_n(u))(\psi_R(x) - \psi_R(y))}{|x - y|^{N+2s}} u_n(y) \, dx dy.$$

$$(9.2.17)$$

Further, the Hölder inequality, the boundedness of (u_n) and Remark 1.4.6 imply that

$$\left| \iint_{\mathbb{R}^{2N}} \frac{(u_n(x) - u_n(u))(\psi_R(x) - \psi_R(y))}{|x - y|^{N+2s}} u_n(y) \, dx dy \right|$$

$$\le \left(\iint_{\mathbb{R}^{2N}} \frac{|u_n(x) - u_n(y)|^2}{|x - y|^{N+2s}} \, dx dy \right)^{\frac{1}{2}} \left(\iint_{\mathbb{R}^{2N}} \frac{|\psi_R(x) - \psi_R(y)|^2}{|x - y|^{N+2s}} u_n^2(y) \, dx dy \right)^{\frac{1}{2}}$$

$$\le C \left(\iint_{\mathbb{R}^{2N}} \frac{|\psi_R(x) - \psi_R(y)|^2}{|x - y|^{N+2s}} u_n^2(y) \, dx dy \right)^{\frac{1}{2}} \le \frac{C}{R^s}.$$

$$(9.2.18)$$

On the other hand, by (9.2.16), for all $n \ge n_0$

$$\int_{\mathbb{R}^N} \frac{\tilde{K}_\varepsilon(u_n)}{\ell_0} V(\varepsilon x) u_n^2 \psi_R \, dx \le \int_{\mathbb{R}^N} \frac{\sup_{n \ge n_0} \|\tilde{K}_\varepsilon(u_n)\|_{L^\infty(\mathbb{R}^N)}}{\ell_0} V(\varepsilon x) u_n^2 \psi_R \, dx$$

$$\le \frac{1}{2} \int_{\mathbb{R}^N} V(\varepsilon x) u_n^2 \psi_R \, dx.$$

$$(9.2.19)$$

Putting together (9.2.17), (9.2.18), (9.2.19) and using the definition of ψ_R, we get

$$\limsup_{n\to\infty} \int_{\mathbb{R}^N\setminus B_R} \left(|(-\Delta)^{\frac{s}{2}} u_n|^2 + V(\varepsilon x) u_n^2 \right) dx \le \frac{C}{R^s},$$

and letting $R\to\infty$ we obtain (9.2.15).

Step 2   Let us prove that $u_n \to u$ in $\mathcal{H}_\varepsilon$ as $n \to \infty$.

Set $\Psi_n = \|u_n - u\|_\varepsilon^2$ and observe that

$$\Psi_n = \langle J_\varepsilon'(u_n), u_n \rangle - \langle J_\varepsilon'(u_n), u \rangle + \int_{\mathbb{R}^N} \tilde{K}_\varepsilon(u_n) g(\varepsilon x, u_n)(u_n - u)\, dx + o_n(1). \qquad (9.2.20)$$

Since $\langle J_\varepsilon'(u_n), u_n \rangle = \langle J_\varepsilon'(u_n), u \rangle = o_n(1)$, to infer that $\Psi_n \to 0$ as $n \to \infty$ we only need to show that

$$\int_{\mathbb{R}^N} \tilde{K}_\varepsilon(u_n) g(\varepsilon x, u_n)(u_n - u)\, dx = o_n(1).$$

We note that $G(\varepsilon x, u_n)$ is bounded in $L^{\frac{2N}{2N-\mu}}(\mathbb{R}^N)$ (since $2 < q < \frac{2(N-\mu)}{N-2s}$), $u_n \to u$ a.e. in $\mathbb{R}^N$, and $G(\cdot, t)$ is continuous in t, so we deduce that

$$G(\varepsilon x, u_n) \rightharpoonup G(\varepsilon x, u) \quad \text{in } L^{\frac{2N}{2N-\mu}}(\mathbb{R}^N). \qquad (9.2.21)$$

By virtue of Theorem 9.2.1, the convolution

$$\frac{1}{|x|^\mu} * h(x) \in L^{\frac{2N}{\mu}}(\mathbb{R}^N), \quad h \in L^{\frac{2N}{2N-\mu}}(\mathbb{R}^N)$$

gives a bounded linear operator from $L^{\frac{2N}{2N-\mu}}(\mathbb{R}^N)$ to $L^{\frac{2N}{\mu}}(\mathbb{R}^N)$, which implies that

$$\tilde{K}_\varepsilon(u_n) = \frac{1}{|x|^\mu} * G(\varepsilon x, u_n) \rightharpoonup \frac{1}{|x|^\mu} * G(\varepsilon x, u) = \tilde{K}_\varepsilon(u) \quad \text{in } L^{\frac{2N}{\mu}}(\mathbb{R}^N). \qquad (9.2.22)$$

Since g has subcritical growth, for any fixed $R > 0$, the local compact embedding in Theorem 1.1.8 and (9.2.22) show that

$$\lim_{n\to\infty} \int_{B_R} \tilde{K}_\varepsilon(u_n) g(\varepsilon x, u_n)(u_n - u)\, dx = 0.$$

By (9.2.4) and the boundedness of $\tilde{K}_\varepsilon(u_n)$, we obtain

$$\int_{\mathbb{R}^N \setminus B_R} \tilde{K}_\varepsilon(u_n) |g(\varepsilon x, u_n) u_n| \, dx \le C \int_{\mathbb{R}^N \setminus B_R} |g(\varepsilon x, u_n) u_n| \, dx$$

$$\le C \int_{\mathbb{R}^N \setminus B_R} (|u_n|^2 + |u_n|^q) \, dx.$$

From Step 1, (V_1), the interpolation inequality in $L^q(\mathbb{R}^N \setminus B_R)$ with $q \in (2, 2_s^*)$, and the boundedness of (u_n) in $L^{2_s^*}(\mathbb{R}^N)$, we see that for every $\eta > 0$ there exists an $R = R_\eta > 0$ such that

$$\limsup_{n \to \infty} \int_{\mathbb{R}^N \setminus B_R} |u_n|^2 \, dx \le C\eta \quad \text{and} \quad \limsup_{n \to \infty} \int_{\mathbb{R}^N \setminus B_R} |u_n|^q \, dx \le C\eta,$$

which implies that

$$\limsup_{n \to \infty} \int_{\mathbb{R}^N \setminus B_R} \tilde{K}_\varepsilon(u_n) |g(\varepsilon x, u_n) u_n| \, dx \le C\eta.$$

On the other hand, by (9.2.4), the boundedness of $\tilde{K}_\varepsilon(u_n)$ and Hölder's inequality we get

$$\int_{\mathbb{R}^N \setminus B_R} \tilde{K}_\varepsilon(u_n) |g(\varepsilon x, u_n) u| \, dx \le C \left[\left(\int_{\mathbb{R}^N \setminus B_R} |u_n|^2 \, dx \right)^{\frac{1}{2}} + \left(\int_{\mathbb{R}^N \setminus B_R} |u_n|^q \, dx \right)^{\frac{q-1}{q}} \right],$$

whence

$$\limsup_{n \to \infty} \int_{\mathbb{R}^N \setminus B_R} \tilde{K}_\varepsilon(u_n) |g(\varepsilon x, u_n) u| \, dx \le C(\eta^{\frac{1}{2}} + \eta^{\frac{q-1}{q}}).$$

Taking into account the above limit inequalities, we infer that

$$\lim_{n \to \infty} \int_{\mathbb{R}^N} \tilde{K}_\varepsilon(u_n) g(\varepsilon x, u_n)(u_n - u) \, dx = 0.$$

$\square$

We also have the following result:

Lemma 9.2.9 *The functional ψ_ε satisfies the* (PS)$_c$ *condition on $\mathbb{S}_\varepsilon$ for any $c \in [c_\varepsilon, \kappa]$.*

Proof Let $(u_n) \subset \mathbb{S}_\varepsilon$ be a Palais–Smale sequence for ψ_ε at the level c. Then $\psi_\varepsilon(u_n) \to c$ and $\|\psi_\varepsilon'(u_n)\|_* \to 0$, where $\| \cdot \|_*$ denotes the norm in the dual space of $(T_{u_n} \mathbb{S}_\varepsilon)^*$. By

Proposition 9.2.6-(c), we infer that $(m_\varepsilon(u_n))$ is a Palais–Smale sequence for J_ε at the level c. In view of Lemma 9.2.8, we see that, up to a subsequence, there exists $u \in \mathbb{S}_\varepsilon$ such that $m_\varepsilon(u_n) \to m_\varepsilon(u)$ in $\mathcal{H}_\varepsilon$. By Lemma 9.2.4-(c), we conclude that $u_n \to u$ in $\mathbb{S}_\varepsilon$.   □

Finally, we establish the existence of a ground state solution to (9.2.2).

Theorem 9.2.10 *For all $\varepsilon > 0$, problem (9.2.2) admits a positive ground state solution.*

Proof Combining Lemmas 9.2.2, 9.2.8, and Theorem 2.2.9, we can see that for every $\varepsilon > 0$ there exists $u_\varepsilon \in \mathcal{H}_\varepsilon$ such that $J_\varepsilon(u_\varepsilon) = c_\varepsilon$ and $J'_\varepsilon(u_\varepsilon) = 0$. Since $g(x,t) = 0$ for $t \leq 0$ and $(x - y)(x^- - y^-) \geq |x^- - y^-|^2$ for all $x, y \in \mathbb{R}$, it is readily verified that $\langle J'_\varepsilon(u_\varepsilon), u_\varepsilon^- \rangle = 0$ implies that $u_\varepsilon \geq 0$ in $\mathbb{R}^N$. Since $u_\varepsilon \in \mathcal{B}$, we can proceed as in the proof of Lemma 9.2.3 to see that $\tilde{K}_\varepsilon(u_\varepsilon) \in L^\infty(\mathbb{R}^N)$. Then, arguing as in the proof of Lemma 9.5.2 below, we have that $u_\varepsilon \in L^\infty(\mathbb{R}^N) \cap C_{loc}^{0,\alpha}(\mathbb{R}^N)$. Finally, using Theorem 1.3.5, we deduce that $u_\varepsilon > 0$ in $\mathbb{R}^N$.   □

9.3   The Autonomous Choquard Problem

In this section we deal with the limit problem associated with (9.2.1), namely

$$\begin{cases} (-\Delta)^s u + V_0 u = \left(\frac{1}{|x|^\mu} * F(u) \right) f(u) & \text{in } \mathbb{R}^N, \\ u \in H^s(\mathbb{R}^N), \ u > 0 & \text{in } \mathbb{R}^N. \end{cases} \tag{P_0}$$

The corresponding functional $J_0 : \mathcal{H}_0 \to \mathbb{R}$ associated is given by

$$J_0(u) = \frac{1}{2}\|u\|_0^2 - \Sigma_0(u),$$

where $\mathcal{H}_0$ is the space $H^s(\mathbb{R}^N)$, endowed with the norm

$$\|u\|_0 = \left([u]_s^2 + \int_{\mathbb{R}^N} V_0 u^2 \, dx \right)^{\frac{1}{2}},$$

and

$$\Sigma_0(u) = \frac{1}{2} \int_{\mathbb{R}^N} \left(\frac{1}{|x|^\mu} * F(u) \right) F(u) \, dx.$$

Consider the corresponding Nehari manifold

$$\mathcal{N}_0 = \{u \in \mathcal{H}_0 \setminus \{0\} : \langle J'_0(u), u \rangle = 0\}$$

and denote by $\mathbb{S}_0$ the unit sphere in $\mathcal{H}_0$. Arguing as in the proofs of Lemma 9.2.4 and Proposition 9.2.6, we can see that the following results hold.

Lemma 9.3.1 *Suppose that f satisfies (f_1)–(f_4). Then, the following assertions hold true:*

(a) *For any $u \in \mathcal{H}_0 \setminus \{0\}$, let $h_u : \mathbb{R}_+ \to \mathbb{R}$ be defined by $h_u(t) = J_0(tu)$. Then, there is a unique $t_u > 0$ such that $h'_u(t) > 0$ in $(0, t_u)$ and $h'_u(t) < 0$ in (t_u, ∞).*
(b) *There is $\tau > 0$, independent of u, such that $t_u \geq \tau$ for every $u \in \mathbb{S}_0$. Moreover, for each compact set $\mathcal{W} \subset \mathbb{S}_0$, there is $C_{\mathcal{W}} > 0$ such that $t_u \leq C_{\mathcal{W}}$ for every $u \in \mathcal{W}$.*
(c) *The map $\hat{m}_0 : \mathcal{H}_0 \setminus \{0\} \to \mathcal{N}_0$ given by $\hat{m}_0(u) = t_u u$ is continuous and $m_0 = \hat{m}|_{\mathbb{S}_0}$ is a homeomorphism between $\mathbb{S}_0$ and $\mathcal{N}_0$. Moreover, $m_0^{-1}(u) = \frac{u}{\|u\|_0}$.*

Let us define the maps $\hat{\psi}_0 : \mathcal{H}_0 \setminus \{0\} \to \mathbb{R}$ by $\hat{\psi}_0(u) = J_0(\hat{m}_0(u))$, and $\psi = \hat{\psi}_0|_{\mathbb{S}_0}$.

Proposition 9.3.2 *Suppose that f satisfies (f_1)–(f_4). Then, one has:*

(a) *$\hat{\psi}_0 \in C^1(\mathcal{H}_0 \setminus \{0\}, \mathbb{R})$ and*

$$\langle \hat{\psi}'_0(u), v \rangle = \frac{\|\hat{m}_0(u)\|_0}{\|u\|_0} \langle J'_0(\hat{m}_0(u)), v \rangle,$$

for every $u \in \mathcal{H}_0 \setminus \{0\}$ and $v \in \mathcal{H}_0$.
(b) *$\psi_0 \in C^1(\mathbb{S}_0, \mathbb{R})$ and $\langle \psi'_0(u), v \rangle = \|m_0(u)\|_0 \langle J'_0(m_0(u)), v \rangle$, for every $v \in T_u \mathbb{S}_0$.*
(c) *If (u_n) is a Palais–Smale sequence for ψ_0, then $(m_0(u_n))$ is a Palais–Smale sequence for J_0. Moreover, if $(u_n) \subset \mathcal{N}_0$ is a bounded Palais–Smale sequence for J_0, then $(m_0^{-1}(u_n))$ is a Palais–Smale sequence for ψ_0.*
(d) *u is a critical point of ψ_0 if and only if $m_0(u)$ is a nontrivial critical point for J_0. Moreover, the corresponding critical values coincide and*

$$\inf_{u \in \mathbb{S}_0} \psi_0(u) = \inf_{u \in \mathcal{N}_0} J_0(u).$$

Furthermore, we have the following characterization of the infimum of J_0 on $\mathcal{N}_0$:

$$c_0 = \inf_{u \in \mathcal{N}_0} J_0(u) = \inf_{u \in \mathcal{H}_0 \setminus \{0\}} \max_{t > 0} J_0(tu) = \inf_{u \in \mathbb{S}_0} \max_{t > 0} J_0(tu). \tag{9.3.1}$$

Arguing as in the proof of Lemma 9.2.2, we see that J_0 has a mountain pass geometry. By the mountain pass theorem without the Palais–Smale condition (see Remark 2.2.10), there exists a Palais–Smale sequence $(u_n) \subset \mathcal{H}_0$ such that

$$J_0(u_n) \to c_0 \quad \text{and} \quad J'_0(u_n) \to 0 \quad \text{in } \mathcal{H}_0^*.$$

The next lemma allows us to assume that the weak limit of a Palais–Smale sequence at the level c_0 is nontrivial.

Lemma 9.3.3 *Let $(u_n) \subset \mathcal{H}_0$ be a Palais–Smale sequence for J_0 at the level c_0 and such that $u_n \rightharpoonup 0$ in $\mathcal{H}_0$. Then*

(a) *either $u_n \to 0$ in $\mathcal{H}_0$, or*
(b) *there exist a sequence $(y_n) \subset \mathbb{R}^N$, and constants $R > 0$ and $\gamma > 0$ such that*

$$\liminf_{n \to \infty} \int_{B_R(y_n)} u_n^2 \, dx \geq \gamma > 0.$$

Proof Suppose that (b) does not hold. Then, for all $R > 0$,

$$\lim_{n \to \infty} \sup_{y \in \mathbb{R}^N} \int_{B_R(y)} u_n^2 \, dx = 0.$$

Since (u_n) is bounded in $\mathcal{H}_0$, Lemma 1.4.4 shows that $u_n \to 0$ in $L^r(\mathbb{R}^N)$ for any $r \in (2, 2_s^*)$. Then, by (f_1)–(f_2) and applying Theorem 9.2.1, we get

$$\int_{\mathbb{R}^N} \left(\frac{1}{|x|^\mu} * F(u_n) \right) f(u_n) u_n \, dx = o_n(1).$$

Taking into account that $\langle J_0'(u_n), u_n \rangle = o_n(1)$, we conclude that $\|u_n\|_0 \to 0$ as $n \to \infty$. $\quad\square$

Now, we prove the following result for the autonomous problem.

Lemma 9.3.4 *Let $(u_n) \subset \mathcal{H}_0$ be a Palais–Smale sequence for J_0 at the level c_0. Then, problem (P_0) has a positive ground state.*

Proof Since

$$J_0(u_n) - \frac{1}{4}\langle J_0'(u_n), u_n \rangle \geq \frac{1}{4}\|u_n\|_0^2,$$

it is readily verified that (u_n) is bounded in $\mathcal{H}_0$. Moreover, arguing as in the proof of Lemma 9.3.3, we can find a sequence $(y_n) \subset \mathbb{R}^N$ and constants $R > 0$ and $\gamma > 0$ such that

$$\liminf_{n \to \infty} \int_{B_R(y_n)} u_n^2 \, dx \geq \gamma > 0.$$

Set $v_n = u_n(\cdot - y_n)$. Since J_0 and J_0' are both invariant under translation,

$$J_0(v_n) \to c_0 \text{ and } J_0'(v_n) \to 0 \text{ in } \mathcal{H}_0^*.$$

We observe that (v_n) is bounded in $\mathcal{H}_0$, so we may assume that $v_n \rightharpoonup v$ in $\mathcal{H}_0$, for some $v \neq 0$. Let us show that v is a weak solution to (P_{V_0}). Fix $\varphi \in C_c^\infty(\mathbb{R}^N)$. Since $\|v_n\|_0 \leq C$ for all $n \in \mathbb{N}$, we can argue as in the proof of Lemma 9.2.3 to deduce that

$$\left\| \frac{1}{|\cdot|^\mu} * F(v_n) \right\|_{L^\infty(\mathbb{R}^N)} \leq C \quad \text{for any } n \in \mathbb{N}.$$

Then, since f is a continuous function with subcritical growth and $v_n \to v$ in $L^r_{\mathrm{loc}}(\mathbb{R}^N)$ for any $r \in [1, 2_s^*)$, the dominated convergence theorem gives

$$\int_{\mathbb{R}^N} \left(\frac{1}{|x|^\mu} * F(v_n) \right) f(v_n)\varphi \, dx \to \int_{\mathbb{R}^N} \left(\frac{1}{|x|^\mu} * F(v) \right) f(v)\varphi \, dx,$$

which combined with the weak convergence of (v_n) in $\mathcal{H}_0$ and the fact that $\langle J_0'(v_n), \varphi \rangle = o_n(1)$, implies that $\langle J_0'(v), \varphi \rangle = 0$. Since $C_c^\infty(\mathbb{R}^N)$ is dense in $\mathcal{H}_0$, we get $\langle J_0'(v), \varphi \rangle = 0$ for all $\varphi \in \mathcal{H}_0$. In particular, $v \in \mathcal{N}_{V_0}$. Using the definition of c_0 together with Fatou's lemma, we also deduce that $J_0(v) = c_0$.

Now, recalling that $f(t) = 0$ for $t \leq 0$ and $(x - y)(x^- - y^-) \geq |x^- - y^-|^2$ for all $x, y \in \mathbb{R}$, it is easy to deduce that $\langle J_0'(v), v^- \rangle = 0$ implies that $v \geq 0$ in $\mathbb{R}^N$.

Proceeding as in the proof of Lemma 9.2.3, we see that $K(v)(x) = \frac{1}{|x|^\mu} * F(v) \in L^\infty(\mathbb{R}^N)$, and arguing as in the proof of Lemma 9.5.2 below, we have that $v \in L^\infty(\mathbb{R}^N)$. Since f has subcritical growth and $K(v)$ is bounded, we obtain that $K(v)f(v) \in L^\infty(\mathbb{R}^N)$, so we can apply Proposition 1.3.2 to infer that $v \in C^{0,\alpha}(\mathbb{R}^N)$ for some $\alpha \in (0, 1)$. Using Theorem 1.3.5 (or Proposition 1.3.11-(ii)), we finally conclude that $v > 0$ in $\mathbb{R}^N$. $\qquad \square$

Remark 9.3.5 As mentioned before, if u is the weak limit of a Palais–Smale sequence (u_n) of J_0 at the level c_0, then we can assume that $u \neq 0$. Otherwise, we would have $u_n \rightharpoonup 0$ in $\mathcal{H}_0$ and, assuming that $u_n \nrightarrow 0$ in $\mathcal{H}_0$, we conclude from Lemma 11.3.4 that there are $(y_n) \subset \mathbb{R}^N$ and $R, \beta > 0$ such that

$$\liminf_{n\to\infty} \int_{B_R(y_n)} u_n^2 \, dx \geq \beta > 0.$$

Set $v_n(x) = u_n(x + y_n)$. Then we see that (v_n) is a Palais–Smale sequence for J_0 at the level c_0, (v_n) is bounded in $\mathcal{H}_0$ and there exists $v \in \mathcal{H}_0$ such that $v_n \rightharpoonup v$ and $v \neq 0$.

Remark 9.3.6 By using a Brezis-Lieb splitting type result for $\Sigma_0(u)$(see [1]), it is easy to see that if (u_n) is a Palais–Smale sequence for J_0 at the level c_0, and $u_n \rightharpoonup u$ in $\mathcal{H}_0$, then

$$J_0(u_n - u) = c_0 - J_0(u) + o_n(1) \quad \text{and} \quad J_0'(u_n - u) = o_n(1).$$

If one assumes that $u \neq 0$, then by Fatou's lemma and (f_3) we have

$$c_0 = \lim_{n\to\infty} \left(J_0(u_n) - \frac{1}{2}\langle J_0'(u_n), u_n\rangle \right) = \liminf_{n\to\infty} \left(\frac{1}{2}\langle \Sigma_0'(u_n), u_n\rangle - \Sigma_0(u_n)\right)$$

$$\geq \left(\frac{1}{2}\langle \Sigma_0'(u), u\rangle - \Sigma_0(u)\right)$$

$$= J_0(u) - \frac{1}{2}\langle J_0'(u), u\rangle \geq c_0,$$

and thus $J_0(u_n - u) = o_n(1)$ and $J_0'(u_n - u) = o_n(1)$. Therefore,

$$\frac{1}{4}\limsup_{n\to\infty} \|u_n - u\|_0^2 \leq \limsup_{n\to\infty} \left(J_0(u_n - u) - \frac{1}{4}\langle J_0'(u_n - u), u_n - u\rangle \right) = 0,$$

which implies that $u_n \to u$ in $\mathcal{H}_0$ as $n \to \infty$. In particular, we deduce that J_0 satisfies the Palais–Smale condition at level c_0.

The next statement is a compactness result for the autonomous problem which will be used later.

Lemma 9.3.7 *Let $(\tilde{v}_n) \subset \mathcal{N}_0$ be such that $J_0(\tilde{v}_n) \to c_0$. Then $(\tilde{v}_n)$ has a convergent subsequence in $\mathcal{H}_0$.*

Proof Since $(\tilde{v}_n) \subset \mathcal{N}_0$ and $J_{V_0}(\tilde{v}_n) \to c_0$, we can apply Lemma 9.3.1-(c) and Proposition 9.3.2-(d) to infer that

$$w_n = m_0^{-1}(\tilde{v}_n) = \frac{\tilde{v}_n}{\|\tilde{v}_n\|_0} \in \mathbb{S}_0$$

and

$$\psi_0(w_n) = J_0(\tilde{v}_n) \to c_0 = \inf_{v\in\mathbb{S}_0} \psi_0(v).$$

Since ψ_0 is a continuous functional bounded below, we can apply Theorem 2.2.1 to find $(\tilde{w}_n) \subset \mathbb{S}_0$ such that $(\tilde{w}_n)$ is a Palais–Smale sequence for ψ_0 at the level c_0 and $\|\tilde{w}_n - w_n\|_0 = o_n(1)$. By Proposition 9.3.2-(c), $(m_0(\tilde{w}_n))$ is a Palais–Smale sequence for J_0 at the level c_0. Taking into account Lemma 9.3.4 and Remark 9.3.6, it follows that there exists $\tilde{w} \in \mathbb{S}_0$ such that $m_0(\tilde{w}_n) \to m_0(\tilde{w})$ in $\mathcal{H}_0$. This fact together with Lemma 9.3.1-(c) shows that $\tilde{v}_n \to \tilde{v}$ in $\mathcal{H}_0$. $\qquad\square$

9.4 Multiplicity Results

In order to study the multiplicity of solutions to (9.1.1), we need to introduce some useful tools. Take $\delta > 0$ such that $M_\delta \subset \Lambda$, where

$$M_\delta = \{x \in \mathbb{R}^N : \operatorname{dist}(x, M) \leq \delta\}.$$

Let η be a smooth nonincreasing cut-off function defined in $[0, \infty)$, such that $\eta = 1$ in $[0, \frac{\delta}{2}]$, $\eta = 0$ in $[\delta, \infty)$, $0 \leq \eta \leq 1$ and $|\eta'| \leq c$ for some $c > 0$. For any $y \in M$, set

$$\Psi_{\varepsilon, y}(x) = \eta(|\varepsilon x - y|) w \left(\frac{\varepsilon x - y}{\varepsilon} \right),$$

where w is a positive ground state solution for J_0 (by Lemma 9.3.4).

Let $t_\varepsilon > 0$ denote the unique positive number satisfying

$$J_\varepsilon(t_\varepsilon \Psi_{\varepsilon, y}) = \max_{t \geq 0} J_\varepsilon(t \Psi_{\varepsilon, y}).$$

Finally, we consider $\Phi_\varepsilon(y) = t_\varepsilon \Psi_{\varepsilon, y}$.

In the next lemma we prove an important relation between Φ_ε and the set M.

Lemma 9.4.1 *The functional Φ_ε has the property that*

$$\lim_{\varepsilon \to 0} J_\varepsilon(\Phi_\varepsilon(y)) = c_0, \quad \text{uniformly in } y \in M.$$

Proof Assume, by contradiction, that there exist $\delta_0 > 0$, $(y_n) \subset M$ and $\varepsilon_n \to 0$ such that

$$|J_{\varepsilon_n}(\Phi_{\varepsilon_n}(y_n)) - c_0| \geq \delta_0. \tag{9.4.1}$$

We first show that $\lim_{n \to \infty} t_{\varepsilon_n} < \infty$. Let us observe that, by using the change of variable $z = \frac{\varepsilon_n x - y_n}{\varepsilon_n}$, if $z \in B_{\frac{\delta}{\varepsilon_n}}$, it follows that $\varepsilon_n z \in B_\delta$ and thus $\varepsilon_n z + y_n \in B_\delta(y_n) \subset M_\delta \subset \Lambda$. Since $G = F$ on $\Lambda \times \mathbb{R}$, we see that

$$J_{\varepsilon_n}(\Phi_{\varepsilon_n}(z_n)) = \frac{t_{\varepsilon_n}^2}{2} \int_{\mathbb{R}^N} |(-\Delta)^{\frac{s}{2}} (\eta(|\varepsilon_n z|) w(z))|^2 \, dz$$

$$+ \frac{t_{\varepsilon_n}^2}{2} \int_{\mathbb{R}^N} V(\varepsilon_n z + y_n)(\eta(|\varepsilon_n z|) w(z))^2 \, dz$$

$$- \Sigma_0(t_{\varepsilon_n} \eta(|\varepsilon_n \cdot|) w).$$

In view of the dominated convergence theorem and Lemma 1.4.8, we obtain that

$$\lim_{n \to \infty} \| \Psi_{\varepsilon_n, y_n} \|_{\varepsilon_n} = \| w \|_0$$

and

$$\lim_{n \to \infty} \Sigma_0(\Psi_{\varepsilon_n, y_n}) = \Sigma_0(w).$$

Using the fact that $t_{\varepsilon_n} \Psi_{\varepsilon_n, y_n} \in \mathcal{N}_{\varepsilon_n}$ and the growth assumptions on f, it is easy to prove that $t_{\varepsilon_n} \to 1$. Indeed, since

$$t_{\varepsilon_n}^2 \| \Psi_{\varepsilon_n, y_n} \|_{\varepsilon_n}^2 = \iint_{\mathbb{R}^{2N}} \frac{F(t_{\varepsilon_n} \Psi_{\varepsilon_n, y_n}) f(t_{\varepsilon_n} \Psi_{\varepsilon_n, y_n}) t_{\varepsilon_n} \Psi_{\varepsilon_n, y_n}}{|x - y|^{\mu}}, \tag{9.4.2}$$

we have that

$$\| w \|_0^2 = \lim_{n \to \infty} \iint_{\mathbb{R}^{2N}} \frac{F(t_{\varepsilon_n} \Psi_{\varepsilon_n, y_n}) f(t_{\varepsilon_n} \Psi_{\varepsilon_n, y_n}) t_{\varepsilon_n} \Psi_{\varepsilon_n, y_n}}{t_{\varepsilon_n}^2 |x - y|^{\mu}}.$$

Using that w is a ground state to (P_0) and condition (f_4), we conclude that $t_{\varepsilon_n} \to 1$. Consequently,

$$\lim_{n \to \infty} \Sigma_0(t_{\varepsilon_n} \eta(| \varepsilon_n \cdot |) w) = \Sigma_0(w)$$

and this yields

$$\lim_{n \to \infty} J_{\varepsilon_n}(\Phi_{\varepsilon_n}(y_n)) = J_0(w) = c_0,$$

which contradicts (9.4.1).   □

Now, take $\delta > 0$ such that $M_\delta \subset \Lambda$, and choose $\rho = \rho(\delta) > 0$ such that $M_\delta \subset B_\delta$. Define $\Upsilon : \mathbb{R}^N \to \mathbb{R}^N$ by setting $\Upsilon(x) = x$ for $|x| \leq \rho$ and $\Upsilon(x) = \frac{\rho x}{|x|}$ for $|x| \geq \rho$. Then we consider the barycenter map $\beta_\varepsilon : \mathcal{N}_\varepsilon \to \mathbb{R}^N$ given by

$$\beta_\varepsilon(u) = \frac{\int_{\mathbb{R}^N} \Upsilon(\varepsilon x) u^2(x) \, dx}{\int_{\mathbb{R}^N} u^2(x) \, dx}.$$

Arguing as in the proof of Lemma 6.3.18 we obtain.

Lemma 9.4.2 *The function β_ε has the property that*

$$\lim_{\varepsilon \to 0} \beta_\varepsilon(\Phi_\varepsilon(y)) = y, \quad \text{uniformly in } y \in M.$$

The next compactness result will play a fundamental role in showing that the solutions of the modified problem are solutions of the original problem.

Lemma 9.4.3 *Let $\varepsilon_n \to 0$ and $(u_n) = (u_{\varepsilon_n}) \subset \mathcal{N}_{\varepsilon_n}$ be such that $J_{\varepsilon_n}(u_n) \to c_0$. Then there exists $(\tilde{y}_n) = (\tilde{y}_{\varepsilon_n}) \subset \mathbb{R}^N$ such that $v_n(x) = u_n(x + \tilde{y}_n)$ has a convergent subsequence in $H^s(\mathbb{R}^N)$. Moreover, up to a subsequence, $y_n = \varepsilon_n \tilde{y}_n \to y_0 \in M$.*

Proof Since $\langle J'_{\varepsilon_n}(u_n), u_n \rangle = 0$ and $J_{\varepsilon_n}(u_n) \to c_0$, we deduce that (u_n) is bounded in $\mathcal{H}_{\varepsilon_n}$. Note that $c_0 > 0$, and since $\|u_n\|_{\varepsilon_n} \to 0$ would imply $J_{\varepsilon_n}(u_n) \to 0$, we can argue as in Lemma 9.3.3 to obtain a sequence $(\tilde{y}_n) \subset \mathbb{R}^N$ and two constants $R, \gamma > 0$ such that

$$\liminf_{n \to \infty} \int_{B_R(\tilde{y}_n)} u_n^2 \, dx \geq \gamma > 0. \tag{9.4.3}$$

Now set $v_n(x) = u_n(x + \tilde{y}_n)$. Then the sequence (v_n) is bounded in $\mathcal{H}_0$, and we may assume that $v_n \rightharpoonup v \neq 0$ in $\mathcal{H}_0$ as $n \to \infty$. Fix $t_n > 0$ such that $\tilde{v}_n = t_n v_n \in \mathcal{N}_0$. Since $u_n \in \mathcal{N}_{\varepsilon_n}$, we see that

$$c_0 \leq J_0(\tilde{v}_n) = J_0(t_n u_n) \leq J_{\varepsilon_n}(t_n u_n) \leq J_{\varepsilon_n}(u_n) = c_0 + o_n(1),$$

and so $J_0(\tilde{v}_n) \to c_0$. In particular, we get $\tilde{v}_n \rightharpoonup \tilde{v}$ in $\mathcal{H}_0$ and $t_n \to t^* > 0$. Then, by the uniqueness of the weak limit, we have $\tilde{v} = t^* v \neq 0$. Using Lemma 9.3.7, we obtain that

$$\tilde{v}_n \to \tilde{v} \quad \text{in } \mathcal{H}_0. \tag{9.4.4}$$

To complete the proof of the lemma, we consider $y_n = \varepsilon_n \tilde{y}_n$ and show that (y_n) admits a subsequence, still denoted by (y_n), such that $y_n \to y_0$ for some $y_0 \in M$. First, we prove that (y_n) is bounded in $\mathbb{R}^N$. We argue by contradiction and assume that, up to a subsequence, $|y_n| \to \infty$ as $n \to \infty$. Since $J_{\varepsilon_n}(u_n) \to c_0$ and $\langle J'_{\varepsilon_n}(u_n), u_n \rangle = 0$, we see that $u_n \in \mathcal{B}$ for all n big enough. Then, in view of Lemma 9.2.3, there exists $C_0 \in (0, \frac{\ell_0}{2}]$ such that for n large enough

$$\|\tilde{K}_{\varepsilon_n}(u_n)\|_{L^\infty(\mathbb{R}^N)} \leq C_0.$$

Fix $R > 0$ such that $\Lambda \subset B_R$. Then, for n large enough, we may assume that $|y_n| > 2R$, and for all $z \in B_{\frac{R}{\varepsilon_n}}$ we have

$$|\varepsilon_n z + y_n| \geq |y_n| - |\varepsilon_n z| > R. \tag{9.4.5}$$

Using the change of variable $x \mapsto z + \tilde{y}_n$ and (9.4.5), we deduce that for n large enough

$$[v_n]_s^2 + \int_{\mathbb{R}^N} V_0 v_n^2 \, dx \leq C_0 \int_{\mathbb{R}^N} g(\varepsilon_n z + y_n, v_n) v_n \, dz$$

$$\leq C_0 \int_{B_{\frac{R}{\varepsilon_n}}} \tilde{f}(v_n) v_n \, dx + C_0 \int_{\mathbb{R}^N \setminus B_{\frac{R}{\varepsilon_n}}} f(v_n) v_n \, dx.$$

Since $v_n \to v$ in $\mathcal{H}_0$ as $n \to \infty$, the dominated convergence theorem implies that

$$\int_{\mathbb{R}^N \setminus B_{\frac{R}{\varepsilon_n}}} f(v_n) v_n \, dx = o_n(1),$$

which combined with the inequality $\tilde{f}(t) \leq \frac{V_0}{\ell_0} t$ shows that

$$\frac{1}{2} \left([v_n]_s^2 + \int_{\mathbb{R}^N} V_0 v_n^2 \, dx \right) \leq C_0 \int_{\mathbb{R}^N \setminus B_{\frac{R}{\varepsilon_n}}} f(v_n) v_n \, dx = o_n(1),$$

which contradicts the fact that $v \not\equiv 0$. Therefore, (y_n) is bounded, and we may assume that $y_n \to y_0 \in \mathbb{R}^N$. Clearly, if $y_0 \notin \overline{\Lambda}$, then we can argue as before to deduce that $v_n \to 0$ in $\mathcal{H}_0$, which is impossible. Hence, $y_0 \in \overline{\Lambda}$. Now, note that if $V(y_0) = V_0$, then $y_0 \notin \partial\Lambda$ in view of (V_2), and consequently $y_0 \in M$. We claim that $V(y_0) = V_0$. Suppose, by contradiction, that $V(y_0) > V_0$. Then, using that $\tilde{v}_n \to \tilde{v}$ in $\mathcal{H}_0$, Fatou's lemma and the invariance of $\mathbb{R}^N$ by translation, we get

$$c_0 = J_0(\tilde{v}) < \frac{1}{2} \left([\tilde{v}]_s^2 + \int_{\mathbb{R}^N} V(y_0) \tilde{v}^2 \, dx \right) - \Xi_0(\tilde{v})$$

$$\leq \liminf_{n \to \infty} \left[\frac{1}{2} [\tilde{v}_n]_s^2 + \frac{1}{2} \int_{\mathbb{R}^N} V(\varepsilon_n x + y_n) \tilde{v}_n^2 \, dx - \Xi_0(\tilde{v}_n) \right]$$

$$\leq \liminf_{n \to \infty} J_{\varepsilon_n}(t_n u_n) \leq \liminf_{n \to \infty} J_{\varepsilon_n}(u_n) = c_0,$$

which gives a contradiction.   □

Define

$$\tilde{\mathcal{N}}_\varepsilon = \{u \in \mathcal{N}_\varepsilon : J_\varepsilon(u) \le c_0 + h(\varepsilon)\},$$

where $h(\varepsilon) = \sup_{y \in M} |J_\varepsilon(\Phi_\varepsilon(y)) - c_0|$. By Lemma 9.4.1, $h(\varepsilon) \to 0$ as $\varepsilon \to 0$. From the definition of $h(\varepsilon)$ we deduce that, for all $y \in M$ and $\varepsilon > 0$, $\Phi_\varepsilon(y) \in \tilde{\mathcal{N}}_\varepsilon$ and so $\tilde{\mathcal{N}}_\varepsilon \ne \emptyset$. Moreover, arguing as in the proof of Lemma 6.3.19, we obtain the following result.

Lemma 9.4.4

$$\lim_{\substack{\varepsilon \to 0 \\ u \in \tilde{\mathcal{N}}_\varepsilon}} \sup \operatorname{dist}(\beta_\varepsilon(u), M_\delta) = 0.$$

9.5 Proof of Theorem 9.1.1

This last section is devoted to the proof of the main result of this chapter. First, arguing as in the proof of Theorem 6.3.22, we obtain the following multiplicity result for the modified problem (9.2.2).

Theorem 9.5.1 *Assume that (V_1)–(V_2) and (f_1)–(f_4) hold. Then, for every $\delta > 0$ such that $M_\delta \subset \Lambda$, there exists $\bar{\varepsilon}_\delta > 0$ such that, for any $\varepsilon \in (0, \bar{\varepsilon}_\delta)$, problem (9.2.2) has at least $\operatorname{cat}_{M_\delta}(M)$ positive solutions.*

Now we show that the solutions obtained in Theorem 9.5.1 satisfy the estimate $u_\varepsilon < a$ in Λ_ε^c for $\varepsilon > 0$ small enough. The following result will be of crucial importance in the study of the behavior of the maximum points of solutions to (9.1.1).

Lemma 9.5.2 *Let (v_n) be the sequence provided by Lemma 9.4.3. Then v_n satisfies the following problem*:

$$\begin{cases} (-\Delta)^s v_n + V_n(x) v_n = \left(\frac{1}{|x|^\mu} * G_n(x, v_n)\right) g_n(x, v_n) & \text{in } \mathbb{R}^N, \\ v_n \in H^s(\mathbb{R}^N), \; v_n > 0 & \text{in } \mathbb{R}^N, \end{cases} \tag{9.5.1}$$

where $V_n(x) = V(\varepsilon_n x + \varepsilon_n \tilde{y}_n)$ and $g_n(x, v_n) = g(\varepsilon_n x + \varepsilon_n \tilde{y}_n, v_n)$, $v_n \in L^\infty(\mathbb{R}^N)$ and there exists $C > 0$ such that $\|v_n\|_{L^\infty(\mathbb{R}^N)} \le C$ for all $n \in \mathbb{N}$. Furthermore,

$$\lim_{|x| \to \infty} v_n(x) = 0 \text{ uniformly in } n \in \mathbb{N}.$$

Proof Mimicking the proof of Lemma 6.3.23 we obtain that

$$
\left(\frac{1}{\beta}\right)^2 S_* \|v_n v_{L,n}^{\beta-1}\|^2_{L^{2_s^*}(\mathbb{R}^N)} + \int_{\mathbb{R}^N} V_n(x) v_n^2 v_{L,n}^{2(\beta-1)} dx
$$
$$
\leq \iint_{\mathbb{R}^{2N}} \frac{(v_n(x) - v_n(y))}{|x-y|^{N+2s}} ((v_n v_{L,n}^{2(\beta-1)})(x) - (v_n v_{L,n}^{2(\beta-1)})(y)) \, dx dy \tag{9.5.2}
$$
$$
+ \int_{\mathbb{R}^N} V_n(x) v_n^2 v_{L,n}^{2(\beta-1)} dx
$$
$$
= \int_{\mathbb{R}^N} \left(\frac{1}{|x|^\mu} * G_n(x, v_n)\right) g_n(x, v_n) v_n v_{L,n}^{2(\beta-1)} dx.
$$

On the other hand, thanks to the boundedness of (v_n) there exists a constant $C_0 > 0$ such that

$$
\sup_{n\in\mathbb{N}} \left\| \frac{1}{|\cdot|^\mu} * G_n(\cdot, v_n) \right\|_{L^\infty(\mathbb{R}^N)} \leq C_0. \tag{9.5.3}
$$

By assumptions (g_1) and (g_2), for every $\xi > 0$ there exists $C_\xi > 0$ such that

$$
|g_n(x, v_n)| \leq \xi |v_n| + C_\xi |v_n|^{2_s^*-1}. \tag{9.5.4}
$$

Taking ξ sufficiently small and using (9.5.3) and (9.5.4), we see that (9.5.2) yields

$$
\|v_n v_{L,n}^{\beta-1}\|^2_{L^{2_s^*}(\mathbb{R}^N)} \leq C\beta^2 \int_{\mathbb{R}^N} |v_n|^{2_s^*} v_{L,n}^{2(\beta-1)} dx.
$$

Now we can argue as in the proof of Lemma 6.3.23 to deduce the L^∞-estimate $\|v_n\|_{L^\infty(\mathbb{R}^N)} \leq C$ for all $n \in \mathbb{N}$. We observe that, if $w_n(x, y) = Ext(v_n) = P_s(x, y) * v_n(x)$ is the s-harmonic extension of v_n, then w_n solves

$$
\begin{cases}
- \operatorname{div}(y^{1-2s} \nabla w_n) = 0 & \text{in } \mathbb{R}^{N+1}_+, \\
w_n(\cdot, 0) = v_n & \text{on } \partial\mathbb{R}^{N+1}_+, \\
\frac{\partial w}{\partial \nu^{1-2s}} = -V_n(x) v_n + \left(\frac{1}{|x|^\mu} * G_n(x, v_n)\right) g_n(x, v_n) & \text{on } \partial\mathbb{R}^{N+1}_+.
\end{cases}
$$

In view of (9.5.3), (9.5.4) and the above L^∞-estimate, we can repeat the same arguments as in Lemma 6.3.23 to show that $v_n(x) \to 0$ as $|x| \to \infty$ uniformly in $n \in \mathbb{N}$ (see also Remark 7.2.10 and note that v_n is a subsolution to $(-\Delta)^s v_n + V_0 v_n = C_0 g_n(x, v_n)$ in $\mathbb{R}^N$). $\qquad\square$

At this point, we are ready to give the proof of our main result.

Proof of Theorem 9.1.1 Take $\delta > 0$ such that $M_\delta \subset \Lambda$. We begin by proving that there exists $\tilde{\varepsilon}_\delta > 0$ such that for any $\varepsilon \in (0, \tilde{\varepsilon}_\delta)$ and any solution $u_\varepsilon \in \tilde{\mathcal{N}}_\varepsilon$ of (9.2.2), it holds

$$\|u_\varepsilon\|_{L^\infty(\mathbb{R}^N \setminus \Lambda_\varepsilon)} < a. \tag{9.5.5}$$

Assume, by contradiction, that there exist $\varepsilon_n \to 0$, $u_n = u_{\varepsilon_n} \in \tilde{\mathcal{N}}_{\varepsilon_n}$ such that $J'_{\varepsilon_n}(u_n) = 0$ and $\|u_n\|_{L^\infty(\mathbb{R}^N \setminus \Lambda_{\varepsilon_n})} \geq a$. Since $J_{\varepsilon_n}(u_n) \leq c_0 + h(\varepsilon_n)$ and $h(\varepsilon_n) \to 0$, we can argue as in the first part of the proof of Lemma 9.4.3 to deduce that $J_{\varepsilon_n}(u_n) \to c_0$. Then, by using Lemma 9.4.3, we can find $(\tilde{y}_n) \subset \mathbb{R}^N$ such that $v_n(x) = u_n(x + \tilde{y}_n) \to v$ converges strongly in $H^s(\mathbb{R}^N)$ and $y_n = \varepsilon_n \tilde{y}_n \to y_0 \in M$.

If we take $r > 0$ such that $B_r(y_0) \subset B_{2r}(y_0) \subset \Lambda$, then $B_{\frac{r}{\varepsilon_n}}(\frac{y_0}{\varepsilon_n}) \subset \Lambda_{\varepsilon_n}$. In particular, for any $y \in B_{\frac{r}{\varepsilon_n}}(\tilde{y}_n)$,

$$\left| y - \frac{y_0}{\varepsilon_n} \right| \leq |y - \tilde{y}_n| + \left| \tilde{y}_n - \frac{y_0}{\varepsilon_n} \right| < \frac{2r}{\varepsilon_n} \qquad \text{for } n \text{ sufficiently large.}$$

Therefore, for these values of n,

$$\mathbb{R}^N \setminus \Lambda_{\varepsilon_n} \subset \mathbb{R}^N \setminus B_{\frac{r}{\varepsilon_n}}(\tilde{y}_n).$$

Now, by Lemma 9.5.2, we obtain that

$$|v_n(x)| \to 0 \quad \text{as } |x| \to \infty, \text{ uniformly in } n \in \mathbb{N},$$

which implies that there exists $R > 0$ such that $v_n(x) < a$ for $|x| \geq R$ and $n \in \mathbb{N}$. Hence,

$$u_n(x) < a \text{ for any } x \in \mathbb{R}^N \setminus B_R(\tilde{y}_n), \ n \in \mathbb{N}.$$

On the other hand, there exists $\nu \in \mathbb{N}$ such that, for any $n \geq \nu$ and $r/\varepsilon_n > R$,

$$\mathbb{R}^N \setminus \Lambda_{\varepsilon_n} \subset \mathbb{R}^N \setminus B_{\frac{r}{\varepsilon_n}}(\tilde{y}_n) \subset \mathbb{R}^N \setminus B_R(\tilde{y}_n),$$

which gives $u_n(x) < a$ for any $x \in \mathbb{R}^N \setminus \Lambda_{\varepsilon_n}$ and $n \geq \nu$, which is impossible.

Let $\bar{\varepsilon}_\delta$ be as given in Theorem 9.5.1 and take $\varepsilon_\delta = \min\{\tilde{\varepsilon}_\delta, \bar{\varepsilon}_\delta\}$. Fix $\varepsilon \in (0, \varepsilon_\delta)$. By Theorem 9.5.1, problem (9.2.2) admits at least $\text{cat}_{M_\delta}(M)$ nontrivial solutions u_ε. Since $u_\varepsilon \in \tilde{\mathcal{N}}_\varepsilon$ satisfies (9.5.5), by the definition of g it follows that u_ε is a solution of (9.2.1). Consequently, for all $\varepsilon \in (0, \varepsilon_\delta)$, $\hat{u}_\varepsilon(x) = u_\varepsilon(\frac{x}{\varepsilon})$ is a solution of (9.1.1).

Now, consider a sequence $\varepsilon_n \to 0$ and a corresponding sequence $(u_n) \subset \mathcal{H}_{\varepsilon_n}$ of solutions to (9.2.2) as above. Let us observe that (g_1) implies that we can find $\gamma \in (0, a)$ such that

$$g(\varepsilon x, t)t \leq \frac{V_0}{\ell_0} t^2 \qquad \text{for any } x \in \mathbb{R}^N, 0 \leq t \leq \gamma. \tag{9.5.6}$$

Arguing as before, we can find $R > 0$ such that

$$\|u_n\|_{L^\infty(B_R^c(\tilde{y}_n))} < \gamma. \tag{9.5.7}$$

Moreover, up to extracting a subsequence, we may assume that

$$\|u_n\|_{L^\infty(B_R(\tilde{y}_n))} \geq \gamma. \tag{9.5.8}$$

Indeed, if (9.5.8) does not hold, then, by (9.5.7), we have $\|u_n\|_{L^\infty(\mathbb{R}^N)} < \gamma$. Then, using that $\langle J'_{\varepsilon_n}(u_n), u_n \rangle = 0$, (9.5.6) and that for all n sufficiently large

$$\|\tilde{K}_{\varepsilon_n}(u_n)\|_{L^\infty(\mathbb{R}^N)} \leq C_0$$

(since $u_n \in \mathcal{B}$ for all n sufficiently large), we infer that

$$\|u_n\|_{\varepsilon_n}^2 \leq C_0 \int_{\mathbb{R}^N} g(\varepsilon_n x, u_n)u_n \, dx \leq \frac{C_0}{\ell_0} \int_{\mathbb{R}^N} V_0 u_n^2 \, dx$$

which together with $\frac{C_0}{\ell_0} \leq \frac{1}{2}$ yields $\|u_n\|_{\varepsilon_n} = 0$, which is a contradiction. Consequently, (9.5.8) holds. Taking into account (9.5.7) and (9.5.8), we deduce that if $p_n \in \mathbb{R}^N$ is a global maximum point of u_n, then $p_n \in B_R(\tilde{y}_n)$. Therefore, $p_n = \tilde{y}_n + q_n$ for some $q_n \in B_R$. Hence, $\eta_n = \varepsilon_n \tilde{y}_n + \varepsilon_n q_n$ is a global maximum point of $\hat{u}_n(x) = u_n(x/\varepsilon_n)$. Since $|q_n| < R$ for any $n \in \mathbb{N}$ and $\varepsilon_n \tilde{y}_n \to y_0 \in M$, it follows from the continuity of V that

$$\lim_{n \to \infty} V(\eta_n) = V(y_0) = V_0.$$

This ends the proof of Theorem 9.1.1.   □

Remark 9.5.3 Proceeding as in the proof of Theorem 7.1.2, we can show that if u_ε is a positive solution of (9.1.1) and x_ε is a global maximum point of u_ε, then there exists a constant $C > 0$ such that

$$0 < u_\varepsilon(x) \leq \frac{C \, \varepsilon^{N+2s}}{\varepsilon^{N+2s} + |x - x_\varepsilon|^{N+2s}} \quad \text{for all } x \in \mathbb{R}^N.$$

Indeed, it is enough to observe that

$$\|\tilde{K}_\varepsilon(u_\varepsilon)\|_{L^\infty(\mathbb{R}^N)} \leq C_0,$$

and use (g_1) to repeat the same arguments in the proof of Theorem 7.1.2 to deduce the desired decay estimate.

A Multiplicity Result for a Fractional Kirchhoff Equation with a General Nonlinearity

10

10.1 Introduction

In this paper we study the multiplicity of weak solutions to the nonlinear fractional Kirchhoff equation

$$\left(p + q(1-s) \iint_{\mathbb{R}^{2N}} \frac{|u(x) - u(y)|^2}{|x-y|^{N+2s}} \, dx \, dy \right) (-\Delta)^s u = g(u) \quad \text{in } \mathbb{R}^N, \tag{10.1.1}$$

where $s \in (0, 1)$, $N \geq 2$, $p > 0$, q is a small positive parameter and g is a nonlinearity which satisfies suitable assumptions.

When $s \to 1^-$ in (10.1.1), from the celebrated Bourgain–Brezis–Mironescu formula [108], we know that

$$\lim_{s \to 1^-} (1-s) \iint_{\mathbb{R}^{2N}} \frac{|u(x) - u(y)|^2}{|x-y|^{N+2s}} \, dx \, dy = \frac{\omega_{N-1}}{2N} \int_{\mathbb{R}^N} |\nabla u|^2 \, dx \quad \text{for all } u \in H^1(\mathbb{R}^N),$$

and thus (10.1.1) becomes a Kirchhoff equation of the form

$$-\left(p + q \int_{\mathbb{R}^N} |\nabla u(x)|^2 \, dx \right) \Delta u = g(u) \text{ in } \mathbb{R}^N, \tag{10.1.2}$$

which has been extensively studied in the last decade. Equation (10.1.2) is related to the stationary analogue of the Kirchhoff equation

$$u_{tt} - \left(p + q \int_{\Omega} |\nabla u(x)|^2 \, dx \right) \Delta u = g(x, u)$$

© The Author(s), under exclusive license to Springer Nature Switzerland AG 2021
V. Ambrosio, *Nonlinear Fractional Schrödinger Equations in* $\mathbb{R}^N$,
Frontiers in Mathematics, https://doi.org/10.1007/978-3-030-60220-8_10

with $\Omega \subset \mathbb{R}^N$ bounded domain, which was proposed by Kirchhoff in 1883 [240] as an extension of the classical D'Alembert's wave equation

$$\rho\, u_{tt} - \left(\frac{P_0}{h} + \frac{E}{2L} \int_0^L |u_x|^2\, dx \right) u_{xx}^2 = g(x, u) \tag{10.1.3}$$

governing the free vibrations of elastic strings. Kirchhoff's model takes into account the changes in length of the string produced by transverse vibrations. Here $u = u(x, t)$ is the transverse string displacement at the space coordinate x and time t, L is the length of the string, h is the area of the cross section, E is the Young modulus of the material, ρ is the mass density and P_0 is the initial tension. We also note that nonlocal boundary value problems like (10.1.2) model several physical and biological systems where u describes a process which depends on the average of itself, as for example, the population density; see [9, 142]. The early classical studies devoted to Kirchhoff equations were carried out by Bernstein [103] and Pohozaev [294]. However, equation (10.1.2) received much attention only after the paper by Lions [253], where a functional analysis approach was proposed to attack it. For more recent results concerning Kirchhoff-type equations we refer to [8, 9, 79, 140, 187, 191, 291, 331].

On the other hand, great attention has been recently given to the study of nonlinear fractional Kirchhoff problems. In [196], Fiscella and Valdinoci proposed a stationary fractional Kirchhoff variational model in a bounded domain $\Omega \subset \mathbb{R}^N$ with homogeneous Dirichlet boundary conditions and involving a critical nonlinearity:

$$\begin{cases} M \left(\int_{\mathbb{R}^N} |(-\Delta)^{\frac{s}{2}} u|^2 dx \right) (-\Delta)^s u = \lambda f(x, u) + |u|^{2_s^* - 2} u & \text{in } \Omega, \\ u = 0 & \text{in } \mathbb{R}^N \setminus \Omega, \end{cases}$$

where M is a continuous Kirchhoff function whose model case is given by $M(t) = a + bt$. The authors gave an interesting interpretation of Kirchhoff's equation in the fractional scenario. In their correction of the early (one-dimensional) model, the tension on the string, which classically has a "nonlocal" nature arising from the average of the kinetic energy $\frac{|u_x|^2}{2}$ on $[0, L]$, possesses a further nonlocal behavior governed by the H^s-norm (or other more general fractional norms) of the function u.

Nyamoradi [283] established the existence of at least three solutions for hemivariational inequalities driven by nonlocal integro-differential operators. Pucci and Saldi in [295] studied the existence and multiplicity of nontrivial non-negative entire solutions for a Kirchhoff-type eigenvalue problem in $\mathbb{R}^N$ involving a critical nonlinearity and the fractional Laplacian. Further results can be found in [55, 62, 72, 73, 195, 262, 273, 296].

In this chapter we study the multiplicity of weak solutions to (10.1.1) with q a small parameter and g a general subcritical nonlinearity. More precisely, we assume that $g :$ $\mathbb{R} \to \mathbb{R}$ satisfies Berestycki–Lions type conditions [100, 101], that is:

(g_1) $g \in C^{1,\alpha}(\mathbb{R}, \mathbb{R})$, with $\alpha > \max\{0, 1 - 2s\}$, and odd;

(g_2) $-\infty < \liminf_{t \to 0} \dfrac{g(t)}{t} \leq \limsup_{t \to 0} \dfrac{g(t)}{t} = -m < 0$;

(g_3) $\lim_{t \to \pm\infty} \dfrac{|g(t)|}{|t|^{2^*_s - 1}} = 0$;

(g_4) there exists $\zeta > 0$ such that $G(\zeta) = \displaystyle\int_0^\zeta g(t)\, dt > 0$.

Let us recall that when $q = 0$ and $p = 1$ in (10.1.1), the papers [39, 139] established the existence and multiplicity of radially symmetric solutions to the fractional scalar field problem

$$(-\Delta)^s u = g(u) \quad \text{in } \mathbb{R}^N. \tag{10.1.4}$$

Now, we aim to study a generalization of (10.1.4), and we look for weak solutions to (10.1.1) with $q > 0$ sufficiently small. Our main result is the following:

Theorem 10.1.1 ([72]) *Assume that* (g$_1$), (g$_2$), (g$_3$) *and* (g$_4$) *hold. Then, for every $h \in \mathbb{N}$ there exists $q(h) > 0$ such that for any $0 < q < q(h)$ Eq.* (10.1.1) *admits at least h couples of solutions in $H^s(\mathbb{R}^N)$ with radial symmetry.*

The proof of the above result is inspired by [83] and combines the mountain pass approach introduced in [218] with the truncation argument of [232].

10.2   The Truncated Problem

In this section we provide an abstract multiplicity result which allows us to prove Theorem 10.1.1. Let us introduce the following functional, defined for $u \in H^s(\mathbb{R}^N)$:

$$\mathcal{F}_q(u) = \frac{1}{2}[u]_s^2 + q\mathcal{R}(u) - \int_{\mathbb{R}^N} G(u)\, dx, \tag{10.2.1}$$

where $q > 0$ is a small parameter and $\mathcal{R} : H^s(\mathbb{R}^N) \to \mathbb{R}$.
We suppose that

$$\mathcal{R} = \sum_{i=1}^k \mathcal{R}_i$$

and, for each $i = 1, \ldots, k$ the functional $\mathcal{R}_i$ satisfies

($\mathcal{R}_1$) $\mathcal{R}_i \in C^1(H^s(\mathbb{R}^N), \mathbb{R})$ is non-negative and even;
($\mathcal{R}_2$) there exists $\delta_i > 0$ such that $\langle \mathcal{R}_i'(u), u \rangle \leq C \|u\|_{H^s(\mathbb{R}^N)}^{\delta_i}$ for all $u \in H^s(\mathbb{R}^N)$;
($\mathcal{R}_3$) if $(u_j) \subset H^s(\mathbb{R}^N)$ converges weakly to $u \in H^s(\mathbb{R}^N)$, then

$$\limsup_{j \to \infty} \langle \mathcal{R}_i'(u_j), u - u_j \rangle \leq 0;$$

($\mathcal{R}_4$) there exist $\alpha_i, \beta_i \geq 0$ such that if $u \in H^s(\mathbb{R}^N)$, $t > 0$ and $u_t = u\left(\frac{\cdot}{t}\right)$, then

$$\mathcal{R}_i(u_t) = t^{\alpha_i} \mathcal{R}_i(t^{\beta_i} u);$$

($\mathcal{R}_5$) $\mathcal{R}_i$ is invariant under the action of the N-dimensional orthogonal group, i.e., $\mathcal{R}_i(u(g\cdot)) = \mathcal{R}_i(u(\cdot))$ for every $g \in \mathcal{O}(N)$.

Let us observe that for any $u \in H^s(\mathbb{R}^N)$, $\mathcal{R}_i(u) - \mathcal{R}_i(0) = \int_0^1 \frac{d}{dt} \mathcal{R}_i(tu)\, dt$, so by assumption ($\mathcal{R}_2$) we have

$$\mathcal{R}_i(u) \leq C_1 + C_2 \|u\|_{H^s(\mathbb{R}^N)}^{\delta_i}. \tag{10.2.2}$$

The main result of this section is

Theorem 10.2.1 *Suppose that* (g_1)–(g_4) *and* $(\mathcal{R}_1)$–$(\mathcal{R}_5)$ *hold. Then, for every* $h \in \mathbb{N}$ *there exists* $q(h) > 0$ *such that for any* $0 < q < q(h)$ *the functional* $\mathcal{F}_q$ *admits at least* h *couples of critical points in* $H_{\mathrm{rad}}^s(\mathbb{R}^N)$.

Let us introduce, for any $t \geq 0$, the functions

$$g_1(t) = (g(t) + mt)^+,$$

$$g_2(t) = g_1(t) - g(t),$$

and extend them as odd functions for $t \leq 0$. Observing that

$$\lim_{t \to 0} \frac{g_1(t)}{t} = 0, \tag{10.2.3}$$

$$\lim_{t \to \infty} \frac{g_1(t)}{t^{2_s^* - 1}} = 0, \tag{10.2.4}$$

$$g_2(t) \geq mt \quad \forall t \geq 0, \tag{10.2.5}$$

we deduce that for any $\varepsilon > 0$ there exists $C_\varepsilon > 0$ such that

$$g_1(t) \le C_\varepsilon t^{2_s^* - 1} + \varepsilon g_2(t) \quad \forall t \ge 0. \tag{10.2.6}$$

Setting

$$G_i(t) = \int_0^t g_i(\tau) \, d\tau \quad i = 1, 2,$$

(10.2.5) immediately implies that

$$G_2(t) \ge \frac{m}{2} t^2 \quad \forall t \in \mathbb{R}, \tag{10.2.7}$$

and, by (10.2.6) we can see that for every $\varepsilon > 0$ there exists $C_\varepsilon > 0$ such that

$$G_1(t) \le C_\varepsilon |t|^{2_s^*} + \varepsilon \, G_2(t) \quad \forall t \in \mathbb{R}. \tag{10.2.8}$$

In view of ($\mathcal{R}_5$), all functionals that we will consider below are invariant under rotations, so, from now on we will directly define our functionals in $H_{\mathrm{rad}}^s(\mathbb{R}^N)$.

Following [232], let $\chi \in C^\infty([0, \infty), \mathbb{R})$ be a cut-off function such that

$$\begin{cases} \chi(t) = 1, & \text{for } t \in [0, 1], \\ 0 \le \chi(t) \le 1, & \text{for } t \in (1, 2), \\ \chi(t) = 0, & \text{for } t \in [2, \infty), \\ \|\chi'\|_{L^\infty(0,\infty)} \le 2, \end{cases}$$

and we set

$$\xi_\Lambda(u) = \chi \left(\frac{\|u\|_{H^s(\mathbb{R}^N)}^2}{\Lambda^2} \right).$$

Then we introduce the truncated functional $\mathcal{F}_q^\Lambda : H_{\mathrm{rad}}^s(\mathbb{R}^N) \to \mathbb{R}$ defined as

$$\mathcal{F}_q^\Lambda(u) = \frac{1}{2} [u]_s^2 + q \, \xi_\Lambda(u) \mathcal{R}(u) - \int_{\mathbb{R}^N} G(u) \, dx.$$

Clearly, a critical point u of $\mathcal{F}_q^\Lambda$ with $\|u\|_{H^s(\mathbb{R}^N)} \le \Lambda$ is a critical point of $\mathcal{F}_q$.

Our first aim is to prove that the truncated functional $\mathcal{F}_q^\Lambda$ has a symmetric mountain pass geometry:

Lemma 10.2.2 *There exist $r_0 > 0$ and $\rho_0 > 0$ such that*

$$
\begin{aligned}
\mathcal{F}_q^\Lambda(u) \geq 0, && \text{for } \|u\|_{H^s(\mathbb{R}^N)} \leq r_0, \\
\mathcal{F}_q^\Lambda(u) \geq \rho_0, && \text{for } \|u\|_{H^s(\mathbb{R}^N)} = r_0.
\end{aligned}
\tag{10.2.9}
$$

Moreover, for any $n \in \mathbb{N}$ there exists an odd continuous map

$$
\gamma_n : \mathbb{S}^{n-1} \longrightarrow H^s_{\mathrm{rad}}(\mathbb{R}^N)
$$

such that

$$
\mathcal{F}_q^\Lambda(\gamma_n(\sigma)) < 0 \quad \text{for all } \sigma \in \mathbb{S}^{n-1},
\tag{10.2.10}
$$

where

$$
\mathbb{S}^{n-1} = \{\sigma = (\sigma_1, \ldots, \sigma_n) \in \mathbb{R}^n : |\sigma| = 1\}.
$$

Proof Taking $\varepsilon = \frac{1}{2}$ in (10.2.8), and using (10.2.7), the positivity of $\mathcal{R}$, and Theorem 1.1.8, we have

$$
\begin{aligned}
\mathcal{F}_q^\Lambda(u) &= \frac{1}{2}[u]_s^2 + \int_{\mathbb{R}^N} G_2(u)\,dx + q\,\xi_\Lambda(u)\mathcal{R}(u) - \int_{\mathbb{R}^N} G_1(u)\,dx \\
&\geq \frac{1}{2}[u]_s^2 + \frac{m}{4}\|u\|_{L^2(\mathbb{R}^N)}^2 - C_{\frac{1}{2}}\|u\|_{L^{2_s^*}(\mathbb{R}^N)}^{2_s^*} \\
&\geq \min\left\{\frac{1}{2}, \frac{m}{4}\right\}\|u\|_{H^s(\mathbb{R}^N)}^2 - C_{\frac{1}{2}} C^* \|u\|_{H^s(\mathbb{R}^N)}^{2_s^*},
\end{aligned}
$$

which easily yields (10.2.9).

Proceeding similarly to Theorem 10 in [101], for any $n \in \mathbb{N}$, there exists an odd continuous map $\pi_n : \mathbb{S}^{n-1} \to H^s_{\mathrm{rad}}(\mathbb{R}^N)$ such that

$$
0 \notin \pi_n(\mathbb{S}^{n-1}),
$$

$$
\int_{\mathbb{R}^N} G(\pi_n(\sigma))\,dx \geq 1 \quad \text{for all } \sigma \in \mathbb{S}^{n-1}.
$$

Let us define

$$
\psi_n^t(\sigma) = \pi_n(\sigma)\left(\frac{\cdot}{t}\right), \quad \text{for } t \geq 1.
$$

Then, for t sufficiently large, we get

$$\mathcal{F}_q^\Lambda(\psi_n^t(\sigma)) = \frac{t^{N-2s}}{2}[\pi_n(\sigma)]_s^2$$

$$+ q \chi \left(\frac{t^{N-2s}[\pi_n(\sigma)]_s^2 + t^N\|\pi_n(\sigma)\|_{L^2(\mathbb{R}^N)}^2}{\Lambda^2} \right) \mathcal{R}(\psi_n^t(\sigma))$$

$$- t^N \int_{\mathbb{R}^N} G(\pi_n(\sigma))\,dx$$

$$\leq t^{N-2s} \left\{ \frac{[\pi_n(\sigma)]_s^2}{2} - t^{2s} \right\} < 0.$$

Therefore, we can choose $\bar{t}$ such that $\mathcal{F}_q^\Lambda(\psi_n^{\bar{t}}(\sigma)) < 0$ for all $\sigma \in \mathbb{S}^{n-1}$, and by setting $\gamma_n(\sigma)(x) = \psi_n^{\bar{t}}(\sigma)(x)$, we see that γ_n satisfies the required properties. $\square$

Now we define the minimax value of $\mathcal{F}_q^\Lambda$ by using the maps $\gamma_n : \partial\mathcal{D}_n \to H_{\text{rad}}^s(\mathbb{R}^N)$ obtained in Lemma 10.2.2. For any $n \in \mathbb{N}$, let

$$b_n = b_n(q, \Lambda) = \inf_{\gamma \in \Gamma_n} \max_{\sigma \in \mathcal{D}_n} \mathcal{F}_q^\Lambda(\gamma(\sigma)),$$

where $\mathcal{D}_n = \{\sigma = (\sigma_1, \ldots, \sigma_n) \in \mathbb{R}^n : |\sigma| \leq 1\}$ and

$$\Gamma_n = \left\{ \gamma \in C(\mathcal{D}_n, H_{\text{rad}}^s(\mathbb{R}^N)) : \begin{array}{l} \gamma(-\sigma) = -\gamma(\sigma) \ \text{ for all } \sigma \in \mathcal{D}_n \\ \gamma(\sigma) = \gamma_n(\sigma) \quad \ \text{ for all } \sigma \in \partial\mathcal{D}_n \end{array} \right\}.$$

Let us introduce the following modified functionals

$$\widetilde{\mathcal{F}_q}(\theta, u) = \mathcal{F}_q(u(\cdot/e^\theta)),$$

$$\widetilde{\mathcal{F}_q^\Lambda}(\theta, u) = \mathcal{F}_q^\Lambda(u(\cdot/e^\theta)),$$

for $(\theta, u) \in \mathbb{R} \times H_{\text{rad}}^s(\mathbb{R}^N)$, and set

$$\widetilde{\mathcal{F}_q'}(\theta, u) = \frac{\partial}{\partial u}\widetilde{\mathcal{F}_q}(\theta, u),$$

$$(\widetilde{\mathcal{F}_q^\Lambda})'(\theta, u) = \frac{\partial}{\partial u}\widetilde{\mathcal{F}_q^\Lambda}(\theta, u),$$

$$\tilde{b}_n = \tilde{b}_n(q, \Lambda) = \inf_{\tilde{\gamma} \in \tilde{\Gamma}_n} \max_{\sigma \in \mathcal{D}_n} \widetilde{\mathcal{F}_q^\Lambda}(\tilde{\gamma}(\sigma)),$$

where

$$
\widetilde{\Gamma}_n = \left\{ \tilde{\gamma} \in C(\mathcal{D}_n, \mathbb{R} \times H^s_{\mathrm{rad}}(\mathbb{R}^N)) : \begin{array}{ll} \tilde{\gamma}(\sigma) = (\theta(\sigma), \eta(\sigma)) \text{ satisfies} & \\ (\theta(-\sigma), \eta(-\sigma)) = (\theta(\sigma), -\eta(\sigma)) & \text{for all } \sigma \in \mathcal{D}_n \\ (\theta(\sigma), \eta(\sigma)) = (0, \gamma_n(\sigma)) & \text{for all } \sigma \in \partial \mathcal{D}_n \end{array} \right\}.
$$

By assumption $(\mathcal{R}_4)$ we get

$$
\widetilde{\mathcal{F}}_q(\theta, u) = \frac{e^{(N-2s)\theta}}{2}[u]_s^2 + q \sum_{i=1}^{k} e^{\alpha_i \theta} \mathcal{R}_i(e^{\beta_i \theta} u) - e^{N\theta} \int_{\mathbb{R}^N} G(u) \, dx,
$$

and

$$
\widetilde{\mathcal{F}}_q^{\Lambda}(\theta, u) = \frac{e^{(N-2s)\theta}}{2}[u]_s^2 + q \chi \left(\frac{e^{(N-2s)\theta}[u]_s^2 + e^{N\theta} \|u\|^2_{L^2(\mathbb{R}^N)}}{\Lambda^2} \right) \sum_{i=1}^{k} e^{\alpha_i \theta} \mathcal{R}_i(e^{\beta_i \theta} u)
$$

$$
- e^{N\theta} \int_{\mathbb{R}^N} G(u) \, dx.
$$

Proceeding as in [39, 218, 298], we can see that the following results hold.

Lemma 10.2.3 *We have that*

(1) *there exists $\bar{b} > 0$ such that $b_n \geq \bar{b}$ for any $n \in \mathbb{N}$;*
(2) *$b_n \to \infty$;*
(3) *$b_n = \tilde{b}_n$ for any $n \in \mathbb{N}$.*

Lemma 10.2.4 *For every $n \in \mathbb{N}$ there exists a sequence $((\theta_j, u_j)) \subset \mathbb{R} \times H^s_{\mathrm{rad}}(\mathbb{R}^N)$ such that*

(i) *$\theta_j \to 0$;*
(ii) *$\widetilde{\mathcal{F}}_q^{\Lambda}(\theta_j, u_j) \to b_n$;*
(iii) *$(\widetilde{\mathcal{F}}_q^{\Lambda})'(\theta_j, u_j) \to 0$ strongly in $(H^s_{\mathrm{rad}}(\mathbb{R}^N))^*$;*
(iv) *$\dfrac{\partial}{\partial \theta} \widetilde{\mathcal{F}}_q^{\Lambda}(\theta_j, u_j) \to 0$.*

Our goal is to prove that, for a suitable choice of Λ and q, the sequence $((\theta_j, u_j))$ given by Lemma 10.2.4 is a bounded Palais–Smale sequence for $\mathcal{F}_q$. We begin by proving the boundedness of (u_j) in $H^s(\mathbb{R}^N)$.

Proposition 10.2.5 *Let $n \in \mathbb{N}$ and let $\Lambda_n > 0$ be sufficiently large. There exists q_n, depending on Λ_n, such that for any $0 < q < q_n$, if $((\theta_j, u_j)) \subset \mathbb{R} \times H^s_{\mathrm{rad}}(\mathbb{R}^N)$ is the sequence given in Lemma 10.2.4, then, up to a subsequence, $\|u_j\|_{H^s(\mathbb{R}^N)} \leq \Lambda_n$, for all $j \in \mathbb{N}$.*

Proof Lemmas 10.2.3 and 10.2.4 imply that

$$N\widetilde{\mathcal{F}_q^\Lambda}(\theta_j, u_j) - \frac{\partial}{\partial\theta}\widetilde{\mathcal{F}_q^\Lambda}(\theta_j, u_j) = Nb_n + o_j(1),$$

which can be written as

$$se^{(N-2s)\theta_j}[u_j]_s^2$$

$$= q\,\chi\left(\frac{\|u_j(\cdot/e^{\theta_j})\|_{H^s(\mathbb{R}^N)}^2}{\Lambda^2}\right)\sum_{i=1}^k (\alpha_i - N)\mathcal{R}_i(u_j(\cdot/e^{\theta_j}))$$

$$+ q\,\chi\left(\frac{\|u_j(\cdot/e^{\theta_j})\|_{H^s(\mathbb{R}^N)}^2}{\Lambda^2}\right)\sum_{i=1}^k e^{\alpha_i\theta_j}\langle\mathcal{R}_i'(e^{\beta_i\theta_j}u_j), \beta_i e^{\beta_i\theta_j}u_j\rangle$$

$$+ q\,\chi'\left(\frac{\|u_j(\cdot/e^{\theta_j})\|_{H^s(\mathbb{R}^N)}^2}{\Lambda^2}\right)\frac{(N-2s)e^{(N-2s)\theta_j}[u_j]_s^2 + Ne^{N\theta_j}\|u_j\|_{L^2(\mathbb{R}^N)}^2}{\Lambda^2}\mathcal{R}(u_j(\cdot/e^{\theta_j}))$$

$$+ Nb_n + o_j(1)$$

$$= \mathcal{I}_j + \mathcal{II}_j + \mathcal{III}_j + Nb_n + o_j(1). \tag{10.2.11}$$

By the definition of b_n, if $\gamma \in \Gamma_n$, we deduce that

$$b_n \le \max_{\sigma\in\mathcal{D}_n}\mathcal{F}_q^\Lambda(\gamma(\sigma))$$

$$\le \max_{\sigma\in\mathcal{D}_n}\left\{\frac{1}{2}[\gamma(\sigma)]_s^2 - \int_{\mathbb{R}^N}G(\gamma(\sigma))\,dx\right\} + \max_{\sigma\in\mathcal{D}_n}(q\,\xi_\Lambda(\gamma(\sigma))\mathcal{R}(\gamma(\sigma)))$$

$$= A_1 + A_2(\Lambda). \tag{10.2.12}$$

Now, if $\|\gamma(\sigma)\|_{H^s(\mathbb{R}^N)}^2 \ge 2\Lambda^2$, then $A_2(\Lambda) = 0$: otherwise, by (10.2.2), we can find $\delta > 0$ such that

$$A_2(\Lambda) \le q\left(C_1 + C_2\|\gamma(\sigma)\|_{H^s(\mathbb{R}^N)}^\delta\right) \le q\left(C_1 + C_2'\Lambda^\delta\right).$$

In addition, we have the following estimates:

$$|\mathcal{I}_j| \le q\left(C_3 + C_4\Lambda^\delta\right), \tag{10.2.13}$$

$$|\mathcal{II}_j| \le C_5\,q\,\Lambda^\delta, \tag{10.2.14}$$

$$|\mathcal{III}_j| \le q\left(C_6 + C_7\Lambda^\delta\right). \tag{10.2.15}$$

In view of (10.2.12), (10.2.13), (10.2.14) and (10.2.15), we deduce from (10.2.11) that

$$[u_j]_s^2 \leq C' + q\left(C_8 + C_9 \Lambda^\delta\right). \tag{10.2.16}$$

On the other hand, by item (iv) of Lemma 10.2.4 and (10.2.8),

$$\frac{(N-2s)e^{(N-2s)\theta_j}}{2}[u_j]_s^2 + q\,\chi\left(\frac{\|u_j(\cdot/e^{\theta_j})\|_{H^s(\mathbb{R}^N)}^2}{\Lambda^2}\right)\sum_{i=1}^k \alpha_i\,\mathcal{R}_i(u_j(\cdot/e^{\theta_j}))$$

$$+\,q\,\chi\left(\frac{\|u_j(\cdot/e^{\theta_j})\|_{H^s(\mathbb{R}^N)}^2}{\Lambda^2}\right)\sum_{i=1}^k e^{\alpha_i\theta_j}\,\langle \mathcal{R}_i'(e^{\beta_i\theta_j}u_j),\,\beta_i e^{\beta_i\theta_j}u_j\rangle$$

$$+\,q\,\chi'\left(\frac{\|u_j(\cdot/e^{\theta_j})\|_{H^s(\mathbb{R}^N)}^2}{\Lambda^2}\right)\frac{(N-2s)e^{(N-2s)\theta_j}[u_j]_s^2 + Ne^{N\theta_j}\|u_j\|_{L^2(\mathbb{R}^N)}^2}{\Lambda^2}\mathcal{R}(u_j(\cdot/e^{\theta_j}))$$

$$+\,Ne^{N\theta_j}\int_{\mathbb{R}^N}G_2(u_j)\,dx$$

$$=\,Ne^{N\theta_j}\int_{\mathbb{R}^N}G_1(u_j)\,dx + o_j(1)$$

$$\leq\,Ne^{N\theta_j}\left(C_\varepsilon\int_{\mathbb{R}^N}|u_j|^{2_s^*}\,dx + \varepsilon\int_{\mathbb{R}^N}G_2(u_j)\,dx\right) + o_j(1). \tag{10.2.17}$$

Then, using (10.2.7), (10.2.14), (10.2.15), (10.2.16), (10.2.17) and Theorem 1.1.8, we infer that

$$\frac{Ne^{N\theta_j}m(1-\varepsilon)}{2}\int_{\mathbb{R}^N}u_j^2\,dx$$

$$\leq (1-\varepsilon)Ne^{N\theta_j}\int_{\mathbb{R}^N}G_2(u_j)\,dx$$

$$\leq Ne^{N\theta_j}C_\varepsilon\int_{\mathbb{R}^N}|u_j|^{2_s^*}\,dx - q\,\chi\left(\frac{\|u_j(\cdot/e^{\theta_j})\|_{H^s(\mathbb{R}^N)}^2}{\Lambda^2}\right)\sum_{i=1}^k e^{\alpha_i\theta_j}\,\langle \mathcal{R}_i'(e^{\beta_i\theta_j}u_j),\,\beta_i e^{\beta_i\theta_j}u_j\rangle$$

$$-\,q\,\chi'\left(\frac{\|u_j(\cdot/e^{\theta_j})\|_{H^s(\mathbb{R}^N)}^2}{\Lambda^2}\right)\frac{(N-2s)e^{(N-2s)\theta_j}[u_j]_s^2 + Ne^{N\theta_j}\|u_j\|_{L^2(\mathbb{R}^N)}^2}{\Lambda^2}\mathcal{R}(u_j(\cdot/e^{\theta_j})) + o_j(1)$$

$$\leq C_{10}\left([u_j]_s^2\right)^{\frac{2_s^*}{2}} + q\left(C_{11} + C_{12}\Lambda^\delta\right) + o_j(1)$$

$$\leq C_{10}\left(C' + q\left(C_8 + C_9\Lambda^\delta\right)\right)^{\frac{2_s^*}{2}} + q\left(C_{11} + C_{12}\Lambda^\delta\right) + o_j(1). \tag{10.2.18}$$

Now, we argue by contradiction. Suppose that there is no subsequence (u_j) that is uniformly bounded by Λ in the H^s-norm. Then we can find $j_0 \in \mathbb{N}$ such that

$$\|u_j\|_{H^s(\mathbb{R}^N)} > \Lambda \text{ for all } j \geq j_0. \tag{10.2.19}$$

Without loss of generality, we may assume that (10.2.19) is true for all u_j. Consequently, using (10.2.16), (10.2.18) and (10.2.19), we deduce that

$$\Lambda^2 < \|u_j\|_{H^s(\mathbb{R}^N)}^2 \leq C_{13} + C_{14}q\, \Lambda^{\frac{2^*_s}{2}\delta}$$

which is not true for Λ large and q small enough. Indeed, to see this, note that one can find Λ_0 such that $\Lambda_0^2 > C_{13} + 1$ and $q_0 = q_0(\Lambda_0)$ such that $C_{14}q\, \Lambda_0^{\frac{2^*_s}{2}\delta} < 1$, for any $q < q_0$, and this gives a contradiction. □

At this point, we prove the following compactness result:

Lemma 10.2.6 *Let $n \in \mathbb{N}$, $\Lambda_n, q_n > 0$ as in Proposition 10.2.5 and $((\theta_j, u_j)) \subset \mathbb{R} \times H^s_{\mathrm{rad}}(\mathbb{R}^N)$ be the sequence given in Lemma 10.2.4. Then (u_j) admits a subsequence which converges in $H^s_{\mathrm{rad}}(\mathbb{R}^N)$ to a nontrivial critical point of $\mathcal{F}_q$ at the level b_n.*

Proof By Proposition 10.2.5, (u_j) is bounded, so, by using Theorem 1.1.11, we can assume, up to a subsequence, that there exists $u \in H^s_{\mathrm{rad}}(\mathbb{R}^N)$ such that

$$u_j \rightharpoonup u \quad \text{in } H^s_{\mathrm{rad}}(\mathbb{R}^N),$$

$$u_j \to u \quad \text{in } L^p(\mathbb{R}^N), \ 2 < p < 2^*_s, \tag{10.2.20}$$

$$u_j \to u \quad \text{a.e. in } \mathbb{R}^N.$$

Thanks to the weak lower semicontinuity,

$$[u]_s^2 \leq \liminf_{j \to \infty}[u_j]_s^2. \tag{10.2.21}$$

Recalling that $\|u_j\|_{H^s(\mathbb{R}^N)} \leq \Lambda_n$ for any $j \in \mathbb{N}$, we see that, for every $v \in H^s_{\mathrm{rad}}(\mathbb{R}^N)$,

$$\langle \widetilde{\mathcal{F}'_q}(\theta_j, u_j), v \rangle = \langle (\widetilde{\mathcal{F}_q^{\Lambda_n}})'(\theta_j, u_j), v \rangle$$

$$= e^{(N-2s)\theta_j} \iint_{\mathbb{R}^{2N}} \frac{(u_j(x) - u_j(y))(v(x) - v(y))}{|x-y|^{N+2s}}\, dx\, dy$$

$$+ q \sum_{i=1}^{k} e^{(\alpha_i + \beta_i)\theta_j} \langle \mathcal{R}'_i(e^{\beta_i \theta_j} u_j), v \rangle$$

$$+ e^{N\theta_j} \int_{\mathbb{R}^N} g_2(u_j) v\, dx - e^{N\theta_j} \int_{\mathbb{R}^N} g_1(u_j) v\, dx. \tag{10.2.22}$$

Taking into account (10.2.22) and item (iii) of Lemma 10.2.4, we have

$$
o_j(1) = \langle \widetilde{\mathcal{F}}'_q(\theta_j, u_j), u \rangle - \langle \widetilde{\mathcal{F}}'_q(\theta_j, u_j), u_j \rangle
$$

$$
= e^{(N-2s)\theta_j} \iint_{\mathbb{R}^{2N}} \frac{(u_j(x) - u_j(y))}{|x - y|^{N+2s}} [(u(x) - u(y)) - (u_j(x) - u_j(y))] \, dx \, dy
$$

$$
+ q \sum_{i=1}^{k} e^{(\alpha_i + \beta_i)\theta_j} \langle \mathcal{R}'_i(e^{\beta_i \theta_j} u_j), u - u_j \rangle
$$

$$
+ e^{N\theta_j} \int_{\mathbb{R}^N} g_2(u_j)(u - u_j) \, dx - e^{N\theta_j} \int_{\mathbb{R}^N} g_1(u_j)(u - u_j) \, dx. \tag{10.2.23}
$$

Now, applying the first part of Lemma 1.4.2 with $P(t) = g_i(t)$, $i = 1, 2$, $Q(t) = |t|^{2^*_s - 1}$, $v_j = u_j$, $v = g_i(u)$, $i = 1, 2$ and $w \in C^\infty_c(\mathbb{R}^N)$, and using (g_3), (10.2.4) and (10.2.20), we see that, as $j \to \infty$,

$$
\int_{\mathbb{R}^N} g_i(u_j) w \, dx \to \int_{\mathbb{R}^N} g_i(u) w \, dx \quad i = 1, 2,
$$

so we obtain

$$
\int_{\mathbb{R}^N} g_i(u_j) u \, dx \to \int_{\mathbb{R}^N} g_i(u) u \, dx \quad i = 1, 2. \tag{10.2.24}
$$

Taking $X = H^s(\mathbb{R}^N)$, $q_1 = 2$, $q_2 = 2^*_s$, $v_j = u_j$, $v = g_1(u)u$ and $P(t) = g_1(t)t$ in Lemma 1.4.3, and using (10.2.3), (10.2.4) and (10.2.20), we deduce that

$$
\int_{\mathbb{R}^N} g_1(u_j) u_j \, dx \to \int_{\mathbb{R}^N} g_1(u) u \, dx. \tag{10.2.25}
$$

On the other hand, (10.2.20) and Fatou's lemma yield

$$
\int_{\mathbb{R}^N} g_2(u) u \, dx \le \liminf_{j \to \infty} \int_{\mathbb{R}^N} g_2(u_j) u_j \, dx. \tag{10.2.26}
$$

Putting together (10.2.23), (10.2.24), (10.2.25), (10.2.26), and using $(\mathcal{R}_3)$ we get

$$
\limsup_{j \to \infty} [u_j]_s^2 = \limsup_{j \to \infty} e^{(N-2s)\theta_j} [u_j]_s^2
$$

$$
= \limsup_{j \to \infty} \left[e^{(N-2s)\theta_j} \iint_{\mathbb{R}^{2N}} \frac{(u_j(x) - u_j(y))(u(x) - u(y))}{|x - y|^{N+2s}} (u(x) - u(y)) \, dx \, dy \right.
$$

$$
+ q \sum_{i=1}^{k} e^{(\alpha_i + \beta_i)\theta_j} \langle \mathcal{R}'_i(e^{\beta_i \theta_j} u_j), u - u_j \rangle
$$

$$+e^{N\theta_j} \int_{\mathbb{R}^N} g_2(u_j)(u-u_j)\,dx - e^{N\theta_j} \int_{\mathbb{R}^N} g_1(u_j)(u-u_j)\,dx \Bigg]$$

$$\leq [u]_s^2. \tag{10.2.27}$$

Therefore (10.2.21) and (10.2.27) give

$$\lim_{j\to\infty} [u_j]_s^2 = [u]_s^2, \tag{10.2.28}$$

which, in view of (10.2.23), yields

$$\lim_{j\to\infty} \int_{\mathbb{R}^N} g_2(u_j)u_j\,dx = \int_{\mathbb{R}^N} g_2(u)u\,dx. \tag{10.2.29}$$

Since $g_2(t)t = mt^2 + h(t)$, where $h(t) = t(g(t)+mt)_-$ is a non-negative and continuous function, it follows from Fatou's lemma that

$$\int_{\mathbb{R}^N} h(u)\,dx \leq \liminf_{j\to\infty} \int_{\mathbb{R}^N} h(u_j)\,dx, \tag{10.2.30}$$

$$\int_{\mathbb{R}^N} u^2\,dx \leq \liminf_{j\to\infty} \int_{\mathbb{R}^N} u_j^2\,dx. \tag{10.2.31}$$

Using (10.2.29) and (10.2.30), we have that

$$\limsup_{j\to\infty} \int_{\mathbb{R}^N} mu_j^2\,dx = \limsup_{j\to\infty} \int_{\mathbb{R}^N} (g_2(u_j)u_j - h(u_j))\,dx$$

$$= \int_{\mathbb{R}^N} g_2(u)u\,dx + \limsup_{j\to\infty} \left(-\int_{\mathbb{R}^N} h(u_j)\,dx\right)$$

$$= \int_{\mathbb{R}^N} (mu^2 + h(u))\,dx - \liminf_{j\to\infty} \int_{\mathbb{R}^N} h(u_j)\,dx$$

$$= \int_{\mathbb{R}^N} mu^2\,dx + \int_{\mathbb{R}^N} h(u)\,dx - \liminf_{j\to\infty} \int_{\mathbb{R}^N} h(u_j)\,dx$$

$$\leq \int_{\mathbb{R}^N} mu^2\,dx$$

which in conjunction with (10.2.31), implies that $u_j \to u$ strongly in $L^2(\mathbb{R}^N)$. We conclude that $u_j \to u$ strongly in $H^s_{\mathrm{rad}}(\mathbb{R}^N)$, and since $b_n > 0$, u is a nontrivial critical point of $\mathcal{F}_q$ at the level b_n. $\qquad\square$

Now we are ready to prove the main result of this section:

Proof of Theorem 10.2.1 Let $h \geq 1$. Since $b_n \to \infty$ (see item (2) of Lemma 10.2.3), up to a subsequence, we can consider that $b_1 < b_2 < \cdots < b_h$. Then, in view of Lemma 10.2.6, we define $q(h) = q_h > 0$ and we get the desired conclusion. □

10.3   A Multiplicity Result

In this section we give the proof of Theorem 10.1.1. Let us introduce the functional

$$\mathcal{F}_q(u) = \frac{1}{2}\left(p + \frac{q}{2}(1-s)[u]^2_{H^s(\mathbb{R}^N)}\right)[u]^2_s - \int_{\mathbb{R}^N} G(u)\,dx.$$

In view of Theorem 10.2.1, it is enough to verify that

$$\mathcal{R}(u) = \frac{1-s}{4}[u]^4_s \tag{10.3.1}$$

satisfies the assumptions $(\mathcal{R}_1)$–$(\mathcal{R}_5)$. Clearly, $\mathcal{R}$ is an even and nonnegative C^1-functional on $H^s(\mathbb{R}^N)$. Since $[u]^2_s \leq \|u\|^2_{H^s(\mathbb{R}^N)}$, we can see that assumptions $(\mathcal{R}_1)$ and $(\mathcal{R}_2)$ are satisfied.

Regarding $(\mathcal{R}_3)$, suppose that $u_j \rightharpoonup u$ weakly in $H^s_{\mathrm{rad}}(\mathbb{R}^N)$ and $[u_j]^2_s \to \ell \geq 0$. If $\ell = 0$, then we are done. Assume that $\ell > 0$. By the weak lower semicontinuity,

$$[u]^2_s \leq \liminf_{j\to\infty}[u_j]^2_s. \tag{10.3.2}$$

Using the following properties of $\liminf$ and $\limsup$ for sequences of real numbers:

$$\limsup_{j\to\infty} a_j b_j = a \limsup_{j\to\infty} b_j \quad \text{if } a_j, b_j \geq 0,\ a_j \to a,$$

$$\limsup_{j\to\infty} (a_j + b_j) = a + \limsup_{j\to\infty} b_j \quad \text{if } a_j \to a,$$

$$\limsup_{j\to\infty} (-a_j) = -\liminf_{j\to\infty} a_j,$$

and applying (10.3.2), we obtain

$$\limsup_{j\to\infty} \langle \mathcal{R}'(u_j), u - u_j \rangle =$$

$$= (1-s)\limsup_{j\to\infty}\left([u_j]^2_s \iint_{\mathbb{R}^{2N}} \frac{(u_j(x) - u_j(y))}{|x-y|^{N+2s}}[(u(x) - u(y)) - (u_j(x) - u_j(y))]\,dx\,dy\right)$$

$$
= (1-s)\ell \limsup_{j\to\infty} \iint_{\mathbb{R}^{2N}} \frac{(u_j(x)-u_j(y))}{|x-y|^{N+2s}}[(u(x)-u(y))-(u_j(x)-u_j(y))]\, dx\, dy
$$

$$
= (1-s)\ell \left(\lim_{j\to\infty} \iint_{\mathbb{R}^{2N}} \frac{(u_j(x)-u_j(y))(u(x)-u(y))}{|x-y|^{N+2s}}\, dx\, dy - \liminf_{j\to\infty}[u_j]_s^2 \right)
$$

$$
= (1-s)\ell \left([u]_s^2 - \liminf_{j\to\infty}[u_j]_s^2 \right) \le 0,
$$

which gives $(\mathcal{R}_3)$.

Now, recalling the definition of u_t and using (10.3.1), we conclude that $(\mathcal{R}_4)$ is verified, because

$$
\mathcal{R}(u_t) = \frac{1-s}{4} \left(\iint_{\mathbb{R}^{2N}} \frac{|u(\frac{x}{t})-u(\frac{y}{t})|^2}{|x-y|^{N+2s}}\, dx\, dy \right)^2
$$

$$
= \frac{(1-s)\, t^{2(N-2s)}}{4} \left(\iint_{\mathbb{R}^{2N}} \frac{|u(x)-u(y)|^2}{|x-y|^{N+2s}}\, dx\, dy \right)^4
$$

$$
= t^{2(N-2s)} \mathcal{R}(u).
$$

Finally, we prove $(\mathcal{R}_5)$. Using a change of variable, we see that, for any $g \in \mathcal{O}(N)$,

$$
\mathcal{R}(u(g\cdot)) = \frac{1-s}{4}[u(g\cdot)]_s^4 = \frac{1-s}{4}[u]_s^4 = \mathcal{R}(u).
$$

Then, by applying Theorem 10.2.1, we infer that for every $h \in \mathbb{N}$ there exists $q(h) > 0$ such that for any $0 < q < q(h)$ the functional $\mathcal{F}_q$ admits at least h couples of critical points in $H^s(\mathbb{R}^N)$ with radial symmetry. This means that (10.1.1) admits at least h couples of weak solutions in $H^s_{\mathrm{rad}}(\mathbb{R}^N)$.

Multiplicity and Concentration of Positive Solutions for a Fractional Kirchhoff Equation   **11**

11.1   Introduction

In this chapter we study the multiplicity and concentration of positive solutions for the following fractional Schrödinger–Kirchhoff type problem

$$
\begin{cases}
M\left(\dfrac{1}{\varepsilon^{3-2s}} \iint_{\mathbb{R}^6} \dfrac{|u(x) - u(y)|^2}{|x - y|^{3+2s}}\, dx dy + \dfrac{1}{\varepsilon^3} \int_{\mathbb{R}^3} V(x) u^2\, dx \right)\left[\varepsilon^{2s}(-\Delta)^s u + V(x)u\right] = f(u) \ \text{ in } \mathbb{R}^3, \\
u \in H^s(\mathbb{R}^3), \ u > 0 \text{ in } \mathbb{R}^3,
\end{cases}
$$

$$(11.1.1)$$

where $\varepsilon > 0$ is a small parameter and $s \in (\frac{3}{4}, 1)$.

We assume that $M : [0, \infty) \to [0, \infty)$ is a continuous function such that:

(M_1) there exists $m_0 > 0$ such that $M(t) \geq m_0$ for any $t \geq 0$;
(M_2) $M(t)$ is nondecreasing;
(M_3) for any $t_1 \geq t_2 > 0$,

$$
\frac{M(t_1)}{t_1} - \frac{M(t_2)}{t_2} \leq m_0 \left(\frac{1}{t_1} - \frac{1}{t_2} \right).
$$

As a model for M, we can take $M(t) = m_0 + bt + \sum_{i=1}^{k} b_i t^{\gamma_i}$ with $b_i \geq 0$ and $\gamma_i \in (0, 1)$ for all $i \in \{1, \dots, k\}$.

Concerning the potential $V : \mathbb{R}^3 \to \mathbb{R}$, we suppose that $V \in C(\mathbb{R}^3, \mathbb{R})$ and fulfills the following hypotheses:

(V_1) there exists $V_0 > 0$ such that $V_0 = \inf_{x \in \mathbb{R}^3} V(x)$;

© The Author(s), under exclusive license to Springer Nature Switzerland AG 2021
V. Ambrosio, *Nonlinear Fractional Schrödinger Equations in* $\mathbb{R}^N$,
Frontiers in Mathematics, https://doi.org/10.1007/978-3-030-60220-8_11

(V_2) there is a bounded set $\Omega \subset \mathbb{R}^3$ such that

$$V_0 < \min_{\partial \Omega} V, \quad \text{and} \quad \Lambda = \{x \in \Omega : V(x) = V_0\} \neq \emptyset.$$

Without loss of generality, we may assume that $0 \in \Lambda$.

Finally, concerning the nonlinear term in (11.1.1), we assume that $f : \mathbb{R} \to \mathbb{R}$ is a continuous function satisfying the following conditions:

(f_1) $\lim\limits_{t \to 0} \dfrac{f(t)}{t^3} = 0$;

(f_2) there is $q \in (4, \frac{6}{3-2s})$ such that $\lim\limits_{t \to \infty} \dfrac{f(t)}{t^{q-1}} = 0$;

(f_3) there is $\vartheta \in (4, q)$ such that $0 < \vartheta F(t) \leq f(t)t$ for all $t > 0$;

(f_4) the function $t \mapsto \dfrac{f(t)}{t^3}$ is increasing in $(0, \infty)$.

A typical example is

$$f(t) = \sum_{i=1}^{k} a_i (t^+)^{q_i - 1}$$

with $a_i \geq 0$ not all identically zero and $q_i \in [\vartheta, \frac{6}{3-2s})$ for all $i \in \{1, \ldots, k\}$.

Since we are interested in positive solutions, we assume that f vanishes in $(-\infty, 0)$.

Now, we are ready to state our main result.

Theorem 11.1.1 *Let $s \in (\frac{3}{4}, 1)$ and assume that (M_1)–(M_3), (V_1)–(V_2) and (f_1)–(f_4) hold. Then, for every $\delta > 0$ such that*

$$\Lambda_\delta = \{x \in \mathbb{R}^3 : \operatorname{dist}(x, \Lambda) \leq \delta\} \subset \Omega,$$

there exists an $\varepsilon_\delta > 0$ such that, for any $\varepsilon \in (0, \varepsilon_\delta)$, problem (11.1.1) has at least $\operatorname{cat}_{\Lambda_\delta}(\Lambda)$ positive solutions. Moreover, if u_ε denotes one of these positive solutions and $\eta_\varepsilon \in \mathbb{R}^3$ is a global maximum point of u_ε, then

$$\lim_{\varepsilon \to 0} V(\eta_\varepsilon) = V_0.$$

The proof of Theorem 11.1.1 relies on variational methods inspired by [191, 321, 322]. In particular, Theorem 11.1.1 corresponds to the fractional counterpart of Theorem 1.1 in [191]. Clearly, the presence of the fractional Laplacian makes our analysis more delicate and intriguing compared to the case $s = 1$, and the results obtained in Chaps. 6 and 7 about fractional Schrödinger equations, will play a fundamental role to overcome our difficulties.

In what follows, we give a sketch of the proof. The lack of informations on the behavior of V at infinity suggest us to use the penalization method introduced by del Pino and Felmer [165]. Since f and M are only continuous, the Nehari manifold associated with the modified problem is not differentiable, so we will use some abstract results obtained by Szulkin and Weth in [321, 322]. After a careful study of the autonomous problem associated to (11.1.1), we deal with the multiplicity of solutions of the modified problem, by invoking the Lusternik–Schnirelman theory. Then, in order to prove that the solutions u_ε of the truncated problem are also solutions to (11.1.1) when $\varepsilon > 0$ is sufficiently small, we argue as in Chap. 7. We point out that the restriction $s \in (\frac{3}{4}, 1)$ is essential in our technical approach in order to guarantee the continuous embedding of the space $H^s(\mathbb{R}^3)$ into the Lebesgue spaces $L^r(\mathbb{R}^N)$ with $4 \le r < \frac{6}{3-2s}$ (see conditions (f_1)–(f_3)).

11.2   The Modified Kirchhoff Problem

This section is devoted to the existence of positive solutions to (11.1.1). After a change of variable, problem (11.1.1) reduces to

$$
\begin{cases}
M\left(\iint_{\mathbb{R}^6} \frac{|u(x)-u(y)|^2}{|x-y|^{3+2s}}\,dxdy + \int_{\mathbb{R}^3} V(\varepsilon x)u^2\,dx\right)\left[(-\Delta)^s u + V(\varepsilon x)u\right] = f(u) \ \text{ in } \mathbb{R}^3, \\
u \in H^s(\mathbb{R}^3), \ u > 0 \text{ in } \mathbb{R}^3.
\end{cases}
$$

$$(11.2.1)$$

Take $K > \frac{2}{m_0}$ and $a > 0$ such that $f(a) = \frac{V_0}{K}a$. Let us define

$$
\tilde{f}(t) = \begin{cases}
f(t), & \text{if } t \le a, \\
\frac{V_0}{K}t, & \text{if } t > a,
\end{cases}
$$

and

$$
g(x, t) = \chi_\Omega(x) f(t) + (1 - \chi_\Omega(x)) \tilde{f}(t).
$$

In view of the assumptions on f we deduce that g is a Carathéodory function and satisfies

(g_1) $\displaystyle\lim_{t \to 0} \frac{g(x, t)}{t^3} = 0$ uniformly in $x \in \mathbb{R}^3$;

(g_2) $\displaystyle\lim_{t \to \infty} \frac{g(x, t)}{t^{q-1}} = 0$ uniformly in $x \in \mathbb{R}^3$;

(g_3) (i) $0 < \vartheta G(x, t) \le g(x, t)t$ for any $x \in \Omega$ and for any $t > 0$,

   (ii) $0 \le 2G(x, t) \le g(x, t)t \le \dfrac{V_0}{K}t^2$ for any $x \in \mathbb{R}^3 \setminus \Omega$ and for any $t > 0$;

(g_4) for each fixed $x \in \Omega$ the function $t \mapsto \dfrac{g(x, t)}{t^3}$ is increasing in $(0, \infty)$ and for each

fixed $x \in \mathbb{R}^3 \setminus \Omega$ the function $t \mapsto \dfrac{g(x, t)}{t^3}$ is increasing in $(0, a)$.

By the definition of g it follows that

$$g(x, t) \leq f(t) \quad \text{for all } t > 0, \text{ for all } x \in \mathbb{R}^3,$$
$$g(x, t) = 0 \qquad \text{for all } t < 0, \text{ for all } x \in \mathbb{R}^3.$$

In what follows, we consider the auxiliary problem

$$M\left(\iint_{\mathbb{R}^6} \frac{|u(x) - u(y)|^2}{|x - y|^{3+2s}} \, dxdy + \int_{\mathbb{R}^3} V(\varepsilon x)u^2 \, dx \right) [(-\Delta)^s u + V(\varepsilon x)u] = g(\varepsilon x, u) \quad \text{in } \mathbb{R}^3.$$
$$(11.2.2)$$

Moreover, we focus our attention on positive solutions to (11.2.2) having the property $u(x) \leq a$ for each $x \in \mathbb{R}^3 \setminus \Omega$.

Therefore, solutions of (11.2.2) can be found as critical points of the following energy functional

$$\mathcal{J}_\varepsilon(u) = \frac{1}{2}\widehat{M}\left(\iint_{\mathbb{R}^6} \frac{|u(x) - u(y)|^2}{|x - y|^{3+2s}} \, dxdy + \int_{\mathbb{R}^3} V(\varepsilon x)u^2 \, dx \right) - \int_{\mathbb{R}^3} G(\varepsilon x, u) \, dx$$

where

$$\widehat{M}(t) = \int_0^t M(\tau) \, d\tau \quad \text{and} \quad G(\varepsilon x, t) = \int_0^t g(\varepsilon x, \tau) \, d\tau,$$

which is well defined on the Hilbert space

$$\mathcal{H}_\varepsilon = \left\{ u \in H^s(\mathbb{R}^3) : \int_{\mathbb{R}^3} V(\varepsilon x)u^2 dx < \infty \right\}$$

endowed with the inner product

$$\langle u, \varphi \rangle_\varepsilon = \langle u, \varphi \rangle_{\mathcal{D}^{s,2}(\mathbb{R}^3)} + \int_{\mathbb{R}^3} V(\varepsilon x) u\varphi \, dx$$

and the corresponding norm

$$\|u\|_\varepsilon = \left([u]_s^2 + \int_{\mathbb{R}^3} V(\varepsilon x) u^2 \, dx \right)^{\frac{1}{2}}.$$

By the assumptions on M and f, and using Theorem 1.1.8, it is easy to check that $\mathcal{J}_\varepsilon$ is well defined, that $\mathcal{J}_\varepsilon \in C^1(\mathcal{H}_\varepsilon, \mathbb{R})$ and that its differential $\mathcal{J}_\varepsilon'$ is given by

$$\langle \mathcal{J}_\varepsilon'(u), \varphi \rangle = M(\|u\|_\varepsilon^2)\langle u, \varphi \rangle_\varepsilon - \int_{\mathbb{R}^3} g(\varepsilon x, u)\varphi \, dx,$$

for any $u, \varphi \in \mathcal{H}_\varepsilon$. Let us introduce the Nehari manifold associated with $\mathcal{J}_\varepsilon$, that is

$$\mathcal{N}_\varepsilon = \{u \in \mathcal{H}_\varepsilon \setminus \{0\} : \langle \mathcal{J}_\varepsilon'(u), u \rangle = 0\}.$$

The main result of this section reads:

Theorem 11.2.1 *Under assumptions* (M_1)–(M_3), (V_1)–(V_2) *and* (f_1)–(f_4), *the auxiliary problem* (11.2.2) *has a positive ground state solution for all* $\varepsilon > 0$.

Let $\Omega_\varepsilon = \{x \in \mathbb{R}^3 : \varepsilon x \in \Omega\}$ and

$$\mathcal{H}_\varepsilon^+ = \{u \in \mathcal{H}_\varepsilon : |\mathrm{supp}(u^+) \cap \Omega_\varepsilon| > 0\} \subset \mathcal{H}_\varepsilon.$$

Let $\mathbb{S}_\varepsilon$ be the unit sphere of $\mathcal{H}_\varepsilon$ and denote $\mathbb{S}_\varepsilon^+ = \mathbb{S}_\varepsilon \cap \mathcal{H}_\varepsilon^+$. We observe that $\mathcal{H}_\varepsilon^+$ is open in $\mathcal{H}_\varepsilon$. Indeed, consider a sequence $(u_n) \subset \mathcal{H}_\varepsilon \setminus \mathcal{H}_\varepsilon^+$ such that $u_n \to u$ in $\mathcal{H}_\varepsilon$ and assume, by contradiction, that $u \in \mathcal{H}_\varepsilon^+$. Now, by the definition of $\mathcal{H}_\varepsilon^+$ it follows that $|\mathrm{supp}(u_n^+) \cap \Omega_\varepsilon| = 0$ for all $n \in \mathbb{N}$ and $u_n^+(x) \to u^+(x)$ a.e. $x \in \Omega_\varepsilon$. So,

$$u^+(x) = \lim_{n \to \infty} u_n^+(x) = 0 \quad \text{a.e. } x \in \Omega_\varepsilon,$$

and this contradicts the fact that $u \in \mathcal{H}_\varepsilon^+$. Therefore, $\mathcal{H}_\varepsilon^+$ is open.

By the definition of $\mathbb{S}_\varepsilon^+$ and the fact that $\mathcal{H}_\varepsilon^+$ is open in $\mathcal{H}_\varepsilon$, it follows that $\mathbb{S}_\varepsilon^+$ is a incomplete $C^{1,1}$-manifold of codimension 1, modeled on $\mathcal{H}_\varepsilon$ and contained in the open $\mathcal{H}_\varepsilon^+$. Then, $\mathcal{H}_\varepsilon = T_u\mathbb{S}_\varepsilon^+ \oplus \mathbb{R}u$ for each $u \in \mathbb{S}_\varepsilon^+$, where

$$T_u\mathbb{S}_\varepsilon^+ = \{v \in \mathcal{H}_\varepsilon : (u, \varphi)_\varepsilon = 0\}.$$

In the next lemma we prove that $\mathcal{J}_\varepsilon$ has a mountain pass geometry [29].

Lemma 11.2.2 *The functional* $\mathcal{J}_\varepsilon$ *satisfies the following conditions:*

(a) *there exist* $\alpha, \rho > 0$ *such that* $\mathcal{J}_\varepsilon(u) \geq \alpha$ *with* $\|u\|_\varepsilon = \rho$;
(b) *there exists* $e \in \mathcal{H}_\varepsilon$ *such that* $\|e\|_\varepsilon > \rho$ *and* $\mathcal{J}_\varepsilon(e) < 0$.

Proof

(a) By (M_1), (g_1), (g_2) and Theorem 1.1.8, for every $\xi > 0$ there exists $C_\xi > 0$ such that

$$\mathcal{J}_\varepsilon(u) = \frac{1}{2}\widehat{M}(\|u\|_\varepsilon^2) - \int_{\mathbb{R}^3} G(\varepsilon x, u)\,dx \geq \frac{m_0}{2}\|u\|_\varepsilon^2 - \xi C\|u\|_\varepsilon^4 - C_\xi C\|u\|_\varepsilon^q.$$

Then, we can find $\alpha, \rho > 0$ such that $\mathcal{J}_\varepsilon(u) \geq \alpha$ with $\|u\|_\varepsilon = \rho$.

(b) In view of (M_3), there exists a positive constant γ such that

$$M(t) \leq \gamma(1 + t) \quad \text{for all } t \geq 0. \tag{11.2.3}$$

Then, by (g_3)-(i), we can see that, for any $u \in \mathcal{H}_\varepsilon^+$ and $t > 0$

$$\begin{aligned}
\mathcal{J}_\varepsilon(tu) &= \frac{1}{2}\widehat{M}(\|tu\|_\varepsilon^2) - \int_{\mathbb{R}^3} G(\varepsilon x, tu)\,dx \\
&\leq \frac{\gamma}{2}t^2\|u\|_\varepsilon^2 + \frac{\gamma}{4}t^4\|u\|_\varepsilon^4 - \int_{\Omega_\varepsilon} G(\varepsilon x, tu)\,dx \\
&\leq \frac{\gamma}{2}t^2\|u\|_\varepsilon^2 + \frac{\gamma}{4}t^4\|u\|_\varepsilon^4 - C_1 t^\vartheta \int_{\Omega_\varepsilon}(u^+)^\vartheta\,dx + C_2|\mathrm{supp}(u^+) \cap \Omega_\varepsilon|,
\end{aligned} \tag{11.2.4}$$

for some positive constants C_1 and C_2.

Taking into account that $\vartheta \in (4, \frac{6}{3-2s})$, we conclude that $\mathcal{J}_\varepsilon(tu) \to -\infty$ as $t \to \infty$. $\qquad\square$

Since f and M are continuous functions, the next results will play a fundamental role in overcoming the non-differentiability of $\mathcal{N}_\varepsilon$ and the incompleteness of $\mathbb{S}_\varepsilon^+$.

Lemma 11.2.3 *Assume that (M_1)–(M_3), (V_1)–(V_2) and (f_1)–(f_4) hold. Then,*

(i) *For each $u \in \mathcal{H}_\varepsilon^+$, let $h : \mathbb{R}_+ \to \mathbb{R}$ be defined by $h_u(t) = \mathcal{J}_\varepsilon(tu)$. Then, there is a unique $t_u > 0$ such that*

$$h_u'(t) > 0 \text{ in } (0, t_u),$$
$$h_u'(t) < 0 \text{ in } (t_u, \infty).$$

(ii) *There exists $\tau > 0$ independent of u such that $t_u \geq \tau$ for all $u \in \mathbb{S}_\varepsilon^+$. Moreover, for each compact set $\mathbb{K} \subset \mathbb{S}_\varepsilon^+$ there is a positive constant $C_\mathbb{K}$ such that $t_u \leq C_\mathbb{K}$ for any $u \in \mathbb{K}$.*

(iii) *The map $\hat{m}_\varepsilon : \mathcal{H}_\varepsilon^+ \to \mathcal{N}_\varepsilon$ given by $\hat{m}_\varepsilon(u) = t_u u$ is continuous and $m_\varepsilon = \hat{m}_\varepsilon|_{\mathbb{S}_\varepsilon^+}$ is a homeomorphism between $\mathbb{S}_\varepsilon^+$ and $\mathcal{N}_\varepsilon$. Moreover, $m_\varepsilon^{-1}(u) = \frac{u}{\|u\|_\varepsilon}$.*

(iv) *If there is a sequence $(u_n) \subset \mathbb{S}_\varepsilon^+$ such that $\mathrm{dist}(u_n, \partial\mathbb{S}_\varepsilon^+) \to 0$, then $\|m_\varepsilon(u_n)\|_\varepsilon \to \infty$ and $\mathcal{J}_\varepsilon(m_\varepsilon(u_n)) \to \infty$.*

Proof

(i) We know that $h_u \in C^1(\mathbb{R}_+, \mathbb{R})$ and, by Lemma 11.2.2, we have that $h_u(0) = 0$, $h_u(t) > 0$ for $t > 0$ small enough and $h_u(t) < 0$ for $t > 0$ sufficiently large. Hence, there exists $t_u > 0$ such that $h_u'(t_u) = 0$, and t_u is a global maximum for h_u.

Then,

$$0 = h_u'(t_u) = \langle \mathcal{J}_\varepsilon'(t_u u), u \rangle = \frac{1}{t_u} \langle \mathcal{J}_\varepsilon'(t_u u), t_u u \rangle,$$

which implies that $t_u u \in \mathcal{N}_\varepsilon$. Next, let us prove the uniqueness of such a t_u. Assume, by contradiction, that there exist $t_1 > t_2 > 0$ such that $h_u'(t_1) = h_u'(t_2) = 0$, or equivalently,

$$t_1 M(\|t_1 u\|_\varepsilon^2) \|u\|_\varepsilon^2 = \int_{\mathbb{R}^3} g(\varepsilon x, t_1 u) u \, dx, \tag{11.2.5}$$

$$t_2 M(\|t_2 u\|_\varepsilon^2) \|u\|_\varepsilon^2 = \int_{\mathbb{R}^3} g(\varepsilon x, t_2 u) u \, dx. \tag{11.2.6}$$

Dividing both members of (11.2.5) by $t_1^3 \|u\|_\varepsilon^4$ we get

$$\frac{M(\|t_1 u\|_\varepsilon^2)}{\|t_1 u\|_\varepsilon^2} = \frac{1}{\|u\|_\varepsilon^4} \int_{\mathbb{R}^3} \frac{g(\varepsilon x, t_1 u)}{(t_1 u)^3} u^4 \, dx;$$

similarly, dividing both members of (11.2.6) by $t_2^3 \|u\|_\varepsilon^4$ we obtain

$$\frac{M(\|t_2 u\|_\varepsilon^2)}{\|t_2 u\|_\varepsilon^2} = \frac{1}{\|u\|_\varepsilon^4} \int_{\mathbb{R}^3} \frac{g(\varepsilon x, t_2 u)}{(t_2 u)^3} u^4 \, dx.$$

Taking the difference of the last two identities and using (M_3) and (g_4), we can see that

$$\frac{m_0}{\|u\|_\varepsilon^2} \left(\frac{1}{t_2^2} - \frac{1}{t_1^2} \right)$$

$$\geq \frac{M(\|t_1 u\|_\varepsilon^2)}{\|t_1 u\|_\varepsilon^2} - \frac{M(\|t_2 u\|_\varepsilon^2)}{\|t_2 u\|_\varepsilon^2}$$

$$= \frac{1}{\|u\|_\varepsilon^4} \int_{\mathbb{R}^3} \left[\frac{g(\varepsilon x, t_1 u)}{(t_1 u)^3} - \frac{g(\varepsilon x, t_2 u)}{(t_2 u)^3} \right] u^4 dx$$

$$= \frac{1}{\|u\|_\varepsilon^4} \int_{\mathbb{R}^3 \setminus \Omega_\varepsilon} \left[\frac{g(\varepsilon x, t_1 u)}{(t_1 u)^3} - \frac{g(\varepsilon x, t_2 u)}{(t_2 u)^3} \right] u^4 dx + \frac{1}{\|u\|_\varepsilon^4} \int_{\Omega_\varepsilon} \left[\frac{g(\varepsilon x, t_1 u)}{(t_1 u)^3} - \frac{g(\varepsilon x, t_2 u)}{(t_2 u)^3} \right] u^4 dx$$

$$\geq \frac{1}{\|u\|_\varepsilon^4} \int_{\mathbb{R}^3 \setminus \Omega_\varepsilon} \left[\frac{g(\varepsilon x, t_1 u)}{(t_1 u)^3} - \frac{g(\varepsilon x, t_2 u)}{(t_2 u)^3} \right] u^4 dx$$

$$= \frac{1}{\|u\|_\varepsilon^4} \int_{(\mathbb{R}^3 \setminus \Omega_\varepsilon) \cap \{t_2 u > a\}} \left[\frac{g(\varepsilon x, t_1 u)}{(t_1 u)^3} - \frac{g(\varepsilon x, t_2 u)}{(t_2 u)^3} \right] u^4 dx$$

$$+ \frac{1}{\|u\|_\varepsilon^4} \int_{(\mathbb{R}^3 \setminus \Omega_\varepsilon) \cap \{t_2 u \leq a < t_1 u\}} \left[\frac{g(\varepsilon x, t_1 u)}{(t_1 u)^3} - \frac{g(\varepsilon x, t_2 u)}{(t_2 u)^3} \right] u^4 dx$$

$$+ \frac{1}{\|u\|_\varepsilon^4} \int_{(\mathbb{R}^3 \setminus \Omega_\varepsilon) \cap \{t_1 u < a\}} \left[\frac{g(\varepsilon x, t_1 u)}{(t_1 u)^3} - \frac{g(\varepsilon x, t_2 u)}{(t_2 u)^3} \right] u^4 dx = I + II + III.$$

First, note that $III \geq 0$ thanks to (g_4) and the assumption that $t_1 > t_2$. In view of the definition of g, we have

$$I \geq \frac{1}{\|u\|_\varepsilon^4} \int_{(\mathbb{R}^3 \setminus \Omega_\varepsilon) \cap \{t_2 u > a\}} \left[\frac{V_0}{K} \frac{1}{(t_1 u)^2} - \frac{V_0}{K} \frac{1}{(t_2 u)^2} \right] u^4 dx$$

$$= \frac{1}{\|u\|_\varepsilon^4} \frac{1}{K} \left(\frac{1}{t_1^2} - \frac{1}{t_2^2} \right) \int_{(\mathbb{R}^3 \setminus \Omega_\varepsilon) \cap \{t_2 u > a\}} V_0 u^2 dx.$$

Regarding II, using again the definition of g, we infer that

$$II \geq \frac{1}{\|u\|_\varepsilon^4} \int_{(\mathbb{R}^3 \setminus \Omega_\varepsilon) \cap \{t_2 u \leq a < t_1 u\}} \left[\frac{V_0}{K} \frac{1}{(t_1 u)^2} - \frac{f(t_2 u)}{(t_2 u)^3} \right] u^4 dx.$$

Therefore,

$$\frac{m_0}{\|u\|_\varepsilon^2} \left(\frac{1}{t_1^2} - \frac{1}{t_2^2} \right) \geq \frac{1}{\|u\|_\varepsilon^4} \frac{1}{K} \left(\frac{1}{t_1^2} - \frac{1}{t_2^2} \right) \int_{(\mathbb{R}^3 \setminus \Omega_\varepsilon) \cap \{t_2 u > a\}} V_0 u^2 dx$$

$$+ \frac{1}{\|u\|_\varepsilon^4} \int_{(\mathbb{R}^3 \setminus \Omega_\varepsilon) \cap \{t_2 u \leq a < t_1 u\}} \left[\frac{V_0}{K} \frac{1}{(t_1 u)^2} - \frac{f(t_2 u)}{(t_2 u)^3} \right] u^4 dx.$$

Multiplying both sides by $\|u\|_\varepsilon^4 \frac{t_1^2 t_2^2}{t_2^2 - t_1^2} < 0$, we have

$$m_0 \|u\|_\varepsilon^2 \leq \frac{1}{K} \int_{(\mathbb{R}^3 \setminus \Omega_\varepsilon) \cap \{t_2 u > a\}} V_0 u^2 dx + \frac{t_1^2 t_2^2}{t_2^2 - t_1^2} \int_{(\mathbb{R}^3 \setminus \Omega_\varepsilon) \cap \{t_2 u \leq a < t_1 u\}} \left[\frac{V_0}{K} \frac{1}{(t_1 u)^2} - \frac{f(t_2 u)}{(t_2 u)^3} \right] u^4 dx$$

$$= \frac{1}{K} \int_{(\mathbb{R}^3 \setminus \Omega_\varepsilon) \cap \{t_2 u > a\}} V_0 u^2 dx$$

$$
-\frac{t_2^2}{t_1^2 - t_2^2} \int_{(\mathbb{R}^3 \setminus \Omega_\varepsilon) \cap \{t_2 u \le a < t_1 u\}} \frac{V_0}{K} u^2 \, dx + \frac{t_1^2}{t_1^2 - t_2^2} \int_{(\mathbb{R}^3 \setminus \Omega_\varepsilon) \cap \{t_2 u \le a < t_1 u\}} \frac{f(t_2 u)}{t_2 u} u^2 \, dx
$$

$$
\le \frac{1}{K} \int_{\mathbb{R}^3 \setminus \Omega_\varepsilon} V_0 u^2 \, dx \le \frac{1}{K} \|u\|_\varepsilon^2.
$$

Then, since $u \ne 0$ and $K > \frac{2}{m_0}$, we get $m_0 \le \frac{1}{K} < m_0$, which is a contradiction.

(ii) Let $u \in \mathbb{S}_\varepsilon^+$. By (i), there exists $t_u > 0$ such that $h_u'(t_u) = 0$, or equivalently

$$
t_u M(t_u^2) = \int_{\mathbb{R}^3} g(\varepsilon x, t_u u) u \, dx.
$$

By assumptions (g_1) and (g_2), given $\xi > 0$ there exists a positive constant C_ξ such that

$$
|g(x, t)| \le \xi |t|^3 + C_\xi |t|^{q-1}, \qquad \text{for all } (x, t) \in \mathbb{R}^3 \times \mathbb{R},
$$

which in conjunction with (M_1) and Theorem 1.1.8 yields

$$
m_0 t_u \le M(t_u^2) t_u = \int_{\mathbb{R}^3} g(\varepsilon x, t_u u) u \, dx
$$

$$
\le \xi C_1 t_u^3 + C_\xi C_2 t_u^{q-1}.
$$

We conclude that there exists $\tau > 0$, independent of u, such that $t_u \ge \tau$. Now, let $\mathbb{K} \subset \mathbb{S}_\varepsilon^+$ be a compact set. Let us show that t_u can be estimated from above by a constant depending on $\mathbb{K}$. Assume, by contradiction, that there exists a sequence $(u_n) \subset \mathbb{K}$ such that $t_n = t_{u_n} \to \infty$. Therefore, there exists $u \in \mathbb{K}$ such that $u_n \to u$ in $\mathcal{H}_\varepsilon$. From (11.2.4) we get

$$
\mathcal{J}_\varepsilon(t_n u_n) \to -\infty. \tag{11.2.7}
$$

Fix $v \in \mathcal{N}_\varepsilon$. Then, using the fact that $\langle \mathcal{J}_\varepsilon'(v), v \rangle = 0$, and assumptions (g_3)-(i) and (g_3)-(ii), we see that

$$
\mathcal{J}_\varepsilon(v) = \mathcal{J}_\varepsilon(v) - \frac{1}{\vartheta} \langle \mathcal{J}_\varepsilon'(v), v \rangle
$$

$$
= \frac{1}{2} \widehat{M}(\|v\|_\varepsilon^2) - \frac{1}{\vartheta} M(\|v\|_\varepsilon^2) \|v\|_\varepsilon^2 + \frac{1}{\vartheta} \int_{\mathbb{R}^3} [g(\varepsilon x, v) v - \vartheta G(\varepsilon x, v)] \, dx
$$

$$
= \frac{1}{2} \widehat{M}(\|v\|_\varepsilon^2) - \frac{1}{\vartheta} M(\|v\|_\varepsilon^2) \|v\|_\varepsilon^2 + \frac{1}{\vartheta} \int_{\mathbb{R}^3 \setminus \Omega_\varepsilon} [g(\varepsilon x, v) v - \vartheta G(\varepsilon x, v)] \, dx
$$

$$+ \frac{1}{\vartheta} \int_{\Omega_\varepsilon} [g(\varepsilon x, v)v - \vartheta G(\varepsilon x, v)] \, dx$$

$$\geq \frac{1}{2} \widehat{M}(\|v\|_\varepsilon^2) - \frac{1}{\vartheta} M(\|v\|_\varepsilon^2) \|v\|_\varepsilon^2 + \frac{1}{\vartheta} \int_{\mathbb{R}^3 \setminus \Omega_\varepsilon} [g(\varepsilon x, v)v - \vartheta G(\varepsilon x, v)] \, dx$$

$$\geq \frac{1}{2} \widehat{M}(\|v\|_\varepsilon^2) - \frac{1}{\vartheta} M(\|v\|_\varepsilon^2) \|v\|_\varepsilon^2 - \left(\frac{\vartheta - 2}{2\vartheta} \right) \frac{1}{K} \int_{\mathbb{R}^3 \setminus \Omega_\varepsilon} V(\varepsilon x) v^2 \, dx$$

$$\geq \frac{1}{2} \widehat{M}(\|v\|_\varepsilon^2) - \frac{1}{\vartheta} M(\|v\|_\varepsilon^2) \|v\|_\varepsilon^2 - \left(\frac{\vartheta - 2}{2\vartheta} \right) \frac{1}{K} \|v\|_\varepsilon^2.$$

Now, by (M_3), we know that

$$\widehat{M}(t) \geq \frac{M(t) + m_0}{2} t \quad \text{for all } t \geq 0. \tag{11.2.8}$$

This together with (M_1) implies that

$$\mathcal{J}_\varepsilon(v) \geq \frac{1}{4} \left[M(\|v\|_\varepsilon^2) + m_0 \right] \|v\|_\varepsilon^2 - \frac{1}{\vartheta} M(\|v\|_\varepsilon^2) \|v\|_\varepsilon^2 - \left(\frac{\vartheta - 2}{2\vartheta} \right) \frac{1}{K} \|v\|_\varepsilon^2$$

$$= \frac{\vartheta - 4}{4\vartheta} M(\|v\|_\varepsilon^2) \|v\|_\varepsilon^2 + \frac{1}{4} m_0 \|v\|_\varepsilon^2 - \left(\frac{\vartheta - 2}{2\vartheta} \right) \frac{1}{K} \|v\|_\varepsilon^2$$

$$\geq \left(\frac{\vartheta - 4}{4\vartheta} + \frac{1}{4} \right) m_0 \|v\|_\varepsilon^2 - \left(\frac{\vartheta - 2}{2\vartheta} \right) \frac{1}{K} \|v\|_\varepsilon^2$$

$$= \left(\frac{\vartheta - 2}{2\vartheta} \right) \left(m_0 - \frac{1}{K} \right) \|v\|_\varepsilon^2. \tag{11.2.9}$$

Taking into account that $(t_{u_n} u_n) \subset \mathcal{N}_\varepsilon$ and $K > \frac{2}{m_0}$, from (11.2.9) we deduce that (11.2.7) cannot hold.

(iii) First, note that $\hat{m}_\varepsilon$, m_ε and m_ε^{-1} are well defined. Indeed, by (i), for each $u \in \mathcal{H}_\varepsilon^+$ there exists a unique $m_\varepsilon(u) \in \mathcal{N}_\varepsilon$. On the other hand, if $u \in \mathcal{N}_\varepsilon$, then $u \in \mathcal{H}_\varepsilon^+$. Otherwise, if $u \notin \mathcal{H}_\varepsilon^+$, then

$$|\text{supp}(u^+) \cap \Omega_\varepsilon| = 0,$$

which together with $(g3)$-(ii) gives

$$0 < M(\|u\|_\varepsilon^2) \|u\|_\varepsilon^2 = \int_{\mathbb{R}^3} g(\varepsilon x, u) u \, dx$$

$$= \int_{\mathbb{R}^3 \setminus \Omega_\varepsilon} g(\varepsilon x, u) u \, dx + \int_{\Omega_\varepsilon} g(\varepsilon x, u) u \, dx$$

$$= \int_{\mathbb{R}^3 \setminus \Omega_\varepsilon} g(\varepsilon x, u^+) u^+ \, dx$$

$$\leq \frac{1}{K} \int_{\mathbb{R}^3 \setminus \Omega_\varepsilon} V(\varepsilon x) u^2 \, dx \leq \frac{1}{K} \|u\|_\varepsilon^2. \qquad (11.2.10)$$

Using (M_1) and (11.2.10) we get

$$0 < m_0 \|u\|_\varepsilon^2 \leq M(\|u\|_\varepsilon^2) \|u\|_\varepsilon^2 \leq \frac{1}{K} \|u\|_\varepsilon^2$$

and this leads to a contradiction because $\frac{1}{K} < \frac{m_0}{2}$. Consequently, $m_\varepsilon^{-1}(u) = \frac{u}{\|u\|_\varepsilon} \in \mathbb{S}_\varepsilon^+$, m_ε^{-1} is well defined and continuous.

Let $u \in \mathbb{S}_\varepsilon^+$. Then,

$$m_\varepsilon^{-1}(m_\varepsilon(u)) = m_\varepsilon^{-1}(t_u u) = \frac{t_u u}{\|t_u u\|_\varepsilon} = \frac{u}{\|u\|_\varepsilon} = u$$

from which m_ε is a bijection. Let us prove that $\hat{m}_\varepsilon$ is a continuous function. Let $(u_n) \subset \mathcal{H}_\varepsilon^+$ and $u \in \mathcal{H}_\varepsilon^+$ such that $u_n \to u$ in $\mathcal{H}_\varepsilon$. Hence,

$$\frac{u_n}{\|u_n\|_\varepsilon} \to \frac{u}{\|u\|_\varepsilon} \quad \text{in } \mathcal{H}_\varepsilon.$$

Set $v_n = \frac{u_n}{\|u_n\|_\varepsilon}$ and $t_n = t_{v_n}$. By (ii), there exists $t_0 > 0$ such that $t_n \to t_0$. Since $t_n v_n \in \mathcal{N}_\varepsilon$ and $\|v_n\|_\varepsilon = 1$, we have

$$M(t_n^2) t_n = \int_{\mathbb{R}^3} g(\varepsilon x, t_n v_n) v_n \, dx$$

or, equivalently,

$$M(t_n^2) t_n = \frac{1}{\|u_n\|_\varepsilon} \int_{\mathbb{R}^3} g(\varepsilon x, t_n v_n) u_n \, dx.$$

Letting $n \to \infty$ we obtain

$$M(t_0^2) t_0 = \frac{1}{\|u\|_\varepsilon} \int_{\mathbb{R}^3} g(\varepsilon x, t_0 v) u \, dx,$$

where $v = \frac{u}{\|u\|_\varepsilon}$, which implies that $t_0 v \in \mathcal{N}_\varepsilon$. By (i), we deduce that $t_v = t_0$, and this shows that

$$\hat{m}_\varepsilon(u_n) = \hat{m}_\varepsilon\left(\frac{u_n}{\|u_n\|_\varepsilon}\right) \to \hat{m}_\varepsilon\left(\frac{u}{\|u\|_\varepsilon}\right) = \hat{m}_\varepsilon(u) \quad \text{in } \mathcal{H}_\varepsilon.$$

Therefore $\hat{m}_\varepsilon$ and m_ε are continuous functions.

(iv) Let $(u_n) \subset \mathbb{S}_\varepsilon^+$ be such that $\mathrm{dist}(u_n, \partial \mathbb{S}_\varepsilon^+) \to 0$. Observing that for each $p \in [2, 2_s^*]$ and $n \in \mathbb{N}$ it holds that

$$\|u_n^+\|_{L^p(\Omega_\varepsilon)} \le \inf_{v \in \partial \mathbb{S}_\varepsilon^+} \|u_n - v\|_{L^p(\Omega_\varepsilon)}$$

$$\le C_p \inf_{v \in \partial \mathbb{S}_\varepsilon^+} \|u_n - v\|_\varepsilon,$$

we use (g_1), (g_2), and (g_3)-(ii) to infer that

$$\int_{\mathbb{R}^3} G(\varepsilon x, t u_n) \, dx = \int_{\mathbb{R}^3 \setminus \Omega_\varepsilon} G(\varepsilon x, t u_n) \, dx + \int_{\Omega_\varepsilon} G(\varepsilon x, t u_n) \, dx$$

$$\le \frac{t^2}{K} \int_{\mathbb{R}^3 \setminus \Omega_\varepsilon} V(\varepsilon x) u_n^2 \, dx + \int_{\Omega_\varepsilon} F(t u_n) \, dx$$

$$\le \frac{t^2}{K} \|u_n\|_\varepsilon^2 + C_1 t^4 \int_{\Omega_\varepsilon} (u_n^+)^4 \, dx + C_2 t^q \int_{\Omega_\varepsilon} (u_n^+)^q \, dx$$

$$\le \frac{t^2}{K} + C_1' t^4 \mathrm{dist}(u_n, \partial \mathbb{S}_\varepsilon^+)^4 + C_2' t^q \mathrm{dist}(u_n, \partial \mathbb{S}_\varepsilon^+)^q.$$

Consequently,

$$\limsup_{n \to \infty} \int_{\mathbb{R}^3} G(\varepsilon x, t u_n) \, dx \le \frac{t^2}{K} \qquad (11.2.11)$$

for all $t > 0$. Keeping in mind the definitions of $m_\varepsilon(u_n)$ and $\widehat{M}(t)$, and by using (11.2.11) and assumption (M_1), we have

$$\liminf_{n \to \infty} \mathcal{J}_\varepsilon(m_\varepsilon(u_n)) \ge \liminf_{n \to \infty} \mathcal{J}_\varepsilon(m_\varepsilon(u_n))$$

$$= \liminf_{n \to \infty} \left[\frac{1}{2} \widehat{M}(\|t u_n\|_\varepsilon^2) - \int_{\mathbb{R}^3} G(\varepsilon x, t u_n) \, dx \right]$$

$$\ge \frac{1}{2} \widehat{M}(t^2) - \frac{t^2}{K}$$

$$\ge \left(\frac{1}{2} m_0 - \frac{1}{K} \right) t^2.$$

Recalling that $K > 2/m_0$, we get

$$\lim_{n \to \infty} \mathcal{J}_\varepsilon(m_\varepsilon(u_n)) = \infty.$$

Moreover, the definition of $\mathcal{J}_\varepsilon(m_\varepsilon(u_n))$ and (11.2.3) imply that

$$\mathcal{J}_\varepsilon(m_\varepsilon(u_n)) \le \frac{1}{2}\widehat{M}(t_{u_n}^2) \le C(t_{u_n}^2 + t_{u_n}^4) \quad \text{for all } n \in \mathbb{N},$$

and so $\|m_\varepsilon(u_n)\|_\varepsilon \to \infty$ as $n \to \infty$.

$\square$

Let us define the maps

$$\hat{\psi}_\varepsilon : \mathcal{H}_\varepsilon^+ \to \mathbb{R} \quad \text{and} \quad \psi_\varepsilon : \mathbb{S}_\varepsilon^+ \to \mathbb{R},$$

by $\hat{\psi}_\varepsilon(u) = \mathcal{J}_\varepsilon(\hat{m}_\varepsilon(u))$ and $\psi_\varepsilon = \hat{\psi}_\varepsilon|_{\mathbb{S}_\varepsilon^+}$.

The next result is a direct consequence of Lemma 11.2.3 and Corollary 2.4.4 (see also [191, 321, 322]).

Proposition 11.2.4 *Assume that (M_1)–(M_3), (V_1)–(V_2) and (f_1)–(f_4) hold. Then,*

(a) $\hat{\psi}_\varepsilon \in C^1(\mathcal{H}_\varepsilon^+, \mathbb{R})$ *and*

$$\langle \hat{\psi}_\varepsilon'(u), v \rangle = \frac{\|\hat{m}_\varepsilon(u)\|_\varepsilon}{\|u\|_\varepsilon} \langle \mathcal{J}_\varepsilon'(\hat{m}_\varepsilon(u)), v \rangle,$$

for every $u \in \mathcal{H}_\varepsilon^+$ and $v \in \mathcal{H}_\varepsilon$.

(b) $\psi_\varepsilon \in C^1(\mathbb{S}_\varepsilon^+, \mathbb{R})$ *and*

$$\langle \psi_\varepsilon'(u), v \rangle = \|m_\varepsilon(u)\|_\varepsilon \langle \mathcal{J}_\varepsilon'(m_\varepsilon(u)), v \rangle,$$

for every $v \in T_u \mathbb{S}_\varepsilon^+$.

(c) *If (u_n) is a Palais–Smale sequence for ψ_ε, then $(m_\varepsilon(u_n))$ is a Palais–Smale sequence for $\mathcal{J}_\varepsilon$. If $(u_n) \subset \mathcal{N}_\varepsilon$ is a bounded Palais–Smale sequence for $\mathcal{J}_\varepsilon$, then $(m_\varepsilon^{-1}(u_n))$ is a Palais–Smale sequence for ψ_ε.*

(d) *u is a critical point of ψ_ε if and only if $m_\varepsilon(u)$ is a nontrivial critical point for $\mathcal{J}_\varepsilon$. Moreover, the corresponding critical values coincide and*

$$\inf_{u \in \mathbb{S}_\varepsilon^+} \psi_\varepsilon(u) = \inf_{u \in \mathcal{N}_\varepsilon} \mathcal{J}_\varepsilon(u).$$

Remark 11.2.5 As in [322], we can see that, thanks to (M_1)–(M_3),

$$c_\varepsilon = \inf_{u \in \mathcal{N}_\varepsilon} \mathcal{J}_\varepsilon(u) = \inf_{u \in \mathcal{H}_\varepsilon^+} \max_{t>0} \mathcal{J}_\varepsilon(tu) = \inf_{u \in \mathbb{S}_\varepsilon^+} \max_{t>0} \mathcal{J}_\varepsilon(tu).$$

It is also easy to check that c_ε coincides with the mountain pass level of $\mathcal{J}_\varepsilon$.

Let us verify that the functional $\mathcal{J}_\varepsilon$ satisfies the Palais–Smale condition.

Lemma 11.2.6 *Let* $(u_n) \subset \mathcal{H}_\varepsilon$ *be a Palais–Smale sequence for* $\mathcal{J}_\varepsilon$ *at the level* $d \in \mathbb{R}$*. Then* (u_n) *is bounded in* $\mathcal{H}_\varepsilon$*.*

Proof Since (u_n) is a Palais–Smale sequence at the level d, we know that

$$\mathcal{J}_\varepsilon(u_n) \to d \quad \text{and} \quad \mathcal{J}'_\varepsilon(u_n) \to 0 \text{ in } \mathcal{H}^*_\varepsilon.$$

Then, arguing as in the proof of Lemma 11.2.3-(ii) (see formula (11.2.9)), we see that

$$C(1 + \|u_n\|_\varepsilon) \geq \mathcal{J}_\varepsilon(u_n) - \frac{1}{\vartheta}\langle \mathcal{J}'_\varepsilon(u_n), u_n \rangle$$

$$\geq \left(\frac{\vartheta - 2}{2\vartheta}\right)\left(m_0 - \frac{1}{K}\right)\|u_n\|^2_\varepsilon.$$

Since $\vartheta > 4$ and $K > 2/m_0$, we conclude that (u_n) is bounded in $\mathcal{H}_\varepsilon$. $\qquad\square$

The next two lemmas are fundamental to obtain compactness of bounded Palais–Smale sequences.

Lemma 11.2.7 *Let* $(u_n) \subset \mathcal{H}_\varepsilon$ *be a Palais–Smale sequence for* $\mathcal{J}_\varepsilon$ *at the level* d*. Then, for each* $\zeta > 0$*, there exists* $R = R(\zeta) > 0$ *such that*

$$\limsup_{n \to \infty} \left[\int_{\mathbb{R}^3 \setminus B_R} dx \int_{\mathbb{R}^3} \frac{|u_n(x) - u_n(y)|^2}{|x - y|^{3+2s}} dy + \int_{\mathbb{R}^3 \setminus B_R} V(\varepsilon x)u_n^2 \, dx\right] < \zeta.$$

Proof For any $R > 0$, let $\eta_R \in C^\infty(\mathbb{R}^3)$ be such that $\eta_R = 0$ in $B_{\frac{R}{2}}$ and $\eta_R = 1$ in $\mathbb{R}^3 \setminus B_R$, with $0 \leq \eta_R \leq 1$ and $\|\nabla \eta_R\|_{L^\infty(\mathbb{R}^3)} \leq \frac{C}{R}$, where C is a constant independent of R. Since $(\eta_R u_n)$ is bounded in $\mathcal{H}_\varepsilon$, it follows that $\langle \mathcal{J}'_\varepsilon(u_n), \eta_R u_n \rangle = o_n(1)$, that is

$$M(\|u_n\|^2_\varepsilon)\left[\iint_{\mathbb{R}^6} \frac{|u_n(x) - u_n(y)|^2}{|x - y|^{3+2s}}\eta_R(x)\,dxdy + \int_{\mathbb{R}^3} V(\varepsilon x)u_n^2\eta_R\,dx\right]$$

$$= o_n(1) + \int_{\mathbb{R}^3} g(\varepsilon x, u_n)u_n\eta_R\,dx - M(\|u_n\|^2_\varepsilon)\iint_{\mathbb{R}^6} \frac{(\eta_R(x) - \eta_R(y))(u_n(x) - u_n(y))}{|x - y|^{3+2s}}u_n(y)\,dxdy.$$

Take $R > 0$ such that $\Omega_\varepsilon \subset B_{\frac{R}{2}}$. Then, using (M_1) and (g_3)-(ii), we have

$$m_0\left[\iint_{\mathbb{R}^6} \frac{|u_n(x) - u_n(y)|^2}{|x - y|^{3+2s}}\eta_R(x)\,dxdy + \int_{\mathbb{R}^3} V(\varepsilon x)u_n^2\eta_R\,dx\right]$$

$$\leq \int_{\mathbb{R}^3} \frac{1}{K}V(\varepsilon x)u_n^2\eta_R\,dx - M(\|u_n\|^2_\varepsilon)\iint_{\mathbb{R}^6} \frac{(\eta_R(x) - \eta_R(y))(u_n(x) - u_n(y))}{|x - y|^{3+2s}}u_n(y)\,dxdy + o_n(1),$$

whence

$$\left(m_0 - \frac{1}{K}\right)\left[\iint_{\mathbb{R}^6} \frac{|u_n(x) - u_n(y)|^2}{|x-y|^{3+2s}}\eta_R(x)\,dxdy + \int_{\mathbb{R}^3} V(\varepsilon x)u_n^2\eta_R\,dx\right]$$

$$\leq -M(\|u_n\|_\varepsilon^2)\iint_{\mathbb{R}^6} \frac{(\eta_R(x) - \eta_R(y))(u_n(x) - u_n(y))}{|x-y|^{3+2s}}u_n(y)\,dxdy + o_n(1).$$

$$(11.2.12)$$

Now note that the boundedness of (u_n) in $\mathcal{H}_\varepsilon$ and assumption (M_2) imply that

$$M(\|u_n\|_\varepsilon^2) \leq C \quad \text{for any } n \in \mathbb{N}.$$

$$(11.2.13)$$

On the other hand, using Hölder's inequality, the boundedness of (u_n) and Lemma 1.4.5 we get

$$\lim_{R \to \infty} \limsup_{n \to \infty} \iint_{\mathbb{R}^6} \frac{(\eta_R(x) - \eta_R(y))(u_n(x) - u_n(y))}{|x-y|^{3+2s}}u_n(y)\,dxdy = 0.$$

$$(11.2.14)$$

Finally, (11.2.12), (11.2.13), (11.2.14) and the definition of η_R show that

$$\lim_{R \to \infty} \limsup_{n \to \infty} \int_{\mathbb{R}^3 \backslash B_R} dx \int_{\mathbb{R}^3} \frac{|u_n(x) - u_n(y)|^2}{|x-y|^{3+2s}}\,dy + \int_{\mathbb{R}^3 \backslash B_R} V(\varepsilon x)u_n^2\,dx = 0,$$

which completes the proof of lemma. $\qquad\square$

Lemma 11.2.8 *Let $(u_n) \subset \mathcal{H}_\varepsilon$ be a Palais–Smale sequence for $\mathcal{J}_\varepsilon$ at the level d and such that $u_n \rightharpoonup u$. Then, for all $R > 0$*

$$\lim_{n \to \infty} \int_{B_R} dx \int_{\mathbb{R}^3} \frac{|u_n(x) - u_n(y)|^2}{|x-y|^{3+2s}}\,dy + \int_{B_R} V(\varepsilon x)u_n^2\,dx$$

$$= \int_{B_R} dx \int_{\mathbb{R}^3} \frac{|u(x) - u(y)|^2}{|x-y|^{3+2s}}\,dy + \int_{B_R} V(\varepsilon x)u^2\,dx.$$

Proof By Lemma 11.2.6, (u_n) is bounded in $\mathcal{H}_\varepsilon$, so we may assume that $u_n \rightharpoonup u$ in $\mathcal{H}_\varepsilon$ and $\|u_n\|_\varepsilon \to t_0$. Then, by the weak lower semicontinuity $\|u\|_\varepsilon \leq t_0$. Moreover, by the continuity of M, we know that $M(\|u_n\|_\varepsilon^2) \to M(t_0)$.

Let $\eta_\rho \in C^\infty(\mathbb{R}^3)$ be such that $\eta_\rho = 1$ in B_ρ and $\eta_\rho = 0$ in $B_{2\rho}^c$, with $0 \leq \eta_\rho \leq 1$. Fix $R > 0$ and choose $\rho > R$. Then we have

$$0 \leq M(\|u_n\|_\varepsilon^2)\left[\int_{B_R} dx \int_{\mathbb{R}^3} \frac{|(u_n(x) - u_n(y)) - (u(x) - u(y))|^2}{|x-y|^{3+2s}}\,dy + \int_{B_R} V(\varepsilon x)(u_n - u)^2\,dx\right]$$

$$\leq M(\|u_n\|_\varepsilon^2)\left[\iint_{\mathbb{R}^6}\frac{|(u_n(x)-u_n(y))-(u(x)-u(y))|^2}{|x-y|^{3+2s}}\eta_\rho(x)\,dxdy+\int_{\mathbb{R}^3}V(\varepsilon x)(u_n-u)^2\eta_\rho\,dx\right]$$

$$\leq M(\|u_n\|_\varepsilon^2)\left[\iint_{\mathbb{R}^6}\frac{|u_n(x)-u_n(y)|^2}{|x-y|^{3+2s}}\eta_\rho(x)\,dxdy+\int_{\mathbb{R}^3}V(\varepsilon x)u_n^2\eta_\rho\,dx\right]$$

$$+M(\|u_n\|_\varepsilon^2)\left[\iint_{\mathbb{R}^6}\frac{|u(x)-u(y)|^2}{|x-y|^{3+2s}}\eta_\rho(x)\,dxdy+\int_{\mathbb{R}^3}V(\varepsilon x)u^2\eta_\rho\,dx\right]$$

$$-2M(\|u_n\|_\varepsilon^2)\left[\iint_{\mathbb{R}^6}\frac{(u_n(x)-u_n(y))(u(x)-u(y))}{|x-y|^{3+2s}}\eta_\rho(x)\,dxdy+\int_{\mathbb{R}^3}V(\varepsilon x)u_nu\eta_\rho\,dx\right]$$

$$=I_{n,\rho}-II_{n,\rho}+III_{n,\rho}+IV_{n,\rho}\leq|I_{n,\rho}|+|II_{n,\rho}|+|III_{n,\rho}|+|IV_{n,\rho}| \qquad (11.2.15)$$

where

$$I_{n,\rho}=M(\|u_n\|_\varepsilon^2)\left[\iint_{\mathbb{R}^6}\frac{|u_n(x)-u_n(y)|^2}{|x-y|^{3+2s}}\eta_\rho(x)\,dxdy+\int_{\mathbb{R}^3}V(\varepsilon x)u_n^2\eta_\rho\,dx\right]-\int_{\mathbb{R}^3}g(\varepsilon x,u_n)u_n\eta_\rho\,dx,$$

$$II_{n,\rho}=M(\|u_n\|_\varepsilon^2)\left[\iint_{\mathbb{R}^6}\frac{(u_n(x)-u_n(y))(u(x)-u(y))}{|x-y|^{3+2s}}\eta_\rho(x)\,dxdy+\int_{\mathbb{R}^3}V(\varepsilon x)u_nu\eta_\rho\,dx\right]$$

$$-\int_{\mathbb{R}^3}g(\varepsilon x,u_n)u\eta_\rho\,dx,$$

$$III_{n,\rho}=-M(\|u_n\|_\varepsilon^2)\left[\iint_{\mathbb{R}^6}\frac{(u_n(x)-u_n(y))(u(x)-u(y))}{|x-y|^{3+2s}}\eta_\rho(x)dxdy+\int_{\mathbb{R}^3}V(\varepsilon x)u_nu\eta_\rho\,dx\right]$$

$$+M(\|u_n\|_\varepsilon^2)\left[\iint_{\mathbb{R}^6}\frac{|u(x)-u(y)|^2}{|x-y|^{3+2s}}\eta_\rho(x)\,dxdy+\int_{\mathbb{R}^3}V(\varepsilon x)u^2\eta_\rho\,dx\right],$$

$$IV_{n,\rho}=\int_{\mathbb{R}^3}g(\varepsilon x,u_n)u_n\eta_\rho\,dx-\int_{\mathbb{R}^3}g(\varepsilon x,u_n)u\eta_\rho\,dx.$$

Let us prove that

$$\lim_{\rho\to\infty}\limsup_{n\to\infty}|I_{n,\rho}|=0. \qquad (11.2.16)$$

First, $I_{n,\rho}$ can be written as

$$I_{n,\rho}=\langle\mathcal{J}_\varepsilon'(u_n),u_n\eta_\rho\rangle-M(\|u_n\|_\varepsilon^2)\iint_{\mathbb{R}^6}\frac{(u_n(x)-u_n(y))(\eta_\rho(x)-\eta_\rho(y))}{|x-y|^{3+2s}}u_n(y)\,dxdy.$$

Since $(u_n\eta_\rho)$ is bounded in $\mathcal{H}_\varepsilon$, we have $\langle\mathcal{J}_\varepsilon'(u_n),u_n\eta_\rho\rangle=o_n(1)$, so

$$I_{n,\rho}=o_n(1)-M(\|u_n\|_\varepsilon^2)\iint_{\mathbb{R}^6}\frac{(u_n(x)-u_n(y))(\eta_\rho(x)-\eta_\rho(y))}{|x-y|^{3+2s}}u_n(y)\,dxdy. \qquad (11.2.17)$$

By Lemma 1.4.5,

$$\lim_{\rho\to\infty}\ \limsup_{n\to\infty}\ \left|M(\|u_n\|_\varepsilon^2)\iint_{\mathbb{R}^6}\frac{(u_n(x)-u_n(y))(\eta_\rho(x)-\eta_\rho(y))}{|x-y|^{3+2s}}u_n(x)\,dxdy\right|=0,$$

which together with (11.2.17) implies that (11.2.16) holds. Next, note that

$$II_{n,\rho}=\langle\mathcal{J}_\varepsilon'(u_n),u\eta_\rho\rangle-M(\|u_n\|_\varepsilon^2)\iint_{\mathbb{R}^6}\frac{(u_n(x)-u_n(y))(\eta_\rho(x)-\eta_\rho(y))}{|x-y|^{3+2s}}u(x)\,dxdy.$$

Similar calculations to the proof of Lemma 1.4.5 show that

$$\lim_{\rho\to\infty}\ \limsup_{n\to\infty}\ \left|M(\|u_n\|_\varepsilon^2)\iint_{\mathbb{R}^6}\frac{(u_n(x)-u_n(y))(\eta_\rho(x)-\eta_\rho(y))}{|x-y|^{3+2s}}u(x)\,dxdy\right|=0,$$

and since $\langle\mathcal{J}_\varepsilon'(u_n),u\eta_\rho\rangle=o_n(1)$, we obtain

$$\lim_{\rho\to\infty}\ \limsup_{n\to\infty}|II_{n,\rho}|=0.\tag{11.2.18}$$

Now let us prove that

$$\lim_{\rho\to\infty}\ \limsup_{n\to\infty}|III_{n,\rho}|=0.\tag{11.2.19}$$

By the weak convergence,

$$III_{n,\rho}=-M(\|u_n\|_\varepsilon^2)\left[\iint_{\mathbb{R}^6}\frac{[(u_n(x)-u(x))-(u_n(y)-u(y))]}{|x-y|^{3+2s}}(u(x)-u(y))\eta_\rho(x)\,dxdy\right.$$

$$\left.+\int_{\mathbb{R}^3}V(\varepsilon x)(u_n-u)u\eta_\rho\,dx\right]$$

$$=-M(\|u_n\|_\varepsilon^2)\left[\iint_{\mathbb{R}^6}\frac{[(u_n(x)-u(x))-(u_n(y)-u(y))]}{|x-y|^{3+2s}}(u(x)\eta_\rho(x)-u(y)\eta_\rho(y))\,dxdy\right.$$

$$\left.+\int_{\mathbb{R}^3}V(\varepsilon x)(u_n-u)u\eta_\rho\,dx\right]$$

$$-M(\|u_n\|_\varepsilon^2)\iint_{\mathbb{R}^6}\frac{[(u_n(x)-u(x))-(u_n(y)-u(y))]}{|x-y|^{3+2s}}(\eta_\rho(x)-\eta_\rho(y))u(y)\,dxdy$$

$$=o_n(1)-M(\|u_n\|_\varepsilon^2)\iint_{\mathbb{R}^6}\frac{[(u_n(x)-u(x))-(u_n(y)-u(y))]}{|x-y|^{3+2s}}(\eta_\rho(x)-\eta_\rho(y))u(y)\,dxdy.$$

Using Hölder's inequality and the boundedness of (u_n) and $M(\|u_n\|_\varepsilon^2)$ we see that

$$\sup_{n\in\mathbb{N}}\left|M(\|u_n\|_\varepsilon^2)\iint_{\mathbb{R}^6}\frac{[(u_n(x)-u(x))-(u_n(y)-u(y))]}{|x-y|^{3+2s}}(\eta_\rho(x)-\eta_\rho(y))u(y)\,dxdy\right|$$

$$\leq C \left(\iint_{\mathbb{R}^6} \frac{|\eta_\rho(x) - \eta_\rho(y)|^2}{|x-y|^{3+2s}} |u(y)|^2 \, dx \, dy \right)^{1/2} \to 0 \quad \text{as } \rho \to \infty,$$

so (11.2.19) holds true. Finally, we deal with the fourth term. By Theorem 1.1.8, $u_n \to u$ in $L^p_{\text{loc}}(\mathbb{R}^3)$ for all $1 \leq p < \frac{6}{3-2s}$. Hence, in view of (g_1) and (g_2), we deduce that, for any $\rho > R$,

$$\lim_{n \to \infty} |IV_{n,\rho}| = 0. \tag{11.2.20}$$

Putting together (11.2.15), (11.2.16), (11.2.18), (11.2.19) and (11.2.20), and recalling that $\|u_n\|_\varepsilon \to t_0$ we complete the proof. □

The previous lemmas enable us to prove the following result.

Proposition 11.2.9 *The functional $\mathcal{J}_\varepsilon$ satisfies the* (PS)$_d$ *condition in $\mathcal{H}_\varepsilon$ at any level $d \in \mathbb{R}$.*

Proof Let $(u_n) \subset \mathcal{H}_\varepsilon$ be a Palais–Smale sequence for $\mathcal{J}_\varepsilon$ at the level d. By Lemma 11.2.6, (u_n) is bounded in $\mathcal{H}_\varepsilon$. Thus, up to a subsequence,

$$u_n \rightharpoonup u \quad \text{in } \mathcal{H}_\varepsilon. \tag{11.2.21}$$

Lemma 11.2.8 implies that

$$\begin{aligned}
\lim_{n \to \infty} \int_{B_R} dx \int_{\mathbb{R}^3} &\frac{|u_n(x) - u_n(y)|^2}{|x-y|^{3+2s}} \, dy + \int_{B_R} V(\varepsilon x) u_n^2 \, dx \\
&= \int_{B_R} dx \int_{\mathbb{R}^3} \frac{|u(x) - u(y)|^2}{|x-y|^{3+2s}} \, dy + \int_{B_R} V(\varepsilon x) u^2 \, dx.
\end{aligned} \tag{11.2.22}$$

Moreover, by Lemma 11.2.7, for each $\zeta > 0$ there exists $R = R(\zeta) > \frac{C}{\zeta}$, with $C > 0$ independent of ζ, such that

$$\limsup_{n \to \infty} \left[\int_{\mathbb{R}^3 \setminus B_R} dx \int_{\mathbb{R}^3} \frac{|u_n(x) - u_n(y)|^2}{|x-y|^{3+2s}} \, dy + \int_{\mathbb{R}^3 \setminus B_R} V(\varepsilon x) u_n^2 \, dx \right] < \zeta. \tag{11.2.23}$$

Putting together (11.2.21), (11.2.22) and (11.2.23) we infer that

$$\|u\|_\varepsilon^2 \leq \liminf_{n \to \infty} \|u_n\|_\varepsilon^2 \leq \limsup_{n \to \infty} \|u_n\|_\varepsilon^2$$

$$= \limsup_{n \to \infty} \left[\int_{B_R} dx \int_{\mathbb{R}^3} \frac{|u_n(x) - u_n(y)|^2}{|x-y|^{3+2s}} \, dy + \int_{B_R} V(\varepsilon x) u_n^2 \, dx \right]$$

$$+ \int_{\mathbb{R}^3 \setminus B_R} dx \int_{\mathbb{R}^3} \frac{|u_n(x) - u_n(y)|^2}{|x - y|^{3+2s}} \, dy + \int_{\mathbb{R}^3 \setminus B_R} V(\varepsilon x) u_n^2 \, dx \Bigg]$$

$$\leq \int_{B_R} dx \int_{\mathbb{R}^3} \frac{|u(x) - u(y)|^2}{|x - y|^{3+2s}} \, dy + \int_{B_R} V(\varepsilon x) u^2 \, dx + \zeta.$$

Taking the limit as $\zeta \to 0$, we have $R \to \infty$, therefore

$$\|u\|_\varepsilon^2 \leq \liminf_{n \to \infty} \|u_n\|_\varepsilon^2 \leq \limsup_{n \to \infty} \|u_n\|_\varepsilon^2 \leq \|u\|_\varepsilon^2,$$

which implies $\|u_n\|_\varepsilon \to \|u\|_\varepsilon$. Since $\mathcal{H}_\varepsilon$ is a Hilbert space, we deduce that $u_n \to u$ in $\mathcal{H}_\varepsilon$. $\quad\square$

Corollary 11.2.10 *The functional ψ_ε satisfies the (PS)$_d$ condition on $\mathbb{S}_\varepsilon^+$ at any level $d \in \mathbb{R}$.*

Proof Let $(u_n) \subset \mathbb{S}_\varepsilon^+$ be a Palais–Smale sequence for ψ_ε at the level d. Then

$$\psi_\varepsilon(u_n) \to d \quad \text{and} \quad \psi_\varepsilon'(u_n) \to 0 \quad \text{in } (T_{u_n} \mathbb{S}_\varepsilon^+)^*.$$

It follows from Proposition 11.2.4-(c) that $(m_\varepsilon(u_n))$ is a Palais–Smale sequence for $\mathcal{J}_\varepsilon$ in $\mathcal{H}_\varepsilon$ at the level d. Then, using Proposition 11.2.9, we see that $\mathcal{J}_\varepsilon$ fulfills the (PS)$_d$ condition in $\mathcal{H}_\varepsilon$, so there exists $u \in \mathbb{S}_\varepsilon^+$ such that, up to a subsequence,

$$m_\varepsilon(u_n) \to m_\varepsilon(u) \quad \text{in } \mathcal{H}_\varepsilon.$$

Applying Lemma 11.2.3-(iii) we conclude that $u_n \to u$ in $\mathbb{S}_\varepsilon^+$. $\quad\square$

At this point, we are able to prove the main result of this section.

Proof of Theorem 11.2.1 In view of Lemma 11.2.2 and Proposition 11.2.9, we can apply Theorem 2.2.9, and thus obtain the existence of a nontrivial critical point u_ε of $\mathcal{J}_\varepsilon$. Now, we show that $u_\varepsilon \geq 0$ in $\mathbb{R}^3$. Since $\langle \mathcal{J}_\varepsilon'(u_\varepsilon), u_\varepsilon^- \rangle = 0$, where $x^- = \min\{x, 0\}$, we see that

$$M(\|u_\varepsilon\|_\varepsilon^2) \left[\iint_{\mathbb{R}^6} \frac{(u_\varepsilon(x) - u_\varepsilon(y))(u_\varepsilon^-(x) - u_\varepsilon^-(y))}{|x - y|^{3-2s}} \, dx dy + \int_{\mathbb{R}^3} V(\varepsilon x) u_\varepsilon u_\varepsilon^- \, dx \right]$$

$$= \int_{\mathbb{R}^3} g(\varepsilon x, u_\varepsilon) u_\varepsilon^- \, dx.$$

Recalling that $(x - y)(x^- - y^-) \geq |x^- - y^-|^2$ and $g(x, t) = 0$ for $t \leq 0$, we deduce that

$$0 \leq M(\|u_\varepsilon\|_\varepsilon^2)\|u_\varepsilon^-\|_\varepsilon^2 \leq 0.$$

Using (M_1), we have $\|u_\varepsilon^-\|_\varepsilon^2 = 0$, that is, $u_\varepsilon \geq 0$ in $\mathbb{R}^3$. Finally, arguing as in the proof of Lemma 11.5.1 below, we can see that $u_\varepsilon \in L^\infty(\mathbb{R}^3) \cap C_{\text{loc}}^{0,\alpha}(\mathbb{R}^3)$, and by Theorem 1.3.5 we deduce that $u_\varepsilon > 0$ in $\mathbb{R}^3$. $\square$

11.3   The Autonomous Kirchhoff Problem

In this section we deal with the limit problem associated with (11.2.1), namely,

$$\begin{cases} M\left(\iint_{\mathbb{R}^6} \frac{|u(x)-u(y)|^2}{|x-y|^{3+2s}} \, dx\, dy + \int_{\mathbb{R}^3} V_0 u^2 \, dx\right)[(-\Delta)^s u + V_0 u] = f(u) \text{ in } \mathbb{R}^3, \\ u \in H^s(\mathbb{R}^3), \ u > 0 \text{ in } \mathbb{R}^3. \end{cases}$$

$$(11.3.1)$$

The Euler–Lagrange functional associated with (11.3.1) is

$$\mathcal{J}_0(u) = \frac{1}{2}\widehat{M}\left(\iint_{\mathbb{R}^6} \frac{|u(x) - u(y)|^2}{|x - y|^{3+2s}} \, dx\, dy + \int_{\mathbb{R}^3} V_0 u^2 \, dx\right) - \int_{\mathbb{R}^3} F(u) \, dx,$$

and so is well defined on the Hilbert space $\mathcal{H}_0 = H^s(\mathbb{R}^3)$ endowed with the inner product

$$\langle u, \varphi \rangle_0 = \langle u, \varphi \rangle_{\mathcal{D}^{s,2}(\mathbb{R}^3)} + \int_{\mathbb{R}^3} V_0 u \varphi \, dx.$$

The norm induced by this inner product is

$$\|u\|_0 = \left([u]_s^2 + V_0 \|u\|_{L^2(\mathbb{R}^3)}^2\right)^{\frac{1}{2}}.$$

The Nehari manifold associated with $\mathcal{J}_0$ is given by

$$\mathcal{N}_0 = \{u \in \mathcal{H}_0 \setminus \{0\} : \langle \mathcal{J}_0'(u), u \rangle = 0\}.$$

Let $\mathcal{H}_0^+$ denote the open subset of $\mathcal{H}_0$ defined as

$$\mathcal{H}_0^+ = \{u \in \mathcal{H}_0 : |\text{supp}(u^+)| > 0\},$$

and let $\mathbb{S}_0^+ = \mathbb{S}_0 \cap \mathcal{H}_0^+$, where $\mathbb{S}_0$ is the unit sphere of $\mathcal{H}_0$. We note that $\mathbb{S}_0^+$ is an incomplete $C^{1,1}$-manifold of codimension 1 modeled on $\mathcal{H}_0$ and contained in $\mathcal{H}_0^+$. Thus $\mathcal{H}_0 = T_u\mathbb{S}_0^+ \oplus$

$\mathbb{R}u$ for each $u \in \mathbb{S}_0^+$, where $T_u\mathbb{S}_0^+ = \{u \in \mathcal{H}_0 : (u, v)_0 = 0\}$. Arguing as in the previous subsection, we can see that the following results hold.

Lemma 11.3.1 *Assume that* (M_1)–(M_3) *and* (f_1)–(f_4) *hold. Then,*

(i) *For each* $u \in \mathcal{H}_0^+$, *let* $h : \mathbb{R}_+ \to \mathbb{R}$ *be defined by* $h_u(t) = \mathcal{J}_0(tu)$. *Then, there is a unique* $t_u > 0$ *such that*

$$h_u'(t) > 0 \text{ in } (0, t_u),$$
$$h_u'(t) < 0 \text{ in } (t_u, \infty).$$

(ii) *There exists* $\tau > 0$ *independent of* u *such that* $t_u \geq \tau$ *for any* $u \in \mathbb{S}_0^+$. *Moreover, for each compact set* $\mathbb{K} \subset \mathbb{S}_0^+$ *there is a positive constant* $C_{\mathbb{K}}$ *such that* $t_u \leq C_{\mathbb{K}}$ *for any* $u \in \mathbb{K}$.

(iii) *The map* $\hat{m}_0 : \mathcal{H}_0^+ \to \mathcal{N}_0$ *given by* $\hat{m}_0(u) = t_u u$ *is continuous and* $m_0 = \hat{m}_0|_{\mathbb{S}_0^+}$ *is a homeomorphism between* $\mathbb{S}_0^+$ *and* $\mathcal{N}_0$. *Moreover* $m_0^{-1}(u) = \frac{u}{\|u\|_0}$.

(iv) *If there is a sequence* $(u_n) \subset \mathbb{S}_0^+$ *such that* $\text{dist}(u_n, \partial\mathbb{S}_0^+) \to 0$ *then* $\|m_0(u_n)\|_0 \to \infty$ *and* $\mathcal{J}_0(m_0(u_n)) \to \infty$.

Let us define the maps

$$\hat{\psi}_0 : \mathcal{H}_0^+ \to \mathbb{R} \quad \text{and} \quad \psi_0 : \mathbb{S}_0^+ \to \mathbb{R},$$

by $\hat{\psi}_0(u) = \mathcal{J}_0(\hat{m}_0(u))$ and $\psi_0 = \hat{\psi}_0|_{\mathbb{S}_0^+}$.

Proposition 11.3.2 *Assume that* (M_1)–(M_3) *and* (f_1)–(f_4) *hold. Then,*

(a) $\hat{\psi}_0 \in C^1(\mathcal{H}_0^+, \mathbb{R})$ *and*

$$\langle \hat{\psi}_0'(u), v \rangle = \frac{\|\hat{m}_0(u)\|_0}{\|u\|_0} \langle \mathcal{J}_0'(\hat{m}_0(u)), v \rangle$$

for every $u \in \mathcal{H}_0^+$ *and* $v \in \mathcal{H}_0$.

(b) $\psi_0 \in C^1(\mathbb{S}_0^+, \mathbb{R})$ *and*

$$\langle \psi_0'(u), v \rangle = \|m_0(u)\|_0 \langle \mathcal{J}_0'(m_0(u)), v \rangle,$$

for every $v \in T_u\mathbb{S}_0^+$.

(c) *If* (u_n) *is a Palais–Smale sequence for* ψ_0, *then* $(m_0(u_n))$ *is a Palais–Smale sequence for* $\mathcal{J}_0$. *If* $(u_n) \subset \mathcal{N}_0$ *is a bounded Palais–Smale sequence for* $\mathcal{J}_0$, *then* $(m_0^{-1}(u_n))$ *is a Palais–Smale sequence for* ψ_0.

(d) *u is a critical point of ψ_0 if and only if $m_0(u)$ is a nontrivial critical point for $\mathcal{J}_0$.*
   Moreover, the corresponding critical values coincide and

$$\inf_{u \in \mathbb{S}_0^+} \psi_0(u) = \inf_{u \in \mathcal{N}_0} \mathcal{J}_0(u).$$

Remark 11.3.3 We have the following equalities:

$$c_0 = \inf_{u \in \mathcal{N}_0} \mathcal{J}_0(u) = \inf_{u \in \mathcal{H}_0^+} \max_{t > 0} \mathcal{J}_0(tu) = \inf_{u \in \mathbb{S}_0^+} \max_{t > 0} \mathcal{J}_0(tu).$$

Arguing as in the proof of Lemma 11.2.2, it is easy to check that $\mathcal{J}_0$ has a mountain pass geometry. Then we can apply a variant of the mountain pass theorem without the Palais–Smale condition (see Remark 2.2.10) to find a Palais–Smale sequence of $\mathcal{J}_0$ at level c_0 (which coincides with the mountain pass level of $\mathcal{J}_0$). The next lemma is very important because it allows us to deduce that the weak limit of a Palais–Smale sequence at the level d is nontrivial.

Lemma 11.3.4 *Let $(u_n) \subset \mathcal{H}_0$ be a Palais–Smale sequence for $\mathcal{J}_0$ at the level d with $u_n \rightharpoonup 0$ in $\mathcal{H}_0$. Then*

(a) *either $u_n \to 0$ in $\mathcal{H}_0$, or*
(b) *there are a sequence $(y_n) \subset \mathbb{R}^3$ and constants $R, \beta > 0$ such that*

$$\liminf_{n \to \infty} \int_{B_R(y_n)} u_n^2 \, dx \geq \beta > 0.$$

Proof Assume that (b) does not hold. Then, for every $R > 0$,

$$\lim_{n \to \infty} \sup_{y \in \mathbb{R}^3} \int_{B_R(y)} u_n^2 \, dx = 0.$$

Since (u_n) is bounded in $\mathcal{H}_0$, Lemma 1.4.4 shows that

$$u_n \to 0 \quad \text{in } L^p(\mathbb{R}^3) \quad \text{for any } 2 < p < 2_s^*.$$

Moreover, by conditions (f_1) and (f_2), we can see that

$$\lim_{n \to \infty} \int_{\mathbb{R}^3} f(u_n) u_n \, dx = 0.$$

Then, using $\langle \mathcal{J}_0'(u_n), u_n \rangle = o_n(1)$ and (M_1), we get

$$0 \le m_0 \|u_n\|_0^2 \le M(\|u_n\|_0^2) \|u_n\|_0^2 = \int_{\mathbb{R}^3} f(u_n) u_n \, dx + o_n(1) = o_n(1).$$

Therefore, (a) holds true.   □

Remark 11.3.5 Let us observe that, if u is the weak limit of a Palais–Smale sequence (u_n) of $\mathcal{J}_0$ at the level c_0, then we can assume that $u \ne 0$. Otherwise, we would have $u_n \rightharpoonup 0$ and, if $u_n \not\to 0$ in $\mathcal{H}_0$, we conclude from Lemma 11.3.4 that there are $(y_n) \subset \mathbb{R}^3$ and $R, \beta > 0$ such that

$$\liminf_{n \to \infty} \int_{B_R(y_n)} u_n^2 \, dx \ge \beta > 0.$$

Set $v_n(x) = u_n(x + y_n)$. Then we see that (v_n) is a Palais–Smale sequence for $\mathcal{J}_0$ at the level c_0, (v_n) is bounded in $\mathcal{H}_0$ and there exists $v \in \mathcal{H}_0$ such that $v_n \rightharpoonup v$ and $v \ne 0$.

Theorem 11.3.6 *Problem* (11.3.1) *admits a positive ground state solution.*

Proof Arguing as in the proof of Lemma 11.2.2, it is easy to check that $\mathcal{J}_0$ has a mountain pass geometry. Then there exists a Palais–Smale sequence $(u_n) \subset \mathcal{H}_0$ for $\mathcal{J}_0$ at the level c_0, that is,

$$\mathcal{J}_0(u_n) \to c_0 \quad \text{and} \quad \mathcal{J}_0'(u_n) \to 0 \quad \text{in } \mathcal{H}_0^*.$$

We claim that (u_n) is a bounded sequence in $\mathcal{H}_0$. Indeed, using (11.2.8) and assumptions (f_3) and (M_1), we have

$$C(1 + \|u_n\|_0) \ge \mathcal{J}_0(u_n) - \frac{1}{\vartheta} \langle \mathcal{J}_0'(u_n), u_n \rangle$$

$$= \frac{1}{2} \widehat{M}(\|u_n\|_0^2) - \frac{1}{\vartheta} M(\|u_n\|_0^2) \|u_n\|_0^2 + \frac{1}{\vartheta} \int_{\mathbb{R}^3} [f(u)u - F(u)] \, dx$$

$$\ge \left(\frac{1}{4} - \frac{1}{\vartheta} \right) M(\|u_n\|_0^2) \|u_n\|_0^2 + \frac{m_0}{4} \|u_n\|_0^2$$

$$\ge \frac{\vartheta - 2}{2\vartheta} m_0 \|u_n\|_0^2,$$

which yields the boundedness of (u_n) because $\vartheta > 4$. Therefore, in view of Theorem 1.1.8 and Remark 11.3.5, we may assume that there exists $u \in \mathcal{H}_0, u \ne 0$, such that

$$u_n \rightharpoonup u \quad \text{in } \mathcal{H}_0, \tag{11.3.2}$$

$$u_n \to u \quad \text{in } L^p_{loc}(\mathbb{R}^3) \quad \text{for all } 2 \le p < \frac{6}{3 - 2s}, \tag{11.3.3}$$

$$\|u_n\|_0 \to t_0. \tag{11.3.4}$$

Recall that $\langle \mathcal{J}'_0(u_n), \varphi \rangle = o_n(1)$ for any $\varphi \in \mathcal{H}_0$, that is

$$M(\|u_n\|_0^2)\langle u_n, \varphi \rangle_0 = \int_{\mathbb{R}^3} f(u_n)\varphi \, dx + o_n(1). \tag{11.3.5}$$

Take $\varphi \in C_c^\infty(\mathbb{R}^3)$. By (11.3.2),

$$\langle u_n, \varphi \rangle_0 \to \langle u, \varphi \rangle_0. \tag{11.3.6}$$

On the other hand, by (f_1), (f_2) and (11.3.3),

$$\int_{\mathbb{R}^3} f(u_n)\varphi \, dx \to \int_{\mathbb{R}^3} f(u)\varphi \, dx. \tag{11.3.7}$$

Since $C_c^\infty(\mathbb{R}^3)$ is dense in $H^s(\mathbb{R}^3)$, (11.3.6) and (11.3.7) hold for all $\varphi \in \mathcal{H}_0$. In order to prove that u is a weak solution to (11.3.1), it remains to show that

$$M(\|u_n\|_0^2) \to M(t_0^2). \tag{11.3.8}$$

In the light of (11.3.4), we note that

$$\|u\|_0^2 \le \liminf_{n \to \infty} \|u_n\|_0^2 = t_0^2,$$

so, by using (M_2), we deduce that $M(\|u\|_0^2) \le M(t_0^2)$. At this point our aim is to prove that $M(\|u\|_0^2) = M(t_0^2)$. Assume by contradiction that $M(\|u\|_0^2) < M(t_0^2)$, then

$$M(\|u\|_0^2)\|u\|_0^2 < M(t_0^2)\|u\|_0^2 = \int_{\mathbb{R}^3} f(u)u \, dx,$$

that is, $\langle \mathcal{J}'_0(u), u \rangle < 0$. Hence there exists $t \in (0, 1)$ such that $tu \in \mathcal{N}_0$. Now, observing that (M_3) implies that $t \mapsto \frac{1}{2}\widehat{M}(t) - \frac{1}{4}M(t)t$ is increasing in $(0, \infty)$, and (f_4) yields that $t \mapsto \frac{1}{4}f(t)t - F(t)$ is increasing for $t > 0$, we have

$$c_0 \le \mathcal{J}_0(tu) = \mathcal{J}_0(tu) - \frac{1}{4}\langle \mathcal{J}'_0(tu), tu \rangle$$

$$= \frac{1}{2}\widehat{M}(\|tu\|_0^2) - \frac{1}{4}M(\|tu\|_0^2)\|tu\|_0^2 + \int_{\mathbb{R}^3}\left[\frac{1}{4}f(tu)tu - F(tu)\right]dx$$

$$< \frac{1}{2}\widehat{M}(\|u\|_0^2) - \frac{1}{4}M(\|u\|_0^2)\|u\|_0^2 + \int_{\mathbb{R}^3}\left[\frac{1}{4}f(u)u - F(u)\right]dx$$

$$\leq \liminf_{n\to\infty}\left\{\frac{1}{2}\widehat{M}(\|u_n\|_0^2) - \frac{1}{4}M(\|u_n\|_0^2)\|u_n\|_0^2 + \int_{\mathbb{R}^3}\left[\frac{1}{4}f(u_n)u_n - F(u_n)\right]dx\right\}$$

$$= \liminf_{n\to\infty}\left\{\mathcal{J}_0(u_n) - \frac{1}{4}\langle\mathcal{J}_0'(u_n), u_n\rangle\right\} = c_0,$$

so we reached a contradiction. Therefore, combining (11.3.5), (11.3.6), (11.3.7) and (11.3.8) we obtain

$$M(\|u\|_0^2)\langle u, \varphi\rangle_0 = \int_{\mathbb{R}^3} f(u)\varphi\, dx \quad \text{for any } \varphi \in C_c^\infty(\mathbb{R}^3).$$

Since $C_c^\infty(\mathbb{R}^3)$ is dense in $\mathcal{H}_0$, we deduce that $\mathcal{J}_0'(u) = 0$. Moreover, a similar calculation as done above with $t = 1$ yields $u \in \mathcal{N}_0$. Let us show that $u > 0$ in $\mathbb{R}^3$. First we prove that $u \geq 0$. Indeed, observing that $\langle\mathcal{J}_0'(u), u^-\rangle = 0$ and using the inequality $(x - y)(x^- - y^-) \geq |x^- - y^-|^2$ and the fact that $f(t) = 0$ for $t \leq 0$, we have

$$0 \leq M(\|u_0\|_0^2)\|u^-\|_0^2 \leq 0,$$

which together with (M_1) implies that $u \geq 0$ and $u \not\equiv 0$. Next, we claim that $u \in C^{1,\alpha}(\mathbb{R}^3)$ for some $\alpha \in (0, 1)$. Let

$$v = \frac{u}{M(\|u\|_0^2)^{\frac{1}{q-2}}}.$$

Then v is a nonnegative solution to

$$(-\Delta)^s v + V_0 v = h(v), \quad \text{in } \mathbb{R}^3$$

where h is the continuous function

$$h(t) = \frac{f(tM(\|u\|_0^2)^{\frac{1}{q-2}})}{M(\|u\|_0^2)^{\frac{q-1}{q-2}}}.$$

Using the conditions (f_1) and (f_3), we can see that

$$\lim_{t\to 0}\frac{h(t)}{t} = \lim_{t\to 0}\frac{f(tM(\|u\|_0^2)^{\frac{1}{q-2}})}{(tM(\|u\|_0^2)^{\frac{1}{q-2}})^3}\frac{(tM(\|u\|_0^2)^{\frac{1}{q-2}})^3}{M(\|u\|_0^2)^{\frac{q-1}{q-2}}} = 0$$

and

$$\lim_{t \to \infty} \frac{h(t)}{t^{q-1}} = \lim_{t \to \infty} \frac{f\left(tM(\|u\|_0^2)^{\frac{1}{q-2}}\right)}{\left(tM(\|u\|_0^2)^{\frac{1}{q-2}}\right)^{q-1}} = 0,$$

so we deduce that $h(t) \leq C(|t| + |t|^{q-1})$ for any $t \in \mathbb{R}$. Using Proposition 3.2.14 we deduce that $v \in L^\infty(\mathbb{R}^3)$. Since $s > \frac{3}{4} > \frac{1}{2}$, Proposition 1.3.2 shows that $v \in C^{1,\alpha}(\mathbb{R}^3)$ for any $\alpha < 2s - 1$. From Proposition 1.3.11-(ii) (or Theorem 1.3.5), we get $v > 0$ in $\mathbb{R}^3$. Therefore, $u \in C^{1,\alpha}(\mathbb{R}^3)$ is a positive solution to (11.3.1) and this ends the proof of theorem. $\square$

The next lemma is a compactness result for the autonomous problem that will be used later.

Lemma 11.3.7 *Let* $(u_n) \subset \mathcal{N}_0$ *be a sequence such that* $\mathcal{J}_0(u_n) \to c_0$. *Then* (u_n) *has a convergent subsequence in* $H^s(\mathbb{R}^3)$.

Proof Since $(u_n) \subset \mathcal{N}_0$ and $\mathcal{J}_0(u_n) \to c_0$, we can apply Lemma 11.3.1-(iii), Proposition 11.3.2-(d) and Remark 11.3.3 to infer that

$$v_n = m^{-1}(u_n) = \frac{u_n}{\|u_n\|_0} \in \mathbb{S}_0^+$$

and

$$\psi_0(v_n) = \mathcal{J}_0(u_n) \to c_0 = \inf_{v \in \mathbb{S}_0^+} \psi_0(v).$$

Let us introduce the map $\mathcal{F} : \overline{\mathbb{S}}_0^+ \to \mathbb{R} \cup \{\infty\}$ defined by

$$\mathcal{F}(u) = \begin{cases} \psi_0(u), & \text{if } u \in \mathbb{S}_0^+, \\ \infty, & \text{if } u \in \partial\mathbb{S}_0^+. \end{cases}$$

We note that

- $(\overline{\mathbb{S}}_0^+, d_0)$, where $d(u, v) = \|u - v\|_0$, is a complete metric space;
- $\mathcal{F} \in C(\overline{\mathbb{S}}_0^+, \mathbb{R} \cup \{\infty\})$, by Lemma 11.3.1-(iii);
- $\mathcal{F}$ is bounded below, by Proposition 11.3.2-(d).

Hence, by applying Theorem 2.2.1 to $\mathcal{F}$, we can find $(\hat{v}_n) \subset \mathbb{S}_0^+$ such that $(\hat{v}_n)$ is a Palais–Smale sequence for ψ_0 at the level c_0 and $\|\hat{v}_n - v_n\|_0 = o_n(1)$. Then, using Proposition

11.3.2, Theorem 11.3.6 and arguing as in the proof of Corollary 11.2.10 we obtain the desired conclusion.   □

Lemma 11.3.8 $\lim\sup_{\varepsilon\to 0} c_\varepsilon = c_0$.

Proof Arguing as in the proof of Lemma 7.2.7 and using the continuity of M, we see that

$$\lim_{\varepsilon\to 0} \sup c_\varepsilon \leq c_0.$$

To complete the proof, note that, by (V_1), $\lim\inf_{\varepsilon\to 0} c_\varepsilon \geq c_0$.   □

11.4   Multiple Solutions for (11.2.2)

In this section, our main purpose is to apply the Lusternik–Schnirelman category theory to prove a multiplicity result for the problem (11.2.2). We begin with several technical results.

Lemma 11.4.1 *Let* $\varepsilon_n \to 0$ *and* $(u_n) = (u_{\varepsilon_n}) \subset \mathcal{N}_{\varepsilon_n}$ *be such that* $\mathcal{J}_{\varepsilon_n}(u_n) \to c_0$. *Then there exists* $(\tilde{y}_n) = (\tilde{y}_{\varepsilon_n}) \subset \mathbb{R}^3$ *such that the translated sequence*

$$\tilde{u}_n(x) = u_n(x + \tilde{y}_n)$$

has a subsequence which converges in $H^s(\mathbb{R}^3)$. *Moreover, up to a subsequence,* $(y_n) = (\varepsilon_n \tilde{y}_n)$ *is such that* $y_n \to y_0 \in \Lambda$.

Proof Since $\langle \mathcal{J}'_{\varepsilon_n}(u_n), u_n \rangle = 0$ and $\mathcal{J}_{\varepsilon_n}(u_n) \to c_0$, it is easy to see that (u_n) is bounded in $\mathcal{H}_{\varepsilon_n}$. Let us observe that $\|u_n\|_{\varepsilon_n} \nrightarrow 0$, since $c_0 > 0$. Therefore, arguing as in Remark 11.3.5, we can find a sequence $(\tilde{y}_n) \subset \mathbb{R}^3$ and constants $R, \alpha > 0$ such that

$$\lim_{n\to\infty}\inf \int_{B_R(\tilde{y}_n)} u_n^2\, dx \geq \alpha.$$

Set $\tilde{u}_n(x) = u_n(x + \tilde{y}_n)$. Then it is clear that $(\tilde{u}_n)$ is bounded in $H^s(\mathbb{R}^3)$, and we may assume that

$$\tilde{u}_n \rightharpoonup \tilde{u} \quad \text{in } H^s(\mathbb{R}^3),$$

for some $\tilde{u} \neq 0$. Let $(t_n) \subset (0, \infty)$ be such that $\tilde{v}_n = t_n \tilde{u}_n \in \mathcal{N}_0$ and set $y_n = \varepsilon_n \, \tilde{y}_n$. Then, using (M_2) and the fact that $g(x, t) \leq f(t)$, we have

$$
\begin{aligned}
c_0 \leq \mathcal{J}_0(\tilde{v}_n) &= \frac{1}{2} \widehat{M}(t_n^2 \|u_n\|_0^2) - \int_{\mathbb{R}^3} F(t_n u_n) \, dx \\
&\leq \frac{1}{2} \widehat{M}(t_n^2 \|u_n\|_{\varepsilon_n}^2) - \int_{\mathbb{R}^3} G(\varepsilon x, t_n u_n) \, dx \\
&= \mathcal{J}_{\varepsilon_n}(t_n u_n) \leq \mathcal{J}_{\varepsilon_n}(u_n) = c_0 + o_n(1),
\end{aligned}
$$

which shows that

$$
\mathcal{J}_0(\tilde{v}_n) \to c_0 \quad \text{and} \quad (\tilde{v}_n) \subset \mathcal{N}_0. \tag{11.4.1}
$$

In particular, (11.4.1) implies that $(\tilde{v}_n)$ is bounded in $H^s(\mathbb{R}^3)$, so we may assume that $\tilde{v}_n \rightharpoonup \tilde{v}$. Obviously, (t_n) is bounded and $t_n \to t_0 \geq 0$. If $t_0 = 0$, then the boundedness of $(\tilde{u}_n)$ implies that $\|\tilde{v}_n\|_0 = t_n \|\tilde{u}_n\|_0 \to 0$, that is $\mathcal{J}_0(\tilde{v}_n) \to 0$, in contrast with the fact $c_0 > 0$. Then, $t_0 > 0$. From the uniqueness of the weak limit, we have $\tilde{v} = t_0 \tilde{u}$ and $\tilde{u} \neq 0$. By (11.4.1) and Lemma 11.3.7 we deduce that

$$
\tilde{v}_n \to \tilde{v} \quad \text{in } H^s(\mathbb{R}^3), \tag{11.4.2}
$$

which implies that $\tilde{u}_n = \dfrac{\tilde{v}_n}{t_n} \to \dfrac{\tilde{v}}{t_0} = \tilde{u}$ in $H^s(\mathbb{R}^3)$ and

$$
\mathcal{J}_0(\tilde{v}) = c_0 \quad \text{and} \quad \langle \mathcal{J}_0'(\tilde{v}), \tilde{v} \rangle = 0.
$$

Let us show that (y_n) has a subsequence such that $y_n \to y_0 \in \Lambda$. Assume, by contradiction, that (y_n) is not bounded, that is, there exists a subsequence, still denoted by (y_n), such that $|y_n| \to \infty$. Since $u_n \in \mathcal{N}_{\varepsilon_n}$, we see that

$$
m_0 \|\tilde{u}_n\|_0^2 \leq \int_{\mathbb{R}^3} g(\varepsilon_n x + y_n, \tilde{u}_n) \tilde{u}_n \, dx.
$$

Take $R > 0$ such that $\Omega \subset B_R$. We may assume that $|y_n| > 2R$, so, for any $x \in B_{R/\varepsilon_n}$ we get $|\varepsilon_n x + y_n| \geq |y_n| - |\varepsilon_n x| > R$.

Then,

$$
m_0 \|\tilde{u}_n\|_0^2 \leq \int_{B_{R/\varepsilon_n}} \tilde{f}(\tilde{u}_n) \tilde{u}_n \, dx + \int_{\mathbb{R}^3 \setminus B_{R/\varepsilon_n}} f(\tilde{u}_n) \tilde{u}_n \, dx.
$$

Since $\tilde{u}_n \to \tilde{u}$ in $H^s(\mathbb{R}^3)$, the dominated convergence theorem shows that

$$\int_{\mathbb{R}^3 \setminus B_{R/\varepsilon_n}} f(\tilde{u}_n)\tilde{u}_n \, dx = o_n(1).$$

Recalling that $\tilde{f}(\tilde{u}_n) \leq \frac{V_0}{K}\tilde{u}_n$, we get

$$m_0 \|\tilde{u}_n\|_0^2 \leq \frac{1}{K} \int_{B_{R/\varepsilon_n}} V_0 \tilde{u}_n^2 \, dx + o_n(1),$$

which yields

$$\left(m_0 - \frac{1}{K}\right) \|\tilde{u}_n\|_0^2 \leq o_n(1).$$

Since $\tilde{u}_n \to \tilde{u} \neq 0$ in $H^s(\mathbb{R}^3)$, we have a contradiction. Thus (y_n) is bounded and, up to a subsequence, we may assume that $y_n \to y_0$. If $y_0 \notin \bar{\Omega}$, then there exists $r > 0$ such that $y_n \in B_{r/2}(y_0) \subset \mathbb{R}^3 \setminus \bar{\Omega}$ for any n large enough. Reasoning as before, we get a contradiction. Hence, $y_0 \in \bar{\Omega}$. Now, we prove that $V(y_0) = V_0$. Assume, by contradiction, that $V(y_0) > V_0$. Taking into account (11.4.2), Fatou's lemma and the translation invariance of $\mathbb{R}^3$, we have

$$c_0 < \liminf_{n \to \infty} \left[\frac{1}{2}\hat{M}\left(\int_{\mathbb{R}^3} \left(|(-\Delta)^{\frac{s}{2}}\tilde{v}_n|^2 + V(\varepsilon_n z + y_n)\tilde{v}_n^2 \right) dz \right) - \int_{\mathbb{R}^3} F(\tilde{v}_n) \, dx \right]$$

$$\leq \liminf_{n \to \infty} \mathcal{J}_{\varepsilon_n}(t_n u_n) \leq \liminf_{n \to \infty} \mathcal{J}_{\varepsilon_n}(u_n) = c_0,$$

which gives a contradiction. Therefore, $V(y_0) = V_0$ and using (V_2) we deduce that $y_0 \in \Lambda$. $\qquad\square$

Our next objective is to relate the number of positive solutions of (11.2.2) to the topology of the set Λ. Take $\delta > 0$ such that

$$\Lambda_\delta = \{x \in \mathbb{R}^3 : \text{dist}(x, \Lambda) \leq \delta\} \subset \Omega.$$

Let η be a smooth nonincreasing cut-off function defined in $[0, \infty)$, such that $\eta = 1$ in $[0, \frac{\delta}{2}]$, $\eta = 0$ in $[\delta, \infty)$, $0 \leq \eta \leq 1$ and $|\eta'| \leq c$ for some $c > 0$. For any $y \in \Lambda$, we define

$$\Psi_{\varepsilon, y}(x) = \eta(|\varepsilon x - y|)w\left(\frac{\varepsilon x - y}{\varepsilon}\right),$$

where $w \in H^s(\mathbb{R}^3)$ is a positive ground state solution to the autonomous problem (11.3.1) (such a solution exists in view of Theorem 11.3.6).

Let $t_\varepsilon > 0$ be the unique number such that

$$\max_{t \geq 0} \mathcal{J}_\varepsilon(t\Psi_{\varepsilon,y}) = \mathcal{J}_\varepsilon(t_\varepsilon \Psi_{\varepsilon,y}).$$

Finally, we consider $\Phi_\varepsilon : \Lambda \to \mathcal{N}_\varepsilon$ defined by

$$\Phi_\varepsilon(y) = t_\varepsilon \Psi_{\varepsilon,y}.$$

Lemma 11.4.2 *The functional Φ_ε has the property that*

$$\lim_{\varepsilon \to 0} \mathcal{J}_\varepsilon(\Phi_\varepsilon(y)) = c_0, \quad \textit{uniformly in } y \in \Lambda.$$

Proof Assume, by contradiction, that there exist $\delta_0 > 0$, $(y_n) \subset \Lambda$ and $\varepsilon_n \to 0$ such that

$$|\mathcal{J}_{\varepsilon_n}(\Phi_{\varepsilon_n}(y_n)) - c_0| \geq \delta_0. \tag{11.4.3}$$

With the change of variable $z = \dfrac{\varepsilon_n x - y_n}{\varepsilon_n}$, if $z \in B_{\frac{\delta}{\varepsilon_n}}$, then $\varepsilon_n z \in B_\delta$ and thus $\varepsilon_n z + y_n \in B_\delta(y_n) \subset \Lambda_\delta \subset \Omega$. Since $G = F$ in $\Omega \times \mathbb{R}$ we have

$$\mathcal{J}_\varepsilon(\Phi_{\varepsilon_n}(y_n)) = \frac{1}{2}\widehat{M}\left(t_{\varepsilon_n}^2 \left(\int_{\mathbb{R}^3} |(-\Delta)^{\frac{s}{2}}(\eta(|\varepsilon_n z|)w(z))|^2 \, dz \right.\right.$$
$$\left.\left.+ \int_{\mathbb{R}^3} V(\varepsilon_n z + y_n)(\eta(|\varepsilon_n z|)w(z))^2 \, dz\right)\right)$$
$$- \int_{\mathbb{R}^3} F(t_{\varepsilon_n}\eta(|\varepsilon_n z|)w(z)) \, dz. \tag{11.4.4}$$

Let us to show that the sequence (t_{ε_n}) is such that $t_{\varepsilon_n} \to 1$ as $n \to \infty$. By the definition of t_{ε_n}, it follows that $\langle \mathcal{J}'_{\varepsilon_n}(\Phi_{\varepsilon_n}(y_n)), \Phi_{\varepsilon_n}(y_n)\rangle = 0$, which gives

$$\frac{M(t_{\varepsilon_n}^2 A_n^2)}{t_{\varepsilon_n}^2 A_n^2} = \frac{1}{A_n^4}\int_{\mathbb{R}^3}\left[\frac{f(t_{\varepsilon_n}\eta(|\varepsilon_n z|)w(z))}{(t_{\varepsilon_n}\eta(|\varepsilon_n z|)w(z))^3}\right](\eta(|\varepsilon_n z|)w(z))^4 \, dz \tag{11.4.5}$$

where

$$A_n^2 = \int_{\mathbb{R}^3}|(-\Delta)^{\frac{s}{2}}(\eta(|\varepsilon_n z|)w(z))|^2 \, dz + \int_{\mathbb{R}^3}V(\varepsilon_n z + y_n)(\eta(|\varepsilon_n z|)w(z))^2 \, dz.$$

Since $\eta(|x|) = 1$ for $x \in B_{\frac{\delta}{2}}$ and $B_{\frac{\delta}{2}} \subset B_{\frac{\delta}{2\varepsilon_n}}$ for all n large enough, (11.4.5) shows that

$$\frac{M(t_{\varepsilon_n}^2 A_n^2)}{t_{\varepsilon_n}^2 A_n^2} \geq \frac{1}{A_n^4}\int_{B_{\frac{\delta}{2}}}\left[\frac{f(t_{\varepsilon_n}w(z))}{(t_{\varepsilon_n}w(z))^3}\right]w^4(z) \, dz.$$

By the continuity of w we can find a vector $\hat{z} \in \mathbb{R}^3$ such that

$$w(\hat{z}) = \min_{z \in \bar{B}_{\frac{\delta}{2}}} w(z) > 0.$$

Then, using (f_4), we deduce that

$$\frac{M(t_{\varepsilon_n}^2 A_n^2)}{t_{\varepsilon_n}^2 A_n^2} \geq \frac{1}{A_n^4} \left[\frac{f(t_{\varepsilon_n} w(\hat{z}))}{(t_{\varepsilon_n} w(\hat{z}))^3} \right] \int_{B_{\frac{\delta}{2}}} w^4(z) \, dz \geq \frac{1}{A_n^4} \left[\frac{f(t_{\varepsilon_n} w(\hat{z}))}{(t_{\varepsilon_n} w(\hat{z}))^3} \right] w^4(\hat{z}) |B_{\frac{\delta}{2}}|.$$

$$(11.4.6)$$

Now, assume by contradiction that $t_{\varepsilon_n} \to \infty$. Let us observe that Lemma 1.4.8 and the dominated convergence theorem yield

$$\|\Psi_{\varepsilon_n, y_n}\|_{\varepsilon_n}^2 = A_n^2 \to \|w\|_0^2 \in (0, \infty).$$

$$(11.4.7)$$

Using that $t_{\varepsilon_n} \to \infty$, (M_3) and (11.4.7), we see that

$$\lim_{n \to \infty} \frac{M(t_{\varepsilon_n}^2 A_n^2)}{t_{\varepsilon_n}^2 A_n^2} \leq \lim_{n \to \infty} \frac{\gamma(1 + t_{\varepsilon_n}^2 A_n^2)}{t_{\varepsilon_n}^2 A_n^2} \leq C < \infty.$$

$$(11.4.8)$$

On the other hand, assumption (f_3) implies that

$$\lim_{n \to \infty} \frac{f(t_{\varepsilon_n} w(\hat{z}))}{(t_{\varepsilon_n} w(\hat{z}))^3} = \infty.$$

$$(11.4.9)$$

Putting together (11.4.6), (11.4.8) and (11.4.9) we get a contradiction. Therefore, (t_{ε_n}) is bounded and, up to subsequence, we may assume that $t_{\varepsilon_n} \to t_0$ for some $t_0 \geq 0$. Let us prove that $t_0 > 0$. Suppose, by contradiction, that $t_0 = 0$. Then, taking into account (11.4.7), (M_1), (f_1) and (f_2), we see that (11.4.5) yields

$$0 < m_0 \|w\|_0^2 \leq \lim_{n \to \infty} \left[C t_{\varepsilon_n}^2 \int_{\mathbb{R}^3} w^4 \, dz + C t_{\varepsilon_n}^{q-2} \int_{\mathbb{R}^3} w^q \, dz \right] = 0,$$

which is impossible. Hence $t_0 > 0$. Passing to the limit as $n \to \infty$ in (11.4.5), it follows from (11.4.7), the continuity of M and the dominated convergence theorem that

$$M(t_0^2 \|w\|_0^2) t_0 \|w\|_0^2 = \int_{\mathbb{R}^3} f(t_0 w) w \, dx.$$

Since $w \in \mathcal{N}_0$, we obtain that

$$\frac{M(t_0^2 \|w\|_0^2)}{t_0^2 \|w\|_0^2} - \frac{M(\|w\|_0^2)}{\|w\|_0^2} = \frac{1}{\|w\|_0^4} \int_{\mathbb{R}^3} \left[\frac{f(t_0 w)}{(t_0 w)^3} - \frac{f(w)}{w^3} \right] w^4 \, dx. \tag{11.4.10}$$

If $t_0 > 1$, then by (M_3) and (f_4), the left-hand side of (11.4.10) is negative and the right-hand side is positive. A similar argument works when $t_0 < 1$. Therefore, $t_0 = 1$. Then, taking the limit as $n \to \infty$ in (11.4.4) and using that $t_{\varepsilon_n} \to 1$, that

$$\int_{\mathbb{R}^3} F(\eta(|\varepsilon_n z|) w(z)) \, dz \to \int_{\mathbb{R}^3} F(w) \, dz,$$

and (11.4.7), we obtain

$$\lim_{n \to \infty} \mathcal{J}_{\varepsilon_n}(\Phi_{\varepsilon_n, y_n}) = \mathcal{J}_0(w) = c_0,$$

which contradicts (11.4.3).   □

For $\delta > 0$, take $\rho = \rho(\delta) > 0$ such that $\Lambda_\delta \subset B_\rho$, and consider the map $\Upsilon : \mathbb{R}^3 \to \mathbb{R}^3$ given by

$$\Upsilon(x) = \begin{cases} x, & \text{if } |x| < \rho, \\ \frac{\rho x}{|x|}, & \text{if } |x| \geq \rho. \end{cases}$$

Now define the barycenter map $\beta_\varepsilon : \mathcal{N}_\varepsilon \to \mathbb{R}^3$ by

$$\beta_\varepsilon(u) = \frac{\displaystyle\int_{\mathbb{R}^3} \Upsilon(\varepsilon x) u^2(x) \, dx}{\displaystyle\int_{\mathbb{R}^3} u^2(x) \, dx}.$$

Arguing as in the proof of Lemma 6.3.18 we can prove the next result.

Lemma 11.4.3 *The function β_ε has the property that*

$$\lim_{\varepsilon \to 0} \beta_\varepsilon(\Phi_\varepsilon(y)) = y, \quad \text{uniformly in } y \in \Lambda.$$

At this point, we introduce a subset $\tilde{\mathcal{N}}_\varepsilon$ of $\mathcal{N}_\varepsilon$ by taking a function $h_1 : \mathbb{R}_+ \to \mathbb{R}_+$ such that $h_1(\varepsilon) \to 0$ as $\varepsilon \to 0$, and setting

$$\tilde{\mathcal{N}}_\varepsilon = \{u \in \mathcal{N}_\varepsilon : \mathcal{J}_\varepsilon(u) \leq c_0 + h_1(\varepsilon)\}.$$

In view of Lemma 11.4.2, $h_1(\varepsilon) = \sup_{y \in \Lambda} |\mathcal{J}_\varepsilon(\Phi_\varepsilon(y)) - c_0| \to 0$ as $\varepsilon \to 0$, so $\tilde{\mathcal{N}}_\varepsilon \neq \emptyset$ for all $\varepsilon > 0$. Furthermore, arguing as in the proof of Lemma 6.3.19, we deduce the following assertion.

Lemma 11.4.4

$$\lim_{\varepsilon \to 0} \sup_{u \in \tilde{\mathcal{N}}_\varepsilon} \mathrm{dist}(\beta_\varepsilon(u), \Lambda_\delta) = 0.$$

Now we are in the position to prove the following multiplicity result.

Theorem 11.4.5 *Assume that (M_1)–(M_3), (V_1)–(V_2) and (f_1)–(f_4) hold. Then, given $\delta > 0$ such that $\Lambda_\delta \subset \Omega$, there exists $\bar{\varepsilon}_\delta > 0$ such that, for any $\varepsilon \in (0, \bar{\varepsilon}_\delta)$, problem (11.2.2) has at least $\mathrm{cat}_{\Lambda_\delta}(\Lambda)$ positive solutions.*

Proof Let $\varepsilon > 0$, and consider the map $\alpha_\varepsilon : \Lambda \to \mathbb{S}_\varepsilon^+$ defined as $\alpha_\varepsilon(y) = m_\varepsilon^{-1}(\Phi_\varepsilon(y))$. Using Lemma 11.4.2, we see that

$$\lim_{\varepsilon \to 0} \psi_\varepsilon(\alpha_\varepsilon(y)) = \lim_{\varepsilon \to 0} \mathcal{J}_\varepsilon(\Phi_\varepsilon(y)) = c_0 \quad \text{uniformly in } y \in \Lambda. \tag{11.4.11}$$

Set

$$\tilde{\mathbb{S}}_\varepsilon^+ = \{w \in \mathbb{S}_\varepsilon^+ : \psi_\varepsilon(w) \leq c_0 + h_1(\varepsilon)\},$$

where $h_1(\varepsilon) = \sup_{y \in \Lambda} |\psi_\varepsilon(\alpha_\varepsilon(y)) - c_0|$. It follows from (11.4.11) that $h_1(\varepsilon) \to 0$ as $\varepsilon \to 0$. Moreover, $\alpha_\varepsilon(y) \in \tilde{\mathbb{S}}_\varepsilon^+$ for all $y \in \Lambda$ and this shows that $\tilde{\mathbb{S}}_\varepsilon^+ \neq \emptyset$ for all $\varepsilon > 0$.

In the light of Lemmas 11.4.2, 11.2.3-(iii), 11.4.4, and 11.4.3, we can find $\bar{\varepsilon} = \bar{\varepsilon}_\delta > 0$ such that the following diagram

$$\Lambda \xrightarrow{\Phi_\varepsilon} \Phi_\varepsilon(\Lambda) \xrightarrow{m_\varepsilon^{-1}} \alpha_\varepsilon(\Lambda) \xrightarrow{m_\varepsilon} \Phi_\varepsilon(\Lambda) \xrightarrow{\beta_\varepsilon} \Lambda_\delta$$

is well defined for any $\varepsilon \in (0, \bar{\varepsilon})$. Thanks to Lemma 11.4.3, and decreasing $\bar{\varepsilon}$ if necessary, we see that $\beta_\varepsilon(\Phi_\varepsilon(y)) = y + \theta(\varepsilon, y)$ for all $y \in \Lambda$, for some function $\theta(\varepsilon, y)$ satisfying $|\theta(\varepsilon, y)| < \frac{\delta}{2}$ uniformly in $y \in \Lambda$ and for all $\varepsilon \in (0, \bar{\varepsilon})$. Define $H(t, y) = y + (1 - t)\theta(\varepsilon, y)$. Then $H : [0, 1] \times \Lambda \to \Lambda_\delta$ is continuous. Clearly, $H(0, y) = \beta_\varepsilon(\Phi_\varepsilon(y))$ and $H(1, y) = y$ for all $y \in \Lambda$. Consequently, $H(t, y)$ is a homotopy between $\beta_\varepsilon \circ \Phi_\varepsilon = (\beta_\varepsilon \circ m_\varepsilon) \circ (m_\varepsilon^{-1} \circ \Phi_\varepsilon)$ and the inclusion map id: $\Lambda \to \Lambda_\delta$. This fact together with Lemma 6.3.21 yields

$$\mathrm{cat}_{\alpha_\varepsilon(\Lambda)}\alpha_\varepsilon(\Lambda) \geq \mathrm{cat}_{\Lambda_\delta}(\Lambda). \tag{11.4.12}$$

Applying Corollary 11.2.10, Lemma 11.3.8, and Theorem 2.4.6 with $c = c_\varepsilon \le c_0 + h_1(\varepsilon) = d$ and $K = \alpha_\varepsilon(\Lambda)$, we infer that ψ_ε has at least $\mathrm{cat}_{\alpha_\varepsilon(\Lambda)}\alpha_\varepsilon(\Lambda)$ critical points on $\widetilde{S}_\varepsilon^+$. Taking into account Proposition 11.2.4-(d) and (11.4.12), we conclude that $\mathcal{J}_\varepsilon$ admits at least $\mathrm{cat}_{\Lambda_\delta}(\Lambda)$ critical points in $\widetilde{\mathcal{N}}_\varepsilon$.   □

11.5   Proof of Theorem 11.1.1

In this last section, we provide the proof of Theorem 11.1.1. First, we establish the following useful L^∞-estimate for the solutions of the modified problem (11.2.2).

Lemma 11.5.1 *Let $\varepsilon_n \to 0$ and $u_n \in \widetilde{\mathcal{N}}_{\varepsilon_n}$ be a solution to (11.2.2). Then, up to a subsequence, $\tilde{u}_n = u_n(\cdot + \tilde{y}_n) \in L^\infty(\mathbb{R}^N)$, and there exists $C > 0$ such that*

$$\|\tilde{u}_n\|_{L^\infty(\mathbb{R}^3)} \le C \quad \text{for all } n \in \mathbb{N}.$$

Moreover, $\tilde{u}_n(x) \to 0$ as $|x| \to \infty$ uniformly in $n \in \mathbb{N}$.

Proof Since $\mathcal{J}_{\varepsilon_n}(u_n) \le c_0 + h(\varepsilon_n)$ with $h(\varepsilon_n) \to 0$ as $n \to \infty$, we can proceed as in the proof of (11.4.1) to deduce that $\mathcal{J}_{\varepsilon_n}(u_n) \to c_0$. Thus, we may invoke Lemma 11.4.1 to find a sequence $(\tilde{y}_n) \subset \mathbb{R}^3$ such that $\varepsilon_n \tilde{y}_n \to y_0 \in \Lambda$ and $\tilde{u}_n = u_n(\cdot + \tilde{y}_n)$ admits a convergent subsequence in $H^s(\mathbb{R}^3)$.

Now, arguing as in the proof of Lemma 6.3.23, and using (M_1), (g_1) and (g_2), we have

$$\left(\frac{1}{\beta}\right)^2 S_*^{-1} \|\tilde{u}_n \tilde{u}_{n,L}^{\beta-1}\|_{L^{2_s^*}(\mathbb{R}^3)}^2 \le \frac{1}{m_0} \int_{\mathbb{R}^3} [g(\varepsilon_n x + \varepsilon_n \tilde{y}_n, \tilde{u}_n) - m_0 V(\varepsilon_n x + \varepsilon_n \tilde{y}_n)]\tilde{u}_n \tilde{u}_{n,L}^{2(\beta-1)} \, dx$$

$$\le C \int_{\mathbb{R}^3} \tilde{u}_n^{2_s^*} \tilde{u}_{n,L}^{2(\beta-1)} \, dx.$$

Hence, we can argue as in the proof of Lemma 6.3.23 to obtain the required L^∞-estimate. We note that the s-harmonic extension $\tilde{w}_n(x, y) = \mathrm{Ext}(\tilde{u}_n) = P_s(x, y) * \tilde{u}_n(x)$ of $\tilde{u}_n$ solves

$$\begin{cases} -\operatorname{div}(y^{1-2s}\nabla \tilde{w}_n) = 0 & \text{in } \mathbb{R}_+^{3+1}, \\ \tilde{w}_n(\cdot, 0) = \tilde{u}_n & \text{on } \partial\mathbb{R}_+^{3+1}, \\ \frac{\partial \tilde{w}_n}{\partial \nu^{1-2s}} = \frac{1}{M(\|u_n\|_{\varepsilon_n})}[-V(\varepsilon_n x + \varepsilon_n \tilde{y}_n)\tilde{u}_n + g(\varepsilon_n x + \varepsilon_n \tilde{y}_n, \tilde{u}_n)] & \text{on } \partial\mathbb{R}_+^{3+1}, \end{cases}$$

and that

$$m_0 \le M(\|u_n\|_{\varepsilon_n}) \le C \quad \text{for all } n \in \mathbb{N}. \tag{11.5.1}$$

Applying Proposition 1.3.11-(iii), we see that $\tilde{u}_n \in L^\infty(\mathbb{R}^3) \cap C^{0,\alpha}_{loc}(\mathbb{R}^3)$. Next, using (11.5.1) and the L^∞-estimate, we argue as in the proof of Lemma 6.3.23 to conclude that $\tilde{u}_n(x) \to 0$ as $|x| \to \infty$ uniformly in $n \in \mathbb{N}$ (see also Remark 7.2.10 and note that $\tilde{u}_n$ is a subsolution to $(-\Delta)^s \tilde{u}_n + V_0 \tilde{u}_n = \frac{1}{m_0} g(\varepsilon_n x + \varepsilon_n \tilde{y}_n, \tilde{u}_n)$ in $\mathbb{R}^3$). $\qquad\qquad\square$

Now, we are able to give the proof of our main result.

Proof of Theorem 11.1.1 Take $\delta > 0$ such that $\Lambda_\delta \subset \Omega$. We begin by proving that there exists $\tilde{\varepsilon}_\delta > 0$ such that for any $\varepsilon \in (0, \tilde{\varepsilon}_\delta)$ and any solution $u_\varepsilon \in \tilde{\mathcal{N}}_\varepsilon$ of (11.2.2),

$$\|u_\varepsilon\|_{L^\infty(\mathbb{R}^3 \setminus \Omega_\varepsilon)} < a. \tag{11.5.2}$$

Suppose, by contradiction, that for some subsequence (ε_n) such that $\varepsilon_n \to 0$, we can find $u_n = u_{\varepsilon_n} \in \tilde{\mathcal{N}}_{\varepsilon_n}$ such that $\mathcal{J}'_{\varepsilon_n}(u_n) = 0$ and

$$\|u_n\|_{L^\infty(\mathbb{R}^3 \setminus \Omega_{\varepsilon_n})} \geq a. \tag{11.5.3}$$

Since $\mathcal{J}_{\varepsilon_n}(u_n) \leq c_0 + h_1(\varepsilon_n)$ and $h_1(\varepsilon_n) \to 0$, we can proceed as in the first part of the proof of Lemma 11.4.1 to deduce that $\mathcal{J}_{\varepsilon_n}(u_n) \to c_0$. Then, by Lemma 11.4.1, there exists $(\tilde{y}_n) \subset \mathbb{R}^3$ such that $\tilde{u}_n = u_n(\cdot + \tilde{y}_n) \to \tilde{u}$ in $H^s(\mathbb{R}^3)$ and $\varepsilon_n \tilde{y}_n \to y_0 \in \Lambda$.

Now, if we choose $r > 0$ such that $B_r(y_0) \subset B_{2r}(y_0) \subset \Omega$, we see that $B_{\frac{r}{\varepsilon_n}}(\frac{y_0}{\varepsilon_n}) \subset \Omega_{\varepsilon_n}$. Moreover, for any $y \in B_{\frac{r}{\varepsilon_n}}(\tilde{y}_n)$,

$$\left|y - \frac{y_0}{\varepsilon_n}\right| \leq |y - \tilde{y}_n| + \left|\tilde{y}_n - \frac{y_0}{\varepsilon_n}\right| < \frac{1}{\varepsilon_n}(r + o_n(1)) < \frac{2r}{\varepsilon_n} \quad \text{for } n \text{ sufficiently large.}$$

Therefore,

$$\mathbb{R}^3 \setminus \Omega_{\varepsilon_n} \subset \mathbb{R}^3 \setminus B_{\frac{r}{\varepsilon_n}}(\tilde{y}_n) \tag{11.5.4}$$

for n sufficiently large. On the other hand, by Lemma 11.5.1,

$$\tilde{u}_n(x) \to 0 \quad \text{as } |x| \to \infty$$

uniformly in $n \in \mathbb{N}$. Therefore, there exists $R > 0$ such that

$$\tilde{u}_n(x) < a \quad \text{for all } |x| \geq R \text{ and all } n \in \mathbb{N}.$$

Hence, $u_n(x) < a$ for all $x \in \mathbb{R}^3 \setminus B_R(\tilde{y}_n)$ and $n \in \mathbb{N}$. On the other hand, there exists $\nu \in \mathbb{N}$ such that for any $n \geq \nu$ and $r/\varepsilon_n > R$, we have

$$\mathbb{R}^3 \setminus \Lambda_{\varepsilon_n} \subset \mathbb{R}^3 \setminus B_{\frac{r}{\varepsilon_n}}(\tilde{y}_n) \subset \mathbb{R}^3 \setminus B_R(\tilde{y}_n),$$

which implies that $u_n(x) < a$ for all $x \in \mathbb{R}^3 \setminus \Omega_{\varepsilon_n}$ and $n \geq \nu$. This contradicts (11.5.3).

Let $\bar{\varepsilon}_\delta > 0$ be given by Theorem 11.4.5, and fix $\varepsilon \in (0, \varepsilon_\delta)$, where $\varepsilon_\delta = \min\{\tilde{\varepsilon}_\delta, \bar{\varepsilon}_\delta\}$. In view of Theorem 11.4.5, problem (11.2.2) admits at least $\operatorname{cat}_{\Lambda_\delta}(\Lambda)$ nontrivial solutions. Let u_ε be one of these solutions. Since $u_\varepsilon \in \tilde{\mathcal{N}}_\varepsilon$ satisfies (11.5.2), by the definition of g it follows that u_ε is a solution of (11.2.1). Then, $\hat{u}_\varepsilon(x) = u(x/\varepsilon)$ is a solution to (11.1.1), and we deduce that (11.1.1) has at least $\operatorname{cat}_{\Lambda_\delta}(\Lambda)$ solutions.

Finally, we study the behavior of the maximum points of solutions to (11.2.1). Take $\varepsilon_n \to 0$ and consider a sequence $(u_n) \subset \mathcal{H}_{\varepsilon_n}$ of solutions to (11.2.1) as above. Condition (g_1) ensurers that we can find $\gamma > 0$ such that

$$g(\varepsilon x, t)t \leq \frac{V_0}{K}t^2 \quad \text{for all } x \in \mathbb{R}^3 \text{ and } 0 \leq t \leq \gamma. \tag{11.5.5}$$

Arguing as before, we can find $R > 0$ such that

$$\|u_n\|_{L^\infty(B_R^c(\tilde{y}_n))} < \gamma. \tag{11.5.6}$$

Moreover, up to extracting a subsequence, we may assume that

$$\|u_n\|_{L^\infty(B_R(\tilde{y}_n))} \geq \gamma. \tag{11.5.7}$$

Indeed, if (11.5.7) does not hold, it follows from (11.5.6) that $\|u_n\|_{L^\infty(\mathbb{R}^3)} < \gamma$. Then, using $\langle \mathcal{J}'_{\varepsilon_n}(u_n), u_n \rangle = 0$ and (11.5.5) we can infer that

$$m_0 \|u_n\|_{\varepsilon_n}^2 \leq \int_{\mathbb{R}^3} g(\varepsilon_n x, u_n)u_n \, dx \leq \frac{V_0}{K} \int_{\mathbb{R}^3} u_n^2 \, dx$$

which yields $\|u_n\|_{\varepsilon_n} = 0$, and this is impossible. Consequently, (11.5.7) holds. Taking into account (11.5.6) and (11.5.7), we deduce that if $p_n \in \mathbb{R}^3$ is a global maximum point of u_n, then $p_n \in B_R(\tilde{y}_n)$. Therefore, $p_n = \tilde{y}_n + q_n$ for some $q_n \in B_R$. Hence, $\eta_n = \varepsilon_n \tilde{y}_n + \varepsilon_n q_n$ is a global maximum point of $\hat{u}_n(x) = u_n(x/\varepsilon_n)$. Since $|q_n| < R$ for all $n \in \mathbb{N}$ and since $\varepsilon_n \tilde{y}_n \to y_0 \in \Lambda$ (in view of Lemma 11.4.1), it follows from the continuity of V that

$$\lim_{n \to \infty} V(\eta_n) = V(y_0) = V_0.$$

$\square$

Remark 11.5.2 In [62] we considered the problem

$$\begin{cases} \varepsilon^{2s} \, M(\varepsilon^{2s-N}[u]_s^2)(-\Delta)^s u + V(x)u = f(u) & \text{in } \mathbb{R}^N, \\ u \in H^s(\mathbb{R}^N), \quad u > 0 & \text{in } \mathbb{R}^N, \end{cases}$$

where $s \in (0, 1)$, $N \geq 2$, $V : \mathbb{R}^N \to \mathbb{R}$ is a continuous potential satisfying the local conditions (V_1)–(V_2) introduced in Chap. 7, the nonlinearity $f : \mathbb{R} \to \mathbb{R}$ is a continuous function such that $f(t) = 0$ for $t \leq 0$ and f fulfills the following Berestycki-Lions type assumptions [100]:

(f_1) $f \in C_{loc}^{0,\alpha}(\mathbb{R})$ for some $\alpha \in (1 - 2s, 1)$ if $s \in (0, \frac{1}{2}]$,
(f_2) $\lim_{t \to 0} \frac{f(t)}{t} = 0$,
(f_3) $\limsup_{t \to \infty} \frac{f(t)}{t^p} < \infty$ for some $p \in (1, 2_s^* - 1)$,
(f_4) there exists $T > 0$ such that $F(T) > \frac{V_0}{2}T^2$, where $F(t) = \int_0^t F(\tau)\, d\tau$,

and the Kirchhoff term $M : [0, \infty) \to \mathbb{R}_+$ is a continuous function such that:

$(M1)$ there exists $m_0 > 0$ such that $M(t) \geq m_0$ for all $t \geq 0$,
$(M2)$ $\liminf_{t \to \infty} \left[\widehat{M}(t) - (1 - \frac{2s}{N})M(t)t \right] = \infty$,
$(M3)$ $M(t)/t^{\frac{2s}{N-2s}} \to 0$ as $t \to \infty$,
$(M4)$ M is nondecreasing in $[0, \infty)$,
$(M5)$ $t \mapsto M(t)/t^{\frac{2s}{N-2s}}$ is nonincreasing in $(0, \infty)$.

Clearly, $M(t) = m_0 + bt$, with $b \geq 0$, satisfies $(M1)$–$(M5)$ when $b = 0$, $N \geq 2$, $s \in (0, 1)$, and when $b > 0$, $N = 3$, $s \in (\frac{3}{4}, 1)$. By combining the penalization approaches in [165] and [121], we established the existence of a family of positive solutions (u_ε) which concentrates at a local minimum of V as $\varepsilon \to 0$.

Concentrating Solutions for a Fractional Kirchhoff Equation with Critical Growth

12

12.1 Introduction

This chapter is devoted to the existence and concentration of positive solutions for the following fractional Kirchhoff type equation with critical nonlinearity:

$$\begin{cases} \left(\varepsilon^{2s} a + \varepsilon^{4s-3} b \int_{\mathbb{R}^3} |(-\Delta)^{\frac{s}{2}} u|^2 \, dx \right) (-\Delta)^s u + V(x) u = f(u) + |u|^{2^*_s - 2} u & \text{in } \mathbb{R}^3, \\ u \in H^s(\mathbb{R}^3), \quad u > 0 \text{ in } \mathbb{R}^3, \end{cases}$$

$$(12.1.1)$$

where $\varepsilon > 0$ is a small parameter, $a, b > 0$ are constants, $s \in (\frac{3}{4}, 1)$ is fixed, $2^*_s = \frac{6}{3-2s}$ is the fractional critical exponent. The potential $V : \mathbb{R}^3 \to \mathbb{R}$ is a continuous function satisfying the following conditions:

(V_1) $V_1 = \inf_{x \in \mathbb{R}^3} V(x) > 0$;
(V_2)   there exists a bounded open set $\Lambda \subset \mathbb{R}^3$ such that

$$0 < V_0 = \inf_{\Lambda} V < \min_{\partial \Lambda} V.$$

The nonlinearity $f : \mathbb{R} \to \mathbb{R}$ is a continuous function which fulfills the following hypotheses:

(f_1) $f(t) = o(t^3)$ as $t \to 0$;
(f_2)   there exist $q, \sigma \in (4, 2^*_s)$, $C_0 > 0$ such that

$$f(t) \geq C_0 t^{q-1} \quad \forall t > 0, \quad \lim_{t \to \infty} \frac{f(t)}{t^{\sigma-1}} = 0;$$

© The Author(s), under exclusive license to Springer Nature Switzerland AG 2021
V. Ambrosio, *Nonlinear Fractional Schrödinger Equations in* $\mathbb{R}^N$,
Frontiers in Mathematics, https://doi.org/10.1007/978-3-030-60220-8_12

(f_3)   there exists $\vartheta \in (4, \sigma)$ such that $0 < \vartheta F(t) \le tf(t)$ for all $t > 0$;
(f_4)   the function $t \mapsto \frac{f(t)}{t^3}$ is increasing in $(0, \infty)$.

Since we will look for positive solutions to (12.1.1), we assume that $f(t) = 0$ for $t \le 0$.

We note that when $a = 1, b = 0$ and $\mathbb{R}^3$ is replaced by $\mathbb{R}^N$, then (12.1.1) reduces to the fractional Schrödinger equation

$$\varepsilon^{2s}(-\Delta)^s u + V(x)u = h(x, u) \quad \text{in } \mathbb{R}^N \tag{12.1.2}$$

studied in Chaps. 6, 7, and 8. On the other hand, if we set $s = \varepsilon = 1$ and we replace $f(u) + |u|^{2^*_s - 2}u$ by a more general nonlinearity $h(x, u)$, then (12.1.1) becomes the well-known classical Kirchhoff equation

$$-\left(a + b\int_{\mathbb{R}^3} |\nabla u|^2 dx\right)\Delta u + V(x)u = h(x, u) \quad \text{in } \mathbb{R}^3. \tag{12.1.3}$$

We recall that He and Zou [215] obtained existence and multiplicity results for small $\varepsilon > 0$ for the perturbed Kirchhoff equation

$$-\left(a\varepsilon^2 + b\varepsilon\int_{\mathbb{R}^3} |\nabla u|^2 dx\right)\Delta u + V(x)u = g(u) \quad \text{in } \mathbb{R}^3, \tag{12.1.4}$$

where the potential V satisfies condition (V) and g is a subcritical nonlinearity. Wang et al. [331] studied the multiplicity and concentration phenomenon for (12.1.4) when $g(u) = \lambda f(u) + |u|^4 u$, f is a continuous subcritical nonlinearity and λ is large. Figueiredo and Santos Junior [191] used the generalized Nehari manifold method to obtain a multiplicity result for a subcritical Kirchhoff equation under conditions (V_1)–(V_2). He et al. [214] dealt with the existence and multiplicity of solutions to (12.1.4), where $g(u) = f(u) + u^5$, $f \in C^1$ is a subcritical nonlinearity which does not satisfies the Ambrosetti–Rabinowitz condition [29] and V fulfills (V_1)–(V_2).

In this chapter we study the existence and concentration behavior of solutions to (12.1.1) under assumptions (V_1)–(V_2) and (f_1)–(f_4). More precisely, our main result can be stated as follows:

Theorem 12.1.1 ([50]) *Assume that (V_1)–(V_2) and (f_1)–(f_4) hold. Then, there exists $\varepsilon_0 > 0$ such that, for each $\varepsilon \in (0, \varepsilon_0)$, problem (12.1.1) has a positive solution u_ε. Moreover, if η_ε denotes a global maximum point of u_ε, then*

$$\lim_{\varepsilon \to 0} V(\eta_\varepsilon) = V_0,$$

and there exists a constant $C > 0$ such that

$$0 < u_\varepsilon(x) \le \frac{C\varepsilon^{3+2s}}{\varepsilon^{3+2s} + |x - \eta_\varepsilon|^{3+2s}} \quad \text{for all } x \in \mathbb{R}^3.$$

The proof of Theorem 12.1.1 will be carried out by means of appropriate variational arguments. After considering the ε-rescaled problem associated with (12.1.1), we use a variant of the penalization technique introduced in [165] (see also [12, 190]). The solutions of the modified problem will be obtained as critical points of the modified energy functional $\mathcal{J}_\varepsilon$ which, in view of the growth assumptions on f and the auxiliary nonlinearity, possesses a mountain pass geometry [29]. In order to recover some compactness properties for $\mathcal{J}_\varepsilon$, we have to circumvent several difficulties which make our study rather delicate. The first one is related to the presence of the Kirchhoff term in (12.1.1) which does not permit to verify in a standard way that if u is the weak limit of a Palais–Smale sequence (u_n) for $\mathcal{J}_\varepsilon$, then u is a weak solution for the modified problem. The second one is due to the lack of compactness caused by the unboundedness of the domain $\mathbb{R}^3$ and the critical Sobolev exponent. Anyway, we will be able to overcome these problems looking for critical points of a suitable functional whose quadratic part involves the limit term of $(a + b[u_n]_s^2)$, and showing that the mountain pass level c_ε of $\mathcal{J}_\varepsilon$ is strictly less than a threshold value related to the best constant of the embedding $H^s(\mathbb{R}^3)$ in $L^{2_s^*}(\mathbb{R}^3)$. Then, applying the mountain pass lemma, we will deduce the existence of a positive solution for the modified problem. Finally, combining a compactness argument with a Moser iteration procedure [278], we prove that the solution of the modified problem is also a solution to the original one for $\varepsilon > 0$ small enough, and that it decays to zero at infinity with polynomial rate.

12.2   The Modified Critical Problem

In order to study (12.1.1), we use the change of variable $x \mapsto \varepsilon x$ and look for solutions to

$$
\begin{cases}
(a + b[u]_s^2)(-\Delta)^s u + V(\varepsilon x)u = f(u) + |u|^{2_s^* - 2}u & \text{in } \mathbb{R}^3, \\
u \in H^s(\mathbb{R}^3), \quad u > 0 \text{ in } \mathbb{R}^3.
\end{cases}
\tag{12.2.1}
$$

Now, we introduce a penalization method in the spirit of [165] which will be crucial for obtaining our main result. First of all, without loss of generality we may assume that

$$
0 \in \Lambda \quad \text{and} \quad V(0) = V_0 = \inf_\Lambda V.
$$

Let $K > \frac{\vartheta}{\vartheta - 2}$ and $a_0 > 0$ be such that

$$
f(a_0) + a_0^{2_s^* - 1} = \frac{V_1}{K} a_0
\tag{12.2.2}
$$

and we define

$$
\tilde{f}(t) =
\begin{cases}
f(t) + (t^+)^{2_s^* - 1}, & \text{if } t \leq a_0, \\
\frac{V_1}{K} t, & \text{if } t > a_0,
\end{cases}
$$

and

$$g(x, t) = \begin{cases} \chi_\Lambda(x)(f(t) + (t^+)^{2_s^*-1}) + (1 - \chi_\Lambda(x))\tilde{f}(t), & \text{if } t > 0, \\ 0, & \text{if } t \le 0. \end{cases}$$

It is easy to check that g enjoys the following properties:

(g_1) $\lim_{t \to 0} \frac{g(x,t)}{t^3} = 0$ uniformly with respect to $x \in \mathbb{R}^3$;

(g_2) $g(x, t) \le f(t) + t^{2_s^*-1}$ for all $x \in \mathbb{R}^3, t > 0$;

(g_3) (i) $0 < \vartheta G(x, t) \le g(x, t)t$ for all $x \in \Lambda$ and $t > 0$,

   (ii) $0 \le 2G(x, t) \le g(x, t)t \le \frac{V_1}{K}t^2$ for all $x \in \mathbb{R}^3 \setminus \Lambda$ and $t > 0$;

(g_4) for each $x \in \Lambda$ the function $t \mapsto \frac{g(x,t)}{t^3}$ is increasing in $(0, \infty)$, and for each $x \in \mathbb{R}^3 \setminus \Lambda$ the function $t \mapsto \frac{g(x,t)}{t^3}$ is increasing in $(0, a_0)$.

Then we consider the following modified problem:

$$\begin{cases} (a + b[u]_s^2)(-\Delta)^s u + V(\varepsilon x)u = g(\varepsilon x, u) & \text{in } \mathbb{R}^3, \\ u \in H^s(\mathbb{R}^3), \quad u > 0 \text{ in } \mathbb{R}^3. \end{cases} \tag{12.2.3}$$

The corresponding energy functional is given by

$$\mathcal{J}_\varepsilon(u) = \frac{1}{2}\|u\|_\varepsilon^2 + \frac{b}{4}[u]_s^4 - \int_{\mathbb{R}^3} G(\varepsilon x, u) \, dx,$$

and is well defined on the space

$$\mathcal{H}_\varepsilon = \left\{ u \in H^s(\mathbb{R}^3) : \int_{\mathbb{R}^3} V(\varepsilon x)u^2 \, dx < \infty \right\}$$

endowed with the norm

$$\|u\|_\varepsilon = \left(a[u]_s^2 + \int_{\mathbb{R}^3} V(\varepsilon x)u^2 \, dx \right)^{\frac{1}{2}}.$$

Clearly, $\mathcal{H}_\varepsilon$ is a Hilbert space with the inner product

$$\langle u, v \rangle_\varepsilon = a\langle u, v \rangle_{\mathcal{D}^{s,2}(\mathbb{R}^3)} + \int_{\mathbb{R}^3} V(\varepsilon x)uv \, dx.$$

It is standard to show that $\mathcal{J}_\varepsilon \in C^1(\mathcal{H}_\varepsilon, \mathbb{R})$ and its differential is given by

$$\langle \mathcal{J}_\varepsilon'(u), v \rangle = \langle u, v \rangle_\varepsilon + b[u]_s^2 \langle u, v \rangle_{\mathcal{D}^{s,2}(\mathbb{R}^3)} - \int_{\mathbb{R}^3} g(\varepsilon x, u)v \, dx$$

for all $u, v \in \mathcal{H}_\varepsilon$. Let us introduce the Nehari manifold associated with (12.2.3), that is,

$$\mathcal{N}_\varepsilon = \left\{ u \in \mathcal{H}_\varepsilon \setminus \{0\} : \langle \mathcal{J}_\varepsilon'(u), u \rangle = 0 \right\}.$$

We begin by proving that $\mathcal{J}_\varepsilon$ possesses a nice geometric structure:

Lemma 12.2.1 *The functional $\mathcal{J}_\varepsilon$ has a mountain pass geometry:*

(a) *there exist $\alpha, \rho > 0$ such that $\mathcal{J}_\varepsilon(u) \geq \alpha$ with $\|u\|_\varepsilon = \rho$;*
(b) *there exists $e \in \mathcal{H}_\varepsilon$ such that $\|e\|_\varepsilon > \rho$ and $\mathcal{J}_\varepsilon(e) < 0$.*

Proof

(a) By assumptions (g_1) and (g_2), for every $\xi > 0$ there exists $C_\xi > 0$ such that

$$\mathcal{J}_\varepsilon(u) \geq \frac{1}{2}\|u\|_\varepsilon^2 - \int_{\mathbb{R}^3} G(\varepsilon x, u)\, dx \geq \frac{1}{2}\|u\|_\varepsilon^2 - \xi C\|u\|_\varepsilon^2 - C_\xi C\|u\|_\varepsilon^{2_s^*}.$$

Then, there exist $\alpha, \rho > 0$ such that $\mathcal{J}_\varepsilon(u) \geq \alpha$ with $\|u\|_\varepsilon = \rho$.
(b) Using (g_3)-(i), we deduce that for any $u \in C_c^\infty(\mathbb{R}^3)$ such that $u \geq 0$, $u \not\equiv 0$ and $\mathrm{supp}(u) \subset \Lambda_\varepsilon$, and

$$\mathcal{J}_\varepsilon(\tau u) = \frac{\tau^2}{2}\|u\|_\varepsilon^2 + b\frac{\tau^4}{4}[u]_s^4 - \int_{\Lambda_\varepsilon} G(\varepsilon x, \tau u)\, dx$$

$$\leq \frac{\tau^2}{2}\|u\|_\varepsilon^2 + b\frac{\tau^4}{4}[u]_s^4 - C_1 \tau^\vartheta \int_{\Lambda_\varepsilon} u^\vartheta\, dx + C_2, \tag{12.2.4}$$

for all $\tau > 0$, with some constants $C_1, C_2 > 0$. Recalling that $\vartheta \in (4, 2_s^*)$, we conclude that $\mathcal{J}_\varepsilon(\tau u) \to -\infty$ as $\tau \to \infty$.

$\square$

In view of Lemma 12.2.1, we can use a variant of the mountain pass theorem without the Palais–Smale condition (see Remark 2.2.10) to deduce the existence of a Palais–Smale sequence $(u_n) \subset \mathcal{H}_\varepsilon$ such that

$$\mathcal{J}_\varepsilon(u_n) \to c_\varepsilon \quad \text{and} \quad \mathcal{J}_\varepsilon'(u_n) \to 0 \text{ in } \mathcal{H}_\varepsilon^*, \tag{12.2.5}$$

where

$$c_\varepsilon = \inf_{\gamma \in \Gamma_\varepsilon} \max_{t \in [0,1]} \mathcal{J}_\varepsilon(\gamma(t)) \quad \text{and} \quad \Gamma_\varepsilon = \{\gamma \in C([0, 1], \mathcal{H}_\varepsilon) : \gamma(0) = 0, \mathcal{J}_\varepsilon(\gamma(1)) < 0\}.$$

$$\tag{12.2.6}$$

As in [299], we use the following equivalent characterization of c_ε that is more appropriate for our aims:

$$c_\varepsilon = \inf_{u \in \mathcal{H}_\varepsilon \setminus \{0\}} \max_{t \geq 0} \mathcal{J}_\varepsilon(tu).$$

Moreover, the monotonicity of g readily implies that for every $u \in \mathcal{H}_\varepsilon \setminus \{0\}$ there exists a unique $t_0 = t_0(u) > 0$ such that

$$\mathcal{J}_\varepsilon(t_0 u) = \max_{t \geq 0} \mathcal{J}_\varepsilon(tu).$$

In the next lemma, we will see that c_ε is less then a threshold value involving the best constant S_* of Sobolev embedding of $\mathcal{D}^{s,2}(\mathbb{R}^3)$ in $L^{2_s^*}(\mathbb{R}^3)$. More precisely:

Lemma 12.2.2 *There exists $T > 0$ such that*

$$0 < c_\varepsilon < \frac{a}{2} S_* T^{3-2s} + \frac{b}{4} S_*^2 T^{6-4s} - \frac{1}{2_s^*} T^3 = c_*.$$

Proof We argue as in [262]. Let $\eta \in C_c^\infty(\mathbb{R}^3)$ be a cut-off function such that $\eta = 1$ in B_ρ, $\mathrm{supp}(\eta) \subset B_{2\rho}$ and $0 \leq \eta \leq 1$, where $\rho > 0$ is such that $B_{2\rho} \subset \Lambda$. For simplicity, we assume that $\rho = 1$. We know (see Remark 1.1.9) that S_* is achieved by $U(x) = \kappa(\mu^2 + |x - x_0|^2)^{-\frac{3-2s}{2}}$, with $\kappa \in \mathbb{R}$, $\mu > 0$ and $x_0 \in \mathbb{R}^3$. Taking $x_0 = 0$, as in [310], we define

$$v_\delta(x) = \eta(\varepsilon x) u_\delta(x) \quad \forall \delta > 0,$$

where

$$u_\delta(x) = \delta^{-\frac{3-2s}{2}} u^*(x/\delta) \quad \text{and} \quad u^*(x) = \frac{U(x/S_*^{\frac{1}{2s}})}{\|U\|_{L^{2_s^*}(\mathbb{R}^3)}}.$$

Then $(-\Delta)^s u_\delta = |u_\delta|^{2_s^*-2} u_\delta$ in $\mathbb{R}^3$ and $[u_\delta]_s^2 = \|u_\delta\|_{L^{2_s^*}(\mathbb{R}^3)}^{2_s^*} = S_*^{\frac{3}{2s}}$. We also recall the following useful estimates (see [310]):

$$A_\delta = [v_\delta]_s^2 = S_*^{\frac{3}{2s}} + O(\delta^{3-2s}), \tag{12.2.7}$$

$$B_\delta = \|v_\delta\|_{L^2(\mathbb{R}^3)}^2 = O(\delta^{3-2s}), \tag{12.2.8}$$

$$C_\delta = \|v_\delta\|^q_{L^q(\mathbb{R}^3)} \geq \begin{cases} O(\delta^{3-\frac{(3-2s)q}{2}}), & \text{if } q > \frac{3}{3-2s}, \\ O(\log(\frac{1}{\delta})\delta^{3-\frac{(3-2s)q}{2}}), & \text{if } q = \frac{3}{3-2s}, \\ O(\delta^{\frac{(3-2s)q}{2}}), & \text{if } q < \frac{3}{3-2s}, \end{cases} \tag{12.2.9}$$

$$D_\delta = \|v_\delta\|^{2^*_s}_{L^{2^*_s}(\mathbb{R}^3)} = S_*^{\frac{3}{2s}} + O(\delta^3). \tag{12.2.10}$$

Note that for all $\delta > 0$ there exists $t_0 > 0$ such that $\mathcal{J}_\varepsilon(\gamma_\delta(t_0)) < 0$, where $\gamma_\delta(t) = v_\delta(\cdot/t)$. Indeed, setting $V_2 = \max_{x \in \bar{\Lambda}} V(x)$, by (f_2) we have

$$\mathcal{J}_\varepsilon(\gamma_\delta(t)) \leq \frac{a}{2} t^{3-2s} [v_\delta]^2_s + \frac{V_2}{2} t^3 \|v_\delta\|^2_{L^2(\mathbb{R}^3)} + \frac{b}{4} t^{6-4s} [v_\delta]^4_s - \frac{t^3}{2^*_s} \|v_\delta\|^{2^*_s}_{L^{2^*_s}(\mathbb{R}^3)} - \frac{t^3}{q} \|v_\delta\|^q_{L^q(\mathbb{R}^3)} C_0$$

$$= \frac{a}{2} t^{3-2s} A_\delta + \frac{b}{4} A_\delta^2 t^{6-4s} + \left(V_2 \frac{B_\delta}{2} - \frac{D_\delta}{2^*_s} - \frac{C_0 C_\delta}{q} \right) t^3. \tag{12.2.11}$$

Since $0 < 6 - 4s < 3$, we use (12.2.8) to deduce that

$$V_2 \frac{B_\delta}{2} - \frac{D_\delta}{2^*_s} \to -\frac{1}{2^*_s} S_*^{\frac{3}{2s}}$$

as $\delta \to 0$. Hence, using (12.2.7), we see that for every $\delta > 0$ sufficiently small, $\mathcal{J}_\varepsilon(\gamma_\delta(t)) \to -\infty$ as $t \to \infty$, that is, there exists $t_0 > 0$ such that $\mathcal{J}_\varepsilon(\gamma_\delta(t_0)) < 0$.

Now, as $t \to 0$, we have

$$[\gamma_\delta(t)]^2_s + \|\gamma_\delta(t)\|^2_{L^2(\mathbb{R}^3)} = t^{3-2s} A_\delta + t^3 B_\delta \to 0 \quad \text{uniformly for } \delta > 0 \text{ small.}$$

We set $\gamma_\delta(0) = 0$. Then $\gamma_\delta(t_0\cdot) \in \Gamma_\varepsilon$, where Γ_ε is defined as in (12.2.6) and we infer that

$$c_\varepsilon \leq \sup_{t \geq 0} \mathcal{J}_\varepsilon(\gamma_\delta(t)).$$

Since $c_\varepsilon > 0$, inequality (12.2.11) shows that exists $t_\delta > 0$ such that

$$\sup_{t \geq 0} \mathcal{J}_\varepsilon(\gamma_\delta(t)) = \mathcal{J}_\varepsilon(\gamma_\delta(t_\delta)).$$

In the light of (12.2.7), (12.2.9) and (12.2.11), we deduce that $\mathcal{J}_\varepsilon(\gamma_\delta(t)) \to 0$ as $t \to 0$ and $\mathcal{J}_\varepsilon(\gamma_\delta(t)) \to -\infty$ as $t \to \infty$ uniformly for $\delta > 0$ small. Then there exist $t_1, t_2 > 0$ (independent of $\delta > 0$) satisfying $t_1 \leq t_\delta \leq t_2$.

Setting

$$H_\delta(t) = \frac{a A_\delta}{2} t^{3-2s} + \frac{b A_\delta^2}{4} t^{6-4s} - \frac{D_\delta}{2^*_s} t^3,$$

we thus have,

$$c_\varepsilon \le \sup_{t \ge 0} H_\delta(t) + \left(\frac{V_2 B_\delta}{2} - \frac{C_0 C_\delta}{q} \right) t_\delta^3.$$

Next, by (12.2.9), for every $q \in (2, 2_s^*)$, we have $C_\delta \ge O(\delta^{3 - \frac{(3-2s)q}{2}})$. Then, by (12.2.8), we infer

$$c_\varepsilon \le \sup_{t \ge 0} H_\delta(t) + O(\delta^{3-2s}) - O(C_0 \delta^{3 - \frac{(3-2s)q}{2}}).$$

Since $3 - 2s > 0$ and $3 - \frac{(3-2s)q}{2} > 0$, we obtain

$$\sup_{t \ge 0} H_\delta(t) \ge \frac{c_\varepsilon}{2} \quad \text{uniformly for } \delta > 0 \text{ small.}$$

Arguing as above, there exist $t_3, t_4 > 0$ (independent of $\delta > 0$) such that

$$\sup_{t \ge 0} H_\delta(t) = \sup_{t \in [t_3, t_4]} H_\delta(t).$$

By (12.2.7) we deduce

$$c_\varepsilon \le \sup_{t \ge 0} K(S_*^{\frac{1}{2s}} t) + O(\delta^{3-2s}) - O(C_0 \delta^{3 - \frac{(3-2s)q}{2}}), \tag{12.2.12}$$

where

$$K(t) = \frac{a S_*}{2} t^{3-2s} + \frac{b S_*^2}{4} t^{6-4s} - \frac{1}{2_s^*} t^3.$$

Let us note that for $t > 0$,

$$K'(t) = \frac{3-2s}{2} a S_* t^{2-2s} + \frac{3-2s}{2} b S_*^2 t^{5-4s} - \frac{3-2s}{2} t^2$$

$$= \frac{(3-2s) t^{2-2s}}{2} \left(a S_* + b S_*^2 t^{3-2s} - t^{2s} \right) = \frac{(3-2s) t^{2-2s}}{2} \tilde{K}(t).$$

Moreover,

$$\tilde{K}'(t) = b S_* (3 - 2s) t^{2-2s} - 2s t^{2s-1} = t^{2-2s} [b S_*^2 (3 - 2s) - 2s t^{4s-3}].$$

Since $4s > 3$, there exists a unique $T > 0$ such that $\tilde{K}(t) > 0$ for $t \in (0, T)$ and $\tilde{K}(t) < 0$ for $t > T$. Then, T is the unique maximum point of $K(t)$. By virtue of (12.2.12), we have,

$$c_\varepsilon \le K(T) + O(\delta^{3-2s}) - O(C_0 \delta^{3 - \frac{(3-2s)q}{2}}). \tag{12.2.13}$$

If $q > \frac{4s}{3-2s}$, then $0 < 3 - \frac{(3-2s)q}{2} < 3 - 2s$, and by (12.2.13), for any fixed $C_0 > 0$, it holds that $c_\varepsilon < K(T)$ for $\delta > 0$ small. If $2 < q < \frac{4s}{3-2s}$, then, for $\delta > 0$ small and $C_0 > \delta^{\frac{(3-2s)q}{2}-2s-1}$, we also have $c_\varepsilon < K(T)$. $\qquad\square$

Lemma 12.2.3 *Every sequence (u_n) satisfying (12.2.5) is bounded in $\mathcal{H}_\varepsilon$.*

Proof In view of (g_3), we deduce that for all $n \in \mathbb{N}$

$$C(1 + \|u_n\|_\varepsilon) \geq \mathcal{J}_\varepsilon(u_n) - \frac{1}{\vartheta}\langle \mathcal{J}_\varepsilon'(u_n), u_n\rangle \tag{12.2.14}$$

$$= \left(\frac{\vartheta - 2}{2\vartheta}\right)\|u_n\|_\varepsilon^2 + b\left(\frac{\vartheta - 4}{4\vartheta}\right)[u_n]^4 + \frac{1}{\vartheta}\int_{\mathbb{R}^3\setminus\Lambda_\varepsilon}[g(\varepsilon x, u_n)u_n - \vartheta G(\varepsilon x, u_n)]\,dx$$

$$\quad + \frac{1}{\vartheta}\int_{\Lambda_\varepsilon}[g(\varepsilon x, u_n)u_n - \vartheta G(\varepsilon x, u_n)]\,dx$$

$$\geq \left(\frac{\vartheta - 2}{2\vartheta}\right)\|u_n\|_\varepsilon^2 + \frac{1}{\vartheta}\int_{\mathbb{R}^3\setminus\Lambda_\varepsilon}[g(\varepsilon x, u_n)u_n - \vartheta G(\varepsilon x, u_n)]\,dx$$

$$\geq \left(\frac{\vartheta - 2}{2\vartheta}\right)\|u_n\|_\varepsilon^2 - \left(\frac{\vartheta - 2}{2\vartheta}\right)\frac{1}{K}\int_{\mathbb{R}^3\setminus\Lambda_\varepsilon}V(\varepsilon x)u_n^2\,dx$$

$$\geq \left(\frac{\vartheta - 2}{2\vartheta}\right)\left(1 - \frac{1}{K}\right)\|u_n\|_\varepsilon^2. \tag{12.2.15}$$

Since $\vartheta > 4$ and $K > 2$, we conclude that (u_n) is bounded in $\mathcal{H}_\varepsilon$. $\qquad\square$

Remark 12.2.4 Arguing as in Remark 5.2.8, we may always suppose that the Palais–Smale sequence (u_n) is non-negative in $\mathbb{R}^3$.

Lemma 12.2.5 *There exist a sequence $(z_n) \subset \mathbb{R}^3$ and constants $R, \beta > 0$ such that*

$$\int_{B_R(z_n)} u_n^2\,dx \geq \beta.$$

Moreover, (z_n) is bounded in $\mathbb{R}^3$.

Proof Assume, by contradiction, that the first conclusion of lemma is not true. Then, by Lemma 1.4.4, we have

$$u_n \to 0 \quad \text{in } L^q(\mathbb{R}^3) \quad \forall q \in (2, 2_s^*),$$

which together with (f_1) and (f_2) shows that

$$\int_{\mathbb{R}^3} F(u_n)\, dx = \int_{\mathbb{R}^3} f(u_n) u_n\, dx = o_n(1) \quad \text{as } n \to \infty.$$

Since (u_n) is bounded in $\mathcal{H}_\varepsilon$, we may assume that $u_n \rightharpoonup u$ in $\mathcal{H}_\varepsilon$.

Now, we observe that

$$\int_{\mathbb{R}^3} G(\varepsilon x, u_n)\, dx \le \frac{1}{2_s^*} \int_{\Lambda_\varepsilon \cup \{u_n \le a_0\}} (u_n^+)^{2_s^*}\, dx + \frac{V_1}{2K} \int_{(\mathbb{R}^3 \setminus \Lambda_\varepsilon) \cap \{u_n > a_0\}} u_n^2\, dx + o_n(1) \tag{12.2.16}$$

and

$$\int_{\mathbb{R}^3} g(\varepsilon x, u_n) u_n\, dx = \int_{\Lambda_\varepsilon \cup \{u_n \le a_0\}} (u_n^+)^{2_s^*}\, dx + \frac{V_1}{K} \int_{(\mathbb{R}^3 \setminus \Lambda_\varepsilon) \cap \{u_n > a_0\}} u_n^2\, dx + o_n(1). \tag{12.2.17}$$

Using the fact that $\langle \mathcal{J}_\varepsilon'(u_n), u_n \rangle = o_n(1)$ and (12.2.17), we have

$$\|u_n\|_\varepsilon^2 - \frac{V_1}{K} \int_{(\mathbb{R}^3 \setminus \Lambda_\varepsilon) \cap \{u_n > a_0\}} u_n^2\, dx + b[u_n]_s^4 = \int_{\Lambda_\varepsilon \cup \{u_n \le a_0\}} (u_n^+)^{2_s^*}\, dx + o_n(1). \tag{12.2.18}$$

Assume that

$$\int_{\Lambda_\varepsilon \cup \{u_n \le a_0\}} (u_n^+)^{2_s^*}\, dx \to \ell^3 \ge 0$$

and

$$[u_n]_s^2 \to B^2.$$

Note that $\ell > 0$: otherwise, (12.2.18) yields $\|u_n\|_\varepsilon \to 0$ as $n \to \infty$ which implies that $\mathcal{J}_\varepsilon(u_n) \to 0$, and this is impossible because $c_\varepsilon > 0$. Then, by (12.2.18) and (1.1.1) we obtain

$$a S_* \left(\int_{\Lambda_\varepsilon \cup \{u_n \le a_0\}} (u_n^+)^{2_s^*}\, dx \right)^{\frac{2}{2_s^*}} + b S_*^2 \left(\int_{\Lambda_\varepsilon \cup \{u_n \le a_0\}} (u_n^+)^{2_s^*}\, dx \right)^{\frac{4}{2_s^*}}$$

$$\le \int_{\Lambda_\varepsilon \cup \{u_n \le a_0\}} (u_n^+)^{2_s^*}\, dx + o_n(1). \tag{12.2.19}$$

Since $\ell > 0$, it follows from (12.2.19) that

$$K'(\ell) = \frac{3 - 2s}{2} \ell^{-1}(aS_*\ell^{3-2s} + bS_*^2\ell^{6-4s} - \ell^3) \leq 0,$$

so we deduce that $\ell \geq T$, where T is the unique maximum of K defined in Lemma 12.2.2. Let us consider the following functional:

$$\mathcal{I}_\varepsilon(u) = \frac{(a + bB^2)}{2}[u]_s^2 + \frac{1}{2}\int_{\mathbb{R}^3} V(\varepsilon x)u^2\,dx - \int_{\mathbb{R}^3} G(\varepsilon x, u)\,dx$$

$$= \mathcal{J}_\varepsilon(u) - \frac{b}{4}[u]_s^4 + \frac{b}{2}B^2[u]_s^2, \tag{12.2.20}$$

and note that (u_n) is a Palais–Smale sequence for $\mathcal{I}_\varepsilon$ at the level $c_\varepsilon + \frac{b}{4}B^4$, that is

$$\mathcal{I}_\varepsilon(u_n) = c_\varepsilon + \frac{b}{4}B^4 + o_n(1), \quad \mathcal{I}_\varepsilon'(u_n) = o_n(1). \tag{12.2.21}$$

Then, since $\ell \geq T$, using (12.2.16), (12.2.21) and (1.1.1) we infer that

$$c_\varepsilon = \mathcal{I}_\varepsilon(u_n) - \frac{b}{4}B^4 + o_n(1)$$

$$\geq \frac{a}{2}[u_n]_s^2 + \frac{bB^2}{2}[u_n]_s^2 - \frac{b}{4}B^4 + \frac{1}{2}\int_{\mathbb{R}^3} V(\varepsilon x)u_n^2\,dx - \frac{V_1}{2K}\int_{(\mathbb{R}^3\setminus\Lambda_\varepsilon)\cap\{u_n>a_0\}} u_n^2\,dx$$

$$- \frac{1}{2_s^*}\int_{\Lambda_\varepsilon\cup\{u_n\leq a_0\}} (u_n^+)^{2_s^*}\,dx + o_n(1)$$

$$\geq \frac{a}{2}[u_n]_s^2 + \frac{b}{4}[u_n]_s^4 - \frac{1}{2_s^*}\int_{\Lambda_\varepsilon\cup\{u_n\leq a_0\}} (u_n^+)^{2_s^*}\,dx + o_n(1)$$

$$\geq \frac{a}{2}S_*\left(\int_{\Lambda_\varepsilon\cup\{u_n\leq a_0\}} (u_n^+)^{2_s^*}\,dx\right)^{\frac{2}{2_s^*}} + \frac{b}{4}S_*^2\left(\int_{\Lambda_\varepsilon\cup\{u_n\leq a_0\}} (u_n^+)^{2_s^*}\,dx\right)^{\frac{4}{2_s^*}}$$

$$- \frac{1}{2_s^*}\int_{\Lambda_\varepsilon\cup\{u_n\leq a_0\}} (u_n^+)^{2_s^*}\,dx + o_n(1)$$

$$= \frac{a}{2}S_*\ell^{3-2s} + \frac{b}{4}S_*^2\ell^{6-4s} - \frac{1}{2_s^*}\ell^3$$

$$\geq \frac{a}{2}S_*T^{3-2s} + \frac{b}{4}S_*^2T^{6-4s} - \frac{1}{2_s^*}T^3 = c_*,$$

and this gives a contradiction by Lemma 12.2.2.

Let us show that (z_n) is bounded in $\mathbb{R}^3$. For any $\rho > 0$, let $\psi_\rho \in C^\infty(\mathbb{R}^3)$ be such that $\psi_\rho = 0$ in B_ρ and $\psi_\rho = 1$ in $\mathbb{R}^3 \setminus B_{2\rho}$, with $0 \leq \psi_\rho \leq 1$ and $\|\nabla\psi_\rho\|_{L^\infty(\mathbb{R}^3)} \leq \frac{C}{\rho}$, where C is a constant independent of ρ. Since $(\psi_\rho u_n)$ is bounded in $\mathcal{H}_\varepsilon$, it follows that $\langle \mathcal{J}'_\varepsilon(u_n), \psi_\rho u_n \rangle = o_n(1)$, that is

$$(a + b[u_n]_s^2) \iint_{\mathbb{R}^6} \frac{|u_n(x) - u_n(y)|^2}{|x-y|^{3+2s}} \psi_\rho(x)\, dx\, dy + \int_{\mathbb{R}^3} V(\varepsilon x) u_n^2 \psi_\rho\, dx$$

$$= o_n(1) + \int_{\mathbb{R}^3} g(\varepsilon x, u_n) u_n \psi_\rho\, dx$$

$$- (a + b[u_n]_s^2) \iint_{\mathbb{R}^6} \frac{(\psi_\rho(x) - \psi_\rho(y))(u_n(x) - u_n(y))}{|x-y|^{3+2s}} u_n(y)\, dx\, dy.$$

Take $\rho > 0$ sufficiently large such that $\Lambda_\varepsilon \subset B_\rho$. Then, using (g_3)-(ii), we get

$$\iint_{\mathbb{R}^6} a \frac{|u_n(x) - u_n(y)|^2}{|x-y|^{3+2s}} \psi_\rho(x)\, dx\, dy + \int_{\mathbb{R}^3} V(\varepsilon x) u_n^2 \psi_\rho\, dx$$

$$\leq \int_{\mathbb{R}^3} \frac{1}{K} V(\varepsilon x) u_n^2 \psi_\rho\, dx$$

$$- (a + b[u_n]_s^2) \iint_{\mathbb{R}^6} \frac{(\psi_\rho(x) - \psi_\rho(y))(u_n(x) - u_n(y))}{|x-y|^{3+2s}} u_n(y)\, dx\, dy + o_n(1)$$

which implies that

$$\left(1 - \frac{1}{K}\right) V_1 \int_{\mathbb{R}^3} u_n^2 \psi_\rho\, dx$$

$$\leq -(a + b[u_n]_s^2) \iint_{\mathbb{R}^6} \frac{(\psi_\rho(x) - \psi_\rho(y))(u_n(x) - u_n(y))}{|x-y|^{3+2s}} u_n(y)\, dx\, dy + o_n(1).$$

$$(12.2.22)$$

Now, by the Hölder inequality and the boundedness on (u_n) in $\mathcal{H}_\varepsilon$,

$$\left| \iint_{\mathbb{R}^6} \frac{(u_n(x) - u_n(y))(\psi_\rho(x) - \psi_\rho(y))}{|x-y|^{3+2s}} u_n(y)\, dx\, dy \right|$$

$$\leq C \left(\iint_{\mathbb{R}^6} \frac{|\psi_\rho(x) - \psi_\rho(y)|^2}{|x-y|^{3+2s}} |u_n(y)|^2\, dx\, dy \right)^{\frac{1}{2}}. \qquad (12.2.23)$$

On the other hand, recalling that $0 \leq \psi_\rho \leq 1$ and $\|\nabla \psi_\rho\|_{L^\infty(\mathbb{R}^3)} \leq C/\rho$, and using polar coordinates, we obtain

$$\iint_{\mathbb{R}^6} \frac{|\psi_\rho(x) - \psi_\rho(y)|^2}{|x-y|^{3+2s}} |u_n(x)|^2\, dx\, dy$$

$$= \int_{\mathbb{R}^3} \int_{|y-x|>\rho} \frac{|\psi_\rho(x) - \psi_\rho(y)|^2}{|x-y|^{3+2s}} |u_n(x)|^2\, dx\, dy + \int_{\mathbb{R}^3} \int_{|y-x|\leq\rho} \frac{|\psi_\rho(x) - \psi_\rho(y)|^2}{|x-y|^{3+2s}} |u_n(x)|^2\, dx\, dy$$

$$\leq C \int_{\mathbb{R}^3} |u_n(x)|^2 \left(\int_{|y-x|>\rho} \frac{dy}{|x-y|^{3+2s}} \right) dx + \frac{C}{\rho^2} \int_{\mathbb{R}^3} |u_n(x)|^2 \left(\int_{|y-x|\leq\rho} \frac{dy}{|x-y|^{3+2s-2}} \right) dx$$

$$\leq C \int_{\mathbb{R}^3} |u_n(x)|^2 \left(\int_{|z|>\rho} \frac{dz}{|z|^{3+2s}} \right) dx + \frac{C}{\rho^2} \int_{\mathbb{R}^3} |u_n(x)|^2 \left(\int_{|z|\leq\rho} \frac{dz}{|z|^{1+2s}} \right) dx$$

$$\leq C \int_{\mathbb{R}^3} |u_n(x)|^2 dx \left(\int_\rho^\infty \frac{d\rho}{\rho^{2s+1}} \right) + \frac{C}{\rho^2} \int_{\mathbb{R}^3} |u_n(x)|^2 dx \left(\int_0^\rho \frac{d\rho}{\rho^{2s-1}} \right)$$

$$\leq \frac{C}{\rho^{2s}} \int_{\mathbb{R}^3} |u_n(x)|^2 dx + \frac{C}{\rho^2} \rho^{-2s+2} \int_{\mathbb{R}^3} |u_n(x)|^2 dx$$

$$\leq \frac{C}{\rho^{2s}} \int_{\mathbb{R}^3} |u_n(x)|^2 dx \leq \frac{C}{\rho^{2s}}$$

where in the last step we used the boundedness of (u_n) in $\mathcal{H}_\varepsilon$. Taking into account (12.2.22), (12.2.23) and the above estimate, we infer that

$$\left(1 - \frac{1}{K}\right) V_1 \int_{B_{2\rho}^c} u_n^2 \, dx \leq \left(1 - \frac{1}{K}\right) V_1 \int_{\mathbb{R}^3} u_n^2 \psi_\rho \, dx \leq \frac{C}{\rho^s} + o_n(1).$$

Now, if by contradiction $|z_n| \to \infty$, then $|z_n| \geq R + 2\rho$ for n large and we obtain

$$\beta \leq \limsup_{n\to\infty} \int_{B_R(z_n)} u_n^2 \, dx \leq \limsup_{n\to\infty} \int_{B_{2\rho}^c} u_n^2 \, dx \leq \frac{C}{\rho^s},$$

that is, $0 < \beta \leq \frac{C}{\rho^s}$, which is not true for ρ large. Hence, (z_n) is bounded in $\mathbb{R}^3$. $\square$

We conclude this section with the proof of the main result of this section:

Theorem 12.2.6 *Assume that* (V_1)–(V_2) *and* $(f1)$–$(f4)$ *hold. Then, for all* $\varepsilon > 0$, *problem* (12.2.3) *admits a positive ground state.*

Proof Using Lemma 12.2.1 and a variant of the mountain pass theorem without the Palais–Smale condition (see Remark 2.2.10), we know that there exists a Palais–Smale sequence (u_n) for $\mathcal{J}_\varepsilon$ at the level c_ε, where $0 < c_\varepsilon < c_*$ by Lemma 12.2.2. By Lemma 12.2.3, (u_n) is bounded in $\mathcal{H}_\varepsilon$, so we may assume that $u_n \rightharpoonup u$ in $\mathcal{H}_\varepsilon$ and $u_n \to u$ in $L^q_{\text{loc}}(\mathbb{R}^3)$ for all $q \in [1, 2_s^*)$. By Lemma 12.2.5, u is nontrivial. Since $\langle \mathcal{J}'_\varepsilon(u_n), \varphi \rangle = o_n(1)$ for all $\varphi \in \mathcal{H}_\varepsilon$, we see that

$$\langle u, \varphi \rangle_\varepsilon + bB^2 \left(\int_{\mathbb{R}^3} (-\Delta)^{\frac{s}{2}} u (-\Delta)^{\frac{s}{2}} \varphi \, dx \right) = \int_{\mathbb{R}^3} g(\varepsilon x, u)\varphi \, dx, \qquad (12.2.24)$$

where $B^2 = \lim_{n\to\infty} [u_n]_s^2$. Note that $B^2 \geq [u]_s^2$ by Fatou's lemma. If, by contradiction, $B^2 > [u]_s^2$, we may use (12.2.24) to deduce that $\langle \mathcal{J}'_\varepsilon(u), u \rangle < 0$. Moreover, conditions

(g_1)–(g_2) imply that $\langle \mathcal{J}_\varepsilon'(\tau u), \tau u \rangle > 0$ for some $0 < \tau << 1$. Then there exists $t_0 \in (\tau, 1)$ such that $t_0 u \in \mathcal{N}_\varepsilon$ and $\langle \mathcal{J}_\varepsilon'(t_0 u), t_0 u \rangle = 0$. Using Fatou's lemma, $t_0 \in (\tau, 1)$ and (g_3) we get

$$c_\varepsilon \leq \mathcal{J}_\varepsilon(t_0 u) - \frac{1}{4}\langle \mathcal{J}_\varepsilon'(t_0 u), t_0 u \rangle < \mathcal{J}_\varepsilon(u) - \frac{1}{4}\langle \mathcal{J}_\varepsilon'(u), u \rangle$$

$$\leq \liminf_{n \to \infty}\left[\mathcal{J}_\varepsilon(u_n) - \frac{1}{4}\langle \mathcal{J}_\varepsilon'(u_n), u_n \rangle \right] = c_\varepsilon, \tag{12.2.25}$$

which gives a contradiction. Therefore, $B^2 = [u]_s^2$ and we deduce that $\mathcal{J}_\varepsilon'(u) = 0$. Hence, $\mathcal{J}_\varepsilon$ admits a nontrivial critical point $u \in \mathcal{H}_\varepsilon$. Since $\langle \mathcal{J}_\varepsilon'(u), u^- \rangle = 0$, where $u^- = \min\{u, 0\}$, and $g(x, t) = 0$ for $t \leq 0$, it is easy to check that $u \geq 0$ in $\mathbb{R}^3$. Moreover, proceeding as in the proof of Lemma 12.4.2 below, one can show that $u \in L^\infty(\mathbb{R}^3)$. By Proposition 1.3.11-(iii) we deduce that $u \in L^\infty(\mathbb{R}^3) \cap C_{loc}^{0,\alpha}(\mathbb{R}^3)$, and applying Theorem 1.3.5 we obtain that $u > 0$ in $\mathbb{R}^3$. Finally, arguing as in (12.2.25) with $t_0 = 1$, we conclude that u is a ground state solution to (12.2.3). □

12.3   The Autonomous Critical Fractional Kirchhoff Problem

Let us consider the following limit problem related to (12.2.3), that is, for $\mu > 0$:

$$\begin{cases} (a + b[u]_s^2)(-\Delta)^s u + \mu u = f(u) + |u|^{2_s^*-2}u & \text{in } \mathbb{R}^3, \\ u \in H^s(\mathbb{R}^3), \quad u > 0 \text{ in } \mathbb{R}^3. \end{cases} \tag{12.3.1}$$

The corresponding Euler–Lagrange functional is given by

$$\mathcal{I}_\mu(u) = \frac{1}{2}\left(a[u]_s^2 + \mu\|u\|_{L^2(\mathbb{R}^3)}^2 \right) + \frac{b}{4}[u]_s^4 - \int_{\mathbb{R}^3}\left[F(u) + \frac{1}{2_s^*}(u^+)^{2_s^*} \right]dx$$

which is well defined on the Hilbert space $\mathbb{X}_\mu = H^s(\mathbb{R}^3)$ endowed with the inner product

$$\langle u, \varphi \rangle_\mu = a\langle u, v \rangle_{\mathcal{D}^{s,2}(\mathbb{R}^3)} + \mu\int_{\mathbb{R}^3} u\varphi\, dx$$

and the induced norm

$$\|u\|_\mu = \left(a[u]_s^2 + \mu\|u\|_{L^2(\mathbb{R}^3)}^2 \right)^{\frac{1}{2}}.$$

We denote by $\mathcal{M}_\mu$ the Nehari manifold associated with $\mathcal{I}_\mu$, that is,

$$\mathcal{M}_\mu = \left\{ u \in \mathbb{X}_\mu \setminus \{0\} : \langle \mathcal{I}_\mu'(u), u \rangle = 0 \right\},$$

and

$$d_\mu = \inf_{u \in \mathcal{M}_\mu} \mathcal{I}_\mu(u),$$

or equivalently

$$d_\mu = \inf_{u \in \mathbb{X}_\mu \setminus \{0\}} \max_{t \geq 0} \mathcal{I}_\mu(tu).$$

Arguing as in the proof of Theorem 12.2.6 it is easy to deduce that:

Theorem 12.3.1 *For all $\mu > 0$, problem (12.3.1) admits a positive ground state solution.*

The following useful relation holds between c_ε and d_{V_0}:

Lemma 12.3.2 $\limsup_{\varepsilon \to 0} c_\varepsilon \leq d_{V_0}$.

Proof For any $\varepsilon > 0$ we set $\omega_\varepsilon(x) = \psi_\varepsilon(x)\omega(x)$, where ω is a positive ground state provided by Theorem 12.3.1 with $\mu = V_0$, and $\psi_\varepsilon(x) = \psi(\varepsilon x)$ with $\psi \in C_c^\infty(\mathbb{R}^3)$, $\psi \in [0, 1]$, $\psi(x) = 1$ if $|x| \leq \frac{1}{2}$ and $\psi(x) = 0$ if $|x| \geq 1$. Here we assume that $\mathrm{supp}(\psi) \subset B_1 \subset \Lambda$. Using Lemma 1.4.8 and the dominated convergence theorem we can see that $\omega_\varepsilon \to \omega$ in $H^s(\mathbb{R}^3)$ and $\mathcal{I}_{V_0}(\omega_\varepsilon) \to \mathcal{I}_{V_0}(\omega) = d_{V_0}$ as $\varepsilon \to 0$. For every $\varepsilon > 0$ there exists $t_\varepsilon > 0$ such that

$$\mathcal{J}_\varepsilon(t_\varepsilon \omega_\varepsilon) = \max_{t \geq 0} \mathcal{J}_\varepsilon(t\omega_\varepsilon).$$

Then, $\frac{d}{dt}[\mathcal{J}_\varepsilon(t\omega_\varepsilon)]_{t=t_\varepsilon} = 0$ and this implies that

$$\frac{1}{t_\varepsilon^2} \int_{\mathbb{R}^3} a|(-\Delta)^{\frac{s}{2}}\omega_\varepsilon|^2 + V(\varepsilon x)\omega_\varepsilon^2 \, dx + b \left(\int_{\mathbb{R}^3} |(-\Delta)^{\frac{s}{2}}\omega_\varepsilon|^2 \, dx \right)^2$$
$$= \int_{\mathbb{R}^3} \frac{f(t_\varepsilon \omega_\varepsilon)}{(t_\varepsilon \omega_\varepsilon)^3} \omega_\varepsilon^4 \, dx + t_\varepsilon^{2^*-4} \int_{\mathbb{R}^3} \omega_\varepsilon^{2_s^*} \, dx. \tag{12.3.2}$$

By (f_1)–(f_4), $\omega \in \mathcal{M}_{V_0}$ and (12.3.2), it follows that $t_\varepsilon \to 1$ as $\varepsilon \to 0$. On the other hand,

$$c_\varepsilon \leq \max_{t \geq 0} \mathcal{J}_\varepsilon(t\omega_\varepsilon) = \mathcal{J}_\varepsilon(t_\varepsilon \omega_\varepsilon) = \mathcal{I}_{V_0}(t_\varepsilon \omega_\varepsilon) + \frac{t_\varepsilon^2}{2} \int_{\mathbb{R}^3} (V(\varepsilon x) - V_0)\omega_\varepsilon^2 \, dx.$$

Since $V(\varepsilon x)$ is bounded on the support of ω_ε, the dominated convergence theorem and the above inequality complete the proof.   □

12.4   Proof of Theorem 12.1.1

This last section is devoted to the proof of the main result of this chapter. First, we prove the following compactness result which will play a fundamental role in showing that the solutions of (12.2.3) are also solutions to (12.2.1) for $\varepsilon > 0$ small enough.

Lemma 12.4.1 *Let* $\varepsilon_n \to 0$ *and* $(u_n) = (u_{\varepsilon_n}) \subset \mathcal{H}_{\varepsilon_n}$ *be such that* $\mathcal{J}_{\varepsilon_n}(u_n) = c_{\varepsilon_n}$ *and* $\mathcal{J}'_{\varepsilon_n}(u_n) = 0$. *Then there exists* $(\tilde{y}_n) = (\tilde{y}_{\varepsilon_n}) \subset \mathbb{R}^3$ *such that* $\tilde{u}_n(x) = u_n(x + \tilde{y}_n)$ *has a convergent subsequence in* $H^s(\mathbb{R}^3)$. *Moreover, up to a subsequence,* $y_n = \varepsilon_n \tilde{y}_n \to y_0$ *for some* $y_0 \in \Lambda$ *such that* $V(y_0) = V_0$.

Proof Using that $\langle \mathcal{J}'_{\varepsilon_n}(u_n), u_n \rangle = 0$ and the conditions (g_1), (g_2), it is easy to see that there is $\gamma > 0$ (independent of ε_n) such that

$$\|u_n\|_{\varepsilon_n} \geq \gamma > 0 \quad \forall n \in \mathbb{N}.$$

Since $\mathcal{J}_{\varepsilon_n}(u_n) = c_{\varepsilon_n}$ and $\langle \mathcal{J}'_{\varepsilon_n}(u_n), u_n \rangle = 0$, Lemma 12.3.2 shows that we can argue as in the proof of Lemma 12.2.3 to deduce that (u_n) is bounded in $\mathcal{H}_{\varepsilon_n}$. Therefore, proceeding as in the proof of Lemma 12.2.5, we can find a sequence $(\tilde{y}_n) \subset \mathbb{R}^3$ and constants $R, \alpha > 0$ such that

$$\liminf_{n \to \infty} \int_{B_R(\tilde{y}_n)} u_n^2 \, dx \geq \alpha.$$

Set $\tilde{u}_n(x) = u_n(x + \tilde{y}_n)$. Then, $(\tilde{u}_n)$ is bounded in $H^s(\mathbb{R}^3)$, and we may assume that

$$\tilde{u}_n \rightharpoonup \tilde{u} \text{ in } H^s(\mathbb{R}^3), \tag{12.4.1}$$

and $[\tilde{u}_n]_s^2 \to B^2$ as $n \to \infty$. Moreover, $\tilde{u} \neq 0$ because

$$\int_{B_R} \tilde{u}^2 \, dx \geq \alpha. \tag{12.4.2}$$

Next, set $y_n = \varepsilon_n \tilde{y}_n$. First, we show that (y_n) is bounded in $\mathbb{R}^3$.

Claim 1 $\lim_{n \to \infty} \text{dist}(y_n, \overline{\Lambda}) = 0$.

If, by contradiction, this is not true, then we can find $\delta > 0$ and a subsequence of (y_n), still denoted (y_n), such that

$$\text{dist}(y_n, \overline{\Lambda}) \geq \delta \quad \text{for all } n \in \mathbb{N}.$$

Therefore, there is $r > 0$ such that $B_r(y_n) \subset \mathbb{R}^3 \setminus \Lambda$ for all $n \in \mathbb{N}$. Since $\tilde{u} \geq 0$ and $C_c^\infty(\mathbb{R}^3)$ is dense in $H^s(\mathbb{R}^3)$, we can approximate $\tilde{u}$ by a sequence $(\psi_j) \subset C_c^\infty(\mathbb{R}^3)$ such that $\psi_j \geq 0$ in $\mathbb{R}^3$ and $\psi_j \to \tilde{u}$ in $H^s(\mathbb{R}^3)$. Fix $j \in \mathbb{N}$ and use $\psi = \psi_j$ as test function in $\langle \mathcal{J}'_{\varepsilon_n}(u_n), \psi \rangle = 0$. We have

$$(a + b[\tilde{u}_n]_s^2) \iint_{\mathbb{R}^6} \frac{(\tilde{u}_n(x) - \tilde{u}_n(y))(\psi_j(x) - \psi_j(y))}{|x-y|^{3+2s}} \, dx \, dy + \int_{\mathbb{R}^3} V(\varepsilon_n x + \varepsilon_n \tilde{y}_n)\tilde{u}_n \psi_j \, dx$$

$$= \int_{\mathbb{R}^3} g(\varepsilon_n x + \varepsilon_n \tilde{y}_n, \tilde{u}_n)\psi_j \, dx. \tag{12.4.3}$$

Since $u_n, \psi_j \geq 0$, the definition of g implies that

$$\int_{\mathbb{R}^3} g(\varepsilon_n x + \varepsilon_n \tilde{y}_n, \tilde{u}_n)\psi_j \, dx = \int_{B_{r/\varepsilon_n}} g(\varepsilon_n x + \varepsilon_n \tilde{y}_n, \tilde{u}_n)\psi_j \, dx$$

$$+ \int_{\mathbb{R}^3 \setminus B_{r/\varepsilon_n}} g(\varepsilon_n x + \varepsilon_n \tilde{y}_n, \tilde{u}_n)\psi_j \, dx$$

$$\leq \frac{V_1}{K} \int_{B_{r/\varepsilon_n}} \tilde{u}_n \psi_j \, dx + \int_{\mathbb{R}^3 \setminus B_{r/\varepsilon_n}} \left(f(\tilde{u}_n)\psi_j + \tilde{u}_n^{2_s^*-1}\psi_j \right) dx.$$

This fact together with (12.4.3) gives

$$(a + b[\tilde{u}_n]_s^2) \iint_{\mathbb{R}^6} \frac{(\tilde{u}_n(x) - \tilde{u}_n(y))(\psi_j(x) - \psi_j(y))}{|x-y|^{3+2s}} \, dx \, dy + A \int_{\mathbb{R}^3} \tilde{u}_n \psi_j \, dx$$

$$\leq \int_{\mathbb{R}^3 \setminus B_{r/\varepsilon_n}} \left(f(\tilde{u}_n)\psi_j + \tilde{u}_n^{2_s^*-1}\psi_j \right) dx, \tag{12.4.4}$$

where $A = V_1(1 - \frac{1}{K})$. Taking into account (12.4.1), that ψ_j has compact support in $\mathbb{R}^3$ and that $\varepsilon_n \to 0$, we infer that as $n \to \infty$

$$\iint_{\mathbb{R}^6} \frac{(\tilde{u}_n(x) - \tilde{u}_n(y))(\psi_j(x) - \psi_j(y))}{|x-y|^{3+2s}} \, dx \, dy$$

$$\to \iint_{\mathbb{R}^6} \frac{(\tilde{u}(x) - \tilde{u}(y))(\psi_j(x) - \psi_j(y))}{|x-y|^{3+2s}} \, dx \, dy$$

and

$$\int_{\mathbb{R}^3 \setminus B_{r/\varepsilon_n}} \left(f(\tilde{u}_n)\psi_j + \tilde{u}_n^{2_s^*-1}\psi_j \right) dx \to 0.$$

The above limits, (12.4.4) and the fact that $[\tilde{u}_n]_s^2 \to B^2$ imply that

$$(a + bB^2) \iint_{\mathbb{R}^6} \frac{(\tilde{u}(x) - \tilde{u}(y))(\psi_j(x) - \psi_j(y))}{|x-y|^{3+2s}} \, dx \, dy + A \int_{\mathbb{R}^3} \tilde{u}\psi_j \, dx \leq 0$$

and letting $j \to \infty$ we obtain

$$(a + bB^2)[\tilde{u}]_s^2 + A\|\tilde{u}\|_{L^2(\mathbb{R}^3)}^2 \leq 0,$$

which contradicts (12.4.2). Hence, there exists a subsequence of (y_n) such that

$$y_n \to y_0 \in \overline{\Lambda}.$$

Claim 2 $y_0 \in \Lambda$.

In view of (g_2) and (12.4.3),

$$(a + b[\tilde{u}_n]_s^2) \iint_{\mathbb{R}^6} \frac{(\tilde{u}_n(x) - \tilde{u}_n(y))(\psi_j(x) - \psi_j(y))}{|x - y|^{3+2s}} \, dx \, dy + \int_{\mathbb{R}^3} V(\varepsilon_n x + \varepsilon_n \tilde{y}_n)\tilde{u}_n \psi_j \, dx$$
$$\leq \int_{\mathbb{R}^3} (f(\tilde{u}_n) + \tilde{u}_n^{2_s^*-1})\psi_j \, dx.$$

Letting $n \to \infty$ we find that

$$(a + bB^2) \iint_{\mathbb{R}^6} \frac{(\tilde{u}(x) - \tilde{u}(y))(\psi_j(x) - \psi_j(y))}{|x - y|^{3+2s}} \, dx \, dy + \int_{\mathbb{R}^3} V(y_0)\tilde{u}\psi_j \, dx$$
$$\leq \int_{\mathbb{R}^3} (f(\tilde{u}) + \tilde{u}^{2_s^*-1})\psi_j \, dx,$$

and then passing to the limit as $j \to \infty$ we obtain

$$(a + bB^2)[\tilde{u}]_s^2 + V(y_0)\|\tilde{u}\|_{L^2(\mathbb{R}^3)}^2 \leq \int_{\mathbb{R}^3} (f(\tilde{u}) + \tilde{u}^{2_s^*-1})\tilde{u} \, dx.$$

Since $B^2 \geq [\tilde{u}]_s^2$ (by Fatou's lemma), this inequality yields

$$(a + b[\tilde{u}]_s^2)[\tilde{u}]_s^2 + V(y_0)\|\tilde{u}\|_{L^2(\mathbb{R}^3)}^2 \leq \int_{\mathbb{R}^3} (f(\tilde{u}) + \tilde{u}^{2_s^*-1})\tilde{u} \, dx.$$

Therefore, we can find $\tau \in (0, 1)$ such that $\tau\tilde{u} \in \mathcal{M}_{V(y_0)}$. Then, by Lemma 12.3.2,

$$d_{V(y_0)} \leq \mathcal{I}_{V(y_0)}(\tau\tilde{u}) \leq \liminf_{n\to\infty} \mathcal{J}_{\varepsilon_n}(u_n) = \liminf_{n\to\infty} c_{\varepsilon_n} \leq d_{V_0},$$

which implies that $V(y_0) \leq V(0) = V_0$. Since $V_0 = \min_{\overline{\Lambda}} V$, we deduce that $V(y_0) = V_0$. This fact together with (V_2) yields $y_0 \notin \partial\Lambda$. Consequently, $y_0 \in \Lambda$.

Claim 3 $\tilde{u}_n \to \tilde{u}$ in $H^s(\mathbb{R}^3)$ as $n \to \infty$.

Let

$$\tilde{\Lambda}_n = \frac{\Lambda - \varepsilon_n \, \tilde{y}_n}{\varepsilon_n}$$

and

$$\tilde{\chi}_n^1(x) = \begin{cases} 1, & \text{if } x \in \tilde{\Lambda}_n, \\ 0, & \text{if } x \in \mathbb{R}^3 \setminus \tilde{\Lambda}_n, \end{cases}$$

$$\tilde{\chi}_n^2(x) = 1 - \tilde{\chi}_n^1(x).$$

Let us also consider the following functions for $x \in \mathbb{R}^3$:

$$h_n^1(x) = \left(\frac{1}{2} - \frac{1}{\vartheta} \right) V(\varepsilon_n x + \varepsilon_n \, \tilde{y}_n) \tilde{u}_n^2(x) \tilde{\chi}_n^1(x),$$

$$h^1(x) = \left(\frac{1}{2} - \frac{1}{\vartheta} \right) V(y_0) \tilde{u}^2(x),$$

$$h_n^2(x) = \left[\left(\frac{1}{2} - \frac{1}{\vartheta} \right) V(\varepsilon_n x + \varepsilon_n \, \tilde{y}_n) \tilde{u}_n^2(x) + \frac{1}{\vartheta} g(\varepsilon_n x + \varepsilon_n \, \tilde{y}_n, \tilde{u}_n(x)) \tilde{u}_n(x) \right.$$

$$\left. - G(\varepsilon_n x + \varepsilon_n \, \tilde{y}_n, \tilde{u}_n(x)) \right] \tilde{\chi}_n^2(x)$$

$$\geq \left(\left(\frac{1}{2} - \frac{1}{\vartheta} \right) - \frac{1}{2K} \right) V(\varepsilon_n x + \varepsilon_n \, \tilde{y}_n) \tilde{u}_n^2(x) \tilde{\chi}_n^2(x),$$

$$h_n^3(x) = \left(\frac{1}{\vartheta} g(\varepsilon_n x + \varepsilon_n \, \tilde{y}_n, \tilde{u}_n(x)) \tilde{u}_n(x) - G(\varepsilon_n x + \varepsilon_n \, \tilde{y}_n, \tilde{u}_n(x)) \right) \tilde{\chi}_n^1(x)$$

$$= \left[\frac{1}{\vartheta} \left(f(\tilde{u}_n(x)) \tilde{u}_n(x) + (\tilde{u}_n(x))^{2_s^*} \right) - \left(F(\tilde{u}_n(x)) + \frac{1}{2_s^*} (\tilde{u}_n(x))^{2_s^*} \right) \right] \tilde{\chi}_n^1(x),$$

$$h^3(x) = \frac{1}{\vartheta} \left(f(\tilde{u}(x)) \tilde{u}(x) + (\tilde{u}(x))^{2_s^*} \right) - \left(F(\tilde{u}(x)) + \frac{1}{2_s^*} (\tilde{u}(x))^{2_s^*} \right).$$

In view of (f_3) and (g_3), all these functions are non-negative. Moreover, by (12.4.1) and Claim 2,

$$\tilde{u}_n(x) \to \tilde{u}(x) \quad \text{a.e. } x \in \mathbb{R}^3,$$

$$y_n = \varepsilon_n \, \tilde{y}_n \to y_0 \in \Lambda,$$

which implies that

$$\tilde{\chi}_n^1(x) \to 1, \; h_n^1(x) \to h^1(x), \; h_n^2(x) \to 0 \text{ and } h_n^3(x) \to h^3(x) \quad \text{a.e. } x \in \mathbb{R}^3.$$

Hence, applying Fatou's lemma and using the translation invariance of $\mathbb{R}^3$, we see that

$$
\begin{aligned}
d_{V_0} &\geq \limsup_{n \to \infty} c_{\varepsilon_n} = \limsup_{n \to \infty} \left(\mathcal{J}_{\varepsilon_n}(u_n) - \frac{1}{\vartheta} \langle \mathcal{J}'_{\varepsilon_n}(u_n), u_n \rangle \right) \\
&\geq \limsup_{n \to \infty} \left[a \left(\frac{1}{2} - \frac{1}{\vartheta} \right) [\tilde{u}_n]_s^2 + \left(\frac{1}{4} - \frac{1}{\vartheta} \right) b [\tilde{u}_n]_s^4 + \int_{\mathbb{R}^3} (h_n^1 + h_n^2 + h_n^3)\, dx \right] \\
&\geq \liminf_{n \to \infty} \left[a \left(\frac{1}{2} - \frac{1}{\vartheta} \right) [\tilde{u}_n]_s^2 + \left(\frac{1}{4} - \frac{1}{\vartheta} \right) b [\tilde{u}_n]_s^4 + \int_{\mathbb{R}^3} (h_n^1 + h_n^2 + h_n^3)\, dx \right] \\
&\geq a \left(\frac{1}{2} - \frac{1}{\vartheta} \right) [\tilde{u}]_s^2 + \left(\frac{1}{4} - \frac{1}{\vartheta} \right) b [\tilde{u}]_s^4 + \int_{\mathbb{R}^3} (h^1 + h^3)\, dx \geq d_{V_0}.
\end{aligned}
$$

Accordingly,

$$
\lim_{n \to \infty} [\tilde{u}_n]_s^2 = [\tilde{u}]_s^2 \tag{12.4.5}
$$

and

$$
h_n^1 \to h^1, \ h_n^2 \to 0 \ \text{and} \ h_n^3 \to h^3 \ \text{in} \ L^1(\mathbb{R}^3).
$$

Then

$$
\lim_{n \to \infty} \int_{\mathbb{R}^3} V(\varepsilon_n x + \varepsilon_n \tilde{y}_n) \tilde{u}_n^2 \, dx = \int_{\mathbb{R}^3} V(y_0) \tilde{u}^2 \, dx,
$$

and we deduce that

$$
\lim_{n \to \infty} \|\tilde{u}_n\|_{L^2(\mathbb{R}^3)}^2 = \|\tilde{u}\|_{L^2(\mathbb{R}^3)}^2. \tag{12.4.6}
$$

Putting together (12.4.1), (12.4.5) and (12.4.6) and using the fact that $H^s(\mathbb{R}^3)$ is a Hilbert space we conclude that

$$
\|\tilde{u}_n - \tilde{u}\|_{V_0} \to 0 \quad \text{as } n \to \infty,
$$

which completes the proof of lemma.   □

The next lemma provides a very useful L^∞-estimate for the solutions of the modified problem (12.2.3).

Lemma 12.4.2 *Let $(\tilde{u}_n)$ be the sequence given in Lemma 12.4.1. Then, $\tilde{u}_n \in L^\infty(\mathbb{R}^3)$ and there exists $C > 0$ such that*

$$\|\tilde{u}_n\|_{L^\infty(\mathbb{R}^3)} \leq C \quad \text{for all } n \in \mathbb{N}.$$

Moreover, $\tilde{u}_n(x) \to 0$ as $|x| \to \infty$, uniformly in $n \in \mathbb{N}$.

Proof Arguing as in the proof of Lemma 6.3.23 we can see that

$$a\left(\frac{1}{\beta}\right)^2 S_* \|\tilde{u}_n \tilde{u}_{L,n}^{\beta-1}\|^2_{L^{2_s^*}(\mathbb{R}^3)} + \int_{\mathbb{R}^3} V_n(x) \tilde{u}_n^2 \tilde{u}_{L,n}^{2(\beta-1)} \, dx$$

$$\leq (a + b[\tilde{u}_n]_s^2) \iint_{\mathbb{R}^6} \frac{(\tilde{u}_n(x) - \tilde{u}_n(y))}{|x - y|^{N+2s}} ((\tilde{u}_n \tilde{u}_{L,n}^{2(\beta-1)})(x) - (\tilde{u}_n \tilde{u}_{L,n}^{2(\beta-1)})(y)) \, dx dy$$

$$+ \int_{\mathbb{R}^3} V_n(x) \tilde{u}_n^2 \tilde{u}_{L,n}^{2(\beta-1)} \, dx$$

$$\leq \int_{\mathbb{R}^3} g_n(x, \tilde{u}_n) \tilde{u}_n \tilde{u}_{L,n}^{2(\beta-1)} \, dx, \tag{12.4.7}$$

where $V_n(x) = V(\varepsilon_n x + \varepsilon_n \tilde{y}_n)$ and $g_n(x, \tilde{u}_n) = g(\varepsilon_n x + \varepsilon_n \tilde{y}_n, \tilde{u}_n)$. By assumptions ($g_1$) and ($g_2$), for every $\xi > 0$ there exists $C_\xi > 0$ such that

$$|g_n(x, \tilde{u}_n)| \leq \xi |\tilde{u}_n| + C_\xi |\tilde{u}_n|^{2_s^*-1}. \tag{12.4.8}$$

Taking $\xi \in (0, V_1)$, and using (12.4.8) and (12.4.7) we get

$$\|\tilde{u}_n \tilde{u}_{L,n}^{\beta-1}\|^2_{L^{2_s^*}(\mathbb{R}^3)} \leq C\beta^2 \int_{\mathbb{R}^3} |\tilde{u}_n|^{2_s^*} \tilde{u}_{L,n}^{2(\beta-1)} \, dx.$$

Then we can proceed as in the proof of Lemma 11.5.1 to complete the proof (see also Remark 7.2.10 and note that $\tilde{u}_n$ is a subsolution to $(-\Delta)^s \tilde{u}_n + \frac{V_1}{bD} \tilde{u}_n = \frac{1}{a} g_n(x, \tilde{u}_n)$ in $\mathbb{R}^3$, where $D > 0$ is such that $a \leq a + b[\tilde{u}_n]_s^2 \leq D$ for all $n \in \mathbb{N}$). □

Now, we give the proof of Theorem 12.1.1.

Proof of Theorem 12.1.1 Firstly, we prove that there exists $\tilde{\varepsilon}_0 > 0$ such that for any $\varepsilon \in (0, \tilde{\varepsilon}_0)$ and any mountain pass solution $u_\varepsilon \in \mathcal{H}_\varepsilon$ of (12.2.3),

$$\|u_\varepsilon\|_{L^\infty(\mathbb{R}^3 \setminus \Lambda_\varepsilon)} < a_0. \tag{12.4.9}$$

Suppose, by contradiction, that for some subsequence (ε_n) so that $\varepsilon_n \to 0$, we can find $u_n = u_{\varepsilon_n} \in \mathcal{H}_{\varepsilon_n}$ such that $\mathcal{J}_{\varepsilon_n}(u_n) = c_{\varepsilon_n}$, $\mathcal{J}'_{\varepsilon_n}(u_n) = 0$ and

$$\|u_n\|_{L^\infty(\mathbb{R}^3 \setminus \Lambda_{\varepsilon_n})} \geq a_0. \tag{12.4.10}$$

By Lemma 12.4.1, there exists $(\tilde{y}_n) \subset \mathbb{R}^3$ such that $\tilde{u}_n = u_n(\cdot + \tilde{y}_n) \to \tilde{u}$ in $H^s(\mathbb{R}^3)$ and $\varepsilon_n \tilde{y}_n \to y_0$ for some $y_0 \in \Lambda$ such that $V(y_0) = V_0$. Now, if we choose $r > 0$ so that $B_r(y_0) \subset B_{2r}(y_0) \subset \Lambda$, we have $B_{\frac{r}{\varepsilon_n}}(\frac{y_0}{\varepsilon_n}) \subset \Lambda_{\varepsilon_n}$. Then, for any $y \in B_{\frac{r}{\varepsilon_n}}(\tilde{y}_n)$,

$$\left| y - \frac{y_0}{\varepsilon_n} \right| \leq |y - \tilde{y}_n| + \left| \tilde{y}_n - \frac{y_0}{\varepsilon_n} \right| < \frac{1}{\varepsilon_n}(r + o_n(1)) < \frac{2r}{\varepsilon_n} \quad \text{for } n \text{ sufficiently large.}$$

Hence, for these values of n, we get

$$\mathbb{R}^3 \setminus \Lambda_{\varepsilon_n} \subset \mathbb{R}^3 \setminus B_{\frac{r}{\varepsilon_n}}(\tilde{y}_n). \tag{12.4.11}$$

By Lemma 12.4.2, we deduce that

$$\tilde{u}_n(x) \to 0 \quad \text{as } |x| \to \infty, \text{ uniformly in } n \in \mathbb{N}. \tag{12.4.12}$$

Therefore, we can find $R > 0$ such that

$$\tilde{u}_n(x) < a_0 \quad \text{for all } |x| \geq R, \ n \in \mathbb{N},$$

which yields $u_n(x) < a_0$ for any $x \in \mathbb{R}^3 \setminus B_R(\tilde{y}_n)$ and $n \in \mathbb{N}$.

On the other hand, there exists $\nu \in \mathbb{N}$ such that for any $n \geq \nu$ and $r/\varepsilon_n > R$, it holds

$$\mathbb{R}^3 \setminus \Lambda_{\varepsilon_n} \subset \mathbb{R}^3 \setminus B_{\frac{r}{\varepsilon_n}}(\tilde{y}_n) \subset \mathbb{R}^3 \setminus B_R(\tilde{y}_n),$$

which gives

$$u_n(x) < a_0 \quad \text{for all } x \in \mathbb{R}^3 \setminus \Lambda_{\varepsilon_n}.$$

contradicting (12.4.10). Thus, (12.4.9) is verified.

Now, let u_ε be a solution to (12.2.3). Since u_ε satisfies (12.4.9) for any $\varepsilon \in (0, \tilde{\varepsilon}_0)$, it follows from the definition of g that u_ε is a solution to (12.2.1), and then $\hat{u}_\varepsilon(x) = u(x/\varepsilon)$ is a solution to (12.1.1) for any $\varepsilon \in (0, \tilde{\varepsilon}_0)$.

Finally, we study the behavior of the maximum points of solutions to problem (12.2.1). Take $\varepsilon_n \to 0$ and consider a sequence $(u_n) \subset \mathcal{H}_{\varepsilon_n}$ of solutions to (12.2.1). We first notice that, by (g_1), there exists $\gamma \in (0, a_0)$ such that

$$g(\varepsilon_n x, t)t = f(t)t + t^{2^*_s} \leq \frac{V_1}{K} t^2 \quad \text{for any } x \in \mathbb{R}^3, \, 0 \leq t \leq \gamma. \tag{12.4.13}$$

The same argument as before shows that, for some $R > 0$,

$$\|u_n\|_{L^\infty(\mathbb{R}^3 \setminus B_R(\tilde{y}_n))} < \gamma. \tag{12.4.14}$$

Moreover, up to extracting a subsequence, we may assume that

$$\|u_n\|_{L^\infty(B_R(\tilde{y}_n))} \geq \gamma. \tag{12.4.15}$$

Indeed, if (12.4.15) does not hold, then (12.4.14) implies that $\|u_n\|_{L^\infty(\mathbb{R}^3)} < \gamma$, and so in view of $\langle \mathcal{J}'_{\varepsilon_n}(u_n), u_n \rangle = 0$ and (12.4.13),

$$\|u_n\|^2_{\varepsilon_n} \leq \|u_n\|^2_{\varepsilon_n} + b[u_n]^4_s = \int_{\mathbb{R}^3} g(\varepsilon_n x, u_n)u_n \, dx \leq \frac{V_1}{K} \int_{\mathbb{R}^3} u_n^2 \, dx.$$

This implies that $\|u_n\|_{\varepsilon_n} = 0$, a contradiction. Hence, (12.4.15) holds true.

Let $p_n \in \mathbb{R}^3$ be a global maximum point of u_n. In the light of (12.4.14) and (12.4.15), we deduce that $p_n \in B_R(\tilde{y}_n)$. Thus $p_n = \tilde{y}_n + q_n$ for some $q_n \in B_R$. Recalling that the solution to (12.1.1) is of the form $\hat{u}_n(x) = u_n(x/\varepsilon_n)$, we conclude that $\eta_n = \varepsilon_n \tilde{y}_n + \varepsilon_n q_n$ is a global maximum point of $\hat{u}_n$. Since $(q_n) \subset B_R$ is bounded and $\varepsilon_n \tilde{y}_n \to y_0$ with $V(y_0) = V_0$, it follows from the continuity of V that

$$\lim_{n \to \infty} V(\eta_n) = V(y_0) = V_0.$$

Next, we give a decay estimate for $\hat{u}_n$. Invoking Lemma 3.2.17, we know that there exists a positive function w such that

$$0 < w(x) \leq \frac{C}{1 + |x|^{3+2s}} \quad \text{for all } x \in \mathbb{R}^3 \tag{12.4.16}$$

and

$$(-\Delta)^s w + \frac{V_1}{2(a + bA_1^2)} w = 0 \quad \text{in } \mathbb{R}^3 \setminus \overline{B}_{R_1}, \tag{12.4.17}$$

for some suitable $R_1 > 0$, with $A_1 > 0$ such that

$$a \leq a + b[u_n]^2_s \leq a + bA_1^2 \quad \text{for all } n \in \mathbb{N}.$$

Using (g_1) and (12.4.12), we can find $R_2 > 0$ sufficiently large such that

$$
(-\Delta)^s \tilde{u}_n + \frac{V_1}{2(a + bA_1^2)} \tilde{u}_n \leq (-\Delta)^s \tilde{u}_n + \frac{V_1}{2(a + b[\tilde{u}_n]^2)} \tilde{u}_n
$$

$$
= \frac{1}{a + b[\tilde{u}_n]_s^2} \left[g_n(x, \tilde{u}_n) - \left(V_n - \frac{V_1}{2} \right) \tilde{u}_n \right]
$$

$$
\leq \frac{1}{a + b[\tilde{u}_n]_s^2} \left[g_n(x, \tilde{u}_n) - \frac{V_1}{2} \tilde{u}_n \right] \leq 0 \text{ in } \mathbb{R}^3 \setminus \overline{B}_{R_2}.
$$

$$(12.4.18)$$

Set $R_3 = \max\{R_1, R_2\} > 0$ and

$$
c = \min_{\overline{B}_{R_3}} w > 0 \quad \text{and} \quad \tilde{w}_n = (d + 1)w - c\tilde{u}_n, \tag{12.4.19}
$$

where $d = \sup_{n \in \mathbb{N}} \|\tilde{u}_n\|_{L^\infty(\mathbb{R}^3)} < \infty$. We claim that

$$
\tilde{w}_n \geq 0 \text{ in } \mathbb{R}^3. \tag{12.4.20}
$$

First observe that (12.4.17), (12.4.18) and (12.4.19) yield

$$
\tilde{w}_n \geq cd + w - cd > 0 \text{ in } \overline{B}_{R_3},
$$

$$
(-\Delta)^s \tilde{w}_n + \frac{V_1}{2(a + bA_1^2)} \tilde{w}_n \geq 0 \text{ in } \mathbb{R}^3 \setminus \overline{B}_{R_3}.
$$

Applying Lemma 1.3.8 we deduce that (12.4.20) is satisfied. Combining (12.4.16) and (12.4.20) we obtain

$$
0 < \tilde{u}_n(x) \leq \frac{\tilde{C}}{1 + |x|^{3+2s}} \quad \text{for all } x \in \mathbb{R}^3, \ n \in \mathbb{N}, \tag{12.4.21}
$$

for some constant $\tilde{C} > 0$. Since $\hat{u}_n(x) = u_n(\frac{x}{\varepsilon_n}) = \tilde{u}_n(\frac{x}{\varepsilon_n} - \tilde{y}_n)$ and $\eta_n = \varepsilon_n \tilde{y}_n + \varepsilon_n q_n$, we can use (12.4.21) to deduce that

$$
0 < \hat{u}_n(x) = u_n \left(\frac{x}{\varepsilon_n} \right) = \tilde{u}_n \left(\frac{x}{\varepsilon_n} - \tilde{y}_n \right)
$$

$$
\leq \frac{\tilde{C}}{1 + |\frac{x}{\varepsilon_n} - \tilde{y}_n|^{3+2s}}
$$

$$= \frac{\tilde{C} \, \varepsilon_n^{3+2s}}{\varepsilon_n^{3+2s} + |x - \varepsilon_n \, \tilde{y}_n|^{3+2s}}$$

$$\leq \frac{\tilde{C} \, \varepsilon_n^{3+2s}}{\varepsilon_n^{3+2s} + |x - \eta_n|^{3+2s}} \qquad \text{for all } x \in \mathbb{R}^3.$$

This ends the proof of Theorem 12.1.1.   □

Remark 12.4.3 In [62] we extended Theorem 12.1.1 for more general nonlinearities with critical growth.

Multiplicity and Concentration Results for a Fractional Schrödinger-Poisson System with Critical Growth

13

13.1 Introduction

In this chapter we focus our attention on the multiplicity and concentration of positive solutions for the following critical fractional nonlinear Schrödinger-Poisson system:

$$\begin{cases} \varepsilon^{2s}(-\Delta)^s u + V(x)u + \phi u = f(u) + |u|^{2^*_s - 2}u & \text{in } \mathbb{R}^3, \\ \varepsilon^{2t}(-\Delta)^t \phi = u^2 & \text{in } \mathbb{R}^3, \\ u \in H^s(\mathbb{R}^3), \ u > 0 \text{ in } \mathbb{R}^3, \end{cases} \tag{13.1.1}$$

where $\varepsilon > 0$ is a small parameter, $s \in (\frac{3}{4}, 1), t \in (0, 1), 2^*_s = \frac{6}{3-2s}$ is the fractional critical Sobolev exponent. Here, the potential $V : \mathbb{R}^3 \to \mathbb{R}$ is a continuous function satisfying the following del Pino-Felmer hypotheses [165]:

(V_1) there exists $V_0 > 0$ such that $V_0 = \inf_{x \in \mathbb{R}^3} V(x)$;
(V_2) there exists a bounded open set $\Lambda \subset \mathbb{R}^3$ such that

$$V_0 < \min_{\partial \Lambda} V \quad \text{and} \quad M = \{x \in \Lambda : V(x) = V_0\} \neq \emptyset.$$

Without loss of generality, we may assume that $0 \in M$.

The nonlinearity $f : \mathbb{R} \to \mathbb{R}$ is assumed to be a continuous function such that $f(t) = 0$ for $t \leq 0$ and to satisfy the following conditions:

(f_1) $f(t) = o(t^3)$ as $t \to 0$;
(f_2) there exist $q, \sigma \in (4, 2^*_s), C_0 > 0$ such that

$$f(t) \geq C_0 t^{q-1} \quad \forall t > 0, \quad \lim_{t \to \infty} \frac{f(t)}{t^{\sigma-1}} = 0;$$

V. Ambrosio, *Nonlinear Fractional Schrödinger Equations in $\mathbb{R}^N$*,
Frontiers in Mathematics, https://doi.org/10.1007/978-3-030-60220-8_13

(f_3)   there exists $\vartheta \in (4, \sigma)$ such that $0 < \vartheta F(t) \leq tf(t)$ for all $t > 0$;

(f_4)   the function $t \mapsto \dfrac{f(t)}{t^3}$ is increasing in $(0, \infty)$.

We note that when $\phi = 0$, then (13.1.1) reduces to a fractional Schrödinger equation of the type

$$\varepsilon^{2s}(-\Delta)^s u + V(x)u = h(x, u) \quad \text{in } \mathbb{R}^3, \tag{13.1.2}$$

which was investigated in depth in Chaps. 6–8.

It $s = t = 1$, then (13.1.1) becomes the classical Schrödinger-Poisson system

$$\begin{cases} -\varepsilon^2 \Delta u + V(x)u + \mu\phi u = g(u) & \text{in } \mathbb{R}^3, \\ -\varepsilon^2 \Delta \phi = u^2 & \text{in } \mathbb{R}^3, \end{cases} \tag{13.1.3}$$

which describes systems of identical charged particles interacting with each other in the case that magnetic field effects can be ignored, and its solution represents, in particular, a standing wave for such a system. For a more detailed physical description of this system we refer to [96]. Concerning some classical existence and multiplicity results for Schrödinger-Poisson systems we refer to [82, 212, 213, 303, 331, 348]. For instance, Ruiz [303] obtained existence and nonexistence results to (13.1.3) when $g(u) = u^p$, $p \in (1, 5)$ and $\mu > 0$. Azzollini et al. [82] investigated the existence of nontrivial solutions when g satisfies Berestycki-Lions type assumptions. Wang et al. [331] considered the existence and concentration of positive solutions to (13.1.3) involving subcritical nonlinearities. He and Li [213] obtained an existence result for a critical Schrödinger-Poisson system, assuming that the potential V satisfies the conditions (V_1)-(V_2).

By contrast, only few results for fractional Schrödinger-Poisson systems are available in literature. Giammetta [204] studied the local and global well-posedness of a fractional Schrödinger-Poisson system in which the fractional diffusion term appears only in the second equation in (13.1.1). Teng [325] analyzed the existence of ground state solutions for (13.1.1) with critical Sobolev exponent, by combining the method of the Pohozaev-Nehari manifold, arguments of Brezis-Nirenberg type [114], the monotonicity trick and a global compactness lemma. In [342] Zhang et al. used a perturbation approach to prove the existence of positive solutions to (13.1.1) when $V(x) = \mu > 0$ and g is a general nonlinearity with subcritical or critical growth. They also investigated the asymptotic behavior of solutions as $\mu \to 0$. Liu and Zhang [264] studied the multiplicity and concentration of solutions to (13.1.1) when the potential V satisfies the global condition due to Rabinowitz [299]. Murcia and Siciliano [281] showed that, for suitably small ε, the number of positive solutions to (13.1.1) is estimated below by the Lusternik–Schnirelman category of the set of minima of the potential.

In this chapter, we study the multiplicity and concentration of solutions to (13.1.1) under the local conditions (V_1)-(V_2) on the potential V and the conditions (f_1)–(f_4) for the nonlinearity f. Then we are able to prove the following result:

Theorem 13.1.1 ([48]) *Assume that (V_1)-(V_2) and (f_1)–(f_4) hold. Then, for any $\delta > 0$ such that*

$$M_\delta = \{x \in \mathbb{R}^3 : \mathrm{dist}(x, M) \leq \delta\} \subset \Lambda,$$

there exists $\varepsilon_\delta > 0$ such that, for any $\varepsilon \in (0, \varepsilon_\delta)$, problem (13.1.1) admits at least $\mathrm{cat}_{M_\delta}(M)$ positive solutions in $\mathcal{H}_\varepsilon \times \mathcal{D}^{t,2}(\mathbb{R}^3)$. Moreover, if $(u_\varepsilon, \phi_\varepsilon)$ denotes one of these solutions and $x_\varepsilon \in \mathbb{R}^3$ is a global maximum point of u_ε, then

$$\lim_{\varepsilon \to 0} V(x_\varepsilon) = V_0,$$

and there exists $C > 0$ such that

$$0 < u_\varepsilon(x) \leq \frac{C \varepsilon^{3+2s}}{\varepsilon^{3+2s} + |x - x_\varepsilon|^{3+2s}} \quad \text{for all } x \in \mathbb{R}^3.$$

The proof of Theorem 13.1.1 is obtained by applying a penalization argument [165], a concentration-compactness lemma and the generalized Nehari method [321, 322]. We emphasize that the presence of two fractional Laplacian operators and the critical Sobolev exponent make our task more complicated and intriguing compared to the ones in the previous chapters and a more careful analysis will be needed.

We conclude this introduction by giving some requisite preliminary results. Let $s, t \in (0, 1)$ be such that $4s + 2t \geq 3$. Using Theorem 1.1.8 we can see that

$$H^s(\mathbb{R}^3) \subset L^{\frac{12}{3+2t}}(\mathbb{R}^3). \tag{13.1.4}$$

For any $u \in H^s(\mathbb{R}^3)$, the linear functional $\mathcal{L}_u : \mathcal{D}^{t,2}(\mathbb{R}^3) \to \mathbb{R}$ given by

$$\mathcal{L}_u(v) = \int_{\mathbb{R}^3} u^2 v \, dx$$

is well defined and continuous thanks to the Hölder inequality and (13.1.4). Indeed,

$$|\mathcal{L}_u(v)| \leq \left(\int_{\mathbb{R}^3} |u|^{\frac{12}{3+2t}} \, dx \right)^{\frac{3+2t}{6}} \left(\int_{\mathbb{R}^3} |v|^{2_t^*} \, dx \right)^{\frac{1}{2_t^*}} \leq C \|u\|^2 \|v\|_{\mathcal{D}^{t,2}},$$

where

$$\|v\|_{\mathcal{D}^{t,2}}^2 = \iint_{\mathbb{R}^6} \frac{|v(x) - v(y)|^2}{|x - y|^{3+2t}} \, dx dy.$$

Then, by the Lax-Milgram Theorem, there exists exactly one $\phi_u^t \in \mathcal{D}^{t,2}(\mathbb{R}^3)$ such that

$$(-\Delta)^t \phi_u^t = u^2 \text{ in } \mathbb{R}^3.$$

Therefore, the following t-Riesz formula holds:

$$\phi_u^t(x) = c_t \int_{\mathbb{R}^3} \frac{u^2(y)}{|x-y|^{3-2t}}\, dy \quad (x \in \mathbb{R}^3), \quad c_t = \pi^{-\frac{3}{2}} 2^{-2t} \frac{\Gamma(\frac{3-2t}{2})}{\Gamma(t)}. \tag{13.1.5}$$

In the sequel, we will omit the constant c_t in order to lighten the notation. Finally, we collect some useful properties of ϕ_u^t which will be used later.

Lemma 13.1.2 *If $s, t \in (0,1)$ and $4s + 2t \geq 3$, then, for all $u \in H^s(\mathbb{R}^3)$ we have:*

(1) $\|\phi_u^t\|_{\mathcal{D}^{t,2}} \leq C\|u\|_{L^{\frac{12}{3+2t}}(\mathbb{R}^3)}^2 \leq C\|u\|_{H^s(\mathbb{R}^3)}^2$ *and* $\int_{\mathbb{R}^3} \phi_u^t u^2 dx \leq C_t \|u\|_{L^{\frac{12}{3+2t}}(\mathbb{R}^3)}^4$.
   Moreover, the mapping $u \in H^s(\mathbb{R}^3) \mapsto \phi_u^t \in \mathcal{D}^{t,2}(\mathbb{R}^3)$ is continuous and sends bounded sets into bounded sets;
(2) $\phi_u^t \geq 0$ *in $\mathbb{R}^3$, and ϕ_u^t is radial if u is radial;*
(3) *if $y \in \mathbb{R}^3$ and $\bar{u}(x) = u(x+y)$, then $\phi_{\bar{u}}^t(x) = \phi_u^t(x+y)$ and $\int_{\mathbb{R}^3} \phi_{\bar{u}}^t \bar{u}^2 dx = \int_{\mathbb{R}^3} \phi_u^t u^2 dx$;*
(4) $\phi_{ru}^t = r^2 \phi_u^t$ *for all $r \in \mathbb{R}$, $\phi_{u_\theta}^t(x) = \theta^{2s} \phi_u^t(\frac{x}{\theta})$ for any $\theta > 0$, where $u_\theta(x) = u(\frac{x}{\theta})$;*
(5) *if $u_n \rightharpoonup u$ in $H^s(\mathbb{R}^3)$, then $\phi_{u_n}^t \rightharpoonup \phi_u^t$ in $\mathcal{D}^{t,2}(\mathbb{R}^3)$;*
(6) *if $u_n \rightharpoonup u$ in $H^s(\mathbb{R}^3)$, then*

$$\int_{\mathbb{R}^3} \phi_{u_n}^t u_n^2 dx = \int_{\mathbb{R}^3} \phi_{(u_n-u)}^t (u_n-u)^2 dx + \int_{\mathbb{R}^3} \phi_u^t u^2 dx + o_n(1);$$

(7) *if $u_n \to u$ in $H^s(\mathbb{R}^3)$, then $\phi_{u_n}^t \to \phi_u^t$ in $\mathcal{D}^{t,2}(\mathbb{R}^3)$ and $\int_{\mathbb{R}^3} \phi_{u_n}^t u_n^2 dx \to \int_{\mathbb{R}^3} \phi_u^t u^2 dx$.*
(8) *Let $4s + 2t > 3$. If $u_n \rightharpoonup u$ in $H^s(\mathbb{R}^3)$ and $u_n \to u$ in $L^{\frac{12}{3+2t}}(\mathbb{R}^3)$, then $\int_{\mathbb{R}^3} \phi_{u_n}^t u_n v\, dx \to \int_{\mathbb{R}^3} \phi_u^t u v\, dx$ for all $v \in H^s(\mathbb{R}^3)$ and $\int_{\mathbb{R}^3} \phi_{u_n}^t u_n^2 dx \to \int_{\mathbb{R}^3} \phi_u^t u^2 dx$.*

Proof The proofs of properties (1)–(7) are similar to those in [303, 348], so we omit them (see [264, 325, 342] for details). Here we only prove (8). Note that $4s + 2t > 3$ and $t \in (0,1)$ imply that $2 < \frac{12}{3+2t} < \frac{6}{3-2s}$. Take $v \in H^s(\mathbb{R}^3)$. From (1) and $u_n \to u$ in $L^{\frac{12}{3+2t}}(\mathbb{R}^3)$, we see that $\phi_{u_n}^t \to \phi_u^t$ in $\mathcal{D}^{t,2}(\mathbb{R}^3)$. Then, by using the Hölder inequality, the embedding $\mathcal{D}^{t,2}(\mathbb{R}^3) \subset L^{\frac{6}{3-2t}}(\mathbb{R}^3)$ and the boundedness of (u_n) in $H^s(\mathbb{R}^3)$, we see that

$$\left| \int_{\mathbb{R}^3} \phi_{u_n}^t u_n v\, dx - \int_{\mathbb{R}^3} \phi_u^t u v\, dx \right| = \left| \int_{\mathbb{R}^3} (\phi_{u_n}^t - \phi_u^t) u_n v\, dx + \int_{\mathbb{R}^3} \phi_u^t (u_n - u) v\, dx \right|$$

$$\leq \|\phi_{u_n}^t - \phi_u^t\|_{L^{\frac{6}{3-2t}}(\mathbb{R}^3)} \|u_n\|_{L^{\frac{12}{3+2t}}(\mathbb{R}^3)} \|v\|_{L^{\frac{12}{3+2t}}(\mathbb{R}^3)}$$

$$+ \|\phi_u^t\|_{L^{\frac{6}{3-2t}}(\mathbb{R}^3)} \|u_n - u\|_{L^{\frac{12}{3+2t}}(\mathbb{R}^3)} \|v\|_{L^{\frac{12}{3+2t}}(\mathbb{R}^3)}$$

$$\leq C\|\phi_{u_n}^t - \phi_u^t\|_{\mathcal{D}^{t,2}} + C\|u_n - u\|_{L^{\frac{12}{3+2t}}(\mathbb{R}^3)} \to 0.$$

In much the same way,

$$\left| \int_{\mathbb{R}^3} \phi_{u_n}^t u_n^2 \, dx - \int_{\mathbb{R}^3} \phi_u^t u^2 \, dx \right| = \left| \int_{\mathbb{R}^3} (\phi_{u_n}^t - \phi_u^t) u_n^2 \, dx + \int_{\mathbb{R}^3} \phi_u^t (u_n - u)(u_n + u) \, dx \right|$$

$$\leq \|\phi_{u_n}^t - \phi_u^t\|_{L^{\frac{6}{3-2t}}(\mathbb{R}^3)} \|u_n\|_{L^{\frac{12}{3+2t}}(\mathbb{R}^3)}^2$$

$$+ \|\phi_u^t\|_{L^{\frac{6}{3-2t}}(\mathbb{R}^3)} \|u_n - u\|_{L^{\frac{12}{3+2t}}(\mathbb{R}^3)} \|u_n + u\|_{L^{\frac{12}{3+2t}}(\mathbb{R}^3)}$$

$$\leq C\|\phi_{u_n}^t - \phi_u^t\|_{\mathcal{D}^{t,2}} + C\|u_n - u\|_{L^{\frac{12}{3+2t}}(\mathbb{R}^3)} \to 0.$$

$\square$

Remark 13.1.3 As in [303], we can see that if $4s + 2t > 3$ and $u_n \rightharpoonup u$ in $H_{\mathrm{rad}}^s(\mathbb{R}^3)$, then $\phi_{u_n}^t \to \phi_u^t$ in $\mathcal{D}^{t,2}(\mathbb{R}^3)$ and $\int_{\mathbb{R}^3} \phi_{u_n}^t u_n^2 \, dx \to \int_{\mathbb{R}^3} \phi_u^t u^2 \, dx$.

13.2 Functional Setting

In order to study (13.1.1), we use the change of variable $x \mapsto \varepsilon x$ and we look for solutions to

$$\begin{cases} (-\Delta)^s u + V(\varepsilon x)u + \phi_u^t u = f(u) + |u|^{2_s^* - 2} u \text{ in } \mathbb{R}^3, \\ u \in H^s(\mathbb{R}^3), \quad u > 0 \text{ in } \mathbb{R}^3, \end{cases} \tag{13.2.1}$$

where ϕ_u^t is given by (13.1.5). In what follows we introduce a useful penalization function [165].

Let $K > 2$ and $a > 0$ such that $f(a) + a^{2_s^* - 1} = \frac{V_0}{K} a$, and define

$$\tilde{f}(t) = \begin{cases} f(t) + (t^+)^{2_s^* - 1}, & \text{if } t \leq a, \\ \frac{V_0}{K} t, & \text{if } t > a, \end{cases}$$

and

$$g(x, t) = \chi_\Lambda(x)(f(t) + (t^+)^{2_s^* - 1}) + (1 - \chi_\Lambda(x))\tilde{f}(t).$$

It is easy to check that g satisfies the following properties:

(g_1) $\lim_{t \to 0} \frac{g(x,t)}{t^3} = 0$ uniformly with respect to $x \in \mathbb{R}^3$;

(g_2) $g(x,t) \leq f(t) + t^{2^*_s - 1}$ for all $x \in \mathbb{R}^3$, $t > 0$;

(g_3) (i) $0 < \vartheta G(x,t) \leq g(x,t)t$ for all $x \in \Lambda$ and $t > 0$,
   (ii) $0 \leq 2G(x,t) \leq g(x,t)t \leq \frac{V_0}{K}t^2$ for all $x \in \mathbb{R}^3 \setminus \Lambda$ and $t > 0$;

(g_4) for each fixed $x \in \Lambda$ the function $\frac{g(x,t)}{t^3}$ is increasing in $(0, \infty)$, and for each fixed
   $x \in \mathbb{R}^3 \setminus \Lambda$ the function $\frac{g(x,t)}{t^3}$ is increasing in $(0, a)$.

Let us consider the following modified problem:

$$\begin{cases} (-\Delta)^s u + V(\varepsilon x)u + \phi^t_u u = g(\varepsilon x, u) \text{ in } \mathbb{R}^3, \\ u \in H^s(\mathbb{R}^3), \quad u > 0 \text{ in } \mathbb{R}^3. \end{cases} \tag{13.2.2}$$

It is clear that weak solutions to (13.2.2) are critical points of the following functional

$$\mathcal{J}_\varepsilon(u) = \frac{1}{2}\|u\|^2_\varepsilon + \frac{1}{4}\int_{\mathbb{R}^3} \phi^t_u u^2 \, dx - \int_{\mathbb{R}^3} G(\varepsilon x, u) \, dx,$$

defined for all $u \in \mathcal{H}_\varepsilon$, where

$$\mathcal{H}_\varepsilon = \left\{ u \in H^s(\mathbb{R}^3) : \int_{\mathbb{R}^3} V(\varepsilon x)u^2 \, dx < \infty \right\}$$

is endowed with the norm

$$\|u\|_\varepsilon = \left([u]^2_s + \int_{\mathbb{R}^3} V(\varepsilon x)u^2 \, dx \right)^{\frac{1}{2}}.$$

Obviously, $\mathcal{H}_\varepsilon$ is a Hilbert space with the inner product

$$\langle u, v \rangle_\varepsilon = \langle u, v \rangle_{\mathcal{D}^{s,2}(\mathbb{R}^3)} + \int_{\mathbb{R}^3} V(\varepsilon x)uv \, dx.$$

We also note that $\mathcal{J}_\varepsilon \in C^1(\mathcal{H}_\varepsilon, \mathbb{R})$ and its differential is given by

$$\langle \mathcal{J}'_\varepsilon(u), v \rangle = \langle u, v \rangle_\varepsilon + \int_{\mathbb{R}^3} \phi^t_u uv \, dx - \int_{\mathbb{R}^3} g(\varepsilon x, u)v \, dx \quad \forall u, v \in \mathcal{H}_\varepsilon.$$

Let us introduce the Nehari manifold associated with (13.2.2), that is,

$$\mathcal{N}_\varepsilon = \{ u \in \mathcal{H}_\varepsilon \setminus \{0\} : \langle \mathcal{J}'_\varepsilon(u), u \rangle = 0 \},$$

and denote

$$\mathcal{H}_\varepsilon^+ = \{u \in \mathcal{H}_\varepsilon : |\mathrm{supp}(u^+) \cap \Lambda_\varepsilon| > 0\}$$

and $\mathbb{S}_\varepsilon^+ = \mathbb{S}_\varepsilon \cap \mathcal{H}_\varepsilon^+$, where $\mathbb{S}_\varepsilon$ is the unit sphere in $\mathcal{H}_\varepsilon$. Then $\mathcal{H}_\varepsilon = T_u \mathbb{S}_\varepsilon^+ \oplus \mathbb{R}u$ and $T_u \mathbb{S}_\varepsilon^+ = \{v \in \mathcal{H}_\varepsilon : \langle u, v\rangle_\varepsilon = 0\}$.

Lemma 13.2.1 *The functional $\mathcal{J}_\varepsilon$ has a mountain pass geometry:*

(a) *there exist $\alpha, \rho > 0$ such that $\mathcal{J}_\varepsilon(u) \geq \alpha$ with $\|u\|_\varepsilon = \rho$;*
(b) *there exists $e \in \mathcal{H}_\varepsilon$ such that $\|e\|_\varepsilon > \rho$ and $\mathcal{J}_\varepsilon(e) < 0$.*

Proof

(a) Indeed, by assumptions (g_1), (g_2), (f_2), for every $\xi > 0$ we can find $C_\xi > 0$ such that

$$\mathcal{J}_\varepsilon(u) \geq \frac{1}{2}\|u\|_\varepsilon^2 - \int_{\mathbb{R}^3} G(\varepsilon x, u)\, dx \geq \frac{1}{2}\|u\|_\varepsilon^2 - \xi C\|u\|_\varepsilon^4 - C_\xi C\|u\|_\varepsilon^{2_s^*}.$$

Then there exist $\alpha, \rho > 0$ such that $\mathcal{J}_\varepsilon(u) \geq \alpha$ with $\|u\|_\varepsilon = \rho$.
(b) In view of (g_3)-(i) and Lemma 13.1.2-(4), for any $u \in \mathcal{H}_\varepsilon^+$ and $\tau > 0$

$$\mathcal{J}_\varepsilon(\tau u) \leq \frac{\tau^2}{2}\|u\|_\varepsilon^2 + \frac{\tau^4}{4}\int_{\mathbb{R}^3} \phi_u^t u^2\, dx - \int_{\mathbb{R}^3} G(\varepsilon x, \tau u)\, dx$$

$$\leq \frac{\tau^2}{2}\|u\|_\varepsilon^2 + \frac{\tau^4}{4}\int_{\mathbb{R}^3} \phi_u^t u^2\, dx - \int_{\Lambda_\varepsilon} G(\varepsilon x, \tau u)\, dx$$

$$\leq \frac{\tau^2}{2}\|u\|_\varepsilon^2 + \frac{\tau^4}{4}\int_{\mathbb{R}^3} \phi_u^t u^2\, dx - C_1 \tau^\vartheta \int_{\Lambda_\varepsilon} (u^+)^\vartheta\, dx + C_2|\mathrm{supp}(u^+) \cap \Lambda_\varepsilon|,$$

$$(13.2.3)$$

for some positive constants C_1 and C_2. Since $\vartheta \in (4, 2_s^*)$, we see that $\mathcal{J}_\varepsilon(\tau u) \to -\infty$ as $\tau \to \infty$. $\qquad \square$

Since f is only continuous, the next results will be crucial for overcoming the non-differentiability of $\mathcal{N}_\varepsilon$ and the incompleteness of $\mathbb{S}_\varepsilon^+$.

Lemma 13.2.2 *Assume that (V_1)-(V_2) and (f_1)–(f_4) hold true. Then,*

(i) *For each $u \in \mathcal{H}_\varepsilon^+$, let $h_u : \mathbb{R}_+ \to \mathbb{R}$ be defined by $h_u(t) = \mathcal{J}_\varepsilon(tu)$. Then, there is a unique $t_u > 0$ such that*

$$h_u'(t) > 0 \quad in \ (0, t_u),$$
$$h_u'(t) < 0 \quad in \ (t_u, \infty).$$

(ii) *There exists $\tau > 0$, independent of u, such that $t_u \geq \tau$ for all $u \in \mathbb{S}_\varepsilon^+$. Moreover, for each compact set $\mathbb{K} \subset \mathbb{S}_\varepsilon^+$ there is a positive constant $C_\mathbb{K}$ such that $t_u \leq C_\mathbb{K}$ for all $u \in \mathbb{K}$.*

(iii) *The map $\hat{m}_\varepsilon : \mathcal{H}_\varepsilon^+ \to \mathcal{N}_\varepsilon$ given by $\hat{m}_\varepsilon(u) = t_u u$ is continuous and $m_\varepsilon = \hat{m}_\varepsilon|_{\mathbb{S}_\varepsilon^+}$ is a homeomorphism between $\mathbb{S}_\varepsilon^+$ and $\mathcal{N}_\varepsilon$. Moreover, $m_\varepsilon^{-1}(u) = \frac{u}{\|u\|_\varepsilon}$.*

(iv) *If there is a sequence $(u_n) \subset \mathbb{S}_\varepsilon^+$ such that $\mathrm{dist}(u_n, \partial \mathbb{S}_\varepsilon^+) \to 0$, then $\|m_\varepsilon(u_n)\|_\varepsilon \to \infty$ and $\mathcal{J}_\varepsilon(m_\varepsilon(u_n)) \to \infty$.*

Proof (i) We note that $h_u \in C^1(\mathbb{R}_+, \mathbb{R})$, and in view of Lemma 13.2.1, we can see that $h_u(0) = 0$, $h_u(t) > 0$ for $t > 0$ small enough and $h_u(t) < 0$ for $t > 0$ sufficiently large. Then there exists $t_u > 0$ such that $h_u'(t_u) = 0$ and t_u is a global maximum for h_u. This implies that $t_u u \in \mathcal{N}_\varepsilon$. Let us show that there is exactly one t_u. Suppose, by contradiction, that there exist $t_1 > t_2 > 0$ such that $h_u'(t_1) = h_u'(t_2) = 0$, that is,

$$t_1 \|u\|_\varepsilon^2 + t_1^3 \int_{\mathbb{R}^3} \phi_u^t u^2 \, dx = \int_{\mathbb{R}^3} g(\varepsilon x, t_1 u) u \, dx \tag{13.2.4}$$

$$t_2 \|u\|_\varepsilon^2 + t_2^3 \int_{\mathbb{R}^3} \phi_u^t u^2 \, dx = \int_{\mathbb{R}^3} g(\varepsilon x, t_2 u) u \, dx. \tag{13.2.5}$$

Using (13.2.4), (13.2.5) and (g_4) we see that

$$\|u\|_\varepsilon^2 \left(\frac{1}{t_1^2} - \frac{1}{t_2^2} \right) = \int_{\mathbb{R}^3} \left[\frac{g(\varepsilon x, t_1 u)}{(t_1 u)^3} - \frac{g(\varepsilon x, t_2 u)}{(t_2 u)^3} \right] u^4 dx$$

$$= \int_{\mathbb{R}^3 \setminus \Lambda_\varepsilon} \left[\frac{g(\varepsilon x, t_1 u)}{(t_1 u)^3} - \frac{g(\varepsilon x, t_2 u)}{(t_2 u)^3} \right] u^4 dx$$

$$+ \int_{\Lambda_\varepsilon} \left[\frac{g(\varepsilon x, t_1 u)}{(t_1 u)^3} - \frac{g(\varepsilon x, t_2 u)}{(t_2 u)^3} \right] u^4 dx$$

$$\geq \int_{\mathbb{R}^3 \setminus \Lambda_\varepsilon} \left[\frac{g(\varepsilon x, t_1 u)}{(t_1 u)^3} - \frac{g(\varepsilon x, t_2 u)}{(t_2 u)^3} \right] u^4 dx$$

$$= \int_{(\mathbb{R}^3 \setminus \Lambda_\varepsilon) \cap \{t_2 u > a\}} \left[\frac{g(\varepsilon x, t_1 u)}{(t_1 u)^3} - \frac{g(\varepsilon x, t_2 u)}{(t_2 u)^3} \right] u^4 dx$$

$$+ \int_{(\mathbb{R}^3 \setminus \Lambda_\varepsilon) \cap \{t_2 u \le a < t_1 u\}} \left[\frac{g(\varepsilon x, t_1 u)}{(t_1 u)^3} - \frac{g(\varepsilon x, t_2 u)}{(t_2 u)^3} \right] u^4 \, dx$$

$$+ \int_{(\mathbb{R}^3 \setminus \Lambda_\varepsilon) \cap \{t_1 u < a\}} \left[\frac{g(\varepsilon x, t_1 u)}{(t_1 u)^3} - \frac{g(\varepsilon x, t_2 u)}{(t_2 u)^3} \right] u^4 \, dx = I + II + III.$$

First, note that $III \ge 0$ because (g_4) holds and $t_1 > t_2$. In view of the definition of g, we have

$$I \ge \int_{(\mathbb{R}^3 \setminus \Lambda_\varepsilon) \cap \{t_2 u > a\}} \left[\frac{V_0}{K} \frac{1}{(t_1 u)^2} - \frac{V_0}{K} \frac{1}{(t_2 u)^2} \right] u^4 \, dx$$

$$= \frac{1}{K} \left(\frac{1}{t_1^2} - \frac{1}{t_2^2} \right) \int_{(\mathbb{R}^3 \setminus \Lambda_\varepsilon) \cap \{t_2 u > a\}} V_0 u^2 \, dx.$$

Concerning II, using again the definition of g, we get

$$II \ge \int_{(\mathbb{R}^3 \setminus \Lambda_\varepsilon) \cap \{t_2 u \le a < t_1 u\}} \left[\frac{V_0}{K} \frac{1}{(t_1 u)^2} - \frac{f(t_2 u) + (t_2 u^+)^{2_s^* - 1}}{(t_2 u)^3} \right] u^4 \, dx.$$

Therefore,

$$\|u\|_\varepsilon^2 \left(\frac{1}{t_1^2} - \frac{1}{t_2^2} \right) \ge \frac{1}{K} \left(\frac{1}{t_1^2} - \frac{1}{t_2^2} \right) \int_{(\mathbb{R}^3 \setminus \Lambda_\varepsilon) \cap \{t_2 u > a\}} V_0 u^2 \, dx$$

$$+ \int_{(\mathbb{R}^3 \setminus \Lambda_\varepsilon) \cap \{t_2 u \le a < t_1 u\}} \left[\frac{V_0}{K} \frac{1}{(t_1 u)^2} - \frac{f(t_2 u) + (t_2 u^+)^{2_s^* - 1}}{(t_2 u)^3} \right] u^4 \, dx.$$

Multiplying both sides by $\frac{t_1^2 t_2^2}{t_2^2 - t_1^2} < 0$ and recalling that $\frac{f(a)}{a} + a^{2_s^* - 2} = \frac{V_0}{K}$, we obtain

$$\|u\|_\varepsilon^2 \le \frac{1}{K} \int_{(\mathbb{R}^3 \setminus \Lambda_\varepsilon) \cap \{t_2 u > a\}} V_0 u^2 \, dx$$

$$+ \frac{t_1^2 t_2^2}{t_2^2 - t_1^2} \int_{(\mathbb{R}^3 \setminus \Lambda_\varepsilon) \cap \{t_2 u \le a < t_1 u\}} \left[\frac{V_0}{K} \frac{1}{(t_1 u)^2} - \frac{f(t_2 u) + (t_2 u^+)^{2_s^* - 1}}{(t_2 u)^3} \right] u^4 \, dx$$

$$\le \frac{1}{K} \int_{(\mathbb{R}^3 \setminus \Lambda_\varepsilon) \cap \{t_2 u > a\}} V_0 u^2 \, dx - \frac{t_2^2}{t_1^2 - t_2^2} \int_{(\mathbb{R}^3 \setminus \Lambda_\varepsilon) \cap \{t_2 u \le a < t_1 u\}} \frac{V_0}{K} u^2 \, dx$$

$$+ \frac{t_1^2}{t_1^2 - t_2^2} \int_{(\mathbb{R}^3 \setminus \Lambda_\varepsilon) \cap \{t_2 u \le a < t_1 u\}} \frac{f(t_2 u) + (t_2 u^+)^{2_s^* - 1}}{t_2 u} u^2 \, dx$$

$$\le \frac{1}{K} \int_{\mathbb{R}^3 \setminus \Lambda_\varepsilon} V_0 u^2 \, dx \le \frac{1}{K} \|u\|_\varepsilon^2.$$

Since $u \neq 0$ and $K > 2$, we reached a contradiction.

(ii) Let $u \in \mathbb{S}_\varepsilon^+$. By (i), there exists $t_u > 0$ such that $h_u'(t_u) = 0$, or equivalently

$$t_u + t_u^3 \int_{\mathbb{R}^3} \phi_u^t u^2 \, dx = \int_{\mathbb{R}^3} g(\varepsilon x, t_u u) u \, dx. \tag{13.2.6}$$

In the light of (g_1) and (g_2), given $\xi > 0$ there exists a positive constant C_ξ such that

$$|g(x,t)| \le \xi |t|^3 + C_\xi |t|^{2_s^* - 1}, \quad \text{for all } (x,t) \in \mathbb{R}^3 \times \mathbb{R}.$$

From (13.2.6) and applying Theorem 1.1.8 we have that

$$t_u \le \xi t_u^3 C_1 + C_\xi t_u^{2_s^* - 1} C_2,$$

which implies that there exists $\tau > 0$, independent of u, such that $t_u \ge \tau$. Now, let $\mathbb{K} \subset \mathbb{S}_\varepsilon^+$ be a compact set. Let us show that t_u can be estimated from above by a constant depending on $\mathbb{K}$. Assume, by contradiction, that there exists a sequence $(u_n) \subset \mathbb{K}$ such that $t_n = t_{u_n} \to \infty$. Therefore, there exists $u \in \mathbb{K}$ such that $u_n \to u$ in $\mathcal{H}_\varepsilon$. In view of (13.2.3),

$$\mathcal{J}_\varepsilon(t_n u_n) \to -\infty. \tag{13.2.7}$$

Fix $v \in \mathcal{N}_\varepsilon$. Then, using the fact that $\langle \mathcal{J}_\varepsilon'(v), v \rangle = 0$, and assumptions (g_3)-(i) and (g_3)-(ii), we can infer that

$$\mathcal{J}_\varepsilon(v) = \mathcal{J}_\varepsilon(v) - \frac{1}{\vartheta} \langle \mathcal{J}_\varepsilon'(v), v \rangle$$

$$= \left(\frac{\vartheta - 2}{2\vartheta} \right) \|v\|_\varepsilon^2 + \left(\frac{\vartheta - 4}{4\vartheta} \right) \int_{\mathbb{R}^3} \phi_v^t v^2 \, dx + \frac{1}{\vartheta} \int_{\mathbb{R}^3 \setminus \Lambda_\varepsilon} [g(\varepsilon x, v)v - \vartheta G(\varepsilon x, v)] \, dx$$

$$+ \frac{1}{\vartheta} \int_{\Lambda_\varepsilon} [g(\varepsilon x, v)v - \vartheta G(\varepsilon x, v)] \, dx$$

$$\ge \left(\frac{\vartheta - 2}{2\vartheta} \right) \|v\|_\varepsilon^2 + \frac{1}{\vartheta} \int_{\mathbb{R}^3 \setminus \Lambda_\varepsilon} [g(\varepsilon x, v)v - \vartheta G(\varepsilon x, v)] \, dx$$

$$\ge \left(\frac{\vartheta - 2}{2\vartheta} \right) \|v\|_\varepsilon^2 - \left(\frac{\vartheta - 2}{2\vartheta} \right) \frac{1}{K} \int_{\mathbb{R}^3 \setminus \Lambda_\varepsilon} V(\varepsilon x) v^2 \, dx$$

$$\ge \left(\frac{\vartheta - 2}{2\vartheta} \right) \left(1 - \frac{1}{K} \right) \|v\|_\varepsilon^2. \tag{13.2.8}$$

Since $(t_{u_n} u_n) \subset \mathcal{N}_\varepsilon$ and $K > 2$, (13.2.8) implies that (13.2.7) does not hold, a contradiction.

(iii) First of all, we observe that $\hat{m}_\varepsilon$, m_ε and m_ε^{-1} are well defined. In fact, by (i), for each $u \in \mathcal{H}_\varepsilon^+$ there exists a unique $\hat{m}_\varepsilon(u) \in \mathcal{N}_\varepsilon$. On the other hand, if $u \in \mathcal{N}_\varepsilon$ then $u \in \mathcal{H}_\varepsilon^+$. Otherwise, if $u \notin \mathcal{H}_\varepsilon^+$, then

$$|\text{supp}(u^+) \cap \Lambda_\varepsilon| = 0,$$

which together with $(g3)$-(ii) gives

$$\|u\|_\varepsilon^2 + \int_{\mathbb{R}^3} \phi_u^t u^2 \, dx = \int_{\mathbb{R}^3 \setminus \Lambda_\varepsilon} g(\varepsilon x, u^+) u^+ \, dx$$

$$\leq \frac{1}{K} \int_{\mathbb{R}^3 \setminus \Lambda_\varepsilon} V(\varepsilon x) u^2 \, dx \leq \frac{1}{K} \|u\|_\varepsilon^2. \tag{13.2.9}$$

Using that $\phi_u^t \geq 0$ and (13.2.9), we get

$$0 < \|u\|_\varepsilon^2 \leq \frac{1}{K} \|u\|_\varepsilon^2$$

which is not possible, because $K > 2$. Accordingly, $m_\varepsilon^{-1}(u) = \frac{u}{\|u\|_\varepsilon} \in \mathbb{S}_\varepsilon^+$, m_ε^{-1} is well defined and it is a continuous mapping. Now, take $u \in \mathbb{S}_\varepsilon^+$; then

$$m_\varepsilon^{-1}(m_\varepsilon(u)) = m_\varepsilon^{-1}(t_u u) = \frac{t_u u}{\|t_u u\|_\varepsilon} = \frac{u}{\|u\|_\varepsilon} = u,$$

which shows that m_ε is a bijection. Next, we show that $\hat{m}_\varepsilon$ is continuous. Let $(u_n) \subset \mathcal{H}_\varepsilon^+$ and $u \in \mathcal{H}_\varepsilon^+$ be such that $u_n \to u$ in $\mathcal{H}_\varepsilon$. Since $\hat{m}_\varepsilon(tu) = \hat{m}_\varepsilon(u)$ for all $t > 0$, we may assume that $\|u_n\|_\varepsilon = \|u\|_\varepsilon = 1$ for all $n \in \mathbb{N}$. Then, in view of (ii), we can find $t_0 > 0$ such that $t_n = t_{u_n} \to t_0$. Since $t_n u_n \in \mathcal{N}_\varepsilon$, we obtain

$$t_n^2 \|u_n\|_\varepsilon^2 + t_n^4 \int_{\mathbb{R}^3} \phi_{u_n}^t u_n^2 \, dx = \int_{\mathbb{R}^3} g(\varepsilon x, t_n u_n) t_n u_n \, dx,$$

and letting $n \to \infty$ we get

$$t_0^2 \|u\|_\varepsilon^2 + t_0^4 \int_{\mathbb{R}^3} \phi_u^t u^2 \, dx = \int_{\mathbb{R}^3} g(\varepsilon x, t_0 u) t_0 u \, dx,$$

which shows that $t_0 u \in \mathcal{N}_\varepsilon$ and $t_u = t_0$. Therefore,

$$\hat{m}_\varepsilon(u_n) \to \hat{m}_\varepsilon(u) \quad \text{in } \mathcal{H}_\varepsilon,$$

and $\hat{m}_\varepsilon$ and m_ε are continuous maps.

(iv) Let $(u_n) \subset \mathbb{S}_\varepsilon^+$ be such that $\mathrm{dist}(u_n, \partial\mathbb{S}_\varepsilon^+) \to 0$. Note that for each $p \in [2, 2_s^*]$ and $n \in \mathbb{N}$,

$$\|u_n^+\|_{L^p(\Lambda_\varepsilon)} \leq \inf_{v \in \partial\mathbb{S}_\varepsilon^+} \|u_n - v\|_{L^p(\Lambda_\varepsilon)}$$

$$\leq C_p \inf_{v \in \partial\mathbb{S}_\varepsilon^+} \|u_n - v\|_\varepsilon,$$

Hence, by (g_1), (g_2), and (g_3)-(ii), we can infer that for all $t > 0$

$$\int_{\mathbb{R}^3} G(\varepsilon x, t u_n)\, dx = \int_{\mathbb{R}^3 \setminus \Lambda_\varepsilon} G(\varepsilon x, t u_n)\, dx + \int_{\Lambda_\varepsilon} G(\varepsilon x, t u_n)\, dx$$

$$\leq \frac{t^2}{K} \int_{\mathbb{R}^3 \setminus \Lambda_\varepsilon} V(\varepsilon x) u_n^2\, dx + \int_{\Lambda_\varepsilon} F(t u_n) + \frac{t^{2_s^*}}{2_s^*}(u_n^+)^{2_s^*}\, dx$$

$$\leq \frac{t^2}{K} \|u_n\|_\varepsilon^2 + C_1 t^4 \int_{\Lambda_\varepsilon} (u_n^+)^4\, dx + C_2 t^{2_s^*} \int_{\Lambda_\varepsilon} (u_n^+)^{2_s^*}\, dx$$

$$\leq \frac{t^2}{K} + C_1' t^4 \mathrm{dist}(u_n, \partial\mathbb{S}_\varepsilon^+)^4 + C_2' t^{2_s^*} \mathrm{dist}(u_n, \partial\mathbb{S}_\varepsilon^+)^{2_s^*}.$$

Therefore, for all $t > 0$,

$$\limsup_{n \to \infty} \int_{\mathbb{R}^3} G(\varepsilon x, t u_n)\, dx \leq \frac{t^2}{K}. \tag{13.2.10}$$

Recalling the definition of $m_\varepsilon(u_n)$ and using (13.2.10) we see that

$$\liminf_{n \to \infty} \left[\frac{1}{2} \|m_\varepsilon(u_n)\|_\varepsilon^2 + \frac{1}{4} \int_{\mathbb{R}^3} \phi_{m_\varepsilon(u_n)}^t (m_\varepsilon(u_n))^2 dx \right]$$

$$\geq \liminf_{n \to \infty} \mathcal{J}_\varepsilon(t u_n)$$

$$\geq \liminf_{n \to \infty} \mathcal{J}_\varepsilon(m_\varepsilon(u_n))$$

$$\geq \liminf_{n \to \infty} \left[\frac{t^2}{2} \|u_n\|_\varepsilon^2 + \frac{t^4}{4} \int_{\mathbb{R}^3} \phi_{u_n}^t u_n^2\, dx - \int_{\mathbb{R}^3} G(\varepsilon x, t u_n)\, dx \right]$$

$$\geq \left(\frac{1}{2} - \frac{1}{K} \right) t^2.$$

Since $t > 0$ is arbitrary and $K > 2$, we conclude that $\mathcal{J}_\varepsilon(m_\varepsilon(u_n)) \to \infty$ and $\|m_\varepsilon(u_n)\|_\varepsilon \to \infty$ as $n \to \infty$, and this ends the proof of Lemma 13.2.2. $\qquad \square$

Let us introduce the maps

$$\hat{\psi}_{\varepsilon} : \mathcal{H}_{\varepsilon}^+ \to \mathbb{R} \quad \text{and} \quad \psi_{\varepsilon} : \mathbb{S}_{\varepsilon}^+ \to \mathbb{R},$$

by $\hat{\psi}_{\varepsilon}(u) = \mathcal{J}_{\varepsilon}(\hat{m}_{\varepsilon}(u))$ and $\psi_{\varepsilon} = \hat{\psi}_{\varepsilon}|_{\mathbb{S}_{\varepsilon}^+}$.

Using Lemma 13.2.2 and arguing as in Chap. 11, we obtain the following result.

Proposition 13.2.3 *Assume that hypotheses* (V_1)-(V_2) *and* (f_1)–(f_4) *hold. Then,*

(a) $\hat{\psi}_{\varepsilon} \in C^1(\mathcal{H}_{\varepsilon}^+, \mathbb{R})$ *and*

$$\langle \hat{\psi}_{\varepsilon}'(u), v \rangle = \frac{\|\hat{m}_{\varepsilon}(u)\|_{\varepsilon}}{\|u\|_{\varepsilon}} \langle \mathcal{J}_{\varepsilon}'(\hat{m}_{\varepsilon}(u)), v \rangle$$

for every $u \in \mathcal{H}_{\varepsilon}^+$ *and* $v \in \mathcal{H}_{\varepsilon}$.

(b) $\psi_{\varepsilon} \in C^1(\mathbb{S}_{\varepsilon}^+, \mathbb{R})$ *and*

$$\langle \psi_{\varepsilon}'(u), v \rangle = \|m_{\varepsilon}(u)\|_{\varepsilon} \langle \mathcal{J}_{\varepsilon}'(m_{\varepsilon}(u)), v \rangle,$$

for every $v \in T_u \mathbb{S}_{\varepsilon}^+$.

(c) *If* (u_n) *is a Palais-Smale sequence for* ψ_{ε}, *then* $(m_{\varepsilon}(u_n))$ *is a Palais-Smale sequence for* $\mathcal{J}_{\varepsilon}$. *If* $(u_n) \subset \mathcal{N}_{\varepsilon}$ *is a bounded Palais-Smale sequence for* $\mathcal{J}_{\varepsilon}$, *then* $(m_{\varepsilon}^{-1}(u_n))$ *is a Palais-Smale sequence for* ψ_{ε}.

(d) *The point* u *is a critical point of* ψ_{ε} *if and only if* $m_{\varepsilon}(u)$ *is a nontrivial critical point for* $\mathcal{J}_{\varepsilon}$. *Moreover, the corresponding critical values coincide and*

$$\inf_{u \in \mathbb{S}_{\varepsilon}^+} \psi_{\varepsilon}(u) = \inf_{u \in \mathcal{N}_{\varepsilon}} \mathcal{J}_{\varepsilon}(u).$$

Remark 13.2.4 As in [322], we see that

$$c_{\varepsilon} = \inf_{u \in \mathcal{N}_{\varepsilon}} \mathcal{J}_{\varepsilon}(u) = \inf_{u \in \mathcal{H}_{\varepsilon}^+} \max_{t > 0} \mathcal{J}_{\varepsilon}(tu) = \inf_{u \in \mathbb{S}_{\varepsilon}^+} \max_{t > 0} \mathcal{J}_{\varepsilon}(tu).$$

It is also easy to check that c_{ε} coincides with the mountain pass level of $\mathcal{J}_{\varepsilon}$.

Remark 13.2.5 Let us note that if $u \in \mathcal{N}_{\varepsilon}$, then using (g_1), (g_2) and taking $\xi \in (0, \frac{1}{2})$ we obtain that

$$0 = \|u\|_{\varepsilon}^2 + \int_{\mathbb{R}^3} \phi_u u^2 \, dx - \int_{\mathbb{R}^3} g(\varepsilon x, u) u \, dx$$

$$\geq \frac{1}{2} \|u\|_{\varepsilon}^2 - C\|u\|_{\varepsilon}^{2_s^*},$$

and so $\|u\|_{\varepsilon} \geq \alpha > 0$ for some α independent of u.

The next lemma gives an upper bound of the minimax level c_ε.

Lemma 13.2.6 *It holds,* $0 < c_\varepsilon < \frac{s}{3} S_*^{\frac{3}{2s}}$.

Proof We follow [264]. We know (see Remark 1.1.9) that S_* is achieved by

$$z_\delta(x) = \frac{\kappa \delta^{-\frac{3-2s}{2}}}{(\mu^2 + |\frac{x}{\delta S_*^{1/2s}}|^2)^{\frac{3-2s}{2}}}$$

for any fixed $\delta > 0$, where $\kappa \in \mathbb{R}$ and $\mu > 0$ are fixed constants. Let $\eta \in C_c^\infty(\mathbb{R}^3)$ be a cut-off function such that $\eta = 1$ in B_ρ, $\text{supp}(\eta) \subset B_{2\rho}$ and $0 \leq \eta \leq 1$, where $B_{2\rho} \subset \Lambda$. For simplicity, we assume that $\rho = 1$. We define

$$Z_\delta(x) = \eta(\varepsilon x) z_\delta(x),$$

and let $v_\delta = \frac{Z_\delta}{\|Z_\delta\|_{L^{2^*_s}(\mathbb{R}^3)}}$ be such that

$$[v_\delta]_s^2 \leq S_* + O(\delta^{3-2s}). \tag{13.2.11}$$

Moreover,

$$\|v_\delta\|_{L^2(\mathbb{R}^3)}^2 = O(\delta^{3-2s}), \tag{13.2.12}$$

$$\|v_\delta\|_{L^p(\mathbb{R}^3)}^q = \begin{cases} O(\delta^{3-\frac{(3-2s)p}{2}}), & \text{if } p > \frac{3}{3-2s}, \\ O(\log(\frac{1}{\delta})\delta^{3-\frac{(3-2s)p}{2}}), & \text{if } p = \frac{3}{3-2s}, \\ O(\delta^{\frac{(3-2s)p}{2}}), & \text{if } p < \frac{3}{3-2s}. \end{cases} \tag{13.2.13}$$

Consider the function

$$y(r) = \frac{r^2}{2}[v_\delta]_s^2 - \frac{r^{2^*_s}}{2^*_s}.$$

Clearly, $y(r)$ achieves its maximum at $r_\delta = [v_\delta]_s^{\frac{2}{2^*_s-2}}$ and

$$y(r_\delta) = \frac{1}{2}[v_\delta]_s^{\frac{4}{2^*_s-2}}[v_\delta]_s^2 - \frac{1}{2^*_s}[v_\delta]_s^{\frac{22^*_s}{2^*_s-2}}.$$

Then, by (13.2.11), $y(r_\delta) \leq \frac{s}{3} S_*^{\frac{3}{2s}} + O(\delta^{3-2s})$. Note that there exists $r' \in (0, 1)$ such that for all $\delta < 1$ we have

$$\max_{r \in [0, r']} \mathcal{J}_\varepsilon(r v_\delta) \leq \max_{r \in [0, r']} \left[\frac{r^2}{2} \left([v_\delta]_s^2 + \|V\|_{L^\infty(\Lambda)} \|v_\delta\|_{L^2(\mathbb{R}^3)}^2 \right) + \frac{r^4}{4} \int_{\mathbb{R}^3} \phi_{v_\delta}^t v_\delta^2 dx \right]$$
$$< \frac{s}{3} S_*^{\frac{3}{2s}},$$
(13.2.14)

where we used Lemma 13.1.2-(4). By (f_2), we see that

$$\mathcal{J}_\varepsilon(r v_\delta) \leq \frac{r^2}{2} \left([v_\delta]_s^2 + \|V\|_{L^\infty(\Lambda)} \|v_\delta\|_{L^2(\mathbb{R}^3)}^2 \right) + \frac{r^4}{4} \int_{\mathbb{R}^3} \phi_{v_\delta}^t v_\delta^2 dx - \frac{C_0 r^q}{q} \|v_\delta\|_{L^q(\mathbb{R}^3)}^q - \frac{r^{2_s^*}}{2_s^*}$$

which combined with (13.2.12) and (13.2.13) implies that there exists $\delta_0 \in (0, 1)$ such that $\mathcal{J}_\varepsilon(r v_\delta) \to -\infty$ as $r \to \infty$, uniformly for $\delta \in (0, \delta_0)$. Then we can find $r'' > 0$ such that, for all $\delta \in (0, \delta_0)$,

$$\max_{r \in [r'', \infty)} \mathcal{J}_\varepsilon(r v_\delta) < \frac{s}{3} S_*^{\frac{3}{2s}}.$$
(13.2.15)

On the other hand, by the definition of $\mathcal{J}_\varepsilon$, Lemma 13.1.2-(1), (f_2), (13.2.12), (13.2.13), we deduce that

$$\max_{r \in [r', r'']} \mathcal{J}_\varepsilon(r v_\delta)$$
$$\leq \max_{r \in (0, \infty)} y(r) + \max_{r \in [r', r'']} \left[\int_{\mathbb{R}^3} \frac{r^4}{4} \phi_{v_\delta}^t v_\delta^2 dx + \|V\|_{L^\infty(\Lambda)} \frac{r^2}{2} \|v_\delta\|_{L^2(\mathbb{R}^3)}^2 - \int_{\mathbb{R}^3} F(r v_\delta) dx \right]$$
$$\leq \frac{s}{3} S_*^{\frac{3}{2s}} + O(\delta^{3-2s}) + C_1 \|v_\delta\|_{L^{\frac{12}{3+2t}}(\mathbb{R}^3)}^4 + C_2 \|v_\delta\|_{L^2(\mathbb{R}^3)}^2 - C_3 \|v_\delta\|_{L^q(\mathbb{R}^3)}^q$$
$$\leq \frac{s}{3} S_*^{\frac{3}{2s}} + O(\delta^{3-2s}) + O(\delta^{2t+4s-3}) - O(\delta^{3-\frac{(3-2s)q}{2}})$$
$$< \frac{s}{3} S_*^{\frac{3}{2s}}$$
(13.2.16)

for small $\delta > 0$, since $3 - \frac{(3-2s)q}{2} < \min\{3 - 2s, 2t + 4s - 3\}$. Then, putting together (13.2.14), (13.2.15) and (13.2.16), we complete the proof. $\square$

13.3   An Existence Result for the Modified Problem

In this section we focus our attention on the existence of positive solutions to (13.2.2) for small $\varepsilon > 0$. We begin by showing that the functional $\mathcal{J}_\varepsilon$ satisfies the Palais-Smale condition at any level $0 < d < \frac{s}{3} S_*^{\frac{3}{2s}}$, where S_* is the best constant of the Sobolev embedding $\mathcal{D}^{s,2}(\mathbb{R}^3)$ into $L^{2_s^*}(\mathbb{R}^3)$. We recall that the existence of Palais-Smale sequences of $\mathcal{J}_\varepsilon$ is justified by Lemma 13.2.1 and a variant of the mountain pass theorem without the Palais-Smale condition (see Remark 2.2.10). Firstly, we note that every Palais-Smale sequence is bounded.

Lemma 13.3.1 *Let $0 < d < \frac{s}{3} S_*^{\frac{3}{2s}}$ and let $(u_n) \subset \mathcal{H}_\varepsilon$ be a Palais-Smale sequence for $\mathcal{J}_\varepsilon$ at the level d. Then (u_n) is bounded in $\mathcal{H}_\varepsilon$.*

Proof Let $(u_n) \subset \mathcal{H}_\varepsilon$ be a Palais-Smale sequence at the level d, that is

$$\mathcal{J}_\varepsilon(u_n) \to d \quad \text{and} \quad \mathcal{J}_\varepsilon'(u_n) \to 0 \text{ in } \mathcal{H}_\varepsilon^*.$$

Arguing as in the proof of Lemma 13.2.2-(ii) (see formula (13.2.8) there), we deduce that

$$C(1 + \|u_n\|_\varepsilon) \geq \mathcal{J}_\varepsilon(u_n) - \frac{1}{\vartheta} \langle \mathcal{J}_\varepsilon'(u_n), u_n \rangle$$

$$\geq \left(\frac{\vartheta - 2}{2\vartheta} \right) \left(1 - \frac{1}{K} \right) \|u_n\|_\varepsilon^2.$$

Since $\vartheta > 4$ and $K > 2$, we conclude that (u_n) is bounded in $\mathcal{H}_\varepsilon$.   □

Remark 13.3.2 Arguing as in Remark 5.2.8, we may always assume that the Palais-Smale sequence (u_n) is non-negative in $\mathbb{R}^N$.

The next result will play a crucial role in establishing compactness of bounded Palais-Smale sequences.

Lemma 13.3.3 *Let $0 < d < \frac{s}{3} S_*^{\frac{3}{2s}}$ and let $(u_n) \subset \mathcal{H}_\varepsilon$ be a Palais-Smale sequence for $\mathcal{J}_\varepsilon$ at the level d. Then, for each $\zeta > 0$, there exists $R = R(\zeta) > 0$ such that*

$$\limsup_{n \to \infty} \left[\int_{\mathbb{R}^3 \setminus B_R} dx \int_{\mathbb{R}^3} \frac{|u_n(x) - u_n(y)|^2}{|x - y|^{3+2s}} \, dy + \int_{\mathbb{R}^3 \setminus B_R} V(\varepsilon x) u_n^2 \, dx \right] < \zeta.$$

Proof For $R > 0$, let $\eta_R \in C^\infty(\mathbb{R}^3)$ be such that $\eta_R = 0$ in $B_{\frac{R}{2}}$ and $\eta_R = 1$ in B_R^c, with $0 \leq \eta_R \leq 1$ and $\|\nabla \eta_R\|_{L^\infty(\mathbb{R}^3)} \leq \frac{C}{R}$, where C is a constant independent of R. Since $(\eta_R u_n)$ is bounded in $\mathcal{H}_\varepsilon$, it follows that $\langle \mathcal{J}_\varepsilon'(u_n), \eta_R u_n \rangle = o_n(1)$, that is

$$\iint_{\mathbb{R}^6} \frac{|u_n(x) - u_n(y)|^2}{|x - y|^{3+2s}} \eta_R(x) \, dx dy + \int_{\mathbb{R}^3} V(\varepsilon x) u_n^2 \eta_R \, dx + \int_{\mathbb{R}^3} \phi_{u_n}^t u_n^2 \eta_R dx$$

$$= o_n(1) + \int_{\mathbb{R}^3} g(\varepsilon x, u_n) u_n \eta_R \, dx - \iint_{\mathbb{R}^6} \frac{(\eta_R(x) - \eta_R(y))(u_n(x) - u_n(y))}{|x - y|^{3+2s}} u_n(y) \, dx dy.$$

Take $R > 0$ such that $\Lambda_\varepsilon \subset B_{\frac{R}{2}}$. Then, using (g_3)-(ii), we see that

$$\iint_{\mathbb{R}^6} \frac{|u_n(x) - u_n(y)|^2}{|x - y|^{3+2s}} \eta_R(x) \, dx dy + \int_{\mathbb{R}^3} V(\varepsilon x) u_n^2 \eta_R \, dx$$

$$\leq \int_{\mathbb{R}^3} \frac{1}{K} V(\varepsilon x) u_n^2 \eta_R \, dx - \iint_{\mathbb{R}^6} \frac{(\eta_R(x) - \eta_R(y))(u_n(x) - u_n(y))}{|x - y|^{3+2s}} u_n(y) \, dx dy + o_n(1),$$

which implies that

$$\iint_{\mathbb{R}^6} \frac{|u_n(x) - u_n(y)|^2}{|x - y|^{3+2s}} \eta_R(x) \, dx dy + \left(1 - \frac{1}{K}\right) \int_{\mathbb{R}^3} V(\varepsilon x) u_n^2 \eta_R \, dx$$

$$\leq -\iint_{\mathbb{R}^6} \frac{(\eta_R(x) - \eta_R(y))(u_n(x) - u_n(y))}{|x - y|^{3+2s}} u_n(y) \, dx dy + o_n(1). \tag{13.3.1}$$

Using Hölder's inequality, the boundedness of (u_n) and Remark 1.4.6, we have

$$\left| \iint_{\mathbb{R}^6} \frac{(\eta_R(x) - \eta_R(y))(u_n(x) - u_n(y))}{|x - y|^{3+2s}} u_n(y) \, dx dy \right| \leq C \left(\iint_{\mathbb{R}^6} \frac{|\eta_R(x) - \eta_R(y)|^2}{|x - y|^{3+2s}} \, dx dy \right)^{\frac{1}{2}}$$

$$\leq \frac{C}{R^s}. \tag{13.3.2}$$

Using (13.3.1), (13.3.2) and the definition of η_R, we obtain that

$$\int_{\mathbb{R}^3 \setminus B_R} dx \int_{\mathbb{R}^3} \frac{|u_n(x) - u_n(y)|^2}{|x - y|^{3+2s}} \, dy + \left(1 - \frac{1}{K}\right) \int_{\mathbb{R}^3 \setminus B_R} V(\varepsilon x) u_n^2 \, dx$$

$$\leq \iint_{\mathbb{R}^6} \frac{|u_n(x) - u_n(y)|^2}{|x - y|^{3+2s}} \eta_R(x) \, dx dy + \left(1 - \frac{1}{K}\right) \int_{\mathbb{R}^3} V(\varepsilon x) u_n^2 \eta_R \, dx \tag{13.3.3}$$

$$\leq \frac{C}{R^s} + o_n(1),$$

from which we deduce the thesis. $\square$

Proposition 13.3.4 *The functional $\mathcal{J}_\varepsilon$ satisfies the* (PS)$_d$ *condition in* $\mathcal{H}_\varepsilon$ *at any level* $0 < d < \frac{s}{3} S_*^{\frac{3}{2s}}$.

Proof Let $(u_n) \subset \mathcal{H}_\varepsilon$ be a Palais-Smale sequence for $\mathcal{J}_\varepsilon$ at the level d. By Lemma 13.3.1, we know that (u_n) is bounded in $\mathcal{H}_\varepsilon$, and, up to a subsequence, we may assume that

$$u_n \rightharpoonup u \quad \text{in } \mathcal{H}_\varepsilon. \tag{13.3.4}$$

In view of Lemma 13.3.3, for each $\zeta > 0$ there exists $R = R_\zeta > 0$ such that $\Lambda_\varepsilon \subset B_{\frac{R}{2}}$ and

$$\limsup_{n \to \infty} \left[\int_{\mathbb{R}^3 \setminus B_R} dx \int_{\mathbb{R}^3} \frac{|u_n(x) - u_n(y)|^2}{|x - y|^{3+2s}} \, dy + \int_{\mathbb{R}^3 \setminus B_R} V(\varepsilon x) u_n^2 \, dx \right] < \zeta. \tag{13.3.5}$$

Using (13.3.5) and the fact that $\mathcal{H}_\varepsilon$ is compactly embedded in $L^2_{\text{loc}}(\mathbb{R}^3)$, it is easy to deduce that $u_n \to u$ in $L^2(\mathbb{R}^3)$. By interpolation, $u_n \to u$ in $L^r(\mathbb{R}^3)$ for all $r \in [2, 2_s^*)$. In particular,

$$u_n \to u \quad \text{in } L^{\frac{12}{3+2t}}(\mathbb{R}^3). \tag{13.3.6}$$

Then, in view of (13.3.4), (13.3.6), the fact that $4s + 2t > 3$ and Lemma 13.1.2-(8),

$$\int_{\mathbb{R}^3} \phi_{u_n}^t u_n^2 \, dx \to \int_{\mathbb{R}^3} \phi_u^t u^2 \, dx, \tag{13.3.7}$$

and

$$\int_{\mathbb{R}^3} \phi_{u_n}^t u_n \psi \, dx \to \int_{\mathbb{R}^3} \phi_u^t u \psi \, dx \quad \text{for all } \psi \in C_c^\infty(\mathbb{R}^3).$$

Now, we know that $\langle \mathcal{J}_\varepsilon'(u_n), \psi \rangle = o_n(1)$ for all $\psi \in C_c^\infty(\mathbb{R}^3)$. Since it is clear that (13.3.4) and (f_1)-(f_2) ensure that

$$(u_n, \psi)_\varepsilon \to (u, \psi)_\varepsilon \quad \text{and} \quad \int_{\mathbb{R}^3} g(\varepsilon x, u_n) \psi \, dx \to \int_{\mathbb{R}^3} g(\varepsilon x, u) \psi \, dx,$$

for all $\psi \in C_c^\infty(\mathbb{R}^3)$, we can use the denseness of $C_c^\infty(\mathbb{R}^3)$ in $\mathcal{H}_\varepsilon$ to deduce that u is a critical point of $\mathcal{J}_\varepsilon$. In particular,

$$\|u\|_\varepsilon^2 + \int_{\mathbb{R}^3} \phi_u^t u^2 \, dx = \int_{\mathbb{R}^3} g(\varepsilon x, u) u \, dx. \tag{13.3.8}$$

Let us show that

$$\lim_{n \to \infty} \int_{\mathbb{R}^3} g(\varepsilon x, u_n) u_n \, dx = \int_{\mathbb{R}^3} g(\varepsilon x, u) u \, dx. \tag{13.3.9}$$

For this purpose, we argue as in the proof of Lemma 7.3.2. By the fractional Sobolev inequality (1.1.1) in Theorem 1.1.8,

$$\left(\int_{\mathbb{R}^3 \setminus B_R} |u_n|^{2_s^*} dx \right)^{\frac{2}{2_s^*}} \leq \left(\int_{\mathbb{R}^3} |u_n \eta_R|^{2_s^*} dx \right)^{\frac{2}{2_s^*}} \leq C [u_n \eta_R]_s^2.$$

Since $0 \leq \eta_R \leq 1$, (13.3.3) and Remark 1.4.6 imply that

$$
\begin{aligned}
[u_n \eta_R]_s^2 &= \iint_{\mathbb{R}^{2N}} \frac{|(u_n(x) - u_n(y))\eta_R(x) + (\eta_R(x) - \eta_R(y))u_n(y)|^2}{|x - y|^{N+2s}} \, dx \, dy \\
&\leq C \left[\iint_{\mathbb{R}^{2N}} \frac{|u_n(x) - u_n(y)|^2}{|x - y|^{N+2s}} \eta_R^2(x) \, dx \, dy + \iint_{\mathbb{R}^{2N}} \frac{|\eta_R(x) - \eta_R(y)|^2}{|x - y|^{N+2s}} |u_n(y)|^2 \, dx \, dy \right] \\
&\leq C \left[\iint_{\mathbb{R}^{2N}} \frac{|u_n(x) - u_n(y)|^2}{|x - y|^{N+2s}} \eta_R(x) \, dx \, dy + \frac{C}{R^{2s}} \right] \\
&\leq \frac{C}{R^s} + o_n(1) + \frac{C}{R^{2s}},
\end{aligned}
$$

and so

$$\lim_{R \to \infty} \limsup_{n \to \infty} \int_{\mathbb{R}^3 \setminus B_R} |u_n|^{2_s^*} dx = 0. \tag{13.3.10}$$

Now, note that by (13.3.5) and (V_1),

$$\lim_{R \to \infty} \limsup_{n \to \infty} \int_{\mathbb{R}^3 \setminus B_R} |u_n|^2 dx = 0, \tag{13.3.11}$$

and using interpolation on L^p-spaces and the boundedness of (u_n) in $L^{2_s^*}(\mathbb{R}^3)$, we also deduce that for every $p \in (2, 2_s^*)$

$$\lim_{R \to \infty} \limsup_{n \to \infty} \int_{\mathbb{R}^3 \setminus B_R} |u_n|^p dx = 0. \tag{13.3.12}$$

By (f_1), (f_2), (g_2), (13.3.10), (13.3.11) and (13.3.12), we see that for every $\zeta > 0$ there exists an $R = R_\zeta > 0$ such that

$$\limsup_{n \to \infty} \int_{\mathbb{R}^3 \setminus B_R} g(\varepsilon x, u_n) u_n \, dx < C\zeta. \tag{13.3.13}$$

On the other hand, choosing R larger if necessary, we may assume that

$$\int_{\mathbb{R}^3 \setminus B_R} g(\varepsilon x, u)u \, dx < \zeta. \tag{13.3.14}$$

Combining (13.3.13) and (13.3.14) we get

$$\limsup_{n \to \infty} \left| \int_{\mathbb{R}^3 \setminus B_R} g(\varepsilon x, u_n)u_n \, dx - \int_{\mathbb{R}^3 \setminus B_R} g(\varepsilon x, u)u \, dx \right| < C\zeta \quad \text{for all } \zeta > 0,$$

and consequently

$$\lim_{n \to \infty} \int_{\mathbb{R}^3 \setminus B_R} g(\varepsilon x, u_n)u_n \, dx = \int_{\mathbb{R}^3 \setminus B_R} g(\varepsilon x, u)u \, dx. \tag{13.3.15}$$

Next, note that, by the definition of g,

$$g(\varepsilon x, u_n)u_n \leq f(u_n)u_n + a^{2_s^*} + \frac{V_0}{K} u_n^2 \quad \text{in } \mathbb{R}^3 \setminus \Lambda_\varepsilon.$$

Since the set $B_R \cap (\mathbb{R}^3 \setminus \Lambda_\varepsilon)$ is bounded, we can use (f_1)-(f_2), the dominated convergence theorem and the strong convergence in $L_{\text{loc}}^r(\mathbb{R}^3)$ for all $r \in [1, 2_s^*)$, to deduce that

$$\lim_{n \to \infty} \int_{B_R \cap (\mathbb{R}^3 \setminus \Lambda_\varepsilon)} g(\varepsilon x, u_n)u_n \, dx = \int_{B_R \cap (\mathbb{R}^3 \setminus \Lambda_\varepsilon)} g(\varepsilon x, u)u \, dx \tag{13.3.16}$$

as $n \to \infty$.

We claim that

$$\lim_{n \to \infty} \int_{\Lambda_\varepsilon} (u_n^+)^{2_s^*} \, dx = \int_{\Lambda_\varepsilon} (u^+)^{2_s^*} \, dx. \tag{13.3.17}$$

If we can prove this, then from Theorem 1.1.8, (g_2), (f_1)-(f_2), (13.3.4) and the dominated convergence theorem it will follow that

$$\lim_{n \to \infty} \int_{\Lambda_\varepsilon \cap B_R} g(\varepsilon x, u_n)u_n \, dx = \int_{\Lambda_\varepsilon \cap B_R} g(\varepsilon x, u)u \, dx. \tag{13.3.18}$$

Putting together (13.3.15), (13.3.16) and (13.3.18), we conclude that (13.3.9) holds. Hence, in view of $\langle \mathcal{J}_\varepsilon'(u_n), u_n \rangle = o_n(1)$, we see that (13.3.7), (13.3.8) and (13.3.9) imply that $\|u_n\|_\varepsilon \to \|u\|_\varepsilon$, and then $u_n \to u$ in $\mathcal{H}_\varepsilon$ (since $\mathcal{H}_\varepsilon$ is a Hilbert space).

Thus, it remains to prove (13.3.17). Since (u_n) is bounded in $\mathcal{H}_\varepsilon$, we may assume that $|(-\Delta)^{\frac{s}{2}} u_n^+|^2 \rightharpoonup \mu$ and $(u_n^+)^{2_s^*} \rightharpoonup \nu$, where μ and ν are two bounded non-negative measures on $\mathbb{R}^3$. Invoking Lemma 1.5.1 we can find an at most countable index set I and sequences $(x_i)_{i \in I} \subset \mathbb{R}^3$, $(\mu_i)_{i \in I}$, $(\nu_i)_{i \in I} \subset (0, \infty)$, such that

$$\mu \geq |(-\Delta)^{\frac{s}{2}} u^+|^2 + \sum_{i \in I} \mu_i \delta_{x_i},$$

$$(13.3.19)$$

$$\nu = (u^+)^{2^*_s} + \sum_{i \in I} \nu_i \delta_{x_i} \quad \text{and} \quad S_* \nu_i^{\frac{2}{2^*_s}} \leq \mu_i$$

for any $i \in I$, where δ_{x_i} is the Dirac mass at the point x_i. Let us show that $(x_i)_{i \in I} \cap \Lambda_\varepsilon = \emptyset$. Assume, by contradiction, that $x_i \in \Lambda_\varepsilon$ for some $i \in I$. For any $\rho > 0$, we define $\psi_\rho(x) = \psi(\frac{x - x_i}{\rho})$ where $\psi \in C_c^\infty(\mathbb{R}^3, [0, 1])$ is such that $\psi = 1$ in B_1, $\psi = 0$ in $\mathbb{R}^3 \setminus B_2$ and $\|\nabla \psi\|_{L^\infty(\mathbb{R}^3)} \leq 2$. We suppose that $\rho > 0$ is such that $\mathrm{supp}(\psi_\rho) \subset \Lambda_\varepsilon$. Since $(\psi_\rho u_n^+)$ is bounded, we have $\langle \mathcal{J}_\varepsilon'(u_n), \psi_\rho u_n^+ \rangle = o_n(1)$, whence

$$\iint_{\mathbb{R}^6} \psi_\rho(y) \frac{|u_n^+(x) - u_n^+(y)|^2}{|x - y|^{3+2s}} \, dx \, dy$$

$$\leq \iint_{\mathbb{R}^6} \psi_\rho(y) \frac{|u_n^+(x) - u_n^+(y)|^2}{|x - y|^{3+2s}} \, dx \, dy + \int_{\mathbb{R}^3} \phi_{u_n}^t (u_n^+)^2 \psi_\rho \, dx + \int_{\mathbb{R}^3} V(\varepsilon x)(u_n^+)^2 \psi_\rho \, dx$$

$$\leq - \iint_{\mathbb{R}^6} u_n^+(x) \frac{(u_n(x) - u_n(y))(\psi_\rho(x) - \psi_\rho(y))}{|x - y|^{3+2s}} \, dx \, dy$$

$$+ \int_{\mathbb{R}^3} u_n^+ \psi_\rho f(u_n) \, dx + \int_{\mathbb{R}^3} \psi_\rho (u_n^+)^{2^*_s} \, dx + o_n(1).$$

$$(13.3.20)$$

Since f has subcritical growth and ψ_ρ has compact support, we obtain that

$$\lim_{\rho \to 0} \lim_{n \to \infty} \int_{\mathbb{R}^3} f(u_n) u_n^+ \psi_\rho \, dx = \lim_{\rho \to 0} \int_{\mathbb{R}^3} \psi_\rho f(u) u^+ \, dx = 0.$$

On the other hand, by Hölder's inequality, the boundedness of (u_n) in $\mathcal{H}_\varepsilon$ and Lemma 1.4.7 we get

$$\lim_{\rho \to 0} \limsup_{n \to \infty} \iint_{\mathbb{R}^6} u_n^+(x) \frac{(u_n(x) - u_n(y))(\psi_\rho(x) - \psi_\rho(y))}{|x - y|^{3+2s}} \, dx \, dy = 0.$$

Then, by (13.3.19) and taking the limit as $\rho \to 0$ and $n \to \infty$ in (13.3.20), we deduce that $\nu_i \geq \mu_i$. In view of the last statement in (13.3.19), we have $\nu_i \geq S^{\frac{3}{2s}}$, and using (g_3), (V_1) and the fact that $K > 2$ we obtain that

$$d = \mathcal{J}_\varepsilon(u_n) - \frac{1}{4} \langle \mathcal{J}_\varepsilon'(u_n), u_n \rangle + o_n(1)$$

$$\geq \frac{1}{4}\|u_n\|_\varepsilon^2 + \int_{\mathbb{R}^3\setminus\Lambda_\varepsilon}\left[\frac{1}{4}u_n g(\varepsilon x, u_n) - G(\varepsilon x, u_n)\right]dx + \frac{4s-3}{12}\int_{\Lambda_\varepsilon}(u_n^+)^{2_s^*}\,dx + o_n(1)$$

$$\geq \frac{1}{4}\left[\int_{\Lambda_\varepsilon}\psi_\rho|(-\Delta)^{\frac{s}{2}}u_n|^2 dx + \int_{\mathbb{R}^3\setminus\Lambda_\varepsilon}V(\varepsilon x)u_n^2\,dx\right] - \frac{1}{4K}\int_{\mathbb{R}^3\setminus\Lambda_\varepsilon}V_0 u_n^2\,dx$$

$$+ \frac{4s-3}{12}\int_{\Lambda_\varepsilon}(u_n^+)^{2_s^*}\,dx + o_n(1)$$

$$\geq \frac{1}{4}\int_{\Lambda_\varepsilon}\psi_\rho|(-\Delta)^{\frac{s}{2}}u_n|^2\,dx$$

$$+ \left(\frac{1}{4} - \frac{1}{4K}\right)\int_{\mathbb{R}^3\setminus\Lambda_\varepsilon}V(\varepsilon x)u_n^2\,dx + \frac{4s-3}{12}\int_{\Lambda_\varepsilon}(u_n^+)^{2_s^*}\,dx + o_n(1)$$

$$\geq \frac{1}{4}\int_{\Lambda_\varepsilon}\psi_\rho|(-\Delta)^{\frac{s}{2}}u_n^+|^2\,dx + \frac{4s-3}{12}\int_{\Lambda_\varepsilon}\psi_\rho(u_n^+)^{2_s^*}\,dx + o_n(1),$$

where in the last inequality we used that $K > 1$, and that $|x^+ - y^+| \leq |x - y|$ for all $x, y \in \mathbb{R}$ yields

$$|u_n^+(x) - u_n^+(y)|^2 \leq |u_n(x) - u_n(y)|^2.$$

Putting together (13.3.19) and the fact that $v_i \geq S^{\frac{3}{2s}}$, and letting $n \to \infty$, we see that

$$d \geq \frac{1}{4}\sum_{\{i\in I:x_i\in\Lambda_\varepsilon\}}\psi_\rho(x_i)\mu_i + \frac{4s-3}{12}\sum_{\{i\in I:x_i\in\Lambda_\varepsilon\}}\psi_\rho(x_i)v_i$$

$$\geq \frac{1}{4}\sum_{\{i\in I:x_i\in\Lambda_\varepsilon\}}\psi_\rho(x_i)S_*v_i^{2/2_s^*} + \frac{4s-3}{12}\sum_{\{i\in I:x_i\in\Lambda_\varepsilon\}}\psi_\rho(x_i)v_i$$

$$\geq \frac{1}{4}S_*^{\frac{3}{2s}} + \frac{4s-3}{12}S_*^{\frac{3}{2s}} = \frac{s}{3}S_*^{\frac{3}{2s}},$$

which gives a contradiction. This means that (13.3.17) holds, which completes the proof.   □

Corollary 13.3.5 *The functional ψ_ε satisfies the* (PS)$_d$ *condition on $\mathbb{S}_\varepsilon^+$ at any level $0 < d < \frac{s}{3}S_*^{\frac{3}{2s}}$.*

Proof Let $(u_n) \subset \mathbb{S}_\varepsilon^+$ be a Palais-Smale sequence for ψ_ε at the level d, that is

$$\psi_\varepsilon(u_n) \to d \quad \text{and} \quad \psi_\varepsilon'(u_n) \to 0 \text{ in } (T_{u_n}\mathbb{S}_\varepsilon^+)^*.$$

Using Proposition 13.2.3-(c), we have that $(m_\varepsilon(u_n))$ is a Palais-Smale sequence for $\mathcal{J}_\varepsilon$ at the level d. Then, by Proposition 13.3.4, $\mathcal{J}_\varepsilon$ satisfies the (PS)$_d$ condition in $\mathcal{H}_\varepsilon$, so there exists $u \in \mathbb{S}_\varepsilon^+$ such that, up to a subsequence,

$$m_\varepsilon(u_n) \to m_\varepsilon(u) \quad \text{in } \mathcal{H}_\varepsilon.$$

Finally, by Lemma 13.2.2-(iii), we infer that $u_n \to u$ in $\mathbb{S}_\varepsilon^+$. $\qquad\square$

We conclude with the proof of the main result of this section:

Theorem 13.3.6 *Assume that (V_1)-(V_2) and $(f1)$-$(f4)$ hold. Then, for all $\varepsilon > 0$, problem (13.2.2) admits a positive ground state.*

Proof By Lemma 13.2.6, $0 < c_\varepsilon < \frac{3}{3}S_*^{\frac{3}{2s}}$. Then, taking into account Lemma 13.2.1, Lemma 13.3.1, Proposition 13.3.4, and applying Theorem 2.2.9, we see that $\mathcal{J}_\varepsilon$ admits a nontrivial critical point $u \in \mathcal{H}_\varepsilon$. Since $\langle \mathcal{J}_\varepsilon'(u), u^- \rangle = 0$, where $u^- = \min\{u, 0\}$, it is easy to check that $u \geq 0$ in $\mathbb{R}^3$. Proceeding as in the proof of Lemma 13.6.1 below, we see that $u \in L^\infty(\mathbb{R}^3)$. In particular,

$$\phi_u^t(x) = \int_{|y-x| \geq 1} \frac{u^2(y)}{|x-y|^{3-2t}} dy + \int_{|y-x| < 1} \frac{u^2(y)}{|x-y|^{3-2t}} dy$$

$$\leq \|u\|_{L^2(\mathbb{R}^3)}^2 + \|u\|_{L^\infty(\mathbb{R}^3)}^2 \int_{|y-x|<1} \frac{1}{|x-y|^{3-2t}} dy \leq C,$$

so that $\phi_u^t u \in L^\infty(\mathbb{R}^3)$. Then, by Proposition 1.3.11-(iii), we deduce that $u \in L^\infty(\mathbb{R}^3) \cap C_{loc}^{0,\alpha}(\mathbb{R}^3)$. Further, Theorem 1.3.5 implies that $u > 0$ in $\mathbb{R}^3$. Finally, we show that u is a ground state solution. Indeed, in view of $(g3)$ and applying Fatou's lemma, we obtain

$$c_\varepsilon \leq \mathcal{J}_\varepsilon(u) - \frac{1}{\vartheta}\langle \mathcal{J}_\varepsilon'(u), u \rangle$$

$$= \left(\frac{1}{2} - \frac{1}{\vartheta}\right)\|u\|_\varepsilon^2 + \left(\frac{1}{4} - \frac{1}{\vartheta}\right)\int_{\mathbb{R}^3} \phi_u^t u^2 \, dx + \int_{\mathbb{R}^3} \frac{1}{\vartheta} g(\varepsilon x, u)u - G(\varepsilon x, u) \, dx$$

$$\leq \liminf_{n \to \infty} \left[\left(\frac{1}{2} - \frac{1}{\vartheta}\right)\|u_n\|_\varepsilon^2 + \left(\frac{1}{4} - \frac{1}{\vartheta}\right)\int_{\mathbb{R}^3} \phi_u^t u_n^2 \, dx\right.$$

$$\left.+ \int_{\mathbb{R}^3} \frac{1}{\vartheta} g(\varepsilon x, u_n)u_n - G(\varepsilon x, u_n) \, dx\right]$$

$$= \liminf_{n \to \infty} \left[\mathcal{J}_\varepsilon(u_n) - \frac{1}{\vartheta}\langle \mathcal{J}_\varepsilon'(u_n), u_n \rangle\right],$$

$$= c_\varepsilon$$

whence $\mathcal{J}_\varepsilon(u) = c_\varepsilon$. $\qquad\square$

13.4   The Autonomous Schrödinger-Poisson Equation

In this section we consider the limit problem associated with (13.2.2). More precisely, we deal with the autonomous problem

$$
\begin{cases}
(-\Delta)^s u + V_0 u + \phi_u^t u = f(u) + |u|^{2_s^*-2} u & \text{in } \mathbb{R}^3, \\
u \in H^s(\mathbb{R}^3), \quad u > 0 & \text{in } \mathbb{R}^3.
\end{cases}
\tag{13.4.1}
$$

The Euler-Lagrange functional associated with (13.4.1) is given by

$$
\mathcal{J}_0(u) = \frac{1}{2} \|u\|_0^2 + \frac{1}{4} \int_{\mathbb{R}^3} \phi_u^t u^2 \, dx - \int_{\mathbb{R}^3} F(u) \, dx - \frac{1}{2_s^*} \int_{\mathbb{R}^3} (u^+)^{2_s^*} \, dx,
$$

which is well defined on $\mathcal{H}_0 = H^s(\mathbb{R}^3)$ endowed with the norm

$$
\|u\|_0 = \left([u]_s^2 + V_0 \|u\|_{L^2(\mathbb{R}^3)}^2 \right)^{\frac{1}{2}}.
$$

Clearly, $\mathcal{H}_0$ is a Hilbert space with the inner product

$$
\langle u, \varphi \rangle_0 = \langle u, \varphi \rangle_{\mathcal{D}^{s,2}(\mathbb{R}^3)} + \int_{\mathbb{R}^3} V_0 u \varphi \, dx.
$$

The Nehari manifold associated with $\mathcal{J}_0$ is given by

$$
\mathcal{N}_0 = \{ u \in \mathcal{H}_0 \setminus \{0\} : \langle \mathcal{J}_0'(u), u \rangle = 0 \}.
$$

Let $\mathcal{H}_0^+$ denote the open subset of $\mathcal{H}_0$ defined as

$$
\mathcal{H}_0^+ = \{ u \in \mathcal{H}_0 : |\text{supp}(u^+)| > 0 \},
$$

and $\mathbb{S}_0^+ = \mathbb{S}_0 \cap \mathcal{H}_0^+$, where $\mathbb{S}_0$ is the unit sphere of $\mathcal{H}_0$. We note that $\mathbb{S}_0^+$ is a incomplete $C^{1,1}$-manifold of codimension 1 modeled on $\mathcal{H}_0$ and contained in $\mathcal{H}_0^+$. Thus, $\mathcal{H}_0 = T_u \mathbb{S}_0^+ \oplus \mathbb{R} u$ for each $u \in \mathbb{S}_0^+$, where $T_u \mathbb{S}_0^+ = \{ u \in \mathcal{H}_0 : (u, v)_0 = 0 \}$.

As in the previous section, we can verify that the following results hold.

Lemma 13.4.1 *Assume that* (f_1)–(f_4) *hold. Then:*

(i) *For each* $u \in \mathcal{H}_0^+$, *let* $h_u : \mathbb{R}_+ \to \mathbb{R}$ *be defined by* $h_u(t) = \mathcal{J}_0(tu)$. *Then, there is a unique* $t_u > 0$ *such that*

$$
h_u'(t) > 0 \text{ in } (0, t_u),
$$

$$
h_u'(t) < 0 \text{ in } (t_u, \infty).
$$

(ii) *There exists $\tau > 0$ independent of u such that $t_u \geq \tau$ for any $u \in \mathbb{S}_0^+$. Moreover, for each compact set $\mathbb{K} \subset \mathbb{S}_0^+$ there is a positive constant $C_{\mathbb{K}}$ such that $t_u \leq C_{\mathbb{K}}$ for any $u \in \mathbb{K}$.*

(iii) *The map $\hat{m}_0 : \mathcal{H}_0^+ \to \mathcal{N}_0$ given by $\hat{m}_0(u) = t_u u$ is continuous and $m_0 = \hat{m}_0|_{\mathbb{S}_0^+}$ is a homeomorphism between $\mathbb{S}_0^+$ and $\mathcal{N}_0$. Moreover $m_0^{-1}(u) = \frac{u}{\|u\|_0}$.*

(iv) *If there is a sequence $(u_n) \subset \mathbb{S}_0^+$ such that $\mathrm{dist}(u_n, \partial\mathbb{S}_0^+) \to 0$, then $\|m_0(u_n)\|_0 \to \infty$ and $\mathcal{J}_0(m_0(u_n)) \to \infty$.*

Let us define the maps

$$\hat{\psi}_0 : \mathcal{H}_0^+ \to \mathbb{R} \quad \text{and} \quad \psi_0 : \mathbb{S}_0^+ \to \mathbb{R},$$

by $\hat{\psi}_0(u) = \mathcal{J}_0(\hat{m}_0(u))$ and $\psi_0 = \hat{\psi}_0|_{\mathbb{S}_0^+}$.

Proposition 13.4.2 *Assume that assumptions (f_1)–(f_4) hold. Then:*

(a) *$\hat{\psi}_0 \in C^1(\mathcal{H}_0^+, \mathbb{R})$ and*

$$\langle \hat{\psi}_0'(u), v \rangle = \frac{\|\hat{m}_0(u)\|_0}{\|u\|_0} \langle \mathcal{J}_0'(\hat{m}_0(u)), v \rangle$$

for every $u \in \mathcal{H}_0^+$ and $v \in \mathcal{H}_0$.

(b) *$\psi_0 \in C^1(\mathbb{S}_0^+, \mathbb{R})$ and*

$$\langle \psi_0'(u), v \rangle = \|m_0(u)\|_0 \langle \mathcal{J}_0'(m_0(u)), v \rangle,$$

for every $v \in T_u \mathbb{S}_0^+$.

(c) *If (u_n) is a Palais-Smale sequence for ψ_0, then $(m_0(u_n))$ is a Palais-Smale sequence for $\mathcal{J}_0$. If $(u_n) \subset \mathcal{N}_0$ is a bounded Palais-Smale sequence for $\mathcal{J}_0$, then $(m_0^{-1}(u_n))$ is a Palais-Smale sequence for ψ_0.*

(d) *u is a critical point of ψ_0 if and only if $m_0(u)$ is a nontrivial critical point for $\mathcal{J}_0$. Moreover, the corresponding critical values coincide and*

$$\inf_{u \in \mathbb{S}_0^+} \psi_0(u) = \inf_{u \in \mathcal{N}_0} \mathcal{J}_0(u).$$

Remark 13.4.3 The following equalities hold:

$$c_0 = \inf_{u \in \mathcal{N}_0} \mathcal{J}_0(u) = \inf_{u \in \mathcal{H}_0^+} \max_{t > 0} \mathcal{J}_0(tu) = \inf_{u \in \mathbb{S}_0^+} \max_{t > 0} \mathcal{J}_0(tu).$$

Moreover, $\mathcal{J}_0$ has a mountain pass geometry and $0 < c_0 < \frac{s}{3} S_*^{\frac{3}{2s}}$.

Now we prove the following useful lemma.

Lemma 13.4.4 *Let $0 < d < \frac{s}{3} S_*^{\frac{3}{2s}}$. Let $(u_n) \subset \mathcal{H}_0$ be a Palais-Smale sequence for $\mathcal{J}_0$ at the level d and $u_n \rightharpoonup 0$ in $\mathcal{H}_0$. Then*

(a) *either $u_n \to 0$ in $\mathcal{H}_0$, or*
(b) *there exist a sequence $(y_n) \subset \mathbb{R}^3$ and constants $R, \beta > 0$ such that*

$$\liminf_{n \to \infty} \int_{B_R(y_n)} u_n^2 \, dx \geq \beta > 0.$$

Proof Assume that (b) does not occur. Then, by Lemma 1.4.4, $u_n \to 0$ in $L^p(\mathbb{R}^3)$ for all $p \in (2, 2_s^*)$. Moreover, by Lemma 13.1.2-(8), $\int_{\mathbb{R}^3} \phi_{u_n}^t u_n^2 \, dx \to 0$. Now we can argue as in the proof of Lemma 6.4.9 to get the thesis. □

Remark 13.4.5 Let us observe that, if (u_n) is a Palais-Smale sequence at the level c_0 for the functional $\mathcal{J}_0$ such that $u_n \rightharpoonup u$, then we may assume that $u \neq 0$. Otherwise, if $u_n \rightharpoonup 0$ and, if $u_n \not\to 0$ in $\mathcal{H}_0$, then in view of Lemma *13.4.4*, there exist a sequence $(y_n) \subset \mathbb{R}^3$ and $R, \beta > 0$ such that

$$\liminf_{n \to \infty} \int_{B_R(y_n)} u_n^2 \, dx \geq \beta > 0.$$

Set $v_n(x) = u_n(x + y_n)$. Then we see that (v_n) is a Palais-Smale sequence for $\mathcal{J}_0$ at the level c_0, (v_n) is bounded in $\mathcal{H}_0$, and there exists $v \in \mathcal{H}_0$ such that $v_n \rightharpoonup v$ with $v \neq 0$.

As a consequence of Lemma 13.4.4, we can prove the following existence result:

Theorem 13.4.6 *Problem (13.4.1) admits a positive ground state solution.*

Proof To establish the existence of a ground state solution u to (13.4.1), it is enough to argue as in the proof of Lemma 6.4.10. We only have to observe that

$$c_0 + o_n(1) = \mathcal{J}_0(u_n) - \frac{1}{4}\langle \mathcal{J}_0'(u_n), u_n \rangle$$

$$= \frac{1}{4}\|u_n\|_0^2 + \int_{\mathbb{R}^3} \left[\frac{1}{4} f(u_n)u_n - F(u_n)\right] dx + \left(\frac{1}{4} - \frac{1}{2_s^*}\right) \int_{\mathbb{R}^3} (u_n^+)^{2_s^*} dx$$

and use assumption (f_3). From $\langle \mathcal{J}_0'(u), u^- \rangle = 0$ we get $u \geq 0$ in $\mathbb{R}^3$. Arguing as in the proof of Lemma 13.6.1, we see that $u \in L^\infty(\mathbb{R}^3)$. In particular,

$$\phi_u^t(x) = \int_{|y-x|\geq 1} \frac{u^2(y)}{|x-y|^{3-2t}} dy + \int_{|y-x|<1} \frac{u^2(y)}{|x-y|^{3-2t}} dy$$

$$\leq \|u\|_{L^2(\mathbb{R}^3)}^2 + \|u\|_{L^\infty(\mathbb{R}^3)}^2 \int_{|y-x|<1} \frac{1}{|x-y|^{3-2t}} dy \leq C,$$

so that $h(x) = f(u^2)u - V_0 u - \phi_u^t u \in L^\infty(\mathbb{R}^3)$. Proposition 1.3.2 and the fact that $s > \frac{3}{4}$ imply that $u \in C^{1,\alpha}(\mathbb{R}^3)$ for all $\alpha < 2s - 1$, and by Theorem 1.3.5 (or Proposition 1.3.11-(ii)) we conclude that $u > 0$ in $\mathbb{R}^3$.   □

Now we give a compactness result for the autonomous problem which we will use later.

Lemma 13.4.7 *Let $(u_n) \subset \mathcal{N}_0$ be a sequence such that $\mathcal{J}_0(u_n) \to c_0$. Then (u_n) has a convergent subsequence in $H^s(\mathbb{R}^3)$.*

Proof Since $(u_n) \subset \mathcal{N}_0$ and $\mathcal{J}_0(u_n) \to c_0$, we can apply Lemma 13.4.1-(iii), Proposition 13.4.2-(d) and the definition of c_0 to infer that

$$v_n = m^{-1}(u_n) = \frac{u_n}{\|u_n\|_0} \in \mathbb{S}_0^+$$

and

$$\psi_0(v_n) = \mathcal{J}_0(u_n) \to c_0 = \inf_{v \in \mathbb{S}_0^+} \psi_0(v).$$

Let us introduce the map $\mathcal{F} : \overline{\mathbb{S}}_0^+ \to \mathbb{R} \cup \{\infty\}$ defined by

$$\mathcal{F}(u) = \begin{cases} \psi_0(u), & \text{if } u \in \mathbb{S}_0^+, \\ \infty, & \text{if } u \in \partial\mathbb{S}_0^+. \end{cases}$$

We note that

- $(\overline{\mathbb{S}}_0^+, d_0)$, where $d_0(u, v) = \|u - v\|_0$, is a complete metric space;
- $\mathcal{F} \in C(\overline{\mathbb{S}}_0^+, \mathbb{R} \cup \{\infty\})$, by Lemma 13.4.1-(iv);
- $\mathcal{F}$ is bounded below, by Proposition 13.4.2-(d).

Hence, applying Theorem 2.2.1 to $\mathcal{F}$, we can find $(\hat{v}_n) \subset \mathbb{S}_0^+$ such that $(\hat{v}_n)$ is a Palais-Smale sequence for ψ_0 at the level c_0 and $\|\hat{v}_n - v_n\|_0 = o_n(1)$. Then, using

Proposition 13.4.2, and Theorem 13.4.6 and arguing as in the proof of Corollary 13.3.5, we complete the proof.   □

Finally, arguing as in the proof of Lemma 12.3.2 and using the fact that $V(\varepsilon \cdot) \geq V_0$, we obtain the following useful relation between c_ε and c_0:

Lemma 13.4.8 $\lim_{\varepsilon \to 0} c_\varepsilon = c_0$.

13.5   Barycenter Map and Multiplicity of Solutions to (13.2.2)

In this section, our main purpose is to apply the Lusternik-Schnirelman category theory to prove a multiplicity result for problem (13.2.2). We begin with the following technical result.

Lemma 13.5.1 *Let $\varepsilon_n \to 0$ and $(u_n) = (u_{\varepsilon_n}) \subset \mathcal{N}_{\varepsilon_n}$ be such that $\mathcal{J}_{\varepsilon_n}(u_n) \to c_0$. Then there exists $(\tilde{y}_n) = (\tilde{y}_{\varepsilon_n}) \subset \mathbb{R}^3$ such that the translated sequence*

$$\tilde{u}_n(x) = u_n(x + \tilde{y}_n)$$

has a subsequence which converges in $H^s(\mathbb{R}^3)$. Moreover, up to a subsequence, $(y_n) = (\varepsilon_n \tilde{y}_n)$ is such that $y_n \to y_0 \in M$.

Proof Since $\langle \mathcal{J}'_{\varepsilon_n}(u_n), u_n \rangle = 0$ and $\mathcal{J}_{\varepsilon_n}(u_n) \to c_0$, it is easy to see that (u_n) is bounded in $\mathcal{H}_{\varepsilon_n}$. Let us observe that $\|u_n\|_{\varepsilon_n} \nrightarrow 0$, since $c_0 > 0$. Therefore, arguing as in the first part of Lemma 7.3.7, we can find a sequence $(\tilde{y}_n) \subset \mathbb{R}^3$ and constants $R, \alpha > 0$ such that

$$\liminf_{n \to \infty} \int_{B_R(\tilde{y}_n)} u_n^2 \, dx \geq \alpha.$$

Set $\tilde{u}_n(x) = u_n(x + \tilde{y}_n)$. Then, $(\tilde{u}_n)$ is bounded in $H^s(\mathbb{R}^3)$, and we may assume that

$$\tilde{u}_n \rightharpoonup \tilde{u} \quad \text{in } H^s(\mathbb{R}^3),$$

for some $\tilde{u} \neq 0$. Let $(t_n) \subset (0, \infty)$ be such that $\tilde{v}_n = t_n \tilde{u}_n \in \mathcal{N}_0$ (see Lemma 13.4.1-(i)), and set $y_n = \varepsilon_n \tilde{y}_n$. Then, using (g_2), Lemma 13.1.2-(4) and the translation invariance of $\mathbb{R}^3$, we see that

$c_0 \leq \mathcal{J}_0(\tilde{v}_n)$

$$\leq \frac{1}{2} \int_{\mathbb{R}^3} |(-\Delta)^{\frac{s}{2}} \tilde{v}_n|^2 + V(\varepsilon_n x + y_n) \tilde{v}_n^2 \, dx + \frac{1}{4} \int_{\mathbb{R}^3} \phi_{\tilde{v}_n}^t \tilde{v}_n^2 \, dx$$

$$- \int_{\mathbb{R}^3} \left(F(\tilde{v}_n) + \frac{1}{2_s^*} (\tilde{v}_n^+)^{2_s^*} \right) dx$$

$$\leq \frac{t_n^2}{2} \int_{\mathbb{R}^3} |(-\Delta)^{\frac{s}{2}} u_n|^2 + V(\varepsilon_n x) u_n^2 \, dx + \frac{t_n^4}{4} \int_{\mathbb{R}^3} \phi_{u_n}^t u_n^2 \, dx - \int_{\mathbb{R}^3} G(\varepsilon_n z, t_n u_n) \, dx$$

$$= \mathcal{J}_{\varepsilon_n}(t_n u_n) \leq \mathcal{J}_{\varepsilon_n}(u_n) = c_0 + o_n(1),$$

and so

$$\mathcal{J}_0(\tilde{v}_n) \to c_0 \quad \text{and} \quad (\tilde{v}_n) \subset \mathcal{N}_0. \tag{13.5.1}$$

In particular, (13.5.1) yields that $(\tilde{v}_n)$ is bounded in $H^s(\mathbb{R}^3)$, so we may assume that $\tilde{v}_n \rightharpoonup \tilde{v}$. Obviously, (t_n) is bounded and we may assume that $t_n \to t_0 \geq 0$. If $t_0 = 0$, then from the boundedness of $(\tilde{u}_n)$ it follows that $\|\tilde{v}_n\|_0 = t_n \|\tilde{u}_n\|_0 \to 0$, that is, $\mathcal{J}_0(\tilde{v}_n) \to 0$, in contrast with the fact $c_0 > 0$. Hence, $t_0 > 0$. By the uniqueness of the weak limit, $\tilde{v} = t_0 \tilde{u}$ and $\tilde{v} \neq 0$. Using (13.5.1) and Lemma 13.4.7 we deduce that

$$\tilde{v}_n \to \tilde{v} \quad \text{in} \quad H^s(\mathbb{R}^3), \tag{13.5.2}$$

which implies that $\tilde{u}_n \to \tilde{u}$ in $H^s(\mathbb{R}^3)$ and

$$\mathcal{J}_0(\tilde{v}) = c_0 \quad \text{and} \quad \langle \mathcal{J}_0'(\tilde{v}), \tilde{v} \rangle = 0.$$

Now let us show that (y_n) admits a subsequence, still denoted by (y_n), such that $y_n \to y_0 \in M$. Assume, by contradiction, that (y_n) is not bounded. Then there exists a subsequence, still denoted by (y_n), such that $|y_n| \to \infty$. Since $u_n \in \mathcal{N}_{\varepsilon_n}$, we see that

$$\|\tilde{u}_n\|_0^2 \leq [\tilde{u}_n]_s^2 + \int_{\mathbb{R}^3} V(\varepsilon_n x + y_n) \tilde{u}_n^2 \, dx + \int_{\mathbb{R}^3} \phi_{\tilde{u}_n}^t \tilde{u}_n^2 \, dx = \int_{\mathbb{R}^3} g(\varepsilon_n x + y_n, \tilde{u}_n) \tilde{u}_n \, dx.$$

Take $R > 0$ such that $\Lambda \subset B_R$, and assume that $|y_n| > 2R$ for n large. Thus, for any $x \in B_{R/\varepsilon_n}$, we get $|\varepsilon_n x + y_n| \geq |y_n| - |\varepsilon_n x| > R$ for all n large enough. Hence, by the definition of g, we deduce that

$$\|\tilde{u}_n\|_0^2 \leq \int_{B_{R/\varepsilon_n}} \tilde{f}(\tilde{u}_n) \tilde{u}_n \, dx + \int_{\mathbb{R}^3 \setminus B_{R/\varepsilon_n}} f(\tilde{u}_n) \tilde{u}_n + \tilde{u}_n^{2_s^*} \, dx.$$

Since $\tilde{u}_n \to \tilde{u}$ in $H^s(\mathbb{R}^3)$, it follows from the dominated convergence theorem that

$$\int_{\mathbb{R}^3 \setminus B_{R/\varepsilon_n}} f(\tilde{u}_n) \tilde{u}_n \, dx = o_n(1).$$

Therefore,

$$\|\tilde{u}_n\|_0^2 \leq \frac{1}{K} \int_{B_{R/\varepsilon_n}} V_0 \tilde{u}_n^2 \, dx + o_n(1),$$

which yields

$$\left(1 - \frac{1}{K}\right) \|\tilde{u}_n\|_0^2 \leq o_n(1).$$

But $\tilde{u}_n \to \tilde{u} \neq 0$ and $K > 2$, so we get a contradiction. Thus (y_n) is bounded and, up to a subsequence, we may assume that $y_n \to y_0$. If $y_0 \notin \overline{\Lambda}$, then there exists $r > 0$ such that $y_n \in B_{r/2}(y_0) \subset \mathbb{R}^3 \setminus \overline{\Lambda}$ for any n large enough. Reasoning as before, we reach a contradiction. Hence, $y_0 \in \overline{\Lambda}$. Now, we show that $V(y_0) = V_0$. Assume, by contradiction, that $V(y_0) > V_0$. Taking into account (13.5.2), Fatou's lemma and the translation invariance of $\mathbb{R}^3$, we have

$$c_0 < \liminf_{n \to \infty} \left[\frac{1}{2} \left(\int_{\mathbb{R}^3} |(-\Delta)^{\frac{s}{2}} \tilde{v}_n|^2 + V(\varepsilon_n x + y_n) \tilde{v}_n^2 \, dx \right) \right.$$

$$\left. + \frac{1}{4} \int_{\mathbb{R}^3} \phi_{\tilde{v}_n}^t \tilde{v}_n^2 \, dx - \int_{\mathbb{R}^3} \left(F(\tilde{v}_n) + \frac{1}{2_s^*} (\tilde{v}_n^+)^{2_s^*} \right) dx \right]$$

$$\leq \liminf_{n \to \infty} \mathcal{J}_{\varepsilon_n}(t_n u_n) \leq \liminf_{n \to \infty} \mathcal{J}_{\varepsilon_n}(u_n) = c_0,$$

which is impossible. Therefore, in view of (V_2), we conclude that $y_0 \in M$. $\qquad \square$

Now, we aim to relate the number of positive solutions of (13.2.2) to the topology of the set Λ. For this reason, we take $\delta > 0$ such that

$$M_\delta = \{x \in \mathbb{R}^3 : \text{dist}(x, M) \leq \delta\} \subset \Lambda,$$

and let η be a smooth nonincreasing cut-off function defined in $[0, \infty)$ such that $\eta = 1$ in $[0, \frac{\delta}{2}]$, $\eta = 0$ in $[\delta, \infty)$, $0 \leq \eta \leq 1$ and $|\eta'| \leq c$ for some $c > 0$. For $y \in \Lambda$, define

$$\Psi_{\varepsilon, y}(x) = \eta(|\varepsilon x - y|) w \left(\frac{\varepsilon x - y}{\varepsilon} \right)$$

where $w \in H^s(\mathbb{R}^3)$ is a positive ground state solution to problem (13.4.1) (see Theorem 13.4.6). Let $t_\varepsilon > 0$ be the unique number such that

$$\max_{t \geq 0} \mathcal{J}_\varepsilon(t \Psi_{\varepsilon, y}) = \mathcal{J}_\varepsilon(t_\varepsilon \Psi_{\varepsilon, y}).$$

Finally, we introduce $\Phi_\varepsilon : M \to \mathcal{N}_\varepsilon$ given by

$$\Phi_\varepsilon(y) = t_\varepsilon \Psi_{\varepsilon,y}.$$

Lemma 13.5.2 *The functional Φ_ε has the property that*

$$\lim_{\varepsilon \to 0} \mathcal{J}_\varepsilon(\Phi_\varepsilon(y)) = c_0, \quad \text{uniformly in } y \in M.$$

Proof Suppose this is not the case, i.e., there exist $\delta_0 > 0$, $(y_n) \subset M$ and $\varepsilon_n \to 0$ such that

$$|\mathcal{J}_{\varepsilon_n}(\Phi_{\varepsilon_n}(y_n)) - c_0| \geq \delta_0. \tag{13.5.3}$$

Then, using the change of variable $z = \dfrac{\varepsilon_n x - y_n}{\varepsilon_n}$, if $z \in B_{\frac{\delta}{\varepsilon_n}}$, we have that $\varepsilon_n z \in B_\delta$ and $\varepsilon_n z + y_n \in B_\delta(y_n) \subset M_\delta \subset \Lambda$. Then, recalling that $G(x,t) = F(t) + \frac{1}{2_s^*}(t^+)^{2_s^*}$ for $(x,t) \in \Lambda \times \mathbb{R}$, we have

$$\mathcal{J}_\varepsilon(\Phi_{\varepsilon_n}(y_n)) = \frac{t_{\varepsilon_n}^2}{2}\left(\int_{\mathbb{R}^3}|(-\Delta)^{\frac{s}{2}}(\eta(|\varepsilon_n z|)w(z))|^2\,dz + \int_{\mathbb{R}^3}V(\varepsilon_n z + y_n)(\eta(|\varepsilon_n z|)w(z))^2\,dz\right)$$
$$+ \frac{t_{\varepsilon_n}^4}{4}\int_{\mathbb{R}^3}\phi_{\eta(|\varepsilon_n z|)}^t(\eta(|\varepsilon_n z|)w(z))^2\,dz - \int_{\mathbb{R}^3}F(t_{\varepsilon_n}\eta(|\varepsilon_n z|)w(z))\,dz$$
$$- \frac{t_{\varepsilon_n}^{2_s^*}}{2_s^*}\int_{\mathbb{R}^3}(\eta(|\varepsilon_n z|)w(z))^{2_s^*}\,dz. \tag{13.5.4}$$

Let us show that the sequence (t_{ε_n}) satisfies $t_{\varepsilon_n} \to 1$ as $\varepsilon_n \to 0$. By the definition of t_{ε_n}, it follows that $\langle \mathcal{J}_{\varepsilon_n}'(\Phi_{\varepsilon_n}(y_n)), \Phi_{\varepsilon_n}(y_n)\rangle = 0$, which gives

$$t_{\varepsilon_n}^2\left(\int_{\mathbb{R}^3}|(-\Delta)^{\frac{s}{2}}(\eta(|\varepsilon_n z|)w(z))|^2 + V(\varepsilon_n z + y_n)(\eta(|\varepsilon_n z|)w(z))^2\,dz\right)$$
$$+ t_{\varepsilon_n}^4\int_{\mathbb{R}^3}\phi_{\eta(|\varepsilon_n z|)}^t(\eta(|\varepsilon_n z|)w(z))^2\,dz$$
$$= \int_{\mathbb{R}^3}g(\varepsilon_n z + y_n, t_{\varepsilon_n}\eta(|\varepsilon_n z|)w(z))t_{\varepsilon_n}\eta(|\varepsilon_n z|)w(z)\,dz. \tag{13.5.5}$$

Since $\eta(|x|) = 1$ for $x \in B_{\frac{\delta}{2}}$ and $B_{\frac{\delta}{2}} \subset B_{\frac{\delta}{2\varepsilon_n}}$ for all n sufficiently large, (13.5.5) yields

$$\frac{1}{t_{\varepsilon_n}^2}\int_{\mathbb{R}^3}|(-\Delta)^{\frac{s}{2}}\Psi_{\varepsilon_n,y_n}|^2 + V(\varepsilon_n x)\Psi_{\varepsilon_n,y_n}^2\,dx + \int_{\mathbb{R}^3}\phi_{\Psi_{\varepsilon_n,y_n}}^t\Psi_{\varepsilon_n,y_n}^2\,dx$$

$$= \int_{\mathbb{R}^3} \frac{f(t_{\varepsilon_n} \Psi_{\varepsilon_n, y_n}) + (t_{\varepsilon_n} \Psi_{\varepsilon_n, y_n})^{2_s^* - 1}}{(t_{\varepsilon_n} \Psi_{\varepsilon_n, y_n})^3} \Psi_{\varepsilon_n, y_n}^4 \, dx$$

$$\geq t_{\varepsilon_n}^{2_s^* - 4} \int_{B_{\frac{\delta}{2}}} |w(z)|^{2_s^*} \, dz.$$

By the continuity of w, there exists a vector $\hat{z} \in \mathbb{R}^3$ such that

$$w(\hat{z}) = \min_{z \in \overline{B}_{\frac{\delta}{2}}} w(z) > 0,$$

which implies that

$$\frac{1}{t_{\varepsilon_n}^2} \int_{\mathbb{R}^3} |(-\Delta)^{\frac{s}{2}} \Psi_{\varepsilon_n, y_n}|^2 + V(\varepsilon_n x) \Psi_{\varepsilon_n, y_n}^2 \, dx + \int_{\mathbb{R}^3} \phi_{\Psi_{\varepsilon_n, y_n}}^t \Psi_{\varepsilon_n, y_n}^2 \, dx$$

$$\geq t_{\varepsilon_n}^{2_s^* - 4} w^{2_s^*}(\hat{z}) |B_{\frac{\delta}{2}}|. \tag{13.5.6}$$

Now, assume by contradiction that $t_{\varepsilon_n} \to \infty$. Lemma 1.4.8, Lemma 13.1.2-(7) and the dominated convergence theorem imply that

$$\|\Psi_{\varepsilon_n, y_n}\|_{\varepsilon_n}^2 \to \|w\|_0^2 \in (0, \infty), \quad \int_{\mathbb{R}^3} \phi_{\Psi_{\varepsilon_n, y_n}}^t \Psi_{\varepsilon_n, y_n}^2 \, dx \to \int_{\mathbb{R}^3} \phi_w^t w^2 \, dx,$$

$$\|\Psi_{\varepsilon_n, y_n}\|_{L^{2_s^*}(\mathbb{R}^3)} \to \|w\|_{L^{2_s^*}(\mathbb{R}^3)}, \quad \int_{\mathbb{R}^3} \frac{f(t_{\varepsilon_n} \Psi_{\varepsilon_n, y_n})}{(t_{\varepsilon_n} \Psi_{\varepsilon_n, y_n})^3} \Psi_{\varepsilon_n, y_n}^4 \, dx \to \int_{\mathbb{R}^3} \frac{f(t_0 w)}{(t_0 w)^3} w^4 \, dx. \tag{13.5.7}$$

Hence, using that $t_{\varepsilon_n} \to \infty$, (13.5.6) and (13.5.7), we obtain

$$\int_{\mathbb{R}^3} \phi_w^t w^2 \, dx = \infty,$$

which is a contradiction. Therefore, the sequence (t_{ε_n}) is bounded and, up to subsequence, we may assume that $t_{\varepsilon_n} \to t_0$ for some $t_0 \geq 0$. Let us prove that $t_0 > 0$. Suppose this is not the case, i.e., $t_0 = 0$. Then, taking into account (13.5.7) and the growth assumptions on g, we see that (13.5.5) gives

$$\|t_{\varepsilon_n} \Psi_{\varepsilon_n, y_n}\|_{\varepsilon_n}^2 \to 0$$

which is impossible because $t_{\varepsilon_n} \Psi_{\varepsilon_n, y_n} \in \mathcal{N}_{\varepsilon_n}$ and because of Remark 13.2.5. Hence, $t_0 > 0$. Then, letting $n \to \infty$ in (13.5.5), we deduce from (13.5.7) and the dominated convergence theorem that

$$\frac{1}{t_0^2} \|w\|_0^2 + \int_{\mathbb{R}^3} \phi_w^t w^2 \, dx = \int_{\mathbb{R}^3} \frac{f(t_0 w) + (t_0 w)^{2_s^* - 1}}{(t_0 w)^3} w^4 \, dx.$$

Since $w \in \mathcal{N}_0$ and (f_5) holds, we infer that $t_0 = 1$. Then, passing to the limit as $n \to \infty$ in (13.5.4), by $t_{\varepsilon_n} \to 1$ and (13.5.7) we obtain

$$\lim_{n \to \infty} \mathcal{J}_{\varepsilon_n}(\Phi_{\varepsilon_n}(y_n)) = \mathcal{J}_0(w) = c_0,$$

which contradicts (13.5.3). $\square$

It is time to define the barycenter map. Take $\rho = \rho(\delta) > 0$ such that $M_\delta \subset B_\rho$, and consider the map $\Upsilon : \mathbb{R}^3 \to \mathbb{R}^3$ given by

$$\Upsilon(x) = \begin{cases} x, & \text{if } |x| < \rho, \\ \frac{\rho x}{|x|}, & \text{if } |x| \geq \rho. \end{cases}$$

Define the barycenter map $\beta_\varepsilon : \mathcal{N}_\varepsilon \to \mathbb{R}^3$ as follows:

$$\beta_\varepsilon(u) = \frac{\displaystyle\int_{\mathbb{R}^3} \Upsilon(\varepsilon x) u^2(x) \, dx}{\displaystyle\int_{\mathbb{R}^3} u^2(x) \, dx}.$$

Arguing as in the proof of Lemma 6.3.18, we see that the function β_ε has the following property:

Lemma 13.5.3

$$\lim_{\varepsilon \to 0} \beta_\varepsilon(\Phi_\varepsilon(y)) = y, \quad \text{uniformly in } y \in M.$$

Next, we introduce a subset $\tilde{\mathcal{N}}_\varepsilon$ of $\mathcal{N}_\varepsilon$ by taking a function $h_1 : \mathbb{R}_+ \to \mathbb{R}_+$ such that $h_1(\varepsilon) \to 0$ as $\varepsilon \to 0$ and setting

$$\tilde{\mathcal{N}}_\varepsilon = \{u \in \mathcal{N}_\varepsilon : \mathcal{J}_\varepsilon(u) \leq c_0 + h_1(\varepsilon)\}.$$

By Lemma 13.5.2, $h_1(\varepsilon) = \sup_{y \in M} |\mathcal{J}_\varepsilon(\Phi_\varepsilon(y)) - c_0| \to 0$ as $\varepsilon \to 0$. Therefore, $\tilde{\mathcal{N}}_\varepsilon \neq \emptyset$ for any $\varepsilon > 0$. Additionally, as in the proof of Lemma 6.3.19, we have:

Lemma 13.5.4 *For every $\delta > 0$,*

$$\lim_{\varepsilon \to 0} \sup_{u \in \tilde{\mathcal{N}}_\varepsilon} \text{dist}(\beta_\varepsilon(u), M_\delta) = 0.$$

Now we prove a multiplicity result for (13.2.2).

Theorem 13.5.5 *Assume that* (V_1)-(V_2) *and* (f_1)–(f_4) *hold. Then, for any* $\delta > 0$ *such that* $M_\delta \subset \Lambda$, *there exists* $\bar{\varepsilon}_\delta > 0$ *such that, for any* $\varepsilon \in (0, \bar{\varepsilon}_\delta)$, *problem* (13.2.2) *has at least* $\mathrm{cat}_{M_\delta}(M)$ *positive solutions.*

Proof For any $\varepsilon > 0$, define the map $\alpha_\varepsilon : M \to \mathbb{S}_\varepsilon^+$ by $\alpha_\varepsilon(y) = m_\varepsilon^{-1}(\Phi_\varepsilon(y))$.
　　By Lemma 13.5.2,

$$\lim_{\varepsilon \to 0} \psi_\varepsilon(\alpha_\varepsilon(y)) = \lim_{\varepsilon \to 0} \mathcal{J}_\varepsilon(\Phi_\varepsilon(y)) = c_0, \quad \text{uniformly in } y \in M. \tag{13.5.8}$$

Set

$$\widetilde{\mathbb{S}}_\varepsilon^+ = \{w \in \mathbb{S}_\varepsilon^+ : \psi_\varepsilon(w) \le c_0 + h_1(\varepsilon)\},$$

where $h_1(\varepsilon) = \sup_{y \in M} |\psi_\varepsilon(\alpha_\varepsilon(y)) - c_0|$. It follows from (13.5.8) that $h_1(\varepsilon) \to 0$ as $\varepsilon \to 0$. By the definition of $h_1(\varepsilon)$, we see that, for all $y \in M$ and $\varepsilon > 0$, $\alpha_\varepsilon(y) \in \widetilde{\mathbb{S}}_\varepsilon^+$. Consequently, $\widetilde{\mathbb{S}}_\varepsilon^+ \ne \emptyset$ for all $\varepsilon > 0$.

　　From the above considerations, together with Lemma 13.2.2-(ii), Lemma 13.5.2, Lemma 13.5.3 and Lemma 13.5.4, we see that there exists $\bar{\varepsilon} = \bar{\varepsilon}_\delta > 0$ such that the following diagram

$$M \xrightarrow{\Phi_\varepsilon} \Phi_\varepsilon(M) \xrightarrow{m_\varepsilon^{-1}} \alpha_\varepsilon(M) \xrightarrow{m_\varepsilon} \Phi_\varepsilon(M) \xrightarrow{\beta_\varepsilon} M_\delta$$

is well defined for any $\varepsilon \in (0, \bar{\varepsilon})$. Thanks to Lemma 13.5.3, and decreasing $\bar{\varepsilon}$ if necessary, we have that $\beta_\varepsilon(\Phi_\varepsilon(y)) = y + \theta(\varepsilon, y)$ for all $y \in M$, for some function $\theta(\varepsilon, y)$ such that $|\theta(\varepsilon, y)| < \frac{\delta}{2}$ uniformly in $y \in M$ and for all $\varepsilon \in (0, \bar{\varepsilon})$. Then, it is easy to check that the map $H : [0, 1] \times M \to M_\delta$ defined by $H(t, y) = y + (1 - t)\theta(\varepsilon, y)$ is a homotopy between $\beta_\varepsilon \circ \Phi_\varepsilon = (\beta_\varepsilon \circ m_\varepsilon) \circ (m_\varepsilon^{-1} \circ \Phi_\varepsilon)$ and the inclusion map id: $M \to M_\delta$. This fact together with Lemma 6.3.21 implies that

$$\mathrm{cat}_{\alpha_\varepsilon(M)} \alpha_\varepsilon(M) \ge \mathrm{cat}_{M_\delta}(M). \tag{13.5.9}$$

Applying Corollary 13.3.5, Lemma 13.4.8 and Theorem 2.4.6 with $c = c_\varepsilon \le c_0 + h_1(\varepsilon) = d$ and $K = \alpha_\varepsilon(M)$, we deduce that ψ_ε has at least $\mathrm{cat}_{\alpha_\varepsilon(M)} \alpha_\varepsilon(M)$ critical points on $\widetilde{\mathbb{S}}_\varepsilon^+$. Therefore, by Proposition 13.2.3-(d) and (13.5.9), we infer that (13.2.2) has at least $\mathrm{cat}_{M_\delta}(M)$ solutions. □

13.6 Proof of Theorem 13.1.1

This last section is devoted to the proof of Theorem 13.1.1, namely, that the solutions of (13.2.2) are indeed solutions of the original problem (13.1.1).

Firstly, we prove the following useful L^∞-estimate for the solutions of the modified problem (13.2.2).

Lemma 13.6.1 *Let $\varepsilon_n \to 0$ and $u_n \in \tilde{\mathcal{N}}_{\varepsilon_n}$ be a solution to (13.2.2). Then, up to a subsequence, $v_n = u_n(\cdot + \tilde{y}_n) \in L^\infty(\mathbb{R}^N)$ and there exists $C > 0$ such that*

$$\|v_n\|_{L^\infty(\mathbb{R}^3)} \leq C \quad \text{for all } n \in \mathbb{N},$$

where $(\tilde{y}_n)$ is given in Lemma 13.5.1. Moreover, $v_n(x) \to 0$ as $|x| \to \infty$ uniformly in $n \in \mathbb{N}$.

Proof Recalling that $\mathcal{J}_{\varepsilon_n}(u_n) \leq c_0 + h_1(\varepsilon_n)$, with $h_1(\varepsilon_n) \to 0$ as $n \to \infty$, we can proceed as in the proof of (13.5.1) to deduce that $\mathcal{J}_{\varepsilon_n}(u_n) \to c_0$. Invoking Lemma 13.5.1, we find a sequence $(\tilde{y}_n) \subset \mathbb{R}^3$ such that $\varepsilon_n \tilde{y}_n \to y_0 \in M$ and $v_n = u_n(\cdot + \tilde{y}_n)$ has a convergent subsequence in $H^s(\mathbb{R}^3)$.

To obtain the L^∞-estimate, we can proceed as in the proof of Lemma 6.3.23. Indeed,

$$\left(\frac{1}{\beta}\right)^2 S_* \|v_n v_{L,n}^{\beta-1}\|_{L^{2^*_s}(\mathbb{R}^3)}^2 + \int_{\mathbb{R}^3} V_n(x) v_n^2 v_{L,n}^{2(\beta-1)}\, dx$$

$$\leq \iint_{\mathbb{R}^6} \left[\frac{(v_n(x) - v_n(y))}{|x-y|^{N+2s}} ((v_n v_{L,n}^{2(\beta-1)})(x) - (v_n v_{L,n}^{2(\beta-1)})(y)) \right] dx\, dy$$

$$+ \int_{\mathbb{R}^3} V_n(x) v_n^2 v_{L,n}^{2(\beta-1)}\, dx$$

$$\leq \int_{\mathbb{R}^3} g_n(x, v_n) v_n v_{L,n}^{2(\beta-1)}\, dx, \tag{13.6.1}$$

where we used the notations $V_n(x) = V(\varepsilon_n x + \varepsilon_n \tilde{y}_n)$ and $g_n(x, v_n) = g(\varepsilon_n x + \varepsilon_n \tilde{y}_n, v_n)$.

Assumptions (g_1) and (g_2) imply that for every $\xi > 0$ there exists $C_\xi > 0$ such that

$$|g_n(x, v_n)| \leq \xi |v_n| + C_\xi |v_n|^{2^*_s - 1}.$$

Taking $\xi \in (0, V_0)$, and using (13.6.1) we obtain

$$\|v_n v_{L,n}^{\beta-1}\|_{L^{2^*_s}(\mathbb{R}^3)}^2 \leq C\beta^2 \int_{\mathbb{R}^3} |v_n|^{2^*_s} v_{L,n}^{2(\beta-1)}\, dx.$$

Therefore, we deduce that $\|v_n\|_{L^\infty(\mathbb{R}^3)} \le C$ for all $n \in \mathbb{N}$. Let now $w_n(x, y) = \text{Ext}(v_n) = P_s(x, y) * v_n(x)$ the s-harmonic extension of v_n and we note that it solves

$$\begin{cases} -\text{div}(y^{1-2s}\nabla w_n) = 0 & \text{in } \mathbb{R}_+^{3+1}, \\ w_n(\cdot, 0) = v_n & \text{on } \partial\mathbb{R}_+^{3+1}, \\ \frac{\partial w}{\partial \nu^{1-2s}} = -V_n(x)v_n - \phi_{v_n}^t v_n + g_n(x, v_n) & \text{on } \partial\mathbb{R}_+^{3+1}. \end{cases}$$

From the uniform boundedness of (v_n) in $L^2(\mathbb{R}^3) \cap L^\infty(\mathbb{R}^3)$, we deduce that $0 \le \phi_{v_n}^t \le C$ for all $n \in \mathbb{N}$ (see the estimate in Theorem 13.4.6). Then we can argue as in the proof of Lemma 6.3.23 to conclude that $v_n(x) \to 0$ as $|x| \to \infty$ uniformly in $n \in \mathbb{N}$ (see also Remark 7.2.10 and note that v_n is a subsolution to $(-\Delta)^s v_n + V_0 v_n = g_n(x, v_n)$ in $\mathbb{R}^3$).

$\square$

Now, we are ready to give the proof of the main result of this chapter.

Proof of Theorem 13.1.1 Take $\delta > 0$ such that $M_\delta \subset \Lambda$. We begin by proving that there exists $\tilde{\varepsilon}_\delta > 0$ such that for every $\varepsilon \in (0, \tilde{\varepsilon}_\delta)$ and every solution $u_\varepsilon \in \tilde{\mathcal{N}}_\varepsilon$ of (13.2.2),

$$\|u_\varepsilon\|_{L^\infty(\mathbb{R}^3 \setminus \Lambda_\varepsilon)} < a. \tag{13.6.2}$$

Assume, by contradiction, that for some subsequence (ε_n) such that $\varepsilon_n \to 0$, there exists $u_n = u_{\varepsilon_n} \in \tilde{\mathcal{N}}_{\varepsilon_n}$ such that $\mathcal{J}'_{\varepsilon_n}(u_n) = 0$ and

$$\|u_n\|_{L^\infty(\mathbb{R}^3 \setminus \Lambda_{\varepsilon_n})} \ge a. \tag{13.6.3}$$

Since $\mathcal{J}_{\varepsilon_n}(u_n) \le c_0 + h_1(\varepsilon_n)$ and $h_1(\varepsilon_n) \to 0$, we can argue as in the first part of the proof of Lemma 13.5.1 to deduce that $\mathcal{J}_{\varepsilon_n}(u_n) \to c_0$. In view of Lemma 13.5.1, there exists a sequence $(\tilde{y}_n) \subset \mathbb{R}^3$ such that $\tilde{u}_n = u_n(\cdot + \tilde{y}_n) \to \tilde{u}$ in $H^s(\mathbb{R}^3)$ and $\varepsilon_n \tilde{y}_n \to y_0 \in M$. Now, if we choose $r > 0$ such that $B_r(y_0) \subset B_{2r}(y_0) \subset \Lambda$, we see that $B_{\frac{r}{\varepsilon_n}}(\frac{y_0}{\varepsilon_n}) \subset \Lambda_{\varepsilon_n}$. Then, for any $y \in B_{\frac{r}{\varepsilon_n}}(\tilde{y}_n)$, we have

$$\left| y - \frac{y_0}{\varepsilon_n} \right| \le |y - \tilde{y}_n| + \left| \tilde{y}_n - \frac{y_0}{\varepsilon_n} \right| < \frac{1}{\varepsilon_n}(r + o_n(1)) < \frac{2r}{\varepsilon_n} \quad \text{for } n \text{ sufficiently large.}$$

Therefore, for these values of n, we get

$$\mathbb{R}^3 \setminus \Lambda_{\varepsilon_n} \subset \mathbb{R}^3 \setminus B_{\frac{r}{\varepsilon_n}}(\tilde{y}_n). \tag{13.6.4}$$

On the other hand, using Lemma 13.6.1 we can verify that

$$\tilde{u}_n(x) \to 0 \quad \text{as } |x| \to \infty \tag{13.6.5}$$

uniformly in $n \in \mathbb{N}$. Consequently, there exists $R > 0$ such that

$$\tilde{u}_n(x) < a \quad \text{for all } |x| \geq R, n \in \mathbb{N}.$$

Hence, $u_n(x) < a$ for any $x \in \mathbb{R}^3 \setminus B_R(\tilde{y}_n)$ and $n \in \mathbb{N}$. On the other hand, there exists $v \in \mathbb{N}$ such that for any $n \geq v$ and $r/\varepsilon_n > R$ we have

$$\mathbb{R}^3 \setminus \Lambda_{\varepsilon_n} \subset \mathbb{R}^3 \setminus B_{\frac{r}{\varepsilon_n}}(\tilde{y}_n) \subset \mathbb{R}^3 \setminus B_R(\tilde{y}_n),$$

which implies that $u_n(x) < a$ for any $x \in \mathbb{R}^3 \setminus \Lambda_{\varepsilon_n}$ and $n \geq v$. This is impossible in view of (13.6.3). Let $\bar{\varepsilon}_\delta > 0$ be given by Theorem 13.5.5, and fix $\varepsilon \in (0, \varepsilon_\delta)$, where $\varepsilon_\delta = \min\{\tilde{\varepsilon}_\delta, \bar{\varepsilon}_\delta\}$. By Theorem 13.5.5, problem (13.2.2) admits at least $\mathrm{cat}_{M_\delta}(M)$ nontrivial solutions. Let u_ε be one of these solutions. Since $u_\varepsilon \in \tilde{\mathcal{N}}_\varepsilon$ satisfies (13.6.2), by the definition of g it follows that u_ε is a solution of (13.2.1). Then $\hat{u}_\varepsilon(x) = u_\varepsilon(x/\varepsilon)$ is a solution to (13.1.1). Therefore, (13.1.1) has at least $\mathrm{cat}_{M_\delta}(M)$ nontrivial solutions.

Finally, let us analyze the behavior of the maximum points of solutions to problem (13.1.1). Take $\varepsilon_n \to 0$ and consider a sequence $(u_n) \subset \mathcal{H}_{\varepsilon_n}$ of solutions to (13.2.1) as above. Note that (g_1) implies that we can find $\gamma \in (0, a)$ sufficiently small such that

$$g(\varepsilon x, t)t \leq \frac{V_0}{K}t^2 \quad \text{for any } x \in \mathbb{R}^3, 0 \leq t \leq \gamma. \tag{13.6.6}$$

Arguing as before, we can find $R > 0$ such that

$$\|u_n\|_{L^\infty(B_R^c(\tilde{y}_n))} < \gamma. \tag{13.6.7}$$

Moreover, up to extracting a subsequence, we may assume that

$$\|u_n\|_{L^\infty(B_R(\tilde{y}_n))} \geq \gamma. \tag{13.6.8}$$

Indeed, if (13.6.8) does not hold, then, in view of (13.6.7), we have $\|u_n\|_{L^\infty(\mathbb{R}^3)} < \gamma$. Further, using that $\langle \mathcal{J}_{\varepsilon_n}'(u_n), u_n \rangle = 0$ and (13.6.6) we infer that

$$\|u_n\|_{\varepsilon_n}^2 \leq \|u_n\|_{\varepsilon_n}^2 + \int_{\mathbb{R}^3} \phi_{u_n}^t u_n^2 \, dx = \int_{\mathbb{R}^3} g(\varepsilon_n x, u_n)u_n \, dx \leq \frac{V_0}{K} \int_{\mathbb{R}^3} u_n^2 \, dx$$

which yields $\|u_n\|_{\varepsilon_n} = 0$, which is impossible. Hence, (13.6.8) holds. Taking into account (13.6.7) and (13.6.8), we deduce that if $p_n \in \mathbb{R}^3$ is a global maximum point of u_n, then it belongs to $B_R(\tilde{y}_n)$. Therefore, $p_n = \tilde{y}_n + q_n$ for some $q_n \in B_R$. Consequently, $\eta_n = \varepsilon_n \tilde{y}_n + \varepsilon_n q_n$ is a global maximum point of $\hat{u}_n(x) = u_n(x/\varepsilon_n)$. Since $|q_n| < R$ for any $n \in \mathbb{N}$ and $\varepsilon_n \tilde{y}_n \to y_0 \in M$ (in view of Lemma 13.5.1), the continuity of V ensures that

$$\lim_{n \to \infty} V(\eta_n) = V(y_0) = V_0.$$

To conclude the proof of Theorem 13.1.1, we estimate the decay of solutions to (13.1.1). By Lemma 3.2.17, we can find a positive function w such that

$$0 < w(x) \leq \frac{C}{1 + |x|^{3+2s}} \quad \text{for all } x \in \mathbb{R}^3, \tag{13.6.9}$$

and

$$(-\Delta)^s w + \frac{V_0}{2} w = 0 \text{ in } \mathbb{R}^3 \setminus \overline{B}_{R_1}, \tag{13.6.10}$$

in the classical sense, for some suitable $R_1 > 0$. By (g_1) and (13.6.5), we can find $R_2 > 0$ sufficiently large such that

$$(-\Delta)^s \tilde{u}_n + \frac{V_0}{2} \tilde{u}_n = g_n(x, \tilde{u}_n) - \left(V_n - \frac{V_0}{2} \right) \tilde{u}_n - \phi_{\tilde{u}_n}^t \tilde{u}_n$$

$$\leq g_n(x, \tilde{u}_n) - \frac{V_0}{2} \tilde{u}_n \leq 0 \text{ in } \mathbb{R}^3 \setminus \overline{B}_{R_2}. \tag{13.6.11}$$

Choose $R_3 = \max\{R_1, R_2\}$, and set

$$a = \min_{\overline{B}_{R_3}} w > 0 \quad \text{and} \quad \tilde{w}_n = (b+1)w - a\tilde{u}_n, \tag{13.6.12}$$

where $b = \sup_{n \in \mathbb{N}} \|\tilde{u}_n\|_{L^\infty(\mathbb{R}^3)} < \infty$. We claim that

$$\tilde{w}_n \geq 0 \text{ in } \mathbb{R}^3. \tag{13.6.13}$$

Indeed, note first that (13.6.10), (13.6.11) and (13.6.12) yield

$$\tilde{w}_n \geq ba + w - ba > 0 \quad \text{in } B_{R_3}, \tag{13.6.14}$$

$$(-\Delta)^s \tilde{w}_n + \frac{V_0}{2} \tilde{w}_n \geq 0 \quad \text{in } \mathbb{R}^3 \setminus \overline{B}_{R_3}. \tag{13.6.15}$$

Then we can use Lemma 1.3.8 to verify that (13.6.13) holds. Using (13.6.9), we see that

$$\tilde{u}_n(x) \leq \frac{\tilde{C}}{1 + |x|^{3+2s}} \quad \text{for all } x \in \mathbb{R}^3, n \in \mathbb{N}, \tag{13.6.16}$$

for some $\tilde{C} > 0$. Since $\hat{u}_n(x) = u_n(\frac{x}{\varepsilon_n}) = \tilde{u}_n(\frac{x}{\varepsilon_n} - \tilde{y}_n)$ and $\eta_n = \varepsilon_n \tilde{y}_n + \varepsilon_n q_n$, inequality (13.6.16) implies that

$$0 < \hat{u}_n(x) = u_n\left(\frac{x}{\varepsilon_n}\right) = \tilde{u}_n\left(\frac{x}{\varepsilon_n} - \tilde{y}_{\varepsilon_n}\right)$$

$$\leq \frac{\tilde{C}}{1 + |\frac{x}{\varepsilon_n} - \tilde{y}_n|^{3+2s}}$$

$$= \frac{\tilde{C}\,\varepsilon_n^{3+2s}}{\varepsilon_n^{3+2s} + |x - \varepsilon_n\,\tilde{y}_n|^{3+2s}}$$

$$\leq \frac{\tilde{C}\,\varepsilon_n^{3+2s}}{\varepsilon_n^{3+2s} + |x - \eta_n|^{3+2s}} \qquad \text{for all } x \in \mathbb{R}^3.$$

This completes the proof of Theorem 13.1.1.   □

Remark 13.6.2 The approach used in this chapter can be easily adapted to deal with the subcritical case (in this situation the Palais-Smale condition holds for all $d \in \mathbb{R}$).

An Existence Result for a Fractional Kirchhoff–Schrödinger–Poisson System

14

14.1 Introduction

In this chapter we deal with the following nonlinear fractional Kirchhoff–Schrödinger–Poisson system

$$\begin{cases} \left(p + q \iint_{\mathbb{R}^6} \frac{|u(x)-u(y)|^2}{|x-y|^{N+2s}}\, dxdy \right)(-\Delta)^s u + \mu\phi u = g(u) & \text{in } \mathbb{R}^3, \\ (-\Delta)^t \phi = \mu u^2 & \text{in } \mathbb{R}^3, \end{cases} \tag{14.1.1}$$

where $s \in (\frac{3}{4}, 1)$, $t \in (0, 1)$, $p > 0$, $q \geq 0$, $\mu > 0$ is a parameter, and $g : \mathbb{R} \to \mathbb{R}$ is an odd $C^{1,\alpha}(\mathbb{R})$ nonlinearity, with $\alpha > \max\{0, 1-2s\}$, satisfying the following Berestycki–Lions type assumptions [100]:

(g_1) $-\infty < \liminf_{\tau \to 0} \frac{g(\tau)}{\tau} \leq \limsup_{\tau \to 0} \frac{g(\tau)}{\tau} = -m < 0$;

(g_2) $-\infty \leq \limsup_{\tau \to \infty} \frac{g(\tau)}{\tau^{2^*_s-1}} \leq 0$, where $2^*_s = \frac{6}{3-2s}$;

(g_3) there exists $\zeta > 0$ such that $G(\zeta) = \int_0^\zeta g(r)\, dr > 0$.

We note that if $\mu = q = 0$ and $p = 1$, problem (14.1.1) reduces to the fractional Schrödinger equation

$$(-\Delta)^s u + V(x)u = g(u) \quad \text{in } \mathbb{R}^3. \tag{14.1.2}$$

On the other hand, when $\mu = 0$, problem (14.1.1) reduces to the fractional Kirchhoff equation

$$\left(p + q \iint_{\mathbb{R}^6} \frac{|u(x)-u(y)|^2}{|x-y|^{N+2s}}\, dxdy \right)(-\Delta)^s u = g(u) \text{ in } \mathbb{R}^3, \tag{14.1.3}$$

© The Author(s), under exclusive license to Springer Nature Switzerland AG 2021
V. Ambrosio, *Nonlinear Fractional Schrödinger Equations in $\mathbb{R}^N$*,
Frontiers in Mathematics, https://doi.org/10.1007/978-3-030-60220-8_14

Finally, when $p = 1$ and $q = 0$, (14.1.1) becomes the fractional Schrödinger-Poisson system

$$
\begin{cases}
(-\Delta)^s u + \mu \phi u = g(u) & \text{in } \mathbb{R}^3, \\
(-\Delta)^t \phi = \mu u^2 & \text{in } \mathbb{R}^3.
\end{cases}
\tag{14.1.4}
$$

In the present chapter, we prove an existence result for (14.1.1) when $\mu > 0$ is small enough. More precisely, we are able to show that:

Theorem 14.1.1 *Let us suppose that (g_1), (g_2), and (g_3) are satisfied. Then, there exists $\mu_0 > 0$ such that for any $0 < \mu < \mu_0$, the system* (14.1.1) *admits a positive solution $(u, \phi) \in H^s(\mathbb{R}^3) \times \mathcal{D}^{t,2}(\mathbb{R}^3)$.*

The proof of our main result is obtained by exploiting suitable variational methods based on a truncation argument and the Struwe's monotonicity trick developed by Jeanjean [231]. It is worth pointing out that our approach for attacking the problem has to take care of the presence of general nonlinearities and combined effects of different nonlocal terms, so an accurate and delicate analysis is required.

14.2   Struwe–Jeanjean Monotonicity Trick for a Perturbed Functional

Since we are dealing with general nonlinearities and we seek positive solutions of (14.2.7), similarly to [100] (see also [82, 139, 342]), we modify the nonlinearity g in a convenient way. Without loss of generality, we assume that

$$
0 < \zeta = \inf\{\tau \in (0, \infty) : G(\tau) > 0\},
$$

where $\zeta > 0$ is given in (g_3). Define $\tilde{g} : \mathbb{R} \to \mathbb{R}$ by

$$
\tilde{g}(\tau) =
\begin{cases}
g(\tau) & \text{for } \tau \in [0, \tau_0], \\
0 & \text{for } \tau \in \mathbb{R} \setminus [0, \tau_0],
\end{cases}
$$

where $\tau_0 = \min\{\tau \in (\zeta, \infty) : g(\tau) = 0\}$ ($\tau_0 = \infty$ if $g(\tau) > 0$ for all $\tau \geq \zeta$). It is easy to check that if u is a nontrivial solution of (14.1.1) with $\tilde{g}$ in the place of g, then $0 \leq u \leq \tau_0$ in $\mathbb{R}^3$, that is, u is a non-negative solution of (14.1.1) with nonlinearity g. Hence, we assume that g is extended as $\tilde{g}$. In particular, g satisfies the assumptions (g_1), (g_3) and

$$
\lim_{t \to \infty} \frac{g(\tau)}{\tau^{2_s^* - 1}} = 0.
\tag{g_2'}
$$

Now set

$$
g_1(\tau) = \begin{cases} (g(\tau) + m\tau)^+ & \text{for } \tau \geq 0, \\ 0 & \text{for } \tau < 0, \end{cases}
$$

$$
g_2(\tau) = g_1(\tau) - g(\tau) \quad \text{for } \tau \in \mathbb{R}.
$$

Note that $g_1, g_2 \geq 0$ in $[0, \infty)$ and

$$
\lim_{\tau \to 0} \frac{g_1(\tau)}{\tau} = 0, \tag{14.2.1}
$$

$$
\lim_{\tau \to \infty} \frac{g_1(\tau)}{\tau^{2_s^* - 1}} = 0, \tag{14.2.2}
$$

$$
g_2(\tau) \geq m\tau \quad \text{for all } \tau \geq 0. \tag{14.2.3}
$$

Consequently, for every $\varepsilon > 0$ there exists $C_\varepsilon > 0$ such that

$$
g_1(\tau) \leq C_\varepsilon \tau^{2_s^* - 1} + \varepsilon g_2(\tau) \quad \text{for all } \tau \geq 0. \tag{14.2.4}
$$

Set

$$
G_i(\tau) = \int_0^\tau g_i(r) \, dr, \quad i = 1, 2.
$$

Then, by (14.2.1)–(14.2.4), we have

$$
G_2(\tau) \geq \frac{m}{2} \tau^2 \quad \text{for all } \tau \in \mathbb{R}, \tag{14.2.5}
$$

and for any $\varepsilon > 0$ there exists $C_\varepsilon > 0$ such that

$$
G_1(\tau) \leq \varepsilon \, G_2(\tau) + C_\varepsilon \, |\tau|^{2_s^*} \quad \text{for all } \tau \in \mathbb{R}. \tag{14.2.6}
$$

As observed in Chap. 13, the system (14.1.1) can be reduced to a single equation. Substituting the expression (13.1.5) of ϕ_u^t in (14.1.1), we obtain the following fractional equation:

$$
(p + q[u]_s^2)(-\Delta)^s u + \mu \phi_u^t u = g(u) \text{ in } \mathbb{R}^3, \tag{14.2.7}
$$

whose solutions can be found by looking for critical points of the functional $\mathcal{J}_\mu : H^s(\mathbb{R}^3) \to \mathbb{R}$ defined by

$$
\mathcal{J}_\mu(u) = \frac{p}{2}[u]_s^2 + \frac{q}{4}[u]_s^4 + \frac{\mu}{4} \int_{\mathbb{R}^3} \phi_u^t u^2 \, dx - \int_{\mathbb{R}^3} G(u) \, dx. \tag{14.2.8}
$$

From the growth assumptions on g and Lemma 13.1.2-(1), it is easy to check that $\mathcal{J}_\mu \in C^1(H^s(\mathbb{R}^3), \mathbb{R})$ and that the critical points of $\mathcal{J}_\mu$ are the weak solutions of (14.2.7).

Definition 14.2.1

(i) We say that $(u, \phi) \in H^s(\mathbb{R}^3) \times \mathcal{D}^{t,2}(\mathbb{R}^3)$ is a weak solution to (14.1.1) if u is a weak solution to (14.2.7).

(ii) We say that $u \in H^s(\mathbb{R}^3)$ is a weak solution to (14.2.7) if

$$\int_{\mathbb{R}^3} \left[(p + q[u]_s^2)(-\Delta)^{\frac{s}{2}} u (-\Delta)^{\frac{s}{2}} v + \mu \phi_u^t uv \right] dx = \int_{\mathbb{R}^3} g(u) v\, dx$$

for any $v \in H^s(\mathbb{R}^3)$.

We will look for critical points of $\mathcal{J}_\mu$ on $H_{\mathrm{rad}}^s(\mathbb{R}^3)$, which is a natural constraint (we remark that, by Lemma 13.1.2-(2), if u is a radial function, then so is ϕ_u^t). Due to the presence of different nonlocal terms and the general nonlinearity, it is not easy to verify the geometric assumptions of the mountain pass theorem and the boundedness of Palais–Smale sequences for $\mathcal{J}_\mu$. Therefore, inspired by [232, 239], we introduce the cut-off function $\chi \in C^\infty([0, \infty), \mathbb{R})$ by

$$\begin{cases} \chi(\tau) = 1 & \text{for } \tau \in [0, 1], \\ 0 \le \chi(\tau) \le 1 & \text{for } \tau \in (1, 2), \\ \chi(\tau) = 0 & \text{for } \tau \in [2, \infty), \\ \|\chi'\|_{L^\infty(0,\infty)} \le 2, \end{cases}$$

and then

$$\xi_k(u) = \chi\left(\frac{\|u\|_{L^\alpha(\mathbb{R}^3)}^\alpha}{k^\alpha} \right) \quad \text{with } \alpha = \frac{12}{3 + 2t}.$$

Now consider the truncated functional $\mathcal{J}_\mu^k : H_{\mathrm{rad}}^s(\mathbb{R}^3) \to \mathbb{R}$ given by

$$\mathcal{J}_\mu^k(u) = \frac{p}{2}[u]_s^2 + \frac{q}{4}[u]_s^4 + \frac{\mu}{4}\xi_k(u) \int_{\mathbb{R}^3} \phi_u^t u^2\, dx - \int_{\mathbb{R}^3} G(u)\, dx.$$

Clearly, if u is a critical point of $\mathcal{J}_\mu^k$ with $\|u\|_{L^\alpha(\mathbb{R}^3)} \le k$, then u is also a critical point of $\mathcal{J}_\mu$. The C^1-functional $\mathcal{J}_\mu^k$ satisfies the geometric assumptions of the mountain pass theorem (note that $s > \frac{3}{4}$) but, since g is a general nonlinearity, we are not able to prove the boundedness of the Palais–Smale sequences. For this reason, we use a slightly different version of Theorem 5.2.2.

Theorem 14.2.2 ([82, 231]) *Let $(X, \| \cdot \|)$ be a Banach space and let $\Lambda \subset \mathbb{R}_+$ be an interval. Consider a family $(\mathcal{J}_\lambda)_{\lambda \in \Lambda}$ of C^1-functionals on X of the form*

$$\mathcal{J}_\lambda(u) = A(u) - \lambda B(u), \quad \text{for } \lambda \in \Lambda,$$

with B nonnegative and either $A(u) \to \infty$ or $B(u) \to \infty$ as $\|u\| \to \infty$, and such that $\mathcal{J}_\lambda(0) = 0$. For $\lambda \in \Lambda$, set

$$\Gamma_\lambda = \{\gamma \in C([0, 1], X) : \gamma(0) = 0, \mathcal{J}_\lambda(\gamma(1)) < 0\}.$$

If for any $\lambda \in \Lambda$ the set Γ_λ is nonempty and

$$c_\lambda = \inf_{\gamma \in \Gamma_\lambda} \max_{t \in [0,1]} \mathcal{J}_\lambda(\gamma(t)) > 0,$$

then, for almost every $\lambda \in \Lambda$, there is a sequence $(u_j) \subset X$ such that

  (i) *(u_j) is bounded;*
 (ii) *$\mathcal{J}_\lambda(u_j) \to c_\lambda$;*
(iii) *$\mathcal{J}'_\lambda(u_j) \to 0$ on X^*.*

Moreover, the function $\lambda \mapsto c_\lambda$ is non-increasing and continuous from the left.

In order to apply Theorem 14.2.2, we consider the following parametrized family of C^1-functionals:

$$\mathcal{J}^k_{\mu,\lambda}(u) = \frac{p}{2}[u]_s^2 + \frac{q}{4}[u]_s^4 + \frac{\mu}{4}\xi_k(u)\int_{\mathbb{R}^3} \phi_u^t u^2 \, dx + \int_{\mathbb{R}^3} G_2(u) \, dx - \lambda \int_{\mathbb{R}^3} G_1(u) \, dx.$$

for $u \in H^s_{\mathrm{rad}}(\mathbb{R}^3)$, with $\lambda \in [\bar{\delta}, 1]$, where $\bar{\delta} \in (0, 1)$ will be chosen later in a suitable way. More precisely, we have the following result.

Lemma 14.2.3 *Under assumptions (g_1), (g'_2), (g_3), the conclusions of Theorem 14.2.2 hold.*

Proof First, we show that there exists $w \in H^s_{\mathrm{rad}}(\mathbb{R}^3)$ such that $\int_{\mathbb{R}^3} G(w) \, dx > 0$. For $R > 1$, we define

$$w_R(x) = \begin{cases} \zeta & \text{for } |x| \leq R, \\ \zeta(R + 1 - |x|) & \text{for } |x| \in [R, R + 1], \\ 0 & \text{for } |x| \geq R + 1. \end{cases}$$

It is clear that $w_R \in H^s_{\mathrm{rad}}(\mathbb{R}^3)$. By the definitions of w_R and g,

$$
\int_{\mathbb{R}^3} G(w_R)\,dx = G(\zeta)|B_R| + \int_{\{R \leq |x| \leq R+1\}} G(\zeta(R+1-|x|))\,dx
$$

$$
\geq G(\zeta)|B_R| - |B_{R+1} - B_R| \max_{t \in [0,\zeta]} |G(t)|
$$

$$
= \frac{\pi^{\frac{3}{2}}}{\Gamma(\frac{3}{2}+1)} [G(\zeta)R^3 - \max_{t \in [0,\zeta]} |G(t)|((R+1)^3 - R^3)]
$$

$$
\geq \frac{\pi^{\frac{3}{2}}}{\Gamma(\frac{3}{2}+1)} \left[G(\zeta) - \max_{t \in [0,\zeta]} |G(t)| \left(\left(1+\frac{1}{R}\right)^3 - 1 \right) \right] R^3,
$$

so there exists $\bar{R} > 0$ such that $\int_{\mathbb{R}^3} G(w_R)\,dx > 0$ for all $R \geq \bar{R}$. Then we take $w = w_R$ with R large enough. Moreover, we can find $\bar{\delta} \in (0,1)$ such that

$$
\bar{\delta} \int_{\mathbb{R}^3} G_1(w)\,dx - \int_{\mathbb{R}^3} G_2(w)\,dx > 0.
$$

Therefore, if we consider the path $\gamma(t) = \bar{w}(\frac{\cdot}{t})$ if $t \in (0,1]$ and $\gamma(0) = 0$, where $\bar{w} = w(\frac{\cdot}{\bar{\theta}})$ and $\bar{\theta} > 0$, we can use the definition of χ, $\lambda \geq \bar{\delta}$ and the above inequality to see that

$$
\mathcal{J}^k_{\mu,\lambda}(\gamma(1)) \leq \frac{p\bar{\theta}^{3-2s}}{2}[w]^2_s + \frac{q\bar{\theta}^{2(3-2s)}}{4}[w]^4_s + \frac{\bar{\theta}^{3+2t}}{4}\mu\chi\left(\frac{\bar{\theta}^3 \|w\|^\alpha_{L^\alpha(\mathbb{R}^3)}}{k^\alpha}\right) \int_{\mathbb{R}^3} \phi^t_w w^2\,dx
$$

$$
+ \bar{\theta}^3 \left(\int_{\mathbb{R}^3} G_2(w)\,dx - \bar{\delta} \int_{\mathbb{R}^3} G_1(w)\,dx \right) \to -\infty \quad \text{as } \bar{\theta} \to \infty,
$$

due to the fact that $6 - 4s < 3$. Now, by using (14.2.5), (14.2.6) with $\varepsilon = \frac{1}{2}$ and Theorem 1.1.8, we have for any $\lambda \in [\bar{\delta}, 1]$

$$
\mathcal{J}^k_{\mu,\lambda}(u) \geq \frac{p}{2}[u]^2_s + \frac{q}{4}[u]^4_s + \frac{\mu}{4}\xi_k(u) \int_{\mathbb{R}^3} \phi^t_u u^2\,dx + \int_{\mathbb{R}^3} G_2(u)\,dx - \int_{\mathbb{R}^3} G_1(u)\,dx
$$

$$
\geq \frac{p}{2}[u]^2_s + \frac{m}{4} \int_{\mathbb{R}^3} u^2\,dx - \frac{C_{1/2}}{2^*_s} \int_{\mathbb{R}^3} |u|^{2^*_s}\,dx
$$

$$
\geq \min\left\{\frac{p}{2}, \frac{m}{4}\right\} \|u\|^2_{H^s(\mathbb{R}^3)} - C^* \|u\|^{2^*_s}_{H^s(\mathbb{R}^3)}.
$$

Then there exists $\rho > 0$ such that for any $\lambda \in [\bar{\delta}, 1]$ and $u \in H^s_{\mathrm{rad}}(\mathbb{R}^3)$ with $u \neq 0$ and $\|u\|_{H^s(\mathbb{R}^3)} \leq \rho$, we get $\mathcal{J}^k_{\mu,\lambda}(u) > 0$. In particular, for any $\|u\|_{H^s(\mathbb{R}^3)} = \rho$, we have that $\mathcal{J}^k_{\mu,\lambda}(u) \geq \bar{c} > 0$, for some $\bar{c} > 0$. Fix $\lambda \in [\bar{\delta}, 1]$ and $\gamma \in \Gamma_\lambda$. Since $\gamma(0) = 0$ and

$\mathcal{J}^k_{\mu,\lambda}(\gamma(1)) < 0$, it is clear that $\|\gamma(1)\|_{H^s(\mathbb{R}^3)} > \rho$. By continuity, there exists $t_\gamma \in (0, 1)$ such that $\|\gamma(t_\gamma)\|_{H^s(\mathbb{R}^3)} = \rho$. Consequently,

$$c_{\mu,\lambda} \geq \inf_{\lambda \in \Gamma_\lambda} \mathcal{J}^k_{\mu,\lambda}(\gamma(t_\gamma)) \geq \bar{c} > 0 \quad \text{for any } \lambda \in [\bar{\delta}, 1].$$

Finally, we are in the position to apply Theorem 14.2.2 with $X = H^s_{\text{rad}}(\mathbb{R}^3)$, $\Lambda = [\bar{\delta}, 1]$,

$$A(u) = \frac{p}{2}[u]^2_s + \frac{q}{4}[u]^4_s + \frac{\mu}{4}\xi_k(u) \int_{\mathbb{R}^3} \phi^t_u u^2 \, dx + \int_{\mathbb{R}^3} G_2(u) \, dx$$

and

$$B(u) = \int_{\mathbb{R}^3} G_1(u) \, dx.$$

Hence, for almost every $\lambda \in [\bar{\delta}, 1]$, there exists a bounded sequence $(u^\lambda_j) \subset H^s_{\text{rad}}(\mathbb{R}^3)$ such that

$$\mathcal{J}^k_{\mu,\lambda}(u^\lambda_j) \to c_{\mu,\lambda} \text{ and } (\mathcal{J}^k_{\mu,\lambda})'(u^\lambda_j) \to 0 \quad \text{in } (H^s_{\text{rad}}(\mathbb{R}^3))^*.$$

□

Next we prove a compactness result.

Lemma 14.2.4 *For any $\lambda \in [\bar{\delta}, 1]$, every bounded Palais–Smale sequence for $\mathcal{J}^k_{\mu,\lambda}$ admits a convergent subsequence.*

Proof Let (u_j) be a bounded Palais–Smale sequence for $\mathcal{J}^k_{\mu,\lambda}$, that is,

$$\mathcal{J}^k_{\mu,\lambda}(u_j) \quad \text{is bounded and} \quad (\mathcal{J}^k_{\mu,\lambda})'(u_j) \to 0 \quad \text{in } (H^s_{\text{rad}}(\mathbb{R}^3))^*. \qquad (14.2.9)$$

Then, using Theorem 1.1.11, we may suppose that, up to a subsequence, there exists $u \in H^s_{\text{rad}}(\mathbb{R}^3)$ such that

$$\begin{aligned}
&[u_j]^2_s \to L \geq 0, \\
&u_j \rightharpoonup u \quad \text{in } H^s_{\text{rad}}(\mathbb{R}^3), \\
&u_j \to u \quad \text{in } L^p(\mathbb{R}^3), \, 2 < p < 2^*_s, \\
&u_j \to u \quad \text{a.e. in } \mathbb{R}^3.
\end{aligned} \qquad (14.2.10)$$

By the weak lower semicontinuity of the fractional Sobolev seminorm,

$$[u]_s^2 \le \liminf_{j \to \infty} [u_j]_s^2. \tag{14.2.11}$$

Applying the first part of Lemma 1.4.2 with $P(t) = g_i(t)$, $i = 1, 2$, $Q(t) = |t|^{2_s^* - 1}$, $v_j = u_j$, $v = g_i(u)$, $i = 1, 2$ and $w \in C_c^\infty(\mathbb{R}^3)$, and using (g_2), (14.2.2) and (14.2.10), we have that

$$\int_{\mathbb{R}^3} g_i(u_j) w \, dx \to \int_{\mathbb{R}^3} g_i(u) w \, dx \quad i = 1, 2$$

as $j \to \infty$. Now (14.2.10) and Remark 13.1.3 imply that, as $j \to \infty$,

$$\xi_k(u_j) \int_{\mathbb{R}^3} \phi_{u_j}^t u_j w \, dx \to \xi_k(u) \int_{\mathbb{R}^3} \phi_u^t u w \, dx$$

$$\chi'\left(\frac{\|u_j\|_{L^\alpha(\mathbb{R}^3)}^\alpha}{k^\alpha} \right) \int_{\mathbb{R}^3} \phi_{u_j}^t u_j^2 \int_{\mathbb{R}^3} |u_j|^{\alpha-2} u_j w \to \chi'\left(\frac{\|u\|_{L^\alpha(\mathbb{R}^3)}^\alpha}{k^\alpha} \right) \int_{\mathbb{R}^3} \phi_u^t u^2 \int_{\mathbb{R}^3} |u|^{\alpha-2} u w.$$

It follows that u satisfies

$$(p + qL)\langle u, w \rangle_{\mathcal{D}^{s,2}(\mathbb{R}^3)} + \mu \xi_k(u) \int_{\mathbb{R}^3} \phi_u^t u w + \frac{\mu \alpha}{4 k^\alpha} \chi'\left(\frac{\|u\|_{L^\alpha(\mathbb{R}^3)}^\alpha}{k^\alpha} \right) \int_{\mathbb{R}^3} \phi_u^t u^2 \int_{\mathbb{R}^3} |u|_\alpha^{\alpha-2} u w$$

$$+ \int_{\mathbb{R}^3} g_2(u) w = \lambda \int_{\mathbb{R}^3} g_1(u) w \tag{14.2.12}$$

for all $w \in C_c^\infty(\mathbb{R}^3)$. Since $C_c^\infty(\mathbb{R}^3)$ is dense in $H_{\mathrm{rad}}^s(\mathbb{R}^3)$, we get

$$(p + qL)[u]_s^2 + \mu \xi_k(u) \int_{\mathbb{R}^3} \phi_u^t u^2 \, dx$$

$$+ \frac{\mu \alpha}{4 k^\alpha} \chi'\left(\frac{\|u\|_{L^\alpha(\mathbb{R}^3)}^\alpha}{k^\alpha} \right) \|u\|_{L^\alpha(\mathbb{R}^3)}^\alpha \int_{\mathbb{R}^3} \phi_u^t u^2 \, dx + \int_{\mathbb{R}^3} g_2(u) u \, dx$$

$$= \lambda \int_{\mathbb{R}^3} g_1(u) u \, dx. \tag{14.2.13}$$

Again, from (14.2.10) and Remark 13.1.3, we obtain that, as $j \to \infty$,

$$\xi_k(u_j) \int_{\mathbb{R}^3} \phi_{u_j}^t u_j^2 \, dx \to \xi_k(u) \int_{\mathbb{R}^3} \phi_u^t u^2 \, dx,$$

$$\chi'\left(\frac{\|u_j\|_{L^\alpha(\mathbb{R}^3)}^\alpha}{k^\alpha} \right) \|u_j\|_{L^\alpha(\mathbb{R}^3)}^\alpha \int_{\mathbb{R}^3} \phi_{u_j}^t u_j^2 \, dx \to \chi'\left(\frac{\|u\|_{L^\alpha(\mathbb{R}^3)}^\alpha}{k^\alpha} \right) \|u\|_{L^\alpha(\mathbb{R}^3)}^\alpha \int_{\mathbb{R}^3} \phi_u^t u^2 \, dx. \tag{14.2.14}$$

Taking $X = H^s_{\mathrm{rad}}(\mathbb{R}^3)$, $q_1 = 2$, $q_2 = 2^*_s$, $v_j = u_j$, $v = g_1(u)u$ and $P(t) = g_1(t)t$ in Lemma 1.4.3, by (14.2.1), (14.2.2) and (14.2.10) we deduce that

$$\int_{\mathbb{R}^3} g_1(u_j)u_j\,dx \to \int_{\mathbb{R}^3} g_1(u)u\,dx. \tag{14.2.15}$$

On the other hand, (14.2.10) and Fatou's lemma yield

$$\int_{\mathbb{R}^3} g_2(u)u\,dx \le \liminf_{j\to\infty} \int_{\mathbb{R}^3} g_2(u_j)u_j\,dx. \tag{14.2.16}$$

Putting together (14.2.13), (14.2.14), (14.2.15), (14.2.16), and using the fact that $\langle (\mathcal{J}^k_{\mu,\lambda})'(u_j), u_j \rangle \to 0$, we see that

$$\limsup_{j\to\infty} (p + qL)[u_j]^2_s$$

$$= \limsup_{j\to\infty} \left[\lambda \int_{\mathbb{R}^3} g_1(u_j)u_j\,dx - \int_{\mathbb{R}^3} g_2(u_j)u_j\,dx - \mu\xi_k(u_j)\int_{\mathbb{R}^3} \phi^t_{u_j}u^2_j\,dx \right.$$

$$\left. - \frac{\mu\alpha}{4k^\alpha}\chi'\left(\frac{\|u_j\|^\alpha_{L^\alpha(\mathbb{R}^3)}}{k^\alpha} \right)\|u_j\|^\alpha_{L^\alpha(\mathbb{R}^3)}\int_{\mathbb{R}^3} \phi^t_{u_j}u^2_j\,dx \right]$$

$$\le \lambda \int_{\mathbb{R}^3} g_1(u)u\,dx - \int_{\mathbb{R}^3} g_2(u)u\,dx - \mu\xi_k(u)\int_{\mathbb{R}^3} \phi^t_u u^2\,dx$$

$$- \frac{\mu\alpha}{4k^\alpha}\chi'\left(\frac{\|u\|^\alpha_{L^\alpha(\mathbb{R}^3)}}{k^\alpha} \right)\|u\|^\alpha_{L^\alpha(\mathbb{R}^3)}\int_{\mathbb{R}^3} \phi^t_u u^2\,dx$$

$$= (p + qL)[u]^2_s. \tag{14.2.17}$$

Now, (14.2.11) and (14.2.17) imply that

$$\lim_{j\to\infty} [u_j]^2_s = [u]^2_s \tag{14.2.18}$$

and thus

$$\lim_{j\to\infty} \int_{\mathbb{R}^3} g_2(u_j)u_j\,dx = \int_{\mathbb{R}^3} g_2(u)u\,dx. \tag{14.2.19}$$

Since $g_2(\tau)\tau = m\tau^2 + h(\tau)$, where $h(\tau) = \tau(g(\tau) + m\tau)_-$ is a non-negative and continuous function, we can apply Fatou's lemma to infer that

$$\int_{\mathbb{R}^3} h(u)\,dx \le \liminf_{j\to\infty} \int_{\mathbb{R}^3} h(u_j)\,dx$$

and

$$\int_{\mathbb{R}^3} u^2 \, dx \le \liminf_{j\to\infty} \int_{\mathbb{R}^3} u_j^2 \, dx.$$

These two inequalities and (14.2.19) imply that

$$\int_{\mathbb{R}^3} mu^2 \, dx \le \liminf_{j\to\infty} \int_{\mathbb{R}^3} mu_j^2 \, dx \le \limsup_{j\to\infty} \int_{\mathbb{R}^3} mu_j^2 \, dx$$

$$= \limsup_{j\to\infty} \int_{\mathbb{R}^3} (g_2(u_j)u_j - h(u_j)) \, dx$$

$$= \int_{\mathbb{R}^3} g_2(u)u \, dx + \limsup_{j\to\infty} \left(- \int_{\mathbb{R}^3} h(u_j) \, dx \right)$$

$$= \int_{\mathbb{R}^3} (mu^2 + h(u)) \, dx - \liminf_{j\to\infty} \int_{\mathbb{R}^3} h(u_j) \, dx$$

$$= \int_{\mathbb{R}^3} mu^2 \, dx + \int_{\mathbb{R}^3} h(u) \, dx - \liminf_{j\to\infty} \int_{\mathbb{R}^3} h(u_j) \, dx$$

$$\le \int_{\mathbb{R}^3} mu^2 \, dx,$$

that is, $u_j \to u$ in $L^2(\mathbb{R}^3)$, which combined with (14.2.18) implies that $u_j \to u$ strongly in $H_{\mathrm{rad}}^s(\mathbb{R}^3)$. $\qquad\square$

Lemma 14.2.3, Lemma 14.2.4 and Theorem 14.2.2 lead to the following result.

Lemma 14.2.5 *For almost every* $\lambda \in [\bar{\delta}, 1]$, *there exists* $u^\lambda \in H_{\mathrm{rad}}^s(\mathbb{R}^3)$, $u^\lambda \ne 0$, *such that* $\mathcal{J}_{\mu,\lambda}^k(u^\lambda) = c_{\mu,\lambda}$ *and* $(\mathcal{J}_{\mu,\lambda}^k)'(u^\lambda) = 0$.

Proof Applying Theorem 5.2.2, we know that for almost every $\lambda \in [\bar{\delta}, 1]$ we can find a bounded sequence $(u_j^\lambda) \subset H_{\mathrm{rad}}^s(\mathbb{R}^3)$ such that

$$\mathcal{J}_{\mu,\lambda}^k(u_j^\lambda) \to c_{\mu,\lambda} \quad \text{and} \quad (\mathcal{J}_{\mu,\lambda}^k)'(u_j^\lambda) \to 0 \text{ in } (H_{\mathrm{rad}}^s(\mathbb{R}^3))^*. \tag{14.2.20}$$

Up to a subsequence, by Lemma 14.2.4, we may assume that there exists $u^\lambda \in H_{\mathrm{rad}}^s(\mathbb{R}^3)$ such that $u_j^\lambda \to u^\lambda$ in $H_{\mathrm{rad}}^s(\mathbb{R}^3)$. By Lemma 14.2.3, we know that $c_{\mu,\lambda} \ge \bar{c} > 0$, which in view of the first relation in (14.2.20) yields $u^\lambda \ne 0$. $\qquad\square$

Therefore, we can find $(\lambda_j) \subset [\bar{\delta}, 1]$, $\lambda_j \to 1$ and $(u_j) \subset H_{\mathrm{rad}}^s(\mathbb{R}^3)$ such that

$$\mathcal{J}_{\mu,\lambda_j}^k(u_j) = c_{\mu,\lambda_j} \quad \text{and} \quad (\mathcal{J}_{\mu,\lambda_j}^k)'(u_j) = 0 \text{ in } (H_{\mathrm{rad}}^s(\mathbb{R}^3))^*.$$

Lemma 14.2.6 *Let u_j be a critical point of $\mathcal{J}^k_{\mu,\lambda_j}$ at the level c_{μ,λ_j}. Then, for every sufficiently large $k > 0$, there exists $\mu_0 = \mu_0(k) > 0$ such that for any $\mu \in (0, \mu_0)$, up to a subsequence, $\|u_j\|_{L^\alpha(\mathbb{R}^3)} \leq k$ for any $j \in \mathbb{N}$.*

Proof Since $(\mathcal{J}^k_{\mu,\lambda_j})'(u_j) = 0$, we can see that u_j is a weak solution to the problem

$$
\begin{cases}
\left(p + q[u]_s^2\right)(-\Delta)^s u + \mu\xi_k(u)\phi u + \frac{\alpha}{k^\alpha}\chi'\left(\frac{\|u\|^\alpha_{L^\alpha(\mathbb{R}^3)}}{k^\alpha}\right)|u|^{\alpha-2}u\int_{\mathbb{R}^3}\phi u\,dx \\
\quad = \lambda g_1(u) - g_2(u) \quad \text{in}\,\mathbb{R}^3 \\
(-\Delta)^t\phi = \mu u^2 \quad \text{in}\,\mathbb{R}^3.
\end{cases}
$$

Therefore, arguing as in the proof of Theorem 3.5.1 (see also [325]), we see that u_j satisfies the following Pohozaev identity:

$$
\frac{3-2s}{2}(p + q[u_j]_s^2)[u_j]_s^2 + \frac{3+2t}{4}\mu\xi_k(u_j)\int_{\mathbb{R}^3}\phi^t_{u_j}u_j^2\,dx
$$

$$
+ \frac{3\mu}{k^\alpha}\chi'\left(\frac{\|u_j\|^\alpha_{L^\alpha(\mathbb{R}^3)}}{k^\alpha}\right)\|u_j\|^\alpha_{L^\alpha(\mathbb{R}^3)}\int_{\mathbb{R}^3}\phi^t_{u_j}u_j^2\,dx
$$

$$
= 3\lambda_j\int_{\mathbb{R}^3}G_1(u_j)\,dx - 3\int_{\mathbb{R}^3}G_2(u_j)\,dx. \tag{14.2.21}
$$

Since $\mathcal{J}^k_{\mu,\lambda_j}(u_j) = c_{\mu,\lambda_j}$, (14.2.21), and Lemma 13.1.2-(1) imply that

$$
s\left(p + \frac{q}{2}[u_j]_s^2\right)[u_j]_s^2
$$

$$
= 3c_{\mu,\lambda_j} + \frac{t}{2}\mu\xi_k(u_j)\int_{\mathbb{R}^3}\phi^t_{u_j}u_j^2\,dx + \frac{3\mu}{k^\alpha}\chi'\left(\frac{\|u_j\|^\alpha_{L^\alpha(\mathbb{R}^3)}}{k^\alpha}\right)\|u_j\|^\alpha_{L^\alpha(\mathbb{R}^3)}\int_{\mathbb{R}^3}\phi^t_{u_j}u_j^2\,dx
$$

$$
\leq 3c_{\mu,\lambda_j} + C_1\mu^2\xi_k(u_j)\|u_j\|^4_{L^\alpha(\mathbb{R}^3)} + C_2\chi'\left(\frac{\|u_j\|^\alpha_{L^\alpha(\mathbb{R}^3)}}{k^\alpha}\right)\frac{\mu^2}{k^\alpha}\|u_j\|^{4+\alpha}_{L^\alpha(\mathbb{R}^3)}. \tag{14.2.22}
$$

Let us estimate the right-hand side of this inequality. Using the min–max definition of c_{μ,λ_j}, we have

$$
c_{\mu,\lambda_j} \leq \max_{\theta>0}\mathcal{J}^k_{\mu,\lambda_j}\left(w\left(\frac{\cdot}{\theta}\right)\right)
$$

$$
\leq \max_{\theta>0}\left\{\frac{\theta^{3-2s}}{2}\left(p + \frac{q\theta^{3-2s}}{2}[w]_s^2\right)[w]_s^2 + \theta^3\left(\int_{\mathbb{R}^3}(G_2(w)\,dx - \bar\delta\int_{\mathbb{R}^3}(G_2(w)\,dx\right)\right\}
$$

$$
+ \max_{\theta>0}\left(\frac{\mu\theta^{3+2t}}{4}\xi_k(\gamma(\sigma))\int_{\mathbb{R}^3}\phi^t_w w^2\,dx\right) = A_1 + A_2(k). \tag{14.2.23}
$$

Now, if $\theta^3 \geq 2k^\alpha / \|w\|_{L^\alpha(\mathbb{R}^3)}^\alpha$ then $A_2(k) = 0$. Otherwise, if $\theta^3 < 2k^\alpha / \|w\|_{L^\alpha(\mathbb{R}^3)}^\alpha$, then by the definition of χ and Lemma 13.1.2-(1), there is $C > 0$ such that

$$A_2(k) \leq \frac{\mu}{4} \left(\frac{2k^\alpha}{\|w\|_{L^\alpha(\mathbb{R}^3)}^\alpha} \right)^{\frac{3+2t}{3}} \int_{\mathbb{R}^3} \phi_w^t w^2 \, dx$$

$$\leq \frac{\mu}{4} \left(\frac{2k^\alpha}{\|w\|_{L^\alpha(\mathbb{R}^3)}^\alpha} \right)^{\frac{3+2t}{3}} C_t \mu \|w\|_{L^\alpha(\mathbb{R}^3)}^4 \leq C_3 \mu^2 k^4.$$

In a similar fashion we can prove the following estimates:

$$C_1 \mu^2 \xi_k(u_j) \|u_j\|_{L^\alpha(\mathbb{R}^3)}^4 \leq C_4 \mu^4 k^4, \tag{14.2.24}$$

$$C_2 \chi' \left(\frac{\|u_j\|_{L^\alpha(\mathbb{R}^3)}^\alpha}{k^\alpha} \right) \frac{\mu^2}{k^\alpha} \|u_j\|_{L^\alpha(\mathbb{R}^3)}^{4+\alpha} \leq C_5 \mu^2 k^4. \tag{14.2.25}$$

Combining (14.2.23), (14.2.24), (14.2.25) and (14.2.22), we obtain

$$sp[u_j]_s^2 \leq s \left(p + \frac{q}{2}[u_j]_s^2 \right) [u_j]_s^2 \leq 3A_1 + C_6 \mu^2 k^4. \tag{14.2.26}$$

On the other hand, by using that $\langle (\mathcal{J}_{\mu,\lambda_j}^k)'(u_j), u_j \rangle = 0$ and (14.2.4), we deduce that

$$(p + q[u_j]_s^2)[u_j]_s^2 + \mu \xi_k(u_j) \int_{\mathbb{R}^3} \phi_{u_j}^t u_j^2 \, dx$$

$$+ \frac{\mu\alpha}{4k^\alpha} \chi' \left(\frac{\|u_j\|_{L^\alpha(\mathbb{R}^3)}^\alpha}{k^\alpha} \right) \|u_j\|_{L^\alpha(\mathbb{R}^3)}^\alpha \int_{\mathbb{R}^3} \phi_{u_j}^t u_j^2 \, dx + \int_{\mathbb{R}^3} g_2(u_j) u_j \, dx$$

$$= \lambda_j \int_{\mathbb{R}^3} g_1(u_j) u_j \, dx \leq C_\varepsilon \|u_j\|_{L^{2_s^*}(\mathbb{R}^3)}^{2_s^*} + \varepsilon \int_{\mathbb{R}^3} g_2(u_j) u_j \, dx. \tag{14.2.27}$$

Hence, using (14.2.3), (14.2.26), (14.2.27), the definition of χ, Lemma 13.1.2 and Theorem 1.1.8, we infer that

$$m(1-\varepsilon) \|u_j\|_{L^2(\mathbb{R}^3)}^2 \leq (1-\varepsilon) \int_{\mathbb{R}^3} g_2(u_j) u_j \, dx$$

$$\leq C_\varepsilon \|u_j\|_{L^{2_s^*}(\mathbb{R}^3)}^{2_s^*} - \frac{\mu\alpha}{4k^\alpha} \chi' \left(\frac{\|u_j\|_{L^\alpha(\mathbb{R}^3)}^\alpha}{k^\alpha} \right) \|u_j\|_{L^\alpha(\mathbb{R}^3)}^\alpha \int_{\mathbb{R}^3} \phi_{u_j}^t u_j^2 \, dx$$

$$\leq C_* C_\varepsilon [u_j]_s^{2_s^*} + C\mu^2 k^4$$

$$\leq \hat{C}(3A_1 + C_6 \mu^2 k^4)^{\frac{3}{3-2s}} + \bar{C}\mu^2 k^4. \tag{14.2.28}$$

Now suppose, by contradiction, that (u_j) admits no subsequence that is uniformly bounded by k in the $L^\alpha(\mathbb{R}^3)$-norm. Then, there exists $\nu \in \mathbb{N}$ such that

$$\|u_j\|_{L^\alpha(\mathbb{R}^3)} > k \quad \text{for any } j \geq \nu. \tag{14.2.29}$$

Without loss of generality, we can assume that (14.2.29) holds for all u_j. Taking into account (14.2.26), (14.2.28) and (14.2.29), we get

$$k^2 < \|u_j\|^2_{L^\alpha(\mathbb{R}^3)} \leq C\|u_j\|^2_{H^s(\mathbb{R}^3)} \leq C_7 + C_8 \mu^{\frac{6}{3-2s}} k^{\frac{12}{3-2s}} + C_9 \mu^2 k^4,$$

which gives a contradiction for k large and μ sufficiently small, because we can find $k_0 > 0$ such that $k_0^2 > C_7 + 1$ and $\bar{\mu} = \bar{\mu}(k_0)$ such that $C_8 \mu^{\frac{6}{3-2s}} k_0^{\frac{12}{3-2s}} + C_9 \mu^2 k_0^4 < 1$ for all $\mu \in (0, \bar{\mu})$. $\qquad \square$

Now, we are ready to give the proof of the main result of this chapter.

Proof of Theorem 14.1.1 Let k and μ_0 be as in Lemma 14.2.6, and fix $\mu \in (0, \mu_0)$. Let u_j be a critical point for $\mathcal{J}^k_{\mu,\lambda_j}$ at the level c_{μ,λ_j}. We claim that (u_j) is a bounded Palais–Smale sequence for $\mathcal{J}_\mu$ at the level $c_{\mu,1}$. Since by Lemma 14.2.6 we know that $\|u_j\|_{L^\alpha(\mathbb{R}^3)} \leq k$, arguments similar to those used to prove (14.2.26) and (14.2.28) imply that (u_j) is bounded in $H^s_{\text{rad}}(\mathbb{R}^3)$. Up to a subsequence, we may assume that $u_j \rightharpoonup u$ in $H^s_{\text{rad}}(\mathbb{R}^3)$. By the definition of χ, it follows that

$$\mathcal{J}^k_{\mu,\lambda_j}(u_j) = \left(\frac{p}{2} + \frac{q}{4}[u_j]_s^2\right)[u_j]_s^2 + \frac{\mu}{4}\int_{\mathbb{R}^3} \phi_{u_j} u_j^2 \, dx + \int_{\mathbb{R}^3} G_2(u_j) \, dx - \lambda_j \int_{\mathbb{R}^3} G_1(u_j) \, dx. \tag{14.2.30}$$

Since $(g_1(u_j))$ is bounded in $(H^s_{\text{rad}}(\mathbb{R}^3))^*$, Lemma 1.4.2 and the fact that

$$\int_{\mathbb{R}^3} g_1(u)\psi \, dx = \int_{\mathbb{R}^3} g_1(u_j)\psi \, dx + o_j(1)$$

for all $\psi \in C_c^\infty(\mathbb{R}^3)$, we obtain that

$$\langle (\mathcal{J}_\mu)'(u_j), \psi \rangle = \langle (\mathcal{J}^k_{\mu,\lambda_j})'(u_j), \psi \rangle + (\lambda_j - 1)\int_{\mathbb{R}^3} g_1(u_j)\psi \, dx \to 0.$$

Since $C_c^\infty(\mathbb{R}^3)$ is dense in $H^s_{\text{rad}}(\mathbb{R}^3)$, we see that $\mathcal{J}'_\mu(u_j) \to 0$ in $(H^s_{\text{rad}}(\mathbb{R}^3))^*$. Further, since the function $\lambda \mapsto c_{\mu,\lambda}$ is left continuous,

$$\mathcal{J}_\mu(u_j) = \mathcal{J}^k_{\mu,\lambda_j}(u_j) + (\lambda_j - 1)\int_{\mathbb{R}^3} g_1(u_j) \, dx = c_{\mu,\lambda_j} + o_j(1) = c_{\mu,1} + o_j(1).$$

Consequently, (u_j) is a bounded Palais–Smale sequence for $\mathcal{J}_\mu$. By Lemma 14.2.4, we deduce that $u_j \rightarrow u$ in $H^s_{\mathrm{rad}}(\mathbb{R}^3)$ and thus $\mathcal{J}_\mu(u) = c_{\mu,1}$ and $\mathcal{J}'_\mu(u) = 0$. From $\langle \mathcal{J}'_\mu(u), u^- \rangle = 0$ we see that $u \geq 0$ in $\mathbb{R}^3$. In light of Lemma 14.2.3, we know that $c_{\mu,1} > 0$, which implies that $u \not\equiv 0$. Since $\phi^t_u \geq 0$, we can argue as in the proof of Lemma 3.2.14 to see that $u \in L^\infty(\mathbb{R}^3)$. Therefore, $\phi^t_u \in L^\infty(\mathbb{R}^3)$. Note that, by (14.2.7), u satisfies

$$(-\Delta)^s u = \frac{1}{(p + q[u]^2_s)}[-\mu\phi^t_u u + g(u)] \quad \text{in } \mathbb{R}^3,$$

and applying Proposition 1.3.2 we obtain that $u \in C^{1,\alpha}(\mathbb{R}^3)$ for any $\alpha < 2s - 1$ (we remark that $s > \frac{3}{4} > \frac{1}{2}$). By using Proposition 1.3.11-(ii) (or Theorem 1.3.5), we have that $u > 0$ in $\mathbb{R}^3$. This completes the proof of Theorem 14.1.1. $\square$

Multiple Positive Solutions for a Non-homogeneous Fractional Schrödinger Equation

15

15.1 Introduction

In this chapter we deal with the existence of positive solutions for the nonlinear fractional equation

$$\begin{cases} (-\Delta)^s u + u = k(x) f(u) + h(x) & \text{in } \mathbb{R}^N, \\ u \in H^s(\mathbb{R}^N), \ u > 0 & \text{in } \mathbb{R}^N, \end{cases} \tag{15.1.1}$$

where $s \in (0, 1)$, $N \geq 2$, k is a bounded positive function, $h \in L^2(\mathbb{R}^N)$, $h \geq 0$, $h \not\equiv 0$, and the nonlinearity $f : \mathbb{R} \to \mathbb{R}$ is a smooth function which can be either asymptotically linear or superlinear at infinity.

Our purpose is to investigate the existence and the multiplicity of positive solutions for the nonhomogeneous equation (15.1.1), subject to a small perturbation $h \in L^2(\mathbb{R}^N)$ and suitable assumptions on the nonlinearity f.

More precisely, we assume that f satisfies the following conditions:

($f1$) $f \in C^1(\mathbb{R}, \mathbb{R}_+)$, $f(0) = 0$ and $f(t) = 0$ for $t \leq 0$;

($f2$) $\lim\limits_{t \to 0} \dfrac{f(t)}{t} = 0$;

($f3$) there exists $p \in (1, 2_s^* - 1)$ such that $\lim\limits_{t \to \infty} \dfrac{f(t)}{t^p} = 0$;

($f4$) there exists $l \in (0, \infty]$ such that $\lim\limits_{t \to \infty} \dfrac{f(t)}{t} = l$.

© The Author(s), under exclusive license to Springer Nature Switzerland AG 2021
V. Ambrosio, *Nonlinear Fractional Schrödinger Equations in* $\mathbb{R}^N$,
Frontiers in Mathematics, https://doi.org/10.1007/978-3-030-60220-8_15

Note that $(f1)$–$(f3)$, yield that for any $\varepsilon > 0$ there exists $C_\varepsilon > 0$ such that

$$|F(t)| \leq \frac{\varepsilon}{2} t^2 + \frac{1}{p+1} C_\varepsilon |t|^{p+1} \qquad \text{for all } t \in \mathbb{R}, \tag{15.1.2}$$

while $(f4)$ implies that f is asymptotically linear if $l < \infty$, or superlinear when $l = \infty$.

Due to the presence of the fractional Laplacian, which is a nonlocal operator, we analyze (15.1.1) by using the s-harmonic extension method [127]. This approach allows us to write a given nonlocal equation in a local way and to apply some known variational techniques to these kind of problems. Hence, instead of (15.1.1), we consider the following degenerate elliptic equation with a nonlinear Neumann boundary condition:

$$\begin{cases} -\operatorname{div}(y^{1-2s} \nabla U) = 0 & \text{in } \mathbb{R}_+^{N+1}, \\ \frac{\partial U}{\partial v^{1-2s}} = \kappa_s [-u + k(x) f(u) + h(x)] & \text{on } \partial \mathbb{R}_+^{N+1}, \end{cases} \tag{15.1.3}$$

where u denotes the trace of U, that is $u = U(\cdot, 0)$. For simplicity, we will assume that $\kappa_s = 1$. Taking into account this fact, we are able to enlist some variational techniques developed in the papers [233, 263, 320, 333], dealing with asymptotically or superlinear classical problems, by introducing the following functional

$$I(U) = \frac{1}{2} \|U\|_{X^s(\mathbb{R}_+^{N+1})}^2 - \int_{\mathbb{R}^N} k(x) F(u)\, dx - \int_{\mathbb{R}^N} h(x) u\, dx$$

where the weighted Sobolev space $X^s(\mathbb{R}_+^{N+1})$ is defined as the completion of $C_c^\infty(\overline{\mathbb{R}_+^{N+1}})$ with respect to the norm

$$\|U\|_{X^s(\mathbb{R}_+^{N+1})} = \left(\iint_{\mathbb{R}_+^{N+1}} y^{1-2s} |\nabla U|^2\, dx dy + \int_{\mathbb{R}^N} u^2\, dx \right)^{\frac{1}{2}} < \infty.$$

Clearly, this functional simplification comes at the price of some additional technical difficulties. For instance, some weighted embedding result will be needed (see Lemma 1.3.9) to obtain convergence results (see Lemma 15.2.1). Moreover, the arguments used in [233, 263] to prove the non-existence of solutions for certain eigenvalue problems have to be handled carefully in order to take care of the trace of the involved functions (see Lemma 15.2.6).

Now, we state our first main result concerning the existence of positive solutions to (15.1.1) in the asymptotically linear case, that is $l < \infty$.

Theorem 15.1.1 ([70]) *Let* $s \in (0, 1)$ *and* $N \geq 2$. *Assume that* $h \in L^2(\mathbb{R}^N)$, $h(x) \geq 0$, $h(x) \not\equiv 0$ *and* $k \in L^\infty(\mathbb{R}^N, \mathbb{R}_+)$ *satisfies the following condition:*

(K) there exists $R_0 > 0$ such that

$$\sup\left\{\frac{f(t)}{t} : t > 0\right\} < \inf\left\{\frac{1}{k(x)} : |x| \geq R_0\right\}. \tag{15.1.4}$$

Suppose that f satisfies the conditions $(f1)–(f4)$ and $\mu^* \in (l, \infty)$, where

$$\mu^* = \inf\left\{\int_{\mathbb{R}^N}(|(-\Delta)^{\frac{s}{2}}u|^2 + u^2)\,dx : u \in H^s(\mathbb{R}^N), \int_{\mathbb{R}^N}k(x)u^2\,dx = 1\right\}. \tag{15.1.5}$$

Assume that

$$\|h\|_{L^2(\mathbb{R}^N)} < m = \max_{t \geq 0}\left[\left(\frac{1}{2} - \frac{\varepsilon}{2}\|k\|_{L^\infty(\mathbb{R}^N)}\right)t - \frac{C_\varepsilon}{p+1}S_*^{p+1}t^p\|k\|_{L^\infty(\mathbb{R}^N)}\right], \tag{15.1.6}$$

where $\varepsilon \in (0, \|k\|_{L^\infty(\mathbb{R}^N)}^{-1})$ is fixed and S_* is the best Sobolev constant of the embedding $H^s(\mathbb{R}^N) \subset L^{2^*_s}(\mathbb{R}^N)$. Let $E : H^s(\mathbb{R}^N) \to \mathbb{R}$ be the energy functional associated with (15.1.1), that is

$$E(u) = \frac{1}{2}\int_{\mathbb{R}^N}(|(-\Delta)^{\frac{s}{2}}u|^2 + u^2)\,dx - \int_{\mathbb{R}^N}k(x)F(u)\,dx - \int_{\mathbb{R}^N}h(x)u\,dx.$$

Then problem (15.1.1) possesses at least two positive solutions $u_1, u_2 \in H^s(\mathbb{R}^N)$ with the property that $E(u_1) < 0 < E(u_2)$.

Remark 15.1.2 The assumption on the size of h is necessary for (15.1.1) to admit a solution. In fact, proceeding as in [130], one can obtain a non-existence result to (15.1.1) when $\|h\|_{L^2(\mathbb{R}^N)}$ is sufficiently large.

The proof of the above theorem goes as follows: under the assumption $l < \infty$, we first use Ekeland's variational principle to prove that for $\|h\|_{L^2(\mathbb{R}^N)}$ small enough, there exists a positive solution to (15.1.3) such that $I(U_0) < 0$. Then, we use a variant of the mountain pass theorem [177] to find a Cerami sequence that converges strongly in $X^s(\mathbb{R}^{N+1}_+)$ to a solution U_1 of (15.1.3) with $I(U_1) > 0$. Clearly, these two solutions U_0 and U_1 are different.

Our second result deals with the existence of positive solutions to (15.1.1) in the superlinear case $l = \infty$.

Theorem 15.1.3 ([70]) *Let $s \in (0, 1)$ and $N \geq 2$. Assume that f fulfills $(f1)–(f4)$ with $l = \infty$. Let $k(x) \equiv 1$, and let $h \in C^1(\mathbb{R}^N) \cap L^2(\mathbb{R}^N)$ be a radial function such that $h(x) \geq 0$, $h(x) \not\equiv 0$ and*

(H) $x \cdot \nabla h(x) \in L^1(\mathbb{R}^N) \cap L^\infty(\mathbb{R}^N)$ *and*

$$x \cdot \nabla h(x) \geq 0 \text{ for all } x \in \mathbb{R}^N.$$

Assume that

$$\|h\|_{L^2(\mathbb{R}^N)} < m_1 = \max_{t \geq 0}\left[\left(\frac{1}{2} - \frac{\varepsilon}{2}\right)t - \frac{C_\varepsilon}{p+1}S_*^{p+1}t^p\right],$$

where $\varepsilon \in (0, 1)$ is fixed. Then, (15.1.1) admits two positive solutions $u_3, u_4 \in H_r^s(\mathbb{R}^N)$ such that $E(u_3) < 0 < E(u_4)$.

    Due to the presence of radial functions $k(x) = 1$ and $h = h(|x|)$, we work in the subspace $X_{\text{rad}}^s(\mathbb{R}_+^{N+1})$ of the weighted space $X^s(\mathbb{R}_+^{N+1})$ that consists of the functions that are radial with respect to $x \in \mathbb{R}^N$. We point out that the methods used to study the asymptotically linear case do not work any more. Indeed, to prove that a Palais-Smale sequence converges to a second solution different from the first one, we have to use the concentration-compactness principle, which seems very hard to apply without requiring further assumptions on $k(x)$ and $f(t)$. This time we use the compactness of $X_{\text{rad}}^s(\mathbb{R}_+^{N+1})$ in $L^q(\mathbb{R}^N)$ for any $q \in (2, 2_s^*)$, and the Ekeland variational principle, to get a first solution to (15.1.3) with negative energy, provided that $\|h\|_{L^2(\mathbb{R}^N)}$ is sufficiently small. The existence of a second solution with positive energy is obtained by combining the Struwe-Jeanjean monotonicity trick in [231], which allows us to prove the existence of bounded Palais-Smale sequences for parametrized functionals, with the Pohozaev identity for the fractional Laplacian and assumption (H), which guarantee the existence of a bounded Palais-Smale sequence for I that converges to a radial positive solution to (15.1.3).

    We point out that in the current literature there are only few papers concerning the existence and the multiplicity of solutions for nonhomogeneous problems in a nonlocal setting [150, 227, 296, 309]; this is rather surprising given that in the classical framework such type of problems have been extensively investigated by many authors [85, 130, 229, 319, 349, 350].

    Now, we provide some examples of functions f, k and h for which our main results are applicable.

Example 15.1.4 Let $R_0 > 0$ and let

$$k(x) = \begin{cases} \frac{1}{1+|x|}, & \text{if } |x| < R_0, \\ \frac{1}{1+R_0}, & \text{if } |x| \geq R_0, \end{cases} \quad \text{and } f(t) = \begin{cases} \frac{R_0 t^2}{1+t}, & \text{if } t > 0, \\ 0, & \text{if } t \leq 0. \end{cases}$$

It is clear that $\|k\|_{L^\infty(\mathbb{R}^N)} = 1$, and f satisfies $(f1)$–$(f3)$ and $(f4)$ with $l = R_0$. Moreover, (K) holds because

$$\sup\left\{\frac{f(t)}{t} : t > 0\right\} = R_0 < R_0 + 1 = \inf\left\{\frac{1}{k(x)} : |x| \geq R_0\right\}.$$

To verify that $l > \mu^*$, we have to choose a special $R_0 > 0$. For $R > 0$, we take $\phi \in C_c^\infty(\mathbb{R}^N)$ such that $\phi(x) = 1$ if $|x| \leq R$, $\phi(x) = 0$ if $|x| \geq 2R$, and $|\nabla\phi(x)| \leq \frac{C}{R}$ for all $x \in \mathbb{R}^N$. Since $\phi \in H^1(\mathbb{R}^N) \subset H^s(\mathbb{R}^N)$, we see that

$$\|\phi\|_{H^s(\mathbb{R}^N)} \leq C\|\phi\|_{H^1(\mathbb{R}^N)}.$$

On the other hand, for any $R_0 > 2R$, we have

$$\frac{\int_{\mathbb{R}^N} \phi^2\, dx}{\int_{\mathbb{R}^N} k(x)\phi^2\, dx} \leq \frac{\int_{\mathbb{R}^N} \phi^2\, dx}{\frac{1}{1+2R}\int_{\mathbb{R}^N} \phi^2\, dx} = 1 + 2R$$

and

$$\frac{\int_{\mathbb{R}^N} |\nabla\phi|^2\, dx}{\int_{\mathbb{R}^N} k(x)\phi^2\, dx} \leq \frac{\frac{C^2}{R^2}|B_{2R}|}{\int_{B_R} k(x)\, dx} \leq \frac{\frac{C^2}{R^2}|B_{2R}|}{\frac{1}{1+R}|B_R|} = C_1\frac{(1+R)}{R^2}.$$

Therefore

$$\frac{\|\phi\|_{H^s(\mathbb{R}^N)}^2}{\int_{\mathbb{R}^N} k(x)\phi^2\, dx} \leq \frac{C\|\phi\|_{H^1(\mathbb{R}^N)}^2}{\int_{\mathbb{R}^N} k(x)\phi^2\, dx} \leq C_2\frac{(1+R)}{R^2} + C_3(1+2R),$$

where $C_2, C_3 > 0$ are constants independent of R.

Choosing $R > 0$ such that $C_2\frac{(1+R)}{R^2} \leq C_3$, we can infer that $\mu^* \leq 2C_3(R+1)$. Then, taking $R_0 = 2C_3(R+1) + 2R$, we have

$$\lim_{t\to\infty}\frac{f(t)}{t} = l = R_0 > \mu^*.$$

Now, fix $\varepsilon \in (0, 1)$, and let $h \in L^2(\mathbb{R}^N)$ be such that

$$\|h\|_{L^2(\mathbb{R}^N)} < m = \max_{t\geq 0}\left[\left(\frac{1}{2} - \frac{\varepsilon}{2}\right)t - \frac{C_\varepsilon}{p+1}S_*^{p+1}t^p\right].$$

Then, all assumptions of Theorem 15.1.1 are satisfied, and we can find at least two positive solutions to (15.1.1).

Example 15.1.5 Fix $\varepsilon \in (0, 1)$, and consider the following functions

$$h(x) = \begin{cases} 0, & \text{if } |x| < \sqrt{3} \vee |x| > 2, \\ C(|x|^2 - 2)^2(|x|^2 - 3)^2(|x|^2 - 4)^2, & \text{if } \sqrt{3} \le |x| \le 2, \end{cases}$$

and

$$f(t) = \begin{cases} t \log(1 + t), & \text{if } t > 0, \\ 0, & \text{if } t \le 0, \end{cases}$$

where $C > 0$ is a constant such that

$$\|h\|_{L^2(\mathbb{R}^N)} < m = \max_{t \ge 0} \left[\left(\frac{1}{2} - \frac{\varepsilon}{2} \right) t - \frac{C_\varepsilon}{p+1} S_*^{p+1} t^p \right].$$

It is clear that f satisfies $(f1)$–$(f3)$ and $(f4)$ with $l = \infty$, and $h \in C^1(\mathbb{R}^N) \cap L^2(\mathbb{R}^N)$. Moreover, for any $\sqrt{3} < |x| < 2$,

$$x \cdot \nabla h = 4C \left[|x|^2(|x|^2 - 2)(|x|^2 - 3)(|x|^2 - 4)(3|x|^4 - 22|x|^2 + 26) \right] \ge 0,$$

so $x \cdot \nabla h \ge 0$ on $\mathbb{R}^N$. In particular, $x \cdot \nabla h \in L^q(\mathbb{R}^N)$ for any $q \in [1, \infty]$. Then, we can apply Theorem 15.1.3 to deduce that the problem (15.1.1) admits at least two positive solutions.

15.2 The Asymptotically Linear Case

In this section we discuss the existence of positive solutions to (15.1.1) under the assumption that f is asymptotically linear. We consider the following degenerate elliptic problem

$$\begin{cases} -\operatorname{div}(y^{1-2s} \nabla U) = 0 & \text{in } \mathbb{R}^{N+1}_+, \\ \frac{\partial U}{\partial v^{1-2s}} = -u + k(x) f(u) + h(x) & \text{on } \mathbb{R}^N, \end{cases} \tag{15.2.1}$$

where $k(x)$ is a bounded positive function, $h \in L^2(\mathbb{R}^N)$, $h \ge 0$ ($h \not\equiv 0$) and f satisfies $(f1)$-$(f4)$ with $l < \infty$.

Since the proof of Theorem 15.1.1 consists of several steps, we first collect some useful lemmas.

Lemma 15.2.1 *Suppose that* $(f1)$–$(f4)$ *with* $l < \infty$ *hold. Let* $h \in L^2(\mathbb{R}^N)$, *let* k *satisfy* (15.1.4), *and let* $(U_n) \subset X^s(\mathbb{R}_+^{N+1})$ *be a bounded Palais-Smale sequence for* I. *Then* (U_n) *has a strongly convergent subsequence in* $X^s(\mathbb{R}_+^{N+1})$.

Proof First, we show that for every $\varepsilon > 0$ there exist an $R(\varepsilon) > R_0$ (where R_0 is given by (K)) and an $n(\varepsilon) > 0$ such that

$$\iint_{\mathbb{R}_+^{N+1} \setminus B_R^+} y^{1-2s} |\nabla U_n|^2 \, dx dy + \int_{\mathbb{R}^N \setminus B_R} u_n^2 \, dx \le \varepsilon, \quad \forall R \ge R(\varepsilon) \text{ and } n \ge n(\varepsilon).$$

$$(15.2.2)$$

Let $\Psi_R \in C^\infty(\mathbb{R}_+^{N+1})$ be a smooth function such that $0 \le \Psi_R \le 1$,

$$\Psi_R(x, y) = \begin{cases} 0 \text{ for } (x, y) \in B_{\frac{R}{2}}^+, \\ 1 \text{ for } (x, y) \notin B_R^+, \end{cases}$$

$$(15.2.3)$$

and

$$|\nabla \Psi_R(x, y)| \le \frac{C}{R} \quad \text{for all } (x, y) \in \mathbb{R}_+^{N+1}$$

$$(15.2.4)$$

for some positive constant C independent of R.

Then, for any $U \in X^s(\mathbb{R}_+^{N+1})$ and all $R \ge 1$, there exists a constant $C_1 > 0$ such that

$$\|\Psi_R U\|_{X^s(\mathbb{R}_+^{N+1})} \le C_1 \|U\|_{X^s(\mathbb{R}_+^{N+1})}.$$

Indeed, using Young's inequality and Lemma 1.3.9-(i), we see that

$$\iint_{\mathbb{R}_+^{N+1}} y^{1-2s} |\nabla(U \Psi_R)|^2 \, dx dy + \int_{\mathbb{R}^N} (u \psi_R)^2 \, dx$$

$$\le 2 \iint_{\mathbb{R}_+^{N+1}} y^{1-2s} |\nabla U|^2 \Psi_R^2 \, dx dy + 2 \iint_{\mathbb{R}_+^{N+1}} y^{1-2s} |\nabla \Psi_R|^2 U^2 \, dx dy + \int_{\mathbb{R}^N} u^2 \, dx$$

$$\le 2 \iint_{\mathbb{R}_+^{N+1}} y^{1-2s} |\nabla U|^2 \, dx dy + \frac{2C}{R^2} \iint_{B_R^+ \setminus B_{\frac{R}{2}}^+} y^{1-2s} U^2 \, dx dy + \int_{\mathbb{R}^N} u^2 \, dx$$

$$\le 2 \iint_{\mathbb{R}_+^{N+1}} y^{1-2s} |\nabla U|^2 \, dx dy + \int_{\mathbb{R}^N} u^2 \, dx$$

$$+ \frac{2C}{R^2} \left(\iint_{B_R^+ \setminus B_{\frac{R}{2}}^+} y^{1-2s} |\nabla U|^{2\gamma} \, dx dy \right)^{\frac{1}{\gamma}} \left(\iint_{B_R^+ \setminus B_{\frac{R}{2}}^+} y^{1-2s} \, dx dy \right)^{\frac{\gamma-1}{\gamma}}$$

$$\le 2(1 + C) \|U\|_{X^s(\mathbb{R}_+^{N+1})}^2 \le C_1 \|U\|_{X^s(\mathbb{R}_+^{N+1})}^2,$$

where we used the facts

$$\iint_{B_R^+ \setminus B_{\frac{R}{2}}^+} y^{1-2s}\, dxdy \le CR^{N+2-2s} \quad \text{and} \quad \frac{\gamma-1}{\gamma} = \frac{2}{N+2-2s}.$$

Since $I'(U_n) \to 0$ as $n \to \infty$ and (U_n) is bounded in $X^s(\mathbb{R}_+^{N+1})$, we know that, for any $\varepsilon > 0$, there exists $n(\varepsilon) > 0$ such that

$$\langle I'(U_n), \Psi_R U_n \rangle \le C_1 \|I'(U_n)\|_* \|U_n\|_{X^s(\mathbb{R}_+^{N+1})} \le \frac{\varepsilon}{4} \quad \text{for } n \ge n(\varepsilon).$$

Equivalently, for all $n \ge n(\varepsilon)$,

$$\iint_{\mathbb{R}_+^{N+1}} y^{1-2s}|\nabla U_n|^2 \Psi_R\, dxdy + \int_{\mathbb{R}^N} u_n^2 \psi_R\, dx$$

$$\le \int_{\mathbb{R}^N} (k(x)f(u_n) + h(x))u_n \psi_R\, dx - \iint_{\mathbb{R}_+^{N+1}} y^{1-2s}\nabla U_n \cdot \nabla \psi_R U_n\, dxdy + \frac{\varepsilon}{4}.$$

$$(15.2.5)$$

Now, by $(f1)$ and (15.1.4), we obtain that there exists $0 < \theta < 1$ such that

$$k(x)f(u_n)u_n \le \theta u_n^2 \quad \text{for } |x| \ge R_0. \tag{15.2.6}$$

Since $h \in L^2(\mathbb{R}^N)$ and $\|U_n\|_{X^s(\mathbb{R}_+^{N+1})} \le C$ for some constant $C > 0$, it follows from (15.2.3) that there exists $R(\varepsilon) > R_0$ such that

$$\int_{\mathbb{R}^N} h(x)u_n \psi_R\, dx \le \|h\psi_R\|_{L^2(\mathbb{R}^N)} \|u_n\|_{L^2(\mathbb{R}^N)} \le \frac{\varepsilon}{4}, \quad \text{for } R \ge R(\varepsilon). \tag{15.2.7}$$

Thanks to the boundedness of (U_n) in $X^s(\mathbb{R}_+^{N+1})$, we may assume that, up to a subsequence, there exists $U \in X^s(\mathbb{R}_+^{N+1})$ such that $U_n \rightharpoonup U$ in $X^s(\mathbb{R}_+^{N+1})$, $u_n \to u$ in $L_{loc}^q(\mathbb{R}^N)$ for any $q \in [1, 2_s^*)$, and $u_n \to u$ a.e. in $\mathbb{R}^N$. Therefore, by (15.2.4), the bound $\|U_n\|_{X^s(\mathbb{R}_+^{N+1})} \le C$, Hölder's inequality and Lemma 1.3.9-(i),

$$\lim_{R\to\infty} \limsup_{n\to\infty} \left| \iint_{\mathbb{R}_+^{N+1}} y^{1-2s}\nabla U_n \cdot \nabla \Psi_R U_n\, dxdy \right|$$

$$\le \lim_{R\to\infty} \limsup_{n\to\infty} \frac{C}{R} \left(\iint_{B_R^+ \setminus B_{\frac{R}{2}}^+} y^{1-2s}|\nabla U_n|^2\, dxdy \right)^{\frac{1}{2}} \left(\iint_{B_R^+ \setminus B_{\frac{R}{2}}^+} y^{1-2s}|U_n|^2\, dxdy \right)^{\frac{1}{2}}$$

$$\leq \lim_{R\to\infty} \frac{C}{R} \left(\iint_{B_R^+\setminus B_{\frac{R}{2}}^+} y^{1-2s} |U|^2 \, dxdy \right)^{\frac{1}{2}}$$

$$\leq \lim_{R\to\infty} \frac{C}{R} \left(\iint_{B_R^+\setminus B_{\frac{R}{2}}^+} y^{1-2s} |U|^{2\gamma} \, dxdy \right)^{\frac{1}{2\gamma}} \left(\iint_{B_R^+\setminus B_{\frac{R}{2}}^+} y^{1-2s} \, dxdy \right)^{\frac{\gamma-1}{2\gamma}}$$

$$\leq C \lim_{R\to\infty} \left(\iint_{B_R^+\setminus B_{\frac{R}{2}}^+} y^{1-2s} |U|^{2\gamma} \, dxdy \right)^{\frac{1}{2\gamma}} = 0. \tag{15.2.8}$$

Then, putting together (15.2.5), (15.2.6), (15.2.7) and (15.2.8), we have for any $R \geq R(\varepsilon)$ and $n \geq n(\varepsilon)$ sufficiently large

$$\iint_{\mathbb{R}_+^{N+1}} y^{1-2s} |\nabla U_n|^2 \Psi_R \, dxdy + \int_{\mathbb{R}^N} (1-\theta) u_n^2 \psi_R \, dx \leq \varepsilon. \tag{15.2.9}$$

From $\theta \in (0,1)$ and (15.2.3), we deduce that (15.2.9) implies (15.2.2).

Now, we exploit the relation (15.2.2) in order to prove the existence of a convergent subsequence for (U_n). Using the fact that $I'(U_n) = 0$ and (U_n) is bounded in $X^s(\mathbb{R}_+^{N+1})$, we see that

$$\langle I'(U_n), U_n \rangle = \iint_{\mathbb{R}_+^{N+1}} y^{1-2s} |\nabla U_n|^2 \, dxdy + \int_{\mathbb{R}^N} u_n^2 \, dx$$

$$- \int_{\mathbb{R}^N} k(x) f(u_n) u_n \, dx - \int_{\mathbb{R}^N} h(x) u_n \, dx = o(1) \tag{15.2.10}$$

and

$$\langle I'(U_n), U \rangle = \iint_{\mathbb{R}_+^{N+1}} y^{1-2s} \nabla U_n \cdot \nabla U \, dxdy + \int_{\mathbb{R}^N} u_n u \, dx$$

$$- \int_{\mathbb{R}^N} k(x) f(u_n) u \, dx - \int_{\mathbb{R}^N} h(x) u \, dx = o(1). \tag{15.2.11}$$

Hence, in order to prove our lemma, it suffices to show that

$$\|U_n\|_{X^s(\mathbb{R}_+^{N+1})} \to \|U\|_{X^s(\mathbb{R}_+^{N+1})} \quad \text{as } n \to \infty.$$

In view of (15.2.10) and (15.2.11), this is equivalent to showing that

$$\int_{\mathbb{R}^N} k(x) f(u_n)(u_n - u) \, dx + \int_{\mathbb{R}^N} h(x)(u_n - u) \, dx = o(1). \tag{15.2.12}$$

Clearly, since $k \in L^\infty(\mathbb{R}^N)$, $h \in L^2(\mathbb{R}^N)$ and $u_n \to u$ in $L^2(B_R)$ for any $R > 0$, we have that

$$\int_{B_R} k(x) f(u_n)(u_n - u)\, dx + \int_{B_R} h(x)(u_n - u)\, dx = o(1). \tag{15.2.13}$$

On the other hand, by (15.2.2), we know that for any $\varepsilon > 0$ there exists $R(\varepsilon) > 0$ such that

$$\int_{|x| \geq R(\varepsilon)} k(x) f(u_n)(u_n - u)\, dx + \int_{\mathbb{R}^N} h(x)(u_n - u)\, dx$$

$$\leq \left(\int_{|x| \geq R(\varepsilon)} k(x) |f(u_n)|^2\, dx \right)^{\frac{1}{2}} \left(\int_{|x| \geq R(\varepsilon)} k(x) |u_n - u|^2\, dx \right)^{\frac{1}{2}}$$

$$+ \left(\int_{|x| \geq R(\varepsilon)} |h(x)|^2\, dx \right)^{\frac{1}{2}} \left(\int_{|x| \geq R(\varepsilon)} |u_n - u|^2\, dx \right)^{\frac{1}{2}}$$

$$\leq C \left(\int_{|x| \geq R(\varepsilon)} |u_n|^2\, dx \right)^{\frac{1}{2}} \left(\int_{|x| \geq R(\varepsilon)} |u_n - u|^2\, dx \right)^{\frac{1}{2}}$$

$$+ \|h\|_{L^2(\mathbb{R}^N)} \left(\int_{|x| \geq R(\varepsilon)} |u_n - u|^2\, dx \right)^{\frac{1}{2}}$$

$$\leq C\varepsilon \tag{15.2.14}$$

for n large enough. Combining (15.2.13) with (15.2.14) we obtain (15.2.12), which completes the proof of lemma. $\square$

In the next lemma we show that I is positive on the boundary of a some ball in $X^s(\mathbb{R}^{N+1}_+)$, as long as $\|h\|_{L^2(\mathbb{R}^N)}$ is sufficiently small. This property will allow us to apply Ekeland's variational principle.

Lemma 15.2.2 *Assume that $(f1)$–$(f3)$ hold, $h \in L^2(\mathbb{R}^N)$ is such that (15.1.6) is satisfied, and $k \in L^\infty(\mathbb{R}^N)$. Then there exist $\rho, \alpha, m > 0$ such that $I(U)|_{\|U\|_{X^s(\mathbb{R}^{N+1}_+)} = \rho} \geq \alpha$ for $\|h\|_{L^2(\mathbb{R}^N)} < m$.*

Proof Fix $\varepsilon \in (0, \|k\|_{L^\infty(\mathbb{R}^N)}^{-1})$. Then, in view of (15.1.2) and (1.2.9),

$$I(U) \geq \frac{1}{2}\|U\|_{X^s(\mathbb{R}^{N+1}_+)}^2 - \frac{\varepsilon}{2}\|k\|_{L^\infty(\mathbb{R}^N)}\|U\|_{X^s(\mathbb{R}^{N+1}_+)}^2 - \frac{C(\varepsilon)}{p+1}\|k\|_{L^\infty(\mathbb{R}^N)}S_*^{p+1}\|U\|_{X^s(\mathbb{R}^{N+1}_+)}^{p+1}$$

$$- \|h\|_{L^2(\mathbb{R}^N)}\|U\|_{X^s(\mathbb{R}^{N+1}_+)}$$

$$= \|U\|_{X^s(\mathbb{R}^{N+1}_+)}\left[\left(\frac{1}{2} - C_1\varepsilon\right)\|U\|_{X^s(\mathbb{R}^{N+1}_+)} - C_2(\varepsilon)\|U\|_{X^s(\mathbb{R}^{N+1}_+)}^p - \|h\|_{L^2(\mathbb{R}^N)}\right],$$

$$(15.2.15)$$

where

$$C_1 = \frac{1}{2}\|k\|_{L^\infty(\mathbb{R}^N)} \quad \text{and} \quad C_2(\varepsilon) = \frac{C(\varepsilon)}{p+1}\|k\|_{L^\infty(\mathbb{R}^N)}S_*^{p+1}.$$

Using (15.1.6) and (15.2.15), we can find $\rho, \alpha > 0$ such that $I(U)|_{\|U\|_{X^s(\mathbb{R}^{N+1}_+)}=\rho} \geq \alpha$, provided that $\|h\|_{L^2(\mathbb{R}^N)} < m$. $\qquad\square$

For ρ given by Lemma 15.2.2, we denote by

$$\mathcal{B}_\rho = \{U \in X^s(\mathbb{R}^{N+1}_+) : \|U\|_{X^s(\mathbb{R}^{N+1}_+)} < \rho\}$$

the ball in $X^s(\mathbb{R}^{N+1}_+)$ with center in 0 and radius ρ. By means of the Ekeland variational principle and Lemma 15.2.1, we can verify that I has a local minimum if $\|h\|_{L^2(\mathbb{R}^N)}$ is small enough.

Theorem 15.2.3 *Assume that $(f1)$–$(f4)$ with $l < \infty$ hold, $h \in L^2(\mathbb{R}^N)$, $h \geq 0$ ($h \not\equiv 0$) and k satisfies (15.1.4). If $\|h\|_{L^2(\mathbb{R}^N)} < m$, where m is given by Lemma 15.2.2, then there exists $U_0 \in X^s(\mathbb{R}^{N+1}_+)$ such that*

$$I(U_0) = \inf\{I(U) : U \in \overline{\mathcal{B}}_\rho\} < 0$$

and U_0 is a nontrivial nonnegative solution of problem (15.2.1).

Proof Since $h(x) \in L^2(\mathbb{R}^N)$, $h \geq 0$ and $h \not\equiv 0$, we can choose a function $V \in X^s(\mathbb{R}^{N+1}_+)$ such that

$$\int_{\mathbb{R}^N} h(x)v\, dx > 0. \qquad (15.2.16)$$

Next, note that

$$I(tV) = \frac{t^2}{2}\left[\int_{\mathbb{R}^{N+1}_+} y^{1-2s}|\nabla V|^2\,dxdy + \int_{\mathbb{R}^N} v^2\,dx\right] - \int_{\mathbb{R}^N} k(x)F(tv)\,dx - t\int_{\mathbb{R}^N} h(x)v\,dx$$

$$\leq \frac{t^2}{2}\|V\|^2_{X^s(\mathbb{R}^{N+1}_+)} - t\int_{\mathbb{R}^N} h(x)v\,dx < 0 \quad \text{for } t > 0 \text{ small enough.}$$

Then

$$c_0 = \inf\{I(U) : U \in \overline{\mathcal{B}}_\rho\} < 0.$$

Note that $\overline{\mathcal{B}}_\rho$ is a complete metric space with the distance

$$\text{dist}(U, V) = \|U - V\|_{X^s(\mathbb{R}^{N+1}_+)}, \quad \text{for any } U, V \in X^s(\mathbb{R}^{N+1}_+).$$

Applying Theorem 2.2.1, there exists $(U_n) \subset \overline{\mathcal{B}}_\rho$ such that

(i) $c_0 \leq I(U_n) < c_0 + \frac{1}{n}$,

(ii) $I(W) \geq I(U_n) - \frac{1}{n}\|W - U_n\|_{X^s(\mathbb{R}^{N+1}_+)}$ for all $W \in \overline{\mathcal{B}}_\rho$.

Let us prove that (U_n) is a bounded Palais-Smale sequence of I. First, we show that $\|U_n\|_{X^s(\mathbb{R}^{N+1}_+)} < \rho$ for a n large enough. If $\|U_n\|_{X^s(\mathbb{R}^{N+1}_+)} = \rho$ for infinitely many n, then we may assume that $\|U_n\|_{X^s(\mathbb{R}^{N+1}_+)} = \rho$ for all $n \geq 1$. Hence, by Lemma 15.2.2, $I(U_n) \geq \alpha > 0$. Taking the limit as $n \to \infty$ and by using (i), we deduce that $0 > c_0 \geq \alpha > 0$, which is a contradiction.

Now, we show that $I'(U_n) \to 0$. Indeed, for any $U \in X^s(\mathbb{R}^{N+1}_+)$ with $\|U\|_{X^s(\mathbb{R}^{N+1}_+)} = 1$, let $W_n = U_n + tU$. For a fixed n, we have $\|W_n\|_{X^s(\mathbb{R}^{N+1}_+)} \leq \|U_n\|_{X^s(\mathbb{R}^{N+1}_+)} + t < \rho$ when t is small enough. Using (ii), we deduce that

$$I(W_n) \geq I(U_n) - \frac{t}{n}\|U\|_{X^s(\mathbb{R}^{N+1}_+)},$$

that is,

$$\frac{I(W_n) - I(U_n)}{t} \geq -\frac{\|U\|_{X^s(\mathbb{R}^{N+1}_+)}}{n} = -\frac{1}{n}.$$

Letting $t \to 0$, we deduce that $\langle I'(U_n), U \rangle \geq -\frac{1}{n}$, which means that $|\langle I'(U_n), U \rangle| \leq \frac{1}{n}$ for any $U \in X^s(\mathbb{R}^{N+1}_+)$ with $\|U\|_{X^s(\mathbb{R}^{N+1}_+)} = 1$. This shows that (U_n) is indeed a bounded Palais-Smale sequence of I. Then, by Lemma 15.2.1, we can find $U_0 \in X^s(\mathbb{R}^{N+1}_+)$ such that $I'(U_0) = 0$ and $I(U_0) = c_0 < 0$. Since $\langle I'(U_0), U_0^- \rangle = 0$ and $h \geq 0$, we get $U_0 \geq 0$ in $\mathbb{R}_+^{N+1}$, $U_0 \not\equiv 0$. $\qquad\square$

In what follows, we show that problem (15.2.1) has a mountain pass type solution. In order to do this, we use Theorem 2.2.15, which furnishes a Cerami sequence (U_n). Since this type of Palais-Smale sequence enjoys some useful properties, we are able to prove its boundedness in the asymptotically linear case.

The lemma below shows that I possesses a mountain pass geometry.

Lemma 15.2.4 *Suppose that* $(f1)$–$(f4)$ *hold and* $\mu^* \in (l, \infty)$ *with* μ^* *given by* (15.1.5). *Then there exists* $V \in X^s(\mathbb{R}_+^{N+1})$ *with* $\|V\|_{X^s(\mathbb{R}_+^{N+1})} > \rho$, ρ *is given by Lemma 15.2.2, such that* $I(V) < 0$.

Proof Since $l > \mu^*$, we can find a non-negative function $W \in X^s(\mathbb{R}_+^{N+1})$ such that

$$\int_{\mathbb{R}^N} k(x) w^2 \, dx = 1 \text{ such that } \int_{\mathbb{R}_+^{N+1}} y^{1-2s} |\nabla W|^2 \, dx dy + \int_{\mathbb{R}^N} w^2 \, dx < l.$$

Using $(f4)$ and Fatou's lemma, we see that

$$\lim_{t \to \infty} \frac{I(tW)}{t^2} = \frac{1}{2} \|W\|_{X^s(\mathbb{R}_+^{N+1})}^2 - \lim_{t \to \infty} \int_{\mathbb{R}^N} k(x) \frac{F(tw)}{t^2} \, dx - \lim_{t \to \infty} \frac{1}{t} \int_{\mathbb{R}^N} h(x) w \, dx$$

$$\leq \frac{1}{2} (\|W\|_{X^s(\mathbb{R}_+^{N+1})}^2 - l) < 0.$$

It remains to take $V = t_0 W$ with t_0 large enough.

$\qquad\square$

Putting together Lemmas 15.2.2 and 15.2.4, we see that the assumptions of Theorem 2.2.15 are satisfied. Then, we can find a sequence $(U_n) \subset X^s(\mathbb{R}_+^{N+1})$ such that

$$I(U_n) \to c > 0 \quad \text{and} \quad \|I'(U_n)\|_* (1 + \|U_n\|_{X^s(\mathbb{R}_+^{N+1})}) \to 0. \tag{15.2.17}$$

Let

$$W_n = \frac{U_n}{\|U_n\|_{X^s(\mathbb{R}_+^{N+1})}}.$$

Obviously, (W_n) is bounded in $X^s(\mathbb{R}^{N+1}_+)$, so there exists a $W \in X^s(\mathbb{R}^{N+1}_+)$ such that, up to a subsequence, we have

$$
\begin{aligned}
W_n &\rightharpoonup W &&\text{in } X^s(\mathbb{R}^{N+1}_+), \\
w_n &\to w &&\text{a.e. in } \mathbb{R}^N, \\
w_n &\to w &&\text{in } L^2_{loc}(\mathbb{R}^N).
\end{aligned}
\tag{15.2.18}
$$

With the notation introduced above, we state.

Lemma 15.2.5 *Assume that $(f1)$–$(f4)$ and (K) hold. Let $h \in L^2(\mathbb{R}^N)$ and $\mu^* \in (l, \infty)$ for μ^* given by (15.1.5). If $\|U_n\|_{X^s(\mathbb{R}^{N+1}_+)} \to \infty$, then W given by (15.2.18) is a nontrivial nonnegative solution of*

$$
\begin{cases}
-\operatorname{div}(y^{1-2s}\nabla W) = 0 & \text{in } \mathbb{R}^{N+1}_+, \\
\dfrac{\partial W}{\partial \nu^{1-2s}} = -w + lk(x)w & \text{on } \mathbb{R}^N.
\end{cases}
\tag{15.2.19}
$$

Proof First, we show that $W \not\equiv 0$. We argue by contradiction and assume that $W \equiv 0$. Then Theorem 1.1.8 shows that $w_n \to 0$ strongly in $L^2(B_{R_0})$, where R_0 is given by condition (K). On the other hand, by $(f1)$, $(f4)$ and $l < \infty$, there is $C > 0$ such that

$$
\frac{f(t)}{t} \le C, \text{ for all } t \in \mathbb{R}.
\tag{15.2.20}
$$

Therefore,

$$
\int_{|x|<R_0} k(x)\frac{f(u_n)}{u_n}w_n^2\,dx \le C\|k\|_{L^\infty(\mathbb{R}^N)}\int_{|x|<R_0} w_n^2\,dx \to 0.
\tag{15.2.21}
$$

By condition (K), we can find $\eta \in (0, 1)$ such that

$$
\sup\left\{\frac{f(t)}{t} : t > 0\right\} < \eta \inf\left\{\frac{1}{k(x)} : |x| \ge R_0\right\},
\tag{15.2.22}
$$

so, for all $n \in \mathbb{N}$, we get

$$
\int_{|x|\ge R_0} k(x)\frac{f(u_n)}{u_n}w_n^2\,dx \le \eta \int_{|x|\ge R_0} w_n^2\,dx \le \eta < 1.
\tag{15.2.23}
$$

Combining (15.2.21) and (15.2.23), we see that

$$
\limsup_{n\to\infty} \int_{\mathbb{R}^N} k(x)\frac{f(u_n)}{u_n}w_n^2\,dx < 1.
\tag{15.2.24}
$$

Next, the fact that $\|U_n\|_{X^s(\mathbb{R}^{N+1}_+)} \to \infty$ and (15.2.17) imply that

$$\frac{\langle I'(U_n), U_n \rangle}{\|U_n\|^2_{X^s(\mathbb{R}^{N+1}_+)}} = o(1),$$

that is

$$o(1) = \|W_n\|^2_{X^s(\mathbb{R}^{N+1}_+)} - \int_{\mathbb{R}^N} k(x) \frac{f(u_n)}{u_n} w_n^2 \, dx = 1 - \int_{\mathbb{R}^N} k(x) \frac{f(u_n)}{u_n} w_n^2 \, dx,$$

which yields a contradiction in view of (15.2.24). Thus, we proved that $W \not\equiv 0$.

In what follows, we show that $W \geq 0$ in $\mathbb{R}^{N+1}_+$. Let $(W_n)_-(x) = \max\{-W_n(x), 0\}$, and we observe that the sequence $((W_n)_-)$ is bounded in $X^s(\mathbb{R}^{N+1}_+)$. Since $\|U_n\|_{X^s(\mathbb{R}^{N+1}_+)} \to \infty$, we obtain that

$$\frac{\langle I'(U_n), (W_n)_- \rangle}{\|U_n\|_{X^s(\mathbb{R}^{N+1}_+)}} = o(1),$$

which gives

$$- \|(W_n)_-\|^2_{X^s(\mathbb{R}^{N+1}_+)} = \int_{\mathbb{R}^N} k(x) \frac{f(u_n)}{\|U_n\|_{X^s(\mathbb{R}^{N+1}_+)}} (w_n)_- \, dx + o(1). \tag{15.2.25}$$

Taking into account $(f1)$, we know that $f(t) \equiv 0$ for all $t \leq 0$, so (15.2.25) implies that

$$\lim_{n \to \infty} \|(W_n)_-\|_{X^s(\mathbb{R}^{N+1}_+)} = 0,$$

which gives $W_- = 0$ in $\mathbb{R}^{N+1}_+$, that is $W \geq 0$.

Finally, we prove that W is a solution to (15.2.19). By (15.2.17) and $\|U_n\|_{X^s(\mathbb{R}^{N+1}_+)} \to \infty$, we get

$$\frac{\langle I'(U_n), \Phi \rangle}{\|U_n\|_{X^s(\mathbb{R}^{N+1}_+)}} = o(1), \quad \text{for all } \Phi \in C_c^\infty(\overline{\mathbb{R}^{N+1}_+}),$$

or explicitly

$$\iint_{\mathbb{R}^{N+1}_+} y^{1-2s} \nabla W_n \cdot \nabla \Phi \, dx \, dy + \int_{\mathbb{R}^N} w_n \phi \, dx = \int_{\mathbb{R}^N} k(x) \frac{f(u_n)}{u_n} w_n \phi \, dx + o(1).$$

$$\tag{15.2.26}$$

where we used the notation $\phi = \Phi(\cdot, 0)$. Since $W_n \rightharpoonup W$ in $X^s(\mathbb{R}^{N+1}_+)$ and $w_n \to w$ in $L^2_{loc}(\mathbb{R}^N)$, we deduce that

$$\iint_{\mathbb{R}^{N+1}_+} y^{1-2s} \nabla W_n \cdot \nabla \Phi \, dx dy + \int_{\mathbb{R}^N} w_n \phi \, dx = \int_{\mathbb{R}^N} k(x) \frac{f(u_n)}{u_n} w_n \phi \, dx + o(1).$$

$$(15.2.27)$$

To prove that W solves (15.2.19), it suffices to show that

$$\int_{\mathbb{R}^N} k(x) \frac{f(u_n)}{u_n} w_n(x) \phi(x) \, dx \to \int_{\mathbb{R}^N} lk(x) w(x) \phi(x) \, dx. \qquad (15.2.28)$$

First, we note that, by (15.2.20) and $\|W_n\|_{X^s(\mathbb{R}^{N+1}_+)} = 1$,

$$\int_{\mathbb{R}^N} \left| \frac{f(u_n)}{u_n} w_n(x) \right|^2 dx \le C \int_{\mathbb{R}^N} w_n^2 \, dx \le C \|W_n\|^2_{X^s(\mathbb{R}^{N+1}_+)} = C,$$

that is, the sequence $(\frac{f(u_n)}{u_n} w_n)$ is bounded in $L^2(\mathbb{R}^N)$.

Now, let us define the following sets

$$\Omega_+ = \{x \in \mathbb{R}^N : w(x) > 0\} \text{ and } \Omega_0 = \{x \in \mathbb{R}^N : w(x) = 0\}.$$

In view of (15.2.18), it is clear that $u_n(x) \to \infty$ a.e. $x \in \Omega_+$. Then, by $(f4)$,

$$\frac{f(u_n)}{u_n} w_n(x) \to lw(x) \quad \text{a.e. } x \in \Omega_+. \qquad (15.2.29)$$

Since $w_n \to 0$ a.e. in Ω_0, it follows from (15.2.20) that

$$\frac{f(u_n)}{u_n} w_n(x) \to 0 \equiv lw(x) \quad \text{a.e. } x \in \Omega_0. \qquad (15.2.30)$$

Now (15.2.29) and (15.2.30) imply that

$$\frac{f(u_n)}{u_n} w_n \rightharpoonup lw \quad \text{in } L^2(\mathbb{R}^N). \qquad (15.2.31)$$

Since $\phi \in C_c^\infty(\mathbb{R}^N)$ and $k \in L^\infty(\mathbb{R}^N)$, we see that $z = k\phi \in L^2(\mathbb{R}^N)$, which combined with (15.2.31) implies that

$$\int_{\mathbb{R}^N} \frac{f(u_n)}{u_n} w_n z \, dx \to \int_{\mathbb{R}^N} lwz \, dx \quad \text{as } n \to \infty,$$

that is, (15.2.28) holds. $\qquad \qquad \square$

Lemma 15.2.6 *Let $k \in L^\infty(\mathbb{R}^N, \mathbb{R}_+)$ and let μ^* be defined by (15.1.5) with $l \in (\mu^*, \infty)$. Then, (15.2.19) has no nontrivial non-negative solution.*

Proof Since $l > \mu^*$, there is a constant $\delta > 0$ such that $\mu^* < \mu^* + \delta < l$. By the definition of μ^*, there exists $V_\delta \in X^s(\mathbb{R}_+^{N+1})$ such that $\int_{\mathbb{R}^N} k(x) v_\delta^2 \, dx = 1$ and

$$\mu^* \leq \|V_\delta\|^2_{X^s(\mathbb{R}_+^{N+1})} < \mu^* + \delta.$$

Since $C_c^\infty(\overline{\mathbb{R}_+^{N+1}})$ is dense in $X^s(\mathbb{R}_+^{N+1})$, we may assume that $V_\delta \in C_c^\infty(\overline{\mathbb{R}_+^{N+1}})$. Let $R > 0$ be such that $\operatorname{supp}(V_\delta) \subset B_R^+$ and define

$$\mu_R = \inf \left\{ \iint_{B_R^+} y^{1-2s} |\nabla U|^2 \, dx dy + \int_{\Gamma_R^0} u^2 \, dx : \int_{\Gamma_R^0} k(x) u^2 \, dx = 1, \, U \in H^1_{\Gamma_R^+}(B_R^+) \right\},$$

where we used the notation

$$H^1_{\Gamma_R^+}(B_R^+) = \{ V \in H^1(B_R^+, y^{1-2s}) : V \equiv 0 \text{ on } \Gamma_R^+ \}.$$

Since $V_\delta \equiv 0$ on Γ_R^+, we infer that $V_\delta \in H^1_{\Gamma_R^+}(B_R^+)$ and

$$\mu_R \leq \|V_\delta\|^2_{X^s(\mathbb{R}_+^{N+1})} < \mu^* + \delta < l. \tag{15.2.32}$$

By the compactness of the embedding $H^1_{\Gamma_R^+}(B_R^+) \subset L^2(\Gamma_R^0)$, it is not difficult to see that there exists $W_R \in H^1_{\Gamma_R^+}(B_R^+) \setminus \{0\}$ with $W_R \geq 0$ and $\int_{\Gamma_R^0} k(x) w_R^2 \, dx = 1$ such that

$$\begin{cases} -\operatorname{div}(y^{1-2s} \nabla W_R) = 0 & \text{in } B_R^+, \\ \dfrac{\partial W_R}{\partial \nu^{1-2s}} = -w_R + \mu_R k(x) w_R & \text{on } \Gamma_R^0, \\ W_R = 0 & \text{on } \Gamma_R^+. \end{cases} \tag{15.2.33}$$

It follows from the strong maximum principle that $W_R > 0$ on B_R^+. Extend W_R by setting $W_R = 0$ in $\mathbb{R}_+^{N+1} \setminus B_R^+$, so that $W_R \in X^s(\mathbb{R}_+^{N+1})$. Therefore, if $U \neq 0$, $U \in X^s(\mathbb{R}_+^{N+1})$ is a non-negative solution of (15.2.19). Then,

$$\mu_R \int_{\Gamma_R^0} k(x) w_R u \, dx = \iint_{B_R^+} y^{1-2s} \nabla W_R \cdot \nabla U \, dx dy + \int_{\Gamma_R^0} w_R u \, dx$$

$$\tag{15.2.34}$$

$$= l \int_{\Gamma_R^0} k(x) u w_R \, dx.$$

Since $u \geq 0$ and $u \neq 0$, we can choose $R > 0$ large enough so that $\int_{\Gamma_R^0} K(x) u w_R \, dx > 0$. Hence, (15.2.34) implies that $\mu_R = l$, which is a contradiction because (15.2.32). □

***Proof of Theorem* 15.1.1** In view of Lemmas 15.2.5 and 15.2.6, it is obvious that the situation $\|U_n\| \to \infty$ cannot occur. Therefore, the sequence (U_n) is bounded in $X^s(\mathbb{R}_+^{N+1})$. Taking into account Lemma 15.2.1, we deduce that problem (15.2.1) admits a nontrivial non-negative solution $U_2 \in X^s(\mathbb{R}_+^{N+1})$ with $I(U_2) > 0$. On the other hand, by Theorem 15.2.3, there exists a nontrivial non-negative solution $U_1 \in X^s(\mathbb{R}_+^{N+1})$ with $I(U_1) < 0$. Applying Theorem 1.3.4, we obtain that $u_1, u_2 > 0$ in $\mathbb{R}^N$. □

15.3 The Superlinear Case

This section is devoted to the proof of Theorem 15.1.3. We consider the following problem:

$$\begin{cases} -\operatorname{div}(y^{1-2s}\nabla U) = 0 & \text{in } \mathbb{R}_+^{N+1}, \\ \frac{\partial U}{\partial \nu^{1-2s}} = -u + f(u) + h(x) & \text{on } \mathbb{R}^N, \end{cases} \tag{15.3.1}$$

where $h(x) = h(|x|) \in C^1(\mathbb{R}^N) \cap L^2(\mathbb{R}^N)$, $h(x) \geq 0$, $h(x) \neq 0$ and f satisfies the conditions $(f1)$–$(f4)$ with $l = \infty$. Since we assume that $k(x) \equiv 1$ and $h(x)$ is radial, it is natural to work on the space of the function belonging to $X^s(\mathbb{R}_+^{N+1})$ which are radial with respect to x, that is

$$X^s_{\text{rad}}(\mathbb{R}_+^{N+1}) = \left\{ U \in X^s(\mathbb{R}_+^{N+1}) : U(x, y) = U(|x|, y) \right\}.$$

We begin by proving the following preliminary result.

Theorem 15.3.1 *Suppose that $h(x) = h(|x|) \in L^2(\mathbb{R}^N)$, $h(x) \geq 0$, $h(x) \neq 0$, and conditions $(f1)$–$(f3)$ hold. Then there exist $m_1 > 0$ and $\tilde{U}_0 \in X^s_{\text{rad}}(\mathbb{R}_+^{N+1})$ such that $I'(\tilde{U}_0) = 0$ and $I(\tilde{U}_0) < 0$ if $\|h\|_{L^2(\mathbb{R}^N)} < m_1$.*

Proof Arguing as in the proof of Theorem 15.2.3, it follows from Theorem 2.2.1 that there exists a bounded Palais-Smale sequence $(\tilde{U}_n) \subset X^s_{\text{rad}}(\mathbb{R}_+^{N+1})$ such that

$$I(\tilde{U}_n) \to \tilde{c}_0 = \inf\{I(U) : U \in X^s_{\text{rad}}(\mathbb{R}_+^{N+1}) \text{ and } \|U\|_{X^s(\mathbb{R}_+^{N+1})} = \rho\} < 0,$$

where ρ is given by Lemma 15.2.2. We claim that this infimum is achieved.

Using Theorem 1.1.11, we may assume that there exists $\tilde{U}_0 \in X^s_{\text{rad}}(\mathbb{R}_+^{N+1})$ such that $\tilde{U}_n \rightharpoonup \tilde{U}_0$ in $X^s_{\text{rad}}(\mathbb{R}_+^{N+1})$, $\tilde{u}_n \to \tilde{u}_0$ in $L^{p+1}(\mathbb{R}^N)$.

Taking into account $(f1)$-$(f3)$, Theorem 1.1.8, and by exploiting the fact that (U_n) is bounded in $X_{\mathrm{rad}}^s(\mathbb{R}_+^{N+1})$, we see that

$$\left| \int_{\mathbb{R}^N} f(\tilde{u}_n)(\tilde{u}_n - \tilde{u}_0)\, dx \right| \leq \varepsilon \|\tilde{u}_n\|_{L^2(\mathbb{R}^N)} \|\tilde{u}_n - \tilde{u}_0\|_{L^2(\mathbb{R}^N)}$$

$$+ C_\varepsilon \|\tilde{u}_n\|_{L^{p+1}(\mathbb{R}^N)}^p \|\tilde{u}_n - \tilde{u}_0\|_{L^{p+1}(\mathbb{R}^N)}$$

$$\leq C\varepsilon + C_\varepsilon C \|\tilde{u}_n - \tilde{u}_0\|_{L^{p+1}(\mathbb{R}^N)}.$$

Hence,

$$\lim_{n \to \infty} \left| \int_{\mathbb{R}^N} f(\tilde{u}_n)(\tilde{u}_n - \tilde{u}_0)\, dx \right| \leq C\varepsilon$$

and so, by the arbitrariness of ε,

$$\int_{\mathbb{R}^N} f(\tilde{u}_n)(\tilde{u}_n - \tilde{u}_0)\, dx \to 0.$$

Using the assumptions $(f1)$–$(f3)$ and Lemma 1.4.3, we obtain that

$$\int_{\mathbb{R}^N} f(\tilde{u}_n)\tilde{u}_n\, dx \to \int_{\mathbb{R}^N} f(\tilde{u}_0)\tilde{u}_0\, dx.$$

Then,

$$\int_{\mathbb{R}^N} (f(\tilde{u}_n) - f(\tilde{u}_0))\tilde{u}_0\, dx = \int_{\mathbb{R}^N} (f(\tilde{u}_n)\tilde{u}_n - f(\tilde{u}_0)\tilde{u}_0)\, dx - \int_{\mathbb{R}^N} f(\tilde{u}_n)(\tilde{u}_n - \tilde{u}_0)\, dx \to 0.$$

On the other hand, $\tilde{u}_n \rightharpoonup \tilde{u}_0$ in $L^2(\mathbb{R}^N)$, and using the fact that $h \in L^2(\mathbb{R}^N)$, we also have

$$\int_{\mathbb{R}^N} h(x)\tilde{u}_n\, dx \to \int_{\mathbb{R}^N} h(x)\tilde{u}_0\, dx.$$

Since $\langle I'(\tilde{U}_n), \tilde{U}_n \rangle \to 0$, and $\langle I'(\tilde{U}_n), \tilde{U}_0 \rangle \to 0$, the above relations imply that $\tilde{U}_n \to \tilde{U}_0$ strongly in $X_{\mathrm{rad}}^s(\mathbb{R}_+^{N+1})$. Therefore,

$$I(\tilde{U}_0) = \tilde{c}_0 < 0 \quad \text{and} \quad I'(\tilde{U}_0) = 0.$$

$\square$

For any $\lambda \in [\frac{1}{2}, 1]$, we introduce the following family of functionals $I_\lambda : X^s_{\mathrm{rad}}(\mathbb{R}^{N+1}_+) \to \mathbb{R}$ defined by

$$I_\lambda(U) = \frac{1}{2} \|U\|^2_{X^s(\mathbb{R}^{N+1}_+)} - \lambda \int_{\mathbb{R}^N} (F(u) + h(x)u)\, dx.$$

for any $U \in X^s_{\mathrm{rad}}(\mathbb{R}^{N+1}_+)$. Our purpose is to show that I_λ satisfies the assumptions of Theorem 5.2.2.

Lemma 15.3.2 *Assume that $(f1)$-$(f4)$ with $l = \infty$ hold. Then,*

(i) *There exists $\bar{V} \in X^s_{\mathrm{rad}}(\mathbb{R}^{N+1}_+) \setminus \{0\}$ such that $I_\lambda(\bar{V}) < 0$ for all $\lambda \in [\frac{1}{2}, 1]$.*

(ii) *For $m_1 > 0$ given in Theorem 15.3.1, if $\|h\|_{L^2(\mathbb{R}^N)} < m_1$, then*

$$c_\lambda = \inf_{\gamma \in \Gamma} \max_{t \in [0,1]} I_\lambda(\gamma(t)) > \max\{I_\lambda(0), I_\lambda(\bar{V})\} \quad \forall \lambda \in \left[\frac{1}{2}, 1\right],$$

where $\Gamma = \{\gamma \in C([0, 1], X^s_r(\mathbb{R}^{N+1}_+))) : \gamma(0) = 0, \gamma(1) = \bar{V}\}$.

Proof

(i) For every $\delta > 0$, we can find $V \in X^s_{\mathrm{rad}}(\mathbb{R}^{N+1}_+) \setminus \{0\}$ and $V \geq 0$ such that

$$\iint_{\mathbb{R}^{N+1}_+} y^{1-2s} |\nabla V|^2\, dxdy < \delta \int_{\mathbb{R}^N} v^2\, dx.$$

This is lawful due to the fact that

$$\inf \left\{ \iint_{\mathbb{R}^{N+1}_+} y^{1-2s} |\nabla U|^2\, dxdy : U \in X^s_{\mathrm{rad}}(\mathbb{R}^{N+1}_+) \text{ and } \|u\|_{L^2(\mathbb{R}^N)} = 1 \right\} = 0$$

(using the Pohozaev identity, one can see that $(-\Delta)^s$ has no eigenvalues in $H^s(\mathbb{R}^N)$). By $(f4)$ with $l = \infty$, and applying Fatou's lemma, we deduce that

$$\lim_{t \to \infty} \int_{\mathbb{R}^N} \frac{F(tv)}{t^2}\, dx \geq (1+\delta) \int_{\mathbb{R}^N} v^2\, dx.$$

Hence, for any $\lambda \in [\frac{1}{2}, 1]$, we get

$$\lim_{t \to \infty} \frac{I_\lambda(tV)}{t^2} \leq \lim_{t \to \infty} \frac{I_{\frac{1}{2}}(tV)}{t^2} \leq \frac{1}{2} \left(\iint_{\mathbb{R}^{N+1}_+} y^{1-2s} |\nabla V|^2\, dxdy - \delta \int_{\mathbb{R}^N} v^2\, dx \right) < 0.$$

Take $t_1 > 0$ large enough such that $I_{\frac{1}{2}}(t_1 V) < 0$, and set $\bar{V} = t_1 V$. Then, $I_\lambda(\bar{V}) \leq I_{\frac{1}{2}}(\bar{V}) < 0$, i.e., (i) holds.

(ii) It is clear that, for any $\lambda \in [\frac{1}{2}, 1]$ and $U \in X^s_{\mathrm{rad}}(\mathbb{R}^{N+1}_+)$, we have

$$I_\lambda(U) \geq \frac{1}{2}\|U\|^2_{X^s(\mathbb{R}^{N+1}_+)} - \int_{\mathbb{R}^N} F(u)\,dx - \|h\|_{L^2(\mathbb{R}^N)}\|u\|_{L^2(\mathbb{R}^N)} = J(U).$$

Proceeding as in the proof of Lemma 15.2.2, we deduce that

$$\inf_{\gamma \in \Gamma} \max_{t \in [0,1]} J(\gamma(t)) > 0,$$

provided that $\|h\|_{L^2(\mathbb{R}^N)} < m_1$, with m_1 given by Theorem 15.3.1. Then, for every $\lambda \in \left[\frac{1}{2}, 1\right]$, it holds

$$c_\lambda = \inf_{\gamma \in \Gamma} \max_{t \in [0,1]} I_\lambda(\gamma(t)) \geq \inf_{\gamma \in \Gamma} \max_{t \in [0,1]} J(\gamma(t)) > \max\{I_\lambda(0), I_\lambda(\bar{V})\}.$$

This ends the proof of the lemma.

$\square$

It follows from Lemma 15.3.2 and Theorem 5.2.2 that there exists $(\lambda_j) \subset [\frac{1}{2}, 1]$ such that

(i) $\lambda_j \to 1$ as $j \to \infty$;
(ii) I_{λ_j} has a bounded Palais-Smale sequence (U_n^j) at the level c_{λ_j}.

In view of Theorem 1.1.11, we deduce that for each $j \in \mathbb{N}$, there exists $U_j \in X^s_{\mathrm{rad}}(\mathbb{R}^{N+1}_+)$ such that $U_n^j \to U_n$ strongly in $X^s_{\mathrm{rad}}(\mathbb{R}^{N+1}_+)$ and U_j is a positive solution of

$$\begin{cases} -\mathrm{div}(y^{1-2s}\nabla U_j) = 0 & \text{in } \mathbb{R}^{N+1}_+, \\ \frac{\partial U_j}{\partial \nu^{1-2s}} = -u_j + \lambda_j[f(u_j) + h(x)] & \text{on } \mathbb{R}^N. \end{cases}$$

Arguing as in Theorem 3.5.1 (see also [307]), it is easy to see that each U_j satisfies the following Pohozaev identity:

$$\frac{N-2s}{2}\iint_{\mathbb{R}^{N+1}_+} y^{1-2s}|\nabla U_j|^2\,dx\,dy + \frac{N}{2}\int_{\mathbb{R}^N} u_j^2\,dx$$

$$= N\lambda_j \int_{\mathbb{R}^N}(F(u_j) + hu_j)\,dx + \lambda_j \int_{\mathbb{R}^N} \nabla h(x)\cdot xu_j\,dx. \tag{15.3.2}$$

In the next lemma, we use condition (H) to prove the boundedness of the sequence (U_j).

Lemma 15.3.3 *Assume that* $(f1)$–$(f4)$ *with* $l = \infty$ *hold, and* h *satisfies* (H) *and* $\|h\|_{L^2(\mathbb{R}^N)} < m_1$, *for* m_1 *given in Theorem 15.3.1. Then,* $(U_j) \subset X^s_{rad}(\mathbb{R}^{N+1}_+)$ *is bounded.*

Proof Using Theorem 5.2.2, we know that the function $\lambda \mapsto c_\lambda$ is continuous from the left. Then, by Lemma 15.3.2-(ii), we deduce that $I_{\lambda_j}(U_j) = c_{\lambda_j} \to c_1 > 0$ as $\lambda_j \to 1$. Hence, we can find a constant $K > 0$ such that $I_{\lambda_j}(U_j) \le K$ for all $j \in \mathbb{N}$. Combining this with (15.3.2), the fact that $u_j > 0$ and (H), we see that

$$\iint_{\mathbb{R}^{N+1}_+} y^{1-2s} |\nabla U_j|^2 \, dx\, dy \le \frac{KN}{s} - \frac{\lambda_j}{s} \int_{\mathbb{R}^N} \nabla h(x) \cdot x\, u_j \, dx \le \frac{KN}{s},$$

which together with the Sobolev inequality (1.2.9) implies that

$$\|u_j\|_{L^{2^*_s}(\mathbb{R}^N)} \le C_* \left(\iint_{\mathbb{R}^{N+1}_+} y^{1-2s} |\nabla U_j|^2 \, dx\, dy \right)^{\frac{1}{2}} \le C. \tag{15.3.3}$$

Now, it follows $I_{\lambda_j}(U_j) \le K$ for all $j \in \mathbb{N}$ that

$$\frac{1}{2} \|U_j\|^2_{X^s(\mathbb{R}^{N+1}_+)} - \lambda_j \int_{\mathbb{R}^N} (F(u_j) + h(x)u_j) \, dx \le K. \tag{15.3.4}$$

On the other hand, by $(f2)$, $(f3)$, there exists a constant $C > 0$ such that

$$\int_{\mathbb{R}^N} F(u_j) \, dx \le \frac{1}{4} \int_{\mathbb{R}^N} u_j^2 \, dx + C \int_{\mathbb{R}^N} |u_j|^{2^*_s} \, dx.$$

Substituting this inequality into (15.3.4) and using (15.3.3), (1.2.9), we deduce that

$$\frac{1}{2} \int_{\mathbb{R}^N} u_j^2 \, dx \le \lambda_j \int_{\mathbb{R}^N} (F(u_j) + h(x)u_j) \, dx + K$$

$$\le \frac{1}{4} \|u_j\|^2_{L^2(\mathbb{R}^N)} + C \|u_j\|^{2^*_s}_{L^{2^*_s}(\mathbb{R}^N)} + \|h\|_{L^2(\mathbb{R}^N)} \|u_j\|_{L^2(\mathbb{R}^N)} + K$$

$$\le \frac{1}{4} \|u_j\|^2_{L^2(\mathbb{R}^N)} + \bar{C} + \|h\|_{L^2(\mathbb{R}^N)} \|u_j\|_{L^2(\mathbb{R}^N)} + K.$$

Then,

$$\frac{1}{4} \|u_j\|^2_{L^2(\mathbb{R}^N)} \le \tilde{C} + \|h\|_{L^2(\mathbb{R}^N)} \|u_j\|_{L^2(\mathbb{R}^N)},$$

that is

$$\|u_j\|_{L^2(\mathbb{R}^N)} \le C \text{ for all } j \in \mathbb{N}, \tag{15.3.5}$$

for some positive constant C independent of j. Putting together (15.3.3) and (15.3.5), we complete the proof.   □

Lemma 15.3.4 *Under the assumptions of Lemma 15.3.3, the above sequence (U_j) is also a Palais-Smale sequence of I.*

Proof By the definitions of I and I_{λ_j} we deduce that

$$I(U_j) = I_{\lambda_j}(U_j) + (\lambda_j - 1) \int_{\mathbb{R}^N} (F(u_j) + h(x)u_j) \, dx. \tag{15.3.6}$$

Using Theorem 5.2.2, we obtain

$$I_{\lambda_j}(U_j) = c_{\lambda_j} \to c_1 > 0 \quad \text{as } \lambda_j \to 1.$$

Hence, applying Lemma 15.3.3 and (15.3.6), we see that $I(U_j) \to c_1 > 0$. Since $I'_{\lambda_j}(U_j) = 0$, we infer that, for any $\Psi \in C_c^\infty(\mathbb{R}_+^{N+1})$,

$$\langle I'(U_j), \Psi \rangle = \langle I'_{\lambda_j}(U_j), \Psi \rangle + (\lambda_j - 1) \int_{\mathbb{R}^N} (f(u_j) + h(x))\psi \, dx \to 0,$$

that is, $I'(U_j) \to 0$ as $j \to \infty$ in the dual space of $X_{\text{rad}}^s(\mathbb{R}_+^{N+1})$.   □

Finally, we give the proof of the main result of this section:

Proof of Theorem 15.1.3 It follows from Theorem 15.3.1 that (15.3.1) admits a nontrivial non-negative solution $U_3 \in X_{\text{rad}}^s(\mathbb{R}_+^{N+1})$ such that $I(U_3) < 0$. On the other hand, by Lemma 15.3.4 and Theorem 1.1.11, we know that problem (15.3.1) has a nontrivial non-negative solution $U_4 \in X_{\text{rad}}^s(\mathbb{R}_+^{N+1})$ with $I(U_4) = c_1 > 0$. Consequently, $\tilde{U}_0 \not\equiv \tilde{U}_1$. By using Theorem 1.3.4, we get $u_3, u_4 > 0$ in $\mathbb{R}^N$ and this ends the proof of Theorem 15.1.3.   □

16.1   Introduction

In this chapter we study the existence of least energy sign-changing (or nodal) solutions for the following nonlinear problem involving the fractional Laplacian:

$$
\begin{cases}
(-\Delta)^s u + V(x)u = K(x)f(u) \ \text{ in } \mathbb{R}^N, \\
u \in \mathcal{D}^{s,2}(\mathbb{R}^N),
\end{cases}
\tag{16.1.1}
$$

with $s \in (0, 1)$, $N > 2s$, V and K are continuous potentials which satisfy suitable assumptions, and f is a continuous nonlinearity.

When $s = 1$, equation in (16.1.1) becomes the classical nonlinear Schrödinger equation

$$
-\Delta u + V(x)u = K(x)f(u) \quad \text{ in } \mathbb{R}^N,
\tag{16.1.2}
$$

which is extensively studied for at least 20 years. We do not intend to review the huge bibliography devoted to equations like (16.1.2); we just emphasize that the potential $V : \mathbb{R}^N \to \mathbb{R}$ has a crucial role concerning the existence and behavior of solutions. For instance, when V is a positive constant, or V is radially symmetric, it is natural to look for radially symmetric solutions, see [318, 340]. On the other hand, after the seminal paper of Rabinowitz [299], where the potential V is assumed to be coercive, several different assumptions were adopted in order to obtain existence and multiplicity results, see [88, 89]. An important class of problems associated with (16.1.2) is the zero mass case, which occurs when

$$
\lim_{|x| \to \infty} V(x) = 0.
$$

© The Author(s), under exclusive license to Springer Nature Switzerland AG 2021
V. Ambrosio, *Nonlinear Fractional Schrödinger Equations in* $\mathbb{R}^N$,
Frontiers in Mathematics, https://doi.org/10.1007/978-3-030-60220-8_16

To study these problems, many authors used several variational methods; see [21,26,30,96, 97] and [100, 107, 192] for problems posed in $\mathbb{R}^N$, as well as [22,91,92,133] for problems in bounded domains with homogeneous boundary conditions.

Motivated by the above papers, our goal is to prove the existence of sign-changing solutions to problem (16.1.1). Before stating our main result, we introduce the basic assumptions on V, K and f. More precisely, we suppose that the functions V, $K : \mathbb{R}^N \to \mathbb{R}$ are continuous on $\mathbb{R}^N$, and we say that $(V, K) \in \mathcal{K}$ if:

(h_1) $V(x)$, $K(x) > 0$ for all $x \in \mathbb{R}^N$ and $K \in L^\infty(\mathbb{R}^N)$;
(h_2) If $(A_n) \subset \mathbb{R}^N$ is a sequence of Borel sets such that the Lebesgue measure $m(A_n) \leq R$, for all $n \in \mathbb{N}$ and some $R > 0$, then

$$\lim_{r \to \infty} \int_{A_n \cap B_r^c} K(x)\, dx = 0,$$

for every $n \in \mathbb{N}$, where $B_r^c = \mathbb{R}^N \setminus B_r$.

Furthermore, one of the following conditions is in force:

(h_3) $K/V \in L^\infty(\mathbb{R}^N)$

or

(h_4) there exists $m \in (2, 2_s^*)$ such that

$$\frac{K(x)}{V(x)^{\frac{2_s^* - m}{2_s^* - 2}}} \to 0 \quad as\ |x| \to \infty.$$

We recall that assumptions (h_1)–(h_4) were introduced for the first time by Alves and Souto in [21]. It is very important to observe that (h_2) is weaker than any one of the conditions listed below that in the above mentioned papers to study zero mass problems:

(a) there are $r \geq 1$ and $\rho \geq 0$ such that $K \in L^r(\mathbb{R}^N \setminus B_\rho)$;
(b) $K(x) \to 0$ as $|x| \to \infty$;
(c) $K = H_1 + H_2$, with H_1 and H_2 satisfying (a) and (b), respectively.

Now, we provide some examples of functions V and K satisfying (h_1)–(h_4). Let (B_n) be a sequence of disjoint open balls in $\mathbb{R}^N$ centered in $\xi_n = (n, 0, \ldots, 0)$ and consider a non-negative function H_3 such that

$$H_3 = 0 \text{ in } \mathbb{R}^N \setminus \bigcup_{n=1}^\infty B_n, \quad H_3(\xi_n) = 1 \text{ and } \int_{B_n} H_3(x)\, dx = 2^{-n}.$$

Then, the pairs (V, K) given by

$$K(x) = V(x) = H_3(x) + \frac{1}{\log(2 + |x|)}$$

and

$$K(x) = H_3(x) + \frac{1}{\log(2 + |x|)} \quad \text{and} \quad V(x) = H_3(x) + \left(\frac{1}{\log(2 + |x|)} \right)^{\frac{2_s^* - 2}{2_s^* - m}}$$

for some $m \in (2, 2_s^*)$, belong to the class $\mathcal{K}$.

On the nonlinearity $f : \mathbb{R} \to \mathbb{R}$ we assume that it is a continuous function and satisfies the following growth conditions at the origin and at infinity:

(f_1) $\lim\limits_{|t| \to 0} \dfrac{f(t)}{|t|} = 0$ if (h_3) holds;

or

$(\tilde{f_1})$ $\lim\limits_{|t| \to 0} \dfrac{f(t)}{|t|^{m-1}} < \infty$ if (h_4) holds for some $m \in (2, 2_s^*)$;

(f_2) f has a quasicritical growth at infinity, namely

$$\lim\limits_{|t| \to \infty} \frac{f(t)}{|t|^{2_s^* - 1}} = 0;$$

(f_3) F has a superquadratic growth at infinity, that is

$$\lim\limits_{|t| \to \infty} \frac{F(t)}{|t|^2} = \infty,$$

where, as usual, we set $F(t) = \displaystyle\int_0^t f(\tau)\, d\tau$;

(f_4) the function $t \mapsto \dfrac{f(t)}{|t|}$ is increasing for every $t \in \mathbb{R} \setminus \{0\}$.

As models for f we can take, for instance, the following nonlinearities

$$f(t) = (t^+)^m \quad \text{and} \quad f(t) = \begin{cases} \log 2(t^+)^m, & \text{if } t \leq 1, \\ t \log(1 + t), & \text{if } t > 1, \end{cases}$$

for some $m \in (2, 2_s^*)$.

Remark 16.1.1 It follows from (f_4) that the function

$$t \mapsto \frac{1}{2} f(t)t - F(t) \quad \text{is increasing for every } t > 0$$

$$\text{and decreasing for every } t < 0.$$

(16.1.3)

Now, we are ready to state the main result of this chapter.

Theorem 16.1.2 ([69]) *Suppose that* $(V, K) \in \mathcal{K}$ *and* $f \in C(\mathbb{R})$ *satisfies either condition* (f_1), *or conditions* ($\tilde{f}_1$) *and* (f_2)–(f_4). *Then, problem* (16.1.1) *possesses a least energy nodal weak solution. In addition, if the nonlinear term* f *is odd, then problem* (16.1.1) *has infinitely many nontrivial weak solutions not necessarily nodals.*

The proof of Theorem 16.1.2 is obtained by enlisting variational arguments. We note that the Euler–Lagrange functional associated with (16.1.1), that is,

$$J(u) = \frac{1}{2} \left([u]_s^2 + \int_{\mathbb{R}^N} V(x)u^2 \, dx \right) - \int_{\mathbb{R}^N} K(x)F(u) \, dx,$$

does not satisfy the following decompositions

$$\langle J'(u), u^{\pm} \rangle = \langle J'(u^{\pm}), u^{\pm} \rangle$$

$$J(u) = J(u^+) + J(u^-),$$

which were fundamental in the application of variational methods to study (16.1.2); see [86, 92, 322, 340].

Anyway, we prove that the geometry of the classical minimization theorem is respected in the nonlocal framework: more precisely, we develop a functional analytical setting that is inspired by (but not equivalent to) the fractional Sobolev spaces, in order to correctly encode the variational formulation of problem (16.1.1). Secondly, the nonlinearity f is only continuous, so to overcome the nondifferentiability of the Nehari manifold associated with J, we use the generalized Nehari manifold method developed by Szulkin and Weth. Of course, also the compactness properties (see Proposition 16.3.2) required by these abstract theorems are satisfied in the nonlocal case, thanks to our functional setting. Then, to obtain nodal solutions, we look for critical points of $J(tu^+ + su^-)$, and, due to the fact that f is only continuous, we do not apply the Miranda's Theorem [270] as in [22, 86], but we use an iterative procedure and the properties of J to prove the existence of a sequence which converges to a critical point of $J(tu^+ + su^-)$ (see Lemma 16.4.1). Finally, we emphasize that Theorem 16.1.2 improves the recent result established in [71], in which the existence of a least energy nodal solution to problem (16.1.1) has been proved under the stronger assumption that $f \in C^1$ and satisfies the Ambrosetti-Rabinowitz condition.

16.2 Preliminary Lemmas

In order to give the weak formulation of problem (16.1.1), we need to work in a special functional space. Indeed, one of the difficulties in treating problem (16.1.1) is related to his variational formulation. With this respect the standard fractional Sobolev spaces are not sufficient in order to study the problem. We overcome this difficulty by working in a suitable functional space, whose definition and basic analytical properties are recalled here. Let us introduce the following functional space

$$\mathbb{X} = \left\{ u \in \mathcal{D}^{s,2}(\mathbb{R}^N) : \int_{\mathbb{R}^N} V(x)u^2 \, dx < \infty \right\}$$

endowed with the norm

$$\|u\| = \left([u]_s^2 + \int_{\mathbb{R}^N} V(x)u^2 \, dx \right)^{\frac{1}{2}}.$$

For $q \in \mathbb{R}$ with $q \geq 1$, we define the Lebesgue space $L_K^q(\mathbb{R}^N)$ as

$$L_K^q(\mathbb{R}^N) = \left\{ u : \mathbb{R}^N \to \mathbb{R} \text{ measurable and } \int_{\mathbb{R}^N} K(x)|u|^q \, dx < \infty \right\},$$

endowed with the norm

$$\|u\|_{L_K^q(\mathbb{R}^N)} = \left(\int_{\mathbb{R}^N} K(x)|u|^q \, dx \right)^{\frac{1}{q}}.$$

We start by giving the following useful results established in [71].

Lemma 16.2.1 *Assume that $(V, K) \in \mathcal{K}$. Then $\mathbb{X}$ is continuously embedded in $L_K^q(\mathbb{R}^N)$ for every $q \in [2, 2_s^*]$ if (h_3) holds. Moreover, $\mathbb{X}$ is continuously embedded in $L_K^m(\mathbb{R}^N)$ if (h_4) holds.*

Proof Assume that (h_3) is true. The proof is trivial if $q = 2$ or $q = 2_s^*$. Fix $q \in (2, 2_s^*)$ and let $\lambda = \dfrac{2_s^* - q}{2_s^* - 2}$. We observe that q can be written as $q = 2\lambda + (1 - \lambda)2_s^*$. Then we have

$$\int_{\mathbb{R}^N} K(x)|u|^q \, dx = \int_{\mathbb{R}^N} K(x)|u|^{2\lambda}|u|^{(1-\lambda)2_s^*} \, dx$$

$$\leq \left(\int_{\mathbb{R}^N} |K(x)|^{\frac{1}{\lambda}}|u|^2 \, dx \right)^{\lambda} \left(\int_{\mathbb{R}^N} |u|^{2_s^*} \, dx \right)^{1-\lambda}$$

$$\leq \left(\sup_{x\in\mathbb{R}^N} \frac{|K(x)|}{|V(x)|^\lambda}\right) \left(\int_{\mathbb{R}^N} V(x)|u|^2\,dx\right)^\lambda \left(\int_{\mathbb{R}^N} |u|^{2_s^*}\,dx\right)^{1-\lambda}$$

$$\leq C \left(\sup_{x\in\mathbb{R}^N} \frac{|K(x)|}{|V(x)|^\lambda}\right) \left(\int_{\mathbb{R}^N} V(x)|u|^2\,dx\right)^\lambda [u]_s^{(1-\lambda)2_s^*}$$

$$\leq C \left(\sup_{x\in\mathbb{R}^N} \frac{|K(x)|}{|V(x)|^\lambda}\right) \|u\|^{2\left(\lambda+\frac{(1-\lambda)2_s^*}{2}\right)}$$

$$= C \left(\sup_{x\in\mathbb{R}^N} \frac{|K(x)|}{|V(x)|^\lambda}\right) \|u\|^q.$$

Since $K \in L^\infty(\mathbb{R}^N)$ and (h_3) hold, we conclude that

$$\|u\|_{L_K^q(\mathbb{R}^N)} \leq C\|u\|.$$

Now, suppose that (h_4) holds. Setting $\lambda_0 = \dfrac{2_s^* - m}{2_s^* - 2}$, we see that m can be written as $m = 2\lambda_0 + (1-\lambda_0)2_s^*$. As above, we have

$$\int_{\mathbb{R}^N} K(x)|u|^m\,dx = \int_{\mathbb{R}^N} K(x)|u|^{2\lambda_0}|u|^{(1-\lambda_0)2_s^*}\,dx$$

$$\leq \left(\int_{\mathbb{R}^N} |K(x)|^{\frac{1}{\lambda_0}}|u|^2\,dx\right)^{\lambda_0} \left(\int_{\mathbb{R}^N} |u|^{2_s^*}\,dx\right)^{1-\lambda_0}$$

$$\leq \left(\sup_{x\in\mathbb{R}^N} \frac{|K(x)|}{|V(x)|^{\lambda_0}}\right) \left(\int_{\mathbb{R}^N} V(x)|u|^2\,dx\right)^{\lambda_0} \left(\int_{\mathbb{R}^N} |u|^{2_s^*}\,dx\right)^{1-\lambda_0}$$

$$\leq C \left(\sup_{x\in\mathbb{R}^N} \frac{|K(x)|}{|V(x)|^\lambda}\right) \|u\|^m.$$

Since $\dfrac{K(x)}{V(x)^{\frac{2_s^*-m}{2_s^*-2}}} \in L^\infty(\mathbb{R}^N)$, we infer that

$$\|u\|_{L_K^m(\mathbb{R}^N)} \leq C\|u\|.$$

This complete the proof of the lemma.   □

Lemma 16.2.2 *Assume that* $(V, K) \in \mathcal{K}$. *Then*

(1) $\mathbb{X}$ *is compactly embedded into* $L_K^q(\mathbb{R}^N)$ *for all* $q \in (2, 2_s^*)$ *if* (h_3) *holds;*

(2) $\mathbb{X}$ *is compactly embedded into* $L_K^m(\mathbb{R}^N)$ *if* (h_4) *holds.*

Proof

(1) Assume that (h_3) holds. Fix $q \in (2, 2_s^*)$ and let $\varepsilon > 0$. Then there exist numbers $0 < t_0 < t_1$ and a positive constant C such that

$$K(x)|t|^q \leq \varepsilon\, C\left[V(x)|t|^2 + |t|^{2_s^*}\right] + C\, K(x)\chi_{[t_0,t_1]}(|t|)|t|^{2_s^*}, \quad \text{for all } t \in \mathbb{R}.$$

Integrating over B_r^c we have, for all $u \in \mathbb{X}$ and $r > 0$,

$$\int_{B_r^c} K(x)|u|^q\, dx \leq \varepsilon\, C \int_{B_r^c}\left[V(x)|u|^2 + |u|^{2_s^*}\right] dx + Ct_1^{2_s^*}\int_{A\cap B_r^c} K(x)\, dx$$

$$= \varepsilon\, C\mathcal{Q}(u) + Ct_1^{2_s^*}\int_{A\cap B_r^c} K(x)\, dx,$$

$$(16.2.1)$$

where we set

$$\mathcal{Q}(u) = \int_{B_r^c}\left[V(x)|u|^2 + |u|^{2_s^*}\right] dx \quad \text{and } A = \left\{x \in \mathbb{R}^N : t_0 \leq |u(x)| \leq t_1\right\}.$$

Now, if $(u_n) \subset \mathbb{X}$ is a sequence such that $u_n \rightharpoonup u$ in $\mathbb{X}$, then there is $M > 0$ such that

$$\|u_n\|^2 \leq M \text{ and } \int_{\mathbb{R}^N} |u_n|^{2_s^*}\, dx \leq M \quad \forall n \in \mathbb{N}. \qquad (16.2.2)$$

This implies that the sequence $(\mathcal{Q}(u_n))$ is bounded from above by a positive constant. Let us denote $A_n = \left\{x \in \mathbb{R}^N : t_0 \leq |u_n| \leq t_1\right\}$. By $(16.2.2)$,

$$t_0^{2_s^*} m(A_n) \leq \int_{A_n} |u_n|^{2_s^*}\, dx \leq M \quad \text{for all } n \in \mathbb{N},$$

which implies that $\sup_{n\in\mathbb{N}} |m(A_n)| < \infty$. Therefore, by (h_2), there exists a positive radius r large enough such that

$$\int_{A_n\cap B_r^c} K(x)\, dx < \frac{\varepsilon}{t_1^{2_s^*}} \quad \text{for all } n \in \mathbb{N}. \qquad (16.2.3)$$

Putting together (16.2.1) and (16.2.3), we see that

$$\int_{B_r^c} K(x)|u_n|^q \, dx \leq \varepsilon \, C \, M + C \, t_1^{2_s^*} \int_{A_n \cap B_r^c} K(x) \, dx$$

$$\leq (C \, M + C) \, \varepsilon \quad \text{for all } n \in \mathbb{N}. \tag{16.2.4}$$

Further, recalling that $q \in (2, 2_s^*)$ and that K is a continuous function, it follows from Theorem 1.1.8 that

$$\lim_{n \to \infty} \int_{B_r} K(x)|u_n|^q \, dx = \int_{B_r} K(x)|u|^q \, dx. \tag{16.2.5}$$

By (16.2.4), for $\varepsilon > 0$ small enough and (16.2.5), we have

$$\lim_{n \to \infty} \int_{\mathbb{R}^N} K(x)|u_n|^q \, dx = \int_{\mathbb{R}^N} K(x)|u|^q \, dx$$

from which we conclude that

$$u_n \to u \text{ in } L_K^q(\mathbb{R}^N), \text{ for every } q \in (2, 2_s^*).$$

(2) Suppose that (h_4) holds. Then we see that for each fixed $x \in \mathbb{R}^N$, the function

$$g(t) = V(x)t^{2-m} + t^{2_s^* - m}, t > 0,$$

has $C_m V(x)^{\frac{2_s^* - m}{2_s^* - 2}}$ as its minimum value, where $C_m = \left(\dfrac{2_s^* - 2}{2_s^* - m} \right) \left(\dfrac{m - 2}{2_s^* - 2} \right)^{\frac{2-m}{2_s^* - 2}}$.

Hence

$$C_m V(x)^{\frac{2_s^* - m}{2_s^* - 2}} \leq V(x)t^{2-m} + t^{2_s^* - m}, \quad \text{for every } x \in \mathbb{R}^N \text{ and } t > 0.$$

Combining this inequality with (h_4), we see that for every $\varepsilon > 0$ we can find a sufficiently large radius r such that

$$K(x)|t|^m \leq \varepsilon C_m' \left[V(x)|t|^2 + |t|^{2_s^*} \right], \quad \text{for every } t \in \mathbb{R} \text{ and } |x| \geq r,$$

where C_m' is the inverse of C_m, and integrating over B_r^c we get

$$\int_{B_r^c} K(x)|u|^m \, dx \leq \varepsilon C_m' \left[\|u\|^2 + \|u\|_{L^{2_s^*}(\mathbb{R}^N)}^{2_s^*} \right], \quad \text{for all } u \in X. \tag{16.2.6}$$

If $(u_n) \subset \mathbb{X}$ is a sequence such that $u_n \rightharpoonup u$ in $\mathbb{X}$, then by (16.2.6) we deduce that

$$\int_{B_r^c} K(x)|u|^m \, dx \le \varepsilon C_m'', \quad \text{for all } n \in \mathbb{N}. \tag{16.2.7}$$

Since $m \in (2, 2_s^*)$ and K is a continuous function, it follows from Theorem 1.1.8 that

$$\lim_{n \to \infty} \int_{B_r} K(x)|u_n|^m \, dx = \int_{B_r} K(x)|u|^m \, dx. \tag{16.2.8}$$

Then, (16.2.7) and (16.2.8) yield

$$\lim_{n \to \infty} \int_{\mathbb{R}^N} K(x)|u_n|^m \, dx = \int_{\mathbb{R}^N} K(x)|u|^m \, dx,$$

from which we deduce that

$$u_n \to u \text{ in } L_K^m(\mathbb{R}^N), \text{ for every } m \in (2, 2_s^*).$$

$\square$

The next lemma is a compactness result related to the nonlinear term (see [71]).

Lemma 16.2.3 *Assume that $(V, K) \in \mathcal{K}$ and f satisfies either (f_1)–(f_2) or $(\tilde{f}_1)$–(f_2). Let (u_n) be a sequence such that $u_n \rightharpoonup u$ in $\mathbb{X}$. Then, up to a subsequence, one has*

$$\lim_{n \to \infty} \int_{\mathbb{R}^N} K(x)F(u_n) \, dx = \int_{\mathbb{R}^N} K(x)F(u) \, dx$$

and

$$\lim_{n \to \infty} \int_{\mathbb{R}^N} K(x)f(u_n)u_n \, dx = \int_{\mathbb{R}^N} K(x)f(u)u \, dx.$$

Proof Assume that (h_3) holds. By (f_1)–(f_2), for fixed $q \in (2, 2_s^*)$ and given $\varepsilon > 0$, there exists $C > 0$ such that

$$|K(x)f(t)t| \le \varepsilon C \left[V(x)|t|^2 + |t|^{2_s^*} \right] + CK(x)|t|^q, \quad \text{for all } t \in \mathbb{R}. \tag{16.2.9}$$

By Proposition 16.2.2,

$$\lim_{n \to \infty} \int_{\mathbb{R}^N} K(x)|u_n|^q \, dx = \int_{\mathbb{R}^N} K(x)|u|^q \, dx,$$

so there exists $r > 0$ such that

$$\int_{B_r^c} K(x)|u_n|^q \, dx < \varepsilon, \qquad \text{for all } n \in \mathbb{N}. \tag{16.2.10}$$

Since $(u_n) \subset \mathbb{X}$ is bounded, there exists a positive constant C' such that

$$\int_{\mathbb{R}^N} V(x)|u_n|^2 \, dx \leq C' \quad \text{and} \quad \int_{\mathbb{R}^N} |u_n|^{2_s^*} \, dx \leq C', \qquad \text{for all } n \in \mathbb{N}, \tag{16.2.11}$$

Next, (16.2.9), (16.2.10), and (16.2.11) imply that

$$\int_{B_r^c} K(x)|u_n|^q \, dx < (2CC' + 1)\varepsilon, \text{ for all } n \in \mathbb{N}.$$

Assume that (h_4) holds. Similarly to the second part of Proposition 16.2.2, given $\varepsilon > 0$ sufficiently small, there exists $r > 0$ large enough such that

$$K(x) \leq \varepsilon C_m' \left[V(x)|t|^{2-m} + |t|^{2_s^*-m} \right], \qquad \text{for all } |t| > 0 \text{ and } |x| > r.$$

Consequently,

$$K(x)|f(t)t| \leq \varepsilon C_m' \left[V(x)|f(t)t||t|^{2-m} + |f(t)t||t|^{2_s^*-m} \right] \qquad \text{for all } |t| > 0 \text{ and } |x| > r.$$

From $(\tilde{f}_1)$ and (f_2), there exist $C, t_0, t_1 > 0$ satisfying

$$K(x)|f(t)t| \leq \varepsilon C \left[V(x)t^2 + |t|^{2_s^*} \right], \text{ for all } t \in I \text{ and } |x| > r,$$

where $I = \{t \in \mathbb{R} : |t| < t_0 \text{ or } |t| > t_1\}$. Therefore, for every $u \in \mathbb{X}$, setting

$$Q(u) = \int_{\mathbb{R}^N} V(x)|u|^2 \, dx + \int_{\mathbb{R}^N} |u|^{2_s^*} \, dx$$

and $A = \{x \in \mathbb{R}^N : t_0 \leq |u(x)| \leq t_1\}$, the following estimate holds true:

$$\int_{B_r^c} K(x)f(u)u \, dx \leq \varepsilon C Q(u) + C \int_{A \cap B_r^c} K(x) \, dx.$$

Due to the boundedness of $(u_n) \subset \mathbb{X}$, we can find $C' > 0$ such that

$$\int_{\mathbb{R}^N} V(x)|u_n|^2 dx \leq C' \quad \text{and} \quad \int_{\mathbb{R}^N} |u_n|^{2_s^*} dx \leq C', \qquad \text{for all } n \in \mathbb{N}.$$

Therefore,

$$\int_{B_r^c} K(x) f(u_n) u_n \, dx \le \varepsilon C'' + C \int_{A_n \cap B_r^c} K(x) \, dx,$$

where $A_n = \{x \in \mathbb{R}^N : t_0 \le |u_n(x)| \le t_1\}$. Arguing as in the proof of Proposition 16.2.2 and using (h_2), we deduce that

$$\int_{A_n \cap B_r^c} K(x) \, dx \to 0 \quad \text{as } r \to +\infty$$

uniformly in $n \in \mathbb{N}$ and, for $\varepsilon > 0$ small enough,

$$\left| \int_{B_r^c} K(x) f(u_n) u_n \, dx \right| < (C'' + 1)\varepsilon.$$

In order to complete the proof, it remains to check that

$$\lim_{n \to \infty} \int_{B_r} K(x) f(u_n) u_n \, dx = \int_{B_r} K(x) f(u) u \, dx$$

which easily follows by Lemma 1.4.2.   □

16.3   The Nehari Manifold Argument

In this section we obtain some preliminary results that serve to overcome the lack of differentiability of the Nehari manifold in which we look for weak solutions to problem (16.1.1).

In the following, we search for a nodal or sign-changing weak solution of problem (16.1.1), that is, a function $u = u^+ + u^- \in X$ such that $u^+ = \max\{u, 0\} \ne 0$, $u^- = \min\{u, 0\} \ne 0$ in $\mathbb{R}^N$ and

$$\iint_{\mathbb{R}^{2N}} \frac{(u(x) - u(y))(\varphi(x) - \varphi(y))}{|x - y|^{N+2s}} \, dx \, dy + \int_{\mathbb{R}^N} V(x) u \varphi \, dx = \int_{\mathbb{R}^N} K(x) f(u) \varphi \, dx,$$

for every $\varphi \in X$. The energy functional associated with (16.1.1) is given by

$$J(u) = \frac{1}{2} \|u\|^2 - \int_{\mathbb{R}^N} K(x) F(u) \, dx.$$

By the assumptions on f, it is clear that $J \in C^1(\mathbb{X}, \mathbb{R})$ and that its differential is given by the formula

$$\langle J'(u), \varphi \rangle = \langle u, \varphi \rangle_{\mathcal{D}^{s,2}(\mathbb{R}^N)} + \int_{\mathbb{R}^N} V(x)u\varphi \, dx - \int_{\mathbb{R}^N} K(x)f(u)\varphi(x) \, dx,$$

for every $u, \varphi \in \mathbb{X}$. Then, the critical points of J are the weak solutions of (16.1.1).

Let us also observe that one has the decompositions

$$J(u) = J(u^+) + J(u^-) - \iint_{\mathbb{R}^{2N}} \frac{u^+(x)u^-(y) + u^-(x)u^+(y)}{|x - y|^{N+2s}} \, dx dy,$$

and

$$\langle J'(u), u^+ \rangle = \langle J'(u^+), u^+ \rangle - \iint_{\mathbb{R}^{2N}} \frac{u^+(x)u^-(y) + u^-(x)u^+(y)}{|x - y|^{N+2s}} \, dx dy.$$

The Nehari manifold associated with the functional J is given by

$$\mathcal{N} = \{u \in \mathbb{X} \setminus \{0\} : \langle J'(u), u \rangle = 0\}.$$

Recalling that a nonzero critical point u of J is a least energy weak solution of problem (16.1.1) if

$$J(u) = \min_{v \in \mathcal{N}} J(v)$$

and, since our purpose is to prove the existence of a least energy sign-changing weak solution of (16.1.1), we look for $u \in \mathcal{M}$ such that

$$J(u) = \min_{v \in \mathcal{M}} J(v),$$

where $\mathcal{M}$ is the subset of $\mathcal{N}$ consisting of all the sign-changing weak solutions of problem (16.1.1), that is

$$\mathcal{M} = \{w \in \mathcal{N} : w^+ \neq 0, w^- \neq 0, \langle J'(w), w^+ \rangle = \langle J'(w), w^- \rangle = 0\}.$$

If we assume that f is only continuous, the following results are crucial, since they allow us to overcome the non-differentiability of $\mathcal{N}$. Below, we denote by $\mathbb{S}$ the unit sphere on $\mathbb{X}$.

Lemma 16.3.1 *Suppose that $(V, K) \in \mathcal{K}$ and f fulfills the conditions (f_1)–(f_4). Then:*

(a) *For each $u \in \mathbb{X} \setminus \{0\}$, let $h_u : \mathbb{R}_+ \to \mathbb{R}$ be defined by $h_u(t) = J(tu)$. Then, there is a unique $t_u > 0$ such that*

$$h'_u(t) > 0 \quad in \ (0, t_u),$$

$$h'_u(t) < 0 \quad in \ (t_u, \infty).$$

(b) *There is $\tau > 0$, independent on u, such that $t_u \geq \tau$ for every $u \in \mathbb{S}$. Moreover, for each compact set $\mathcal{W} \subset \mathbb{S}$, there is $C_{\mathcal{W}} > 0$ such that $t_u \leq C_{\mathcal{W}}$ for every $u \in \mathcal{W}$.*

(c) *The map $\hat{\eta} : \mathbb{X} \setminus \{0\} \to \mathcal{N}$ given by $\hat{\eta}(u) = t_u u$ is continuous and $\eta = \hat{\eta}|_{\mathbb{S}}$ is a homeomorphism between $\mathbb{S}$ and $\mathcal{N}$. Moreover,*

$$\eta^{-1}(u) = \frac{u}{\|u\|}.$$

Proof

(a) We distinguish two cases.

　　Let us assume that (h_3) is satisfied. Using assumptions (f_1) and (f_2), given $\varepsilon > 0$ there exists a positive constant C_ε such that

$$|F(t)| \leq \varepsilon |t|^2 + C_\varepsilon |t|^{2^*_s}, \quad \text{for every } t \in \mathbb{R},$$

which in conjunction with the Sobolev embedding result implies that

$$
\begin{aligned}
J(tu) &= \frac{t^2}{2}[u]^2_s + \frac{t^2}{2} \int_{\mathbb{R}^N} V(x) u^2 \, dx - \int_{\mathbb{R}^N} K(x) F(tu) \, dx \\
&\geq \frac{t^2}{2} \|u\|^2 - \varepsilon \int_{\mathbb{R}^N} K(x) t^2 u^2 \, dx - C_\varepsilon \int_{\mathbb{R}^N} K(x) t^{2^*_s} |u|^{2^*_s} \, dx \\
&\geq \frac{t^2}{2} \|u\|^2 - \varepsilon \|K/V\|_{L^\infty(\mathbb{R}^N)} t^2 \|u\|^2 - C_\varepsilon C' \|K\|_{L^\infty(\mathbb{R}^N)} t^{2^*_s} \|u\|^{2^*_s}.
\end{aligned}
$$

$$(16.3.1)$$

Taking $0 < \varepsilon < \dfrac{1}{2\|K/V\|_{L^\infty(\mathbb{R}^N)}}$, we get a $t_0 > 0$ sufficiently small such that

$$0 < h_u(t) = J(tu), \quad \text{for all } t < t_0. \tag{16.3.2}$$

On the other hand, suppose that (h_4) holds. Then there exists a constant $C_m > 0$ such that, for each $\varepsilon \in (0, C_m)$, one can find an $R > 0$ such that

$$\int_{B_R^c} K(x)|u|^m \, dx \le \varepsilon \int_{B_R^c} (V(x)|u|^2 + |u|^{2_s^*}) \, dx, \tag{16.3.3}$$

for all $u \in \mathbb{X}$. Now, using $(\tilde{f}_1)$ and (f_2), Theorem 1.1.8, (16.3.3) and the Hölder inequality, we have that

$$J(tu) \ge \frac{t^2}{2}\|u\|^2 - C_1 \int_{\mathbb{R}^N} K(x)t^m |u|^m \, dx - C_2 \int_{\mathbb{R}^N} K(x)t^{2_s^*} |u|^{2_s^*} \, dx$$

$$\ge \frac{t^2}{2}\|u\|^2 - C_1 t^m \varepsilon \int_{B_R^c} (V(x)|u|^2 + |u|^{2_s^*}) \, dx - C_1 t^m \int_{B_R} K(x)|u|^m \, dx$$

$$\quad - C_2 t^{2_s^*} \|K\|_{L^\infty(\mathbb{R}^N)} \int_{\mathbb{R}^N} |u|^{2_s^*} \, dx$$

$$\ge \frac{t^2}{2}\|u\|^2 - C_1 t^m \varepsilon \int_{B_R^c} (V(x)|u|^2 + |u|^{2_s^*}) \, dx$$

$$\quad - C_1 t^m \|K\|_{L^{\frac{2_s^*}{2_s^* - m}}(B_R)} \left(\int_{B_R} K(x)|u|^m \, dx \right)^{\frac{m}{2_s^*}}$$

$$\quad - C_2 t^{2_s^*} \|K\|_{L^\infty(\mathbb{R}^N)} \int_{\mathbb{R}^N} |u|^{2_s^*} \, dx$$

$$\ge \frac{t^2}{2}\|u\|^2 - C_1 t^m \left(\varepsilon \|u\|^2 + \varepsilon C \|u\|^{2_s^*} + C \|K\|_{L^{\frac{2_s^*}{2_s^* - m}}(B_R)} \|u\|^m \right)$$

$$\quad - C_2 C t^{2_s^*} \|K\|_{L^\infty(\mathbb{R}^N)} \|u\|^{2_s^*}. \tag{16.3.4}$$

This shows that (16.3.2) holds also in this case. Moreover, since $F(t) \ge 0$ for all $t \in \mathbb{R}$, we have

$$J(tu) \le \frac{t^2}{2}\|u\|^2 - \int_A K(x)F(tu) \, dx,$$

where $A \subset \operatorname{supp}(u)$ is a measurable set with finite positive measure. Consequently,

$$\limsup_{t \to \infty} \frac{J(tu)}{\|tu\|^2} \le \frac{1}{2} - \liminf_{t \to \infty} \left\{ \int_A K(x) \left[\frac{F(tu)}{(tu)^2} \right] \left(\frac{u}{\|u\|} \right)^2 dx \right\}.$$

By (f_3) and Fatou's lemma, it follows that

$$\limsup_{t\to\infty} \frac{J(tu)}{\|tu\|^2} \leq -\infty. \tag{16.3.5}$$

Thus, there exists $R > 0$ sufficiently large such that

$$h_u(R) = J(Ru) < 0. \tag{16.3.6}$$

By the continuity of h_u and (f_4), there is $t_u > 0$ which is a global maximum of h_u with $t_u u \in \mathcal{N}$.

The next objective is to prove that t_u is the unique critical point of h_u. Arguing by contradiction, let us assume that there are critical points t_1, t_2 of h_u with $t_1 > t_2 > 0$. Then we have $h'_u(t_1) = h'_u(t_2) = 0$, or equivalently,

$$\|u\|^2 - \int_{\mathbb{R}^N} K(x) \frac{f(t_1 u)u}{t_1} \, dx = 0,$$

$$\|u\|^2 - \int_{\mathbb{R}^N} K(x) \frac{f(t_2 u)u}{t_2} \, dx = 0.$$

Subtracting and taking into account Remark 16.1.1, we obtain

$$0 = \int_{\mathbb{R}^N} K(x) \left[\frac{f(t_1 u)}{t_1 u} - \frac{f(t_2 u)}{t_2 u} \right] u^2 \, dx > 0$$

which leads a contradiction.

(b) By (a), there exists $t_u > 0$ such that

$$t_u^2 \|u\|^2 = \int_{\mathbb{R}^N} K(x) f(t_u u) t_u u \, dx. \tag{16.3.7}$$

Then, estimating the right-hand side of (16.3.7) similarly to (16.3.1) and (16.3.4), we see that there exists $\tau > 0$, independent of u, such that $t_u \geq \tau$. On the other hand, let $\mathcal{W} \subset \mathbb{S}$ be a compact set. Assume, by contradiction, that there exists $(u_n) \subset \mathcal{W}$ such that $t_n = t_{u_n} \to \infty$. Therefore, there exists $u \in \mathcal{W}$ such that $u_n \to u$ in $\mathbb{X}$. From (16.3.5), we have

$$J(t_n u_n) \to -\infty \text{ in } \mathbb{R}. \tag{16.3.8}$$

Next, by Remark 16.1.1,

$$
\begin{aligned}
J(v) &= J(v) - \frac{1}{2}\langle J'(v), v\rangle \\
&= \int_{\mathbb{R}^N} K(x)\left(\frac{1}{2}f(v)v - F(v)\right) dx \geq 0,
\end{aligned}
\tag{16.3.9}
$$

for each $v \in \mathcal{N}$. Taking into account that $(t_{u_n} u_n) \subset \mathcal{N}$, we conclude from (16.3.8) that (16.3.9) is not true, which is a contradiction.

(c) Since $J \in C^1(\mathbb{X}, \mathbb{R})$, $J(0) = 0$ and satisfies (a) and (b), the assertion follows by Proposition 2.4.2. The proof is now complete.

□

Let us define the maps $\hat{\psi} : \mathbb{X} \to \mathbb{R}$ and $\psi : \mathbb{S} \to \mathbb{R}$ by $\hat{\psi}(u) = J(\hat{\eta}(u))$ and $\psi = \hat{\psi}|_{\mathbb{S}}$. The next result is a consequence of Lemma 16.3.1, Proposition 2.4.3 and Corollary 2.4.4.

Proposition 16.3.2 *Suppose that $(V, K) \in \mathcal{K}$ and f satisfies (f_1)–(f_4). Then, one has:*

(a) $\hat{\psi} \in C^1(\mathbb{X} \setminus \{0\}, \mathbb{R})$ *and*

$$
\langle \hat{\psi}'(u), v\rangle = \frac{\|\hat{\eta}(u)\|}{\|u\|}\langle J'(\hat{\eta}(u)), v\rangle,
$$

for every $u \in \mathbb{X} \setminus \{0\}$ and $v \in \mathbb{X}$.

(b) $\psi \in C^1(\mathbb{S}, \mathbb{R})$ *and $\langle \psi'(u), v\rangle = \|\eta(u)\|\langle J'(\eta(u)), v\rangle$ for all $v \in T_u\mathbb{S}$.*

(c) *If (u_n) is a Palais–Smale sequence for ψ, then $(\eta(u_n))$ is a Palais–Smale sequence for J. Moreover, if $(u_n) \subset \mathcal{N}$ is a bounded Palais–Smale sequence for J, then $(\eta^{-1}(u_n))$ is a Palais–Smale sequence for the functional ψ.*

(d) *u is a critical point of ψ if and only if $\eta(u)$ is a nontrivial critical point for J. Moreover, the corresponding critical values coincide and*

$$
\inf_{u\in\mathbb{S}} \psi(u) = \inf_{u\in\mathcal{N}} J(u).
$$

Remark 16.3.3 We notice that the following equalities hold:

$$
d_\infty = \inf_{u\in\mathcal{N}} J(u) = \inf_{u\in\mathbb{X}\setminus\{0\}} \max_{t>0} J(tu) = \inf_{u\in\mathbb{S}} \max_{t>0} J(tu).
\tag{16.3.10}
$$

In particular, relations (16.3.1), (16.3.5) and (16.3.10) imply that

$$
d_\infty > 0.
\tag{16.3.11}
$$

16.4   Technical Results

The aim of this section is to prove some technical lemmas related to the existence of a least energy nodal solution.

For each $u \in \mathbb{X}$ with $u^{\pm} \not\equiv 0$, consider the function $h^u : [0, \infty) \times [0, \infty) \to \mathbb{R}$ given by

$$h^u(t, \tau) = J(tu^+ + \tau u^-). \tag{16.4.1}$$

Its gradient $\Phi^u : [0, \infty) \times [0, \infty) \to \mathbb{R}^2$ is given by

$$
\begin{aligned}
\Phi^u(t, \tau) &= \left(\Phi_1^u(t, \tau), \Phi_2^u(t, \tau) \right) \\
&= \left(\frac{\partial h^u}{\partial t}(t, \tau), \frac{\partial h^u}{\partial s}(t, \tau) \right) \\
&= \left(\langle J'(tu^+ + \tau u^-), u^+ \rangle, \langle J'(tu^+ + \tau u^-), u^- \rangle \right).
\end{aligned}
\tag{16.4.2}
$$

Lemma 16.4.1 *Suppose that $(V, K) \in \mathcal{K}$ and f satisfies (f_1)–(f_4). Then:*

(i) *The pair (t, τ) is a critical point of h^u with $t, \tau > 0$ if and only if $tu^+ + \tau u^- \in \mathcal{M}$.*
(ii) *The map h^u has a unique critical point (t_+, τ_-), with $t_+ = t_+(u) > 0$ and $\tau_- = \tau_-(u) > 0$, which is the unique global maximum point of h^u.*
(iii) *The maps $a_+(r) = \Phi_1^u(r, \tau_-)r$ and $a_-(r) = \Phi_2^u(t_+, r)r$ are such that*

$$
\begin{aligned}
a_+(r) > 0 \ if \ r \in (0, t_+) \quad &and \quad a_+(r) < 0 \ if \ r \in (t_+, \infty) \\
a_-(r) > 0 \ if \ r \in (0, \tau_-) \quad &and \quad a_-(r) < 0 \ if \ r \in (\tau_-, \infty).
\end{aligned}
\tag{16.4.3}
$$

Proof

(i) By (16.4.2), we have

$$\Phi^u(t, \tau) = \left(\frac{1}{t}\langle J'(tu^+ + \tau u^-), tu^+ \rangle, \frac{1}{\tau}\langle J'(tu^+ + \tau u^-), \tau u^+ \rangle \right),$$

for all $t, \tau > 0$. Therefore, $\Phi^u(t, \tau) = 0$ if and only if

$$\langle J'(tu^+ + \tau u^-), tu^+ \rangle = 0 \quad and \quad \langle J'(tu^+ + \tau u^-), su^- \rangle = 0,$$

and this implies that $tu^+ + \tau u^- \in \mathcal{M}$.
(ii) Firstly we show that h^u has a critical point. For each $u \in \mathbb{X}$ such that $u^{\pm} \not\equiv 0$ and fixed τ_0, we define the function $h_1 : [0, \infty) \to [0, \infty)$ by $h_1(t) = h^u(t, \tau_0)$. Following

the lines of Lemma 16.3.1-(a), we can infer that h_1 has a positive maximum point. Moreover, there exists a unique $t_0 = t_0(u, \tau_0) > 0$ such that

$$h_1'(t) > 0 \quad \text{if } t \in (0, t_0),$$
$$h_1'(t_0) = 0,$$
$$h_1'(t) < 0 \quad \text{if } t \in (t_0, \infty).$$

Thus, the map $\phi_1 : [0, \infty) \to [0, \infty)$ given by $\phi_1(s) = t(u, s)$, where $t(u, s)$ satisfies the properties just mentioned with τ in place of τ_0, is well defined.

By the definition of h_1,

$$h_1'(\phi_1(\tau)) = \Phi_1''(\phi_1(\tau), \tau) = 0 \quad \text{for all } \tau \geq 0, \tag{16.4.4}$$

that is,

$$0 = |\phi_1(\tau)|^2 \|u^+\|^2 - \tau \phi_1(\tau) \iint_{\mathbb{R}^{2N}} \frac{u^+(x)u^-(y) + u^-(x)u^+(y)}{|x - y|^{N+2s}} \, dxdy$$

$$- \int_{\mathbb{R}^N} K(x) f(\phi_1(\tau)u^+) \, \phi_1(\tau)u^+ \, dx. \tag{16.4.5}$$

Now, we establish several properties of ϕ_1.

(a) The map ϕ_1 is continuous.

Let $\tau_n \to \tau_0$ as $n \to \infty$ in $\mathbb{R}$. We want to prove that $(\phi_1(\tau_n))$ is bounded. Assume, by contradiction, that there is a subsequence, denoted again by (τ_n), such that $\phi_1(\tau_n) \to \infty$ as $n \to \infty$. So, $\phi_1(\tau_n) \geq \tau_n$ for n large. By (16.4.5), we have

$$\|u^+\|^2 - \frac{\tau_n}{\phi_1(\tau_n)} \iint_{\mathbb{R}^{2N}} \frac{u^+(x)u^-(y) + u^-(x)u^+(y)}{|x - y|^{N+2s}} \, dxdy \quad = \int_{\mathbb{R}^N} K(x) \frac{f(\phi_1(\tau_n)u^+)}{\phi_1(\tau_n)u^+}(u^+)^2 \, dx. \tag{16.4.6}$$

Since $\tau_n \to \tau_0$, $\phi_1(\tau_n) \to \infty$ as $n \to \infty$, assumptions (f_3)–(f_4) and Fatou's lemma imply that

$$\|u^+\|^2 = \liminf_{n \to \infty} \int_{\mathbb{R}^N} K(x) \frac{f(\phi_1(\tau_n)u^+)}{\phi_1(\tau_n)u^+}(u^+)^2 \, dx \geq \infty,$$

so we reached a contradiction. Hence the sequence $(\phi_1(\tau_n))$ is bounded. Consequently, there exists $t_0 \geq 0$ such that $\phi_1(\tau_n) \to t_0$. Consider (16.4.5) with $\tau = \tau_n$. Letting $n \to \infty$ we have

$$t_0^2 \|u^+\|^2 - \tau_0 t_0 \iint_{\mathbb{R}^{2N}} \frac{u^+(x)u^-(y) + u^-(x)u^+(y)}{|x-y|^{N+2s}} \, dxdy = \int_{\mathbb{R}^N} K(x)f(\phi_1(t_0)u^+)\phi_1(t_0)u^+ \, dx,$$

that is $h_1'(t_0) = \Phi_1^u(t_0, \tau_0) = 0$. Accordingly, $t_0 = \phi_1(\tau_0)$, i.e., ϕ_1 is continuous.

(b) $\phi_1(0) > 0$.

Assume that there exists a sequence (τ_n) such that $\phi_1(\tau_n) \to 0$ and $\tau_n \to 0$ as $n \to \infty$. By assumption (f_1) we get

$$\|u^+\|^2 \leq \|u^+\|^2 - \frac{\tau_n}{\phi_1(\tau_n)} \iint_{\mathbb{R}^{2N}} \frac{u^+(x)u^-(y) + u^-(x)u^+(y)}{|x-y|^{N+2s}} dxdy$$

$$= \int_{\mathbb{R}^N} K(x) \frac{f(\phi_1(\tau_n)u^+)}{\phi_1(\tau_n)u^+}(u^+)^2 \, dx \to 0, \quad \text{as } n \to \infty$$

so we reached a contradiction. Therefore $\phi_1(0) > 0$.

(c) Now we show that $\phi_1(\tau) \leq \tau$ for τ large.

As a matter of fact, proceeding as in the first part of the proof of a), we can see that there is no sequence (τ_n) such that $\tau_n \to \infty$ and $\phi_1(\tau_n) \geq \tau_n$ for all $n \in \mathbb{N}$. This implies that $\phi_1(\tau) \leq \tau$ for τ large.

Analogously, for every $t_0 \geq 0$ we define $h_2(\tau) = h^u(t_0, \tau)$ and, consequently, we can find a map ϕ_2 such that

$$h_2'(\phi_2(t)) = \Phi_2^u(t, \phi_2(t)), \quad \forall t \geq 0 \tag{16.4.7}$$

and (a), (b) and (c) hold.

By (c), there is a positive constant C_1 such that $\phi_1(\tau) \leq \tau$ and $\phi_2(t) \leq t$ for every $t, \tau \geq C_1$.

Let

$$C_2 = \max \left\{ \max_{\tau \in [0, C_1]} \phi_1(\tau), \max_{t \in [0, C_1]} \phi_2(t) \right\}$$

and $C = \max\{C_1, C_2\}$. We define $T : [0, C] \times [0, C] \to \mathbb{R}^2$ by $T(t, \tau) = (\phi_1(\tau), \phi_2(t))$. Let us note that

$$T([0, C] \times [0, C]) \subset [0, C] \times [0, C].$$

Indeed, for every $t \in [0, C]$, we have that

$$
\begin{cases}
\phi_2(t) \leq t \leq C, & \text{if } t \geq C_1 \\
\phi_2(t) \leq \max_{t \in [0, C_1]} \phi_2(t) \leq C_2, & \text{if } t \leq C_1.
\end{cases}
$$

Similarly, we see that $\phi_1(\tau) \leq C$ for all $\tau \in [0, C]$. Moreover, since ϕ_i are continuous for $i = 1, 2$, it is clear that T is a continuous map. Then, by Brouwer fixed point theorem, there exists $(t_+, \tau_-) \in [0, C] \times [0, C]$ such that

$$
(\phi_1(\tau_-), \phi_2(t_+)) = (t_+, \tau_-).
$$

Owing to this fact and recalling that $\phi_i > 0$, we have $t_+ > 0$ and $\tau_- > 0$. By (16.4.4) and (16.4.7),

$$
\Phi_1^u(t_+, \tau_-) = \Phi_2^u(t_+, \tau_-) = 0,
$$

that is (t_+, τ_-) is a critical point of h^u. Next we aim to prove the uniqueness of (t_+, τ_-). Assuming that $w \in \mathcal{M}$, we have

$$
\begin{aligned}
\Phi^w(1, 1) &= \left(\Phi_1^w(1, 1), \Phi_2^w(1, 1) \right) \\
&= \left(\frac{\partial h^w}{\partial t}(1, 1), \frac{\partial h^w}{\partial \tau}(1, 1) \right) \\
&= \left(\langle J'(w^+ + w^-), w^+ \rangle, \langle J'(w^+ + w^-), w^- \rangle \right) = (0, 0),
\end{aligned}
$$

which implies that $(1, 1)$ is a critical point of h^w. Now, assume that (t_0, τ_0) is a critical point of h^w, with $0 < t_0 \leq \tau_0$. This means that

$$
\langle J'(t_0 w^+ + \tau_0 w^-), t_0 w^+ \rangle = 0 \text{ and } \langle J'(t_0 w^+ + \tau_0 w^-), \tau_0 w^- \rangle = 0,
$$

or equivalently

$$
t_0^2 \|w^+\|^2 - \tau_0 t_0 \iint_{\mathbb{R}^{2N}} \frac{w^+(x)w^-(y) + w^-(x)w^+(y)}{|x - y|^{N+2s}} \, dx dy = \int_{\mathbb{R}^N} K(x) f(t_0 w^+) t_0 w^+ \, dx
$$

$$(16.4.8)$$

$$
\tau_0^2 \|w^-\|^2 - \tau_0 t_0 \iint_{\mathbb{R}^{2N}} \frac{w^+(x)w^-(y) + w^-(x)w^+(y)}{|x - y|^{N+2s}} \, dx dy = \int_{\mathbb{R}^N} K(x) f(\tau_0 w^-) \tau_0 w^- \, dx.
$$

$$(16.4.9)$$

Dividing by $\tau_0^2 > 0$ in (16.4.9), we have

$$\|w^-\|^2 - \frac{t_0}{\tau_0} \iint_{\mathbb{R}^{2N}} \frac{w^+(x)w^-(y) + w^-(x)w^+(y)}{|x-y|^{N+2s}} \, dxdy = \int_{\mathbb{R}^N} K(x) \frac{f(s_0 w^-)}{\tau_0 w^-} (w^-)^2 \, dx,$$

and using the fact that $0 < t_0 \le \tau_0$ we see that

$$\|w^-\|^2 - \iint_{\mathbb{R}^{2N}} \frac{w^+(x)w^-(y) + w^-(x)w^+(y)}{|x-y|^{N+2s}} \, dxdy \ge \int_{\mathbb{R}^N} K(x) \frac{f(\tau_0 w^-)}{\tau_0 w^-} (w^-)^2 \, dx.$$

(16.4.10)

Since $w \in \mathcal{M}$, we also have

$$\|w^-\|^2 - \iint_{\mathbb{R}^{2N}} \frac{w^+(x)w^-(y) + w^-(x)w^+(y)}{|x-y|^{N+2s}} \, dxdy = \int_{\mathbb{R}^N} K(x) \frac{f(w^-)}{w^-} (w^-)^2 \, dx.$$

(16.4.11)

Putting together (16.4.10) and (16.4.11), we get

$$0 \ge \int_{\mathbb{R}^N} K(x) \left[\frac{f(\tau_0 w^-)}{\tau_0 w^-} (w^-)^2 - \frac{f(w^-)}{w^-} (w^-)^2 \right] dx.$$

The above relation and assumption (f_4) ensures that $0 < t_0 \le \tau_0 \le 1$.

Let us prove that $t_0 \ge 1$. Dividing by $t_0^2 > 0$ in (16.4.8), we have

$$\|w^+\|^2 - \frac{\tau_0}{t_0} \iint_{\mathbb{R}^{2N}} \frac{w^+(x)w^-(y) + w^-(x)w^+(y)}{|x-y|^{N+2s}} \, dxdy = \int_{\mathbb{R}^N} K(x) \frac{f(t_0 w^+)}{t_0 w^+} (w^+)^2 \, dx$$

and since $0 < t_0 \le \tau_0$ we deduce that

$$\|w^+\|^2 - \iint_{\mathbb{R}^{2N}} \frac{w^+(x)w^-(y) + w^-(x)w^+(y)}{|x-y|^{N+2s}} \, dxdy \le \int_{\mathbb{R}^N} K(x) \frac{f(t_0 w^+)}{t_0 w^+} (w^+)^2 \, dx.$$

(16.4.12)

Further, since $w \in \mathcal{M}$, we also have

$$\|w^+\|^2 - \iint_{\mathbb{R}^{2N}} \frac{w^+(x)w^-(y) + w^-(x)w^+(y)}{|x-y|^{N+2s}} \, dxdy = \int_{\mathbb{R}^N} K(x) \frac{f(w^+)}{w^+} (w^+)^2 \, dx.$$

(16.4.13)

Putting together (16.4.12) and (16.4.13) we get

$$0 \geq \int_{\mathbb{R}^N} K(x) \left[\frac{f(w^+)}{w^+}(w^+)^2 - \frac{f(t_0 w^+)}{t_0 w^+}(w^+)^2 \right] dx.$$

It follows from (f_4) that $t_0 \geq 1$. Consequently, $t_0 = \tau_0 = 1$, and this proves that $(1, 1)$ is the unique critical point of h^w with positive coordinates.

Let $u^\pm \in X$ be such that $u^\pm \neq 0$, and let (t_1, τ_1), (t_2, τ_2) be critical points of h^u with positive coordinates. By (i) it follows that

$$w_1 = t_1 u^+ + \tau_1 u^- \in \mathcal{M} \quad \text{and} \quad w_2 = t_2 u^+ + \tau_2 u^- \in \mathcal{M}.$$

We notice that w_2 can be written as

$$w_2 = \left(\frac{t_2}{t_1}\right) t_1 u^+ + \left(\frac{\tau_2}{\tau_1}\right) \tau_1 u^- = \frac{t_2}{t_1} w_1^+ + \frac{\tau_2}{\tau_1} w_1^- \in \mathcal{M}.$$

Since $w_1 \in X$ is such that $w_1^\pm \neq 0$, we have that $(t_2/t_1, \tau_2/\tau_1)$ is a critical point for h^{w_1} with positive coordinates. On the other hand, since $w_1 \in \mathcal{M}$, we conclude that $t_2/t_1 = \tau_2/\tau_1 = 1$, which gives $t_1 = t_2$ and $\tau_1 = \tau_2$.

Finally, we prove that h^u has a maximum global point $(\bar{t}, \bar{\tau}) \in (0, \infty) \times (0, \infty)$. Let $A^+ \subset \text{supp}(u^+)$ and $A^- \subset \text{supp}(u^-)$ positive with finite measure. By assumption (f_3) and the fact that $F(t) \geq 0$ for every $t \in \mathbb{R}$,

$$h^u(t, \tau) \leq \frac{1}{2} \|tu^+ + \tau u^-\|^2 - \int_{A^+} K(x) F(tu^+) \, dx - \int_{A^-} K(x) F(\tau u^-) \, dx$$

$$\leq \frac{t^2}{2} \|u^+\|^2 + \frac{\tau^2}{2} \|u^-\|^2 - t\tau \iint_{\mathbb{R}^{2N}} \frac{u^+(x)u^-(y) + u^+(y)u^-(x)}{|x - y|^{N+2s}} \, dx \, dy$$

$$- \int_{A^+} K(x) F(tu^+) \, dx - \int_{A^-} K(x) F(\tau u^-) \, dx.$$

Let us suppose that $|t| \geq |\tau| > 0$. Then, using the fact that $F(t) \geq 0$ for every $t \in \mathbb{R}$, we see that

$$h^u(t, \tau) \leq (t^2 + \tau^2) \left[\frac{1}{2} \|u^+\|^2 + \frac{1}{2} \|u^-\|^2 - \frac{1}{2} \iint_{\mathbb{R}^{2N}} \frac{u^+(x)u^-(y) + u^+(y)u^-(x)}{|x - y|^{N+2s}} \, dx \, dy \right]$$

$$- t^2 \int_{A^+} K(x) \frac{F(tu^+)}{(tu^+)^2} (u^+)^2 \, dx.$$

By (f_3), Fatou's lemma and the fact that $0 < t^2 + \tau^2 \leq 2t^2$, we obtain that

$$\limsup_{|(t,\tau)| \to \infty} \frac{h^u(t,\tau)}{t^2 + \tau^2} \leq C(u^+, u^-) - \frac{1}{2} \liminf_{|t| \to \infty} \int_{A^+} K(x) \frac{F(tu^+)}{(tu^+)^2} (u^+)^2 \, dx = -\infty,$$

where $C(u^+, u^-) > 0$ is a constant depending only on u^+ and u^-.

Therefore,

$$\lim_{|(t,\tau)| \to \infty} h^u(t,\tau) = -\infty, \tag{16.4.14}$$

which upon recalling that h^u is a continuous function implies that h^u has a maximum global point $(\bar{t}, \bar{\tau}) \in (0, \infty) \times (0, \infty)$.

The linearity of F and the positivity of K yield

$$\int_{\mathbb{R}^N} K(x)(F(tu^+) + F(\tau u^-)) \, dx = \int_{\mathbb{R}^N} K(x) F(tu^+ + \tau u^-) \, dx. \tag{16.4.15}$$

By (16.4.15), for every $u \in \mathbb{X}$ such that $u^{\pm} \neq 0$ and for every $t, \tau \geq 0$,

$$J(tu^+) + J(\tau u^-) \leq J(tu^+ + \tau u^-).$$

So, for every $u \in \mathbb{X}$ such that $u^{\pm} \neq 0$, one has

$$h^u(t, 0) + h^u(0, \tau) \leq h^u(t, \tau),$$

for every $t, \tau \geq 0$. Then,

$$\max_{t \geq 0} h^u(t, 0) < \max_{t, \tau > 0} h^u(t, \tau) \quad \text{and} \quad \max_{\tau \geq 0} h^u(0, \tau) < \max_{t, \tau > 0} h^u(t, \tau),$$

and this proves that $(\bar{t}, \bar{\tau}) \in (0, \infty) \times (0, \infty)$.

(iii) By Lemma 16.3.1-(a) we easily have that

$$\Phi_1^u(r, \tau_-) = \frac{\partial h^u}{\partial t}(r, \tau_-) > 0 \text{ if } r \in (0, t_+),$$

$$\Phi_1^u(t_+, \tau_-) = \frac{\partial h^u}{\partial t}(t_+, \tau_-) = 0,$$

$$\Phi_1^u(r, \tau_-) = \frac{\partial h^u}{\partial t}(r, \tau_-) > 0 \text{ if } r \in (t_+, \infty).$$

Therefore, (16.4.3) holds true. The proof of Lemma 16.4.1 is now complete. □

Lemma 16.4.2 *If $(u_n) \subset \mathcal{M}$ and $u_n \rightharpoonup u$ in $\mathbb{X}$, then $u \in \mathbb{X}$ and $u^{\pm} \neq 0$.*

Proof First, note that there is $\beta > 0$ such that

$$\beta \leq \|v^{\pm}\| \quad \forall v \in \mathcal{M}. \tag{16.4.16}$$

Indeed, if $v \in \mathcal{M}$, then

$$\|v^{\pm}\|^2 \leq \int_{\mathbb{R}^N} K(x) f(v^{\pm}) v^{\pm} \, dx.$$

Assume that (h_3) holds true. Then, using (f_1), (f_2), and (1.1.1), we see that given $\varepsilon > 0$ there exists a positive constant C_{ε} such that

$$\|v^{\pm}\|^2 \leq \int_{\mathbb{R}^N} K(x) f(v^{\pm}) v^{\pm} \, dx$$

$$\leq \varepsilon \|K/V\|_{L^{\infty}(\mathbb{R}^N)} \int_{\mathbb{R}^N} V(x)(v^{\pm})^2 \, dx + C_{\varepsilon} C_* \|K\|_{L^{\infty}(\mathbb{R}^N)} \|v^{\pm}\|^{2_s^*} \tag{16.4.17}$$

$$\leq \varepsilon \|K/V\|_{L^{\infty}(\mathbb{R}^N)} \|v^{\pm}\|^2 + C_{\varepsilon} C_* \|K\|_{L^{\infty}(\mathbb{R}^N)} \|v^{\pm}\|^{2_s^*}.$$

Choosing

$$\varepsilon \in \left(0, \frac{1}{\|K/V\|_{L^{\infty}(\mathbb{R}^N)}}\right),$$

guarantees that there exists a positive constant β_1 such that $\|v^{\pm}\| > \beta_1$.

Analogously, assuming that (h_4) holds, conditions $(\tilde{f}_1)$, (f_2), Theorem 1.1.8 and Hölder's inequality imply that

$$\|v^{\pm}\|^2 \leq \int_{\mathbb{R}^N} K(x) f(v^{\pm}) v^{\pm} \, dx$$

$$\leq C_1 \varepsilon \|v^{\pm}\|^2 + C_1 C_* (\varepsilon + C_2 \|K\|_{L^{\infty}(\mathbb{R}^N)}) \|v^{\pm}\|^{2_s^*} + \|K\|_{L^{\frac{2_s^*}{2_s^* - m}}(B_R)} C_* \|v^{\pm}\|^m. \tag{16.4.18}$$

Since $m \in (2, 2_s^*)$, we can choose ε sufficiently small so that it is possible to find a positive constant β_2 such that $\|v^{\pm}\| > \beta_2$. Hence, if we set $\beta = \min\{\beta_1, \beta_2\}$, then (16.4.16) holds.

Then, if $(u_n) \subset \mathcal{M}$, we have

$$\beta^2 \leq \int_{\mathbb{R}^N} K(x) f(u_n^{\pm}) u_n^{\pm} \, dx, \quad \forall n \in \mathbb{N}. \tag{16.4.19}$$

Since $u_n \rightharpoonup u$ in $\mathbb{X}$, using Lemma 16.2.2, we can let $n \to \infty$ in (16.4.19). More precisely, using Lemma 16.2.3, it follows that

$$0 < \beta^2 \le \int_{\mathbb{R}^N} K(x) f(u^{\pm}) u^{\pm} \, dx.$$

Thus $u \in \mathbb{X}$ and $u^{\pm} \ne 0$. The proof is now complete.   □

Let us denote

$$c_\infty = \inf_{u \in \mathcal{M}} J(u).$$

Since $\mathcal{M} \subset \mathcal{N}$, we deduce

$$c_\infty \ge d_\infty > 0. \tag{16.4.20}$$

16.5   Existence and Multiplicity Results

In this section we prove the existence of least energy nodal weak solutions by using minimization arguments and a variant of deformation lemma. We start by proving the existence of a minimum point of the functional J in $\mathcal{M}$.

Let $(u_n) \subset \mathcal{M}$ be such that

$$J(u_n) \to c_\infty \quad \text{in } \mathbb{R}. \tag{16.5.1}$$

We claim that (u_n) is bounded in $\mathbb{X}$. Indeed, assume, by contradiction, that that there exists a subsequence, denoted again by (u_n), such that $\|u_n\| \to \infty$ as $n \to \infty$. Set

$$v_n = \frac{u_n}{\|u_n\|}, \quad n \in \mathbb{N}.$$

Since (v_n) is bounded in $\mathbb{X}$, due to the reflexivity of $\mathbb{X}$, there exists $v \in \mathbb{X}$ such that

$$v_n \rightharpoonup v \quad \text{in } \mathbb{X}. \tag{16.5.2}$$

Moreover, by virtue of Lemma 16.2.1,

$$v_n(x) \to v(x) \quad \text{a.e. in } \mathbb{R}^N. \tag{16.5.3}$$

By Lemma 16.4.1-(i) and the fact that $(u_n) \subset \mathcal{M}$, we have that $t_+(v_n) = s_-(v_n) = \|u_n\|$ and

$$
\begin{aligned}
J(u_n) = J(\|u_n\| v_n) &\geq J(t v_n) \\
&= \frac{t^2}{2} \|v_n\|^2 - \int_{\mathbb{R}^N} K(x) F(t v_n) \, dx \\
&= \frac{t^2}{2} - \int_{\mathbb{R}^N} K(x) F(t v_n) \, dx,
\end{aligned}
\tag{16.5.4}
$$

for every $t > 0$ and $n \in \mathbb{N}$.

Suppose that $v = 0$. Taking into account (16.5.2) and Lemma 16.2.3, we get

$$
\int_{\mathbb{R}^N} K(x) F(t v_n) \, dx \to 0, \quad \forall t > 0.
\tag{16.5.5}
$$

Passing to the limit in (16.5.4) as $n \to \infty$, and combining (16.5.1) and (16.5.5) we have

$$
c_\infty \geq \frac{t^2}{2}, \quad \forall t > 0,
$$

which gives a contradiction.

Hence, $v \neq 0$. Taking into account definitions of J and (v_n), we have

$$
\frac{J(u_n)}{\|u_n\|^2} = \frac{1}{2} - \int_{\mathbb{R}^N} K(x) \frac{F(v_n \|u_n\|)}{(v_n \|u_n\|)^2} (v_n)^2 \, dx.
\tag{16.5.6}
$$

Now, since $v \neq 0$ and $\|u_n\| \to \infty$, using (16.5.3) in addition to (f_3), the Fatou lemma ensures that

$$
\int_{\mathbb{R}^N} K(x) \frac{F(v_n \|u_n\|)}{(v_n \|u_n\|)^2} v_n^2 \, dx \to \infty
\tag{16.5.7}
$$

which in conjunction with (16.5.1) show that by passing to the limit in (16.5.6) as $n \to \infty$, we reach a contradiction.

Therefore, $(u_n) \subset \mathbb{X}$ is a bounded subsequence. Consequently, there exists $u \in \mathbb{X}$ such that

$$
u_n \rightharpoonup u \quad \text{in } \mathbb{X}.
\tag{16.5.8}
$$

By Lemma 16.4.2, it follows that $u^\pm \neq 0$. Moreover, by Lemma 16.4.1, there are two constants $t_+, \tau_- > 0$ such that

$$
t_+ u^+ + \tau_- u^- \in \mathcal{M}.
\tag{16.5.9}
$$

Now, our aim is to prove that $t_+, \tau_- \in (0, 1]$. By (16.5.8) and Lemma 16.2.3,

$$\int_{\mathbb{R}^N} K(x) f(u_n^\pm) u_n^\pm \, dx \to \int_{\mathbb{R}^N} K(x) f(u^\pm) u^\pm \, dx \tag{16.5.10}$$

and

$$\int_{\mathbb{R}^N} K(x) F(u_n^\pm) \, dx \to \int_{\mathbb{R}^N} K(x) F(u^\pm) \, dx. \tag{16.5.11}$$

Recalling that $(u_n) \subset \mathcal{M}$, using (16.5.8) and (16.5.10), and applying the Fatou lemma one obtains that

$$\langle J'(u), u^\pm \rangle$$
$$= \|u^\pm\|^2 - \iint_{\mathbb{R}^{2N}} \frac{u^+(x)u^-(y) + u^-(x)u^+(y)}{|x-y|^{N+2s}} \, dx\,dy - \int_{\mathbb{R}^N} K(x) f(u^\pm) u^\pm dx$$
$$\le \liminf_{n \to \infty} \langle J'(u_n), u_n^\pm \rangle = 0. \tag{16.5.12}$$

Let us assume $0 < t_+ < \tau_-$. By (16.5.9),

$$\tau_-^2 \|u^-\|^2 - t_+ \tau_- \iint_{\mathbb{R}^{2N}} \frac{u^+(x)u^-(y) + u^-(x)u^+(y)}{|x-y|^{N+2s}} \, dx\,dy = \int_{\mathbb{R}^N} K(x) f(\tau_- u^-) \tau_- u^- \, dx,$$

and since $t_+ < \tau_-$ we obtain

$$\|u^-\|^2 - \iint_{\mathbb{R}^{2N}} \frac{u^-(x)u^+(y) + u^-(y)u^+(x)}{|x-y|^{N+2s}} \, dx\,dy \ge \int_{\text{supp}(u^-)} K(x) \frac{f(\tau_- u^-)}{\tau_- u^-} (u^-)^2 \, dx. \tag{16.5.13}$$

In view of (16.5.12), we have

$$\|u^-\|^2 - \iint_{\mathbb{R}^{2N}} \frac{u^-(x)u^+(y) + u^-(y)u^+(x)}{|x-y|^{N+2s}} \, dx\,dy \le \int_{\text{supp}(u^-)} K(x) \frac{f(u^-)}{u^-} (u^-)^2 \, dx. \tag{16.5.14}$$

Putting together (16.5.13) and (16.5.14), we deduce that

$$0 \ge \int_{\text{supp}(u^-)} K(x) \left[\frac{f(\tau_- u^-)}{\tau_- u^-} - \frac{f(u^-)}{u^-} \right] (u^-)^2 \, dx, \tag{16.5.15}$$

which yields $\tau_- \in (0, 1]$ in virtue of (f_4). Similarly, we can show that $t_+ \in (0, 1]$.
Now, we prove that

$$J(t_+ u^+ + \tau_- u^-) = c_\infty. \tag{16.5.16}$$

Using the definition of $c_\infty, t_+, \tau_- \in (0, 1]$, (f_4), and taking into account relations (16.5.9), (16.5.10) and (16.5.11), we get

$$c_\infty \leq J(t_+ u^+ + \tau_- u^-)$$

$$= J(t_+ u^+ + \tau_- u^-) - \frac{1}{2}\langle J'(t_+ u^+ + \tau_- u^-), t_+ u^+ + \tau_- u^-\rangle$$

$$= \int_{\mathbb{R}^N} K(x) \left[\frac{1}{2} f(t_+ u^+ + \tau_- u^-)(t_+ u^+ + \tau_- u^-) - F(t_+ u^+ + \tau_- u^-) \right] dx$$

$$= \int_{\mathbb{R}^N} K(x) \left[\frac{1}{2} f(t_+ u^+)(t_+ u^+) - F(t_+ u^+) \right] dx$$

$$+ \int_{\mathbb{R}^N} K(x) \left[\frac{1}{2} f(\tau_- u^-)(\tau_- u^-) - F(\tau_- u^-) \right] dx$$

$$\leq \int_{\mathbb{R}^N} K(x) \left[\frac{1}{2} f(u^+)(u^+) - F(u^+) \right] dx + \int_{\mathbb{R}^N} K(x) \left[\frac{1}{2} f(u^-)(u^-) - F(u^-) \right] dx$$

$$= \int_{\mathbb{R}^N} K(x) \left[\frac{1}{2} f(u)u - F(u) \right] dx$$

$$= \lim_{n\to\infty} \int_{\mathbb{R}^N} K(x) \left[\frac{1}{2} f(u_n)u_n - F(u_n) \right] dx$$

$$= \lim_{n\to\infty} \left[J(u_n) - \frac{1}{2}\langle J'(u_n), u_n\rangle \right] = c_\infty.$$

Hence, (16.5.16) holds true. Furthermore, the above calculations imply that $t_+ = \tau_- = 1$.

Next we show that $u = u^+ + u^-$ is a critical point of the functional J, arguing by contradiction. Thus, suppose that $J'(u) \neq 0$. By continuity, there exist $\delta, \mu > 0$ such that

$$\mu \leq |J'(v)|, \quad \text{since that } \|v - u\| \leq 3\delta. \tag{16.5.17}$$

Define $D = [\frac{1}{2}, \frac{3}{2}] \times [\frac{1}{2}, \frac{3}{2}]$ and $g : D \to \mathbb{X}^\pm$ by

$$g(t, \tau) = t u^+ + \tau u^-,$$

where $\mathbb{X}^\pm = \{u \in \mathbb{X} : u^\pm \neq 0\}$. By Lemma 16.4.1, we deduce that

$$J(g(1, 1)) = c_\infty,$$

$$J(g(t, \tau)) < c_\infty \quad \text{in } D \setminus \{(1, 1)\}.$$

Therefore,

$$\beta = \max_{(t,\tau) \in \partial D} J(g(t,\tau)) < c_\infty. \tag{16.5.18}$$

Now, we apply Lemma 2.2.5 with

$$S = \tilde{S} = \{v \in \mathbb{X} : \|v - u_\pm\| \le \delta\},$$

and $c = c_\infty$. Choosing $\varepsilon = \min\left\{\dfrac{c_\infty - \beta}{4}, \dfrac{\mu\delta}{8}\right\}$, we deduce that there exists a deformation $\eta \in C([0,1] \times \mathbb{X}, \mathbb{X})$ such that:

(a) $\eta(t,v) = v$ if $v \in J^{-1}([c_\infty - 2\varepsilon, c_\infty + 2\varepsilon])$;
(b) $J(\eta(1,v)) \le c_\infty - \varepsilon$ for each $v \in \mathbb{X}$ with $\|v - u_\pm\| \le \delta$ and $J(v) \le c_\infty + \varepsilon$;
(c) $J(\eta(1,v)) \le J(v)$ for all $v \in \mathbb{X}$.

It follows from (b) and (c) that

$$\max_{(t,\tau) \in \partial D} J(\eta(1, g(t,\tau))) < c_\infty. \tag{16.5.19}$$

To complete the proof, it suffices to prove that

$$\eta(1, g(D)) \cap \mathcal{M} \ne \emptyset. \tag{16.5.20}$$

Indeed, the definition of c_∞ and (16.5.20) contradict (16.5.19). Hence, let us define the maps

$$h(t,\tau) = \eta(1, g(t,\tau)),$$

$$\psi_0(t,\tau) = \left(J'(g(t,1))tu^+, J'(g(1,\tau))\tau u^-\right),$$

$$\psi_1(t,\tau) = \left(\frac{1}{t}J'(h(t,1))h(t,1)^+, \frac{1}{\tau}J'(h(1,\tau))h(1,\tau)^-\right).$$

By Lemma 16.4.1-(iii), the C^1-function $\gamma_+(t) = h^u(t,1)$ has a unique global maximum point $t = 1$ (note that $t\gamma_+'(t) = \langle J'(g(t,1)), tu^+ \rangle$). By denseness, given $\varepsilon > 0$ small enough, there is $\gamma_{+,\varepsilon} \in C^\infty([\frac{1}{2}, \frac{3}{2}])$ such that $\|\gamma_+ - \gamma_{+,\varepsilon}\|_{C^1([\frac{1}{2}, \frac{3}{2}])} < \varepsilon$. Therefore, $\|\gamma_+' - \gamma_{+,\varepsilon}'\|_{C([\frac{1}{2}, \frac{3}{2}])} < \varepsilon$, $\gamma_{+,\varepsilon}'(1) = 0$ and $\gamma_{+,\varepsilon}''(1) < 0$. Analogously, there exists $\gamma_{-,\varepsilon} \in C^\infty([\frac{1}{2}, \frac{3}{2}])$ such that $\|\gamma_-' - \gamma_{-,\varepsilon}'\|_{C([\frac{1}{2}, \frac{3}{2}])} < \varepsilon$, $\gamma_{+,\varepsilon}'(1) = 0$ and $\gamma_{+,\varepsilon}''(1) < 0$, where $\gamma_-(\tau) = h^u(1,\tau)$.

Define $\psi_\varepsilon \in C^\infty(D)$ by $\psi_\varepsilon(t, \tau) = (t\gamma'_{+,\varepsilon}(t), \tau\gamma'_{-,\varepsilon}(\tau))$ and note that $\|\psi_\varepsilon - \psi_0\|_{C(D)} < \frac{3\sqrt{2}}{2}\varepsilon$, $(0, 0) \notin \psi_\varepsilon(\partial D)$, and, $(0, 0)$ is a regular value of ψ_ε in D. On the other hand, $(1, 1)$ is the unique solution of $\psi_\varepsilon(t, \tau) = (0, 0)$ in D. By the definition of Brouwer's degree [27, 137], we conclude that

$$\deg(\psi_0, D, (0, 0)) = \deg(\psi_\varepsilon, D, (0, 0)) = \mathrm{sgn}\,\mathrm{Jac}(\psi_\varepsilon)(1, 1),$$

for ε small enough.

Since

$$\mathrm{Jac}(\psi_\varepsilon)(1, 1) = [\gamma'_{+,\varepsilon}(1) + \gamma''_{+,\varepsilon}(1)] \times [\gamma'_{-,\varepsilon}(1) + \gamma''_{-,\varepsilon}(1)] = \gamma''_{+,\varepsilon}(1) \times \gamma''_{-,\varepsilon}(1) > 0$$

we obtain that

$$\deg(\psi_0, D, (0, 0)) = \mathrm{sgn}[\gamma''_{+,\varepsilon}(1) \times \gamma''_{-,\varepsilon}(1)] = 1,$$

where $\mathrm{Jac}(\psi_\varepsilon)$ is the Jacobian determinant of ψ_ε and sgn denotes the sign function.

On the other hand, by (16.5.18) we have

$$J(g(t, \tau)) \leq \beta < \frac{\beta + c_\infty}{2}$$

$$= c_\infty - 2\left(\frac{c_\infty - \beta}{4}\right) \qquad (16.5.21)$$

$$\leq c_\infty - 2\varepsilon, \quad \forall (t, \tau) \in \partial D.$$

By (16.5.21) and (a), it follows that $g = h$ on ∂D. Therefore, $\psi_1 = \psi_0$ on ∂D and consequently

$$\deg(\psi_1, D, (0, 0)) = \deg(\psi_0, D, (0, 0)) = 1, \qquad (16.5.22)$$

which shows that $\psi_1(t, \tau) = (0, 0)$ for some $(t, \tau) \in D$.

Now, in order to show that (16.5.20) holds true, we prove that

$$\psi_1(1, 1) = \left(J'(h(t, 1))h(1, 1)^+, J'(h(1, 1))h(1, 1)^-\right) = 0. \qquad (16.5.23)$$

As a matter of fact, (16.5.23) and the fact that $(1, 1) \in D$, yield $h(1, 1) = \eta(1, g(1, 1)) \in \mathcal{M}$.

We argue as follows. If the zero (t, τ) of ψ_1 obtained above is equal to $(1, 1)$, there is nothing to do. If, however, $(t, \tau) \neq (1, 1)$, then we take $0 < \delta_1 < \min\{|t - 1|, |\tau - 1|\}$ and consider

$$
D_1 = \left[1 - \frac{\delta_1}{2}, 1 + \frac{\delta_1}{2}\right] \times \left[1 - \frac{\delta_1}{2}, 1 + \frac{\delta_1}{2}\right].
$$

Then, $(t, \tau) \in D \setminus D_1$. Hence, we can repeat for D_1 the argument used for D, and obtain a couple $(t_1, \tau_1) \in D_1$ such that $\psi_1(t_1, \tau_1) = 0$. If $(t_1, \tau_1) = (1, 1)$, there is nothing to prove. Otherwise, we can continue with this procedure and find in the n-th step that (16.5.23) holds, or produce a sequence $(t_n, \tau_n) \in D_{n-1} \setminus D_n$ which converges to $(1, 1)$ and such that

$$
\psi_1(t_n, \tau_n) = 0, \qquad \text{for every } n \in \mathbb{N}.
$$

Letting here $n \to \infty$ and using the continuity of ψ_1 we get (16.5.23). Therefore, $u = u^+ + u^-$ is a critical point of J.

Finally, we consider the case when f is odd. Clearly, the functional ψ is even. From (16.3.11) and (16.4.20) we have that ψ is bounded from below in $\mathbb{S}$. Taking into account Lemmas 16.2.2 and 16.2.3, we infer that ψ satisfies the Palais–Smale condition on $\mathbb{S}$. Then, by Proposition 16.3.2 and Theorem 2.3.4, we conclude that the functional J has infinitely many critical points. $\qquad \square$

Fractional Schrödinger Equations with Magnetic Fields   17

17.1   Introduction

In this chapter we deal with the following fractional problem

$$\begin{cases} \varepsilon^{2s}(-\Delta)^s_{\frac{A}{\varepsilon}} u + V(x)u = f(|u|^2)u \text{ in } \mathbb{R}^N, \\ |u| \in H^s(\mathbb{R}^N, \mathbb{R}), \end{cases} \tag{17.1.1}$$

where $\varepsilon > 0$ is a parameter, $s \in (0, 1)$, $N \geq 3$, $V \in C(\mathbb{R}^N, \mathbb{R})$ and $A \in C^{0,\alpha}(\mathbb{R}^N, \mathbb{R}^N)$, $\alpha \in (0, 1]$, are the electric and magnetic potentials, respectively, and $f : \mathbb{R} \to \mathbb{R}$. The fractional magnetic Laplacian is defined by

$$(-\Delta)^s_A u(x) = C(N, s) \lim_{r \to 0} \int_{B^c_r(x)} \frac{u(x) - e^{\iota(x-y)\cdot A(\frac{x+y}{2})}u(y)}{|x - y|^{N+2s}} \, dy, \quad C(N, s) = s \frac{4^s \Gamma\left(\frac{N+2s}{2}\right)}{\pi^{N/2}\Gamma(1 - s)}. \tag{17.1.2}$$

This nonlocal operator was defined in [157] as a fractional extension (for an arbitrary $s \in (0, 1)$) of the magnetic pseudorelativistic operator or Weyl pseudodifferential operator defined with mid-point prescription,

$$\begin{aligned} \mathscr{H}_A u(x) &= \frac{1}{(2\pi)^3} \iint_{\mathbb{R}^6} e^{\iota(x-y)\cdot\xi} \sqrt{\left|\xi - A\left(\frac{x+y}{2}\right)\right|^2} u(y) \, dy d\xi \\ &= \frac{1}{(2\pi)^3} \iint_{\mathbb{R}^6} e^{\iota(x-y)\cdot\left(\xi + A\left(\frac{x+y}{2}\right)\right)} \sqrt{|\xi|^2} u(y) \, dy d\xi, \end{aligned}$$

© The Author(s), under exclusive license to Springer Nature Switzerland AG 2021
V. Ambrosio, *Nonlinear Fractional Schrödinger Equations in* $\mathbb{R}^N$,
Frontiers in Mathematics, https://doi.org/10.1007/978-3-030-60220-8_17

introduced in [225] by Ichinose and Tamura, through oscillatory integrals. Observe that
for smooth functions u,

$$\mathcal{H}_A u(x) = -\lim_{\varepsilon \searrow 0} \int_{B_\varepsilon^c} \left[e^{-\imath y \cdot A\left(x + \frac{y}{2}\right)} u(x + y) - u(x) - 1_{\{|y| < 1\}}(y) y \cdot (\nabla - \imath A(x)) u(x) \right] d\mu$$

$$= \lim_{\varepsilon \searrow 0} \int_{B_\varepsilon^c(x)} \left[u(x) - e^{\imath (x - y) \cdot A\left(\frac{x + y}{2}\right)} u(y) \right] \mu(y - x) \, dy,$$

where

$$d\mu = \mu(y) dy = \frac{\Gamma\left(\frac{N+1}{2}\right)}{\pi^{\frac{N+1}{2}} |y|^{N+1}} dy.$$

More precisely, from a physical point of view, when $s = \frac{1}{2}$, the operator in (17.1.2)
takes inspiration from the definition of a quantized operator corresponding to the classical
relativistic Hamiltonian symbol for a relativistic particle of mass $m \geq 0$, that is

$$\sqrt{(\xi - A(x))^2 + m^2} + V(x), \quad (\xi, x) \in \mathbb{R}^N \times \mathbb{R}^N,$$

which is the sum of the kinetic energy term involving the magnetic vector potential $A(x)$
and the potential energy term given by the electric scalar potential $V(x)$. For the sake
of completeness, we emphasize that in the literature there are three kinds of quantum
relativistic Hamiltonians, depending on how the kinetic energy term $\sqrt{(\xi - A(x))^2 + m^2}$
is quantized. As explained in [224], these three nonlocal operators are in general different
from each other, but coincide when the vector potential A is assumed to be linear, so in
particular, in the case of constant magnetic fields.

In the light of the results in [292, 314], when $s \to 1$, the equation in (17.1.1) is related
to the study of solutions $u : \mathbb{R}^N \to \mathbb{C}$ of the following nonlinear Schrödinger equation
with magnetic field

$$\left(\frac{\varepsilon}{\imath} \nabla - A(x)\right)^2 u + V(x) u = f(|u|^2) u \quad \text{in } \mathbb{R}^N, \tag{17.1.3}$$

where $\left(\frac{\varepsilon}{\imath} \nabla - A(x)\right)^2$ is the magnetic Laplacian given by

$$\left(\frac{\varepsilon}{\imath} \nabla - A(x)\right)^2 u = -\varepsilon^2 \Delta u - \frac{2\varepsilon}{\imath} A(x) \cdot \nabla u + |A(x)|^2 u - \frac{\varepsilon}{\imath} u \operatorname{div}(A(x)).$$

In this context, when $N = 3$, the magnetic field B is exactly the curl of A, while for
higher dimensions $N \geq 4$, B should be thought of as the 2-form given by $B_{ij} = \partial_j A_k - \partial_k A_j$; see [81, 249, 301]. Equation (17.1.3) arises in the investigation of standing wave

solutions $\psi(x, t) = u(x)e^{-\iota \frac{E}{\varepsilon} t}$, with $E \in \mathbb{R}$, for the following time-dependent nonlinear Schrödinger equation

$$\iota \varepsilon \frac{\partial \psi}{\partial t} = \left(\frac{\varepsilon}{\iota}\nabla - A(x)\right)^2 \psi + W(x)\psi - f(|\psi|^2)\psi \quad \text{in } (x, t) \in \mathbb{R}^N \times \mathbb{R}$$

where $W(x) = V(x) + E$. An important class of solutions of (17.1.3) are the so-called semi-classical states which concentrate and develop a spike shape around one, or more, particular points in $\mathbb{R}^N$, while vanishing elsewhere as $\varepsilon \to 0$. This interest in this problem is due to the well-known fact that the transition from Quantum Mechanics to Classical Mechanics can be formally performed by sending $\varepsilon \to 0$. For this reason, Eq. (17.1.3) has been widely studied by many authors [18, 77, 144, 145, 148, 179, 243].

However, in the nonlocal fractional magnetic setting, only few recent papers [157, 194, 272, 343] deal with the existence and multiplicity of solutions.

Here, we are interested in the existence and multiplicity of solutions to (17.1.1) when the potential V fulfills the following condition introduced by Rabinowitz in [299]:

$$V_\infty = \liminf_{|x| \to \infty} V(x) > V_0 = \inf_{x \in \mathbb{R}^N} V(x) > 0. \tag{V}$$

In this context, the presence of the nonlocal operator (17.1.2) makes our analysis more complicated and intriguing, and new techniques are needed to overcome the difficulties that appear.

Before to state our results, we introduce the assumptions on the nonlinearity. We suppose that $f : \mathbb{R} \to \mathbb{R}$ satisfies the following conditions:

$(f1)$ $f \in C^1(\mathbb{R}, \mathbb{R})$;
$(f2)$ $f(t) = 0$ for $t \leq 0$;
$(f3)$ there exists $q \in (2, 2_s^*)$, where $2_s^* = 2N/(N-2s)$, such that $\lim_{t \to \infty} f(t)/t^{\frac{q-2}{2}} = 0$;
$(f4)$ there exists $\theta > 2$ such that $0 < \frac{\theta}{2}F(t) \leq tf(t)$ for any $t > 0$, where $F(t) = \int_0^t f(\tau)d\tau$;
$(f5)$ there exists $\sigma \in (2, 2_s^*)$ such that $f'(t) \geq C_\sigma t^{\frac{\sigma-4}{2}}$ for any $t > 0$.

A first result we get is the following.

Theorem 17.1.1 ([66]) *Assume that (V) and $(f1)$–$(f5)$ hold. Then there exists $\varepsilon_0 > 0$ such that problem (17.1.1) admits a ground state solution for any $\varepsilon \in (0, \varepsilon_0)$.*

Now, let us introduce the sets

$$M = \{x \in \mathbb{R}^N : V(x) = V_0\} \quad \text{and} \quad M_\delta = \{x \in \mathbb{R}^N : \text{dist}(x, M) \leq \delta\} \text{ for } \delta > 0. \tag{17.1.4}$$

Secondly, we obtain a multiplicity result for (17.1.1) by using the Lusternik-Schnirelman category.

Theorem 17.1.2 ([66]) *Assume that (V) and $(f1)$–$(f5)$ hold. Then, for any $\delta > 0$ there exists $\varepsilon_\delta > 0$ such that, for any $\varepsilon \in (0, \varepsilon_\delta)$, problem (17.3.1) has at least $\mathrm{cat}_{M_\delta}(M)$ nontrivial solutions.*

The proof of the above theorems is based on variational methods. In the study of our problem, we will use the diamagnetic inequality established in [157] and the power-type decay at infinity of positive solutions to the limit problem associated with (17.3.1). These facts combined with the Hölder continuity assumption on the magnetic potential, will play an essential role in deriving some useful estimates needed to obtain the existence of solutions and to implement the barycenter machinery.

17.2 A Multiplicity Result Under the Rabinowitz Condition

17.2.1 The Variational Setting

Using the change of variable $x \mapsto \varepsilon x$, we see that problem (17.1.1) is equivalent to

$$
\begin{cases}
(-\Delta)^s_{A_\varepsilon} u + V(\varepsilon x)u = f(|u|^2)u \text{ in } \mathbb{R}^N, \\
|u| \in H^s(\mathbb{R}^N, \mathbb{R}),
\end{cases}
\tag{17.2.1}
$$

where we set $A_\varepsilon(x) = A(\varepsilon x)$.

For a function $u : \mathbb{R}^N \to \mathbb{C}$, let us denote (for simplicity, we set $\frac{C(N,s)}{2} = 1$)

$$
[u]^2_A = \iint_{\mathbb{R}^{2N}} \frac{|u(x) - e^{\iota(x-y)\cdot A(\frac{x+y}{2})}u(y)|^2}{|x-y|^{N+2s}}\, dx dy,
$$

and consider the space

$$
\mathcal{D}^{s,2}_A(\mathbb{R}^N, \mathbb{C}) = \left\{ u \in L^{2^*_s}(\mathbb{R}^N, \mathbb{C}) : [u]^2_A < \infty \right\}.
$$

Then we introduce the Hilbert space

$$
H^s_\varepsilon = \left\{ u \in \mathcal{D}^{s,2}_{A_\varepsilon}(\mathbb{R}^N, \mathbb{C}) : \int_{\mathbb{R}^N} V(\varepsilon x)|u|^2\, dx < \infty \right\}
$$

endowed with the scalar product

$$\langle u, v \rangle_\varepsilon = \Re \iint_{\mathbb{R}^{2N}} \frac{(u(x) - e^{\iota(x-y)\cdot A_\varepsilon(\frac{x+y}{2})}u(y))\overline{(v(x) - e^{\iota(x-y)\cdot A_\varepsilon(\frac{x+y}{2})}v(y))}}{|x-y|^{N+2s}} \, dx \, dy$$

$$+ \Re \int_{\mathbb{R}^N} V(\varepsilon x) u \bar{v} \, dx$$

and the noun $\|u\|_\varepsilon = \sqrt{\langle u, u \rangle_\varepsilon}$.

If $u \in H_\varepsilon^s$, let

$$\hat{u}_j(x) = \varphi_j(x) u(x) \tag{17.2.2}$$

where $j \in \mathbb{N}$ and $\varphi_j(x) = \varphi(2x/j)$ with $\varphi \in C_c^\infty(\mathbb{R}^N, \mathbb{R})$, $0 \le \varphi \le 1$, $\varphi(x) = 1$ if $|x| \le 1$, and $\varphi(x) = 0$ if $|x| \ge 2$. Note that $\hat{u}_j \in H_\varepsilon^s$ and $\hat{u}_j$ has compact support.

Proceeding as in the proof of Lemma 3.2 in [343], we get the following useful result.

Lemma 17.2.1 *For any $\varepsilon > 0$, it holds that $\|\hat{u}_j - u\|_\varepsilon \to 0$ as $j \to \infty$.*

The space H_ε^s enjoys the following fundamental properties.

Lemma 17.2.2 *H_ε^s is complete and $C_c^\infty(\mathbb{R}^N, \mathbb{C})$ is dense in H_ε^s.*

Proof To prove that H_ε^s is a complete space, consider a Cauchy sequence (u_n) in $H_{A_\varepsilon}^s$. In particular, $(\sqrt{V(\varepsilon \cdot)}u_n)$ is a Cauchy sequence in $L^2(\mathbb{R}^N, \mathbb{C})$, and since $V(\varepsilon \cdot) \ge V_0$ in $\mathbb{R}^N$, there exists $u \in L^2(\mathbb{R}^N, \mathbb{C})$ such that $\sqrt{V(\varepsilon \cdot)}u_n \to \sqrt{V(\varepsilon \cdot)}u$ in $L^2(\mathbb{R}^N, \mathbb{C})$ and a.e. in $\mathbb{R}^N$. By Fatou's lemma, $u_n \to u$ in H_ε^s.

To prove that $C_c^\infty(\mathbb{R}^N, \mathbb{C})$ is dense in H_ε^s, fix $u \in H_\varepsilon^s$ and consider the sequence $\hat{u}_j(x) = u(x)\varphi(x/j)$ defined as in (17.2.2). In view of Lemma 17.2.1, $\|\hat{u}_j - u\|_\varepsilon \to 0$ as $j \to \infty$ and so it is enough to prove the denseness for compactly supported functions in H_ε^s.

Now, we consider $u \in H_\varepsilon^s$ with compact support, and assume that $\mathrm{supp}(u) \subset B_R$. Taking into account that

$$|u(x) - u(y)|^2 \le 2|u(x) - u(y)e^{\iota A_\varepsilon(\frac{x+y}{2})\cdot(x-y)}|^2 + 2|u(y)|^2|e^{\iota A_\varepsilon(\frac{x+y}{2})\cdot(x-y)} - 1|^2$$

and $|e^{\iota t} - 1|^2 \le 4$ and $|e^{\iota t} - 1|^2 \le t^2$, we have

$$\int_{B_R} |u(y)|^2 \left(\int_{\mathbb{R}^N} \frac{|e^{\iota A_\varepsilon(\frac{x+y}{2})\cdot(x-y)} - 1|^2}{|x-y|^{N+2s}} \, dx \right) dy$$

$$\leq C \left[\int_{B_R} |u(y)|^2 \left(\int_{|x-y|>1} \frac{1}{|x-y|^{N+2s}} \, dx \right) dy \right.$$

$$\left. + \int_{B_R} |u(y)|^2 \left(\int_{|x-y|\leq 1} \frac{\max_{|z|\leq \frac{2R+1}{2}} |A_\varepsilon(z)|^2}{|x-y|^{N+2s-2}} \, dx \right) dy \right]$$

$$< \infty,$$

and since $V(\varepsilon \cdot) \geq V_0$ in $\mathbb{R}^N$, we see that $u \in H^s(\mathbb{R}^N, \mathbb{C})$.

Then, it makes sense to define $u_\varepsilon = \rho_\varepsilon * u \in C_c^\infty(\mathbb{R}^N, \mathbb{C})$, where ρ_ε is a mollifier with $\mathrm{supp}(\rho_\varepsilon) \subset B_\varepsilon$. Arguing as in the proof of Theorem 3.24 in [211], we have that $u_\varepsilon \to u$ in $H^s(\mathbb{R}^N, \mathbb{C})$ as $\varepsilon \to 0$. Moreover there exists $K > 0$ such that $\mathrm{supp}(u_\varepsilon - u) \subset B_K$ for all $\varepsilon > 0$ small enough, and arguing as before, we obtain

$$[u_\varepsilon - u]_{A_\varepsilon}^2 \leq 2[u_\varepsilon - u]_s^2 + 2 \iint_{\mathbb{R}^{2N}} |(u_\varepsilon - u)(y)|^2 \frac{|e^{\iota A_\varepsilon(\frac{x+y}{2})\cdot(x-y)} - 1|^2}{|x-y|^{N+2s}} \, dx dy$$

$$\leq 2[u_\varepsilon - u]_s^2 + C \left[\int_{B_K} |(u_\varepsilon - u)(y)|^2 \left(\int_{|x-y|>1} \frac{1}{|x-y|^{N+2s}} \, dx \right) dy \right.$$

$$\left. + \int_{B_K} |(u_\varepsilon - u)(y)|^2 \left(\int_{|x-y|\leq 1} \frac{(\max_{|z|\leq \frac{2K+1}{2}} |A_\varepsilon(z)|)^2}{|x-y|^{N+2s-2}} \, dx \right) dy \right]$$

$$\leq 2[u_\varepsilon - u]_s^2 + C \int_{B_K} |(u_\varepsilon - u)(y)|^2 \, dy \to 0 \quad \text{as } \varepsilon \to 0.$$

$\square$

Now we recall the following fractional diamagnetic inequality:

Lemma 17.2.3 ([157]) *For any $u \in \mathcal{D}_A^{s,2}(\mathbb{R}^N, \mathbb{C})$, we have that $|u| \in \mathcal{D}^{s,2}(\mathbb{R}^N, \mathbb{R})$ and*

$$[|u|]_s \leq [u]_A. \tag{17.2.3}$$

We also have the following pointwise diamagnetic inequality

$$||u(x)| - |u(y)|| \leq |u(x) - u(y)e^{\iota A(\frac{x+y}{2})\cdot(x-y)}| \quad a.e. \ x, y \in \mathbb{R}^N.$$

Proof The proof is very simple. Indeed, since $|e^{\iota t}| = 1$ for all $t \in \mathbb{R}$ and $\Re z \leq |z|$ for all $z \in \mathbb{C}$, we have for a.e. $x, y \in \mathbb{R}^N$

$$|u(x) - u(y)e^{\iota A(\frac{x+y}{2})\cdot(x-y)}|^2 = |u(x)|^2 + |u(y)|^2 - 2\Re(u(x)\overline{u(y)}e^{-\iota A(\frac{x+y}{2})\cdot(x-y)})$$

$$\geq |u(x)|^2 + |u(y)|^2 - 2|u(x)||u(y)| = ||u(x)| - |u(y)||^2$$

which immediately gives desired inequality.   □

Then, using Theorem 1.1.8, Lemmas 17.2.3 and 6.2.3, we get

Lemma 17.2.4 *The space H_ε^s is continuously embedded in $L^r(\mathbb{R}^N, \mathbb{C})$ for $r \in [2, 2_s^*]$, and compactly embedded in $L_{loc}^r(\mathbb{R}^N, \mathbb{C})$ for $r \in [1, 2_s^*)$. Moreover, if $V_\infty = \infty$, then, for any bounded sequence (u_n) in H_ε^s, we have that, up to a subsequence, $(|u_n|)$ is strongly convergent in $L^r(\mathbb{R}^N, \mathbb{R})$ for $r \in [2, 2_s^*)$.*

For compact supported functions in $H^s(\mathbb{R}^N, \mathbb{R})$, we prove the following result.

Lemma 17.2.5 *If $u \in H^s(\mathbb{R}^N, \mathbb{R})$ and u has compact support, then $w = e^{\iota A(0) \cdot x} u \in H_\varepsilon^s$.*

Proof Assume that $\text{supp}(u) \subset B_R$. Since V is continuous, it is clear that

$$\int_{\mathbb{R}^N} V(\varepsilon x) |w|^2 \, dx = \int_{B_R} V(\varepsilon x) |u|^2 \, dx \leq C \|u\|_{L^2(\mathbb{R}^N)}^2 < \infty.$$

Therefore, it is enough to show that $[w]_{A_\varepsilon} < \infty$.

Recalling that A is continuous, $|e^{\iota t} - 1|^2 \leq 4$ and $|e^{\iota t} - 1|^2 \leq t^2$ for all $t \in \mathbb{R}$, we have

$$[w]_{A_\varepsilon}^2 = \iint_{\mathbb{R}^{2N}} \frac{|e^{\iota A(0) \cdot x} u(x) - e^{\iota A(0) \cdot y} e^{\iota A_\varepsilon(\frac{x+y}{2}) \cdot (x-y)} u(y)|^2}{|x-y|^{N+2s}} \, dx dy$$

$$\leq 2[u]_s^2 + 2 \iint_{\mathbb{R}^{2N}} \frac{u^2(y) |e^{\iota[A_\varepsilon(\frac{x+y}{2}) - A(0)] \cdot (x-y)} - 1|^2}{|x-y|^{N+2s}} \, dx dy$$

$$\leq 2[u]_s^2 + 2 \int_{B_R} u^2(y) \left[\int_{|x-y| \geq 1} \frac{4}{|x-y|^{N+2s}} \, dx + \int_{|x-y| < 1} \frac{|A_\varepsilon(\frac{x+y}{2}) - A(0)|^2}{|x-y|^{N+2s-2}} \, dx \right] dy$$

$$\leq 2[u]_s^2 + 2 \int_{B_R} u^2(y) \left[\int_{|x-y| \geq 1} \frac{4}{|x-y|^{N+2s}} \, dx + \int_{|x-y| < 1} \frac{(\max_{|z| \leq \frac{2R+1}{2}} [|A_\varepsilon(z)| + |A(0)|])^2}{|x-y|^{N+2s-2}} \, dx \right] dy$$

$$\leq 2[u]_s^2 + C \int_{B_R} u^2(y) \, dy \left[\int_1^\infty \frac{1}{\rho^{2s+1}} \, d\rho + \int_0^1 \frac{1}{\rho^{2s-1}} \, d\rho \right] < \infty,$$

because $u \in H^s(\mathbb{R}^N, \mathbb{R})$ and $s \in (0, 1)$.   □

Arguing as in the proof of Lemma 6.2.4-(B) and taking into account Lemma 17.2.4, we deduce the following result:

Lemma 17.2.6 *Let $\tau \in [2, 2_s^*)$ and $(u_n) \subset H_\varepsilon^s$ be a bounded sequence. Then one can extract a subsequence $(u_{n_j}) \subset H_\varepsilon^s$ such that for any $\sigma > 0$ there exists $r_{\sigma, \tau} > 0$ such that*

$$\limsup_{j \to \infty} \int_{B_j \setminus B_r} |u_{n_j}|^\tau \, dx \leq \sigma \tag{17.2.4}$$

for all $r \geq r_\sigma$.

We conclude this section with some properties on the nonlinearity that will be useful in the proofs of our results.

Lemma 17.2.7 *The nonlinearity f satisfies the following properties:*

(*i*) *for every $\xi > 0$ there exists $C_\xi > 0$ such that*

$$\frac{\theta}{2} F(t^2) \leq f(t^2)t^2 \leq \xi t^2 + C_\xi |t|^q \quad \text{for all } t \in \mathbb{R};$$

(*ii*) *there exist $C_1, C_2 > 0$ such that, $F(t^2) \geq C_1 |t|^\theta - C_2 \quad$ for all $t \in \mathbb{R}$;*

(*iii*) *if $u_{n_j} \rightharpoonup u$ in H_ε^s and $\hat{u}_j$ is defined as in (17.2.2), then*

$$\int_{\mathbb{R}^N} F(|u_{n_j}|^2) - F(|u_{n_j} - \hat{u}_j|^2) - F(|\hat{u}_j|^2) \, dx = o_j(1) \quad \text{as } j \to \infty;$$

(*iv*) *if $(u_n) \subset H_\varepsilon^s$ is bounded, (u_{n_j}) a subsequence as in Lemma 17.2.6 such that $u_{n_j} \rightharpoonup u$ in H_ε^s and $\hat{u}_j$ is defined as in (17.2.2), then*

$$\int_{\mathbb{R}^N} [f(|u_{n_j}|^2)u_{n_j} - f(|u_{n_j} - \hat{u}_j|^2)(u_{n_j} - \hat{u}_j) - f(|\hat{u}_j|^2)\hat{u}_j]\phi dx \to 0 \quad \text{as } j \to \infty$$

uniformly with respect to $\phi \in H_\varepsilon^s$ with $\|\phi\|_\varepsilon \leq 1$.

Proof Properties (i) and (ii) are easy consequences of $(f1)$–$(f4)$.

Let us prove (iii). Recalling that $\hat{u}_j = \varphi_j u$ with $\varphi_j \in [0, 1]$, it follows from (i) and the Young inequality that

$$|F(|u_{n_j}|^2) - F(|u_{n_j} - \hat{u}_j|^2)| \leq 2 \int_0^1 |f(|u_{n_j} - t\hat{u}_j|^2)||u_j - t\hat{u}_j||\hat{u}_j| \, dt$$

$$\leq C \left[(|u_{n_j}| + |u|)|u| + (|u_{n_j}| + |u|)^{q-1}|u| \right]$$

$$\leq \xi (|u_{n_j}|^2 + |u_{n_j}|^q) + C(|u|^2 + |u|^q)$$

for any $\xi > 0$. Then

$$|F(|u_{n_j}|^2) - F(|u_{n_j} - \hat{u}_j|^2) - F(|\hat{u}_j|^2)| \leq \xi (|u_{n_j}|^2 + |u_{n_j}|^q) + C(|u|^2 + |u|^q).$$

Now, let

$$G_j^\xi = \max\left\{|F(|u_{n_j}|^2) - F(|u_{n_j} - \hat{u}_j|^2) - F(|\hat{u}_j|^2)| - \xi(|u_{n_j}|^2 + |u_{n_j}|^q), 0\right\}.$$

Note that $G_j^\xi \to 0$ as $j \to \infty$ a.e. in $\mathbb{R}^N$ and $0 \le G_j^\xi \le C(|u|^2 + |u|^q) \in L^1(\mathbb{R}^N, \mathbb{R})$. Then, applying the dominated convergence theorem, we deduce that

$$\int_{\mathbb{R}^N} G_j^\xi \, dx \to 0 \quad \text{as } j \to \infty.$$

On the other hand, by the definition of G_j^ξ,

$$|F(|u_{n_j}|^2) - F(|u_{n_j} - \hat{u}_j|^2) - F(|\hat{u}_j|^2)| \le \xi(|u_j|^2 + |u_j|^{2_s^*}) + G_j^\xi.$$

Hence, since (u_{n_j}) is bounded in H_ε^s, we have

$$\limsup_{j \to \infty} \int_{\mathbb{R}^N} |F(|u_{n_j}|^2) - F(|u_{n_j} - \hat{u}_j|^2) - F(|\hat{u}_j|^2)| \, dx \le C\xi$$

and, from the arbitrariness of ξ, we get the thesis. To prove (iv), let us consider $\phi \in H_\varepsilon^s$ such that $\|\phi\|_\varepsilon \le 1$ and $\sigma > 0$. Note that, for any $r \ge \max\{r_{\sigma,2}, r_{\sigma,q}\}$, where $r_{\sigma,\tau}$ has been introduced in Lemma 17.2.4,

$$\left|\int_{\mathbb{R}^N}[f(|u_{n_j}|^2)u_{n_j} - f(|u_{n_j} - \hat{u}_j|^2)(u_{n_j} - \hat{u}_j) - f(|\hat{u}_j|^2)\hat{u}_j]\phi \, dx\right|$$

$$\le \int_{B_r} |f(|u_{n_j}|^2)u_{n_j} - f(|v_j|^2)v_j - f(|\hat{u}_j|^2)\hat{u}_j||\phi| \, dx$$

$$+ \int_{B_r^c} |f(|u_{n_j}|^2)u_{n_j} - f(|v_j|^2)v_j - f(|\hat{u}_j|^2)\hat{u}_j||\phi| \, dx$$

$$= D_j + E_j.$$

Taking into account Lemmas 17.2.4 and 17.2.1, we can apply the dominated convergence theorem to obtain that $D_j \to 0$ uniformly in $\phi \in H_\varepsilon^s$ with $\|\phi\|_\varepsilon \le 1$.

On the other hand, by (i) in Lemma 17.2.7 and $\hat{u}_j = 0$ in B_j^c for any $j \ge 1$, we deduce that, for j large enough,

$$E_j = \int_{B_j \setminus B_r} |f(|u_{n_j}|^2)u_{n_j} - f(|u_{n_j} - \hat{u}_j|^2)(u_{n_j} - \hat{u}_j) - f(|\hat{u}_j|^2)\hat{u}_j||\phi| \, dx$$

$$\le C \int_{B_j \setminus B_r} (|u_{n_j}| + |\hat{u}_j| + |u_{n_j}|^{q-1} + |\hat{u}_j|^{q-1})|\phi| \, dx.$$

Since $\|\phi\|_\varepsilon \leq 1$, Hölder's inequality and Lemma 17.2.4 imply that

$$\int_{B_j \setminus B_r} (|u_{n_j}| + |u_{n_j}|^{q-1})|\phi| \, dx \leq C \left[\left(\int_{B_j \setminus B_r} |u_{n_j}|^2 \, dx \right)^{\frac{1}{2}} + \left(\int_{B_j \setminus B_r} |u_{n_j}|^q \, dx \right)^{\frac{q-1}{q}} \right]$$

which combined with Lemma 17.2.4 yields

$$\limsup_{j \to \infty} \int_{B_j \setminus B_r} (|u_{n_j}| + |u_{n_j}|^{q-1})|\phi| \, dx \leq C(\sigma^{\frac{1}{2}} + \sigma^{\frac{q-1}{q}}).$$

Moreover, Lemmas 17.2.4 and 17.2.1 ensure that $\hat{u}_j \to u$ in $L^2(\mathbb{R}^N, \mathbb{C}) \cap L^q(\mathbb{R}^N, \mathbb{C})$ as $j \to \infty$. This and the Hölder inequality give

$$\limsup_{j \to \infty} \int_{B_j \setminus B_r} (|\hat{u}_j| + |\hat{u}_j|^{q-1})|\phi| \, dx = \int_{B_r^c} (|u| + |u|^{q-1})|\phi| \, dx \leq C(\sigma^{\frac{1}{2}} + \sigma^{\frac{q-1}{q}})$$

for r large enough. The arbitrariness of $\sigma > 0$ implies that $E_j \to 0$ as $j \to \infty$ uniformly with respect to ϕ, $\|\phi\|_\varepsilon \leq 1$. The proof of Lemma 17.2.7 is thus complete. $\qquad \square$

17.2.2 A First Existence Result

The goal of this section is to prove Theorem 17.1.1. We want to find solutions of (17.2.1) in the sense of the following definition.

Definition 17.2.8 We say that $u \in H_\varepsilon^s$ is a weak solution to (17.2.1) if for any $v \in H_\varepsilon^s$

$$\langle u, v \rangle_\varepsilon = \Re \left(\int_{\mathbb{R}^N} f(|u|^2) u \bar{v} \, dx \right).$$

Such solutions can be found as critical points of the functional $J_\varepsilon : H_\varepsilon^s \to \mathbb{R}$ defined as

$$J_\varepsilon(u) = \frac{1}{2} \|u\|_\varepsilon^2 - \frac{1}{2} \int_{\mathbb{R}^N} F(|u|^2) \, dx.$$

Using Lemmas 17.2.4 and 17.2.7, we see that J_ε is well defined and that $J_\varepsilon \in C^1(H_\varepsilon^s, \mathbb{R})$.

Let us show that for any $\varepsilon > 0$ the functional J_ε satisfies the geometrical assumptions of the mountain pass theorem [29].

Lemma 17.2.9 *The functional J_ε satisfies the following conditions:*

(i) there exist $\alpha, \rho > 0$ such that $J_\varepsilon(u) \geq \alpha$ with $\|u\|_\varepsilon = \rho$;

(ii) *there exists $e \in H_\varepsilon^s$ such that $\|e\|_\varepsilon > \rho$ and $J_\varepsilon(e) < 0$.*

Proof In view of item (i) in Lemma 17.2.7, Lemma 17.2.4, and the condition (V), we have that for $\xi \in (0, V_0)$

$$J_\varepsilon(u) \geq \frac{1}{2}[u]_{A_\varepsilon}^2 + \frac{1}{2}\left(1 - \frac{\xi}{V_0}\right)\int_{\mathbb{R}^N} V(\varepsilon x)|u|^2 \, dx - \frac{C_\xi}{2}\int_{\mathbb{R}^N} |u|^q \, dx \geq C_1\|u\|_\varepsilon^2 - C_2\|u\|_\varepsilon^q,$$

from which we deduce the first statement.

To prove (ii), we observe that by item (ii) in Lemma 17.2.7 and taking $\varphi \in C_c^\infty(\mathbb{R}^N, \mathbb{C})$ such that $\varphi \not\equiv 0$ we have

$$J_\varepsilon(t\varphi) \leq \frac{t^2}{2}\|\varphi\|_\varepsilon^2 - t^\theta C_1 \|\varphi\|_{L^\theta(\mathbb{R}^N)}^\theta + C_2|\mathrm{supp}(\varphi)| \to -\infty \quad \text{as } t \to \infty$$

since $\theta > 2$. $\square$

Using a variant of mountain pass theorem without the Palais–Smale condition (see Remark 2.2.10), we deduce that there exists a sequence $(u_n) \subset H_\varepsilon^s$ such that

$$J_\varepsilon(u_n) \to c_\varepsilon \quad \text{and} \quad J_\varepsilon'(u_n) \to 0, \tag{17.2.5}$$

where c_ε is the minimax level of the mountain pass theorem, namely

$$c_\varepsilon = \inf_{\gamma \in \Gamma} \max_{t \in [0,1]} J_\varepsilon(\gamma(t))$$

with $\Gamma = \{\gamma \in H([0, 1], H_\varepsilon^s) : \gamma(0) = 0, J_\varepsilon(\gamma(1)) < 0\}$.

The sequence (u_n) is bounded in H_ε^s. Indeed, by (17.2.5) and $(f4)$,

$$C(1 + \|u_n\|_\varepsilon) \geq J_\varepsilon(u_n) - \frac{1}{\theta}\langle J_\varepsilon'(u_n), u_n \rangle$$

$$= \left(\frac{1}{2} - \frac{1}{\theta}\right)\|u_n\|_\varepsilon^2 + \int_{\mathbb{R}^N}\left[\frac{1}{\theta}f(|u_n|^2)|u_n|^2 - \frac{1}{2}F(|u_n|^2)\right]dx$$

$$\geq \left(\frac{1}{2} - \frac{1}{\theta}\right)\|u_n\|_\varepsilon^2.$$

Moreover, it is standard to verify the characterization

$$c_\varepsilon = \inf_{u \in H_\varepsilon^s \setminus \{0\}} \sup_{t \geq 0} J_\varepsilon(tu) = \inf_{u \in \mathcal{N}_\varepsilon} J_\varepsilon(u),$$

where

$$\mathcal{N}_\varepsilon = \{u \in H_\varepsilon^s \setminus \{0\} : \langle J_\varepsilon'(u), u \rangle = 0\}$$

is the usual Nehari manifold associated with J_ε.

Lemma 17.2.10 *We have:*

(i) *there exists $K > 0$ such that, for all $u \in \mathcal{N}_\varepsilon$, $\|u\|_\varepsilon \geq K$;*
(ii) *for every $u \in H_\varepsilon^s \setminus \{0\}$ there exists a unique $t_0 = t_0(u)$ such that $J_\varepsilon(t_0 u) = \max_{t \geq 0} J_\varepsilon(tu)$ and then $t_0 u \in \mathcal{N}_\varepsilon$.*

Proof Property (i) follows easily from item (i) in Lemmas 17.2.7 and 17.2.4, since, if $u \in \mathcal{N}_\varepsilon$, then, for all $\xi > 0$

$$\|u\|_\varepsilon^2 = \int_{\mathbb{R}^N} f(|u|^2)|u|^2 \, dx \leq \xi \|u\|_\varepsilon^2 + C\|u\|_\varepsilon^q.$$

To prove (ii), fix $u \in H_\varepsilon^s \setminus \{0\}$ and consider the smooth function $h(t) = J_\varepsilon(tu)$ for $t \geq 0$. Arguing as in Lemma 17.2.9, we get that

$$J_\varepsilon(tu) \geq C_1 t^2 \|u\|_\varepsilon^2 - C_2 t^q \|u\|_\varepsilon^q$$

and

$$J_\varepsilon(tu) \leq \frac{t^2}{2} \|u\|_\varepsilon^2 - t^\theta C_1 \int_\Omega |u|^\theta \, dx + C_2|\Omega| \to -\infty \quad \text{as } t \to \infty,$$

where Ω is a compact subset of $\operatorname{supp}(u)$ with $|\Omega| > 0$. Then there exists a maximum point of h. To prove the uniqueness, let $0 < t_1 < t_2$ be two maximum points of h. Then, since $h'(t_1) = h'(t_2) = 0$,

$$\|u\|_\varepsilon^2 = \int_{\mathbb{R}^N} f(|t_1 u|^2)|u|^2 \, dx = \int_{\mathbb{R}^N} f(|t_2 u|^2)|u|^2 \, dx,$$

which contradicts the strict monotonicity of f assumed in $(f5)$. $\qquad\square$

To prove the compactness of the Palais–Smale sequences at the level d, for suitable $d \in \mathbb{R}$, we will use the following preliminary result.

Lemma 17.2.11 *Let $d \in \mathbb{R}$ and $(u_n) \subset H_\varepsilon^s$ be a Palais–Smale sequence for J_ε at the level d and such that $u_n \rightharpoonup 0$ in H_ε^s. Then, one of the following alternatives occurs:*

(a) $u_n \to 0$ in H_ε^s;

(b) there are a sequence $(y_n) \subset \mathbb{R}^N$ and constants $R, \beta > 0$ such that

$$\liminf_{n\to\infty} \int_{B_R(y_n)} |u_n|^2 \, dx \geq \beta > 0.$$

Proof Assume that (b) does not hold. Then, for every $R > 0$,

$$\lim_{n\to\infty} \sup_{y\in\mathbb{R}^N} \int_{B_R(y)} |u_n|^2 \, dx = 0.$$

Since (u_n) is bounded in H_ε^s, it follows from (17.2.3) that $(|u_n|)$ is bounded in $H^s(\mathbb{R}^N, \mathbb{R})$, and applying Lemma 1.4.4 we deduce that $\|u_n\|_{L^q(\mathbb{R}^N)} \to 0$.

Then, using the fact that (u_n) is a Palais–Smale sequence for J_ε at the level d, and (i) in Lemma 17.2.7, we see that, for every $\xi > 0$,

$$0 \leq \|u_n\|_\varepsilon^2 = \int_{\mathbb{R}^N} f(|u_n|^2)|u_n|^2 \, dx + o_n(1)$$

$$\leq \xi \|u_n\|_{L^2(\mathbb{R}^N)}^2 + C_\xi \|u_n\|_{L^q(\mathbb{R}^N)}^q + o_n(1)$$

$$\leq \frac{\xi}{V_0} \|u_n\|_\varepsilon^2 + C_\xi \|u_n\|_{L^q(\mathbb{R}^N)}^q + o_n(1).$$

Taking ξ small enough, we get (a). $\square$

To continue our argument, we need to consider the following family of scalar limit problems associated with

$$\begin{cases} (-\Delta)^s u + \mu u = f(u^2)u \text{ in } \mathbb{R}^N, \\ u \in H^s(\mathbb{R}^N, \mathbb{R}), \end{cases} \tag{P_μ}$$

with $\mu > 0$, whose corresponding C^1 functional $I_\mu : H^s(\mathbb{R}^N, \mathbb{R}) \to \mathbb{R}$ is given by

$$I_\mu(u) = \frac{1}{2}\|u\|_\mu^2 - \frac{1}{2}\int_{\mathbb{R}^N} F(u^2) \, dx,$$

where

$$\|u\|_\mu = \left([u]_s^2 + V_0\|u\|_{L^2(\mathbb{R}^N)}^2 \right)^{\frac{1}{2}}.$$

Even in this case we can introduce the Nehari manifold

$$\mathcal{M}_\mu = \{u \in H^s(\mathbb{R}^N, \mathbb{R}) : \langle I_\mu'(u), u \rangle = 0\},$$

and then

$$m_\mu = \inf_{\gamma \in \Xi_\mu} \max_{t \in [0,1]} I_\mu(\gamma(t)) = \inf_{u \in H^s(\mathbb{R}^N,\mathbb{R})\setminus\{0\}} \sup_{t \geq 0} I_\mu(tu) = \inf_{u \in \mathcal{M}_\mu} I_\mu(u)$$

with $\Xi_\mu = \{\gamma \in C([0,1], H^s(\mathbb{R}^N, \mathbb{R})) : \gamma(0) = 0, I_\mu(\gamma(1)) < 0\}$.

We will call ground state for (P_μ) each minimum of I_μ in $\mathcal{M}_\mu$, wich is also a solution of (P_μ).

Remark 17.2.12 Arguing as in Lemma 17.2.10, we can prove that for every fixed $\mu > 0$ there exists $K > 0$ such that, for all $u \in \mathcal{M}_\mu$, $\|u\|_\varepsilon \geq K$ and that for any $u \in H^s(\mathbb{R}^N, \mathbb{R}) \setminus \{0\}$ there exists a unique $t_0 = t_0(u)$ such that $I_\mu(t_0 u) = \max_{t \geq 0} I_\mu(tu)$, and then $t_0 u \in \mathcal{M}_\mu$.

Following the same arguments as in Theorem 6.3.11, we obtain the following result:

Lemma 17.2.13 *Let $(w_n) \subset \mathcal{M}_\mu$ be a sequence satisfying $I_\mu(w_n) \to m_\mu$. Then (w_n) is bounded in $H^s(\mathbb{R}^N, \mathbb{R})$ and, up to a subsequence, $w_n \rightharpoonup w$ in $H^s(\mathbb{R}^N, \mathbb{R})$. If $w \neq 0$, then $w_n \to w \in \mathcal{M}_\mu$ in $H^s(\mathbb{R}^N, \mathbb{R})$ and w is a ground state for (P_μ). If $w = 0$, then there exist $(\tilde{y}_n) \subset \mathbb{R}^N$ and $\tilde{w} \in H^s(\mathbb{R}^N, \mathbb{R}) \setminus \{0\}$ such that, up to a subsequence, $w_n(\cdot + \tilde{y}_n) \to \tilde{w} \in \mathcal{M}_\mu$ in $H^s(\mathbb{R}^N, \mathbb{R})$ and $\tilde{w}$ is a ground state for (P_μ).*

Remark 17.2.14 In view of Lemma 17.2.13 and arguing as in the proof of Theorem 3.2.11, we can see that (P_μ) admits a positive ground state $w \in H^s(\mathbb{R}^N, \mathbb{R})$ that is Hölder continuous and has a power-type decay at infinity, more precisely

$$0 < w(x) \leq \frac{C}{|x|^{N+2s}} \quad \text{if } |x| > 1.$$

Now we prove a fundamental property on the Palais–Smale sequences for J_ε in the noncoercive case ($V_\infty < \infty$).

Lemma 17.2.15 *Let $d \in \mathbb{R}$. Assume that $V_\infty < \infty$ and let (v_n) be a Palais–Smale sequence for J_ε at the level d with $v_n \rightharpoonup 0$ in H_ε^s. If $v_n \not\to 0$ in H_ε^s, then $d \geq m_{V_\infty}$.*

Proof Let $(t_n) \subset (0, \infty)$ be such that $(t_n|v_n|) \subset \mathcal{M}_{V_\infty}$.

First we prove that $\limsup_n t_n \leq 1$. Assume, by contradiction, that there exist $\delta > 0$ and a subsequence, still denoted by (t_n), such that

$$t_n \geq 1 + \delta \quad \forall n \in \mathbb{N}. \tag{17.2.6}$$

Since (v_n) is a Palais–Smale sequence for J_ε at the level d,

$$[v_n]_{A_\varepsilon}^2 + \int_{\mathbb{R}^N} V(\varepsilon x)|v_n|^2\, dx = \int_{\mathbb{R}^N} f(|v_n|^2)|v_n|^2\, dx + o_n(1). \qquad (17.2.7)$$

On the other hand, $t_n|v_n| \in \mathcal{M}_{V_\infty}$. Thus we get

$$[|v_n|]_s^2 + V_\infty \|v_n\|_{L^2(\mathbb{R}^N)}^2 = \int_{\mathbb{R}^N} f(t_n^2|v_n|^2)|v_n|^2\, dx. \qquad (17.2.8)$$

Putting together (17.2.7), (17.2.8) and using (17.2.3), we conclude that

$$\int_{\mathbb{R}^N} \left[f(t_n^2|v_n|^2) - f(|v_n|^2) \right]|v_n|^2\, dx \le \int_{\mathbb{R}^N} (V_\infty - V(\varepsilon x))\,|v_n|^2\, dx + o_n(1).$$
$$\qquad (17.2.9)$$

Now, by condition (V), for every $\zeta > 0$ there exists $R = R(\zeta) > 0$ such that

$$V_\infty - V(\varepsilon x) \le \zeta \quad \text{for any } |x| \ge R. \qquad (17.2.10)$$

Combining (17.2.10) with the fact that, by Lemma 17.2.4, $v_n \to 0$ in $L^2(B_R, \mathbb{C})$, so that $|v_n| \to 0$ in $L^2(B_R)$, and with the boundedness of (v_n) in H_ε^s, we get

$$\int_{\mathbb{R}^N} (V_\infty - V(\varepsilon x))\,|v_n|^2\, dx = \int_{B_R} (V_\infty - V(\varepsilon x))\,|v_n|^2\, dx + \int_{B_R^c} (V_\infty - V(\varepsilon x))\,|v_n|^2\, dx$$

$$\le V_\infty \int_{B_R} |v_n|^2 dx + \zeta \int_{B_R^c} |v_n|^2\, dx$$

$$\le o_n(1) + \frac{\zeta}{V_0}\|v_n\|_\varepsilon^2 \le o_n(1) + \zeta C.$$

Then, in view of (17.2.9), we deduce that

$$\int_{\mathbb{R}^N} \left[f(t_n^2|v_n|^2) - f(|v_n|^2) \right]|v_n|^2\, dx \le \zeta C + o_n(1). \qquad (17.2.11)$$

Since $v_n \nrightarrow 0$, we can apply Lemma 17.2.11 to deduce the existence of a sequence $(y_n) \subset \mathbb{R}^N$, and the existence of two positive numbers $\bar{R}, \beta$, such that

$$\int_{B_{\bar{R}}(y_n)} |v_n|^2\, dx \ge \beta > 0. \qquad (17.2.12)$$

Now, consider the functions $w_n = |v_n|(\cdot + y_n)$. Thanks to condition (V), (17.2.3), and the boundedness of (v_n) in H_ε^s, we have that

$$\|w_n\|_{V_0}^2 = \||v_n|\|_{V_0}^2 \le \|v_n\|_\varepsilon^2 \le C.$$

Therefore, $w_n \rightharpoonup w$ in $H^s(\mathbb{R}^N, \mathbb{R})$ and $w_n \to w$ in $L_{\text{loc}}^r(\mathbb{R}^N, \mathbb{R})$ for all $r \in [2, 2_s^*)$. By (17.2.12)

$$\int_{B_{\bar{R}}} w^2 \, dx = \lim_{n \to \infty} \int_{B_{\bar{R}}} w_n^2 \, dx \ge \beta,$$

and so there exists a set $\Omega \subset \mathbb{R}^N$ of positive measure such that $w \neq 0$ in Ω. It follows from (17.2.6) and (17.2.11) that

$$\int_\Omega \left(f((1+\delta)^2 w_n^2) - f(w_n^2) \right) w_n^2 \, dx \le \zeta C + o_n(1).$$

Applying Fatou's lemma and using $(f5)$ we obtain

$$0 < \int_\Omega \left(f((1+\delta)^2 w^2) - f(w^2) \right) w^2 \, dx \le \zeta C,$$

so by the arbitrariness of $\zeta > 0$ we reached a contradiction.

Now, two cases can occur.

Case 1 $\limsup_{n \to \infty} t_n = 1$.

In this case there exists a subsequence, still denoted by (t_n), such that $t_n \to 1$. Taking into account that (v_n) is a Palais–Smale sequence for J_ε at the level d, m_{V_∞} is the minimax level of I_{V_∞}, and (17.2.3), we have

$$
\begin{aligned}
d + o_n(1) = J_\varepsilon(v_n) \\
\ge J_\varepsilon(v_n) - I_{V_\infty}(t_n|v_n|) + m_{V_\infty} \\
\ge \frac{1 - t_n^2}{2} [|v_n|]^2 + \frac{1}{2} \int_{\mathbb{R}^N} \left(V(\varepsilon x) - t_n^2 V_\infty \right) |v_n|^2 \, dx \\
+ \frac{1}{2} \int_{\mathbb{R}^N} \left[F(t_n^2 |v_n|^2) - F(|v_n|^2) \right] dx + m_{V_\infty}.
\end{aligned}
\tag{17.2.13}
$$

Since $(|v_n|)$ is bounded in $H^s(\mathbb{R}^N, \mathbb{R})$ and $t_n \to 1$, we see that

$$\frac{(1 - t_n^2)}{2} [|v_n|]_s^2 = o_n(1).
\tag{17.2.14}$$

Now, using (V), we deduce that for every $\zeta > 0$ there exists $R = R(\zeta) > 0$ such that, for every $|x| > R$,

$$V(\varepsilon x) - t_n^2 V_\infty = (V(\varepsilon x) - V_\infty) + (1 - t_n^2)V_\infty \geq -\zeta + (1 - t_n^2)V_\infty.$$

Thus, since (v_n) is bounded in H_ε^s, $|v_n| \to 0$ in $L^p(B_R)$, and $t_n \to 1$, we get

$$\int_{\mathbb{R}^N} \left(V(\varepsilon x) - t_n^2 V_\infty \right) |v_n|^2 \, dx = \int_{B_R} \left(V(\varepsilon x) - t_n^2 V_\infty \right) |v_n|^2 \, dx$$

$$+ \int_{B_R^c} \left(V(\varepsilon x) - t_n^2 V_\infty \right) |v_n|^2 \, dx$$

$$\geq (V_0 - t_n^2 V_\infty) \int_{B_R} |v_n|^2 \, dx - \zeta \int_{B_R^c} |v_n|^2 \, dx$$

$$+ V_\infty (1 - t_n^2) \int_{B_R^c} |v_n|^2 \, dx$$

$$\geq o_n(1) - \frac{C}{V_0} \zeta.$$

(17.2.15)

Finally, using the mean value theorem, item (i) in Lemma 17.2.7, the fact that $t_n \to 1$, and the boundedness of $(|v_n|)$, we get

$$\left| \int_{\mathbb{R}^N} \left[F(t_n^2 |v_n|^2) - F(|v_n|^2) \right] dx \right| \leq \int_{\mathbb{R}^N} |f(\theta_n |v_n|^2)| |t_n^2 - 1| |v_n|^2 \, dx$$

$$\leq (C_1 \|v_n\|_{L^2(\mathbb{R}^N)}^2 + C_2 \|v_n\|_{L^q(\mathbb{R}^N)}^q)|t_n^2 - 1| = o_n(1).$$

(17.2.16)

Combining (17.2.13), (17.2.14), (17.2.15) and (17.2.16) we can infer that

$$d + o_n(1) \geq o_n(1) - \zeta C + m_{V_\infty},$$

and taking the limit as $n \to \infty$ we get $d \geq m_{V_\infty}$.

Case 2 $\limsup_{n \to \infty} t_n = t_0 < 1$.

In this case there exists a subsequence, still denoted by (t_n), such that $t_n \to t_0$ and $t_n < 1$ for all $n \in \mathbb{N}$. Since (v_n) is a bounded Palais–Smale sequence for J_ε at the level d, we have

$$d + o_n(1) = J_\varepsilon(v_n) - \frac{1}{2}\langle J_\varepsilon'(v_n), v_n \rangle = \frac{1}{2} \int_{\mathbb{R}^N} [f(|v_n|^2)|v_n|^2 - F(|v_n|^2)]dx.$$

(17.2.17)

Observe that, by $(f5)$, the function $t \mapsto f(t)t - F(t)$ is increasing for $t > 0$.

Hence, since $t_n|v_n| \in \mathcal{M}_{V_\infty}$ and $t_n < 1$, from (17.2.17), we obtain

$$
\begin{aligned}
m_{V_\infty} &\leq I_{V_\infty}(t_n|v_n|) \\
&= I_{V_\infty}(t_n|v_n|) - t_n \frac{1}{2} \langle I'_{V_\infty}(t_n|v_n|), |v_n| \rangle \\
&= \frac{1}{2} \int_{\mathbb{R}^N} \left(f(t_n^2|v_n|^2)t_n^2|v_n|^2 - F(t_n^2|v_n|^2) \right) dx \\
&\leq \frac{1}{2} \int_{\mathbb{R}^N} \left(f(|v_n|^2)|v_n|^2 - F(|v_n|^2) \right) dx \\
&= d + o_n(1).
\end{aligned}
$$

Passing to the limit as $n \to \infty$, we get $d \geq m_{V_\infty}$.   □

Thus we are ready to give conditions on the levels c which ensure that J_ε satisfies the $(PS)_c$ condition.

Proposition 17.2.16 *The functional J_ε satisfies the $(PS)_c$ condition at any level $c < m_{V_\infty}$ if $V_\infty < \infty$, and at any level $c \in \mathbb{R}$ if $V_\infty = \infty$.*

Proof Let (u_n) be a Palais–Smale sequence for J_ε at the level c. Then (u_n) is bounded in H_ε^s and, up to a subsequence, $u_n \rightharpoonup u$ in H_ε^s and $u_n \to u$ in $L_{loc}^q(\mathbb{R}^N, \mathbb{C})$ for any $q \in [1, 2_s^*)$. Using $(f1)$–$(f3)$, it is easy to deduce that $J'_\varepsilon(u) = 0$, which combined with $(f4)$ gives

$$
J_\varepsilon(u) = J_\varepsilon(u) - \frac{1}{2}\langle J'_\varepsilon(u), u \rangle = \frac{1}{2} \int_{\mathbb{R}^N} [f(|u|^2)|u|^2 - F(|u|^2)] \, dx \geq 0. \qquad (17.2.18)
$$

In view of Lemma 17.2.6, we can find a subsequence $(u_{n_j}) \subset H_\varepsilon^s$ satisfying (17.2.4).

Now, let $v_j = u_{n_j} - \hat{u}_j$, where $\hat{u}_j$ is defined as in (17.2.2). We claim that

$$
J_\varepsilon(v_j) = c - J_\varepsilon(u) + o_j(1) \qquad (17.2.19)
$$

and

$$
J'_\varepsilon(v_j) = o_j(1). \qquad (17.2.20)
$$

To prove (17.2.19), note that

$$J_\varepsilon(v_j) - J_\varepsilon(u_{n_j}) + J_\varepsilon(\hat{u}_j) = [\|\hat{u}_j\|_\varepsilon^2 - \langle u_{n_j}, \hat{u}_j\rangle_\varepsilon] + \int_{\mathbb{R}^N} [F(|u_{n_j}|^2) - F(|v_j|^2) - F(|\hat{u}_j|^2)]\, dx$$

$$= A_j + B_j.$$

In view of the weak convergence of (u_{n_j}) to u in H_ε^s and Lemma 17.2.1, we can see that $A_j \to 0$ as $j \to \infty$. Moreover, by (iii) in Lemma 17.2.7, we have that $B_j \to 0$ as $j \to \infty$.

To show (17.2.20), we observe that

$$\left|\langle J_\varepsilon'(v_j) - J_\varepsilon'(u_{n_j}) + J_\varepsilon'(\hat{u}_j), \phi\rangle\right| = \left|\Re \int_{\mathbb{R}^N} [f(|u_{n_j}|^2)u_{n_j} - f(|v_j|^2)v_j - f(|\hat{u}_j|^2)\hat{u}_j]\bar{\phi}\, dx\right|$$

$$\leq \int_{\mathbb{R}^N} |f(|u_{n_j}|^2)u_{n_j} - f(|v_j|^2)v_j - f(|\hat{u}_j|^2)\hat{u}_j||\phi|\, dx$$

and applying (iv) in Lemma 17.2.7 we get that $\langle J_\varepsilon'(v_j) - J_\varepsilon'(u_{n_j}) + J_\varepsilon'(\hat{u}_j), \phi\rangle \to 0$ for any $\phi \in H_\varepsilon^s$ such that $\|\phi\|_\varepsilon \leq 1$. Then, since $J_\varepsilon'(u_{n_j}) \to 0$ and $J_\varepsilon'(\hat{u}_j) \to J_\varepsilon'(u) = 0$, we can infer that (17.2.20) is satisfied.

Let us assume that $V_\infty < \infty$ and $c < m_{V_\infty}$. By (17.2.19) and (17.2.18), we have that $c - J_\varepsilon(u) \leq c < m_{V_\infty}$. Since (v_j) is a Palais–Smale sequence for J_ε at the level $c - J_\varepsilon(u)$ and $v_j \rightharpoonup 0$ in H_ε^s, it follows from Lemma 17.2.15 that $v_j \to 0$ in H_ε^s. Hence, Lemma 17.2.1 implies that $u_{n_j} \to u$ in H_ε^s as $j \to \infty$.

If $V_\infty = \infty$, then, by Lemma 17.2.4, $v_j \to 0$ in $L^r(\mathbb{R}^N, \mathbb{C})$ for any $r \in [2, 2_s^*)$ and by (17.2.20) and item (i) in Lemma 17.2.7, we deduce that

$$\|v_j\|_\varepsilon^2 = \int_{\mathbb{R}^N} f(|v_j|^2)|v_j|^2\, dx + o_j(1) = o_j(1).$$

Hence, as before, $u_{n_j} \to u$ in H_ε^s as $j \to \infty$ which completes the proof. $\qquad\square$

Now we show that $\mathcal{N}_\varepsilon$ is a natural constraint, namely that the constrained critical points of the functional J_ε on $\mathcal{N}_\varepsilon$ are critical points of J_ε in H_ε^s.

Proposition 17.2.17 *The functional J_ε restricted to $\mathcal{N}_\varepsilon$ satisfies the $(PS)_c$ condition at any level $c < m_{V_\infty}$ if $V_\infty < \infty$, and at any level $c \in \mathbb{R}$ if $V_\infty = \infty$.*

Proof Let $(u_n) \subset \mathcal{N}_\varepsilon$ be a Palais–Smale sequence at the level c for J_ε restricted to $\mathcal{N}_\varepsilon$. Then, by Proposition 2.3.10, $J_\varepsilon(u_n) \to c$ as $n \to \infty$ and there exists $(\lambda_n) \subset \mathbb{R}$ such that

$$J_\varepsilon'(u_n) = \lambda_n T_\varepsilon'(u_n) + o_n(1), \tag{17.2.21}$$

where $T_\varepsilon : H_\varepsilon^s \to \mathbb{R}$ is defined by

$$T_\varepsilon(u) = \|u\|_\varepsilon^2 - \int_{\mathbb{R}^N} f(|u|^2)|u|^2 dx.$$

Thanks to $(f5)$,

$$\langle T_\varepsilon'(u_n), u_n \rangle = 2\|u_n\|_\varepsilon^2 - 2 \int_{\mathbb{R}^N} f(|u_n|^2)|u_n|^2 \, dx - 2 \int_{\mathbb{R}^N} f'(|u_n|^2)|u_n|^4 \, dx$$

$$= -2 \int_{\mathbb{R}^N} f'(|u_n|^2)|u_n|^4 \, dx \le -2C_\sigma \|u_n\|_{L^\sigma(\mathbb{R}^N)}^\sigma < 0.$$

Up to a subsequence, we may assume that $\langle T_\varepsilon'(u_n), u_n \rangle \to \ell \le 0$.
     If $\ell = 0$, then

$$o_n(1) = |\langle T_\varepsilon'(u_n), u_n \rangle| \ge C\|u_n\|_{L^\sigma(\mathbb{R}^N)}^\sigma,$$

so we obtain that $u_n \to 0$ in $L^\sigma(\mathbb{R}^N, \mathbb{C})$. Observe that, since $(u_n) \subset \mathcal{N}_\varepsilon$ and $J_\varepsilon(u_n) \to c$ as $n \to \infty$, the sequence (u_n) is bounded in H_ε^s. By interpolation, we also have $u_n \to 0$ in $L^q(\mathbb{R}^N, \mathbb{C})$. Hence, by (i) in Lemma 17.2.7, we get

$$\|u_n\|_\varepsilon^2 = \int_{\mathbb{R}^N} f(|u_n|^2)|u_n|^2 \, dx \le \frac{\xi}{V_0}\|u_n\|_\varepsilon^2 + C_\xi \|u_n\|_{L^q(\mathbb{R}^N)}^q = \frac{\xi}{V_0}\|u_n\|_\varepsilon^2 + o_n(1),$$

which implies that $u_n \to 0$ in H_ε^s. This is impossible in view of item (i) of Lemma 17.2.10. Therefore, $\ell < 0$ and by (17.2.21) we deduce that $\lambda_n = o_n(1)$. Moreover, by the assumptions on f,

$$|\langle T_\varepsilon'(u_n), \phi \rangle| \le 2\|u_n\|_\varepsilon\|\phi\|_\varepsilon + 2 \int_{\mathbb{R}^N} |f(|u_n|^2)||u_n||\phi| \, dx + 2 \int_{\mathbb{R}^N} |f'(|u_n|^2)||u_n|^3|\phi| \, dx$$

$$\le C\|u_n\|_\varepsilon(1 + \|u_n\|_\varepsilon^{q-2})\|\phi\|_\varepsilon \text{ for every } \phi \in H_\varepsilon^s.$$

Then, the boundedness of (u_n) implies the boundedness of $T_\varepsilon'(u_n)$ and so, by (17.2.21), we infer that $J_\varepsilon'(u_n) = o_n(1)$, that is, (u_n) is a Palais–Smale sequence for J_ε at the level c. Hence, it remains to apply Proposition 17.2.16. $\square$

Consequently, we have the following result.

Corollary 17.2.18 *The constrained critical points of the functional J_ε on $\mathcal{N}_\varepsilon$ are critical points of J_ε in H_ε^s.*

Now we are ready to give the proof of the main result of this section.

Proof of Theorem 17.1.1 By Lemma 17.2.9, we know that J_ε has a mountain pass geometry. Then, by a variant of mountain pass theorem without the Palais–Smale condition (see Remark 2.2.10), there exists a Palais–Smale sequence $(u_n) \subset H_\varepsilon^s$ for J_ε at the level c_ε.

If $V_\infty = \infty$, then by Lemma 17.2.4 and Proposition 17.2.16, we deduce that $J_\varepsilon(u) = c_\varepsilon$ and $J_\varepsilon'(u) = 0$, where $u \in H_\varepsilon^s$ is the weak limit of (u_n).

Now, suppose that $V_\infty < \infty$. In view of Proposition 17.2.16, it is enough to show that $c_\varepsilon < m_{V_\infty}$. Without loss of generality, we suppose that

$$V(0) = V_0 = \inf_{x \in \mathbb{R}^N} V(x).$$

Let $\mu \in (V_0, V_\infty)$. Clearly, $m_{V_0} < m_\mu < m_{V_\infty}$. Let $w \in H^s(\mathbb{R}^N, \mathbb{R})$ be a positive ground state to the autonomous problem (P_μ) and $\eta \in C_c^\infty(\mathbb{R}^N, \mathbb{R})$ be a cut-off function such that $\eta = 1$ in B_1 and $\eta = 0$ in B_2^c. Define $w_r(x) = \eta_r(x)w(x)e^{\iota A(0)\cdot x}$, with $\eta_r(x) = \eta(x/r)$ for $r > 0$, and observe that $|w_r| = \eta_r w$ and $w_r \in H_\varepsilon^s$ in view of Lemma 17.2.5. Take $t_r > 0$ such that

$$I_\mu(t_r|w_r|) = \max_{t \geq 0} I_\mu(t|w_r|)$$

Let us prove that there exists a sufficiently large r such that $I_\mu(t_r|w_r|) < m_{V_\infty}$.

If, by contradiction, $I_\mu(t_r|w_r|) \geq m_{V_\infty}$ for any $r > 0$, then by using the fact that $|w_r| \to w$ in $H^s(\mathbb{R}^N, \mathbb{R})$ as $r \to \infty$ (see Lemma 1.4.8), we have $t_r \to 1$ and

$$m_{V_\infty} \leq \liminf_{r \to \infty} I_\mu(t_r|w_r|) = I_\mu(w) = m_\mu,$$

which contradicts the inequality $m_{V_\infty} > m_\mu$. Hence, there exists $r > 0$ such that

$$I_\mu(t_r|w_r|) = \max_{\tau \geq 0} I_\mu(\tau(t_r|w_r|)) \text{ and } I_\mu(t_r|w_r|) < m_{V_\infty}. \qquad (17.2.22)$$

Now, we show that

$$\lim_{\varepsilon \to 0} [w_r]_{A_\varepsilon}^2 = [\eta_r w]_s^2. \qquad (17.2.23)$$

First, note that

$$[w_r]_{A_\varepsilon}^2 = \iint_{\mathbb{R}^{2N}} \frac{|e^{\iota A(0)\cdot x}\eta_r(x)w(x) - e^{\iota A_\varepsilon(\frac{x+y}{2})\cdot(x-y)}e^{\iota A(0)\cdot y}\eta_r(y)w(y)|^2}{|x - y|^{N+2s}} \, dx dy$$

$$= [\eta_r w]_s^2 + \iint_{\mathbb{R}^{2N}} \frac{\eta_r^2(y)w^2(y)|e^{\iota[A_\varepsilon(\frac{x+y}{2})-A(0)]\cdot(x-y)} - 1|^2}{|x - y|^{N+2s}} \, dx dy$$

$$+ 2\Re \iint_{\mathbb{R}^{2N}} \frac{(\eta_r(x)w(x) - \eta_r(y)w(y))\eta_r(y)w(y)(1 - e^{-\imath[A_\varepsilon(\frac{x+y}{2}) - A(0)]\cdot(x-y)})}{|x-y|^{N+2s}}\, dx\, dy$$

$$= [\eta_r w]_s^2 + X_\varepsilon + 2Y_\varepsilon.$$

Since $|Y_\varepsilon| \leq [\eta_r w]_s \sqrt{X_\varepsilon}$, it suffices to show that $X_\varepsilon \to 0$ as $\varepsilon \to 0$ to deduce that (17.2.23) holds. Observe that, for $0 < \beta < \alpha/(1+\alpha-s)$,

$$X_\varepsilon \leq \int_{\mathbb{R}^N} w^2(y) \left(\int_{|x-y|\geq\varepsilon^{-\beta}} \frac{|e^{\imath[A_\varepsilon(\frac{x+y}{2}) - A(0)]\cdot(x-y)} - 1|^2}{|x-y|^{N+2s}}\, dx \right) dy$$

$$+ \int_{\mathbb{R}^N} w^2(y) \left(\int_{|x-y|<\varepsilon^{-\beta}} \frac{|e^{\imath[A_\varepsilon(\frac{x+y}{2}) - A(0)]\cdot(x-y)} - 1|^2}{|x-y|^{N+2s}}\, dx \right) dy \qquad (17.2.24)$$

$$= X_\varepsilon^1 + X_\varepsilon^2.$$

Since $|e^{\imath t} - 1|^2 \leq 4$, recalling that $w \in H^s(\mathbb{R}^N, \mathbb{R})$, we see that

$$X_\varepsilon^1 \leq C \int_{\mathbb{R}^N} w^2(y)\, dy \int_{\varepsilon^{-\beta}}^\infty \rho^{-1-2s}\, d\rho \leq C\, \varepsilon^{2\beta s} \to 0. \qquad (17.2.25)$$

Concerning X_ε^2, since $|e^{\imath t} - 1|^2 \leq t^2$ for all $t \in \mathbb{R}$, $A \in C^{0,\alpha}(\mathbb{R}^N, \mathbb{R}^N)$ for $\alpha \in (0, 1]$, and $|x+y|^2 \leq 2(|x-y|^2 + 4|y|^2)$, we have

$$X_\varepsilon^2 \leq \int_{\mathbb{R}^N} w^2(y) \left(\int_{|x-y|<\varepsilon^{-\beta}} \frac{|A_\varepsilon\left(\frac{x+y}{2}\right) - A(0)|^2}{|x-y|^{N+2s-2}}\, dx \right) dy$$

$$\leq C\, \varepsilon^{2\alpha} \int_{\mathbb{R}^N} w^2(y) \left(\int_{|x-y|<\varepsilon^{-\beta}} \frac{|x+y|^{2\alpha}}{|x-y|^{N+2s-2}}\, dx \right) dy$$

$$\leq C\, \varepsilon^{2\alpha} \left[\int_{\mathbb{R}^N} w^2(y) \left(\int_{|x-y|<\varepsilon^{-\beta}} \frac{1}{|x-y|^{N+2s-2-2\alpha}}\, dx \right) dy \right. \qquad (17.2.26)$$

$$\left. + \int_{\mathbb{R}^N} |y|^{2\alpha} w^2(y) \left(\int_{|x-y|<\varepsilon^{-\beta}} \frac{1}{|x-y|^{N+2s-2}}\, dx \right) dy \right]$$

$$= C\, \varepsilon^{2\alpha}(X_\varepsilon^{2,1} + X_\varepsilon^{2,2}).$$

Then,

$$X_\varepsilon^{2,1} = C \int_{\mathbb{R}^N} w^2(y)\, dy \int_0^{\varepsilon^{-\beta}} \rho^{1+2\alpha-2s}\, d\rho \leq C\varepsilon^{-2\beta(1+\alpha-s)}. \qquad (17.2.27)$$

On the other hand, using Remark 17.2.14, we infer that

$$
\begin{aligned}
X_\varepsilon^{2,2} &\leq C \int_{\mathbb{R}^N} |y|^{2\alpha} w^2(y)\, dy \int_0^{\varepsilon^{-\beta}} \rho^{1-2s}\, d\rho \\
&\leq C\,\varepsilon^{-2\beta(1-s)} \left[\int_{B_1} w^2(y)\, dy + \int_{B_1^c} \frac{1}{|y|^{2(N+2s)-2\alpha}}\, dy \right] \\
&\leq C\,\varepsilon^{-2\beta(1-s)}.
\end{aligned}
\tag{17.2.28}
$$

Taking into account (17.2.24), (17.2.25), (17.2.26), (17.2.27) and (17.2.28), we conclude that $X_\varepsilon \to 0$.

Now, in view of condition (V), there exists $\varepsilon_0 > 0$ such that

$$
V(\varepsilon x) \leq \mu \quad \text{for all } x \in \mathrm{supp}(|w_r|),\ \varepsilon \in (0, \varepsilon_0).
\tag{17.2.29}
$$

Therefore, putting together (17.2.22), (17.2.23) and (17.2.29), we deduce that

$$
\limsup_{\varepsilon \to 0} c_\varepsilon \leq \limsup_{\varepsilon \to 0} \left[\max_{\tau \geq 0} J_\varepsilon(\tau t_r w_r) \right] \leq \max_{\tau \geq 0} I_\mu(\tau t_r |w_r|) = I_\mu(t_r |w_r|) < m_{V_\infty},
$$

which implies that $c_\varepsilon < m_{V_\infty}$ for any $\varepsilon > 0$ sufficiently small. $\qquad\square$

17.2.3 A Multiplicity Result via Lusternik–Schnirelman Category

In this section, our main purpose is to apply the Lusternik-Schnirelman category theory to prove a multiplicity result for problem (17.2.1). In order to obtain our main result, we first give some useful preliminary lemmas.

Let $\delta > 0$ be fixed and $\omega \in H^s(\mathbb{R}^N, \mathbb{R})$ be a ground state solution of the problem (P_μ) for $\mu = V_0$ given by Lemma 17.2.13 (see also Remark 17.2.14).

Let ψ be a smooth nonincreasing cut-off function defined in $[0, \infty)$ such that $\psi = 1$ in $[0, \delta/2]$, $\psi = 0$ in $[\delta, \infty)$, $0 \leq \psi \leq 1$ and $|\psi'| \leq c$ for some $c > 0$. For each fixed $y \in M$, we introduce the function

$$
\Psi_{\varepsilon,y}(x) = \psi(|\varepsilon x - y|)\omega\left(\frac{\varepsilon x - y}{\varepsilon}\right) e^{\iota \tau_y\left(\frac{\varepsilon x - y}{\varepsilon}\right)},
$$

where M is defined in (17.1.4) and $\tau_y(x) = \sum_{j=1}^N A_j(y)x_j$.

By Lemma 17.2.10, let $t_\varepsilon > 0$ be the unique positive number such that

$$
J_\varepsilon(t_\varepsilon \Psi_{\varepsilon,y}) = \max_{t \geq 0} J_\varepsilon(t_\varepsilon \Psi_{\varepsilon,y}).
$$

and let us introduce the map $\Phi_\varepsilon : M \to \mathcal{N}_\varepsilon$ by setting $\Phi_\varepsilon(y) = t_\varepsilon \Psi_{\varepsilon,y}$. By construction, $\Phi_\varepsilon(y)$ has compact support for any $y \in M$. We begin by proving the following result.

Lemma 17.2.19 $\|\Psi_{\varepsilon,y}\|_\varepsilon^2 \to \|\omega\|_{V_0}^2$ as $\varepsilon \to 0$, uniformly with respect to $y \in M$.

Proof Applying the dominated convergence theorem we easily see that

$$\int_{\mathbb{R}^N} V(\varepsilon x)|\Psi_{\varepsilon,y}(x)|^2\,dx \to V_0 \int_{\mathbb{R}^N} \omega^2(x)\,dx.$$

Thus, we only need to prove that, as $\varepsilon \to 0$

$$\iint_{\mathbb{R}^{2N}} \frac{|\Psi_{\varepsilon,y}(x_1) - \Psi_{\varepsilon,y}(x_2)e^{\imath(x_1-x_2)\cdot A_\varepsilon(\frac{x_1+x_2}{2})}|^2}{|x_1 - x_2|^{N+2s}}\,dx_1 dx_2 \to \iint_{\mathbb{R}^{2N}} \frac{|\omega(x_1) - \omega(x_2)|^2}{|x_1 - x_2|^{N+2s}}\,dx_1 dx_2.$$

Using the change of variable $\varepsilon x_i - y = \varepsilon z_i$ $(i = 1, 2)$, we can write

$$\iint_{\mathbb{R}^{2N}} \frac{|\Psi_{\varepsilon,y}(x_1) - \Psi_{\varepsilon,y}(x_2)e^{\imath(x_1-x_2)\cdot A_\varepsilon(\frac{x_1+x_2}{2})}|^2}{|x_1 - x_2|^{N+2s}}\,dx_1 dx_2$$

$$= \iint_{\mathbb{R}^{2N}} \frac{|\psi(|\varepsilon z_1|)\omega(z_1)e^{\imath \tau_y(z_1)} - \psi(|\varepsilon z_2|)\omega(z_2)e^{\imath \tau_y(z_2)}e^{\imath(z_1-z_2)\cdot A(\varepsilon\frac{z_1+z_2}{2}+y)}|^2}{|z_1 - z_2|^{N+2s}}\,dz_1 dz_2$$

$$= \iint_{\mathbb{R}^{2N}} \frac{|\psi(|\varepsilon z_1|)\omega(z_1) - \psi(|\varepsilon z_2|)\omega(z_2)|^2}{|z_1 - z_2|^{N+2s}}\,dz_1 dz_2$$

$$+ 2\iint_{\mathbb{R}^{2N}} \frac{\psi^2(|\varepsilon z_2|)\omega^2(z_2)\left(1 - \cos\left\{(z_1 - z_2)\cdot[A(\varepsilon(\frac{z_1+z_2}{2}) + y) - A(y)]\right\}\right)}{|z_1 - z_2|^{N+2s}}\,dz_1 dz_2$$

$$+ 2\Re \iint_{\mathbb{R}^{2N}} \frac{[\psi(|\varepsilon z_1|)\omega(z_1) - \psi(|\varepsilon z_2|)\omega(z_2)]\psi(|\varepsilon z_2|)\omega(z_2)\left[1 - e^{\imath(z_2-z_1)\cdot[A(\varepsilon(\frac{z_1+z_2}{2})+y) - A(y)]}\right]}{|z_1 - z_2|^{N+2s}}\,dz_1 dz_2$$

$$= X_\varepsilon + Y_\varepsilon + 2Z_\varepsilon.$$

Since $\psi(|x|) = 1$ for $x \in B_{\delta/2}$, we can use Lemma 1.4.8 to get

$$X_\varepsilon = \iint_{\mathbb{R}^{2N}} \frac{|\psi(|\varepsilon z_1|)\omega(z_1) - \psi(|\varepsilon z_2|)\omega(z_2)|^2}{|z_1 - z_2|^{N+2s}}\,dz_1 dz_2 \to \iint_{\mathbb{R}^{2N}} \frac{|\omega(z_1) - \omega(z_2)|^2}{|z_1 - z_2|^{N+2s}}\,dz_1 dz_2$$

as $\varepsilon \to 0$.

On the other hand, by the Hölder inequality,

$$|Z_\varepsilon| \leq \sqrt{X_\varepsilon}\sqrt{Y_\varepsilon}.$$

Therefore, it is enough to show that $Y_\varepsilon \to 0$ as $\varepsilon \to 0$.

Since $\psi = 0$ in B_δ^c, we have

$$Y_\varepsilon = 2 \int_{B_{\delta/\varepsilon}(0)} \psi^2(|\varepsilon z_2|)\omega^2(z_2) \left\{ \int_{|z_1-z_2|<\varepsilon^{-\beta}} \frac{1 - \cos\left\{(z_1 - z_2) \cdot [A(\varepsilon(\frac{z_1+z_2}{2}) + y) - A(y)]\right\}}{|z_1 - z_2|^{N+2s}} \, dz_1 \right.$$

$$\left. + \int_{|z_1-z_2|\geq \varepsilon^{-\beta}} \frac{1 - \cos\left\{(z_1 - z_2) \cdot [A(\varepsilon(\frac{z_1+z_2}{2}) + y) - A(y)]\right\}}{|z_1 - z_2|^{N+2s}} \, dz_1 \right\} \, dz_2 = Y_\varepsilon^1 + Y_\varepsilon^2,$$

$$(17.2.30)$$

where $0 < \beta < \frac{\alpha}{1+\alpha-s}$.

Taking into account that $|z_1 + z_2|^{2\alpha} \leq C(|z_1 - z_2|^{2\alpha} + |z_2|^{2\alpha})$ for any $z_1, z_2 \in \mathbb{R}^N$, $2(1 - \cos t) \leq t^2$ in $\mathbb{R}$, the assumptions on A, and recalling that $0 \leq \psi \leq 1$, we can see that

$$Y_\varepsilon^1 \leq C \varepsilon^{2\alpha} \int_{B_{\delta/\varepsilon}} \omega^2(z_2) \left\{ \int_{|z_1-z_2|<\varepsilon^{-\beta}} \frac{dz_1}{|z_1 - z_2|^{N+2s-2-2\alpha}} + \int_{|z_1-z_2|<\varepsilon^{-\beta}} \frac{|z_2|^{2\alpha}}{|z_1 - z_2|^{N+2s-2}} \, dz_1 \right\} dz_2$$

$$= C \varepsilon^{2\alpha}[Y_\varepsilon^{1,1} + Y_\varepsilon^{1,2}].$$

$$(17.2.31)$$

We have

$$Y_\varepsilon^{1,1} \leq C \int_{\mathbb{R}^N} \omega^2(z_2) \, dz_2 \int_0^{\varepsilon^{-\beta}} \rho^{1+2\alpha-2s} \, d\rho = C \varepsilon^{-2\beta(1+\alpha-s)} \qquad (17.2.32)$$

and, taking into account Remark 17.2.14 and that $N \geq 3$,

$$Y_\varepsilon^{1,2} \leq C \int_{\mathbb{R}^N} |z_2|^{2\alpha}\omega^2(z_2) \, dz_2 \int_0^{\varepsilon^{-\beta}} \rho^{1-2s} \, d\rho$$

$$\leq C \varepsilon^{-2\beta(1-s)} \left[\int_{|z_2|>1} \frac{1}{|z_2|^{2(N+2s)-2\alpha}} \, dz_2 + \int_{|z_2|<1} \omega^2(z_2) \, dz_2 \right] \qquad (17.2.33)$$

$$\leq C \varepsilon^{-2\beta(1-s)}$$

Putting together (17.2.31), (17.2.32) and (17.2.33), we infer that

$$Y_\varepsilon^1 \to 0 \quad \text{as } \varepsilon \to 0. \qquad (17.2.34)$$

Finally, using that $0 \leq \psi \leq 1$ and $0 \leq 1 - \cos t \leq 1$ in $\mathbb{R}$, we have

$$Y_\varepsilon^2 \leq C \int_{\mathbb{R}^N} \omega^2(z_2) \, dz_2 \int_{\varepsilon^{-\beta}}^\infty \frac{1}{\rho^{2s+1}} \, d\rho \leq C \varepsilon^{2s\beta}. \qquad (17.2.35)$$

Taking into account (17.2.30), (17.2.34) and (17.2.35) we can conclude. $\qquad \square$

The next result will be very useful to define a map from M to a suitable sublevel in the Nehari manifold.

Lemma 17.2.20 *The functional* Φ_ε *has the property that*

$$\lim_{\varepsilon \to 0} J_\varepsilon(\Phi_\varepsilon(y)) = m_{V_0}, \quad uniformly\ in\ y \in M.$$

Proof Assume, by contradiction, that there there exist $\kappa > 0$, $(y_n) \subset M$ and $\varepsilon_n \to 0$ such that

$$|J_{\varepsilon_n}(\Phi_{\varepsilon_n}(y_n)) - m_{V_0}| \geq \kappa.$$

Since $\langle J'_{\varepsilon_n}(\Phi_{\varepsilon_n}(y_n)), \Phi_{\varepsilon_n}(y_n)\rangle = 0$, using the change of variable $z = (\varepsilon_n x - y_n)/\varepsilon_n$, condition $(f5)$, and the fact that, if $z \in B_{\delta/\varepsilon_n}$, then $\varepsilon_n z + y_n \in B_\delta(y_n) \subset M_\delta$, we deduce that

$$
\begin{aligned}
\|\Psi_{\varepsilon_n, y_n}\|_{\varepsilon_n}^2 &= \int_{\mathbb{R}^N} f(|t_{\varepsilon_n}\Psi_{\varepsilon_n}|^2)|\Psi_{\varepsilon_n}|^2\, dx \\
&= \int_{\mathbb{R}^N} f((t_{\varepsilon_n}\psi(|\varepsilon_n z|)\omega(z))^2)(\psi(|\varepsilon_n z|)\omega(z))^2\, dz \\
&\geq \int_{B_{\delta/2}} f((t_{\varepsilon_n}\omega(z))^2)\omega^2(z)\, dz \\
&\geq f((t_n\alpha)^2) \int_{B_{\delta/2}} \omega^2(z)\, dz
\end{aligned}
$$

for all $n \geq n_0$, with $n_0 \in \mathbb{N}$ such that $B_{\frac{\delta}{2}} \subset B_{\frac{\delta}{2\varepsilon_n}}$ and $\alpha = \min\{\omega(z) : |z| \leq \frac{\delta}{2}\}$.

Hence, if $t_{\varepsilon_n} \to \infty$, by $(f4)$ we deduce that $\|\Psi_{\varepsilon_n, y_n}\|^2 \to \infty$ which contradicts Lemma 17.2.19. Therefore, up to a subsequence, we may assume that $t_{\varepsilon_n} \to t_0 \geq 0$. In fact, taking into account Lemma 17.2.19 and passing to the limit as $n \to \infty$ in

$$\|\Psi_{\varepsilon_n, y_n}\|_{\varepsilon_n}^2 = \int_{\mathbb{R}^N} f((t_{\varepsilon_n}\psi(|\varepsilon_n z|)\omega(z))^2)(\psi(|\varepsilon_n z|)\omega(z))^2\, dz$$

it is easy to check that $t_0 > 0$.

Moreover,

$$[t_0\omega]_s^2 + \int_{\mathbb{R}^N} V_0(t_0\omega)^2\, dx = \int_{\mathbb{R}^N} f((t_0\omega)^2)(t_0\omega)^2\, dx,$$

that is, $t_0\omega \in \mathcal{M}_{V_0}$. Since $\omega \in \mathcal{M}_{V_0}$, we get that $t_0 = 1$.

Then,

$$\lim_{n\to\infty} \int_{\mathbb{R}^N} F(|\Phi_{\varepsilon_n}(y_n)|^2)\, dx = \int_{\mathbb{R}^N} F(\omega^2)\, dx,$$

and consequently

$$\lim_{n\to\infty} J_{\varepsilon_n}(\Phi_{\varepsilon_n}(y_n)) = I_{V_0}(\omega) = m_{V_0},$$

which gives a contradiction. □

Now, we introduce the suitable barycenter map. Take $\rho = \rho(\delta) > 0$ such that $M_\delta \subset B_\rho$ and consider $\Upsilon : \mathbb{R}^N \to \mathbb{R}^N$ given by

$$\Upsilon(x) = \begin{cases} x, & \text{if } |x| < \rho, \\ \rho x/|x|, & \text{if } |x| \ge \rho. \end{cases}$$

Define the barycenter map $\beta_\varepsilon : \mathcal{N}_\varepsilon \to \mathbb{R}^N$ by the formula

$$\beta_\varepsilon(u) = \frac{\displaystyle\int_{\mathbb{R}^N} \Upsilon(\varepsilon x)|u(x)|^2\, dx}{\displaystyle\int_{\mathbb{R}^N} |u(x)|^2\, dx}.$$

Lemma 17.2.21 *The function Φ_ε has the property that*

$$\lim_{\varepsilon\to 0} \beta_\varepsilon(\Phi_\varepsilon(y)) = y, \quad \text{uniformly in } y \in M.$$

Proof Suppose, by contradiction, that there exist $\kappa > 0$, $(y_n) \subset M$ and $\varepsilon_n \to 0$ such that

$$|\beta_{\varepsilon_n}(\Phi_{\varepsilon_n}(y_n)) - y_n| \ge \kappa. \tag{17.2.36}$$

Using the change of variable $z = (\varepsilon_n x - y_n)/\varepsilon_n$, we see that

$$\beta_{\varepsilon_n}(\Psi_{\varepsilon_n}(y_n)) = y_n + \frac{\int_{\mathbb{R}^N}[\Upsilon(\varepsilon_n z + y_n) - y_n](\psi(|\varepsilon_n z|))^2(\omega(z))^2\, dz}{\int_{\mathbb{R}^N}(\psi(|\varepsilon_n z|))^2(\omega(z))^2\, dz}.$$

Taking into account $(y_n) \subset M \subset M_\delta \subset B_\rho$ and applying the dominated convergence theorem, we infer that

$$|\beta_{\varepsilon_n}(\Phi_{\varepsilon_n}(y_n)) - y_n| = o_n(1),$$

which contradicts (17.2.36).   □

Next, we prove the following useful compactness result.

Proposition 17.2.22 *Let $\varepsilon_n \to 0$ and $(u_n) = (u_{\varepsilon_n}) \subset \mathcal{N}_{\varepsilon_n}$ be such that $J_{\varepsilon_n}(u_n) \to m_{V_0}$. Then there exists $(\tilde{y}_n) = (\tilde{y}_{\varepsilon_n}) \subset \mathbb{R}^N$ such that the translated sequence*

$$v_n(x) = |u_n|(x + \tilde{y}_n)$$

admits a subsequence that converges in $H^s(\mathbb{R}^N, \mathbb{R})$. Moreover, up to a subsequence, $(y_n) = (\varepsilon_n \tilde{y}_n)$ is such that $y_n \to y \in M$.

Proof Since $\langle J'_{\varepsilon_n}(u_n), u_n \rangle = 0$ and $J_{\varepsilon_n}(u_n) \to m_{V_0}$, we easily conclude that there exists $C > 0$ such that $\|u_n\|_{\varepsilon_n} \le C$ for all $n \in \mathbb{N}$. Let us observe that $\|u_n\|_{\varepsilon_n} \nrightarrow 0$ since $m_{V_0} > 0$. Therefore, as in the proof of Lemma 17.2.11, we can find a sequence $(\tilde{y}_n) \subset \mathbb{R}^N$ and constants $R, \beta > 0$ such that

$$\liminf_{n \to \infty} \int_{B_R(\tilde{y}_n)} |u_n|^2 \, dx \ge \beta. \tag{17.2.37}$$

Set

$$v_n(x) = |u_n|(x + \tilde{y}_n).$$

The diamagnetic inequality (17.2.3) implies that the sequence $(|u_n|)$ is bounded in $H^s(\mathbb{R}^N, \mathbb{R})$ and, using (17.2.37), we may suppose that $v_n \rightharpoonup v$ in $H^s(\mathbb{R}^N, \mathbb{R})$ for some $v \neq 0$.

Let $(t_n) \subset (0, \infty)$ be such that $w_n = t_n v_n \in \mathcal{M}_{V_0}$, and set $y_n = \varepsilon_n \tilde{y}_n$.

Using (17.2.3), we see that

$$m_{V_0} \le I_{V_0}(w_n) \le \max_{t \ge 0} J_{\varepsilon_n}(t u_n) = J_{\varepsilon_n}(u_n) = m_{V_0} + o_n(1),$$

which implies that $I_{V_0}(w_n) \to m_{V_0}$.

Now, the sequence (t_n) is bounded because, by Lemma 17.2.13, (v_n) and (w_n) are bounded in $H^s(\mathbb{R}^N, \mathbb{R})$ and $v_n \nrightarrow 0$ in $H^s(\mathbb{R}^N, \mathbb{R})$. Therefore, up to a subsequence, we may assume that $t_n \to t_0 \ge 0$.

We claim that $t_0 > 0$. In fact, if $t_0 = 0$, then the boundedness of (v_n), we get $w_n = t_n v_n \to 0$ in $H^s(\mathbb{R}^N, \mathbb{R})$, that is $I_{V_0}(w_n) \to 0$ in contrast with the fact $m_{V_0} > 0$.

Thus, up to a subsequence, we may assume that $w_n \rightharpoonup w = t_0 v \neq 0$ in $H^s(\mathbb{R}^N, \mathbb{R})$.

From Lemma 17.2.13, we can deduce that $w_n \to w$ in $H^s(\mathbb{R}^N, \mathbb{R})$, which gives $v_n \to v$ in $H^s(\mathbb{R}^N, \mathbb{R})$.

Now we show that (y_n) has a subsequence such that $y_n \to y \in M$.

Assume by contradiction that (y_n) is not bounded, that is there exists a subsequence, still denoted by (y_n), such that $|y_n| \to \infty$.

First, we deal with the case $V_\infty = \infty$.

Taking into account (17.2.3), we obtain that

$$\int_{\mathbb{R}^N} V(\varepsilon_n x + y_n) |v_n|^2 \, dx \le [|v_n|]_s^2 + \int_{\mathbb{R}^N} V(\varepsilon_n x + y_n) |v_n|^2 \, dx \le \|u_n\|_{\varepsilon_n}^2 \le C.$$

On the other hand, by Fatou's lemma,

$$\liminf_{n \to \infty} \int_{\mathbb{R}^N} V(\varepsilon_n x + y_n) |v_n|^2 \, dx = \infty,$$

so we get a contradiction.

Now, suppose $V_\infty < \infty$.

Using that $w_n \to w$ strongly in $H^s(\mathbb{R}^N, \mathbb{R})$, that $V_0 < V_\infty$, the translation invariance of $\mathbb{R}^N$ and (17.2.3), we obtain

$$m_{V_0} = I_{V_0}(w) < I_{V_\infty}(w)$$

$$\le \liminf_{n \to \infty} \left[\frac{1}{2} [w_n]_s^2 + \frac{1}{2} \int_{\mathbb{R}^N} V(\varepsilon_n x + y_n) w_n^2 \, dx - \frac{1}{2} \int_{\mathbb{R}^N} F(w_n^2) \, dx \right]$$

$$= \liminf_{n \to \infty} \left[\frac{t_n^2}{2} [|u_n|]_s^2 + \frac{t_n^2}{2} \int_{\mathbb{R}^N} V(\varepsilon_n x) |u_n|^2 \, dx - \frac{1}{2} \int_{\mathbb{R}^N} F(t_n^2 |u_n|^2) \, dx \right]$$

$$\le \liminf_{n \to \infty} J_{\varepsilon_n}(t_n u_n) \le \liminf_{n \to \infty} J_{\varepsilon_n}(u_n) = m_{V_0},$$

$$(17.2.38)$$

so again we reached a contradiction.

Then, (y_n) is bounded and, up to a subsequence, we may assume that $y_n \to y$. If $y \notin M$, then $V_0 < V(y)$ and we can argue as in (17.2.38) to get a contradiction and thus complete the proof. $\square$

At this point, we introduce a subset $\tilde{\mathcal{N}}_\varepsilon$ of $\mathcal{N}_\varepsilon$ by

$$\tilde{\mathcal{N}}_\varepsilon = \{u \in \mathcal{N}_\varepsilon : J_\varepsilon(u) \le m_{V_0} + h(\varepsilon)\},$$

where $h : \mathbb{R}_+ \to \mathbb{R}_+$ is such that $h(\varepsilon) \to 0$ as $\varepsilon \to 0$.

It follows from Lemma 17.2.20 that $h(\varepsilon) = \sup_{y \in M} |J_\varepsilon(\Phi_\varepsilon(y)) - m_{V_0}| \to 0$ as $\varepsilon \to 0$. Hence, $\Phi_\varepsilon(y) \in \tilde{\mathcal{N}}_\varepsilon$ and $\tilde{\mathcal{N}}_\varepsilon \ne \emptyset$ for any $\varepsilon > 0$. Moreover, the following relation holds between $\tilde{\mathcal{N}}_\varepsilon$ and the barycenter map.

Lemma 17.2.23 *We have*

$$\lim_{\varepsilon \to 0} \sup_{u \in \tilde{\mathcal{N}}_\varepsilon} \text{dist}(\beta_\varepsilon(u), M_\delta) = 0.$$

Proof Let $\varepsilon_n \to 0$ as $n \to \infty$. For any $n \in \mathbb{N}$, there exists $(u_n) \in \tilde{\mathcal{N}}_{\varepsilon_n}$ such that

$$\sup_{u \in \tilde{\mathcal{N}}_{\varepsilon_n}} \inf_{y \in M_\delta} |\beta_{\varepsilon_n}(u) - y| = \inf_{y \in M_\delta} |\beta_{\varepsilon_n}(u_n) - y| + o_n(1).$$

Therefore, it is suffices to prove that there exists $(y_n) \subset M_\delta$ such that

$$\lim_{n \to \infty} |\beta_{\varepsilon_n}(u_n) - y_n| = 0. \tag{17.2.39}$$

Using the diamagnetic inequality (17.2.3), we see that $I_{V_0}(t|u_n|) \leq J_{\varepsilon_n}(tu_n)$ for all $t \geq 0$. Therefore, recalling that $(u_n) \subset \tilde{\mathcal{N}}_{\varepsilon_n} \subset \mathcal{N}_{\varepsilon_n}$, we deduce that

$$m_{V_0} \leq \max_{t \geq 0} I_{V_0}(t|u_n|) \leq \max_{t \geq 0} J_{\varepsilon_n}(tu_n) = J_{\varepsilon_n}(u_n) \leq m_{V_0} + h(\varepsilon_n)$$

which implies that $J_{\varepsilon_n}(u_n) \to c_{V_0}$, because $h(\varepsilon_n) \to 0$ as $n \to \infty$.

It follows from Proposition 17.2.22 that there exists $(\tilde{y}_n) \subset \mathbb{R}^N$ such that $y_n = \varepsilon_n \tilde{y}_n \in M_\delta$ for n sufficiently large. Accordingly,

$$\beta_{\varepsilon_n}(u_n) = y_n + \frac{\int_{\mathbb{R}^N} [\Upsilon(\varepsilon_n z + y_n) - y_n] |u_n(z + \tilde{y}_n)|^2 \, dz}{\int_{\mathbb{R}^N} |u_n(z + \tilde{y}_n)|^2 \, dz}.$$

Since, up to a subsequence, $|u_n|(\cdot + \tilde{y}_n)$ converges strongly in $H^s(\mathbb{R}^N, \mathbb{R})$ and $\varepsilon_n z + y_n \to y \in M$ for any $z \in \mathbb{R}^N$, we deduce (17.2.39). $\square$

Now, we are ready to present the proof of our multiplicity result.

Proof of Theorem 17.1.2 Given $\delta > 0$, we can apply Lemmas 17.2.20, 17.2.21 and 17.2.23 and argue as in [146] to find $\varepsilon_\delta > 0$ such that for any $\varepsilon \in (0, \varepsilon_\delta)$, the diagram

$$M \xrightarrow{\Phi_\varepsilon} \tilde{\mathcal{N}}_\varepsilon \xrightarrow{\beta_\varepsilon} M_\delta$$

is well defined and $\beta_\varepsilon \circ \Phi_\varepsilon$ is homotopically equivalent to the embedding $\iota : M \to M_\delta$. This fact and Lemma 6.3.21 yield

$$\text{cat}_{\tilde{\mathcal{N}}_\varepsilon}(\tilde{\mathcal{N}}_\varepsilon) \geq \text{cat}_{M_\delta}(M).$$

By the definition of $\widetilde{\mathcal{N}}_\varepsilon$ and Proposition 17.2.17, J_ε fulfills the Palais–Smale condition in $\widetilde{\mathcal{N}}_\varepsilon$ (taking ε_δ smaller if necessary), so we can apply standard Lusternik-Schnirelman theory for C^1 functionals (see Theorem 2.3.12) to obtain at least $\mathrm{cat}_{M_\delta}(M)$ critical points of J_ε restricted to $\mathcal{N}_\varepsilon$. From Corollary 17.2.18, we deduce that J_ε has at least $\mathrm{cat}_{M_\delta}(M)$ critical points in H_ε^s.   □

Remark 17.2.24 If we assume that f is only continuous, one can adapt the arguments in Chap. 6, based on the Nehari manifold approach in [322], to deduce that the conclusion of Theorem 17.1.2 holds again.

17.3   An Existence Result Under del Pino-Felmer Conditions

In the second part of this chapter we study the nonlinear fractional Schrödinger equation

$$\begin{cases} \varepsilon^{2s}(-\Delta)_{\frac{A}{\varepsilon}}^s u + V(x)u = f(|u|^2)u & \text{in } \mathbb{R}^N, \\ |u| \in H^s(\mathbb{R}^N, \mathbb{R}), \end{cases} \tag{17.3.1}$$

where $\varepsilon > 0$ is a small parameter, $s \in (0, 1)$, $N \geq 3$ and $A : \mathbb{R}^N \to \mathbb{R}^N$ is a $C^{0,\alpha}$ magnetic potential, with $\alpha \in (0, 1]$. Here we assume the potential $V : \mathbb{R}^N \to \mathbb{R}$ in (17.3.1) is a continuous function satisfying the following conditions due to del Pino and Felmer [165]:

(V_1) $\inf_{x \in \mathbb{R}^N} V(x) = V_1 > 0$;
(V_2) there exists a bounded open set $\Lambda \subset \mathbb{R}^N$ such that

$$0 < V_0 = \inf_{x \in \Lambda} V(x) < \min_{x \in \partial\Lambda} V(x).$$

Concerning the nonlinearity $f : \mathbb{R} \to \mathbb{R}$, we suppose that f is continuous and that the following conditions are fulfilled:

(f_1) $f(t) = 0$ for $t \leq 0$;
(f_2) there exists $q \in (2, 2_s^*)$ such that $\lim_{t \to \infty} f(t)/t^{\frac{q-2}{2}} = 0$;
(f_3) there exists $\theta > 2$ such that $0 < \frac{\theta}{2}F(t) \leq tf(t)$ for any $t > 0$, where $F(t) = \int_0^t f(\tau)d\tau$;
(f_4) $f(t)$ is increasing for $t > 0$.

The main result of this chapter is the following:

Theorem 17.3.1 ([44, 57]) *Suppose that V satisfies (V_1)–(V_2) and f satisfies (f_1)–(f_4). Then there exists $\varepsilon_0 > 0$ such that, for any $\varepsilon \in (0, \varepsilon_0)$, problem (17.3.1) has a nontrivial* *solution u_ε. Moreover, if $\eta_\varepsilon \in \mathbb{R}^N$ is a global maximum point of $|u_\varepsilon|$, then*

$$\lim_{\varepsilon \to 0} V(\eta_\varepsilon) = V_0,$$

and there exists $C > 0$ such that

$$|u_\varepsilon(x)| \leq \frac{C\,\varepsilon^{N+2s}}{\varepsilon^{N+2s} + |x - \eta_\varepsilon|^{N+2s}} \quad \textit{for all } x \in \mathbb{R}^N.$$

The proof of Theorem 17.3.1 is obtained by using suitable variational methods. More precisely, inspired by [18, 165], we modify the nonlinearity f outside the set Λ in such way that the energy functional of the modified problem satisfies the Palais–Smale condition (see Lemma 17.3.3). In order to prove that the solutions of the modified problem also satisfy (17.3.1) for $\varepsilon > 0$ small enough, we use in an appropriate way a Moser iteration scheme [278] and some recent results established in [183]. We note that, differently from the case $A = 0$, we cannot adapt directly the techniques used in Chap. 7 because the presence of the magnetic fractional Laplacian $(-\Delta)_A^s$ creates several technical difficulties. Indeed, the estimates on the modulus of solutions are considerably more delicate and a more careful analysis is essential for proving that the translated sequence $(\tilde{u}_n)$ of solutions (u_n) of the modified problem is such that $|\tilde{u}_n(x)| \to 0$ as $|x| \to \infty$ uniformly with respect to $n \in \mathbb{N}$. To achieve our purpose, we first prove a nonlocal version of the distributional Kato inequality [238] $-\Delta|u| \leq \Re\left(\text{sign}(\bar{u})\left(\frac{\nabla}{i} - A(x)\right)^2 u\right)$ for the solutions of $(-\Delta)_A^s u = f \in L^1_{\text{loc}}(\mathbb{R}^N, \mathbb{C})$, namely, in distributional sense,

$$(-\Delta)^s|u| \leq \Re(\text{sign}(\bar{u})\,f).$$

The main idea is to use $\dfrac{u}{u_\delta}\varphi$ as test function in the problem solved by u, where $u_\delta = \sqrt{|u|^2 + \delta^2}$ and φ is a real smooth non-negative function with compact support in $\mathbb{R}^N$, and then take the limit as $\delta \to 0$. Then, each $|\tilde{u}_n|$ satisfies

$$(-\Delta)^s|\tilde{u}_n| + V_1|\tilde{u}_n| \leq g_n(x, |\tilde{u}_n|^2)|\tilde{u}_n| \quad \text{in } \mathbb{R}^N, \tag{17.3.2}$$

and by combining a comparison argument with the techniques developed in Chap. 7, we are able to show that $|\tilde{u}_n(x)| \to 0$ as $|x| \to \infty$ uniformly with respect to $n \in \mathbb{N}$. We point out that our approach is completely different from the one used in the local case considered in [18], where the authors used a suitable Moser iteration argument to prove that the solutions of the modified problem are also solutions of the original one. However, the iteration in [18] does not seem to be easy to adapt in our framework. Finally, we also establish a power-type decay estimate for $|u_n|$. In the next sections we follow the approach in [44, 57].

17.3.1 The Modified Problem

Using the change of variable $u(x) \mapsto u(\varepsilon x)$, we see that (17.3.1) is equivalent to the problem

$$\begin{cases} (-\Delta)^s_{A_\varepsilon} u + V(\varepsilon x)u = f(|u|^2)u \text{ in } \mathbb{R}^N, \\ |u| \in H^s(\mathbb{R}^N, \mathbb{R}), \end{cases} \tag{17.3.3}$$

where we set $A_\varepsilon(x) = A(\varepsilon x)$. Without loss of generality, we assume that

$$0 \in \Lambda \quad \text{and} \quad V_0 = V(0) = \inf_\Lambda V.$$

Fix $k > \frac{\theta}{\theta-2}$ and $a > 0$ such that $f(a) = \frac{V_0}{k}$, and we introduce the functions

$$\tilde{f}(t) = \begin{cases} f(t), & \text{if } t \le a, \\ \frac{V_1}{k}, & \text{if } t > a, \end{cases}$$

and

$$g(x, t) = \chi_\Lambda(x) f(t) + (1 - \chi_\Lambda(x)) \tilde{f}(t),$$

where χ_Λ is the characteristic function of Λ, and we write $G(x, t) = \int_0^t g(x, \tau) \, d\tau$.

Assumptions (f_1)–(f_4) imply that g verifies the following conditions:

(g_1) $\lim\limits_{t \to 0} g(x, t) = 0$ uniformly in $x \in \mathbb{R}^N$;

(g_2) $\lim_{t \to \infty} \dfrac{g(x, t)}{t^{\frac{q-2}{2}}} = 0$ uniformly in $x \in \mathbb{R}^N$;

(g_3) (i) $0 < \frac{\theta}{2} G(x, t) \le g(x, t)t$ for any $x \in \Lambda$ and $t > 0$,

(ii) $0 \le G(x, t) \le g(x, t)t \le \frac{V(x)}{k}t$ for any $x \in \mathbb{R}^N \setminus \Lambda$ and $t > 0$;

(g_4) $t \mapsto g(x, t)$ is increasing for $t > 0$.

Now consider the auxiliary problem

$$\begin{cases} (-\Delta)^s_{A_\varepsilon} u + V(\varepsilon x)u = g(\varepsilon x, |u|^2)u \text{ in } \mathbb{R}^N, \\ |u| \in H^s(\mathbb{R}^N, \mathbb{R}). \end{cases} \tag{17.3.4}$$

Note that if u is a solution of (17.3.4) such that

$$|u(x)| \leq \sqrt{a} \text{ for all } x \in \mathbb{R}^N \setminus \Lambda_\varepsilon, \qquad (17.3.5)$$

where $\Lambda_\varepsilon = \{x \in \mathbb{R}^N : \varepsilon x \in \Lambda\}$, then u is also a solution of the original problem (17.3.3).

It is clear that weak solutions to (17.3.4) can be found as critical points of the Euler–Lagrange functional

$$J_\varepsilon(u) = \frac{1}{2}\|u\|_\varepsilon^2 - \frac{1}{2}\int_{\mathbb{R}^N} G(\varepsilon x, |u|^2)\, dx,$$

which is well defined for any function $u : \mathbb{R}^N \to \mathbb{C}$ belonging to the space

$$H_\varepsilon^s = \left\{ u \in \mathcal{D}_{A_\varepsilon}^{s,2}(\mathbb{R}^N, \mathbb{C}) : \int_{\mathbb{R}^N} V(\varepsilon x)|u|^2\, dx < \infty \right\}$$

endowed with the norm

$$\|u\|_\varepsilon^2 = \left([u]_{A_\varepsilon}^2 + \|\sqrt{V(\varepsilon \cdot)}|u|\|_{L^2(\mathbb{R}^N)}^2 \right)^{\frac{1}{2}}.$$

Consider also the scalar autonomous problem associated with (17.3.4), that is

$$(-\Delta)^s u + V_0 u = f(u^2)u \text{ in } \mathbb{R}^N, \qquad (17.3.6)$$

and denote by $I_{V_0} : H^s(\mathbb{R}^N, \mathbb{R}) \to \mathbb{R}$ the corresponding energy functional, namely

$$I_{V_0}(u) = \frac{1}{2}\|u\|_{V_0}^2 - \frac{1}{2}\int_{\mathbb{R}^N} F(u^2)\, dx$$

where we used the notation

$$\|u\|_{V_0} = \left([u]_s^2 + V_0\|u\|_{L^2(\mathbb{R}^N)}^2 \right)^{\frac{1}{2}}$$

which is a norm in $H^s(\mathbb{R}^N, \mathbb{R})$ equivalent to the standard one.

In what follows, we verify that J_ε fulfills the assumptions of the mountain pass theorem.

Lemma 17.3.2

(i) $J_\varepsilon \in C^1(H_\varepsilon^s, \mathbb{R})$ and $J_\varepsilon(0) = 0$;

(ii) there exist $\alpha, \rho > 0$ such that $J_\varepsilon(u) \geq \alpha$ for any $u \in H_\varepsilon^s$ such that $\|u\|_\varepsilon = \rho$;

(iii) there exists $e \in H_\varepsilon^s$ such that $\|e\|_\varepsilon > \rho$ and $J_\varepsilon(e) < 0$.

Proof Using (g_1)–(g_2) and Theorem 1.1.8, we can see that for any $\delta > 0$ there exists $C_\delta > 0$ such that

$$J_\varepsilon(u) \geq \frac{1}{2}\|u\|_\varepsilon^2 - \delta\|u\|_\varepsilon^2 - C_\delta\|u\|_\varepsilon^q.$$

Choosing $\delta > 0$ sufficiently small, we conclude that (i) holds. Regarding (ii), we note that, in view of (g_3), for any $u \in H_\varepsilon^s$, $u \not\equiv 0$ with $\mathrm{supp}(u) \subset \Lambda_\varepsilon$ and $t > 0$ we have

$$J_\varepsilon(tu) \leq \frac{t^2}{2}\|u\|_\varepsilon^2 - \frac{1}{2}\int_{\Lambda_\varepsilon} G(\varepsilon x, t^2|u|^2)\,dx$$

$$\leq \frac{t^2}{2}\|u\|_\varepsilon^2 - Ct^\theta \int_{\Lambda_\varepsilon} |u|^\theta\,dx + C,$$

which implies that $J_\varepsilon(tu) \to -\infty$ as $t \to \infty$. $\qquad\square$

Lemma 17.3.3 *The functional J_ε satisfies the Palais–Smale condition at any level $c \in \mathbb{R}$.*

Proof Let $(u_n) \subset H_\varepsilon^s$ be a Palais–Smale sequence for J_ε at the level c. Then (u_n) is bounded. Indeed, using (g_3), we have

$$C(1 + \|u_n\|_\varepsilon) \geq J_\varepsilon(u_n) - \frac{1}{\theta}\langle J_\varepsilon'(u_n), u_n\rangle$$

$$\geq \left(\frac{1}{2} - \frac{1}{\theta}\right)\|u_n\|_\varepsilon^2 + \frac{1}{\theta}\int_{\mathbb{R}^N \setminus \Lambda_\varepsilon}\left[g(\varepsilon x, |u_n|^2)|u_n|^2 - \frac{\theta}{2}G(\varepsilon x, |u_n|^2)\right]dx$$

$$\geq \frac{1}{2}\left(\frac{\theta - 2}{\theta} - \frac{1}{k}\right)\|u_n\|_\varepsilon^2,$$

and recalling that $k > \frac{\theta}{\theta - 2}$ we get the boundedness. We will show that for every $\xi > 0$ there exists $R = R_\xi > 0$ such that

$$\limsup_{n \to \infty}\int_{\mathbb{R}^N \setminus B_R}\int_{\mathbb{R}^N}\frac{|u_n(x) - u_n(y)e^{\iota A_\varepsilon(\frac{x+y}{2})\cdot(x-y)}|^2}{|x - y|^{N+2s}}\,dx\,dy + \int_{\mathbb{R}^N \setminus B_R} V(\varepsilon x)|u_n|^2\,dx \leq \xi.$$

$$(17.3.7)$$

Assuming that this fact is established, let us show how it can be used to conclude the proof of lemma. We know that $u_n \rightharpoonup u$ in H_ε^s. Since H_ε^s is compactly embedded in $L_{\mathrm{loc}}^r(\mathbb{R}^N, \mathbb{C})$ and g has subcritical growth, it is easy to prove that $J_\varepsilon'(u) = 0$. In particular,

$$\|u\|_\varepsilon^2 = \int_{\mathbb{R}^N} g(\varepsilon x, |u|^2)|u|^2\,dx.$$

Recalling that $\langle J'_\varepsilon(u_n), u_n \rangle = o_n(1)$, we infer that

$$\|u_n\|_\varepsilon^2 = \int_{\mathbb{R}^N} g(\varepsilon x, |u_n|^2)|u_n|^2 \, dx + o_n(1).$$

In view of the above relations and recalling that H_ε^s is a Hilbert space, it is enough to show that

$$\lim_{n \to \infty} \int_{\mathbb{R}^N} g(\varepsilon x, |u_n|^2)|u_n|^2 \, dx = \int_{\mathbb{R}^N} g(\varepsilon x, |u|^2)|u|^2 \, dx.$$

Using (17.3.7) and (V_1), we see that for every $\xi > 0$ there exists $R = R_\xi > 0$ such that

$$\limsup_{n \to \infty} \int_{\mathbb{R}^N \setminus B_R} |u_n|^2 \, dx \leq C\xi,$$

which together with the interpolation inequality in $L^q(\mathbb{R}^N \setminus B_R)$ and the boundedness of $(|u_n|)$ in $L^{2^*_s}(\mathbb{R}^N)$ implies that

$$\limsup_{n \to \infty} \int_{\mathbb{R}^N \setminus B_R} |u_n|^q \, dx \leq C\xi^{\frac{\tau q}{2}},$$

where $\tau \in (0, 1)$ is such that $\frac{1}{q} = \frac{\tau}{2} + \frac{1-\tau}{2^*_s}$. Therefore, by using (g_1), (g_2) and the above inequalities, we deduce that

$$\limsup_{n \to \infty} \int_{\mathbb{R}^N \setminus B_R} g(\varepsilon x, |u_n|^2)|u_n|^2 \, dx \leq C(\xi + \xi^{\frac{\tau q}{2}}).$$

On the other hand, since $g(\varepsilon x, |u|^2)|u|^2 \in L^1(\mathbb{R}^N, \mathbb{R})$, we may assume, eventually taking R larger, that

$$\int_{\mathbb{R}^N \setminus B_R} g(\varepsilon x, |u|^2)|u|^2 \, dx \leq \xi.$$

Thanks to the compactness of H_ε^s in $L^q(B_R, \mathbb{C})$ for all $q \in [1, 2^*_s)$,

$$\lim_{n \to \infty} \int_{B_R} g(\varepsilon x, |u_n|^2)|u_n|^2 \, dx = \int_{B_R} g(\varepsilon x, |u|^2)|u|^2 \, dx.$$

Consequently,

$$\limsup_{n \to \infty} \left| \int_{\mathbb{R}^N} g(\varepsilon x, |u_n|^2)|u_n|^2 \, dx - \int_{\mathbb{R}^N} g(\varepsilon x, |u|^2)|u|^2 \, dx \right| \leq C(\xi + \xi^{\frac{\tau q}{2}}) \quad \text{for all } \xi > 0,$$

and sending $\xi \to 0$ we obtain the desired result.

Now, as promised, we show (17.3.7). Let $\eta_R \in C^\infty(\mathbb{R}^N, \mathbb{R})$ be such that $0 \le \eta_R \le 1$, $\eta_R = 0$ in $B_{\frac{R}{2}}$, $\eta_R = 1$ in $\mathbb{R}^N \setminus B_R$ and $\|\nabla \eta_R\|_{L^\infty(\mathbb{R}^N)} \le \frac{C}{R}$ for some $C > 0$ independent of R. Since $\langle J_\varepsilon'(u_n), \eta_R u_n \rangle = o_n(1)$, we have

$$
\Re\left(\iint_{\mathbb{R}^{2N}} \frac{\overline{(u_n(x) - u_n(y)e^{\iota A_\varepsilon(\frac{x+y}{2})\cdot(x-y)})}(u_n(x)\eta_R(x) - u_n(y)\eta_R(y)e^{\iota A_\varepsilon(\frac{x+y}{2})\cdot(x-y)})}{|x-y|^{N+2s}} \, dxdy \right)
$$

$$
+ \int_{\mathbb{R}^N} V(\varepsilon x)\eta_R |u_n|^2 \, dx = \int_{\mathbb{R}^N} g(\varepsilon x, |u_n|^2)|u_n|^2 \eta_R \, dx + o_n(1).
$$

Fix $R > 0$ such that $\Lambda_\varepsilon \subset B_{R/2}$. Taking into account that

$$
\Re\left(\iint_{\mathbb{R}^{2N}} \frac{\overline{(u_n(x) - u_n(y)e^{\iota A_\varepsilon(\frac{x+y}{2})\cdot(x-y)})}(u_n(x)\eta_R(x) - u_n(y)\eta_R(y)e^{\iota A_\varepsilon(\frac{x+y}{2})\cdot(x-y)})}{|x-y|^{N+2s}} \, dxdy \right)
$$

$$
= \Re\left(\iint_{\mathbb{R}^{2N}} \overline{u_n(y)}e^{-\iota A_\varepsilon(\frac{x+y}{2})\cdot(x-y)} \frac{(u_n(x) - u_n(y)e^{\iota A_\varepsilon(\frac{x+y}{2})\cdot(x-y)})(\eta_R(x) - \eta_R(y))}{|x-y|^{N+2s}} \, dxdy \right)
$$

$$
+ \iint_{\mathbb{R}^{2N}} \eta_R(x) \frac{|u_n(x) - u_n(y)e^{\iota A_\varepsilon(\frac{x+y}{2})\cdot(x-y)}|^2}{|x-y|^{N+2s}} \, dxdy,
$$

and using (g_3)-(ii) we have

$$
\iint_{\mathbb{R}^{2N}} \eta_R(x) \frac{|u_n(x) - u_n(y)e^{\iota A_\varepsilon(\frac{x+y}{2})\cdot(x-y)}|^2}{|x-y|^{N+2s}} \, dxdy + \int_{\mathbb{R}^N} V(\varepsilon x)\eta_R |u_n|^2 \, dx
$$

$$
\le -\Re\left(\iint_{\mathbb{R}^{2N}} \overline{u_n(y)}e^{-\iota A_\varepsilon(\frac{x+y}{2})\cdot(x-y)} \frac{(u_n(x) - u_n(y)e^{\iota A_\varepsilon(\frac{x+y}{2})\cdot(x-y)})(\eta_R(x) - \eta_R(y))}{|x-y|^{N+2s}} \, dxdy \right)
$$

$$
+ \frac{1}{k} \int_{\mathbb{R}^N} V(\varepsilon x)\eta_R |u_n|^2 \, dx + o_n(1). \tag{17.3.8}
$$

It follows from the Hölder inequality and the boundedness of (u_n) in H_ε^s that

$$
\left| \Re\left(\iint_{\mathbb{R}^{2N}} \overline{u_n(y)}e^{-\iota A_\varepsilon(\frac{x+y}{2})\cdot(x-y)} \frac{(u_n(x) - u_n(y)e^{\iota A_\varepsilon(\frac{x+y}{2})\cdot(x-y)})(\eta_R(x) - \eta_R(y))}{|x-y|^{N+2s}} \, dxdy \right) \right|
$$

$$
\le \left(\iint_{\mathbb{R}^{2N}} \frac{|u_n(x) - u_n(y)e^{\iota A_\varepsilon(\frac{x+y}{2})\cdot(x-y)}|^2}{|x-y|^{N+2s}} \, dxdy \right)^{\frac{1}{2}} \left(\iint_{\mathbb{R}^{2N}} |u_n(y)|^2 \frac{|\eta_R(x) - \eta_R(y)|^2}{|x-y|^{N+2s}} \, dxdy \right)^{\frac{1}{2}}
$$

$$
\le C \left(\iint_{\mathbb{R}^{2N}} |u_n(y)|^2 \frac{|\eta_R(x) - \eta_R(y)|^2}{|x-y|^{N+2s}} \, dxdy \right)^{\frac{1}{2}}. \tag{17.3.9}
$$

Since $|u_n| \in H^s(\mathbb{R}^N, \mathbb{R})$ thanks to Lemma 17.2.3, we can apply Lemma 1.4.5 to deduce that

$$\lim_{R \to \infty} \limsup_{n \to \infty} \iint_{\mathbb{R}^{2N}} |u_n(y)|^2 \frac{|\eta_R(x) - \eta_R(y)|^2}{|x - y|^{N+2s}} \, dx \, dy = 0. \tag{17.3.10}$$

Then, putting together (17.3.8), (17.3.9) and (17.3.10) we obtain that

$$\lim_{R \to \infty} \limsup_{n \to \infty} \left[\int_{\mathbb{R}^N \setminus B_R} \int_{\mathbb{R}^N} \frac{|u_n(x) - u_n(y) e^{\imath A_\varepsilon (\frac{x+y}{2}) \cdot (x-y)}|^2}{|x - y|^{N+2s}} \, dx \, dy + \left(1 - \frac{1}{k}\right) \int_{\mathbb{R}^N \setminus B_R} V(\varepsilon x) |u_n|^2 \, dx \right] = 0$$

which implies that (17.3.7) holds true.   □

Taking into account Lemma 17.3.2, we can define the mountain pass level

$$c_\varepsilon = \inf_{\gamma \in \Gamma_\varepsilon} \max_{t \in [0,1]} J_\varepsilon(\gamma(t)),$$

where

$$\Gamma_\varepsilon = \{\gamma \in C([0, 1], H_\varepsilon^s) : \gamma(0) = 0 \text{ and } J_\varepsilon(\gamma(1)) < 0\}.$$

Applying the mountain pass theorem [29], for every $\varepsilon > 0$ there exists $u_\varepsilon \in H_\varepsilon^s \setminus \{0\}$ such that $J_\varepsilon(u_\varepsilon) = c_\varepsilon$ and $J_\varepsilon'(u_\varepsilon) = 0$. Let us now introduce the Nehari manifold associated with J_ε, namely

$$\mathcal{N}_\varepsilon = \{u \in H_\varepsilon^s \setminus \{0\} : \langle J_\varepsilon'(u), u \rangle = 0\}.$$

It is standard to verify that c_ε can be characterized as follows:

$$c_\varepsilon = \inf_{u \in H_\varepsilon^s \setminus \{0\}} \sup_{t \geq 0} J_\varepsilon(tu) = \inf_{u \in \mathcal{N}_\varepsilon} J_\varepsilon(u);$$

see [340] for more details. In a similar fashion one can prove that I_{V_0} has a mountain pass geometry, and denoting by $\mathcal{M}_{V_0}$ the Nehari manifold associated with I_{V_0}, we obtain that $m_{V_0} = \inf_{\mathcal{M}_{V_0}} I_{V_0}$ coincides with the mountain pass level of I_{V_0}. Next, we prove a very interesting relation between c_ε and m_{V_0}.

Lemma 17.3.4 *The numbers c_ε and m_{V_0} satisfy the following inequality*

$$\limsup_{\varepsilon \to 0} c_\varepsilon \leq m_{V_0}.$$

Proof Let $w \in H^s(\mathbb{R}^N, \mathbb{R})$ be a positive ground state of the autonomous problem (17.3.6), so that $I'_{V_0}(w) = 0$ and $I_{V_0}(w) = m_{V_0}$, and let $\eta \in C^\infty_c(\mathbb{R}^N, [0, 1])$ be a cut-off function such that $\eta = 1$ in $B_{\frac{\delta}{2}}$ and $\mathrm{supp}(\eta) \subset B_\delta \subset \Lambda$ for some $\delta > 0$. Recall (see Remark 17.2.14) that $w \in C^{0,\gamma}(\mathbb{R}^N, \mathbb{R})$, for some $\gamma > 0$, and

$$0 < w(x) \le \frac{C}{|x|^{N+2s}} \qquad \text{for all } |x| > 1.$$

Set $w_\varepsilon(x) = \eta_\varepsilon(x)w(x)e^{\imath A(0) \cdot x}$, with $\eta_\varepsilon(x) = \eta(\varepsilon x)$ for $\varepsilon > 0$, and note that $|w_\varepsilon| = \eta_\varepsilon w$ and $w_\varepsilon \in H^s_\varepsilon$ in light of Lemma 17.2.5. Let us prove that

$$\lim_{\varepsilon \to 0} \|w_\varepsilon\|^2_\varepsilon = \|w\|^2_{V_0} \in (0, \infty). \tag{17.3.11}$$

Since it is clear that

$$\int_{\mathbb{R}^N} V(\varepsilon x)|w_\varepsilon|^2 \, dx \to \int_{\mathbb{R}^N} V_0|w|^2 \, dx,$$

it remains to show that

$$\lim_{\varepsilon \to 0} [w_\varepsilon]^2_{A_\varepsilon} = [w]^2_s. \tag{17.3.12}$$

By Lemma 1.4.8,

$$[\eta_\varepsilon w]_s \to [w]_s \qquad \text{as } \varepsilon \to 0. \tag{17.3.13}$$

On the other hand

$$[w_\varepsilon]^2_{A_\varepsilon} = \iint_{\mathbb{R}^{2N}} \frac{|e^{\imath A(0) \cdot x}\eta_\varepsilon(x)w(x) - e^{\imath A_\varepsilon(\frac{x+y}{2}) \cdot (x-y)}e^{\imath A(0) \cdot y}\eta_\varepsilon(y)w(y)|^2}{|x - y|^{N+2s}} \, dxdy$$

$$= [\eta_\varepsilon w]^2_s + \iint_{\mathbb{R}^{2N}} \frac{\eta^2_\varepsilon(y)w^2(y)|e^{\imath [A_\varepsilon(\frac{x+y}{2}) - A(0)] \cdot (x-y)} - 1|^2}{|x - y|^{N+2s}} \, dxdy$$

$$+ 2\Re \iint_{\mathbb{R}^{2N}} \frac{(\eta_\varepsilon(x)w(x) - \eta_\varepsilon(y)w(y))\eta_\varepsilon(y)w(y)(1 - e^{-\imath[A_\varepsilon(\frac{x+y}{2}) - A(0)] \cdot (x-y)})}{|x - y|^{N+2s}} \, dxdy$$

$$= [\eta_\varepsilon w]^2_s + X_\varepsilon + 2Y_\varepsilon.$$

Then, in view of the inequality $|Y_\varepsilon| \leq [\eta_\varepsilon w]_s \sqrt{X_\varepsilon}$ and (17.3.13), to deduce (17.3.12), it suffices to prove that $X_\varepsilon \to 0$ as $\varepsilon \to 0$. For $0 < \beta < \alpha/(1+\alpha-s)$, we have

$$X_\varepsilon \leq \int_{\mathbb{R}^N} w^2(y) \left(\int_{|x-y| \geq \varepsilon^{-\beta}} \frac{|e^{i[A_\varepsilon(\frac{x+y}{2})-A(0)]\cdot(x-y)} - 1|^2}{|x-y|^{N+2s}} \, dx \right) dy$$

$$+ \int_{\mathbb{R}^N} w^2(y) \left(\int_{|x-y| < \varepsilon^{-\beta}} \frac{|e^{i[A_\varepsilon(\frac{x+y}{2})-A(0)]\cdot(x-y)} - 1|^2}{|x-y|^{N+2s}} \, dx \right) dy$$

$$= X_\varepsilon^1 + X_\varepsilon^2.$$

Arguing as in the proof of Theorem 17.1.1, we conclude that (17.3.12) holds.

Next, let $t_\varepsilon > 0$ be the unique number such that

$$J_\varepsilon(t_\varepsilon w_\varepsilon) = \max_{t \geq 0} J_\varepsilon(t w_\varepsilon).$$

Then t_ε satisfies

$$\|w_\varepsilon\|_\varepsilon^2 = \int_{\mathbb{R}^N} g(\varepsilon x, t_\varepsilon^2 |w_\varepsilon|^2)|w_\varepsilon|^2 \, dx = \int_{\mathbb{R}^N} f(t_\varepsilon^2 |w_\varepsilon|^2)|w_\varepsilon|^2 \, dx, \qquad (17.3.14)$$

where we used that $\operatorname{supp}(\eta) \subset \Lambda$ and $g = f$ on Λ. Let us prove that $t_\varepsilon \to 1$ as $\varepsilon \to 0$. Using that $\eta = 1$ in $B_{\frac{\delta}{2}}$ and w is a continuous positive function, we can see that (f_4) yields

$$\|w_\varepsilon\|_\varepsilon^2 \geq f(t_\varepsilon^2 \alpha_0^2) \int_{B_{\frac{\delta}{2}}} w^2 \, dx,$$

where $\alpha_0 = \min_{\bar{B}_{\delta/2}} w > 0$. So, if $t_\varepsilon \to \infty$ as $\varepsilon \to 0$, then we use (f_3) and (17.3.11) to deduce that $\|w\|_{V_0}^2 = \infty$, which gives a contradiction. On the other hand, if $t_\varepsilon \to 0$ as $\varepsilon \to 0$, by the growth assumptions on f and (17.3.11) we infer that $\|w\|_{V_0}^2 = 0$, which is impossible. In conclusion, $t_\varepsilon \to t_0 \in (0, \infty)$ as $\varepsilon \to 0$. Now, taking the limit as $\varepsilon \to 0$ in (17.3.14) and using (17.3.11), we see that

$$\|w\|_{V_0}^2 = \int_{\mathbb{R}^N} f(t_0^2 w^2) w^2 \, dx. \qquad (17.3.15)$$

It follows from $w \in \mathcal{M}_{V_0}$ and (f_4) that $t_0 = 1$. Then, using (17.3.11), the fact that $t_\varepsilon \to 1$ and applying the dominated convergence theorem, we obtain that $\lim_{\varepsilon \to 0} J_\varepsilon(t_\varepsilon w_\varepsilon) = I_{V_0}(w) = m_{V_0}$. Since $c_\varepsilon \leq \max_{t \geq 0} J_\varepsilon(t w_\varepsilon) = J_\varepsilon(t_\varepsilon w_\varepsilon)$, we conclude that $\limsup_{\varepsilon \to 0} c_\varepsilon \leq m_{V_0}$. $\qquad \square$

17.3.2 Proof of Theorem 17.3.1: A Kato's Inequality for $(-\Delta)_A^s$

In this subsection we give the proof of Theorem 17.3.1. We start by proving a fractional Kato's inequality in the spirit of [238] for the solutions of fractional magnetic problems.

Theorem 17.3.5 ([65]) *Let $u \in \mathcal{D}_A^{s,2}(\mathbb{R}^N, \mathbb{C})$ and $f \in L_{loc}^1(\mathbb{R}^N, \mathbb{C})$ be such that*

$$\Re \left(\iint_{\mathbb{R}^{2N}} \frac{(u(x) - u(y)e^{\imath A(\frac{x+y}{2}) \cdot (x-y)})}{|x-y|^{N+2s}} \overline{(\psi(x) - \psi(y)e^{\imath A(\frac{x+y}{2}) \cdot (x-y)})} \, dx \, dy \right)$$

$$= \Re \left(\int_{\mathbb{R}^N} f \bar{\psi} \, dx \right) \tag{17.3.16}$$

for all $\psi : \mathbb{R}^N \to \mathbb{C}$ measurable with compact support and such that $[\psi]_A < \infty$. Then $(-\Delta)^s |u| \le \Re(sign(\bar{u}) f)$ in $\mathcal{D}'$, that is

$$\iint_{\mathbb{R}^{2N}} \frac{(|u(x)| - |u(y)|)(\varphi(x) - \varphi(y))}{|x-y|^{N+2s}} \, dx \, dy \le \Re \left(\int_{\mathbb{R}^N} sign(\bar{u}) f \varphi \, dx \right) \tag{17.3.17}$$

for all $\varphi \in C_c^\infty(\mathbb{R}^N, \mathbb{R})$ such that $\varphi \ge 0$, where

$$sign(\bar{u})(x) = \begin{cases} \dfrac{\overline{u(x)}}{|u(x)|}, & if\ u(x) \ne 0, \\ 0, & if\ u(x) = 0. \end{cases}$$

Proof Take $\varphi \in C_c^\infty(\mathbb{R}^N, \mathbb{R})$ such that $\varphi \ge 0$, and for all $\delta > 0$ take

$$\psi_\delta(x) = \frac{u(x)}{\sqrt{|u(x)|^2 + \delta^2}} \varphi(x) = \frac{u(x)}{u_\delta(x)} \varphi(x)$$

as test function in (17.3.16). First, we show that ψ_δ is admissible. It is clear that ψ_δ has compact support. On the other hand,

$$\psi_\delta(x) - \psi_\delta(y) e^{\imath A(\frac{x+y}{2}) \cdot (x-y)} = \left(\frac{u(x)}{u_\delta(x)} \right) \varphi(x) - \left(\frac{u(y)}{u_\delta(y)} \right) \varphi(y) e^{\imath A(\frac{x+y}{2}) \cdot (x-y)}$$

$$= \left[u(x) - u(y) e^{\imath A(\frac{x+y}{2}) \cdot (x-y)} \right] \frac{\varphi(x)}{u_\delta(x)}$$

$$+ \left[\frac{\varphi(x)}{u_\delta(x)} - \frac{\varphi(y)}{u_\delta(y)} \right] u(y) e^{\imath A(\frac{x+y}{2}) \cdot (x-y)}$$

$$= \left[u(x) - u(y) e^{\imath A(\frac{x+y}{2}) \cdot (x-y)} \right] \frac{\varphi(x)}{u_\delta(x)}$$

$$+\left[\frac{1}{u_\delta(x)} - \frac{1}{u_\delta(y)}\right]\varphi(x)u(y)e^{\iota A(\frac{x+y}{2})\cdot(x-y)}$$

$$+[\varphi(x) - \varphi(y)]\frac{u(y)}{u_\delta(y)}e^{\iota A(\frac{x+y}{2})\cdot(x-y)},$$

which implies that

$$|\psi_\delta(x) - \psi_\delta(y)e^{\iota A(\frac{x+y}{2})\cdot(x-y)}|^2$$

$$\le \frac{4}{\delta^2}|u(x) - u(y)e^{\iota A(\frac{x+y}{2})\cdot(x-y)}|^2\|\varphi\|^2_{L^\infty(\mathbb{R}^N)}$$

$$+4\left|\frac{u(y)}{u_\delta(y)}\right|^2\frac{1}{|u_\delta(x)|^2}\|\varphi\|^2_{L^\infty(\mathbb{R}^N)}|u_\delta(y) - u_\delta(x)|^2 + 4|\varphi(x) - \varphi(y)|^2$$

$$\le \frac{4}{\delta^2}|u(x) - u(y)e^{\iota A(\frac{x+y}{2})\cdot(x-y)}|^2\|\varphi\|^2_{L^\infty(\mathbb{R}^N)} + \frac{4}{\delta^2}||u(x)| - |u(y)||^2\|\varphi\|^2_{L^\infty(\mathbb{R}^N)}$$

$$+4|\varphi(x) - \varphi(y)|^2;$$

here we used the elementary inequalities

$$|z + w + k|^2 \le 4(|z|^2 + |w|^2 + |k|^2) \quad \text{for all } z, w, k \in \mathbb{C},$$

$$\left|\sqrt{|z|^2 + \delta^2} - \sqrt{|w|^2 + \delta^2}\right| \le ||z| - |w|| \quad \text{for all } z, w \in \mathbb{C},$$

and the fact that $|e^{\iota t}| = 1$ for all $t \in \mathbb{R}$, $u_\delta \ge \delta$, $|\frac{u}{u_\delta}| \le 1$. Since $u \in \mathcal{D}^{s,2}_A(\mathbb{R}^N, \mathbb{C})$, $|u| \in \mathcal{D}^{s,2}(\mathbb{R}^N, \mathbb{R})$ (by Lemma 17.2.3) and $\varphi \in C^\infty_c(\mathbb{R}^N, \mathbb{R})$, we deduce that $[\psi_\delta]_A < \infty$. Then we have

$$\Re\left[\iint_{\mathbb{R}^{2N}} \frac{(u(x) - u(y)e^{\iota A(\frac{x+y}{2})\cdot(x-y)})}{|x-y|^{N+2s}}\left(\frac{\overline{u(x)}}{u_\delta(x)}\varphi(x) - \frac{\overline{u(y)}}{u_\delta(y)}\varphi(y)e^{-\iota A(\frac{x+y}{2})\cdot(x-y)}\right)dxdy\right]$$

$$= \Re\left(\int_{\mathbb{R}^N} f\frac{\overline{u}}{u_\delta}\varphi\, dx\right). \tag{17.3.18}$$

Now, since $\Re(z) \le |z|$ for all $z \in \mathbb{C}$ and $|e^{\iota t}| = 1$ for all $t \in \mathbb{R}$, we see that

$$\Re\left[(u(x) - u(y)e^{\iota A(\frac{x+y}{2})\cdot(x-y)})\left(\frac{\overline{u(x)}}{u_\delta(x)}\varphi(x) - \frac{\overline{u(y)}}{u_\delta(y)}\varphi(y)e^{-\iota A(\frac{x+y}{2})\cdot(x-y)}\right)\right]$$

$$= \Re\left[\frac{|u(x)|^2}{u_\delta(x)}\varphi(x) + \frac{|u(y)|^2}{u_\delta(y)}\varphi(y) - \frac{u(x)\overline{u(y)}}{u_\delta(y)}\varphi(y)e^{-\iota A(\frac{x+y}{2})\cdot(x-y)} - \frac{u(y)\overline{u(x)}}{u_\delta(x)}\varphi(x)e^{\iota A(\frac{x+y}{2})\cdot(x-y)}\right]$$

$$\ge \left[\frac{|u(x)|^2}{u_\delta(x)}\varphi(x) + \frac{|u(y)|^2}{u_\delta(y)}\varphi(y) - |u(x)|\frac{|u(y)|}{u_\delta(y)}\varphi(y) - |u(y)|\frac{|u(x)|}{u_\delta(x)}\varphi(x)\right]. \tag{17.3.19}$$

Let us note that

$$
\frac{|u(x)|^2}{u_\delta(x)}\varphi(x) + \frac{|u(y)|^2}{u_\delta(y)}\varphi(y) - |u(x)|\frac{|u(y)|}{u_\delta(y)}\varphi(y) - |u(y)|\frac{|u(x)|}{u_\delta(x)}\varphi(x)
$$

$$
= \frac{|u(x)|}{u_\delta(x)}(|u(x)| - |u(y)|)\varphi(x) - \frac{|u(y)|}{u_\delta(y)}(|u(x)| - |u(y)|)\varphi(y)
$$

$$
= \left[\frac{|u(x)|}{u_\delta(x)}(|u(x)| - |u(y)|)\varphi(x) - \frac{|u(x)|}{u_\delta(x)}(|u(x)| - |u(y)|)\varphi(y)\right]
$$

$$
+ \left(\frac{|u(x)|}{u_\delta(x)} - \frac{|u(y)|}{u_\delta(y)}\right)(|u(x)| - |u(y)|)\varphi(y)
$$

$$
= \frac{|u(x)|}{u_\delta(x)}(|u(x)| - |u(y)|)(\varphi(x) - \varphi(y)) + \left(\frac{|u(x)|}{u_\delta(x)} - \frac{|u(y)|}{u_\delta(y)}\right)(|u(x)| - |u(y)|)\varphi(y)
$$

$$
\geq \frac{|u(x)|}{u_\delta(x)}(|u(x)| - |u(y)|)(\varphi(x) - \varphi(y)) \tag{17.3.20}
$$

where in the last inequality we used the fact that

$$
\left(\frac{|u(x)|}{u_\delta(x)} - \frac{|u(y)|}{u_\delta(y)}\right)(|u(x)| - |u(y)|)\varphi(y) \geq 0
$$

because

$$
h(t) = \frac{t}{\sqrt{t^2 + \delta^2}} \quad \text{is increasing for } t \geq 0 \quad \text{and} \quad \varphi \geq 0 \text{ in } \mathbb{R}^N.
$$

Since

$$
\frac{\left|\frac{|u(x)|}{u_\delta(x)}(|u(x)| - |u(y)|)(\varphi(x) - \varphi(y))\right|}{|x - y|^{N+2s}} \leq \frac{\big||u(x)| - |u(y)|\big|}{|x - y|^{\frac{N+2s}{2}}}\frac{|\varphi(x) - \varphi(y)|}{|x - y|^{\frac{N+2s}{2}}} \in L^1(\mathbb{R}^{2N}, \mathbb{R}),
$$

and $\frac{|u(x)|}{u_\delta(x)} \to 1$ a.e. in $\mathbb{R}^N$ as $\delta \to 0$, we can use (17.3.19), (17.3.20) and the dominated convergence theorem to deduce that

$$
\liminf_{\delta \to 0} \Re\left[\iint_{\mathbb{R}^{2N}} \frac{(u(x) - u(y)e^{\iota A(\frac{x+y}{2})\cdot(x-y)})}{|x - y|^{N+2s}}\left(\frac{\overline{u(x)}}{u_\delta(x)}\varphi(x) - \frac{\overline{u_n(y)}}{u_\delta(y)}\varphi(y)e^{-\iota A(\frac{x+y}{2})\cdot(x-y)}\right)dxdy\right]
$$

$$
\geq \liminf_{\delta \to 0} \iint_{\mathbb{R}^{2N}} \frac{|u(x)|}{u_\delta(x)}\frac{(|u(x)| - |u(y)|)(\varphi(x) - \varphi(y))}{|x - y|^{N+2s}}dxdy
$$

$$
= \iint_{\mathbb{R}^{2N}} \frac{(|u(x)| - |u(y)|)(\varphi(x) - \varphi(y))}{|x - y|^{N+2s}}dxdy. \tag{17.3.21}
$$

On the other hand, since that $|f\frac{\bar{u}}{u_\delta}\varphi| \leq |f\varphi| \in L^1(\mathbb{R}^N, \mathbb{R})$ and $f\frac{\bar{u}}{u_\delta}\varphi \to f\operatorname{sign}(\bar{u})\varphi$ a.e. in $\mathbb{R}^N$ as $\delta \to 0$, we can invoke the dominated convergence theorem to infer that, as $\delta \to 0$,

$$\Re\left(\int_{\mathbb{R}^N} f\frac{\bar{u}}{u_\delta}\varphi\, dx\right) \to \Re\left(\int_{\mathbb{R}^N} f\operatorname{sign}(\bar{u})\varphi\, dx\right). \tag{17.3.22}$$

Putting together (17.3.18), (17.3.21) and (17.3.22), we see that (17.3.17) holds true. □

Remark 17.3.6 We also have the following pointwise Kato's inequality (see [66])

$$(-\Delta)^s|u| \leq \Re(\operatorname{sign}(\bar{u})(-\Delta)^s_A u)$$

for every sufficiently enough $u : \mathbb{R}^N \to \mathbb{C}$, with $s \in (0, 1)$. Indeed, for all $x \in \mathbb{R}^N$ such that $u(x) \neq 0$, we have

$$(-\Delta)^s|u|(x) = C(N, s)\lim_{r\to 0}\int_{B^c_r(x)}\frac{|u(x)| - |u(y)|}{|x - y|^{N+2s}}\, dy$$

$$= C(N, s)\lim_{r\to 0}\int_{B^c_r(x)}\frac{\frac{|u(x)|^2}{|u(x)|} - |u(y)|\frac{|\bar{u}(x)|}{|u(x)|}}{|x - y|^{N+2s}}\, dy$$

$$= C(N, s)\lim_{r\to 0}\int_{B^c_r(x)}\frac{\frac{|u(x)|^2}{|u(x)|} - \frac{|\bar{u}(x)u(y)e^{\iota A(\frac{x+y}{2})\cdot(x-y)}|}{|u(x)|}}{|x - y|^{N+2s}}\, dy$$

$$\leq \Re\left(C(N, s)\lim_{r\to 0}\int_{B^c_r(x)}\frac{\frac{|u(x)|^2}{|u(x)|} - \frac{\bar{u}(x)u(y)e^{\iota A(\frac{x+y}{2})\cdot(x-y)}}{|u(x)|}}{|x - y|^{N+2s}}\, dy\right)$$

$$= \Re\left(\frac{\bar{u}(x)}{|u(x)|}C(N, s)\lim_{r\to 0}\int_{B^c_r(x)}\left[\frac{u(x) - u(y)e^{\iota A(\frac{x+y}{2})\cdot(x-y)}}{|x - y|^{N+2s}}\right]dy\right)$$

$$= \Re(\operatorname{sign}(\bar{u})(-\Delta)^s_A u)(x).$$

We recall that in [219] the authors proved a Kato inequality for the fractional magnetic operator $((-\iota\nabla - A(x))^2 + m^2)^{\frac{\alpha}{2}}$ with $\alpha \in (0, 1]$ and $m > 0$, or $\alpha = 1$ and $m = 0$, borrowing some arguments used in [238]. As observed in [157], when $\alpha = 1$ and $m = 0$, this operator coincides with $(-\Delta)^{\frac{1}{2}}_A$.

The next compactness result will play a fundamental role in obtaining the main result of this work.

Lemma 17.3.7 *Let* $\varepsilon_n \to 0$ *and* $(u_n) = (u_{\varepsilon_n}) \subset H^s_{\varepsilon_n}$ *be such that* $J_{\varepsilon_n}(u_n) = c_{\varepsilon_n}$ *and* $J'_{\varepsilon_n}(u_n) = 0$. *Then there exists* $(\tilde{y}_n) = (\tilde{y}_{\varepsilon_n}) \subset \mathbb{R}^N$ *such that* $v_n(x) = |u_n|(x + \tilde{y}_n)$ *has a convergent subsequence in* $H^s(\mathbb{R}^N, \mathbb{R})$. *Moreover, up to a subsequence,* $y_n = \varepsilon_n \tilde{y}_n \to y_0$ *for some* $y_0 \in \Lambda$ *such that* $V(y_0) = V_0$.

Proof By $\langle J'_{\varepsilon_n}(u_n), u_n \rangle = 0$, $J_{\varepsilon_n}(u_n) = c_{\varepsilon_n}$ and Lemma 17.3.4, we deduce that (u_n) is bounded in $H^s_{\varepsilon_n}$. Then there exists $C > 0$ (independent of n) such that $\|u_n\|_{\varepsilon_n} \leq C$ for all $n \in \mathbb{N}$. Moreover, by Lemma 17.2.3, we have that $(|u_n|)$ is bounded in $H^s(\mathbb{R}^N, \mathbb{R})$.

Next, let us show that there exist $(\tilde{y}_n) \subset \mathbb{R}^N$ and $R, \gamma > 0$ such that

$$\liminf_{n\to\infty} \int_{B_R(\tilde{y}_n)} |u_n|^2 \, dx \geq \gamma > 0. \tag{17.3.23}$$

Assume, by contradiction, that (17.3.23) is not satisfied. Hence, for all $R > 0$ we get

$$\lim_{n\to\infty} \sup_{y\in\mathbb{R}^N} \int_{B_R(y)} |u_n|^2 \, dx = 0.$$

Using the boundedness of $(|u_n|)$ and Lemma 1.4.4, we obtain that $|u_n| \to 0$ in $L^q(\mathbb{R}^N, \mathbb{R})$ for any $q \in (2, 2^*_s)$. Then, by the growth assumptions on g, we deduce

$$\lim_{n\to\infty} \int_{\mathbb{R}^N} g(\varepsilon_n x, |u_n|^2)|u_n|^2 \, dx = 0 = \lim_{n\to\infty} \int_{\mathbb{R}^N} G(\varepsilon_n x, |u_n|^2) \, dx. \tag{17.3.24}$$

It follows from $\langle J'_{\varepsilon_n}(u_n), u_n \rangle = 0$ and (17.3.24) that $\|u_n\|_{\varepsilon_n} \to 0$ as $n \to \infty$. This gives a contradiction, because $u_n \in \mathcal{N}_{\varepsilon_n}$, (g_1) and (g_2) yield $\|u_n\|^2_{\varepsilon_n} \geq \alpha_0$ for all $n \in \mathbb{N}$, for some $\alpha_0 > 0$. Let us define $\tilde{u}_n(x) = u_n(x + \tilde{y}_n)$ and $v_n = |\tilde{u}_n|$. Then, (v_n) is bounded in $H^s(\mathbb{R}^N, \mathbb{R})$, and we may assume that

$$v_n \rightharpoonup v \quad \text{in } H^s(\mathbb{R}^N, \mathbb{R}). \tag{17.3.25}$$

Moreover, $v \neq 0$ thanks to (17.3.23), which gives

$$\int_{B_R} v^2 \, dx \geq \gamma > 0. \tag{17.3.26}$$

Next, we show that $(y_n) = (\varepsilon_n \tilde{y}_n)$ is a bounded sequence in $\mathbb{R}^N$. To this end, it is enough to prove the following claim.

Claim 1 $\lim_{n\to\infty} \operatorname{dist}(y_n, \overline{\Lambda}) = 0$.

Suppose that this is not the case. Then there exist $\delta > 0$ and a subsequence (y_n), still denoted (y_n), such that

$$\text{dist}(y_n, \overline{\Lambda}) \geq \delta \quad \text{for all } n \in \mathbb{N}.$$

Therefore, we can find $r > 0$ such that $B_r(y_n) \subset \Lambda^c$ for all $n \in \mathbb{N}$. Since $C_c^\infty(\mathbb{R}^N, \mathbb{R})$ is dense in $H^s(\mathbb{R}^N, \mathbb{R})$ and $v \geq 0$, we can find a sequence $(\psi_j) \subset C_c^\infty(\mathbb{R}^N, \mathbb{R})$ such that $\psi_j \geq 0$ and $\psi_j \to v$ in $H^s(\mathbb{R}^N, \mathbb{R})$.

Now, we note that $\tilde{u}_n$ solves

$$(-\Delta)_{\tilde{A}_n}^s \tilde{u}_n + \tilde{V}_n(x)\tilde{u}_n = \tilde{g}_n(x, v_n)\tilde{u}_n \quad \text{in } \mathbb{R}^N, \tag{17.3.27}$$

where

$$\tilde{A}_n(x) = A_{\varepsilon_n}(x + \tilde{y}_n),$$

$$\tilde{V}_n(x) = V(\varepsilon_n x + y_n),$$

and

$$\tilde{g}_n(x, v_n) = g(\varepsilon_n x + y_n, v_n^2(x)).$$

Using Theorem 17.3.5, we deduce that v_n satisfies (in distributional sense)

$$(-\Delta)^s v_n + \tilde{V}_n(x)v_n \leq \tilde{g}_n(x, v_n)v_n \quad \text{in } \mathbb{R}^N. \tag{17.3.28}$$

Then, fixing $j \in \mathbb{N}$ and taking ψ_j as test function in (17.3.28), we get

$$\iint_{\mathbb{R}^{2N}} \frac{(v_n(x) - v_n(y))}{|x - y|^{N+2s}}(\psi_j(x) - \psi_j(y)) \, dx dy + \int_{\mathbb{R}^N} \tilde{V}_n(x)v_n\psi_j \, dx \leq \int_{\mathbb{R}^N} \tilde{g}_n(x, v_n)v_n\psi_j \, dx. \tag{17.3.29}$$

Since $v_n, \psi_j \geq 0$ and using (g2), that $B_r(y_n) \subset \Lambda^c$ and that $g(x, t) = \tilde{f}(t) \leq \frac{V_1}{k}$ for $(x, t) \in \Lambda^c \times \mathbb{R}$, we have

$$\int_{\mathbb{R}^N} \tilde{g}_n(x, v_n)v_n\psi_j \, dx = \int_{B_{r/\varepsilon_n}} \tilde{g}_n(x, v_n)v_n\psi_j \, dx + \int_{\mathbb{R}^N \setminus B_{r/\varepsilon_n}} \tilde{g}_n(x, v_n)v_n\psi_j \, dx$$

$$\leq \frac{V_1}{k} \int_{B_{r/\varepsilon_n}} v_n\psi_j \, dx + \int_{\mathbb{R}^N \setminus B_{r/\varepsilon_n}} f(v_n^2)v_n\psi_j \, dx.$$

which together with (17.3.29) implies that

$$\iint_{\mathbb{R}^{2N}} \frac{(v_n(x) - v_n(y))}{|x - y|^{N+2s}} (\psi_j(x) - \psi_j(y)) \, dx \, dy + \mu_0 \int_{\mathbb{R}^N} v_n \psi_j \, dx \le \int_{\mathbb{R}^N \setminus B_{r/\varepsilon_n}} f(v_n^2) v_n \psi_j \, dx$$

(17.3.30)

where $\mu_0 = V_1(1 - \frac{1}{k})$. By (17.3.25), the fact that ψ_j has compact support in $\mathbb{R}^N$ and since $\varepsilon_n \to 0$, we infer that, as $n \to \infty$,

$$\iint_{\mathbb{R}^{2N}} \frac{(v_n(x) - v_n(y))}{|x - y|^{N+2s}} (\psi_j(x) - \psi_j(y)) \, dx \, dy \to \iint_{\mathbb{R}^{2N}} \frac{(v(x) - v(y))}{|x - y|^{N+2s}} (\psi_j(x) - \psi_j(y)) \, dx \, dy$$

and

$$\int_{\mathbb{R}^N \setminus B_{r/\varepsilon_n}} f(v_n^2) v_n \psi_j \, dx \to 0.$$

The above limits and (17.3.30) show that

$$\iint_{\mathbb{R}^{2N}} \frac{(v(x) - v(y))}{|x - y|^{N+2s}} (\psi_j(x) - \psi_j(y)) \, dx \, dy + \mu_0 \int_{\mathbb{R}^N} v \psi_j \, dx \le 0,$$

and sending $j \to \infty$ we obtain that

$$\|v\|_{\mu_0}^2 = [v]_s^2 + \mu_0 \|v\|_{L^2(\mathbb{R}^N)}^2 \le 0,$$

which contradicts (17.3.26). This ends the proof of Claim 1.

Hence, there exist a subsequence of (y_n) and $y_0 \in \overline{\Lambda}$ such that

$$\lim_{n \to \infty} y_n = \lim_{n \to \infty} \varepsilon_n \tilde{y}_n = y_0.$$

Claim 2 $y_0 \in \Lambda$. Using (g_2) and (17.3.29), we see that

$$\iint_{\mathbb{R}^{2N}} \frac{(v_n(x) - v_n(y))}{|x - y|^{N+2s}} (\psi_j(x) - \psi_j(y)) \, dx \, dy + \int_{\mathbb{R}^N} \tilde{V}_n(x) v_n \psi_j \, dx \le \int_{\mathbb{R}^N} f(v_n^2) v_n \psi_j \, dx.$$

Letting $n \to \infty$ we get

$$\iint_{\mathbb{R}^{2N}} \frac{(v(x) - v(y))}{|x - y|^{N+2s}} (\psi_j(x) - \psi_j(y)) \, dx \, dy + \int_{\mathbb{R}^N} V(y_0) v \psi_j \, dx \le \int_{\mathbb{R}^N} f(v^2) v \psi_j \, dx,$$

and passing to the limit as $j \to \infty$ we have

$$[v]_s^2 + V(y_0)\|v\|_{L^2(\mathbb{R}^N)}^2 \leq \int_{\mathbb{R}^N} f(v^2)v^2 \, dx.$$

Accordingly, there exists $\tau \in (0, 1)$ such that

$$\tau v \in \mathcal{N}_{V(y_0)} = \{u \in H^s(\mathbb{R}^N, \mathbb{R}) \setminus \{0\} : \langle I'_{V(y_0)}(u), u \rangle = 0\}.$$

Denoting by $c_{V(y_0)}$ the mountain pass level associated with $I_{V(y_0)}$, and using Lemmas 17.2.3 and 17.3.4, we obtain

$$c_{V(y_0)} \leq I_{V(y_0)}(\tau u) \leq \liminf_{n\to\infty} J_{\varepsilon_n}(u_n) = \liminf_{n\to\infty} c_{\varepsilon_n} \leq c_{V_0}.$$

This in turn implies that $V(y_0) \leq V(0) = V_0$. Since $V_0 = \inf_{\overline{\Lambda}} V$, we infer that $V(y_0) = V_0$. By (V_2), it follows that $y_0 \notin \partial\Lambda$, that is $y_0 \in \Lambda$.

Claim 3 $v_n \to v$ in $H^s(\mathbb{R}^N, \mathbb{R})$ as $n \to \infty$.

Put

$$\tilde{\Lambda}_n = \frac{\Lambda - \varepsilon_n \, \tilde{y}_n}{\varepsilon_n},$$

and define

$$\tilde{\chi}_n^1(x) = \begin{cases} 1, & \text{if } x \in \tilde{\Lambda}_n, \\ 0, & \text{if } x \in \mathbb{R}^N \setminus \tilde{\Lambda}_n, \end{cases}$$

$$\tilde{\chi}_n^2(x) = 1 - \tilde{\chi}_n^1(x).$$

Let us introduce the following functions for all $x \in \mathbb{R}^N$:

$$h_n^1(x) = \left(\frac{1}{2} - \frac{1}{\theta}\right) \tilde{V}_n(x) v_n^2(x) \tilde{\chi}_n^1(x),$$

$$h^1(x) = \left(\frac{1}{2} - \frac{1}{\theta}\right) V(y_0) v^2(x),$$

$$h_n^2(x) = \left[\left(\frac{1}{2} - \frac{1}{\theta}\right) \tilde{V}_n(x) v_n^2(x) + \frac{1}{\theta} \tilde{g}_n(x, v_n) v_n^2(x) - \frac{1}{2} \tilde{G}_n(x, v_n)\right] \tilde{\chi}_n^2(x)$$

$$\geq \left(\left(\frac{1}{2} - \frac{1}{\theta}\right) - \frac{1}{2k}\right) \tilde{V}_n(x) v_n^2(x) \tilde{\chi}_n^2(x),$$

$$h_n^3(x) = \left(\frac{1}{\theta} \tilde{g}_n(x, v_n) v_n^2(x) - \frac{1}{2} \tilde{G}_n(x, v_n) \right) \tilde{\chi}_n^1(x)$$

$$= \left(\frac{1}{\theta} f(v_n^2(x)) v_n^2(x) - \frac{1}{2} F(v_n^2(x)) \right) \tilde{\chi}_n^1(x),$$

$$h^3(x) = \frac{1}{\theta} f(v^2(x)) v^2(x) - \frac{1}{2} F(v^2(x)),$$

where $\tilde{G}_n(x, v_n) = G(\varepsilon_n x + \varepsilon_n \tilde{y}_n, v_n^2(x))$. In view of (f_3) and (g_3), all the above functions are non-negative in $\mathbb{R}^N$. Moreover, using (17.3.25) and Claim 2, we have

$$v_n(x) \to v(x) \quad \text{a.e. } x \in \mathbb{R}^N,$$

$$\varepsilon_n \tilde{y}_n \to y_0 \in \Lambda,$$

which implies that

$$\tilde{\chi}_n^1(x) \to 1, \ h_n^1(x) \to h^1(x), \ h_n^2(x) \to 0 \quad \text{and} \quad h_n^3(x) \to h^3(x) \text{ a.e. } x \in \mathbb{R}^N.$$

Hence, by Fatou's lemma, Lemma 17.2.3 and Lemma 17.3.4 we get

$$c_{V_0} \geq \limsup_{n \to \infty} c_{\varepsilon_n} = \limsup_{n \to \infty} \left(J_{\varepsilon_n}(u_n) - \frac{1}{\theta} \langle J'_{\varepsilon_n}(u_n), u_n \rangle \right)$$

$$\geq \limsup_{n \to \infty} \left[\left(\frac{1}{2} - \frac{1}{\theta} \right) [v_n]_s^2 + \int_{\mathbb{R}^N} (h_n^1 + h_n^2 + h_n^3) \, dx \right]$$

$$\geq \liminf_{n \to \infty} \left[\left(\frac{1}{2} - \frac{1}{\theta} \right) [v_n]_s^2 + \int_{\mathbb{R}^N} (h_n^1 + h_n^2 + h_n^3) \, dx \right]$$

$$\geq \left(\frac{1}{2} - \frac{1}{\theta} \right) [v]_s^2 + \int_{\mathbb{R}^N} (h^1 + h^3) \, dx \geq c_{V_0},$$

which yields that

$$\lim_{n \to \infty} [v_n]_s^2 = [v]_s^2 \tag{17.3.31}$$

and

$$h_n^1 \to h^1, \ h_n^2 \to 0 \quad \text{and} \quad h_n^3 \to h^3 \quad \text{in } L^1(\mathbb{R}^N, \mathbb{R}).$$

Therefore,

$$\lim_{n \to \infty} \int_{\mathbb{R}^N} \tilde{V}_n(x) v_n^2 \, dx = \int_{\mathbb{R}^N} V(y_0) v^2 \, dx,$$

and thus we have

$$\lim_{n \to \infty} \|v_n\|^2_{L^2(\mathbb{R}^N)} = \|v\|^2_{L^2(\mathbb{R}^N)}. \tag{17.3.32}$$

Putting together (17.3.31) and (17.3.32) and using the fact that $H^s(\mathbb{R}^N, \mathbb{R})$ is a Hilbert space, we obtain

$$\|v_n - v\|^2_{V_0} = \|v_n\|^2_{V_0} - \|v\|^2_{V_0} + o_n(1) = o_n(1).$$

This completes the proof of lemma.   □

Now we prove the following lemma that will play a key role in establishing that the solutions of (17.3.4) are indeed solutions of (17.3.1).

Lemma 17.3.8 *Let (v_n) be the translated sequence given in Lemma* 17.3.7. *Then,* $v_n \in L^\infty(\mathbb{R}^N, \mathbb{R})$ *and there exists* $C > 0$ *such that*

$$\|v_n\|_{L^\infty(\mathbb{R}^N)} \le C \quad \textit{for all } n \in \mathbb{N}.$$

Moreover,

$$\lim_{|x| \to \infty} v_n(x) = 0 \quad \textit{uniformly in } n \in \mathbb{N}.$$

Proof As observed in Lemma 17.3.7, the functions $v_n = |u_n|(\cdot + \tilde{y}_n) \in H^s(\mathbb{R}^N, \mathbb{R})$ solve

$$(-\Delta)^s v_n + \tilde{V}_n(x) v_n \le \tilde{g}_n(x, v_n) v_n = h_n \text{ in } \mathbb{R}^N,$$

where

$$\tilde{V}_n(x) = V(\varepsilon_n x + \varepsilon_n \tilde{y}_n), \quad \tilde{g}_n(x, v_n) = g(\varepsilon_n x + \varepsilon_n \tilde{y}_n, v_n^2),$$

and $v_n \to v \ne 0$ in $H^s(\mathbb{R}^N, \mathbb{R})$. Then, we can argue as in the proof of Lemma 6.3.23 to see that

$$\|v_n\|_{L^\infty(\mathbb{R}^N)} \le K \text{ for all } n \in \mathbb{N}. \tag{17.3.33}$$

By interpolation, $v_n \to v$ converges strongly in $L^r(\mathbb{R}^N, \mathbb{R})$ for all $r \in [2, \infty)$. In view of the growth assumptions on g, we can also see that $h_n \to f(v^2) v$ in $L^r(\mathbb{R}^N, \mathbb{R})$ and $\|h_n\|_{L^\infty(\mathbb{R}^N)} \le C$ for all $n \in \mathbb{N}$. Note that, by (V_1), we have

$$(-\Delta)^s v_n + V_1 v_n \le h_n \quad \text{in } \mathbb{R}^N. \tag{17.3.34}$$

Let $z_n \in H^s(\mathbb{R}^N, \mathbb{R})$ denote the unique solution to

$$(-\Delta)^s z_n + V_1 z_n = h_n \quad \text{in } \mathbb{R}^N. \tag{17.3.35}$$

Then $z_n = \mathcal{K} * h_n$, where $\mathcal{K}(x) = \mathcal{F}^{-1}((|k|^{2s} + V_1)^{-1})$, and we deduce that $|z_n(x)| \to 0$ as $|x| \to \infty$, uniformly with respect to $n \in \mathbb{N}$ (see Remark 7.2.10). Since v_n satisfies (17.3.34) and z_n solves (17.3.35), a comparison readily shows that $0 \le v_n \le z_n$ in $\mathbb{R}^N$ and for all $n \in \mathbb{N}$. In particular, we infer that $v_n(x) \to 0$ as $|x| \to \infty$ uniformly with respect to $n \in \mathbb{N}$. $\qquad\square$

Remark 17.3.9 For more details on the boundedness of solutions for fractional magnetic Schrödinger equations one can consult [43].

Remark 17.3.10 To prove only the boundedness of (u_n) we can also modify in as suitable way the proof of Lemma 6.3.23. For the reader's convenience, we give the details. We follow [44]. For each $n \in \mathbb{N}$ and $L > 0$, we define $u_{L,n} = \min\{|u_n|, L\} \ge 0$ and $v_{L,n} = u_{L,n}^{2(\beta-1)} u_n$, where $\beta > 1$ will be chosen later. Taking $v_{L,n}$ as test function in (17.3.4) we see that

$$\Re\left(\iint_{\mathbb{R}^{2N}} \frac{(u_n(x) - u_n(y)e^{\imath A_{\varepsilon n}(\frac{x+y}{2})\cdot(x-y)})}{|x-y|^{N+2s}} \overline{(u_n(x)u_{L,n}^{2(\beta-1)}(x) - u_n(y)u_{L,n}^{2(\beta-1)}(y)e^{\imath A_{\varepsilon n}(\frac{x+y}{2})\cdot(x-y)})} \, dx\, dy\right)$$

$$= \int_{\mathbb{R}^N} g(\varepsilon_n x, |u_n|^2)|u_n|^2 u_{L,n}^{2(\beta-1)} \, dx - \int_{\mathbb{R}^N} V(\varepsilon_n x)|u_n|^2 u_{L,n}^{2(\beta-1)} \, dx. \tag{17.3.36}$$

Note that

$$\Re\left[(u_n(x) - u_n(y)e^{\imath A_{\varepsilon n}(\frac{x+y}{2})\cdot(x-y)})\overline{(u_n(x)u_{L,n}^{2(\beta-1)}(x) - u_n(y)u_{L,n}^{2(\beta-1)}(y)e^{\imath A_{\varepsilon n}(\frac{x+y}{2})\cdot(x-y)})}\right]$$

$$= \Re\left[|u_n(x)|^2 u_{L,n}^{2(\beta-1)}(x) - u_n(x)\overline{u_n(y)}u_{L,n}^{2(\beta-1)}(y)e^{-\imath A_{\varepsilon n}(\frac{x+y}{2})\cdot(x-y)}\right.$$

$$\left. - u_n(y)\overline{u_n(x)}u_{L,n}^{2(\beta-1)}(x)e^{\imath A_{\varepsilon n}(\frac{x+y}{2})\cdot(x-y)} + |u_n(y)|^2 u_{L,n}^{2(\beta-1)}(y)\right]$$

$$\ge (|u_n(x)|^2 u_{L,n}^{2(\beta-1)}(x) - |u_n(x)||u_n(y)|u_{L,n}^{2(\beta-1)}(y) - |u_n(y)||u_n(x)|u_{L,n}^{2(\beta-1)}(x) + |u_n(y)|^2 u_{L,n}^{2(\beta-1)}(y)$$

$$= (|u_n(x)| - |u_n(y)|)(|u_n(x)|u_{L,n}^{2(\beta-1)}(x) - |u_n(y)|u_{L,n}^{2(\beta-1)}(y)),$$

so we have

$$\Re\left(\iint_{\mathbb{R}^{2N}} \frac{(u_n(x) - u_n(y)e^{\imath A_{\varepsilon n}(\frac{x+y}{2})\cdot(x-y)})}{|x-y|^{N+2s}}\right.$$

$$\left. \overline{(u_n(x)u_{L,n}^{2(\beta-1)}(x) - u_n(y)u_{L,n}^{2(\beta-1)}(y)e^{\imath A_{\varepsilon n}(\frac{x+y}{2})\cdot(x-y)})} \, dx\, dy\right)$$

$$\geq \iint_{\mathbb{R}^{2N}} \frac{(|u_n(x)| - |u_n(y)|)}{|x-y|^{N+2s}} (|u_n(x)|u_{L,n}^{2(\beta-1)}(x) - |u_n(y)|u_{L,n}^{2(\beta-1)}(y)) \, dx dy.$$

$$(17.3.37)$$

For all $t \geq 0$, we define

$$\gamma(t) = \gamma_{L,\beta}(t) = tt_L^{2(\beta-1)},$$

where $t_L = \min\{t, L\}$. Since γ is a nondecreasing function,

$$(a - b)(\gamma(a) - \gamma(b)) \geq 0 \quad \text{for any } a, b \geq 0.$$

Consider the function

$$\Gamma(t) = \int_0^t (\gamma'(\tau))^{\frac{1}{2}} d\tau,$$

and note that

$$(a - b)(\gamma(a) - \gamma(b)) \geq |\Gamma(a) - \Gamma(b)|^2 \quad \text{for any } a, b \geq 0.$$

Hence,

$$|\Gamma(|u_n(x)|) - \Gamma(|u_n(y)|)|^2 \leq (|u_n(x)| - |u_n(y)|)((|u_n|u_{L,n}^{2(\beta-1)})(x) - (|u_n|u_{L,n}^{2(\beta-1)})(y)).$$

$$(17.3.38)$$

Then, in view of (17.3.37) and (17.3.38), we obtain

$$\Re\left(\iint_{\mathbb{R}^{2N}} \frac{(u_n(x) - u_n(y)e^{\iota A_{\varepsilon_n}(\frac{x+y}{2})\cdot(x-y)})}{|x-y|^{N+2s}} \overline{(u_n(x)u_{L,n}^{2(\beta-1)}(x) - u_n(y)u_{L,n}^{2(\beta-1)}(y)e^{\iota A_{\varepsilon_n}(\frac{x+y}{2})\cdot(x-y)})} \, dx dy\right)$$

$$\geq [\Gamma(|u_n|)]_s^2. \qquad (17.3.39)$$

Since $\Gamma(|u_n|) \geq \frac{1}{\beta}|u_n|u_{L,n}^{\beta-1}$, using the fractional Sobolev embedding $\mathcal{D}^{s,2}(\mathbb{R}^N, \mathbb{R}) \subset L^{2_s^*}(\mathbb{R}^N, \mathbb{R})$ we deduce that

$$[\Gamma(|u_n|)]_s^2 \geq S_* \|\Gamma(|u_n|)\|_{L^{2_s^*}(\mathbb{R}^N)}^2 \geq \left(\frac{1}{\beta}\right)^2 S_* \||u_n|u_{L,n}^{\beta-1}\|_{L^{2_s^*}(\mathbb{R}^N)}^2. \qquad (17.3.40)$$

Combining (17.3.36), (17.3.39) and (17.3.40), we infer that

$$\left(\frac{1}{\beta}\right)^2 S_* \||u_n|u_{L,n}^{\beta-1}\|_{L^{2_s^*}(\mathbb{R}^N)}^2 + \int_{\mathbb{R}^N} V(\varepsilon_n x)|u_n|^2 u_{L,n}^{2(\beta-1)} \, dx \leq$$

$$\int_{\mathbb{R}^N} g(\varepsilon_n x, |u_n|^2)|u_n|^2 u_{L,n}^{2(\beta-1)}\, dx. \tag{17.3.41}$$

On the other hand, by assumptions (g_1) and (g_2), for every $\xi > 0$ there exists $C_\xi > 0$ such that

$$g(\varepsilon_n x, t^2)t^2 \le \xi|t|^2 + C_\xi|t|^{2_s^*} \quad \text{for all } (x,t) \in \mathbb{R}^N \times \mathbb{R}. \tag{17.3.42}$$

Taking $\xi \in (0, V_1)$ and using (17.3.41) and (17.3.42) we see that

$$\||u_n|u_{L,n}^{\beta-1}\|_{L^{2_s^*}(\mathbb{R}^N)}^2 \le C\beta^2 \int_{\mathbb{R}^N} |u_n|^{2_s^*} u_{L,n}^{2(\beta-1)}\, dx. \tag{17.3.43}$$

Now, we take $\beta = \frac{2_s^*}{2}$ and fix $R > 0$. Recalling that $0 \le u_{L,n} \le |u_n|$ and applying the Hölder inequality we have

$$\int_{\mathbb{R}^N} |u_n|^{2_s^*} u_{L,n}^{2(\beta-1)}\, dx$$

$$= \int_{\mathbb{R}^N} |u_n|^{2_s^*-2}|u_n|^2 u_{L,n}^{2_s^*-2}\, dx$$

$$= \int_{\mathbb{R}^N} |u_n|^{2_s^*-2}(|u_n|u_{L,n}^{\frac{2_s^*-2}{2}})^2\, dx$$

$$\le \int_{\{|u_n|<R\}} R^{2_s^*-2}|u_n|^{2_s^*}\, dx + \int_{\{|u_n|>R\}} |u_n|^{2_s^*-2}(|u_n|u_{L,n}^{\frac{2_s^*-2}{2}})^2\, dx$$

$$\le \int_{\{|u_n|<R\}} R^{2_s^*-2}|u_n|^{2_s^*}\, dx + \left(\int_{\{|u_n|>R\}} |u_n|^{2_s^*}\, dx\right)^{\frac{2_s^*-2}{2_s^*}} \left(\int_{\mathbb{R}^N} (|u_n|u_{L,n}^{\frac{2_s^*-2}{2}})^{2_s^*}\, dx\right)^{\frac{2}{2_s^*}}. \tag{17.3.44}$$

Since $v_n = |u_n|(\cdot + \tilde{y}_n)$ converges strongly in $H^s(\mathbb{R}^N, \mathbb{R})$, we see that for any sufficiently large R,

$$\left(\int_{\{|u_n|>R\}} |u_n|^{2_s^*}\, dx\right)^{\frac{2_s^*-2}{2_s^*}} \le \frac{1}{2C\beta^2}. \tag{17.3.45}$$

Putting together (17.3.43), (17.3.44) and (17.3.45) we get

$$\left(\int_{\mathbb{R}^N} (|u_n|u_{L,n}^{\frac{2_s^*-2}{2}})^{2_s^*}\, dx\right)^{\frac{2}{2_s^*}} \le C\beta^2 \int_{\mathbb{R}^N} R^{2_s^*-2}|u_n|^{2_s^*}\, dx < \infty,$$

and taking the limit as $L \to \infty$ we see that $|u_n| \in L^{\frac{(2_s^*)^2}{2}}(\mathbb{R}^N, \mathbb{R})$. Now, using $0 \le u_{L,n} \le |u_n|$ and letting $L \to \infty$ in (17.3.43), we have

$$\|u_n\|_{L^{\beta 2_s^*}(\mathbb{R}^N)}^{2\beta} \le C\beta^2 \int_{\mathbb{R}^N} |u_n|^{2_s^* + 2(\beta - 1)}\, dx,$$

from which we deduce that

$$\left(\int_{\mathbb{R}^N} |u_n|^{2_s^*\beta}\, dx \right)^{\frac{1}{2_s^*(\beta-1)}} \le (C\beta)^{\frac{1}{\beta-1}} \left(\int_{\mathbb{R}^N} |u_n|^{2_s^* + 2(\beta-1)}\, dx \right)^{\frac{1}{2(\beta-1)}}.$$

For $m \ge 1$ we define β_{m+1} inductively so that $2_s^* + 2(\beta_{m+1} - 1) = 2_s^*\beta_m$ and $\beta_1 = \frac{2_s^*}{2}$. Then we have

$$\left(\int_{\mathbb{R}^N} |u_n|^{2_s^*\beta_{m+1}}\, dx \right)^{\frac{1}{2_s^*(\beta_{m+1}-1)}} \le (C\beta_{m+1})^{\frac{1}{\beta_{m+1}-1}} \left(\int_{\mathbb{R}^N} |u_n|^{2_s^*\beta_m}\, dx \right)^{\frac{1}{2_s^*(\beta_m-1)}}.$$

Let us define

$$D_m = \left(\int_{\mathbb{R}^N} |u_n|^{2_s^*\beta_m}\, dx \right)^{\frac{1}{2_s^*(\beta_m-1)}}.$$

Using an iteration argument, we can find $C_0 > 0$ independent of m such that

$$D_{m+1} \le \prod_{k=1}^{m} (C\beta_{k+1})^{\frac{1}{\beta_{k+1}-1}} D_1 \le C_0 D_1.$$

Taking the limit as $m \to \infty$ we get

$$\|u_n\|_{L^\infty(\mathbb{R}^N)} \le C_0 D_1 = K \quad \text{for all } n \in \mathbb{N}.$$

We end this section with the proof of Theorem 17.3.1.

Proof of Theorem 17.3.1 We begin by showing that there exists $\tilde{\varepsilon}_0 > 0$ such that for any $\varepsilon \in (0, \tilde{\varepsilon}_0)$ and any solution $u_\varepsilon \in H_\varepsilon^s$ of (17.3.4),

$$\|u_\varepsilon\|_{L^\infty(\mathbb{R}^N \setminus \Lambda_\varepsilon)} < \sqrt{a}. \tag{17.3.46}$$

Assume, by contradiction, that for some subsequence (ε_n) such that $\varepsilon_n \to 0$, we can find $u_n = u_{\varepsilon_n} \in H_{\varepsilon_n}^s$ such that $J_{\varepsilon_n}(u_n) = c_{\varepsilon_n}$, $J_{\varepsilon_n}'(u_n) = 0$ and

$$\|u_n\|_{L^\infty(\mathbb{R}^N \setminus \Lambda_{\varepsilon_n})} \ge \sqrt{a}. \tag{17.3.47}$$

In view of Lemma 17.3.7, there is a sequence $(\tilde{y}_n) \subset \mathbb{R}^N$ such that $v_n = |u_n|(\cdot + \tilde{y}_n) \to v$ in $H^s(\mathbb{R}^N, \mathbb{R})$ and $\varepsilon_n \tilde{y}_n \to y_0$ for some $y_0 \in \Lambda$ such that $V(y_0) = V_0$.

Now, if we choose $r > 0$ such that $B_r(y_0) \subset B_{2r}(y_0) \subset \Lambda$, we see that $B_{\frac{r}{\varepsilon_n}}(\frac{y_0}{\varepsilon_n}) \subset \Lambda_{\varepsilon_n}$. Then, for any $y \in B_{\frac{r}{\varepsilon_n}}(\tilde{y}_n)$ it holds

$$\left| y - \frac{y_0}{\varepsilon_n} \right| \leq |y - \tilde{y}_n| + \left| \tilde{y}_n - \frac{y_0}{\varepsilon_n} \right| < \frac{1}{\varepsilon_n}(r + o_n(1)) < \frac{2r}{\varepsilon_n} \qquad \text{for } n \text{ sufficiently large.}$$

Hence,

$$\mathbb{R}^N \setminus \Lambda_{\varepsilon_n} \subset \mathbb{R}^N \setminus B_{\frac{r}{\varepsilon_n}}(\tilde{y}_n) \tag{17.3.48}$$

for any large enough n. Using Lemma 17.3.8, we obtain that

$$v_n(x) \to 0 \quad \text{as } |x| \to \infty, \tag{17.3.49}$$

uniformly in $n \in \mathbb{N}$. Therefore, there exists $R > 0$ such that

$$v_n(x) < \sqrt{a} \quad \text{for all } |x| \geq R, n \in \mathbb{N}.$$

Consequently, $|u_n(x)| < \sqrt{a}$ for any $x \in \mathbb{R}^N \setminus B_R(\tilde{y}_n)$ and $n \in \mathbb{N}$. On the other hand, (17.3.48) implies that there exists $\nu \in \mathbb{N}$ such that, for any $n \geq \nu$ and $r/\varepsilon_n > R$,

$$\mathbb{R}^N \setminus \Lambda_{\varepsilon_n} \subset \mathbb{R}^N \setminus B_{\frac{r}{\varepsilon_n}}(\tilde{y}_n) \subset \mathbb{R}^N \setminus B_R(\tilde{y}_n),$$

which yields $|u_n(x)| < \sqrt{a}$ for any $x \in \mathbb{R}^N \setminus \Lambda_{\varepsilon_n}$ and $n \geq \nu$, and this contradicts (17.3.47). Since $u_\varepsilon \in H^s_\varepsilon$ satisfies (17.3.46), it follows from the definition of g that u_ε is a solution of (17.3.3). Thus, $\hat{u}_\varepsilon(x) = u_\varepsilon(x/\varepsilon)$ is a nontrivial solution to (17.3.1) for $\varepsilon > 0$ sufficiently small. Finally, we study the behavior of the maximum points of solutions to problem (17.3.1). Take $\varepsilon_n \to 0$ and consider a sequence $(u_n) \subset H^s_{\varepsilon_n}$ of solutions to (17.3.4) as above. Observe that (g_1) implies that we can find $\gamma \in (0, \sqrt{a})$ small such that

$$g(\varepsilon x, t^2)t^2 \leq \frac{V_1}{k}t^2 \quad \text{for any } x \in \mathbb{R}^N, |t| \leq \gamma. \tag{17.3.50}$$

Arguing as before, there is $R > 0$ such that

$$\|u_n\|_{L^\infty(\mathbb{R}^N \setminus B_R(\tilde{y}_n))} < \gamma. \tag{17.3.51}$$

Moreover, up to extracting a subsequence, we may assume that

$$\|u_n\|_{L^\infty(B_R(\tilde{y}_n))} \geq \gamma. \tag{17.3.52}$$

Indeed, if (17.3.52) does not hold, then, by (17.3.51), $\|u_n\|_{L^\infty(\mathbb{R}^N)} < \gamma$. Then, using $\langle J'_{\varepsilon_n}(u_n), u_n \rangle = 0$ and (17.3.50), we infer that

$$\|u_n\|_{\varepsilon_n}^2 = \int_{\mathbb{R}^N} g(\varepsilon_n x, |u_n|^2)|u_n|^2 \, dx \leq \frac{V_1}{K} \int_{\mathbb{R}^N} |u_n|^2 \, dx,$$

which yields $\|u_n\|_{\varepsilon_n} = 0$, and this is impossible. Hence (17.3.52) holds true.

Taking into account (17.3.51) and (17.3.52), we deduce that if $p_n \in \mathbb{R}^N$ denotes a global maximum point of $|u_n|$, then p_n belongs to $B_R(\tilde{y}_n)$. Therefore, $p_n = \tilde{y}_n + q_n$ for some $q_n \in B_R$. Consequently, $\eta_n = \varepsilon_n \tilde{y}_n + \varepsilon_n q_n$ is a global maximum point of $|\hat{u}_n|(x) = |u_n|(x/\varepsilon_n)$. Since $|q_n| < R$ for any $n \in \mathbb{N}$ and $\varepsilon_n \tilde{y}_n \to y_0$, the continuity of V implies that

$$\lim_{n \to \infty} V(\eta_n) = V(y_0) = V_0.$$

Next we give a decay estimate for $|\hat{u}_n|$. First, we recall that by virtue of Lemma 3.2.17 there exists a continuous function w such that

$$0 < w(x) \leq \frac{C}{1 + |x|^{N+2s}} \quad \text{for all } x \in \mathbb{R}^N, \tag{17.3.53}$$

and

$$(-\Delta)^s w + \frac{V_1}{2} w = 0 \text{ in } \mathbb{R}^N \setminus \overline{B}_{R_1}, \tag{17.3.54}$$

in the classical sense, for some suitable $R_1 > 0$. Using Lemma 17.3.8, we know that $v_n(x) \to 0$ as $|x| \to \infty$ uniformly in $n \in \mathbb{N}$, so there exists $R_2 > 0$ such that

$$h_n = g(\varepsilon_n x + \varepsilon_n \tilde{y}_n, v_n^2)v_n \leq \frac{V_1}{2} v_n \quad \text{in } \mathbb{R}^N \setminus \overline{B}_{R_2}. \tag{17.3.55}$$

Let w_n denote the unique solution to

$$(-\Delta)^s w_n + V_1 w_n = h_n \quad \text{in } \mathbb{R}^N.$$

Then $w_n(x) \to 0$ as $|x| \to \infty$, uniformly in $n \in \mathbb{N}$ and, by comparison, $0 \leq v_n \leq w_n$ in $\mathbb{R}^N$. Moreover, in view of (17.3.55),

$$(-\Delta)^s w_n + \frac{V_1}{2} w_n = h_n - \frac{V_1}{2} w_n \leq 0 \text{ in } \mathbb{R}^N \setminus \overline{B}_{R_2}.$$

Choose $R_3 = \max\{R_1, R_2\}$ and set

$$c = \min_{\overline{B}_{R_3}} w > 0 \quad \text{and} \quad \tilde{w}_n = (b+1)w - cw_n, \qquad (17.3.56)$$

where $b = \sup_{n \in \mathbb{N}} \|w_n\|_{L^\infty(\mathbb{R}^N)} < \infty$. Our goal is to show that

$$\tilde{w}_n \geq 0 \quad \text{in } \mathbb{R}^N. \qquad (17.3.57)$$

First, we observe that

$$\tilde{w}_n \geq bc + w - bc > 0 \quad \text{in } \overline{B}_{R_3},$$

$$(-\Delta)^s \tilde{w}_n + \frac{V_1}{2} \tilde{w}_n \geq 0 \quad \text{in } \mathbb{R}^N \setminus \overline{B}_{R_3}.$$

Then we can apply Lemma 1.3.8 to deduce that (17.3.57) holds true. Next, from (17.3.53), (17.3.57) and the fact that $v_n \leq w_n$ in $\mathbb{R}^N$, we get that

$$0 \leq v_n(x) \leq w_n(x) \leq \frac{(b+1)}{c} w(x) \leq \frac{\tilde{C}}{1+|x|^{N+2s}} \quad \text{for all } x \in \mathbb{R}^N, n \in \mathbb{N},$$

for some constant $\tilde{C} > 0$. Recalling the definition of v_n, we infer that

$$|\hat{u}_n|(x) = |u_n|\left(\frac{x}{\varepsilon_n}\right) = v_n\left(\frac{x}{\varepsilon_n} - \tilde{y}_n\right)$$

$$\leq \frac{\tilde{C}}{1 + |\frac{x}{\varepsilon_n} - \tilde{y}_n|^{N+2s}}$$

$$= \frac{\tilde{C}\,\varepsilon_n^{N+2s}}{\varepsilon_n^{N+2s} + |x - \varepsilon_n \tilde{y}_n|^{N+2s}}$$

$$\leq \frac{\tilde{C}\,\varepsilon_n^{N+2s}}{\varepsilon_n^{N+2s} + |x - \eta_n|^{N+2s}} \quad \text{for all } x \in \mathbb{R}^N.$$

This ends the proof of Theorem 17.3.1.   □

Remark 17.3.11 The approaches used in this chapter are very flexible and permit to obtain existence and multiplicity results when we consider nonlinearities with critical or supercritical growth as in [44]. Moreover, we can also deal with other classes of fractional magnetic problems, such as the fractional magnetic Choquard equations [45], the fractional magnetic Kirchhoff problems [53,58], and the fractional magnetic Schrödinger-Poisson equations [51,59,60], under local or global conditions on the potential V.

Remark 17.3.12 If we assume that f is only continuous, then one can combine the arguments in Chap. 9 (see also Chaps. 11 and 13), based on the Nehari manifold approach in [322], and the arguments developed in the previous section, to obtain a multiplicity result for (17.3.1). In this case, one works with $(H_\varepsilon^s)^+ = \{u \in H_\varepsilon^s : |\text{supp}(|u|) \cap \Lambda_\varepsilon| > 0\}$ and $\mathbb{S}_\varepsilon^+ = \mathbb{S}_\varepsilon \cap (H_\varepsilon^s)^+$ for $\varepsilon > 0$; see [65] for more details.

17.4 A Multiplicity Result for a Fractional Magnetic Schrödinger Equation with Exponential Critical Growth

In this section we are interested in the existence, multiplicity and concentration of nontrivial solutions for the following fractional magnetic problem

$$
\begin{cases}
\varepsilon(-\Delta)_{\frac{A}{\varepsilon}}^{\frac{1}{2}} u + V(x)u = f(|u|^2)u & \text{in } \mathbb{R}, \\
|u| \in H^{\frac{1}{2}}(\mathbb{R}, \mathbb{R}),
\end{cases}
\tag{17.4.1}
$$

where $\varepsilon > 0$ is a parameter, $A : \mathbb{R} \to \mathbb{R}$ is a magnetic potential belonging to $C^{0,\alpha}(\mathbb{R}, \mathbb{R})$ for some $\alpha \in (0, 1]$, $V : \mathbb{R} \to \mathbb{R}$ is an electric potential, $f : \mathbb{R} \to \mathbb{R}$ is a continuous nonlinearity, and $(-\Delta)_A^{\frac{1}{2}}$ is the $\frac{1}{2}$-magnetic Laplacian given by

$$
(-\Delta)_A^{\frac{1}{2}} u(x) = \frac{1}{\pi} \lim_{r \to 0} \int_{\mathbb{R} \setminus (x-r, x+r)} \frac{u(x) - e^{i(x-y)A(\frac{x+y}{2})}u(y)}{|x-y|^2} \, dy,
$$

for any $u : \mathbb{R} \to \mathbb{C}$ sufficiently smooth.

We note that if $A = 0$, $N = 1$ and $s = \frac{1}{2}$, then the equation in (17.4.1) reduces to

$$
\varepsilon(-\Delta)^{\frac{1}{2}} u + V(x)u = f(x, u) \quad \text{in } \mathbb{R},
\tag{17.4.2}
$$

to which only few papers were devoted; see [11, 125, 162, 173, 199, 223]. Indeed, one of the main difficulties in the study of this class of problems is related to the fact that the embedding $H^{\frac{1}{2}}(\mathbb{R}, \mathbb{R}) \subset L^q(\mathbb{R}, \mathbb{R})$ is continuous for all $q \in [2, \infty)$, but that is no longer true for $L^\infty(\mathbb{R}, \mathbb{R})$. This means that the maximal growth allowed for the nonlinearity f so as to be able to deal with (17.4.2) via variational methods in a suitable subspace of $H^{\frac{1}{2}}(\mathbb{R}, \mathbb{R})$, is given by $e^{\alpha_0 |u|^2}$ as $|u| \to \infty$ for some $\alpha_0 > 0$. This is a consequence of the following fractional Moser-Trudinger inequality established by Ozawa in [285]:

Theorem 17.4.1 ([285]) *There exists $\omega \in (0, \pi]$ such that, for all $\alpha \in (0, \omega)$ there exists $C_\alpha > 0$ such that*

$$
\int_{\mathbb{R}} (e^{\alpha |u|^2} - 1) \, dx \le C_\alpha \|u\|_{L^2(\mathbb{R})}^2,
$$

for all $u \in H^{\frac{1}{2}}(\mathbb{R}, \mathbb{R})$ with $\|(-\Delta)^{\frac{1}{4}} u\|^2_{L^2(\mathbb{R})} \leq 1$.

It is worth pointing out that Moser-Trudinger type inequalities [279,327] have been widely used in the study of several bidimensional elliptic problems with exponential critical growth; see, for instance, [5, 10, 15, 129, 160, 175] and the references therein.

In this section, we will assume that $V \in C(\mathbb{R}, \mathbb{R})$ verifies the following del Pino-Felmer type conditions [165]:

(V_1) $\inf_{x \in \mathbb{R}} V(x) = V_0 > 0$;
(V_2) there exists a bounded open set $\Lambda \subset \mathbb{R}$ such that

$$V_0 < \min_{x \in \partial \Lambda} V(x) \quad \text{and} \quad M = \{x \in \Lambda : V(x) = V_0\} \neq \emptyset.$$

Without loss of generality, we will assume that $0 \in M$.

The function $f : \mathbb{R} \to \mathbb{R}$ satisfies the following assumptions:

(f_1) $f \in C^1(\mathbb{R}, \mathbb{R})$ and $f(t) = 0$ for $t \leq 0$;
(f_2) there exist $\omega \in (0, \pi]$ and $\alpha_0 \in (0, \omega)$ such that

$$\lim_{t \to \infty} \frac{f(t)\sqrt{t}}{e^{\alpha t}} = 0 \quad \text{for any } \alpha > \alpha_0, \quad \lim_{t \to \infty} \frac{f(t)\sqrt{t}}{e^{\alpha t}} = \infty \quad \text{for any } \alpha < \alpha_0;$$

(f_3) there exists $\theta > 2$ such that $0 < \frac{\theta}{2} F(t) \leq t f(t)$ for any $t > 0$, where $F(t) = \int_0^t f(\tau) d\tau$;
(f_4) $f(t)$ is increasing for $t > 0$;
(f_5) there exists $p > 2$ and $C_p > 0$ such that

$$f(t) \geq C_p t^{\frac{p-2}{2}},$$

where

$$C_p > \left[\beta_p \left(\frac{2\theta}{\theta - 2} \right) \frac{1}{\min\{1, V_0\}} \right]^{\frac{p-2}{2}},$$

with

$$\beta_p = \inf_{\mathcal{P}_0} \mathcal{E}_0, \quad \mathcal{P}_0 = \{u \in X_0 \setminus \{0\} : \langle \mathcal{E}'_0(u), u \rangle = 0\},$$

and

$$\mathcal{E}_0(u) = \frac{1}{2}\left(\frac{1}{2\pi}[u]_{\frac{1}{2}}^2 + \int_{\mathbb{R}} V_0|u|^2\,dx\right) - \frac{1}{p}\int_{\mathbb{R}}|u|^p\,dx;$$

(f_6) there exist $\sigma \in (2, \infty)$ and $C_\sigma > 0$ such that

$$f'(t) \geq C_\sigma t^{\frac{\sigma-4}{2}} \qquad \text{for any } t > 0.$$

Our main result is the following:

Theorem 17.4.2 ([49]) *Suppose that V satisfies (V_1)–(V_2) and f satisfies (f_1)–(f_6). Then, for any $\delta > 0$ such that*

$$M_\delta = \{x \in \mathbb{R} : dist(x, M) \leq \delta\} \subset \Lambda,$$

there exists $\varepsilon_\delta > 0$ such that, for any $\varepsilon \in (0, \varepsilon_\delta)$, problem (17.4.1) has at least $cat_{M_\delta}(M)$ nontrivial solutions. Moreover, if u_ε denotes one of these solutions and $\eta_\varepsilon \in \mathbb{R}$ is a global maximum point of $|u_\varepsilon|$, then

$$\lim_{\varepsilon \to 0} V(\eta_\varepsilon) = V_0,$$

and there exists $C > 0$ such that

$$|u_\varepsilon(x)| \leq \frac{C\,\varepsilon^2}{\varepsilon^2 + |x - \eta_\varepsilon|^2} \qquad \text{for all } x \in \mathbb{R}.$$

We note that when $A = 0$ in (17.4.1), and V and f fulfill (V_1)–(V_2) and (f_1)–(f_4) respectively, Alves et al. [11], inspired by [175], established the existence of a positive solution which concentrates around a local minima of $V(x)$ as $\varepsilon \to 0$. Therefore, Theorem 17.4.2 can be seen as a generalization in the fractional magnetic setting of the results in [11] and [175]. Moreover, our results complement and improve them, because here we also consider the issue of the multiplicity of solutions to (17.4.1) for $\varepsilon > 0$ small enough, which is not considered in the above mentioned papers, and we introduce a magnetic field. We emphasize that the presence of the magnetic field creates substantial difficulties that make the study of (17.4.1) rather difficult compared to [11, 175], and some appropriate arguments will be needed. More precisely, after considering a modified problem in the spirit of [165] (see also [18]), we will make use of the diamagnetic inequality [157], the Moser-Trudinger inequality for the modulus of functions belonging to the fractional magnetic space $H_\varepsilon^{\frac{1}{2}}$, and the Hölder continuity of the magnetic field to deduce some fundamental estimates which allow us to deduce the existence of a nontrivial solution for the modified problem when $\varepsilon > 0$ is small enough. After that, we apply a Nehari manifold argument and Lusternik-Schnirelman category theory to obtain a multiplicity result for the auxiliary problem. Finally, we prove that, for $\varepsilon > 0$ sufficiently small, the solutions of the modified problem are also solutions of the original one. This goal is achieved by proving some L^∞-estimates, independent of ε, obtained by combining a Moser iteration procedure [278], which takes care of the exponential critical growth of the nonlinearity, with a Kato's type inequality [238] which allows us to show that the modulus of each solution of the modified problem is a subsolution of a certain fractional problem in $\mathbb{R}$ involving $(-\Delta)^{\frac{1}{2}}$.

17.4.1 Preliminary Results

Let $L^2(\mathbb{R}, \mathbb{C})$ be the space of complex-valued square-integrable functions, endowed with the real scalar product

$$\langle u, v \rangle_{L^2} = \Re \left(\int_\mathbb{R} u \bar{v} \, dx \right)$$

for all $u, v \in L^2(\mathbb{R}, \mathbb{C})$. We define the fractional magnetic Sobolev space

$$H_A^{\frac{1}{2}}(\mathbb{R}, \mathbb{C}) = \{ u \in L^2(\mathbb{R}, \mathbb{C}) : [u]_A < \infty \},$$

where

$$[u]_A = \left(\frac{1}{2\pi} \iint_{\mathbb{R}^2} \frac{|u(x) - u(y)e^{\imath A(\frac{x+y}{2})(x-y)}|^2}{|x - y|^2} \, dx dy \right)^{\frac{1}{2}},$$

endowed with the norm

$$\|u\|_A = \left([u]_A^2 + \|u\|_{L^2(\mathbb{R})}^2 \right)^{\frac{1}{2}}.$$

When $A = 0$ and $u \in H^{\frac{1}{2}}(\mathbb{R}, \mathbb{R})$, we set

$$[u]^2 = \frac{1}{2\pi} [u]_{\frac{1}{2}}^2 = \frac{1}{2\pi} \iint_{\mathbb{R}^2} \frac{|u(x) - u(y)|^2}{|x - y|^2} \, dx dy.$$

By Lemma 17.2.3 and Theorem 1.1.8 we deduce that:

Theorem 17.4.3 *The space $H_A^{\frac{1}{2}}(\mathbb{R}, \mathbb{C})$ is continuously embedded in $L^r(\mathbb{R}, \mathbb{C})$ for any $r \in [2, \infty)$ and compactly embedded in $L^r(K, \mathbb{C})$ for any $r \in [1, \infty)$ and any compact $K \subset \mathbb{R}$.*

We prove the following version of Lemma 1.4.4 in the case $N = 1$ and $s = \frac{1}{2}$.

Lemma 17.4.4 *If* (u_n) *is a bounded sequence in* $H^{\frac{1}{2}}(\mathbb{R}, \mathbb{R})$ *and if*

$$\lim_{n \to \infty} \sup_{y \in \mathbb{R}} \int_{y-R}^{y+R} |u_n|^2 \, dx = 0 \tag{17.4.3}$$

for some $R > 0$, *then* $u_n \to 0$ *in* $L^t(\mathbb{R}^N)$ *for all* $t \in (2, \infty)$.

Proof Let $t \in (2, \infty)$. For any $r > t$, by standard interpolation, we have

$$\|u_n\|^t_{L^t(B_R(y))} \leq \|u_n\|^{1-\lambda}_{L^2(B_R(y))} \|u_n\|^{\lambda}_{L^r(B_R(y))},$$

where $\lambda \in (0, 1)$ is such that

$$\frac{1 - \lambda}{2} + \frac{\lambda}{r} = \frac{1}{t}.$$

Covering $\mathbb{R}$ by balls of radius R in such a way that each point of $\mathbb{R}$ is contained in at most two balls, we obtain

$$\|u_n\|^t_{L^t(\mathbb{R})} \leq C \sup_{y \in \mathbb{R}} \left(\int_{y-R}^{y+R} |u_n|^2 \, dx \right)^{\frac{(1-\lambda)t}{2}} \|u\|^{\lambda t}_{L^r(\mathbb{R})}.$$

Using the continuous embedding $H^{\frac{1}{2}}(\mathbb{R}, \mathbb{R}) \subset L^r(\mathbb{R})$, the boundedness of (u_n) in $H^{\frac{1}{2}}(\mathbb{R}, \mathbb{R})$, and (17.4.3), we deduce that

$$\|u_n\|^t_{L^t(\mathbb{R})} \leq C \sup_{y \in \mathbb{R}} \left(\int_{y-R}^{y+R} |u_n|^2 \, dx \right)^{\frac{(1-\lambda)t}{2}} \to 0 \quad \text{as } n \to \infty,$$

which completes the proof of lemma.   □

Next, we prove a number of technical results that will be used later.

Lemma 17.4.5 *Let* $(u_n) \subset H^{\frac{1}{2}}_A(\mathbb{R}, \mathbb{C})$ *be a bounded sequence, and set* $M = \sup_{n \in \mathbb{N}} \|u_n\|_A$. *Then,*

$$\sup_{n \in \mathbb{N}} \int_{\mathbb{R}} (e^{\alpha |u_n|^2} - 1) \, dx < \infty, \quad \text{for every } \alpha \in \left(0, \frac{\omega}{M^2} \right).$$

In particular, if $M \in (0, 1)$, there exists $\alpha_M > \omega$ such that

$$\sup_{n \in \mathbb{N}} \int_{\mathbb{R}} (e^{\alpha_M |u_n|^2} - 1) \, dx < \infty.$$

Proof Fix $\alpha \in \left(0, \frac{\omega}{M^2}\right)$. Using Lemma 17.2.3 we can see that

$$\left\| (-\Delta)^{\frac{1}{4}} \frac{|u_n|}{\|u_n\|_A} \right\|_{L^2(\mathbb{R})}^2 = \frac{\|(-\Delta)^{\frac{1}{4}} |u_n|\|_{L^2(\mathbb{R})}^2}{\|u_n\|_A^2} = \frac{[|u_n|]^2}{\|u_n\|_A^2} \leq 1.$$

By Theorem 17.4.1, we deduce that

$$\int_{\mathbb{R}} (e^{\alpha |u_n|^2} - 1) \, dx \leq \int_{\mathbb{R}} \left(e^{\alpha M^2 \left(\frac{|u_n|}{\|u_n\|_A} \right)^2} - 1 \right) dx \leq C_{\alpha M^2} \frac{\|u_n\|_{L^2(\mathbb{R})}^2}{\|u_n\|_A^2} \leq C_{\alpha M^2}.$$

$$(17.4.4)$$

If $M \in (0, 1)$, then $\omega < \frac{\omega}{M^2}$, so we can find $\alpha_M \in \left(0, \frac{\omega}{M^2}\right)$. Hence, we can use the above inequality with α replaced by α_M. $\quad\square$

Lemma 17.4.6 *Let $(u_n) \subset H_A^{\frac{1}{2}}(\mathbb{R}, \mathbb{C})$ be a sequence with*

$$\limsup_{n \to \infty} \|u_n\|_A^2 < 1. \qquad (17.4.5)$$

Then, there exists $t > 1$ sufficiently close to 1 and $C > 0$ such that

$$\int_{\mathbb{R}} \left(e^{\omega |u_n|^2} - 1 \right)^t dx \leq C \quad \text{for all } n \in \mathbb{N}.$$

Proof In view of (17.4.5), we can find $m \in (0, 1)$ and $n_0 \in \mathbb{N}$ such that

$$\|u_n\|_A^2 < m < 1 \quad \text{for all } n \geq n_0.$$

Take $t > 1$ sufficiently close to 1 and $\beta > t$ such that $\beta m < 1$. Observing that

$$\left(e^{\omega s^2} - 1 \right)^t \leq C(e^{\omega \beta s^2} - 1) \quad \text{for all } s \in \mathbb{R},$$

for some $C = C_\beta > 0$, we can deduce that

$$\int_{\mathbb{R}} \left(e^{\omega |u_n|^2} - 1 \right)^t dx \leq C \int_{\mathbb{R}} \left(e^{\beta m \omega \left(\frac{|u_n|}{\|u_n\|_A} \right)^2} - 1 \right) dx \quad \text{for all } n \geq n_0.$$

Applying Lemma 17.4.5 we can see that

$$\int_{\mathbb{R}} \left(e^{\omega |u_n|^2} - 1\right)^t dx \leq C_0 \quad \text{for all } n \geq n_0,$$

for some $C_0 > 0$. Taking

$$C = \max \left\{ C_0, \int_{\mathbb{R}} \left(e^{\omega |u_1|^2} - 1\right)^t dx, \ldots, \int_{\mathbb{R}} \left(e^{\omega |u_{n_0}|^2} - 1\right)^t dx \right\}$$

conclude the proof. $\qquad\qquad\qquad\qquad\qquad\qquad\qquad\qquad\qquad\qquad\qquad\qquad\square$

Lemma 17.4.7 *Let* $(u_n) \subset H_A^{\frac{1}{2}}(\mathbb{R}, \mathbb{C})$ *be a sequence satisfying* (17.4.5) *and let* $R > 0$. *If* $u_n \rightharpoonup u$ *in* $H_A^{\frac{1}{2}}(\mathbb{R}, \mathbb{C})$, *then, up to a subsequence, we have*

$$\int_{-R}^{R} F(|u_n|^2) \, dx \to \int_{-R}^{R} F(|u|^2) \, dx, \tag{17.4.6}$$

$$\int_{-R}^{R} f(|u_n|^2) |u_n|^2 \, dx \to \int_{-R}^{R} f(|u|^2) |u|^2 \, dx, \tag{17.4.7}$$

$$\Re \int_{-R}^{R} f(|u_n|^2) u_n \overline{\phi} \, dx \to \Re \int_{-R}^{R} f(|u|^2) u \overline{\phi} \, dx \quad \text{for all } \phi \in H_A^{\frac{1}{2}}. \tag{17.4.8}$$

Proof Using (f_1) and (f_2), we can see that for every $\beta > 1$ and $\alpha > \alpha_0$ there exists $C > 0$ such that

$$|F(t^2)| \leq C \left(|t|^2 + (e^{\alpha\beta|t|^2} - 1)\right) \quad \text{for all } t \in \mathbb{R},$$

which implies that

$$|F(|u_n|^2)| \leq C \left(|u_n|^2 + (e^{\alpha\beta|u_n|^2} - 1)\right) \quad \text{for all } n \in \mathbb{N}. \tag{17.4.9}$$

Put

$$\varphi_n(x) = C \left(e^{\alpha\beta|u_n|^2} - 1\right)$$

and fix $\beta, q > 1$ sufficiently close to 1 and α sufficiently close to α_0 such that $\varphi_n \in L^q(\mathbb{R}, \mathbb{R})$ and $\sup_{n \in \mathbb{N}} \|\varphi_n\|_{L^q(\mathbb{R})} < \infty$, in view of Lemma 17.4.6. Up to a subsequence,

$$\varphi_n \rightharpoonup \varphi = C(e^{\alpha\beta|u|^2} - 1) \quad \text{in } L^q(\mathbb{R}, \mathbb{R}).$$

Now, we show that

$$\varphi_n \to \varphi \quad \text{in } L^1(-R, R) \quad \text{for any } R > 0. \tag{17.4.10}$$

Denoting by χ_R the characteristic function of the interval $(-R, R)$, and noticing that $\chi_R \in L^{q'}(\mathbb{R}, \mathbb{R})$, we can see that

$$\int_{\mathbb{R}} \varphi_n \chi_R \, dx \to \int_{\mathbb{R}} \varphi \chi_R \, dx,$$

that is,

$$\int_{-R}^{R} \varphi_n \, dx \to \int_{-R}^{R} \varphi \, dx.$$

Since $\varphi_n, \varphi \geq 0$, this means that $\|\varphi_n\|_{L^1(-R,R)} \to \|\varphi\|_{L^1(-R,R)}$. Observing that $\varphi_n \to \varphi$ a.e. in $\mathbb{R}$, we can use the Brezis-Lieb lemma [113] to conclude that $\|\varphi_n - \varphi\|_{L^1(-R,R)} \to 0$. Then, recalling that $|u_n| \to |u|$ in $L^2(-R, R)$, and using (17.4.9) and (17.4.10), we can apply the dominated convergence theorem to deduce that (17.4.6) is verified. Similar arguments show that (17.4.7) and (17.4.8) hold. $\qquad \square$

Lemma 17.4.8 *Let* $(u_n) \subset H_A^{\frac{1}{2}}(\mathbb{R}, \mathbb{C})$ *be a sequence satisfying* (17.4.5). *If there exists* $R > 0$ *such that*

$$\lim_{n \to \infty} \sup_{y \in \mathbb{R}} \int_{y-R}^{y+R} |u_n|^2 dx = 0,$$

then

$$\lim_{n \to \infty} \int_{\mathbb{R}} F(|u_n|^2) \, dx = \lim_{n \to \infty} \int_{\mathbb{R}} f(|u_n|^2)|u_n|^2 \, dx = 0.$$

Proof By Lemma 17.4.4,

$$|u_n| \to 0 \quad \text{in } L^q(\mathbb{R}, \mathbb{R}) \text{ for any } q \in (2, \infty). \tag{17.4.11}$$

Since (u_n) satisfies (17.4.5), we can use Lemma 17.4.6 to find $t > 1$ sufficiently close to 1 and $C > 0$ such that

$$\int_{\mathbb{R}} \left(e^{\omega |u_n|^2} - 1 \right)^t dx \leq C \quad \text{for all } n \in \mathbb{N}.$$

Then, from the growth assumptions on f and applying the Hölder inequality we have

$$\int_{\mathbb{R}} f(|u_n|^2)|u_n|^2 dx \leq \delta \int_{\mathbb{R}} |u_n|^2 \, dx + C_\delta \int_{\mathbb{R}} |u_n|(e^{\omega|u_n|^2} - 1) \, dx$$

$$\leq C\delta + C\|u_n\|_{L^{t'}(\mathbb{R})} \quad \text{for any } \delta > 0,$$

which together with (17.4.11) implies that

$$\int_{\mathbb{R}} f(|u_n|^2)|u_n|^2 \, dx \to 0 \quad \text{as } n \to \infty.$$

Consequently, using (f_3), we deduce that

$$\int_{\mathbb{R}} F(|u_n|^2) \, dx \to 0 \quad \text{as } n \to \infty.$$

$\square$

Finally, we give a simple variant of Lemma 1.4.8 in the one-dimensional case.

Lemma 17.4.9 *Let* $u \in H^{\frac{1}{2}}(\mathbb{R}, \mathbb{R})$ *and* $\phi \in C_c^\infty(\mathbb{R}, \mathbb{R})$ *such that* $0 \leq \phi \leq 1$, $\phi = 1$ *in* $(-1, 1)$ *and* $\phi = 0$ *in* $\mathbb{R} \setminus (-2, 2)$. *Set* $\phi_r(x) = \phi(\frac{x}{r})$ *for* $r > 0$. *Then*

$$\lim_{r \to \infty} [u\phi_r - u]_{\frac{1}{2}} = 0 \quad \text{and} \quad \lim_{r \to \infty} \|u\phi_r - u\|_{L^2(\mathbb{R})} = 0.$$

Proof Since $\phi_r u \to u$ a.e. in $\mathbb{R}$ as $r \to \infty$, $0 \leq \phi \leq 1$ and $u \in L^2(\mathbb{R}, \mathbb{R})$, we can use the dominated convergence theorem to see that $\lim_{r \to \infty} \|u\phi_r - u\|_{L^2(\mathbb{R})} = 0$.

Therefore, we only need to show that the first limit is equal to zero. Note that

$$[u\phi_r - u]_{\frac{1}{2}}^2 \leq 2 \left(\iint_{\mathbb{R}^2} |u(x)|^2 \frac{|\phi_r(x) - \phi_r(y)|^2}{|x - y|^2} \, dx dy + \iint_{\mathbb{R}^2} \frac{|\phi_r(x) - 1|^2 |u(x) - u(y)|^2}{|x - y|^2} \, dx dy \right)$$

$$= 2(A_r + B_r).$$

Taking into account that $|\phi_r(x) - 1| \leq 2$, $|\phi_r(x) - 1| \to 0$ a.e. in $\mathbb{R}$ and $u \in H^{\frac{1}{2}}(\mathbb{R}, \mathbb{R})$, and applying the dominated convergence theorem, we see that

$$B_r \to 0 \quad \text{as } r \to \infty.$$

Now we show that

$$A_r \to 0 \quad \text{as } r \to \infty.$$

We have

$$A_r = \int_{\mathbb{R}} |u(x)|^2 \left(\int_{|y-x|>r} \frac{|\phi_r(x) - \phi_r(y)|^2}{|x-y|^2} \, dy + \int_{|y-x|\le r} \frac{|\phi_r(x) - \phi_r(y)|^2}{|x-y|^2} \, dy \right) dx$$

$$\le \int_{\mathbb{R}} |u(x)|^2 \left(\int_{|y-x|>r} \frac{4\|\phi\|_{L^\infty(\mathbb{R})}^2}{|x-y|^2} \, dy + \int_{|y-x|\le r} r^{-2}\|\phi'\|_{L^\infty(\mathbb{R})}^2 \, dy \right) dx$$

$$= \frac{\left(4\|\phi\|_{L^\infty(\mathbb{R})}^2 + 2\|\phi'\|_{L^\infty(\mathbb{R})}^2 \right)}{r} \int_{\mathbb{R}} |u(x)|^2 dx$$

$$\le \frac{C}{r} \to 0 \quad \text{as } r \to \infty.$$

This ends the proof of lemma. $\qquad\qquad\qquad\qquad\qquad\qquad\qquad\qquad\qquad\qquad$ □

17.4.2 Variational Framework and Modified Problem

Using the change of variable $u(x) \mapsto u(\varepsilon x)$, instead of (17.4.1), we deal with the following equivalent problem

$$\begin{cases} (-\Delta)^{\frac{1}{2}}_{A_\varepsilon} u + V(\varepsilon x)u = f(|u|^2)u & \text{in } \mathbb{R}, \\ |u| \in H^{\frac{1}{2}}(\mathbb{R}, \mathbb{R}), \end{cases} \qquad (17.4.12)$$

where we set $A_\varepsilon(x) = A(\varepsilon x)$.

In that follows, we use a penalization argument inspired by [18, 165, 175] to study (17.4.12). Fix $k > \frac{\theta}{\theta-2}$ and $a > 0$ such that $f(a) = \frac{V_0}{k}$, and we consider the function

$$\hat{f}(t) = \begin{cases} f(t), & \text{if } t \le a, \\ \frac{V_0}{k}, & \text{if } t > a. \end{cases}$$

Let $t_a, T_a > 0$ be such that $t_a < a < T_a$ and take $\xi \in C_c^\infty(\mathbb{R}, \mathbb{R})$ such that

(ξ_1) $\xi(t) \le \hat{f}(t)$ for all $t \in [t_a, T_a]$,
(ξ_2) $\xi(t_a) = \hat{f}(t_a), \xi(T_a) = \hat{f}(T_a), \xi'(t_a) = \hat{f}'(t_a)$ and $\xi'(T_a) = \hat{f}'(T_a)$,
(ξ_3) the function $t \mapsto \xi(t)$ is nondecreasing for all $t \in [t_a, T_a]$.

Let us define $\tilde{f} \in C^1(\mathbb{R}, \mathbb{R})$ as follows:

$$\tilde{f}(t) = \begin{cases} \hat{f}(t), & \text{if } t \notin [t_a, T_a], \\ \xi(t), & \text{if } t \in [t_a, T_a]. \end{cases}$$

We introduce the penalized nonlinearity $g : \mathbb{R} \times \mathbb{R} \to \mathbb{R}$ by

$$g(x, t) = \chi_\Lambda(x) f(t) + (1 - \chi_\Lambda(x)) \tilde{f}(t),$$

where χ_Λ is the characteristic function of Λ, and we write $G(x, t) = \int_0^t g(x, \tau) \, d\tau$.

In view of (f_1)–(f_4) and (ξ_1)–(ξ_3), the function g enjoys the following properties:

(g_1) $\lim_{t \to 0} g(x, t) = 0$ uniformly in $x \in \mathbb{R}$;

(g_2) given $q \geq 2$, for any $\delta > 0$ and $\alpha > \alpha_0$ there exists $C = C(\delta, \alpha, q) > 0$ such that

$$g(x, t) \leq \delta + Ct^{\frac{q-2}{2}} \left(e^{\alpha t} - 1 \right), \qquad \text{for any } t \geq 0,$$

and

$$G(x, t) \leq \delta t + Ct^{\frac{q}{2}} (e^{\alpha t} - 1), \qquad \text{for any } t \geq 0,$$

(g_3) (i) $0 < \frac{\theta}{2} G(x, t) \leq g(x, t)t$ for any $x \in \Lambda$ and $t > 0$,

(ii) $0 \leq G(x, t) \leq g(x, t)t \leq \frac{V(x)}{k} t$ for any $x \in \Lambda^c$ and $t > 0$;

(g_4) $t \mapsto g(x, t)$ is increasing for $t > 0$.

From now on, we focus our study on the modified problem

$$\begin{cases} (-\Delta)^{\frac{1}{2}}_{A_\varepsilon} u + V(\varepsilon x)u = g(\varepsilon x, |u|^2)u & \text{in } \mathbb{R}, \\ |u| \in H^{\frac{1}{2}}(\mathbb{R}, \mathbb{R}). \end{cases} \tag{17.4.13}$$

Note that if u is a solution of (17.4.13) such that

$$|u(x)| \leq \sqrt{t_a} \quad \text{for all } x \in \mathbb{R} \setminus \Lambda_\varepsilon, \tag{17.4.14}$$

where $\Lambda_\varepsilon = \{x \in \mathbb{R} : \varepsilon x \in \Lambda\}$, then u is also a solution of the original problem (17.4.12).

To study weak solutions to (17.4.13), we look for the critical points of the Euler–Lagrange functional

$$J_\varepsilon(u) = \frac{1}{2} \|u\|_\varepsilon^2 - \frac{1}{2} \int_\mathbb{R} G(\varepsilon x, |u|^2) \, dx$$

which is well defined for any function u belonging to the space

$$H_\varepsilon^{\frac{1}{2}} = \overline{C_c^\infty(\mathbb{R}, \mathbb{C})}^{\|\cdot\|_\varepsilon}$$

endowed with the norm

$$\|u\|_\varepsilon = \left([u]_{A_\varepsilon}^2 + \|\sqrt{V(\varepsilon\,\cdot)}\,|u|\|_{L^2(\mathbb{R})}^2 \right)^{\frac{1}{2}}.$$

Using (V_1), it is easy to see that the embedding $H_\varepsilon^{\frac{1}{2}} \subset L^q(\mathbb{R}, \mathbb{C})$ is continuous for all $q \in [2, \infty)$ and that $H_\varepsilon^{\frac{1}{2}} \subset L^q(K, \mathbb{C})$ for all $q \in [1, \infty)$ and $K \subset \mathbb{R}$ compact.

As we will see later, it will be crucial to consider the following family of autonomous problem associated with (17.4.12), with $\mu \in \mathbb{R}_+$:

$$\begin{cases} (-\Delta)^{\frac{1}{2}}u + \mu u = f(u^2)u & \text{in } \mathbb{R}, \\ u \in H^{\frac{1}{2}}(\mathbb{R}, \mathbb{R}). \end{cases} \tag{17.4.15}$$

We denote by $I_\mu : H_\mu^{\frac{1}{2}}(\mathbb{R}, \mathbb{R}) \to \mathbb{R}$ the corresponding energy functional

$$I_\mu(u) = \frac{1}{2}\|u\|_\mu^2 - \frac{1}{2}\int_{\mathbb{R}} F(u^2)\,dx,$$

where $H_\mu^{\frac{1}{2}}(\mathbb{R}, \mathbb{R})$ is the space $H^{\frac{1}{2}}(\mathbb{R}, \mathbb{R})$ equipped with the norm

$$\|u\|_\mu = \left([u]^2 + \mu\|u\|_{L^2(\mathbb{R})}^2 \right)^{\frac{1}{2}}.$$

As a first result, we verify that J_ε possesses a mountain pass structure [29].

Lemma 17.4.10

(i) $J_\varepsilon(0) = 0$;

(ii) *there exist* $\sigma, \rho > 0$ *such that* $J_\varepsilon(u) \geq \sigma$ *for any* $u \in H_\varepsilon^{\frac{1}{2}}$ *such that* $\|u\|_\varepsilon = \rho$;

(iii) *there exists* $e \in H_\varepsilon^{\frac{1}{2}}$ *with* $\|e\|_\varepsilon > \rho$, *such that* $J_\varepsilon(e) < 0$.

Proof Fix $u \in H_\varepsilon^{\frac{1}{2}}$ such that $\|u\|_\varepsilon = \rho < 1$ and take $\alpha \in (\omega, \frac{\omega}{\rho^2})$. By the growth assumptions on g, there exists $r > 1$ close to 1 such that $r\alpha < \frac{\omega}{\rho^2}, q > 2$ and $C > 0$ with

$$G(\varepsilon x, t^2) \leq \frac{V_0}{4}t^2 + C\left(e^{r\alpha t^2} - 1 \right)^{\frac{1}{r}} t^q \quad \text{for all } x \in \mathbb{R}, t \geq 0.$$

Therefore, applying the Hölder inequality and using (17.4.4) we get

$$J_\varepsilon(u) \geq \frac{1}{2}\|u\|_\varepsilon^2 - \frac{V_0}{4}\|u\|_{L^2(\mathbb{R})}^2 - C \int_\mathbb{R} \left(e^{r\alpha|u|^2} - 1\right)^{\frac{1}{r}} |u|^q \, dx$$

$$\geq \frac{1}{4}\|u\|_\varepsilon^2 - C \left(\int_\mathbb{R} (e^{r\alpha|u|^2} - 1)\, dx\right)^{\frac{1}{r}} \left(\int_\mathbb{R} |u|^{r'q}\, dx\right)^{\frac{1}{r'}}$$

$$\geq \frac{1}{4}\|u\|_\varepsilon^2 - C\|u\|_\varepsilon^q = \frac{1}{4}\rho^2 - C\rho^q = \sigma > 0,$$

for every ρ sufficiently small.

Regarding (iii), we note that in view of (g_3), for any $u \in H_\varepsilon^{\frac{1}{2}} \setminus \{0\}$ with $\text{supp}(u) \subset \Lambda_\varepsilon$ and $t > 0$, it holds that

$$J_\varepsilon(tu) \leq \frac{t^2}{2}\|u\|_\varepsilon^2 - \frac{1}{2}\int_{\Lambda_\varepsilon} G(\varepsilon x, t^2|u|^2)\, dx$$

$$\leq \frac{t^2}{2}\|u\|_\varepsilon^2 - Ct^\theta \int_{\Lambda_\varepsilon} |u|^\theta \, dx + C$$

which together with $\theta > 2$ implies that $J_\varepsilon(tu) \to -\infty$ as $t \to \infty$. □

Remark 17.4.11 Similar arguments show that I_μ has a mountain pass geometry (concerning condition (iii), we take $u \in C_c^\infty(\mathbb{R}, \mathbb{R}) \setminus \{0\}$).

Taking into account Lemma 17.4.10, we define the mountain pass level

$$c_\varepsilon = \inf_{\gamma \in \Gamma_\varepsilon} \max_{t \in [0,1]} J_\varepsilon(\gamma(t)),$$

where

$$\Gamma_\varepsilon = \{\gamma \in C([0, 1], H_\varepsilon^{\frac{1}{2}}) : \gamma(0) = 0 \text{ and } J_\varepsilon(\gamma(1)) < 0\}.$$

We also introduce the Nehari manifold associated with J_ε, that is

$$\mathcal{N}_\varepsilon = \{u \in H_\varepsilon^{\frac{1}{2}} \setminus \{0\} : \langle J_\varepsilon'(u), u \rangle = 0\}.$$

Note that for all $u \in \mathcal{N}_\varepsilon$, the growth assumptions on g ensure that we can find $r^* > 0$ (independent of u) such that

$$\|u\|_\varepsilon \geq r^* > 0. \tag{17.4.16}$$

It is easy to verify that c_ε can be characterized as follows:

$$c_\varepsilon = \inf_{u \in H_\varepsilon^{\frac{1}{2}} \setminus \{0\}} \sup_{t \geq 0} J_\varepsilon(tu) = \inf_{u \in \mathcal{N}_\varepsilon} J_\varepsilon(u).$$

Then we denote by $\mathcal{M}_\mu$ the Nehari manifold associated with I_μ and we set $m_\mu = \inf_{u \in \mathcal{M}_\mu} I_\mu$. The number m_μ and the manifold $\mathcal{M}_\mu$ have properties similar to those of c_ε and $\mathcal{N}_\varepsilon$.

Remark 17.4.12 The minimax level m_{V_0} satisfies the following upper bound:

$$0 < m_{V_0} < \left(\frac{1}{2} - \frac{1}{\theta} \right) \min\{1, V_0\}.$$

Indeed, let $w_0 \in H^{\frac{1}{2}}(\mathbb{R}, \mathbb{R})$ be such that $\mathcal{E}_0(w_0) = \beta_p$ and $\mathcal{E}_0'(w_0) = 0$. We note that the existence of w_0 can be obtained in a standard way by using a minimization argument and Lemma 17.4.4; see, for instance, [10, 162]. By the characterization of m_{V_0},

$$m_{V_0} \leq \max_{t \geq 0} I_{V_0}(tw_0).$$

By using (f_5) we deduce that

$$
\begin{aligned}
m_{V_0} &\leq \max_{t \geq 0} \left[\frac{t^2}{2} \|w_0\|_{V_0}^2 - \frac{C_p t^p}{p} \|w_0\|_{L^p(\mathbb{R})}^p \right] \\
&= \max_{t \geq 0} \left[\frac{t^2}{2} - \frac{C_p t^p}{p} \right] \|w_0\|_{L^p(\mathbb{R})}^p \\
&= C_p^{\frac{2}{2-p}} \left(\frac{1}{2} - \frac{1}{p} \right) \|w_0\|_{L^p(\mathbb{R})}^p \\
&= C_p^{\frac{2}{2-p}} \beta_p \\
&< \left(\frac{1}{2} - \frac{1}{\theta} \right) \min\{1, V_0\},
\end{aligned}
$$

which gives the desired estimate. Hereafter, we will assume that k is large enough so that

$$0 < m_{V_0} < \left(\frac{1}{2} - \frac{1}{\theta} - \frac{1}{2k} \right) \min\{1, V_0\} < \left(\frac{1}{2} - \frac{1}{\theta} \right) \min\{1, V_0\}.$$

Now we prove the following compactness result for the autonomous problem (17.4.15):

Lemma 17.4.13 *Let $(u_n) \subset \mathcal{M}_{V_0}$ be a sequence such that $I_{V_0}(u_n) \to m_{V_0}$. Then*

(i) *either (u_n) strongly converges in $H_{V_0}^{\frac{1}{2}}(\mathbb{R}, \mathbb{R})$, or*
(ii) *there exists a sequence $(\tilde{y}_n) \subset \mathbb{R}$ such that, up to a subsequence, $v_n(x) = u_n(x + \tilde{y}_n)$ converges strongly in $H_{V_0}^{\frac{1}{2}}(\mathbb{R}, \mathbb{R})$.*

In particular, there exists a positive minimizer w of I_{V_0} in $\mathcal{M}_{V_0}$.

Proof By Theorem 2.2.1, we may assume that (u_n) is a Palais–Smale sequence for I_{V_0} at the level m_{V_0}. Using (f_3) we obtain that (u_n) is bounded in $H_{V_0}^{\frac{1}{2}}$. Indeed,

$$
I_{V_0}(u_n) - \frac{1}{\theta} \langle I_{V_0}'(u_n), u_n \rangle = \left(\frac{1}{2} - \frac{1}{\theta} \right) \|u_n\|_{V_0}^2 + \int_{\mathbb{R}} \left[\frac{1}{\theta} f(u^2) u^2 - \frac{1}{2} F(u^2) \right] dx
$$

$$
\geq \left(\frac{1}{2} - \frac{1}{\theta} \right) \|u_n\|_{V_0}^2. \tag{17.4.17}
$$

Hence, up to a subsequence, $u_n \rightharpoonup u$ in $H_{V_0}^{\frac{1}{2}}$ for some $u \in H_{V_0}^{\frac{1}{2}}$. Now we distinguish two cases. First, assume that $u \neq 0$. Note that, by (17.4.17) and Remark 17.4.12,

$$
\min\{1, V_0\} \left(\frac{1}{2} - \frac{1}{\theta} \right) > m_{V_0} \geq \left(\frac{1}{2} - \frac{1}{\theta} \right) \limsup_{n \to \infty} \|u_n\|_{V_0}^2,
$$

that is, $\limsup_{n \to \infty} \|u_n\|_{V_0}^2 < 1$. Then, arguing as in the proof of Lemma 17.4.7 (see (17.4.8)) and using the denseness of $C_c^\infty(\mathbb{R}, \mathbb{R})$ in $H_{V_0}^{\frac{1}{2}}$, we deduce that $\langle I_{V_0}'(u), \phi \rangle = 0$ for all $\phi \in H_{V_0}^{\frac{1}{2}}$. In particular, $u \in \mathcal{M}_{V_0}$ and $I_{V_0}(u) \geq m_{V_0}$. Using Fatou's lemma, the weak lower semicontinuity of $\| \cdot \|_{V_0}$ and (f_3), it is easy to prove that $I_{V_0}(u) \leq m_{V_0}$. Consequently, u is a minimizer of I_{V_0} in $\mathcal{M}_{V_0}$. Since $\langle I_{V_0}'(u), u^- \rangle = 0$ and $f(t) = 0$ when $t \leq 0$, we have that $u \geq 0$ in $\mathbb{R}$. Using a suitable Moser iteration argument (see the proof of Lemma 17.4.24), we can show that $u \in L^\infty(\mathbb{R}, \mathbb{R})$. This implies that $f(u^2) u \in L^\infty(\mathbb{R}, \mathbb{R})$. Applying Proposition 1.3.2 we can deduce that $u \in C^{0,\alpha}(\mathbb{R}, \mathbb{R})$ for any $\alpha \in (0, 1)$. Since $f \in C^1$ and $u \in C^{0,\alpha}(\mathbb{R}, \mathbb{R})$, we obtain that $f(u^2) u \in C^{0,\alpha}(\mathbb{R}, \mathbb{R})$, and by Proposition 1.3.1, $u \in C^{1,\alpha}(\mathbb{R}, \mathbb{R})$. Using Theorem 1.3.5 (or Proposition 1.3.11-(ii)), it is easy to deduce that $u > 0$ in $\mathbb{R}$. Now, suppose that $u = 0$. In this case, since $(u_n) \subset \mathcal{M}_{V_0}$, similar argument as the one used for (17.4.16) shows that $\|u_n\|_{V_0} \not\to 0$. Moreover, we can find $(\tilde{y}_n) \subset \mathbb{R}$, $R, \beta > 0$ such that

$$
\liminf_{n \to \infty} \int_{y_n - R}^{y_n + R} |u_n|^2 \, dx \geq \beta.
$$

Otherwise, we can argue as in the proof of Lemma 17.4.8 to deduce that

$$\int_{\mathbb{R}} f(u_n^2) u_n^2 \, dx \to 0$$

which combined with the fact that $(u_n) \subset \mathcal{M}_{V_0}$ yields $\|u_n\|_{V_0} \to 0$, which is a contradiction. Define $v_n(x) = u_n(x + \tilde{y}_n)$. Clearly, (v_n) is a bounded Palais–Smale sequence for I_{V_0} at the level m_{V_0}, and we may assume that $v_n \rightharpoonup v$ in $H_{V_0}^{\frac{1}{2}}$ for some $v \in H_{V_0}^{\frac{1}{2}}$, $v \neq 0$. Then we can repeat the arguments used in the case $u \neq 0$ to get the desired conclusion.   □

Next, we show an interesting relation between c_ε and m_{V_0}.

Lemma 17.4.14 *The numbers c_ε and m_{V_0} satisfy the inequality*

$$\limsup_{\varepsilon \to 0} c_\varepsilon \leq m_{V_0}.$$

Proof By Lemma 17.4.13, we can find a positive ground state $w \in H_{V_0}^{\frac{1}{2}}(\mathbb{R}, \mathbb{R})$ for the autonomous problem (17.4.15) with $\mu = V_0$, so that $I'_{V_0}(w) = 0$ and $I_{V_0}(w) = m_{V_0}$. Since $w \in C^{1,\alpha}(\mathbb{R}, \mathbb{R}) \cap L^2(\mathbb{R}, \mathbb{R})$, we deduce that $w(x) \to 0$ as $|x| \to \infty$. In what follows, we prove the following decay estimate for w:

$$0 < w(x) \leq \frac{C}{|x|^2} \quad \text{for all } |x| > 1. \tag{17.4.18}$$

Since $w(x) \to 0$ as $|x| \to \infty$ and since (f_1) holds, we can see that there exists $R_1 > 0$ such that

$$(-\Delta)^{\frac{1}{2}} w + \frac{V_0}{2} w \leq 0 \quad \text{in } \mathbb{R} \setminus [-R_1, R_1]. \tag{17.4.19}$$

By Lemma 3.2.17 and Remark 3.2.18, there exists a continuous and positive function H such that

$$0 < H(x) \leq \frac{C}{1 + |x|^2} \quad \text{for all } x \in \mathbb{R}, \tag{17.4.20}$$

and

$$(-\Delta)^{\frac{1}{2}} H + \frac{V_0}{2} H = 0 \quad \text{in } \mathbb{R} \setminus [-R_2, R_2], \tag{17.4.21}$$

for some $R_2 > 0$. Let $R = \max\{R_1, R_2\}$ and $C_0 = \|w\|_{L^\infty(\mathbb{R})}c^{-1}$, where $c = \min_{[-R,R]} H$. Then the continuous function $z(x) = C_0 H(x) - w(x)$ satisfies $z(x) \geq 0$ in $[-R, R]$, and using (17.4.19) and (17.4.21), we have $(-\Delta)^{\frac{1}{2}}z + \frac{V_0}{2}z \geq 0$ in $\mathbb{R} \setminus [-R, R]$. By Lemma 1.3.8, we obtain that $z \geq 0$ in $\mathbb{R} \setminus [-R, R]$, and thus $z \geq 0$ in $\mathbb{R}$. This fact combined with (17.4.20) implies that (17.4.18) holds true.

Now, take a cut-off function $\eta \in C_c^\infty(\mathbb{R}, [0, 1])$ such that $\eta = 1$ in a neighborhood of zero, say $(-\frac{\delta}{2}, \frac{\delta}{2})$, and $\text{supp}(\eta) \subset (-\delta, \delta) \subset \Lambda$ for some $\delta > 0$. Set $w_\varepsilon(x) = \eta_\varepsilon(x)w(x)e^{\iota A(0)x}$, with $\eta_\varepsilon(x) = \eta(\varepsilon x)$ for $\varepsilon > 0$, and observe that $|w_\varepsilon| = \eta_\varepsilon w$ and $w_\varepsilon \in H_\varepsilon^{\frac{1}{2}}$ in view of Lemma 17.2.5. In what follows, we prove that

$$\lim_{\varepsilon \to 0} \|w_\varepsilon\|_\varepsilon^2 = \|w\|_{V_0}^2 \in (0, \infty). \tag{17.4.22}$$

Clearly, by the dominated convergence theorem, we have

$$\int_\mathbb{R} V(\varepsilon x)|w_\varepsilon|^2\, dx \to \int_\mathbb{R} V_0|w|^2\, dx.$$

Next, we show that

$$\lim_{\varepsilon \to 0} [w_\varepsilon]_{A_\varepsilon}^2 = [w]^2. \tag{17.4.23}$$

We have that

$$[w_\varepsilon]_{A_\varepsilon}^2 = \frac{1}{2\pi} \iint_{\mathbb{R}^2} \frac{|e^{\iota A(0)x}\eta_\varepsilon(x)w(x) - e^{\iota A_\varepsilon(\frac{x+y}{2})(x-y)}e^{\iota A(0)\cdot y}\eta_\varepsilon(y)w(y)|^2}{|x-y|^2}\, dxdy$$

$$= [\eta_\varepsilon w]^2 + \frac{1}{2\pi} \iint_{\mathbb{R}^2} \frac{\eta_\varepsilon^2(y)w^2(y)|e^{\iota[A_\varepsilon(\frac{x+y}{2})-A(0)](x-y)} - 1|^2}{|x-y|^2}\, dxdy$$

$$+ \frac{1}{\pi}\mathfrak{R} \iint_{\mathbb{R}^2} \frac{(\eta_\varepsilon(x)w(x) - \eta_\varepsilon(y)w(y))\eta_\varepsilon(y)w(y)(1 - e^{-\iota[A_\varepsilon(\frac{x+y}{2})-A(0)](x-y)})}{|x-y|^2}\, dxdy$$

$$= [\eta_\varepsilon w]^2 + X_\varepsilon + \frac{1}{\pi}Y_\varepsilon,$$

and

$$|Y_\varepsilon| \leq (2\pi)[\eta_\varepsilon w]\sqrt{X_\varepsilon}.$$

On the other hand, by Lemma 17.4.9, it follows that

$$[\eta_\varepsilon w]_{\frac{1}{2}} \to [w]_{\frac{1}{2}} \quad \text{as } \varepsilon \to 0. \tag{17.4.24}$$

Arguing as in the proof of Corollary 17.2.18 and using (17.4.18), we can verify that $X_\varepsilon \to 0$ as $\varepsilon \to 0$, which implies that (17.4.23) holds.

Let $t_\varepsilon > 0$ be the unique number such that

$$J_\varepsilon(t_\varepsilon w_\varepsilon) = \max_{t \geq 0} J_\varepsilon(t w_\varepsilon).$$

Then t_ε verifies

$$\|w_\varepsilon\|_\varepsilon^2 = \int_{\mathbb{R}} g(\varepsilon x, t_\varepsilon^2 |w_\varepsilon|^2) |w_\varepsilon|^2 \, dx = \int_{\mathbb{R}} f(t_\varepsilon^2 |w_\varepsilon|^2) |w_\varepsilon|^2 \, dx, \tag{17.4.25}$$

where we used that $\mathrm{supp}(\eta) \subset \Lambda$ and $g = f$ on Λ. Let us prove that $t_\varepsilon \to 1$ as $\varepsilon \to 0$. Since $\eta = 1$ in $(-\frac{\delta}{2}, \frac{\delta}{2})$ and recalling that w is a continuous positive function on $\mathbb{R}$, we can see that (f_4) yields

$$\|w_\varepsilon\|_\varepsilon^2 \geq f(t_\varepsilon^2 \sigma_0^2) \int_{-\frac{\delta}{2}}^{\frac{\delta}{2}} |w|^2 \, dx,$$

where $\sigma_0 = \min_{[-\frac{\delta}{2}, \frac{\delta}{2}]} w > 0$.

If $t_\varepsilon \to \infty$ as $\varepsilon \to 0$, then we can use (f_3) and (17.4.22) to deduce that $\|w\|_{V_0}^2 = \infty$ which gives a contradiction. On the other hand, if $t_\varepsilon \to 0$ as $\varepsilon \to 0$, then by the growth assumptions on g and (17.4.22), we infer that $\|w\|_{V_0}^2 = 0$, which is impossible.

In conclusion, $t_\varepsilon \to t_0 \in (0, \infty)$ as $\varepsilon \to 0$. Hence, taking the limit as $\varepsilon \to 0$ in (17.4.25) and using (17.4.22), we can see that

$$\|w\|_{V_0}^2 = \int_{\mathbb{R}} f(t_0^2 |w|^2) |w|^2 \, dx. \tag{17.4.26}$$

Since $w \in \mathcal{M}_{V_0}$ and (f_4) holds, we deduce that $t_0 = 1$. Applying the dominated convergence theorem, we obtain that $\lim_{\varepsilon \to 0} J_\varepsilon(t_\varepsilon w_\varepsilon) = I_{V_0}(w) = m_{V_0}$. Finally, since $c_\varepsilon \leq \max_{t \geq 0} J_\varepsilon(t w_\varepsilon) = J_\varepsilon(t_\varepsilon w_\varepsilon)$, we conclude that $\limsup_{\varepsilon \to 0} c_\varepsilon \leq m_{V_0}$. □

In the next lemma we show a compactness condition for J_ε.

Lemma 17.4.15 *Let $c \in \mathbb{R}$ be such that $0 < c < \left(\frac{1}{2} - \frac{1}{\theta} - \frac{1}{2k}\right) \min\{1, V_0\}$. Then J_ε satisfies the Palais–Smale condition at the level c.*

Proof Let $(u_n) \subset H_\varepsilon^{\frac{1}{2}}$ be a Palais–Smale sequence for J_ε at the level c. Then, using (g_3), we have

$$C(1 + \|u_n\|_\varepsilon) \geq J_\varepsilon(u_n) - \frac{1}{\theta} \langle J_\varepsilon'(u_n), u_n \rangle$$

$$\geq \left(\frac{1}{2} - \frac{1}{\theta}\right) [u_n]_{A_\varepsilon}^2 + \left(\left(\frac{1}{2} - \frac{1}{\theta}\right) - \frac{1}{2k}\right) \int_{\mathbb{R}} V(\varepsilon x) |u_n|^2 \, dx$$

$$\geq \left(\left(\frac{1}{2} - \frac{1}{\theta} \right) - \frac{1}{2k} \right) \|u_n\|_\varepsilon^2.$$

Recalling that $k > \frac{\theta}{\theta-2}$, we can deduce that (u_n) is bounded in $H_\varepsilon^{\frac{1}{2}}$. Moreover, since $c < \left(\frac{1}{2} - \frac{1}{\theta} - \frac{1}{2k} \right) \min\{1, V_0\}$, we can see that

$$\limsup_{n\to\infty} \|u_n\|_\varepsilon^2 < 1. \tag{17.4.27}$$

Since $H_\varepsilon^{\frac{1}{2}}$ is a reflexive space, we can find a subsequence, still denoted by (u_n), and $u \in H_\varepsilon^{\frac{1}{2}}$ such that

$$\begin{aligned}
u_n &\rightharpoonup u \quad \text{in } H_\varepsilon^{\frac{1}{2}} \text{ as } n \to \infty, \\
u_n &\to u \quad \text{in } L_{loc}^q(\mathbb{R}, \mathbb{C}) \text{ for all } q \in [2, \infty) \text{ as } n \to \infty, \\
|u_n| &\to |u| \quad \text{a.e. in } \mathbb{R} \text{ as } n \to \infty.
\end{aligned} \tag{17.4.28}$$

By (17.4.27) and arguing as in the proof of Lemma 17.4.7 we can infer that

$$\lim_{n\to\infty} \Re \int_\mathbb{R} g(\varepsilon x, |u_n|^2) u_n \overline{\phi} \, dx = \Re \int_\mathbb{R} g(\varepsilon x, |u|^2) u \overline{\phi} \, dx \quad \text{for all } \phi \in C_c^\infty(\mathbb{R}, \mathbb{C}). \tag{17.4.29}$$

Taking into account (17.4.28), (17.4.29) and the density of $C_c^\infty(\mathbb{R}, \mathbb{C})$ in $H_\varepsilon^{\frac{1}{2}}$, we deduce that

$$\langle J_\varepsilon'(u), \phi \rangle = 0 \quad \text{for all } \phi \in H_\varepsilon^{\frac{1}{2}},$$

that is, u is a critical point for J_ε. Consequently, $\langle J_\varepsilon'(u), u \rangle = 0$, or equivalently

$$[u]_{A_\varepsilon}^2 + \int_{\Lambda_\varepsilon} V(\varepsilon x)|u|^2 \, dx + \int_{\Lambda_\varepsilon^c} \mathcal{C}(\varepsilon x, |u|^2) \, dx = \int_{\Lambda_\varepsilon} f(|u|^2)|u|^2 \, dx, \tag{17.4.30}$$

where $\mathcal{C}(x, t) = V(x)t - g(x, t)t$. Note that, by (g_3)-(ii),

$$V(x)t \geq \mathcal{C}(x, t) \geq \left(1 - \frac{1}{k} \right) V(x)t \geq 0 \quad \text{for all } x \in \Lambda^c, \, t \geq 0. \tag{17.4.31}$$

Recalling that $\langle J_\varepsilon'(u_n), u_n \rangle = o_n(1)$, we also know that

$$[u_n]_{A_\varepsilon}^2 + \int_{\Lambda_\varepsilon} V(\varepsilon x)|u_n|^2 \, dx + \int_{\Lambda_\varepsilon^c} \mathcal{C}(\varepsilon x, |u_n|^2) \, dx = \int_{\Lambda_\varepsilon} f(|u_n|^2)|u_n|^2 \, dx + o_n(1). \tag{17.4.32}$$

Since Λ_ε is bounded, by (17.4.27) and arguing as in the proof of Lemma 17.4.7, we have

$$\lim_{n\to\infty} \int_{\Lambda_\varepsilon} f(|u_n|^2)|u_n|^2 \, dx = \int_{\Lambda_\varepsilon} f(|u|^2)|u|^2 \, dx, \tag{17.4.33}$$

and using the compact embedding in Theorem 17.4.3 it holds that

$$\lim_{n\to\infty} \int_{\Lambda_\varepsilon} V(\varepsilon x)|u_n|^2 \, dx = \int_{\Lambda_\varepsilon} V(\varepsilon x)|u|^2 \, dx. \tag{17.4.34}$$

Putting together (17.4.30), (17.4.32), (17.4.33) and (17.4.34), we deduce that

$$\limsup_{n\to\infty} \left([u_n]_{A_\varepsilon}^2 + \int_{\Lambda_\varepsilon^c} \mathcal{C}(\varepsilon x, |u_n|^2) \, dx \right) = [u]_{A_\varepsilon}^2 + \int_{\Lambda_\varepsilon^c} \mathcal{C}(\varepsilon x, |u|^2) \, dx.$$

Next, by (17.4.31) and Fatou's lemma,

$$\liminf_{n\to\infty} \left([u_n]_{A_\varepsilon}^2 + \int_{\Lambda_\varepsilon^c} \mathcal{C}(\varepsilon x, |u_n|^2) \, dx \right) \geq [u]_{A_\varepsilon}^2 + \int_{\Lambda_\varepsilon^c} \mathcal{C}(\varepsilon x, |u|^2) \, dx.$$

Hence,

$$\lim_{n\to\infty} [u_n]_{A_\varepsilon}^2 = [u]_{A_\varepsilon}^2, \tag{17.4.35}$$

and

$$\lim_{n\to\infty} \int_{\Lambda_\varepsilon^c} \mathcal{C}(\varepsilon x, |u_n|^2) \, dx = \int_{\Lambda_\varepsilon^c} \mathcal{C}(\varepsilon x, |u|^2) \, dx.$$

The last limit, Fatou's lemma and (17.4.31) imply that

$$\int_{\Lambda_\varepsilon^c} V(\varepsilon x)|u|^2 \, dx \leq \liminf_{n\to\infty} \int_{\Lambda_\varepsilon^c} V(\varepsilon x)|u_n|^2 \, dx$$

$$\leq \limsup_{n\to\infty} \int_{\Lambda_\varepsilon^c} V(\varepsilon x)|u_n|^2 \, dx$$

$$\leq \left(\frac{k}{k-1} \right) \limsup_{n\to\infty} \int_{\Lambda_\varepsilon^c} \mathcal{C}(\varepsilon x, |u_n|^2) \, dx$$

$$= \left(\frac{k}{k-1} \right) \int_{\Lambda_\varepsilon^c} \mathcal{C}(\varepsilon x, |u|^2) \, dx \leq \left(\frac{k}{k-1} \right) \int_{\Lambda_\varepsilon^c} V(\varepsilon x)|u|^2 \, dx \quad \text{for all } k > \frac{\theta}{\theta-2}.$$

Letting here $k \to \infty$, we find that

$$\lim_{n \to \infty} \int_{\Lambda_\varepsilon^c} V(\varepsilon x) |u_n|^2 \, dx = \int_{\Lambda_\varepsilon^c} V(\varepsilon x) |u|^2 \, dx,$$

which combined with (17.4.34) gives

$$\lim_{n \to \infty} \int_{\mathbb{R}} V(\varepsilon x) |u_n|^2 \, dx = \int_{\mathbb{R}} V(\varepsilon x) |u|^2 \, dx. \tag{17.4.36}$$

Putting together (17.4.35) and (17.4.36), we obtain that

$$\lim_{n \to \infty} \|u_n\|_\varepsilon^2 = \|u\|_\varepsilon^2.$$

Since $H_\varepsilon^{\frac{1}{2}}$ is a Hilbert space and $u_n \rightharpoonup u$ in $H_\varepsilon^{\frac{1}{2}}$ as $n \to \infty$, we conclude that $u_n \to u$ in $H_\varepsilon^{\frac{1}{2}}$. $\qquad\square$

In order to obtain multiple critical points of J_ε, we will consider J_ε constrained on $\mathcal{N}_\varepsilon$. Therefore, it is needed to prove the following result.

Proposition 17.4.16 *Let $c \in \mathbb{R}$ be such that $c < \left(\frac{1}{2} - \frac{1}{\theta} - \frac{1}{2k}\right) \min\{1, V_0\}$. Then, the functional J_ε restricted to $\mathcal{N}_\varepsilon$ satisfies the $(PS)_c$ condition at the level c.*

Proof Let $(u_n) \subset \mathcal{N}_\varepsilon$ be such that $J_\varepsilon(u_n) \to c$ and $\|J_\varepsilon'(u_n)_{|\mathcal{N}_\varepsilon}\|_* = o_n(1)$. Then there exists a sequence $(\lambda_n) \subset \mathbb{R}$ such that

$$J_\varepsilon'(u_n) = \lambda_n T_\varepsilon'(u_n) + o_n(1) \tag{17.4.37}$$

where $T_\varepsilon : H_\varepsilon^{\frac{1}{2}} \to \mathbb{R}$ is given by

$$T_\varepsilon(u) = \|u\|_\varepsilon^2 - \int_{\mathbb{R}} g(\varepsilon x, |u|^2) |u|^2 \, dx.$$

Taking into account that $\langle J_\varepsilon'(u_n), u_n \rangle = 0$, that $g_\varepsilon(x, |u|^2)$ is constant on $\Lambda_\varepsilon^c \cap \{|u|^2 > T_a\}$, the definition of g, the monotonicity of ξ and condition (f_6), we can see that

$$\langle T_\varepsilon'(u_n), u_n \rangle = 2\|u_n\|_\varepsilon^2 - 2 \int_{\mathbb{R}} g'(\varepsilon x, |u_n|^2) |u_n|^4 \, dx - 2 \int_{\mathbb{R}} g(\varepsilon x, |u_n|^2) |u_n|^2 \, dx$$

$$= -2 \int_{\mathbb{R}} g'(\varepsilon x, |u_n|^2) |u_n|^4 \, dx$$

$$\leq -2 \int_{\Lambda_\varepsilon \cup \{|u_n|^2 < t_a\}} f'(|u_n|^2)|u_n|^4 \, dx$$

$$\leq -2C_\sigma \int_{\Lambda_\varepsilon \cup \{|u_n|^2 < t_a\}} |u_n|^\sigma \, dx$$

$$\leq -2C_\sigma \int_{\Lambda_\varepsilon} |u_n|^\sigma dx < 0.$$

By the boundedness of (u_n) in $H_\varepsilon^{\frac{1}{2}}$, we can assume that $\langle T_\varepsilon'(u_n), u_n \rangle \to \ell \leq 0$.

If $\ell = 0$, then $|u_n| \to 0$ in $L^\sigma(\Lambda_\varepsilon, \mathbb{R})$. By interpolation, $|u_n| \to 0$ in $L^r(\Lambda_\varepsilon, \mathbb{R})$ for all $r \geq \sigma$. Now, we note that $u_n \in \mathcal{N}_\varepsilon$ and $J_\varepsilon(u_n) \to c < \left(\frac{1}{2} - \frac{1}{\theta} - \frac{1}{2k}\right)\min\{1, V_0\}$ imply that $\limsup_{n\to\infty} \|u_n\|_\varepsilon^2 < 1$. Then, using Lemma 17.4.7, we deduce that

$$\int_{\Lambda_\varepsilon} g(\varepsilon x, |u_n|^2)|u_n|^2 \, dx = \int_{\Lambda_\varepsilon} f(|u_n|^2)|u_n|^2 \, dx \to 0.$$

Therefore,

$$\|u_n\|_\varepsilon^2 = \int_{\Lambda_\varepsilon^c} g(\varepsilon x, |u_n|^2)|u_n|^2 \, dx + o_n(1) \leq \frac{1}{k} \int_{\Lambda_\varepsilon^c} V(\varepsilon x)|u_n|^2 \, dx + o_n(1),$$

and so $\|u_n\|_\varepsilon \to 0$, which is a contradiction because $\|u\|_\varepsilon \geq r > 0$ for all $u \in \mathcal{N}_\varepsilon$.

Consequently, $\ell < 0$ and taking into account (17.4.37) we conclude $\lambda_n \to 0$, that is, (u_n) is a Palais–Smale sequence at the level c for the unconstrained functional. The desired assertion follows from Lemma 17.4.15. □

Arguing as in the previous lemma, it is easy to check that:

Corollary 17.4.17 *The critical points of the functional J_ε on $\mathcal{N}_\varepsilon$ are critical points of J_ε.*

We conclude this section by the following existence result for (17.4.13):

Theorem 17.4.18 *There exists $\varepsilon_0 > 0$ such that, for any $\varepsilon \in (0, \varepsilon_0)$, problem (17.4.13) admits a nontrivial solution.*

Proof Taking into account Lemmas 17.4.10, 17.4.14, 17.4.15 and applying the mountain pass theorem [29], we can deduce that, for all $\varepsilon > 0$ sufficiently small, problem (17.4.13) admits a nontrivial solution. □

17.4.3 Multiple Solutions for the Modified Problem

In this subsection, we show that it is possible to relate the number of nontrivial solutions of (17.4.13) to the topology of the set Λ. To this aim, take $\delta > 0$ such that

$$M_\delta = \{x \in \mathbb{R} : \text{dist}(x, M) \le \delta\} \subset \Lambda,$$

and choose a smooth nonincreasing cut-off function η defined in $[0, \infty)$ such that $\eta = 1$ in $[0, \frac{\delta}{2}]$, $\eta = 0$ in $[\delta, \infty)$, $0 \le \eta \le 1$ and $|\eta'| \le c$ for some $c > 0$.

For $y \in \Lambda$, let

$$\Psi_{\varepsilon, y}(x) = \eta(|\varepsilon x - y|) w\left(\frac{\varepsilon x - y}{\varepsilon}\right) e^{i \tau_y \left(\frac{\varepsilon x - y}{\varepsilon}\right)},$$

where $\tau_y(x) = A(y)x$ and $w \in H_{V_0}^{\frac{1}{2}}(\mathbb{R}, \mathbb{R})$ is a positive ground state solution to the autonomous problem (17.4.15) with $\mu = V_0$ (see Lemma 17.4.13). Let $t_\varepsilon > 0$ be the unique number such that

$$\max_{t \ge 0} J_\varepsilon(t \Psi_{\varepsilon, y}) = J_\varepsilon(t_\varepsilon \Psi_{\varepsilon, y}).$$

Noting that $t_\varepsilon \Psi_{\varepsilon, y} \in \mathcal{N}_\varepsilon$, we can define $\Phi_\varepsilon : M \to \mathcal{N}_\varepsilon$ as

$$\Phi_\varepsilon(y) = t_\varepsilon \Psi_{\varepsilon, y}.$$

Lemma 17.4.19 *The functional Φ_ε has the property that*

$$\lim_{\varepsilon \to 0} J_\varepsilon(\Phi_\varepsilon(y)) = m_{V_0}, \qquad \text{uniformly in } y \in M.$$

Proof Assume, by contradiction, that there exist $\delta_0 > 0$, $(y_n) \subset M$ and $\varepsilon_n \to 0$ such that

$$|J_{\varepsilon_n}(\Phi_{\varepsilon_n}(y_n)) - m_{V_0}| \ge \delta_0. \tag{17.4.38}$$

To simplify the notation, we write Φ_n, Ψ_n and t_n for $\Phi_{\varepsilon_n}(y_n)$, $\Psi_{\varepsilon_n, y_n}$ and t_{ε_n}, respectively. Arguing as in the proof of Lemma 17.2.19 and applying the dominated convergence theorem we get

$$\|\Psi_n\|_{\varepsilon_n}^2 \to \|w\|_{V_0}^2 \in (0, \infty). \tag{17.4.39}$$

On the other hand, since $\langle J'_{\varepsilon_n}(t_n \Psi_n), t_n \Psi_n \rangle = 0$, the change of variable

$$z = \frac{\varepsilon_n x - y_n}{\varepsilon_n},$$

leads to

$$t_n^2 \| \Psi_n \|_{\varepsilon_n}^2 = \int_{\mathbb{R}} g(\varepsilon_n z + y_n, |t_n \eta(|\varepsilon_n z|) w(z)|^2) |t_{\varepsilon_n} \eta(|\varepsilon_n z|) w(z)|^2 \, dz.$$

If $z \in (-\frac{\delta}{\varepsilon_n}, \frac{\delta}{\varepsilon_n})$, then $\varepsilon_n z + y_n \in (y_n - \delta, y_n + \delta) \subset M_\delta \subset \Lambda$. Noting that $g(x, t) = f(t)$ for $(x, t) \in \Lambda \times \mathbb{R}$, we have

$$\| \Psi_n \|_{\varepsilon_n}^2 = \int_{\mathbb{R}} f(|t_n \eta(|\varepsilon_n z|) w(z)|^2) |\eta(|\varepsilon_n z|) w(z)|^2 \, dz. \tag{17.4.40}$$

Since $\eta = 1$ in the interval $(-\frac{\delta}{2}, \frac{\delta}{2}) \subset (-\frac{\delta}{2\varepsilon_n}, \frac{\delta}{2\varepsilon_n})$ for all n large enough, from (17.4.40) and (f_4) we get

$$\| \Psi_n \|_{\varepsilon_n}^2 \geq \int_{-\frac{\delta}{2}}^{\frac{\delta}{2}} f(|t_n w(z)|^2) |w(z)|^2 \, dz$$

$$\geq f(|t_n \sigma_0|^2) \int_{-\frac{\delta}{2}}^{\frac{\delta}{2}} |w(z)|^2 \, dz, \tag{17.4.41}$$

where

$$\sigma_0 = \min_{|z| \leq \frac{\delta}{2}} w(z) > 0.$$

Now, if $t_n \to \infty$, we can use (17.4.41), (17.4.39) and (f_3) to reach a contradiction. Therefore, (t_n) is bounded and, up to subsequence, we may assume that $t_n \to t_0$ for some $t_0 \geq 0$. Let us prove that $t_0 > 0$. Otherwise, if $t_0 = 0$, we can use (17.4.39), the growth assumptions on g and (17.4.40) to conclude that

$$\| \Psi_n \|_{\varepsilon_n}^2 \to 0,$$

which is impossible because $t_\varepsilon \Psi_{\varepsilon, y} \in \mathcal{N}_\varepsilon$ and (17.4.16) holds. Hence, $t_0 > 0$. Taking the limit as $n \to \infty$ in (17.4.40), we deduce that

$$\| w \|_{V_0}^2 = \int_{\mathbb{R}} f((t_0 w)^2) \, w^2 \, dx.$$

Since $w \in \mathcal{M}_{V_0}$ and (f_4) holds, we can deduce that $t_0 = 1$. This and the dominated convergence theorem imply that

$$\int_{\mathbb{R}} F(|t_n \Psi_n|^2) \, dx \rightarrow \int_{\mathbb{R}} F(|w|^2) \, dx.$$

Hence, letting $n \rightarrow \infty$ in

$$J_{\varepsilon_n}(\Phi_n) = \frac{t_n^2}{2} \|\Psi_n\|_{\varepsilon_n}^2 - \frac{1}{2} \int_{\mathbb{R}} F(|t_n \Psi_n|^2) \, dx,$$

we can conclude that

$$\lim_{n \rightarrow \infty} J_{\varepsilon_n}(\Phi_{\varepsilon_n}(y_n)) = I_{V_0}(w) = m_{V_0},$$

which contradicts (17.4.38). $\qquad\qquad\qquad\qquad\qquad\qquad\qquad\qquad\qquad\qquad\qquad\quad$ □

For any $\delta > 0$, take $\rho = \rho(\delta) > 0$ such that $M_\delta \subset (-\rho, \rho)$. Let $\Upsilon : \mathbb{R} \rightarrow \mathbb{R}$ be defined as

$$\Upsilon(x) = x, \quad \text{if } |x| < \rho \quad \text{and} \quad \Upsilon(x) = \frac{\rho x}{|x|}, \quad \text{if } |x| \geq \rho.$$

Finally, consider the barycenter map $\beta_\varepsilon : \mathcal{N}_\varepsilon \rightarrow \mathbb{R}$ given by

$$\beta_\varepsilon(u) = \frac{\displaystyle\int_{\mathbb{R}} \Upsilon(\varepsilon x) |u(x)|^2 \, dx}{\displaystyle\int_{\mathbb{R}} |u(x)|^2 \, dx}.$$

Arguing as in the proof of Lemma 17.2.21, it is easy to see that the function β_ε has the following property:

Lemma 17.4.20

$$\lim_{\varepsilon \rightarrow 0} \beta_\varepsilon(\Phi_\varepsilon(y)) = y, \quad \text{uniformly in } y \in M.$$

The next compactness result is crucial for showing that the solutions of the modified problem are solutions of the original problem.

Lemma 17.4.21 *Let $\varepsilon_n \rightarrow 0$ and $u_n = u_{\varepsilon_n} \in \mathcal{N}_{\varepsilon_n}$ for all $n \in \mathbb{N}$ be such that $J_{\varepsilon_n}(u_n) \rightarrow m_{V_0}$. Then there exists $(\tilde{y}_n) = (\tilde{y}_{\varepsilon_n}) \subset \mathbb{R}$ such that $v_n(x) = |u_n|(x + \tilde{y}_n)$ has a convergent subsequence in $H^{\frac{1}{2}}(\mathbb{R}, \mathbb{R})$. Moreover, up to a subsequence, $y_n = \varepsilon_n \tilde{y}_n \rightarrow y_0$ for some $y_0 \in M$.*

Proof Taking into account that $\langle J'_{\varepsilon_n}(u_n), u_n \rangle = 0$, $J_{\varepsilon_n}(u_n) = m_{V_0} + o_n(1)$ and Lemma 17.4.13, we can argue as in the first part of Lemma 17.4.15 to see that (u_n) is bounded in $H^{\frac{1}{2}}_{\varepsilon_n}$ and $\limsup_{n\to\infty} \|u_n\|^2_{\varepsilon_n} < 1$. Moreover, by Lemma 17.2.3 and assumption (V_1), $(|u_n|)$ is bounded in $H^{\frac{1}{2}}_{V_0}(\mathbb{R}, \mathbb{R})$.

Now, we claim that there exist a sequence $(\tilde{y}_n) \subset \mathbb{R}$ and constants $R > 0$ and $\gamma > 0$ such that

$$\liminf_{n\to\infty} \int_{\tilde{y}_n - R}^{\tilde{y}_n + R} |u_n|^2 \, dx \geq \gamma > 0. \tag{17.4.42}$$

Indeed, suppose, by contradiction, that (17.4.42) does not hold. Then for all $R > 0$ we get

$$\limsup_{n\to\infty} \sup_{y\in\mathbb{R}} \int_{y-R}^{y+R} |u_n|^2 \, dx = 0.$$

The boundedness $(|u_n|)$ and Lemma 17.4.4 imply that $|u_n| \to 0$ in $L^q(\mathbb{R}, \mathbb{R})$ for any $q \in (2, \infty)$. Arguing as in the proof of Lemma 17.4.8 we see that

$$\lim_{n\to\infty} \int_{\mathbb{R}} g(\varepsilon_n x, |u_n|^2)|u_n|^2 \, dx = 0 = \lim_{n\to\infty} \int_{\mathbb{R}} G(\varepsilon_n x, |u_n|^2) \, dx. \tag{17.4.43}$$

Taking into account $\langle J'_{\varepsilon_n}(u_n), u_n \rangle = 0$ and (17.4.43), we infer that $\|u_n\|_{\varepsilon_n} \to 0$ as $n \to \infty$, and then $J_{\varepsilon_n}(u_n) \to 0$, which is a contradiction because $c_{V_0} > 0$.

Let $v_n(x) = |u_n|(x + \tilde{y}_n)$. Then (v_n) is bounded in $H^{\frac{1}{2}}_{V_0}(\mathbb{R}, \mathbb{R})$, and we may assume that $v_n \rightharpoonup v \neq 0$ in $H^{\frac{1}{2}}_{V_0}(\mathbb{R}, \mathbb{R})$ as $n \to \infty$. Fix $t_n > 0$ such that $\tilde{v}_n = t_n v_n \in \mathcal{M}_{V_0}$. Using Lemma 17.2.3 and the fact that $u_n \in \mathcal{N}_{\varepsilon_n}$, we have

$$m_{V_0} \leq I_{V_0}(\tilde{v}_n) \leq \max_{t\geq 0} J_{\varepsilon_n}(t u_n) = J_{\varepsilon_n}(u_n) = m_{V_0} + o_n(1)$$

which implies that $I_{V_0}(\tilde{v}_n) \to m_{V_0}$. In particular, $\tilde{v}_n \nrightarrow 0$ in $H^{\frac{1}{2}}_{V_0}(\mathbb{R}, \mathbb{R})$. Since (v_n) and $(\tilde{v}_n)$ are bounded in $H^{\frac{1}{2}}_{V_0}(\mathbb{R}, \mathbb{R})$ and $\tilde{v}_n \nrightarrow 0$ in $H^{\frac{1}{2}}_{V_0}(\mathbb{R}, \mathbb{R})$, we deduce that $t_n \to t^* \geq 0$.

Indeed, $t^* > 0$ since $\tilde{v}_n \nrightarrow 0$ in $H^{\frac{1}{2}}_{V_0}(\mathbb{R}, \mathbb{R})$. The uniqueness of the weak limit implies that $\tilde{v}_n \rightharpoonup \tilde{v} = t^* v \neq 0$ in $H^{\frac{1}{2}}_{V_0}(\mathbb{R}, \mathbb{R})$. This combined with Lemma 17.4.13 yields

$$\tilde{v}_n \to \tilde{v} \quad \text{in } H^{\frac{1}{2}}_{V_0}(\mathbb{R}, \mathbb{R}). \tag{17.4.44}$$

Consequently, $v_n \to v$ in $H^{\frac{1}{2}}_{V_0}(\mathbb{R}, \mathbb{R})$ as $n \to \infty$.

Now, we set $y_n = \varepsilon_n \tilde{y}_n$ and we show that (y_n) admits a subsequence, still denoted (y_n), such that $y_n \to y_0$ for some $y_0 \in M$. Firstly, we prove that (y_n) is bounded. Assume, by contradiction, that, up to a subsequence, $|y_n| \to \infty$ as $n \to \infty$. Take $R > 0$ such that $\Lambda \subset (-R, R)$. Since we may suppose that $|y_n| > 2R$, for any $z \in (-\frac{R}{\varepsilon_n}, \frac{R}{\varepsilon_n})$

$$|\varepsilon_n z + y_n| \geq |y_n| - |\varepsilon_n z| > R.$$

Now, using that $u_n \in \mathcal{N}_{\varepsilon_n}$ for all $n \in \mathbb{N}$, (V_1), Lemma 17.2.3 and the change of variable $x \mapsto z + \tilde{y}_n$ we observe that

$$[v_n]^2 + \int_{\mathbb{R}} V_0 v_n^2 \, dx \leq \int_{\mathbb{R}} g(\varepsilon_n x + y_n, |v_n|^2) |v_n|^2 \, dx$$

$$\leq \int_{-\frac{R}{\varepsilon_n}}^{\frac{R}{\varepsilon_n}} \tilde{f}(|v_n|^2) |v_n|^2 \, dx + \int_{\mathbb{R} \setminus (-\frac{R}{\varepsilon_n}, \frac{R}{\varepsilon_n})} f(|v_n|^2) |v_n|^2 \, dx.$$

$$(17.4.45)$$

Then, recalling that $v_n \to v$ in $H_{V_0}^{\frac{1}{2}}(\mathbb{R}, \mathbb{R})$ as $n \to \infty$ and $\tilde{f}(t) \leq \frac{V_0}{k}$, we can see that (17.4.45) yields

$$\min\left\{1, V_0\left(1 - \frac{1}{k}\right)\right\} \left([v_n]^2 + \int_{\mathbb{R}} |v_n|^2 \, dx\right) = o_n(1),$$

that is, $v_n \to 0$ in $H_{V_0}^{\frac{1}{2}}(\mathbb{R}, \mathbb{R})$, which is a contradiction. Therefore, (y_n) is bounded and we may assume that $y_n \to y_0 \in \mathbb{R}$. If $y_0 \notin \overline{\Lambda}$, then we can argue as above to infer that $v_n \to 0$ in $H_{V_0}^{\frac{1}{2}}(\mathbb{R}, \mathbb{R})$, which is impossible. Hence $y_0 \in \overline{\Lambda}$. Note that if $V(y_0) = V_0$, then condition (V_2) implies that $y_0 \notin \partial\Lambda$. Therefore, it is enough to verify that $V(y_0) = V_0$. Suppose that, on the contrary, $V(y_0) > V_0$. Using (17.4.44), Fatou's Lemma, the translation invariance of $\mathbb{R}$ and Lemma 17.2.3 we get

$$m_{V_0} = I_{V_0}(\tilde{v}) < \frac{1}{2}[\tilde{v}]^2 + \frac{1}{2}\int_{\mathbb{R}} V(y_0)\tilde{v}^2 \, dx - \frac{1}{2}\int_{\mathbb{R}} F(|\tilde{v}|^2) \, dx$$

$$\leq \liminf_{n \to \infty} \left[\frac{1}{2}[\tilde{v}_n]^2 + \frac{1}{2}\int_{\mathbb{R}} V(\varepsilon_n x + y_n)|\tilde{v}_n|^2 \, dx - \frac{1}{2}\int_{\mathbb{R}} F(|\tilde{v}_n|^2) \, dx\right]$$

$$\leq \liminf_{n \to \infty} \left[\frac{t_n^2}{2}[|u_n|]^2 + \frac{t_n^2}{2}\int_{\mathbb{R}} V(\varepsilon_n x)|u_n|^2 \, dx - \frac{1}{2}\int_{\mathbb{R}} F(|t_n u_n|^2) \, dz\right]$$

$$\leq \liminf_{n \to \infty} J_{\varepsilon_n}(t_n u_n) \leq \liminf_{n \to \infty} J_{\varepsilon_n}(u_n) = m_{V_0},$$

a contradiction.   □

Now, consider the subset $\widetilde{\mathcal{N}}_\varepsilon$ of $\mathcal{N}_\varepsilon$ defined as

$$\widetilde{\mathcal{N}}_\varepsilon = \left\{ u \in \mathcal{N}_\varepsilon : J_\varepsilon(u) \leq m_{V_0} + h(\varepsilon) \right\},$$

where $h : \mathbb{R}^+ \to \mathbb{R}^+$ is such that $h(\varepsilon) \to 0$ as $\varepsilon \to 0$. Using Lemma 17.4.19 we infer that $h(\varepsilon) = \sup_{y \in M} |J_\varepsilon(\Phi_\varepsilon(y)) - m_{V_0}| \to 0$ as $\varepsilon \to 0$. Thus, $\Phi_\varepsilon(y) \in \widetilde{\mathcal{N}}_\varepsilon$, and $\widetilde{\mathcal{N}}_\varepsilon \neq \emptyset$ for any $\varepsilon > 0$. Moreover, proceeding as in the proof of Lemma 17.2.23 we have:

Lemma 17.4.22 *For every $\delta > 0$ we have*

$$\lim_{\varepsilon \to 0} \sup_{u \in \widetilde{\mathcal{N}}_\varepsilon} \ dist(\beta_\varepsilon(u), M_\delta) = 0.$$

We end this section with the proof of a multiplicity result for (17.4.13).

Theorem 17.4.23 *For every $\delta > 0$ such that $M_\delta \subset \Lambda$, there exists $\tilde{\varepsilon}_\delta > 0$ such that, for any $\varepsilon \in (0, \tilde{\varepsilon}_\delta)$, problem (17.4.13) has at least $cat_{M_\delta}(M)$ nontrivial solutions.*

Proof Given $\delta > 0$ such that $M_\delta \subset \Lambda$, we can use Lemmas 17.4.19, 17.4.20, 17.4.22 and argue as in [146] to deduce the existence of $\tilde{\varepsilon}_\delta > 0$ such that, for any $\varepsilon \in (0, \varepsilon_\delta)$, the diagram

$$M \xrightarrow{\Phi_\varepsilon} \widetilde{\mathcal{N}}_\varepsilon \xrightarrow{\beta_\varepsilon} M_\delta$$

is well defined and $\beta_\varepsilon \circ \Phi_\varepsilon$ is homotopically equivalent to the embedding $\iota : M \to M_\delta$. Thus, $cat_{\widetilde{\mathcal{N}}_\varepsilon}(\widetilde{\mathcal{N}}_\varepsilon) \geq cat_{M_\delta}(M)$. It follows from Proposition 17.4.16 and standard Lusternik-Schnirelman theory that J_ε possesses at least $cat_{\widetilde{\mathcal{N}}_\varepsilon}(\widetilde{\mathcal{N}}_\varepsilon)$ critical points on $\mathcal{N}_\varepsilon$. Applying Corollary 17.4.17, we can deduce that (17.4.13) has at least $cat_{M_\delta}(M)$ nontrivial solutions.   □

17.4.4 Proof of Theorem 17.4.2

In this last subsection we prove that the solutions obtained in Theorem 17.4.23 verify $|u_\varepsilon(x)| \leq \sqrt{t_a}$ for any $x \in \mathbb{R} \setminus \Lambda_\varepsilon$ and ε small.

We begin with the following lemma which will play a fundamental role in the study of the behavior of the maximum points of solutions.

Lemma 17.4.24 *Let $\varepsilon_n \to 0$ and $u_n \in \widetilde{\mathcal{N}}_{\varepsilon_n}$ be a solution to (17.4.13) such that*

$$m = \limsup_{n \to \infty} \|u_n\|_{\varepsilon_n}^2 < 1.$$

Then $v_n = |u_n|(\cdot + \tilde{y}_n)$ belongs to $L^\infty(\mathbb{R}, \mathbb{R})$ and there exists $C > 0$ such that

$$\|v_n\|_{L^\infty(\mathbb{R})} \leq C \quad \text{for all } n \in \mathbb{N},$$

where $(\tilde{y}_n)$ is given by Lemma 17.4.21. Moreover

$$\lim_{|x| \to \infty} v_n(x) = 0, \quad \text{uniformly in } n \in \mathbb{N}.$$

Proof Recalling that $J_{\varepsilon_n}(u_n) \leq m_{V_0} + h(\varepsilon_n)$ with $h(\varepsilon_n) \to 0$ as $n \to \infty$, we can proceed as in the first part of the proof of Lemma 17.4.21 to deduce that $J_{\varepsilon_n}(u_n) \to m_{V_0}$. Invoking Lemma 17.4.21, we find a sequence $(\tilde{y}_n) \subset \mathbb{R}$ such that $\varepsilon_n \tilde{y}_n \to y_0 \in M$ and $v_n = |u_n|(\cdot + \tilde{y}_n)$ has a convergent subsequence in $H^{\frac{1}{2}}(\mathbb{R}, \mathbb{R})$.

Now we use a Moser iteration argument. For each $n \in \mathbb{N}$ and $L > 0$, we set $u_{L,n} = \min\{|u_n|, L\}$ and $v_{L,n} = u_{L,n}^{2(\beta-1)} u_n$, where $\beta > 1$ will be chosen later. Taking $v_{L,n}$ as test function in (17.4.13) we can see that

$$\frac{1}{2\pi} \Re \left(\iint_{\mathbb{R}^2} \frac{(u_n(x) - u_n(y)e^{\iota A_{\varepsilon_n}(\frac{x+y}{2})(x-y)})}{|x-y|^2} \overline{(u_n u_{L,n}^{2(\beta-1)}(x) - u_n u_{L,n}^{2(\beta-1)}(y)e^{\iota A_{\varepsilon_n}(\frac{x+y}{2})(x-y)})} \, dx \, dy \right)$$

$$= \int_{\mathbb{R}} g(\varepsilon_n x, |u_n|^2)|u_n|^2 u_{L,n}^{2(\beta-1)} \, dx - \int_{\mathbb{R}} V(\varepsilon_n x)|u_n|^2 u_{L,n}^{2(\beta-1)} \, dx. \qquad (17.4.46)$$

By assumptions (g_1) and (g_2), for every $\xi > 0$ there exists $C_\xi > 0$ such that

$$g(\varepsilon_n x, |u_n|^2)|u_n|^2 \leq \xi|u_n|^2 + C_\xi|u_n|^2 B(u_n) \quad \text{for all } n \in \mathbb{N}, \qquad (17.4.47)$$

where $B(u_n) = (e^{\omega \tau |u_n|^2} - 1) \in L^q(\mathbb{R})$ for some $q > 1$ close to 1, $\tau > 1$ such that $\tau q m < 1$, and

$$\|B(u_n)\|_{L^q(\mathbb{R})} \leq C \quad \text{for any } n \in \mathbb{N}. \qquad (17.4.48)$$

On the other hand, the same calculations performed in Remark 17.3.10 show that

$$\Re \left(\iint_{\mathbb{R}^2} \frac{(u_n(x) - u_n(y)e^{\iota A_{\varepsilon_n}(\frac{x+y}{2})(x-y)})}{|x-y|^2} \overline{(u_n(x)u_{L,n}^{2(\beta-1)}(x) - u_n(y)u_{L,n}^{2(\beta-1)}(y)e^{\iota A_{\varepsilon_n}(\frac{x+y}{2})(x-y)})} \, dx \, dy \right)$$

$$\geq [\Gamma(|u_n|)]_{\frac{1}{2}}^2. \qquad (17.4.49)$$

Since

$$\Gamma(|u_n|) \geq \frac{1}{\beta} |u_n| u_{L,n}^{\beta-1},$$

and recalling that $H^{\frac{1}{2}}(\mathbb{R}, \mathbb{R}) \subset L^r(\mathbb{R}, \mathbb{R})$ for all $r \in [2, \infty)$, we can see that

$$[\Gamma(|u_n|)]_{\frac{1}{2}}^2 \geq C \|\Gamma(|u_n|)\|_{L^\gamma(\mathbb{R})}^2 \geq \frac{1}{\beta^2} C \||u_n| u_{L,n}^{\beta-1}\|_{L^\gamma(\mathbb{R})}^2, \tag{17.4.50}$$

where $\gamma > 2q'$. Combining (17.4.46), (17.4.49) and (17.4.50) we can infer that

$$\left(\frac{1}{\beta}\right)^2 C \||u_n| u_{L,n}^{\beta-1}\|_{L^\gamma(\mathbb{R})}^2 + \int_{\mathbb{R}} V(\varepsilon_n x) |u_n|^2 u_{L,n}^{2(\beta-1)} \, dx \leq \int_{\mathbb{R}} g(\varepsilon_n x, |u_n|^2) |u_n|^2 u_{L,n}^{2(\beta-1)} \, dx. \tag{17.4.51}$$

Taking $\xi \in (0, V_0)$ and using (17.4.47) and (17.4.51) we have

$$\|w_{L,n}\|_{L^\gamma(\mathbb{R})}^2 \leq C\beta^2 \int_{\mathbb{R}} B(u_n) |u_n|^2 u_{L,n}^{2(\beta-1)} \, dx = C\beta^2 \int_{\mathbb{R}} B(u_n) w_{L,n}^2 \, dx, \tag{17.4.52}$$

where $w_{L,n} = |u_n| u_{L,n}^{\beta-1}$. We derive from (17.4.48) and Hölder's inequality that

$$\|w_{L,n}\|_{L^\gamma(\mathbb{R})}^2 \leq C\beta^2 \left(\int_{\mathbb{R}} (B(u_n))^q \, dx\right)^{\frac{1}{q}} \left(\int_{\mathbb{R}} w_{L,n}^{2q'} \, dx\right)^{\frac{1}{q'}} \leq C\beta^2 \|w_{L,n}\|_{L^{2q'}(\mathbb{R})}^2,$$

and letting $L \to \infty$ we obtain

$$\|u_n\|_{L^{\beta\gamma}(\mathbb{R})} \leq C^{\frac{1}{2\beta}} \beta^{\frac{1}{\beta}} \|u_n\|_{L^{2q'\beta}(\mathbb{R})}.$$

Since $\gamma > 2q'$, we can use an iteration argument to deduce that, for all $m \in \mathbb{N}$,

$$\|u_n\|_{L^{\tau k(m+1)}(\mathbb{R})} \leq C^{\sum_{i=1}^m \frac{1}{2k^i}} k^{\sum_{i=1}^m \frac{i}{k^i}} \|u_n\|_{L^\gamma(\mathbb{R})}$$

where $k = \frac{\gamma}{2q'}$ and $\tau = 2q'$, which together with the boundedness of (u_n) in $H_\varepsilon^{\frac{1}{2}}$ implies that

$$\|u_n\|_{L^\infty(\mathbb{R})} \leq K \quad \text{for all } n \in \mathbb{N}. \tag{17.4.53}$$

Arguing as in the proof of Lemma 17.3.8 and using (V_1), we see that $v_n = |u_n|(\cdot + \tilde{y}_n)$ solves

$$(-\Delta)^{\frac{1}{2}} v_n + V_0 v_n \leq h_n \quad \text{in } \mathbb{R}, \tag{17.4.54}$$

where

$$h_n(x) = g(\varepsilon_n x + \varepsilon_n \tilde{y}_n, v_n^2) v_n.$$

Since (17.4.53) yields $\|v_n\|_{L^\infty(\mathbb{R})} \leq C$ for all $n \in \mathbb{N}$, by interpolation we know that $v_n \to v$ converges strongly in $L^r(\mathbb{R}, \mathbb{R})$ for all $r \in [2, \infty)$, for some $v \in L^r(\mathbb{R}, \mathbb{R})$. Furthermore, in view of the growth assumptions on f, we also have that $h_n \to f(v^2)v$ in $L^r(\mathbb{R}, \mathbb{R})$ and that $\|h_n\|_{L^\infty(\mathbb{R})} \leq C$ for all $n \in \mathbb{N}$. Let $z_n \in H^{\frac{1}{2}}(\mathbb{R}, \mathbb{R})$ be the unique solution of

$$(-\Delta)^{\frac{1}{2}} z_n + V_0 z_n = h_n \quad \text{in } \mathbb{R}. \tag{17.4.55}$$

Then, $z_n = \mathcal{K} * h_n$, where $\mathcal{K}(x) = \mathcal{F}^{-1}((|k| + V_0)^{-1})$ satisfies the following properties (see [105, 198, 199] and Remark 3.2.18):

(1) $\mathcal{K}$ is positive and even in $\mathbb{R} \setminus \{0\}$;
(2) there exists a positive constant $C > 0$ such that $\mathcal{K}(x) \leq \frac{K_1}{|x|^2}$ for any $x \in \mathbb{R} \setminus \{0\}$;
(3) $\mathcal{K} \in L^q(\mathbb{R}^N)$ for any $q \in [1, \infty]$.

Arguing as in Remark 7.2.10, we obtain that $|z_n(x)| \to 0$ as $|x| \to \infty$ uniformly in $n \in \mathbb{N}$. Since v_n satisfies (17.4.54) and z_n solves (17.4.55), a comparison argument shows that $0 \leq v_n \leq z_n$ in $\mathbb{R}$ and for all $n \in \mathbb{N}$. Furthermore, we can infer that $v_n(x) \to 0$ as $|x| \to \infty$, uniformly in $n \in \mathbb{N}$. □

Now, we are ready to present the proof of the main result of this section.

Proof of Theorem 17.4.2 Let $\delta > 0$ be such that $M_\delta \subset \Lambda$, and we show that there exists $\hat{\varepsilon}_\delta > 0$ such that, for every $\varepsilon \in (0, \hat{\varepsilon}_\delta)$ and every solution $u \in \tilde{\mathcal{N}}_\varepsilon$ of (17.4.13),

$$\|u\|_{L^\infty(\mathbb{R} \setminus \Lambda_\varepsilon)} < \sqrt{t_a}. \tag{17.4.56}$$

Assume, by contradiction, that for some sequence $\varepsilon_n \to 0$ we can find a sequence $(u_n) \subset \tilde{\mathcal{N}}_{\varepsilon_n}$ such that $J'_{\varepsilon_n}(u_n) = 0$ and

$$\|u_n\|_{L^\infty(\mathbb{R} \setminus \Lambda_\varepsilon)} \geq \sqrt{t_a}. \tag{17.4.57}$$

Since $J_{\varepsilon_n}(u_n) \leq m_{V_0} + h_1(\varepsilon_n)$, we can argue as in the first part of the proof of Lemma 17.4.21 to show that $J_{\varepsilon_n}(u_n) \to m_{V_0}$. This fact together with $J'_{\varepsilon_n}(u_n) = 0$ implies that

$$\limsup_{n \to \infty} \|u_n\|_{\varepsilon_n}^2 < 1.$$

Using Lemma 17.4.21 there exists $(\tilde{y}_n) \subset \mathbb{R}$ such that $\varepsilon_n \tilde{y}_n \to y_0$ for some $y_0 \in M$. Now, we can find $r > 0$ such that, for some subsequence still denoted $(\tilde{y}_n)$, we have that $(\tilde{y}_n - r, \tilde{y}_n + r) \subset \Lambda$ for all $n \in \mathbb{N}$. Hence, $(\tilde{y}_n - \frac{r}{\varepsilon_n}, \tilde{y}_n + \frac{r}{\varepsilon_n}) \subset \Lambda_{\varepsilon_n}$ for all $n \in \mathbb{N}$, which implies that

$$\mathbb{R} \setminus \Lambda_{\varepsilon_n} \subset \mathbb{R} \setminus \left(\tilde{y}_n - \frac{r}{\varepsilon_n}, \tilde{y}_n + \frac{r}{\varepsilon_n} \right) \qquad \text{for all } n \in \mathbb{N}.$$

Invoking Lemma 17.4.24, there exists $R > 0$ such that

$$v_n(x) < \sqrt{t_a} \quad \text{for } |x| \geq R, \ n \in \mathbb{N},$$

where $v_n(x) = |u_n|(x + \tilde{y}_n)$. Hence $|u_n(x)| < \sqrt{t_a}$ for any $x \in \mathbb{R} \setminus (\tilde{y}_n - R, \tilde{y}_n + R)$ and $n \in \mathbb{N}$. On the other hand, we can find $\nu \in \mathbb{N}$ such that for any $n \geq \nu$ and $r/\varepsilon_n > R$ it holds that

$$\mathbb{R} \setminus \Lambda_{\varepsilon_n} \subset \mathbb{R} \setminus \left(\tilde{y}_n - \frac{r}{\varepsilon_n}, \tilde{y}_n + \frac{r}{\varepsilon_n} \right) \subset \mathbb{R} \setminus (\tilde{y}_n - R, \tilde{y}_n + R).$$

Then $|u_n(x)| < \sqrt{t_a}$ for any $x \in \mathbb{R} \setminus \Lambda_{\varepsilon_n}$ and $n \geq \nu$, and this contradicts (17.4.57).

Let $\tilde{\varepsilon}_\delta > 0$ be given by Theorem 17.4.23 and set $\varepsilon_\delta = \min\{\tilde{\varepsilon}_\delta, \hat{\varepsilon}_\delta\}$. Theorem 17.4.23 provides at least $\mathrm{cat}_{M_\delta}(M)$ nontrivial solutions to (17.4.13). If $u \in H_\varepsilon^{\frac{1}{2}}$ is one of these solutions, then $u \in \tilde{\mathcal{N}}_\varepsilon$, and, in view of (17.4.56) and the definition of g, we can infer that u is also a solution to (17.4.13). Since $\hat{u}_\varepsilon(x) = u_\varepsilon(x/\varepsilon)$ is a solution to (17.4.1), we can deduce that (17.4.1) has at least $\mathrm{cat}_{M_\delta}(M)$ nontrivial solutions.

Finally, we study the behavior of the maximum points of $|\hat{u}_n|$. Take $\varepsilon_n \to 0$ and (u_n) a sequence of solutions to (17.4.13) as above. We first note that (g_1) implies that there exists $\gamma \in (0, \sqrt{t_a})$ small such that

$$g(\varepsilon x, t^2)t^2 \leq \frac{V_0}{2}t^2 \quad \text{for all } x \in \mathbb{R}, |t| \leq \gamma. \tag{17.4.58}$$

Arguing as above, we can take $R > 0$ such that

$$\|u_n\|_{L^\infty(\mathbb{R} \setminus (\tilde{y}_n - R, \tilde{y}_n + R))} < \gamma. \tag{17.4.59}$$

Up to a subsequence, we may also assume that

$$\|u_n\|_{L^\infty(\tilde{y}_n-R,\tilde{y}_n+R)} \geq \gamma. \tag{17.4.60}$$

Indeed, if (17.4.60) is not true, by (17.4.59) we get $\|u_n\|_{L^\infty(\mathbb{R})} < \gamma$, and then from the fact that $J'_{\varepsilon_n}(u_n) = 0$, (17.4.58) and Lemma 17.2.3 we obtain

$$[|u_n|]^2 + \int_\mathbb{R} V_0|u_n|^2\, dx \leq \|u_n\|^2_{\varepsilon_n} = \int_\mathbb{R} g(\varepsilon_n x, |u_n|^2)|u_n|^2\, dx \leq \frac{V_0}{2} \int_\mathbb{R} |u_n|^2\, dx,$$

whence $\||u_n|\|_{V_0} = 0$, which is a contradiction. Hence (17.4.60) holds.

Taking into account (17.4.59) and (17.4.60), we can infer that if p_n is a global maximum point of $|u_n|$, then $p_n \in (\tilde{y}_n - R, \tilde{y}_n + R)$, that is $p_n = \tilde{y}_n + q_n$ for some $q_n \in (-R, R)$. Recalling that the associated solution of (17.4.1) is of the form $\hat{u}_n(x) = u_n(x/\varepsilon_n)$, we can see that $\eta_n = \varepsilon_n \tilde{y}_n + \varepsilon_n q_n$ is a global maximum point of $|\hat{u}_n|$. Since $q_n \in (-R, R)$, $\varepsilon_n \tilde{y}_n \to y_0$ and $V(y_0) = V_0$, the continuity of V implies that

$$\lim_{n\to\infty} V(\eta_n) = V_0.$$

Next we study the decay properties of $|\hat{u}_n|$. By Lemma 3.2.17 and Remark 3.2.18, we can find a function w such that

$$0 < w(x) \leq \frac{C}{1 + |x|^2} \quad \text{for all } x \in \mathbb{R}, \tag{17.4.61}$$

and

$$(-\Delta)^{\frac{1}{2}}w + \frac{V_0}{2}w \geq 0 \quad \text{in } \mathbb{R} \setminus [-R_1, R_1], \tag{17.4.62}$$

for some suitable $R_1 > 0$. Using Lemma 17.4.24, we know that $v_n(x) \to 0$ as $|x| \to \infty$ uniformly in $n \in \mathbb{N}$, so there exists $R_2 > 0$ such that

$$h_n(x) = g(\varepsilon_n x + \varepsilon_n \tilde{y}_n, v_n^2)v_n \leq \frac{V_0}{2}v_n \quad \text{in } \mathbb{R} \setminus [-R_2, R_2]. \tag{17.4.63}$$

Let w_n be the unique solution to

$$(-\Delta)^{\frac{1}{2}}w_n + V_0 w_n = h_n \quad \text{in } \mathbb{R}.$$

Then $w_n(x) \to 0$ as $|x| \to \infty$ uniformly in $n \in \mathbb{N}$, and, by comparison, $0 \leq v_n \leq w_n$ in $\mathbb{R}$. Moreover, by (17.4.63),

$$(-\Delta)^{1/2} w_n + \frac{V_0}{2} w_n = h_n - \frac{V_0}{2} w_n \leq 0 \quad \text{in } \mathbb{R} \setminus [-R_2, R_2]. \tag{17.4.64}$$

Choose $R_3 = \max\{R_1, R_2\}$ and set

$$d = \min_{[-R_3, R_3]} w > 0 \quad \text{and} \quad \tilde{w}_n = (b+1)w - d\, w_n. \tag{17.4.65}$$

where $b = \sup_{n \in \mathbb{N}} \|w_n\|_{L^\infty(\mathbb{R})} < \infty$. Clearly, by (17.4.64) and (17.4.65), we get

$$\tilde{w}_n \geq bd + w - b\, d > 0 \quad \text{in } [-R_3, R_3],$$

$$(-\Delta)^{\frac{1}{2}} \tilde{w}_n + \frac{V_0}{2} \tilde{w}_n \geq 0 \quad \text{in } \mathbb{R} \setminus [-R_3, R_3].$$

Applying Lemma 1.3.8 we deduce that

$$\tilde{w}_n \geq 0 \quad \text{in } \mathbb{R},$$

which combined with (17.4.61) and $v_n \leq w_n$ shows that

$$0 \leq v_n(x) \leq w_n(x) \leq \frac{(b+1)}{d} w(x) \leq \frac{\tilde{C}}{1 + |x|^2} \quad \text{for all } n \in \mathbb{N}, x \in \mathbb{R},$$

for some constant $\tilde{C} > 0$. Recalling the definition of v_n, we have

$$|\hat{u}_n|(x) = |u_n| \left(\frac{x}{\varepsilon_n}\right) = v_n \left(\frac{x}{\varepsilon_n} - \tilde{y}_n\right)$$

$$\leq \frac{\tilde{C}}{1 + |\frac{x}{\varepsilon_n} - \tilde{y}_n|^2}$$

$$= \frac{\tilde{C}\, \varepsilon_n^2}{\varepsilon_n^2 + |x - \varepsilon_n\, \tilde{y}_n|^2}$$

$$\leq \frac{\tilde{C}\, \varepsilon_n^2}{\varepsilon_n^2 + |x - \eta_n|^2} \quad \text{for all } x \in \mathbb{R}.$$

The proof of Theorem 17.4.2 is now complete.   □

Bibliography

1. N. Ackermann, On a periodic Schrödinger equation with nonlocal superlinear part. Math. Z. **248**(2), 423–443 (2004)
2. R.A. Adams, *Sobolev Spaces*. Pure and Applied Mathematics, vol. 65 (Academic Press, New York-London, 1975), xviii+268 pp
3. F.J. Almgren, E.H. Lieb, Symmetric decreasing rearrangement is sometimes continuous. J. Am. Math. Soc. **2**(4), 683–773 (1989)
4. C.O. Alves, Local mountain pass for a class of elliptic system. J. Math. Anal. Appl. **335**(1), 135–150 (2007)
5. C.O. Alves, Existence of a positive solution for a nonlinear elliptic equation with saddle–like potential and nonlinearity with exponential critical growth in $\mathbb{R}^2$. Milan J. Math. **84**(1), 1–22 (2016)
6. C.O. Alves, V. Ambrosio, A multiplicity result for a nonlinear fractional Schrödinger equation in $\mathbb{R}^N$ without the Ambrosetti-Rabinowitz condition. J. Math. Anal. Appl. **466**(1), 498–522 (2018)
7. C. O. Alves, V. Ambrosio, T. Isernia, *Existence, multiplicity and concentration for a class of fractional p&q Laplacian problems in $\mathbb{R}^N$*. Commun. Pure Appl. Anal. **18**(4), 2009–2045 (2019)
8. C.O. Alves, F.J.S.A. Corrêa, G.M. Figueiredo, On a class of nonlocal elliptic problems with critical growth. Differ. Equ. Appl. **2**(3), 409–417 (2010)
9. C.O. Alves, F.J.S.A. Corrêa, T.F. Ma, Positive solutions for a quasi-linear elliptic equation of Kirchhoff type. Comput. Math. Appl. **49**(1), 85–93 (2005)
10. C.O. Alves, J.M. do Ó, O.H. Miyagaki, On nonlinear perturbations of a periodic elliptic problem in $\mathbb{R}^2$ involving critical growth. Nonlinear Anal. **56**(5), 781–791 (2004)
11. C.O. Alves, J.M. do Ó, O.H. Miyagaki, Concentration phenomena for fractional elliptic equations involving exponential critical growth. Adv. Nonlinear Stud. **16**(4), 843–861 (2016)
12. C.O. Alves, J.M. do Ó, M.A.S. Souto, Local mountain-pass for a class of elliptic problems in $\mathbb{R}^N$ involving critical growth. Nonlinear Anal. Ser. A: Theory Methods **46**(4), 495–510 (2001)
13. C.O. Alves, G.M. Figueiredo, Multiplicity of positive solutions for a quasilinear problem in $\mathbb{R}^N$ via penalization method. Adv. Nonlinear Stud. **5**(4), 551–572 (2005)
14. C.O. Alves, G.M. Figueiredo, Existence and multiplicity of positive solutions to a p-Laplacian equation in $\mathbb{R}^N$. Differ. Integr. Equ. **19**(2), 143–162 (2006)
15. C.O. Alves, G.M. Figueiredo, On multiplicity and concentration of positive solutions for a class of quasilinear problems with critical exponential growth in $\mathbb{R}^N$. J. Differ. Equ. **246**(3), 1288–1311 (2009)

© The Author(s), under exclusive license to Springer Nature Switzerland AG 2021
V. Ambrosio, *Nonlinear Fractional Schrödinger Equations in $\mathbb{R}^N$*,
Frontiers in Mathematics, https://doi.org/10.1007/978-3-030-60220-8

16. C.O. Alves, G.M. Figueiredo, M.F. Furtado, Multiplicity of solutions for elliptic systems via local mountain pass method. Commun. Pure Appl. Anal. **8**(6), 1745–1758 (2009)

17. C.O. Alves, G.M. Figueiredo, M.F. Furtado, Multiple solutions for critical elliptic systems via penalization method. Differ. Integr. Equ. **23**(7–8), 703–723 (2010)

18. C.O. Alves, G.M. Figueiredo, M.F. Furtado, Multiple solutions for a nonlinear Schrödinger equation with magnetic fields. Commun. Partial Differ. Equ. **36**(9), 1565–1586 (2011)

19. C.O. Alves, O.H. Miyagaki, Existence and concentration of solution for a class of fractional elliptic equation in $\mathbb{R}^N$ via penalization method. Calc. Var. Partial Differ. Equ. **55**(3), Art. 47, 19 pp (2016)

20. C.O. Alves, S.H.M. Soares, Existence and concentration of positive solutions for a class of gradient systems. NoDEA Nonlinear Differ. Equ. Appl. **12**(4), 437–457 (2005)

21. C.O. Alves, M.A.S. Souto, Existence of solutions for a class of nonlinear Schrödinger equations with potential vanishing at infinity. J. Differ. Equ. **254**(4), 1977–1991 (2013)

22. C.O. Alves, M.A.S. Souto, Existence of least energy nodal solution for a Schrödinger-Poisson system in bounded domains. Z. Angew. Math. Phys. **65**(6), 1153–1166 (2014)

23. C.O. Alves, M.A.S. Souto, M. Montenegro, Existence of a ground state solution for a nonlinear scalar field equation with critical growth. Calc. Var. Partial Differ. Equ. **43**(3–4), 537–554 (2012)

24. C.O. Alves, M.A.S. Souto, M. Montenegro, Existence of solution for two classes of elliptic problems in $\mathbb{R}^N$ with zero mass. J. Differ. Equ. **252**(10), 5735–5750 (2012)

25. C.A. Alves, M. Yang, Investigating the multiplicity and concentration behaviour of solutions for a quasilinear Choquard equation via penalization method. Proc. R. Soc. Edinburgh Sect. A **146**(1), 23–58 (2016)

26. A. Ambrosetti, V. Felli, A. Malchiodi, Ground states of nonlinear Schrödinger equations with potentials vanishing at infinity. J. Eur. Math. Soc. (JEMS) **7**(1), 117–144 (2005)

27. A. Ambrosetti, A. Malchiodi, *Nonlinear Analysis and Semilinear Elliptic Problems*. Cambridge Studies in Advanced Mathematics, vol. 104 (Cambridge University Press, Cambridge, 2007), xii+316 pp

28. A. Ambrosetti, A. Malchiodi, S. Secchi, Multiplicity results for some nonlinear Schrödinger equations with potentials. Arch. Ration. Mech. Anal. **159**(3), 253–271 (2001)

29. A. Ambrosetti, P.H. Rabinowitz, Dual variational methods in critical point theory and applications. J. Funct. Anal. **14**, 349–381 (1973)

30. A. Ambrosetti, Z.-Q. Wang, Nonlinear Schrödinger equations with vanishing and decaying potentials. Differ. Integr. Equ. **18**(12), 1321–1332 (2005)

31. V. Ambrosio, Periodic solutions for a pseudo-relativistic Schrödinger equation. Nonlinear Anal. **120**, 262–284 (2015)

32. V. Ambrosio, Ground states for superlinear fractional Schrödinger equations in $\mathbb{R}^N$. Ann. Acad. Sci. Fenn. Math. **41**(2), 745–756 (2016)

33. V. Ambrosio, Multiple solutions for a fractional p-Laplacian equation with sign-changing potential. Electron. J. Differ. Equ. **151**, 12 (2016)

34. V. Ambrosio, Ground states solutions for a non-linear equation involving a pseudo-relativistic Schrödinger operator. J. Math. Phys. **57**(5), 051502, 18 pp (2016)

35. V. Ambrosio, Multiplicity of positive solutions for a class of fractional Schrödinger equations via penalization method. Ann. Mat. Pura Appl. (4) **196**(6), 2043–2062 (2017)

36. V. Ambrosio, Ground states for a fractional scalar field problem with critical growth. Differ. Integr. Equ. **30**(1–2), 115–132 (2017)

37. V. Ambrosio, Periodic solutions for a superlinear fractional problem without the Ambrosetti-Rabinowitz condition. Discret. Contin. Dyn. Syst. **37**(5), 2265–2284 (2017)

38. V. Ambrosio, Zero mass case for a fractional Berestycki-Lions type problem. Adv. Nonlinear Anal. **7**(3), 365–374 (2018)
39. V. Ambrosio, Mountain pass solutions for the fractional Berestycki-Lions problem. Adv. Differ. Equ. **23**(5–6), 455–488 (2018)
40. V. Ambrosio, Concentration phenomena for critical fractional Schrödinger systems. Commun. Pure Appl. Anal. **17**(5), 2085–2123 (2018)
41. V. Ambrosio, Periodic solutions for critical fractional problems. Calc. Var. Partial Differ. Equ. **57**(2), Art. 45, 31 pp (2018)
42. V. Ambrosio, Multiple solutions for superlinear fractional problems via theorems of mixed type. Adv. Nonlinear Stud. **18**(4), 799–817 (2018)
43. V. Ambrosio, Boundedness and decay of solutions for some fractional magnetic Schrödinger equations in $\mathbb{R}^N$. Milan J. Math. **86**(2), 125–136 (2018)
44. V. Ambrosio, Existence and concentration results for some fractional Schrödinger equations in $\mathbb{R}^N$ with magnetic fields. Commun. Partial Differ. Equ. **44**(8), 637–680 (2019)
45. V. Ambrosio, Concentration phenomena for a fractional Choquard equation with magnetic field. Dyn. Partial Differ. Equ. **16**(2), 125–149 (2019)
46. V. Ambrosio, Multiplicity and concentration results for a fractional Choquard equation via penalization method. Potential Anal. **50**(1), 55–82 (2019)
47. V. Ambrosio, Concentrating solutions for a class of nonlinear fractional Schrödinger equations in $\mathbb{R}^N$. Rev. Mat. Iberoam. **35**(5), 1367–1414 (2019)
48. V. Ambrosio, Multiplicity and concentration results for a class of critical fractional Schrödinger-Poisson systems via penalization method. Commun. Contemp. Math. **22**(1), 1850078, 45 pp (2020)
49. V. Ambrosio, On a fractional magnetic Schrödinger equation in $\mathbb{R}$ with exponential critical growth. Nonlinear Anal. **183**, 117–148 (2019)
50. V. Ambrosio, Concentrating solutions for a fractional Kirchhoff equation with critical growth. Asymptot. Anal. **116**(3-4), 249–278 (2020)
51. V. Ambrosio, Multiplicity and concentration results for a fractional Schrödinger-Poisson type equation with magnetic field. Proc. R. Soc. Edinburgh Sect. A **150**(2), 655–694 (2020)
52. V. Ambrosio, Multiplicity of solutions for fractional Schrödinger systems in $\mathbb{R}^N$. Complex Var. Elliptic Equ. **65**(5), 856–885 (2020)
53. V. Ambrosio, Multiplicity and concentration of solutions for a fractional Kirchhoff equation with magnetic field and critical growth. Ann. Henri Poincaré **20**(8), 2717–2766 (2019)
54. V. Ambrosio, Multiplicity and concentration of solutions for fractional Schrödinger systems via penalization method. Atti Accad. Naz. Lincei Rend. Lincei Mat. Appl. **30**(3), 543–581 (2019)
55. V. Ambrosio, Infinitely many periodic solutions for a class of fractional Kirchhoff problems. Monatsh. Math. **190**(4), 615–639 (2019)
56. V. Ambrosio, On the multiplicity and concentration of positive solutions for a p-fractional Choquard equation in $\mathbb{R}^N$. Comput. Math. Appl. **78**(8), 2593–2617 (2019)
57. V. Ambrosio, A local mountain pass approach for a class of fractional NLS equations with magnetic fields. Nonlinear Anal. **190**, 111622, 14 pp (2020)
58. V. Ambrosio, Multiple concentrating solutions for a fractional Kirchhoff equation with magnetic fields. Discret. Contin. Dyn. Syst. **40**(2), 781–815 (2020)
59. V. Ambrosio, Multiplicity and concentration results for fractional Schrödinger-Poisson equations with magnetic fields and critical growth. Potential Anal. **52**(4), 565–600 (2020)
60. V. Ambrosio, Existence and concentration of nontrivial solutions for a fractional magnetic Schrödinger-Poisson type equation. Ann. Sc. Norm. Super. Pisa Cl. Sci. (5) **XXI**, 1043–1081 (2020)

61. V. Ambrosio, Fractional $p \& q$ Laplacian problems in $\mathbb{R}^N$ with critical growth. Z. Anal. Anwend. **39**(3), 289–314 (2020)

62. V. Ambrosio, Concentration phenomena for a class of fractional Kirchhoff equations in $\mathbb{R}^N$ with general nonlinearities. Nonlinear Anal. **195**, 111761, 39 pp (2020)

63. V. Ambrosio, The nonlinear fractional relativistic Schrödinger equation: existence, multiplicity, decay and concentration results. submitted

64. V. Ambrosio, *On the fractional relativistic Schrödinger operator*. submitted

65. V. Ambrosio, *Concentration phenomena for fractional magnetic NLS*. submitted

66. V. Ambrosio, P. d'Avenia, Nonlinear fractional magnetic Schrödinger equation: existence and multiplicity. J. Differ. Equ. **264**(5), 3336–3368 (2018)

67. V. Ambrosio, G.M. Figueiredo, Ground state solutions for a fractional Schrödinger equation with critical growth. Asymptot. Anal. **105**(3–4), 159–191 (2017)

68. V. Ambrosio, G.M. Figueiredo, T. Isernia, Existence and concentration of positive solutions for p-fractional Schrödinger equations. Ann. Mat. Pura Appl. (4) **199**(1), 317–344 (2020)

69. V. Ambrosio, G.M. Figueiredo, T. Isernia, G. Molica Bisci, Sign-changing solutions for a class of zero mass nonlocal Schrödinger equations. Adv. Nonlinear Stud. **19**(1), 113–132 (2019)

70. V. Ambrosio, H. Hajaiej, Multiple solutions for a class of nonhomogeneous fractional Schrödinger equations in $\mathbb{R}^N$. J. Dynam. Differ. Equ. **30**(3), 1119–1143 (2018)

71. V. Ambrosio, T. Isernia, Sign-changing solutions for a class of fractional Schrödinger equations with vanishing potentials. Atti Accad. Naz. Lincei Rend. Lincei Mat. Appl. **29**(1), 127–152 (2018)

72. V. Ambrosio, T. Isernia, A multiplicity result for a fractional Kirchhoff equation in $\mathbb{R}^N$ with a general nonlinearity. Commun. Contemp. Math. **20**(5), 1750054, 17 pp (2018)

73. V. Ambrosio, T. Isernia, Concentration phenomena for a fractional Schrödinger-Kirchhoff type problem. Math. Methods Appl. Sci. **41**(2), 615–645 (2018)

74. V. Ambrosio, T. Isernia, Multiplicity and concentration results for some nonlinear Schrödinger equations with the fractional p-Laplacian. Discret. Contin. Dyn. Syst. **38**(11), 5835–5881 (2018)

75. V. Ambrosio, T. Isernia, On the multiplicity and concentration for p-fractional Schrödinger equations. Appl. Math. Lett. **95**, 13–22 (2019)

76. D. Applebaum, *Lévy Processes and Stochastic Calculus*. Cambridge Studies in Advanced Mathematics, vol. 93. (Cambridge University Press, Cambridge, 2004), xxiv+384 pp

77. G. Arioli, A. Szulkin, A semilinear Schrödinger equation in the presence of a magnetic field. Arch. Ration. Mech. Anal. **170**(4), 277–295 (2003)

78. N. Aronszajn, K.T. Smith, Theory of Bessel potentials. I. Ann. Inst. Fourier (Grenoble) **11**, 385–475 (1961)

79. A. Arosio, S. Panizzi, On the well-posedness of the Kirchhoff string. Trans. Am. Math. Soc. **348**(1), 305–330 (1996)

80. A.I. Ávila, J. Yang, Multiple solutions of nonlinear elliptic systems. NoDEA Nonlinear Differ. Equ. Appl. **12**(4), 459–479 (2005)

81. J. Avron, I. Herbst, B. Simon, Schrödinger operators with magnetic fields. I. General interactions. Duke Math. J. **45**(4), 847–883 (1978)

82. A. Azzollini, P. d'Avenia, A. Pomponio, On the Schrödinger-Maxwell equations under the effect of a general nonlinear term. Ann. Inst. H. Poincaré Anal. Non Linéaire **27**(2), 779–791 (2010)

83. A. Azzollini, P. d'Avenia, A. Pomponio, Multiple critical points for a class of nonlinear functionals. Ann. Mat. Pura Appl. **190**(3), 507–523 (2011)

84. A. Azzollini, A. Pomponio, On a "zero mass" nonlinear Schrödinger equation. Adv. Nonlinear Stud. **7**(4), 599–627 (2007)

85. A. Bahri, H. Berestycki, A perturbation method in critical point theory and applications. Trans. Am. Math. Soc. **267**(1), 1–32 (1981)

86. S. Barile, G.M. Figueiredo, Existence of least energy positive, negative and nodal solutions for a class of $p\&q$-problems with potentials vanishing at infinity. J. Math. Anal. Appl. **427**(2), 1205–1233 (2015)

87. B. Barrios, E. Colorado, A. de Pablo, U. Sánchez, On some critical problems for the fractional Laplacian operator. J. Differ. Equ. **252**(11), 6133–6162 (2012)

88. T. Bartsch, Z. Liu, T. Weth, Sign changing solutions of superlinear Schrödinger equations. Commun. Partial Differ. Equ. **29**(1–2), 25–42 (2004)

89. T. Bartsch, A. Pankov, Z.-Q. Wang, Nonlinear Schrödinger equations with steep potential well. Commun. Contemp. Math. **4**(4), 549–569 (2001)

90. T. Bartsch, Z.-Q. Wang, Existence and multiplicity results for some superlinear elliptic problems on $\mathbb{R}^N$. Commun. Partial Differ. Equ. **20**(9–10), 1725–1741 (1995)

91. T. Bartsch, T. Weth, Three nodal solutions of singularly perturbed elliptic equations on domains without topology. Ann. Inst. H. Poincaré Anal. Non Linéaire **22**(3), 259–281 (2005)

92. T. Bartsch, T. Weth, M. Willem, Partial symmetry of least energy nodal solutions to some variational problems. J. Anal. Math. **96**, 1–18 (2005)

93. P. Belchior, H. Bueno, O.H. Miyagaki, G.A. Pereira, Remarks about a fractional Choquard equation: ground state, regularity and polynomial decay. Nonlinear Anal. **164**, 38–53 (2017)

94. A.K. Ben-Naoum, C. Troestler, M. Willem, Extrema problems with critical Sobolev exponents on unbounded domains. Nonlinear Anal. **26**(4), 823–833 (1996)

95. V. Benci, G. Cerami, Multiple positive solutions of some elliptic problems via the Morse theory and the domain topology. Calc. Var. Partial Differ. Equ. **2**(1), 29–48 (1994)

96. V. Benci, D. Fortunato, An eigenvalue problem for the Schrödinger-Maxwell equations. Topol. Methods Nonlinear Anal. **11**(2), 283–293 (1998)

97. V. Benci, C.R. Grisanti, A.M. Micheletti, Existence of solutions for the nonlinear Schrödinger equations with $V(\infty) = 0$, in *Contributions to Non-linear Analysis*. Progress in Nonlinear Differential Equations and Their Applications, vol. 66 (Birkhäuser, Basel, 2006), pp. 53–65

98. V. Benci, A.M. Micheletti, Solutions in exterior domains of null mass nonlinear field equations. Adv. Nonlinear Stud. **6**(2), 171–198 (2006)

99. H. Berestycki, T. Gallouët, O. Kavian, Equations de Champs scalaires euclidiens non linéaires dans le plan. C. R. Acad. Sci. Paris Sér. I Math. **297**(5), 307–310 (1983)

100. H. Berestycki, P.L. Lions, Nonlinear scalar field equations. I. Existence of a ground state. Arch. Ration. Mech. Anal. **82**(4), 313–345 (1983)

101. H. Berestycki, P.L. Lions, Nonlinear scalar field equations. II. Existence of infinitely many solutions. Arch. Ration. Mech. Anal. **82**(4), 347–375 (1983)

102. H. Berestycki, P.L. Lions, Existence d'états multiples dans des équations de champs scalaires non linéaires dans le cas de masse nulle. C. R. Acad. Sci. Paris Sér. I Math. **297**(4), 267–270 (1983)

103. S. Bernstein, Sur une classe d'équations fonctionnelles aux dérivées partielles. Bull. Acad. Sci. URSS. Sér. Math. [Izvestia Akad. Nauk SSSR] **4**, 17–26 (1940)

104. G. Bianchi, J. Chabrowski, A. Szulkin, On symmetric solutions of an elliptic equation with a nonlinearity involving critical Sobolev exponent. Nonlinear Anal. **25**(1), 41–59 (1995)

105. R.M. Blumenthal, R.K. Getoor, Some theorems on stable processes. Trans. Am. Math. Soc. **95**, 263–273 (1960)

106. K. Bogdan, T. Byczkowski, T. Kulczycki, M. Ryznar, R. Song, Z. Vondraček, *Potential Analysis of Stable Processes and its Extensions*, ed. by P. Graczyk, A. Stos. Lecture Notes in Mathematics, vol. 1980 (Springer, Berlin, 2009), x+187 pp

107. D. Bonheure, J. Van Schaftingen, Ground states for the nonlinear Schrödinger equation with potential vanishing at infinity. Ann. Mat. Pura Appl. (4) **189**(2), 273–301 (2010)

108. J. Bourgain, H. Brezis, P. Mironescu, Another look at Sobolev spaces, in *Optimal Control and Partial Differential Equations* (IOS, Amsterdam, 2001)

109. C. Brändle, E. Colorado, A. de Pablo, U. Sánchez, A concave-convex elliptic problem involving the fractional Laplacian. Proc. R. Soc. Edinburgh Sect. A **143**(1), 39–71 (2013)

110. L. Brasco, S. Mosconi, M. Squassina, Optimal decay of extremals for the fractional Sobolev inequality. Calc. Var. Partial Differ. Equ. **55**(2), Art. 23, 32 pp (2016)

111. H Brezis, J.-M. Coron, L. Nirenberg, Free vibrations for a nonlinear wave equation and a theorem of P. Rabinowitz. Commun. Pure Appl. Math. **33**(5), 667–684 (1980)

112. H. Brezis, T. Kato, Remarks on the Schrödinger operator with singular complex potentials. J. Math. Pures Appl. (9) **58**(2), 137–151 (1979)

113. H. Brezis, E.H. Lieb, A relation between pointwise convergence of functions and convergence of functionals. Proc. Am. Math. Soc. **88**(3), 486–490 (1983)

114. H. Brezis, L. Nirenberg, Positive solutions of nonlinear elliptic equations involving critical Sobolev exponents. Commun. Pure Appl. Math. **36**(4), 437–477 (1983)

115. C. Bucur, E. Valdinoci, *Nonlocal Diffusion and Applications*. Lecture Notes of the Unione Matematica Italiana, vol. 20 (Springer, [Cham]; Unione Matematica Italiana, Bologna, 2016), xii+155 pp

116. H. Bueno, A.H.S. Medeiros, G.A. Pereira, Pohozaev-type identities for a pseudo-relativistic Schrödinger operator and applications. preprint arXiv:1810.07597

117. H. Bueno, O.H. Miyagaki, G.A. Pereira, Remarks about a generalized pseudo-relativistic Hartree equation. J. Differ. Equ. **266**(1), 876–909 (2019)

118. J. Busca, B. Sirakov, Symmetry results for semilinear elliptic systems in the whole space. J. Differ. Equ. **163**(1), 41–56 (2000)

119. T. Byczkowski, J. Malecki, M. Ryznar, Bessel potentials, hitting distributions and Green functions. Trans. Am. Math. Soc. **361**(9), 4871–4900 (2009)

120. J. Byeon, Singularly perturbed nonlinear Dirichlet problems with a general nonlinearity. Trans. Am. Math. Soc. **362**(4), 1981–2001 (2010)

121. J. Byeon, L. Jeanjean, Standing waves for nonlinear Schrödinger equations with a general nonlinearity. Arch. Ration. Mech. Anal. **185**(2), 185–200 (2007)

122. J. Byeon, O. Kwon, J. Seok, Nonlinear scalar field equations involving the fractional Laplacian. Nonlinearity **30**(4), 1659–1681 (2017)

123. J. Byeon, Z.-Q Wang, Standing waves with a critical frequency for nonlinear Schrödinger equations. II. Calc. Var. Partial Differ. Equ. **18**(2), 207–219 (2003)

124. X. Cabré, Y. Sire, Nonlinear equations for fractional Laplacians I: regularity, maximum principles, and Hamiltonian estimates. Ann. Inst. H. Poincaré Anal. Non Linéaire **31**(1), 23–53 (2014)

125. X. Cabré, J. Solá-Morales, Layer solutions in a half-space for boundary reactions. Commun. Pure Appl. Math. **58**(12), 1678–1732 (2005)

126. X. Cabré, J. Tan, Positive solutions of nonlinear problems involving the square root of the Laplacian. Adv. Math. **224**(5), 2052–2093 (2010)

127. L.A. Caffarelli, L.Silvestre, An extension problem related to the fractional Laplacian. Commun. Partial Differ. Equ. **32**(7–9), 1245–1260 (2007)

128. A.-P. Calderón, Lebesgue spaces of differentiable functions and distributions, in *1961 Proceedings of Symposia in Pure Mathematics*, vol. IV (American Mathematical Society, Providence, 1961), pp. 33–49

129. D.M. Cao, Nontrivial solution of semilinear elliptic equation with critical exponent in $\mathbb{R}^2$. Commun. Partial Differ. Equ. **17**(3–4), 407–435 (1992)

130. D.-M.M. Cao, H.-S. Zhou, Multiple positive solutions of nonhomogeneous semilinear elliptic equations in $\mathbb{R}^N$. Proc. R. Soc. Edinburgh Sect. A **126**(2), 443–463 (1996)
131. A. Capella, J. Davila, L. Dupaigne, Y. Sire, Regularity of radial extremals solutions for some non-local semilinear equation. Commun. Partial Differ. Equ. **36**(8), 1353–1384 (2011)
132. R. Carmona, W.C. Masters, B. Simon, Relativistic Schrödinger operators: asymptotic behavior of the eigenfunctions. J. Func. Anal **91**(1), 117–142 (1990)
133. A. Castro, J. Cossio, J. Neuberger, A sign-changing solution for a superlinear Dirichlet problem. Rocky Mountain J. Math. **27**(4), 1041–1053 (1997)
134. G. Cerami, An existence criterion for the critical points on unbounded manifolds. Istit. Lombardo Accad. Sci. Lett. Rend. A **112**(2), 332–336 (1978, 1979)
135. J. Chabrowski, Concentration-compactness principle at infinity and semilinear elliptic equations involving critical and subcritical Sobolev exponents. Calc. Var. Partial Differ. Equ. **3**(4), 493–512 (1995)
136. J. Chabrowski, J. Yang, Existence theorems for elliptic equations involving supercritical Sobolev exponent. Adv. Differ. Equ. **2**(2), 231–256 (1997)
137. K.-C. Chang, *Methods in Nonlinear Analysis*. Springer Monographs in Mathematics (Springer, Berlin, 2005), x+439 pp
138. W. Chen, C. Li, B. Ou, Classification of solutions for an integral equation. Commun. Pure Appl. Math. **59**(3), 330–343 (2006)
139. X.J. Chang, Z.Q. Wang, Ground state of scalar field equations involving fractional Laplacian with general nonlinearity. Nonlinearity **26**(2), 479–494 (2013)
140. C. Chen, Y. Kuo, T. Wu, The Nehari manifold for a Kirchhoff type problem involving sign-changing weight functions. J. Differ. Equ. **250**(4), 1876–1908 (2011)
141. G. Chen, Y. Zheng, Concentration phenomenon for fractional nonlinear Schrödinger equations. Commun. Pure Appl. Anal. **13**(6), 2359–2376 (2014)
142. M. Chipot, B. Lovat, Some remarks on nonlocal elliptic and parabolic problems. Nonlinear Anal. **30**(7), 4619–4627 (1997)
143. W. Choi, J. Seok, Nonrelativistic limit of standing waves for pseudo-relativistic nonlinear Schrödinger equations. J. Math. Phys. **57**(2), 021510, 15 pp (2016)
144. S. Cingolani, Semiclassical stationary states of nonlinear Schrödinger equations with an external magnetic field. J. Differ. Equ. **188**(1), 52–79 (2003)
145. S. Cingolani, M. Clapp, S. Secchi, Multiple solutions to a magnetic nonlinear Choquard equation. Z. Angew. Math. Phys. **63**(2), 233–248 (2012)
146. S. Cingolani, M. Lazzo, Multiple semiclassical standing waves for a class of nonlinear Schrödinger equations. Topol. Methods Nonlinear Anal. **10**(1), 1–13 (1997)
147. S. Cingolani, M. Lazzo, Multiple positive solutions to nonlinear Schrödinger equations with competing potential functions. J. Differ. Equ. **160**, 118–138 (2000)
148. S. Cingolani, S. Secchi, Semiclassical states for NLS equations with magnetic potentials having polynomial growths. J. Math. Phys. **46**(5), 053503, 19 pp (2005)
149. C.V. Coffman, A minimum-maximum principle for a class of non-linear integral equations. J. Analyse Math. **22**, 391–419 (1969)
150. E. Colorado, A. de Pablo, U. Sánchez, Perturbations of a critical fractional equation. Pac. J. Math. **271**(1), 65–85 (2014)
151. D.G. Costa, C.A. Magalhaes, Variational elliptic problems which are nonquadratic at infinity. Nonlinear Anal. **23**(11), 1401–1412 (1994)
152. V. Coti Zelati, M. Nolasco, Existence of ground states for nonlinear, pseudo-relativistic Schrödinger equations. Atti Accad. Naz. Lincei Rend. Lincei Mat. Appl. **22**(1), 51–72 (2011)
153. V. Coti Zelati, M. Nolasco, Ground states for pseudo-relativistic Hartree equations of critical type. Rev. Mat. Iberoam. **29**(4), 1421–1436 (2013)

154. V. Coti Zelati, P.H. Rabinowitz, Homoclinic type solutions for a semilinear elliptic PDE on $\mathbb{R}^n$. Commun. Pure Appl. Math. **45**(10), 1217–1269 (1992)
155. A. Cotsiolis, N.K. Tavoularis, Best constants for Sobolev inequalities for higher order fractional derivatives. J. Math. Anal. Appl. **295**(1), 225–236 (2004)
156. P. d'Avenia, G. Siciliano, M. Squassina, On fractional Choquard equations. Math. Models Methods Appl. Sci. **25**(8), 1447–1476 (2015)
157. P. d'Avenia, M. Squassina, Ground states for fractional magnetic operators. ESAIM Control Optim. Calc. Var. **24**(1), 1–24 (2018)
158. J. Dávila, M. del Pino, S. Dipierro, E. Valdinoci, Concentration phenomena for the nonlocal Schrödinger equation with Dirichlet datum. Anal. PDE **8**(5), 1165–1235 (2015)
159. J. Dávila, M. del Pino, J. Wei, Concentrating standing waves for the fractional nonlinear Schrödinger equation. J. Differ. Equ. **256**(2), 858–892 (2014)
160. D.G. de Figueiredo, O.H. Miyagaki, B. Ruf, Elliptic equations in $\mathbb{R}^2$ with nonlinearities in the critical growth range. Calc. Var. Partial Differ. Equ. **3**(2), 139–153 (1995)
161. D.C. de Morais Filho, M.A.S. Souto, Systems of p-Laplacian equations involving homogeneous nonlinearities with critical Sobolev exponent degrees. Commun. Partial Differ. Equ. **24**(7–8), 1537–1553 (1999)
162. M. de Souza, Y.L. Araújo, On nonlinear perturbatiie of a periodic fractional Schrödinger equation with critical exponential growth. Math. Nachr. **289**(5–6), 610–625 (2016)
163. L.M. Del Pezzo, A. Quaas, A Hopf's lemma and a strong minimum principle for the fractional p-Laplacian. J. Differ. Equ. **263**(1), 765–778 (2017)
164. L.M. Del Pezzo, A. Quaas, Spectrum of the fractional p-Laplacian in $\mathbb{R}^N$ and decay estimate for positive solutions of a Schrödinger equation. Nonlinear Anal. **193**, 111479 (2020)
165. M. del Pino, P.L. Felmer, Local mountain passes for semilinear elliptic problems in unbounded domains. Calc. Var. Partial Differ. Equ. **4**(2), 121–137 (1996)
166. F. Demengel, G. Demengel, *Functional Spaces for the Theory of Elliptic Partial Differential Equations*. Universitext (Springer, London; EDP Sciences, Les Ulis, 2012), xviii+465 pp
167. A. Di Castro, T. Kuusi, G. Palatucci, Local behavior of fractional p-minimizers. Ann. Inst. H. Poincaré Anal. Non Linéaire, **33**(5), 1279–1299 (2016)
168. E. Di Nezza, G. Palatucci, E. Valdinoci, Hitchhiker's guide to the fractional Sobolev spaces. Bull. Sci. math. **136**(5), 521–573 (2012)
169. Y. Ding, *Variational Methods for Strongly Indefinite Problems*. Interdisciplinary Mathematical Sciences, vol. 7 (World Scientific Publishing Co. Pte. Ltd., Hackensack, 2007), viii+168 pp
170. S. Dipierro, M. Medina, E. Valdinoci, Fractional elliptic problems with critical growth in the whole of $\mathbb{R}^n$. Appunti. Scuola Normale Superiore di Pisa (Nuova Serie) [Lecture Notes. Scuola Normale Superiore di Pisa (New Series)], vol. 15 (Edizioni della Normale, Pisa, 2017), viii+152 pp
171. S. Dipierro, L. Montoro, I. Peral, B. Sciunzi, Qualitative properties of positive solutions to nonlocal critical problems involving the Hardy-Leray potential. Calc. Var. Partial Differ. Equ. **55**(4), Paper No. 99, 29 pp (2016)
172. S. Dipierro, G. Palatucci, E. Valdinoci, Existence and symmetry results for a Schrödinger type problem involving the fractional Laplacian. Matematiche (Catania) **68**(1), 201–216 (2013)
173. J.M. do Ó, O.H. Miyagaki, M. Squassina, Nonautonomous fractional problems with exponential growth. NoDEA Nonlinear Differ. Equ. Appl. **22**(5), 1395–1410 (2015)
174. J.M. do Ó, O.H. Miyagaki, M. Squassina, Critical and subcritical fractional problems with vanishing potentials. Commun. Contemp. Math. **18**(6), 1550063, 20 pp (2016)
175. J.M. do Ó, M.A.S. Souto, On a class of nonlinear Schrödinger equations in $\mathbb{R}^2$ involving critical growth. J. Differ. Equ. **174**(2), 289–311 (2001)
176. I. Ekeland, On the variational principle. J. Math. Anal. Appl. **47**, 324–353 (1974)

177. I. Ekeland, *Convexity Methods in Hamiltonian Mechanics.* (Springer-Verlag, Berlin, 1990), x+247 pp

178. A. Erdelyi, *Higher Trascendental Functions* (McGraw-Hill, New York, 1953)

179. M. Esteban, P.L. Lions, Stationary solutions of nonlinear Schrödinger equations with an external magnetic field, in *Partial Differential Equations and the Calculus of Variations.* Progress in Nonlinear Differential Equations, 1, vol. I (Birkhäuser Boston, Boston, 1989), pp. 401–449

180. E.B. Fabes, C.E. Kenig, R.P. Serapioni, The local regularity of solutions of degenerate elliptic equations. Commun. Partial Differ. Equ. **7**(1), 77–116 (1982)

181. M. Fall, V. Felli, Unique continuation properties for relativistic Schrödinger operators with a singular potential. Discret. Contin. Dyn. Syst. **35**(12), 5827–5867 (2015)

182. M.M. Fall, F. Mahmoudi, E. Valdinoci, Ground states and concentration phenomena for the fractional Schrödinger equation. Nonlinearity **28**(6), 1937–1961 (2015)

183. P. Felmer, A. Quaas, J. Tan, Positive solutions of the nonlinear Schrödinger equation with the fractional Laplacian. Proc. R. Soc. Edinburgh Sect. A **142**(6), 1237–1262 (2012)

184. P. Felmer, I. Vergara, Scalar field equation with non-local diffusion. NoDEA Nonlinear Differ. Equ. Appl. **22**(5), 1411–1428 (2015)

185. P. Felmer, Y. Wang, Radial symmetry of positive solutions to equations involving the fractional Laplacian. Commun. Contemp. Math. **16**(1), 1350023, 24 pp (2014)

186. G.M. Figueiredo, Existence, multiplicity and concentration of positive solutions for a class of quasilinear problems with critical growth. Commun. Appl. Nonlinear Anal. **13**(4), 79–99 (2006)

187. G.M. Figueiredo, Existence of a positive solution for a Kirchhoff problem type with critical growth via truncation argument. J. Math. Anal. Appl. **401**(2), 706–713 (2013)

188. G.M. Figueiredo, M. Furtado, Positive solutions for some quasilinear equations with critical and supercritical growth. Nonlinear Anal. **66**(7), 1600–1616 (2007)

189. G.M. Figueiredo, M.F. Furtado, Multiple positive solutions for a quasilinear system of Schrödinger equations. NoDEA Nonlinear Differ. Equ. Appl. **15**(3), 309–333 (2008)

190. G.M. Figueiredo, M. Furtado, Positive solutions for a quasilinear Schrödinger equation with critical growth. J. Dyn. Differ. Equ. **24**(1), 13–28 (2012)

191. G.M. Figueiredo, J.R. Santos Júnior, Multiplicity and concentration behavior of positive solutions for a Schrödinger-Kirchhoff type problem via penalization method. ESAIM Control Optim. Calc. Var. **20**(2), 389–415 (2014)

192. G.M. Figueiredo, J.R. Santos Júnior, Existence of a least energy nodal solution for a Schrödinger-Kirchhoff equation with potential vanishing at infinity. J. Math. Phys. **56**, 051506 18 pp (2015)

193. G.M. Figueiredo, G. Siciliano, A multiplicity result via Ljusternick-Schnirelmann category and Morse theory for a fractional Schrödinger equation in $\mathbb{R}^N$. NoDEA Nonlinear Differ. Equ. Appl. **23**(2), Art. 12, 22 pp (2016)

194. A. Fiscella, A. Pinamonti, E. Vecchi, Multiplicity results for magnetic fractional problems. J. Differ. Equ. **263**(8), 4617–4633 (2017)

195. A. Fiscella, P. Pucci, Kirchhoff-Hardy fractional problems with lack of compactness. Adv. Nonlinear Stud. **17**(3), 429–456 (2017)

196. A. Fiscella, E. Valdinoci, A critical Kirchhoff type problem involving a nonlocal operator. Nonlinear Anal. **94**, 156–170 (2014)

197. A. Floer, A. Weinstein, Nonspreading wave packets for the cubic Schrödinger equation with a bounded potential. J. Funct. Anal. **69**(3), 397–408 (1986)

198. R.L. Frank, E. Lenzmann, Uniqueness of non-linear ground states for fractional Laplacians in $\mathbb{R}$. Acta Math. **210**(2), 261–318 (2013)

199. R.L. Frank, E. Lenzmann, L. Silvestre, Uniqueness of radial solutions for the fractional Laplacian. Commun. Pure Appl. Math. **69**(9), 1671–1726 (2016)

200. R.L. Frank, R. Seiringer, Non-linear ground state representations and sharp Hardy inequalities. J. Funct. Anal. **255**(12), 3407–3430 (2008)

201. G. Franzina, G. Palatucci, Fractional p-eigenvalues. Riv. Math. Univ. Parma (N.S.) **5**(2), 373–386 (2014)

202. J. Fröhlich, B. Jonsson, G. Lars, E. Lenzmann, Boson stars as solitary waves. Commun. Math. Phys. **274**(1), 1–30 (2007)

203. N. Garofalo, Fractional thoughts, in *New Developments in the Analysis of Nonlocal Operators*. Contemporary Mathematics, vol. 723 (American Mathematical Society, Providence, 2019), pp. 1–135

204. A.R. Giammetta, Fractional Schrödinger-Poisson-Slater system in one dimension. preprint arXiv:1405.2796

205. B. Gidas, Bifurcation phenomena in mathematical physics and related topics, in *Proceedings of the NATO Advanced Study Institute held at Cargèse*, June 24–July 7, 1979, ed. by C. Bardos, D. Bessis, NATO Advanced Study Institute Series. Ser. C Mathematical and Physical Sciences 54, 1980

206. B. Gidas, W.M. Ni, L. Nirenberg, Symmetry and related properties via the maximum principle. Commun. Math. Phys. **68**(3), 209–243 (1979)

207. B. Gidas, W.M. Ni, L. Nirenberg, Symmetry of positive solutions of nonlinear elliptic equations in $\mathbb{R}^n$, in *Mathematical Analysis and Applications, Part A*, Adv. in Math. Suppl. Stud., 7a (Academic Press, New York/London, 1981), pp. 369–402

208. D. Gilbarg, N. Trudinger, *Elliptic Partial Differential Equations of Second Order*. Reprint of the 1998 edition. Classics in Mathematics (Springer, Berlin, 2001), xiv+517 pp

209. L. Grafakos, *Modern Fourier Analysis*. Graduate Texts in Mathematics, vol. 250, 3rd edn. (Springer, New York, 2014), xvi+624 pp

210. Z. Guo, S. Luo, W. Zou, On critical systems involving fractional Laplacian. J. Math. Anal. Appl. **446**(1), 681–706 (2017)

211. D.D. Haroske, H. Triebel, *Distributions, Sobolev Spaces, Elliptic Equations*, EMS Textbooks in Mathematics (European Mathematical Society (EMS), Zürich, 2008)

212. X. He, Multiplicity and concentration of positive solutions for the Schrödinger-Poisson equations. Z. Angew. Math. Phys. **62**(5), 869–889 (2011)

213. Y. He, G. Li, Standing waves for a class of Schrödinger-Poisson equations in $\mathbb{R}^3$ involving critical Sobolev exponents. Ann. Acad. Sci. Fenn. Math. **40**(2), 729–766 (2015)

214. Y. He, G. Li, S. Peng, Concentrating bound states for Kirchhoff type problems in $\mathbb{R}^3$ involving critical Sobolev exponents. Adv. Nonlinear Stud. **14**(2), 483–510 (2014)

215. X. He, W. Zou, Existence and concentration behavior of positive solutions for a Kirchhoff equation in $\mathbb{R}^3$. J. Differ. Equ. **252**(2), 1813–1834 (2012)

216. X. He, W. Zou, Existence and concentration result for the fractional Schrödinger equations with critical nonlinearities. Calc. Var. Partial Differ. Equ. **55**(4), Art. 91, 39 pp (2016)

217. I.W. Herbst, Spectral theory of the operator $(p^2 + m^2)^{1/2} - Ze^2/r$. Commun. Math. Phys. **53**(3), 285–294 (1977)

218. J. Hirata, N. Ikoma, K. Tanaka, Nonlinear scalar field equations in $\mathbb{R}^N$: mountain pass and symmetric mountain pass approaches. Topol. Methods Nonlinear Anal. **35**(2), 253–276 (2010)

219. F. Hiroshima, T. Ichinose, J. Lőrinczi, Kato's inequality for Magnetic Relativistic Schrödinger Operators. Publ. Res. Inst. Math. Sci. **53**(1), 79–117 (2017)

220. L. Hörmander, The analysis of linear partial differential operators. III. Pseudodifferential operators, in *Grundlehren der Mathematischen Wissenschaften* [Fundamental Principles of Mathematical Sciences], vol. 274 (Springer, Berlin, 1985), viii+525 pp

221. L. Hörmander, The analysis of linear partial differential operators. IV. Fourier integral operators, in *Grundlehren der Mathematischen Wissenschaften* [Fundamental Principles of Mathematical Sciences], vol. 275. (Springer, Berlin, 1985), vii+352 pp

222. A. Iannizzotto, S. Mosconi, M. Squassina, Global Hölder regularity for the fractional p-Laplacian. Rev. Mat. Iberoam. **32**(4), 1353–1392 (2016)

223. A. Iannizzotto, M. Squassina, 1/2-Laplacian problems with exponential nonlinearity. J. Math. Anal. Appl. **414**(1), 372–385 (2014)

224. T. Ichinose, Magnetic relativistic Schrödinger operators and imaginary-time path integrals, in *Mathematical Physics, Spectral Theory and Stochastic Analysis*. Operator Theory: Advances and Applications, vol. 232 (Birkhäuser/Springer Basel AG, Basel, 2013), pp. 247–297

225. T. Ichinose, H. Tamura, Imaginary-time path integral for a relativistic spinless particle in an electromagnetic field. Commun. Math. Phys. **105**(2), 239–257 (1986)

226. N. Ikoma, Existence of solutions of scalar field equations with fractional operator. J. Fixed Point Theory Appl. **19**(1), 649–690 (2017)

227. T. Isernia, Positive solution for nonhomogeneous sublinear fractional equations in $\mathbb{R}^N$. Complex Var. Elliptic Equ. **63**(5), 689–714 (2018)

228. S. Jarohs, T. Weth, On the strong maximum principle for nonlocal operators. Math. Z. **293**(1–2), 81–111 (2019)

229. L. Jeanjean, Two positive solutions for a class of nonhomogeneous elliptic equations. Differ. Integr. Equ. **10**(4), 609–624 (1997)

230. L. Jeanjean, Existence of solutions with prescribed norm for semilinear elliptic equations. Nonlinear Anal. **28**(10), 1633–1659 (1997)

231. J. Jeanjean, On the existence of bounded Palais-Smale sequences and application to a Landesman-Lazer-type problem set on $\mathbb{R}^N$. Proc. R. Soc. Edinburgh Sect. A **129**(4), 787–809 (1999)

232. L. Jeanjean, S. Le Coz, An existence and stability result for standing waves of nonlinear Schrödinger equations. Adv. Differ. Equ. **11**(7), 813–840 (2006)

233. L. Jeanjean, K. Tanaka, A positive solution for an asymptotically linear elliptic problem on $\mathbb{R}^N$ autonomous at infinity. ESAIM Control Optim. Calc. Var. **7**, 597–614 (2002)

234. L. Jeanjean, K. Tanaka, A remark on least energy solutions in $\mathbb{R}^N$. Proc. Am. Math. Soc. **131**(8), 2399–2408 (2003)

235. L. Jeanjean, K. Tanaka, Singularly perturbed elliptic problems with superlinear or asymptotically linear nonlinearities. Calc. Var. **21**(3), 287–318 (2004)

236. L. Jeanjean, K. Tanaka, A positive solution for a nonlinear Schrödinger equation on $\mathbb{R}^N$. Indiana Univ. Math. J. **54**(2), 443–464 (2005)

237. T. Jin, Y. Li, J. Xiong, On a fractional Nirenberg problem, part I: blow up analysis and compactness of solutions. J. Eur. Math. Soc. (JEMS) **16**(6), 1111–1171 (2014)

238. T. Kato, Schrödinger operators with singular potentials. Israel J. Math. **13**, 135–148 (1972, 1973)

239. H. Kikuchi, Existence and stability of standing waves for Schrödinger-Poisson-Slater equation. Adv. Nonlinear Stud. **7**(3), 403–437 (2007)

240. G. Kirchhoff, *Mechanik* (Teubner, Leipzig, 1883)

241. S.G. Krantz, Lipschitz spaces, smoothness of functions, and approximation theory. Exposition. Math. **1**(3), 193–260 (1983)

242. M.A. Krasnoselski, *Topological Methods in the Theory of Nonlinear Integral Equations*, Translated by A.H. Armstrong; translation ed. by J. Burlak (A Pergamon Press Book The Macmillan Co., New York, 1964), xi + 395 pp

243. K. Kurata, Existence and semi-classical limit of the least energy solution to a nonlinear Schrödinger equation with electromagnetic fields. Nonlinear Anal. Ser. A: Theory Methods **41**(5–6), 763–778 (2000)

244. N.S. Landkof, *Foundations of Modern Potential Theory*, Translated from the Russian by A.P. Doohovskoy. Die Grundlehren der mathematischen Wissenschaften, Band 180 (Springer, New York/Heidelberg, 1972), x+424 pp

245. N. Laskin, Fractional quantum mechanics and Lèvy path integrals. Phys. Lett. A, **268**(4–6), 298–305 (2000)

246. N. Laskin, Fractional Schrödinger equation. Phys. Rev. E (3) **66**(5), 056108, 7 pp (2002)

247. N. Laskin, *Fractional Quantum Mechanics* (World Scientific Publishing Co. Pte. Ltd., Hackensack, 2018), xv+341 pp

248. E.H. Lieb, Existence and uniqueness of the minimizing solution of Choquard's nonlinear equation. Stud. Appl. Math. **57**(2), 93–105 (1976/1977)

249. E.H. Lieb, M. Loss, *Analysis*. Graduate Studies in Mathematics, vol. 14 (American Mathematical Society, Providence, 1997), xviii+278 pp

250. E.H. Lieb, H.T. Yau, The Chandrasekhar theory of stellar collapse as the limit of quantum mechanics. Commun. Math. Phys. **112**(1), 147–174 (1987)

251. E.H. Lieb, H.T. Yau, The stability and instability of relativistic matter. Commun. Math. Phys. **118**(2), 177–213 (1988)

252. E. Lindgren, P. Lindqvist, Fractional eigenvalues. Calc. Var. Partial Differ. Equ. **49**(1–2), 795–826 (2014)

253. J.L. Lions, On some questions in boundary value problems of mathematical physics, in *Contemporary Developments in Continuum Mechanics and Partial Differential Equations* (Proc. Internat. Sympos., Inst. Mat., Univ. Fed. Rio de Janeiro, Rio de Janeiro, 1977), (North-Holland Math. Stud., 30, North-Holland, Amsterdam-New York, 1978), pp. 284–346

254. P.L. Lions, The Choquard equation and related questions. Nonlinear Anal. **4**(6), 1063–1072 (1980)

255. P.L. Lions, Symetrié et compacité dans les espaces de Sobolev. J. Funct. Anal. **49**(3), 315–334 (1982)

256. P.L. Lions, The concentration-compactness principle in the calculus of variations. The locally compact case. II. Ann. Inst. H. Poincaré Anal. Non Linéaire **1**(4), 223–283 (1984)

257. P.L. Lions, The concentration-compactness principle in the calculus of variations. The limit case. Part *I*. Rev. Mat. Iberoamericana **1**(1), 145–201 (1985)

258. J.-L. Lions, E. Magenes, *Non-Homogeneous Boundary Value Problems and Applications*, vol. I. Translated from the French by P. Kenneth. Die Grundlehren der mathematischen Wissenschaften, Band 181 (Springer, New York/Heidelberg, 1972), xvi+357 pp

259. S.B. Liu, On ground states of superlinear p-Laplacian equations in $\mathbb{R}^N$. J. Math. Anal. Appl. **361**(1), 48–58 (2010)

260. Z.S. Liu, S.J. Guo, On ground state solutions for the Schrödinger-Poisson equations with critical growth. J. Math. Anal. Appl. **412**(1), 435–448 (2014)

261. B. Liu, L. Ma, Radial symmetry results for fractional Laplacian systems. Nonlinear Anal. **146**, 120–135 (2016)

262. Z. Liu, M. Squassina, J. Zhang, Ground states for fractional Kirchhoff equations with critical nonlinearity in low dimension. NoDEA Nonlinear Differ. Equ. Appl. **24**(4), Art. 50, 32 pp (2017)

263. C. Liu, Z. Wang, H.S. Zhou, Asymptotically linear Schrödinger equation with potential vanishing at infinity. J. Differ. Equ. **245**(1), 201–222 (2008)

264. Z. Liu, J. Zhang, Multiplicity and concentration of positive solutions for the fractional Schrödinger-Poisson systems with critical growth. ESAIM Control Optim. Calc. Var. **23**(4), 1515–1542 (2017)
265. L. Ljusternik, L. Schnirelmann, *Méthodes topologique dans les problèmes variationnels* (Hermann and Cie, Paris, 1934)
266. L. Ma, L. Zhao, Classification of positive solitary solutions of the nonlinear Choquard equation. Arch. Ration. Mech. Anal. **195**(2), 455–467 (2010)
267. J. Mawhin, M. Willem, *Critical Point Theory and Hamiltonian Systems*, (Springer-Verlag, New York, 1989), pp. xiv+277
268. V.G. Maz'ja, *Sobolev Spaces*, Translated from the Russian by T.O. Shaposhnikova. Springer Series in Soviet Mathematics (Springer, Berlin, 1985), xix+486 pp
269. C. Mercuri, M. Willem, A global compactness result for the p-Laplacian involving critical nonlinearities. Discret. Contin. Dyn. Syst. **28**(2), 469–493 (2010)
270. C. Miranda, Un'osservazione sul teorema di Brouwer. Boll. Unione Mat. Ital. **3**(2), 5–7 (1940)
271. O.H. Miyagaki, M.A.S. Souto, Superlinear problems without Ambrosetti and Rabinowitz growth condition. J. Differ. Equ. **245**(12), 3628–3638 (2008)
272. X. Mingqi, P. Pucci, M. Squassina, B. Zhang, Nonlocal Schrödinger-Kirchhoff equations with external magnetic field. Discret. Contin. Dyn. Syst. A **37**(3), 503–521 (2017)
273. G. Molica Bisci, V.D. Rădulescu, R. Servadei, *Variational Methods for Nonlocal Fractional Problems*, vol. 162 (Cambridge University Press, Cambridge, 2016)
274. V. Moroz, J. Van Schaftingen, Ground states of nonlinear Choquard equations: existence, qualitative properties and decay asymptotics. J. Funct. Anal. **265**(2), 153–84 (2013)
275. V. Moroz, J. Van Schaftingen, Existence of groundstates for a class of nonlinear Choquard equations. Trans. Am. Math. Soc. **367**(9), 6557–6579 (2015)
276. V. Moroz, J. Van Schaftingen, A guide to the Choquard equation. J. Fixed Point Theory Appl. **19**(1), 773–813 (2017)
277. S. Mosconi, K. Perera, M. Squassina, Y. Yang, The Brezis-Nirenberg problem for the fractional p-Laplacian. Calc. Var. Partial Differ. Equ. **55**(4), Art. 105, 25 pp (2016)
278. J. Moser, A new proof of De Giorgi's theorem concerning the regularity problem for elliptic differential equations. Commun. Pure Appl. Math. **13**, 457–468 (1960)
279. J. Moser, A sharp form of an inequality by N. Trudinger. Indiana Univ. Math. J. **20**, 1077–1092 (1970/1971)
280. D. Motreanu, V.V. Motreanu, N. Papageorgiou, *Topological and Variational Methods with Applications to Nonlinear Boundary Value Problems* (Springer, New York, 2014), xii+459 pp
281. E. Murcia, G. Siciliano, Positive semiclassical states for a fractional Schrödinger-Poisson system. Differ. Integr. Equ. **30**(3–4), 231–258 (2017)
282. R. Musina, A.I. Nazarov, Strong maximum principles for fractional Laplacians. Proc. R. Soc. Edinburgh Sect. A **149**(5), 1223–1240 (2019)
283. N. Nyamoradi, Existence of three solutions for Kirchhoff nonlocal operators of elliptic type. Math. Commun. **18**(2), 489–502 (2013)
284. Y.G. Oh, Existence of semiclassical bound states of nonlinear Schrödinger equations with potentials of the class $(V)_a$. Commun. Partial Differ. Equ. **13**(12), 1499–1519 (1988)
285. T. Ozawa, On critical cases of Sobolev's inequalities. J. Funct. Anal. **127**(2), 259–269 (1995)
286. R.S. Palais, Lusternik-Schnirelman theory on Banach manifolds. Topology **5**, 115–132 (1966)
287. R.S. Palais, S. Smale, A generalized Morse theory. Bull. Am. Math. Soc. **70**, 165–172 (1964)
288. G. Palatucci, A. Pisante, Improved Sobolev embeddings, profile decomposition, and concentration-compactness for fractional Sobolev spaces. Calc. Var. Partial Differ. Equ. **50**(3–4), 799–829 (2014)
289. S. Pekar, *Untersuchung uber die Elektronentheorie der Kristalle* (Akademie, Berlin, 1954).

290. R. Penrose, Quantum computation, entanglement and state reduction. R. Soc. Lond. Philos. Trans. Ser. A Math. Phys. Eng. Sci. **356**(1743), 1927–1939 (1998)
291. K. Perera, Z.T. Zhang, Nontrivial solutions of Kirchhoff-type problems via the Yang index. J. Differ. Equ. **221**(1), 246–255 (2006)
292. A. Pinamonti, M. Squassina, E. Vecchi, Magnetic BV functions and the Bourgain-Brezis-Mironescu formula. Adv. Calc. Var. **12**(3), 225–252 (2019)
293. S.I. Pohožaev, On the eigenfunctions of the equation $\Delta u + \lambda f(u) = 0$. Dokl. Akad. Nauk SSSR **165**, 36–39 (1965)
294. S.I. Pohožaev, A certain class of quasilinear hyperbolic equations. Mat. Sb. **96**(138), 152–166, 168 (1975)
295. P. Pucci, S. Saldi, Critical stationary Kirchhoff equations in $\mathbb{R}^N$ involving nonlocal operators. Rev. Mat. Iberoam. **32**(1), 1–22 (2016)
296. P. Pucci, M. Xiang, B. Zhang, Multiple solutions for nonhomogeneous Schrödinger-Kirchhoff type equations involving the fractional p-Laplacian in $\mathbb{R}^N$. Calc. Var. Partial Differ. Equ. **54**(3), 2785–2806 (2015)
297. P.H. Rabinowitz, Variational methods for nonlinear elliptic eigenvalue problems. Indiana Univ. Math. J. **23**, 729–754 (1973/1974)
298. P.H. Rabinowitz, Minimax methods in critical point theory with applications to differential equations, in *CBMS Regional Conference Series in Mathematics*, vol. 65, 1986
299. P.H. Rabinowitz, On a class of nonlinear Schrödinger equations. Z. Angew. Math. Phys. **43**(2), 270–291 (1992)
300. M. Ramos, Z.-Q. Wang, M. Willem, Positive solutions for elliptic equations with critical growth in unbounded domains, in *Calculus of Variations and Differential Equations* (Haifa, 1998) Chapman & Hall/CRC Res. Notes Math., vol. 410 (Chapman & Hall/CRC, Boca Raton, 2000), pp. 192–199
301. M. Reed, B. Simon, *Methods of Modern Mathematical Physics, I, Functional Analysis* (Academic Press, New York/London, 1972), xvii+325 pp
302. X. Ros-Oton, J. Serra, The Pohozaev identity for the fractional Laplacian. Arch. Ration. Mech. Anal. **213**(2), 587–628 (2014)
303. D. Ruiz, The Schrödinger-Poisson equation under the effect of a nonlinear local term. J. Funct. Anal. **237**(2), 655–674 (2006)
304. M. Ryznar, Estimate of Green function for relativistic α-stable processes. Potential Anal. **17**(1), 1–23 (2002)
305. M. Schechter, W. Zou, Superlinear problems. Pac. J. Math. **214**(1), 145–160 (2004)
306. S. Secchi, Ground state solutions for nonlinear fractional Schrödinger equations in $\mathbb{R}^N$. J. Math. Phys. **54**(3), 031501, 17 pp (2013)
307. S. Secchi, On fractional Schrödinger equations in $\mathbb{R}^N$ without the Ambrosetti-Rabinowitz condition. Topol. Methods Nonlinear Anal. **47**(1), 19–41 (2016)
308. S. Secchi, On some nonlinear fractional equations involving the Bessel potential. J. Dyn. Differ. Equ. **29**(3), 1173–1193 (2017)
309. R. Servadei, Infinitely many solutions for fractional Laplace equations with subcritical nonlinearity, in *Recent Trends in Nonlinear Partial Differential Equations. II. Stationary Problems*, Contemporary Mathematics, vol. 595 (American Mathematical Society, Providence, 2013), pp. 317–340
310. R. Servadei, E. Valdinoci, The Brezis-Nirenberg result for the fractional Laplacian. Trans. Am. Math. Soc. **367**(1), 67–102 (2015)
311. X. Shang, J. Zhang, Y. Yang, On fractional Schrödinger equations with critical growth. J. Math. Phys. **54**(12), 121502, 20 pp (2013)

312. Z. Shen, F. Gao, M. Yang, Ground states for nonlinear fractional Choquard equations with general nonlinearities. Math. Methods Appl. Sci. **39**(14), 4082–4098 (2016)

313. L. Silvestre, Regularity of the obstacle problem for a fractional power of the Laplace operator. Commun. Pure Appl. Math. **60**(1), 67–112 (2007)

314. M. Squassina, B. Volzone, Bourgain-Brezis-Mironescu formula for magnetic operators. C. R. Math. Acad. Sci. Paris **354**(8), 825–831 (2016)

315. E. Stein, *Singular Integrals and Differentiability Properties of Functions*. Princeton Mathematical Series, vol. 30 (Princeton University Press, Princeton, 1970), xiv+290 pp

316. P. R. Stinga, *User's guide to the fractional Laplacian and the method of semigroups*. Handbook of fractional calculus with applications, vol. 2, De Gruyter, Berlin, 235–265 (2019)

317. P.R. Stinga, J.L. Torrea, Extension problem and Harnack's inequality for some fractional operators. Commun. Partial Differ. Equ. **35**(11), 2092–2122 (2010)

318. W.A. Strauss, Existence of solitary waves in higher dimensions. Commun. Math. Phys. **55**(2), 149–162 (1977)

319. M. Struwe, *Variational Methods. Applications to Nonlinear Partial Differential Equations and Hamiltonian Systems*, vol. 34, 4th edn. (Springer, Berlin, 2008), xx+302 pp

320. C.A. Stuart, H.S. Zhou, Applying the mountain pass theorem to an asymptotically linear elliptic equation on $\mathbb{R}^N$. Commun. Partial Differ. Equ. **24**(9–10), 1731–1758 (1999)

321. A. Szulkin, T. Weth, Ground state solutions for some indefinite variational problems. J. Funct. Anal. **257**(12), 3802–3822 (2009)

322. A. Szulkin, T. Weth, *The Method of Nehari Manifold*. Handbook of Nonconvex Analysis and Applications (Int. Press, Somerville, MA, 2010), pp. 597–632

323. M.H. Taibleson, On the theory of Lipschitz spaces of distributions on Euclidean n-space. I. Principal properties. J. Math. Mech. **13**, 407–479 (1964)

324. J. Tan, The Brezis-Nirenberg type problem involving the square root of the Laplacian. Calc. Var. Partial Differ. Equ. **42**(1–2), 21–41 (2011)

325. K. Teng, Existence of ground state solutions for the nonlinear fractional Schrödinger-Poisson system with critical Sobolev exponent. J. Differ. Equ. **261**(6), 3061–3106 (2016)

326. H. Triebel, *Theory of Function Spaces*. Monographs in Mathematics, vol. 78 (Birkhäuser Verlag, Basel, 1983), 284 pp

327. N.S. Trudinger, On imbeddings into Orlicz spaces and some applications. J. Math. Mech. **17**, 473–483 (1967)

328. B.O. Turesson, *Nonlinear Potential Theory and Weighted Sobolev Spaces*. Lecture Notes in Mathematics, vol. 1736 (Springer, Berlin, 2000), xiv+173 pp

329. J.L. Vázquez, The Dirichlet problem for the fractional p-Laplacian evolution equation. J. Differ. Equ. **260**(7), 6038–6056 (2016)

330. X. Wang, On concentration of positive bound states of nonlinear Schrödinger equations. Commun. Math. Phys. **53**(2), 229–244 (1993)

331. J. Wang, L. Tian, J. Xu, F. Zhang, Multiplicity and concentration of positive solutions for a Kirchhoff type problem with critical growth. J. Differ. Equ. **253**(7), 2314–2351 (2012)

332. K. Wang, J. Wei, On the uniqueness of solutions of a nonlocal elliptic system. Math. Ann. **365**(1–2), 105–153 (2016)

333. Z. Wang, H.S. Zhou, Positive solutions for a nonhomogeneous elliptic equation on $\mathbb{R}^N$ without (AR) condition. J. Math. Anal. Appl. **353**(1), 470–479 (2009)

334. M. Warma, The fractional relative capacity and the fractional Laplacian with Neumann and Robin boundary conditions on open sets. Potential Anal. **42**(2), 499–547 (2015)

335. M. Warma, The fractional Neumann and Robin type boundary conditions for the regional fractional p-Laplacian. NoDEA Nonlinear Differ. Equ. Appl. **23**(1), Art. 1, 46 pp (2016)

336. G.N. Watson, *A Treatise on the Theory of Bessel Functions* (Cambridge University Press, Cambridge, England; The Macmillan Company, New York, 1944), vi+804 pp

337. R.A. Weder, Spectral properties of one-body relativistic spin-zero Hamiltonians. Ann. Inst. H. Poincaré Sect. A (N.S.) **20**, 211–220 (1974)

338. R.A. Weder, Spectral analysis of pseudodifferential operators. J. Funct. Anal. **20**(4), 319–337 (1975)

339. J. Wei, M. Winter, Strongly interacting bumps for the Schrödinger-Newton equation. J. Math. Phys. **50**(1), 012905, 22 pp (2009)

340. M. Willem, *Minimax Theorems*. Progress in Nonlinear Differential Equations and their Applications, vol. 24 (Birkhäuser Boston, Inc., Boston, 1996), x+162 pp

341. J. Zhang, Z. Chen, W. Zou, Standing waves for nonlinear Schrödinger equations involving critical growth. J. Lond. Math. Soc. (2) **90**(3), 827–844 (2014)

342. J. Zhang, M. do Ó, M. Squassina, Fractional Schrödinger-Poisson Systems with a General Subcritical or Critical Nonlinearity. Adv. Nonlinear Stud. **16**(1), 15–30 (2016)

343. B. Zhang, M. Squassina, X. Zhang, Fractional NLS equations with magnetic field, critical frequency and critical growth. Manuscripta Math. **155**(1–2), 115–140 (2018)

344. X. Zhang, B. Zhang, D. Repovš, Existence and symmetry of solutions for critical fractional Schrödinger equations with bounded potentials. Nonlinear Anal. **142**, 48–68 (2016)

345. J. Zhang, W. Zou, A Berestycki-Lions theorem revisited. Commun. Contemp. Math. **14**(5), 1250033, 14 pp (2012)

346. J. Zhang, W. Zou, The critical case for a Berestycki-Lions theorem. Sci. China Math. **57**(3), 541–554 (2014)

347. J. Zhang, W. Zou, Solutions concentrating around the saddle points of the potential for critical Schrödinger equations. Calc. Var. Partial Differ. Equ. **54**(4), 4119–4142 (2015)

348. L. Zhao, F. Zhao, Positive solutions for Schrödinger-Poisson equations with a critical exponent. Nonlinear Anal. **70**(6), 2150–2164 (2009)

349. X.P. Zhu, A perturbation result on positive entire solutions of a semilinear elliptic equation. J. Differ. Equ. **92**(2), 163–178 (1991)

350. X.P. Zhu, H.S. Zhu, Existence of multiple positive solutions of inhomogeneous semilinear elliptic problems in unbounded domains. Proc. R. Soc. Edinburgh Sect. A **115**(3–4), 301–318 (1990)

351. W. Zou, M. Schechter, *Critical Point Theory and its Applications* (Springer, New York, 2006), xii+318 pp

Index

© The Author(s), under exclusive license to Springer Nature Switzerland AG 2021
V. Ambrosio, *Nonlinear Fractional Schrödinger Equations in* $\mathbb{R}^N$,
Frontiers in Mathematics, https://doi.org/10.1007/978-3-030-60220-8

Printed in the United States
by Baker & Taylor Publisher Services